Coastal Lagoons

Critical Habitats of Environmental Change

Marine Science Series

The CRC Marine Science Series is dedicated to providing state-of-the-art coverage of important topics in marine biology, marine chemistry, marine geology, and physical oceanography. The series includes volumes that focus on the synthesis of recent advances in marine science.

CRC MARINE SCIENCE SERIES

Series Editor

Michael J. Kennish, Ph.D.

Published Titles

Artificial Reef Evaluation with Application to Natural Marine Habitats, William Seaman, Jr.
The Biology of Sea Turtles, Volume I, Peter L. Lutz and John A. Musick
Chemical Oceanography, Third Edition, Frank J. Millero
Coastal Ecosystem Processes, Daniel M. Alongi
Coastal Pollution: Effects on Living Resources and Humans, Carl J. Sindermann
Ecology of Estuaries: Anthropogenic Effects, Michael J. Kennish
Ecology of Marine Bivalves: An Ecosystem Approach, Richard F. Dame
Ecology of Marine Invertebrate Larvae, Larry McEdward
Ecology of Seashores, George A. Knox
Environmental Oceanography, Second Edition, Tom Beer
Estuarine Indicators, Stephen A. Bortone
Estuarine Research, Monitoring, and Resource Protection, Michael J. Kennish
Estuary Restoration and Maintenance: The National Estuary Program, Michael J. Kennish
Eutrophication Processes in Coastal Systems: Origin and Succession of Plankton Blooms and Effects on Secondary Production in Gulf Coast Estuaries, Robert J. Livingston
Handbook of Marine Mineral Deposits, David S. Cronan
Handbook for Restoring Tidal Wetlands, Joy B. Zedler
Intertidal Deposits: River Mouths, Tidal Flats, and Coastal Lagoons, Doeke Eisma
Marine Chemical Ecology, James B. McClintock and Bill J. Baker
Ocean Pollution: Effects on Living Resources and Humans, Carl J. Sindermann
Physical Oceanographic Processes of the Great Barrier Reef, Eric Wolanski
The Physiology of Fishes, *Second Edition*, David H. Evans
Pollution Impacts on Marine Biotic Communities, Michael J. Kennish
Practical Handbook of Estuarine and Marine Pollution, Michael J. Kennish
Practical Handbook of Marine Science, *Third Edition,* Michael J. Kennish
Restoration of Aquatic Systems, Robert J. Livingston
Seagrasses: Monitoring, Ecology, Physiology, and Management, Stephen A. Bortone
Trophic Organization in Coastal Systems, Robert J. Livingston

COASTAL LAGOONS

Critical Habitats of Environmental Change

Edited by

MICHAEL J. KENNISH

Rutgers University
New Brunswick, New Jersey, U.S.A.

HANS W. PAERL

University of North Carolina
Morehead City, U.S.A.

CRC Press is an imprint of the
Taylor & Francis Group, an **informa** business

Cover image of Hurricane Floyd Aftermath by GeoEye. Reprinted with permission.

CRC Press
Taylor & Francis Group
6000 Broken Sound Parkway NW, Suite 300
Boca Raton, FL 33487-2742

Printed in the United States of America on acid-free paper
10 9 8 7 6 5 4 3 2 1

International Standard Book Number: 978-1-4200-8830-4 (Hardback)

Library of Congress Cataloging-in-Publication Data

Coastal lagoons : critical habitats of environmental change / editors, Michael J. Kennish, Hans W. Paerl.
p. cm. -- (Marine science series)
"A CRC title."
Includes bibliographical references and index.
ISBN 978-1-4200-8830-4 (hardcover : alk. paper)
1. Lagoon ecology. 2. Coast changes. I. Kennish, Michael J. II. Paerl, Hans W. III. Title. IV. Series.

QH541.5.L27C625 2010
577.7'8--dc22 2010017670

Visit the Taylor & Francis Web site at
http://www.taylorandfrancis.com

and the CRC Press Web site at
http://www.crcpress.com

Contents

Preface....................ix
The Editors....................xi
Contributors....................xiii

Chapter 1 Coastal Lagoons: Critical Habitats of Environmental Change....................1

Michael J. Kennish and Hans W. Paerl

Chapter 2 Assessing the Response of the Pamlico Sound, North Carolina, USA to Human and Climatic Disturbances: Management Implications....................17

Hans W. Paerl, Robert R. Christian, J. D. Bales, B. L. Peierls, N. S. Hall, A. R. Joyner, and S. R. Riggs

Chapter 3 Sources and Fates of Nitrogen in Virginia Coastal Bays....................43

Iris C. Anderson, Jennifer W. Stanhope, Amber K. Hardison, and Karen J. McGlathery

Chapter 4 Ecosystem Health Indexed through Networks of Nitrogen Cycling....................73

Robert R. Christian, Christine M. Voss, Cristina Bondavalli, Pierluigi Viaroli, Mariachiara Naldi, A. Christy Tyler, Iris C. Anderson, Karen J. McGlathery, Robert E. Ulanowicz, and Victor Camacho-Ibar

Chapter 5 Blooms in Lagoons:: Different from Those of River-Dominated Estuaries....................91

Patricia M. Glibert, Joseph N. Boyer, Cynthia A. Heil, Christopher J. Madden, Brian Sturgis, and Catherine S. Wazniak

Chapter 6 Relationship between Macroinfaunal Diversity and Community Stability, and a Disturbance Caused by a Persistent Brown Tide Bloom in Laguna Madre, Texas.. 115

Paul A. Montagna, Mary F. Conley, Richard D. Kalke, and Dean A. Stockwell

Chapter 7 The Choptank Basin in Transition: Intensifying Agriculture, Slow Urbanization, and Estuarine Eutrophication....................135

Thomas R. Fisher, Thomas E. Jordan, Kenneth W. Staver, Anne B. Gustafson, Antti I. Koskelo, Rebecca J. Fox, Adrienne J. Sutton, Todd Kana, Kristen A. Beckert, Joshua P. Stone, Gregory McCarty, and Megan Lang

Chapter 8 Seagrass Decline in New Jersey Coastal Lagoons: A Response to Increasing Eutrophication....................167

Michael J. Kennish, Scott M. Haag, and Gregg P. Sakowicz

Chapter 9 Controls Acting on Benthic Macrophyte Communities in a Temperate and a Tropical Estuary 203

Sophia E. Fox, Ylra S. Olsen, Mirta Teichberg, and Ivan Valiela

Chapter 10 Phase Shifts, Alternative Stable States, and the Status of Southern California Lagoons 227

Peggy Fong and Rachel L. Kennison

Chapter 11 Lagoons of the Nile Delta 253

Autumn J. Oczkowski and Scott W. Nixon

Chapter 12 Origins and Fate of Inorganic Nitrogen from Land to Coastal Ocean on the Yucatan Peninsula, Mexico 283

Troy Mutchler, Rae F. Mooney, Sarah Wallace, Larissa Podsim, Stein Fredriksen, and Kenneth H. Dunton

Chapter 13 Subtropical Karstic Coastal Lagoon Assessment, Southeast Mexico: The Yucatan Peninsula Case 307

Jorge A. Herrera-Silveira and Sara M. Morales-Ojeda

Chapter 14 Seasonal and Interannual Variability of Planktonic Microbes in a Mesotidal Coastal Lagoon (Ria Formosa, SE Portugal): Impact of Climatic Changes and Local Human Influences 335

Ana B. Barbosa

Chapter 15 A Comparison of Eutrophication Processes in Three Chinese Subtropical Semi-Enclosed Embayments with Different Buffering Capacities 367

Kedong Yin, Jie Xu, and Paul J. Harrison

Chapter 16 The Wadden Sea: A Coastal Ecosystem under Continuous Change 399

Katja J. M. Philippart and Eric G. Epping

Chapter 17 The Patos Lagoon Estuary, Southern Brazil: Biotic Responses to Natural and Anthropogenic Impacts in the Last Decades (1979–2008) 433

Clarisse Odebrecht, Paulo C. Abreu, Carlos E. Bemvenuti, Margareth Copertino, José H. Muelbert, João P. Vieira, and Ulrich Seeliger

Chapter 18 Structure and Function of Warm Temperate East Australian Coastal Lagoons: Implications for Natural and Anthropogenic Changes 457

Bradley D. Eyre and Damien Maher

Chapter 19 Response of the Venice Lagoon Ecosystem to Natural and Anthropogenic Pressures over the Last 50 Years 483

Cosimo Solidoro, Vinko Bandelj, Fabrizio Aubry Bernardi, Elisa Camatti, Stefano Ciavatta, Gianpiero Cossarini, Chiara Facca, Piero Franzoi, Simone Libralato, Donata Melaku Canu, Roberto Pastres, Fabio Pranovi, Sasa Raicevich, Giorgio Socal, Adriano Sfriso, Marco Sigovini, Davide Tagliapietra, and Patrizia Torricelli

Chapter 20 Effect of Freshwater Inflow on Nutrient Loading and Macrobenthos Secondary Production in Texas Lagoons 513

Paul A. Montagna, and Jian Li

Index 541

Preface

Coastal Lagoons: Critical Habitats of Environmental Change is the outgrowth of a special session of the 19th Biennial Conference of the Estuarine Research Federation held in Providence, Rhode Island, from November 4 to 8, 2007. This special session titled *Human and Climatic Factors Affecting Eutrophication of Coastal Lagoonal Ecosystems* focused on the escalating eutrophication problems in coastal lagoons worldwide, the indicators of eutrophic conditions in these shallow water bodies, the influences of natural factors, and the resulting biotic and ecosystem impairments. Biogeochemical and ecological responses to nutrient enrichment in lagoonal estuaries were compared to those observed in other estuarine types. A synthesis component also examined how anthropogenic nutrient loading and natural influences cause change in the coastal zone.

The chapters comprising this volume represent a wide array of studies on natural and anthropogenic drivers of change in coastal lagoons located in different regions of the world. Detailed descriptions of the physicochemical and biotic characteristics of these diverse coastal lagoonal ecosystems are provided which address environmental factors, forcing features, and stressors affecting hydrologic, biogeochemical, and trophic properties of these important water bodies. They also recount the innovative tools and approaches used for assessing ecological change in the context of anthropogenically and climatically mediated factors. These include such approaches and methods of analysis as biotic indicators, trend analysis, and modeling.

The following individuals are acknowledged for their insightful peer reviews of the manuscripts comprising the chapters of this book: Merryl Alber, Scott Ator, Evelia Rivera Arriaga, Richard Batiuk, Justus van Beusekom, David Borkman, Steven Bouillon, Federico Brandini, Henry Briceno, Edward Buskey, Timothy Carruthers, Franciscus Colijn, Francisco Comin, Michael Connor, Daniele Curiel, Roberto Danovaro, Edward Dettmann, Victor de Jonge, Jeffrey Gaeckle, Anne Giblin, Ralph R. Haese, James Hagy, Marianne Holmer, Peter Herman, David Johnson, Bastiaan Knoppers, Ann E. Krause, Paulo Lana, Brian Lapointe, Harold Marshall, Ernest Matson, Jack Meisinger, Daryl Moorhead, Margaret Mulholland, Robert Nuzzi, Adina Paytan, Edward Phlips, Karsten Reise, Tammi L. Richardson, Gilbert Rowe, Steven Rumrill, Jose Saldivar-Comenges, John Schramski, Kevin Sellner, Karline Soetaert, Martha Sutula, Carol Thornber, Pierluigi Viaroli, Christy Tyler, Ian Webster, Kim Withers, W. J. (Wim) Wolff, Dana Woodruff, Meng Xia, and Joy Zedler.

The Editors

Dr. Michael J. Kennish, Ph.D., is a research professor on the faculty of the Institute of Marine and Coastal Sciences at Rutgers University, New Brunswick, New Jersey. He is also the research director of the Jacques Cousteau National Estuarine Research Reserve in Tuckerton, New Jersey.

Dr. Kennish has conducted biological and geological research on estuarine, coastal ocean, and deep-sea environments for more than 30 years, and has taught marine science classes at Rutgers for many years. Much of his research has involved the development and application of innovative methods to determine the condition and overall ecosystem health of aquatic systems. He is widely known for his work on human impacts on estuarine and marine environments, and has served on environmental panels and workgroups assessing these problems in New Jersey, the mid-Atlantic region, and nationwide, while concomitantly collaborating extensively with state and federal government agencies to remediate degraded habitats. Most notably, he has been heavily engaged in integrative ecosystem assessment, particularly investigations of impairment and remediation of impacted estuarine and coastal marine environments. These include studies of the natural and anthropogenic stressors that effect change in coastal ecosystems, and the dynamics of environmental forcing factors that generate imbalances in biotic community structure and ecosystem function.

Dr. Kennish's research is multidisciplinary in scope, addressing a wide range of nationally significant problems such as the effects of watershed development on coastal bays and near-shore ocean waters, wastewater discharges, habitat loss and alteration of aquatic systems, nutrient enrichment and eutrophication, hypoxia and anoxia, organic pollution, chemical contaminants, climate change, sea-level rise, overfishing, invasive species, watercraft effects, dredging and dredged material disposal, freshwater diversions, calefaction of estuarine waters, and entrainment and impingement of electric generating stations. He has also examined the effects of construction and operation of industrial facilities, maintenance of shorelines and waterways, and human use of coastal space and aquatic systems. Much of his basic research has entailed investigations of benthic communities and habitats, as well as seafloor mapping and habitat characterization. He has also studied the biology and geology of mid-ocean ridge and hydrothermal vent systems as a member of the Center for Deep-Sea Ecology and Biotechnology at Rutgers. As the co-chair of the Coastal Climate Change Group in Rutgers' Climate and Environmental Change Initiative, he is involved in the study of long-term climate change impacts on the New Jersey coast. While maintaining a wide range of research interests in marine ecology and marine geology, Dr. Kennish has been most actively involved in leading research teams investigating estuarine and coastal marine environments in New Jersey.

Dr. Kennish has published 12 books and more than 150 research articles in science journals and books. In addition, he has edited six compendium journal volumes on various topics in marine science.

Dr. Hans W. Paerl, Ph.D., is Kenan Professor of Marine and Environmental Sciences at the University of North Carolina-Chapel Hill Institute of Marine Sciences, Morehead City, North Carolina, where he has been on the faculty for 30 years. During this time he has supervised 41 graduate students, 16 post-doctoral researchers, 27 technicians, and more than 25 undergraduates.

Dr. Paerl has designed and directed numerous water quality and environmental research programs in North Carolina, nationally, and internationally, including those involving nutrient cycling and production dynamics of aquatic ecosystems, environmental controls of algal production, and assessment of the causes and consequences of eutrophication, notably harmful algal blooms and hypoxia. His studies have focused on identifying the importance and ecological impacts of atmospheric nitrogen deposition in estuarine and coastal environments and the development and application of bio-indicators used to assess human- and climate-induced change in aquatic ecosystems. He is involved in the development and application of microbial and biogeochemical indicators of aquatic ecosystem condition and change in response to human and climatic perturbations. He is co-principal investigator of the ferry-based water quality monitoring program, FerryMon (www.ferrymon.org), and the Neuse River Estuary Modeling and Monitoring Program, ModMon (http://www.unc.edu/ims/neuse/modmon), which employ environmental sensors and various microbial indicators to assess near real-time ecological condition of the Pamlico Sound System, the second largest estuarine complex in the United States. Results of these monitoring and assessment programs form the basis for evaluating national and local nutrient management strategies aimed at reducing the unwanted effects of excessive nutrient loading (i.e., eutrophication, harmful algal blooms, and hypoxia). Recently, he has been working with Chinese colleagues to help formulate nutrient reduction approaches for arresting and controlling toxic cyanobacterial blooms in China's third largest lake, Taihu, Jiangsu Province, which provides drinking water and fisheries resources for approximately 40 million people living in the Taihu basin.

In 2003, Dr. Paerl was awarded the G. Evelyn Hutchinson research achievement award by the American Society of Limnology and Oceanography for "contributing to the understanding of aquatic microbial processes, for documenting linkages among the atmospheric deposition of nitrogen, coastal eutrophication, and harmful algal blooms, and for crossing traditional research boundaries to system-level perspectives within freshwater, estuarine, and marine ecosystems." His work, which includes over 240 peer-reviewed publications and book chapters, plays a central role in coastal water quality and fisheries issues on local, regional, and global scales.

Contributors

Paulo C. Abreu
Institute of Oceanography, Federal University of Rio Grande
Rio Grande, Brazil

Iris C. Anderson
College of William and Mary, Virginia Institute of Marine Science, Department of Biological Studies
Gloucester Point, Virginia

J. D. Bales
U.S. Geological Survey
Raleigh, North Carolina

Vinko Bandelj
Istituto Nazionale di Oceanografia e di Geofisica Sperimentale (OGS)
Sgonico, Italy

Ana B. Barbosa
Centre for Marine and Environmental Research (CIMA), University of Algarve
Faro, Portugal

Kristen A. Beckert
Horn Point Laboratory, Center for Environmental Science, University of Maryland
Cambridge, Maryland

Carlos E. Bemvenuti
Institute of Oceanography, Federal University of Rio Grande
Rio Grande, Brazil

Fabrizio Aubry Bernardi
Consiglio Nazionale delle Ricerche, Istituto di Scienze Marine
Venice, Italy

Cristina Bondavalli
University of Parma, Department of Environmental Sciences, Viale Usberti
Parma, Italy

Joseph N. Boyer
Southeast Environmental Research Center, Florida International University
Miami, Florida

Victor Camacho-Ibar
Universidad Autónoma de Baja California, Instituto de Investigaciones Oceanológicas
Ensenada, Mexico

Elisa Camatti
Consiglio Nazionale delle Ricerche, Istituto di Scienze Marine
Venice, Italy

Donata Melaku Canu
Istituto Nazionale di Oceanografia e di Geofisica Sperimentale (OGS)
Sgonico, Italy

Robert R. Christian
Department of Biology, East Carolina University
Greenville, North Carolina

Stefano Ciavatta
Plymouth Marine Laboratory Prospect Place, The Hoe
Plymouth, United Kingdom
and
Università di Venezia Ca' Foscari Dorsoduro
Venice, Italy

Mary F. Conley
The Nature Conservancy
Charleston, South Carolina

Margareth Copertino
Institute of Oceanography, Federal University of Rio Grande
Rio Grande, Brazil

Gianpiero Cossarini
Istituto Nazionale di Oceanografia e di Geofisica Sperimentale (OGS)
Sgonico, Italy

Kenneth H. Dunton
The University of Texas at Austin Marine Science Institute
Port Aransas, Texas

Eric G. Epping
Department of Geology, Royal Netherlands Institute for Sea Research
Den Burg, the Netherlands

Bradley D. Eyre
Centre for Coastal Biogeochemistry, Southern Cross University
Lismore, Australia

Chiara Facca
Università di Venezia Ca' Foscari Dorsoduro
Venice, Italy

Thomas R. Fisher
Horn Point Laboratory, Center for Environmental Science, University of Maryland
Cambridge, Maryland

Peggy Fong
Department of Ecology and Evolutionary Biology, University of California Los Angeles
Los Angeles, California

Rebecca J. Fox
Horn Point Laboratory, Center for Environmental Science, University of Maryland
Cambridge, Maryland

Sophia E. Fox
The Ecosystems Center, Marine Biological Laboratory
Woods Hole, Massachusetts
and
Cape Cod National Seashore
National Park Service
Wellfleet, Massachusetts

Piero Franzoi
Università di Venezia Ca' Foscari Castello
Venice, Italy

Stein Fredriksen
The University of Oslo, Department of Biology
Oslo, Norway

Patricia M. Glibert
University of Maryland Center for Environmental Sciences
Horn Point Laboratory
Cambridge, Maryland

Anne B. Gustafson
Horn Point Laboratory, Center for Environmental Science, University of Maryland
Cambridge, Maryland

Scott M. Haag
Center for Remote Sensing and Spatial Analysis, Rutgers University
New Brunswick, New Jersey

N. S. Hall
Institute of Marine Sciences, University of North Carolina at Chapel Hill
Morehead City, North Carolina

Amber K. Hardison
Virginia Institute of Marine Science, College of William and Mary
Gloucester Point, Virginia

Paul J. Harrison
Division of Environment
Hong Kong University of Science and Technology
Hong Kong SAR, China

Cynthia A. Heil
Florida Fish and Wildlife Conservation Commission, Fish and Wildlife Research Institute
St. Petersburg, Florida

Jorge A. Herrera-Silveira
Centro de Investigacion y Estudios Avanzados del Instituto Politecnico Nacional
Merida, Yucatan Mexico

Thomas E. Jordan
Smithsonian Environmental Research Center
Edgewater, Maryland

A. R. Joyner
Institute of Marine Sciences, University of North Carolina at Chapel Hill
Morehead City, North Carolina

Richard D. Kalke
Harte Research Institute for Gulf of Mexico Studies, Texas A&M University
Corpus Christi, Texas

Todd Kana
Horn Point Laboratory, Center for Environmental Science, University of Maryland
Cambridge, Maryland

Michael J. Kennish
Institute of Marine and Coastal Sciences, Rutgers University
New Brunswick, New Jersey

Rachel L. Kennison
Department of Ecology and Evolutionary Biology, University of California Los Angeles
Los Angeles, California

Antti I. Koskelo
Horn Point Laboratory, Center for Environmental Science, University of Maryland
Cambridge, Maryland

Megan Lang
U.S. Department of Agriculture
Beltsville, Maryland

Jian Li
CTC Fishing, Inc.
Tamuning, Guam

Simone Libralato
Istituto Nazionale di Oceanografia e di Geofisica Sperimentale
Zgonik, Italy

Christopher J. Madden
South Florida Water Management District, Coastal Ecosystems Division
West Palm Beach, Florida

Damien Maher
Centre for Coastal Biogeochemistry, Southern Cross University
Lismore, Australia

Gregory McCarty
U.S. Department of Agriculture
Beltsville, Maryland

Karen J. McGlathery
University of Virginia, Department of Environmental Sciences
Charlottesville, Virginia

Paul A. Montagna
Harte Research Institute for Gulf of Mexico Studies, Texas A&M University–Corpus Christi
Corpus Christi, Texas

Rae F. Mooney
The University of Texas at Austin Marine Science Institute
Port Aransas, Texas

Sara M. Morales-Ojeda
Centro de Investigacion y Estudios Avanzados del Instituto Politecnico Nacional
Merida, Yucatan, Mexico

José H. Muelbert
Institute of Oceanography, Federal University of Rio Grande
Rio Grande, Brazil

Troy Mutchler
Kennesaw State University, Biology and Physics Department
Kennesaw, Georgia

Mariachiara Naldi
University of Parma, Department of Environmental Sciences, Viale Usberti
Parma, Italy

Scott W. Nixon
Graduate School of Oceanography, University of Rhode Island
Narragansett, Rhode Island

Autumn J. Oczkowski
Graduate School of Oceanography, University of Rhode Island
Narragansett, Rhode island

Clarisse Odebrecht
Institute of Oceanography, Federal University of Rio Grande
Rio Grande, Brazil

Ylra S. Olsen
School of Ocean Sciences, University of Bangor, Menai Bridge
Anglesey, UK

Hans W. Paerl
Institute of Marine Sciences, University of North Carolina at Chapel Hill
Morehead City, North Carolina

Roberto Pastres
Università di Venezia Ca' Foscari
Venice, Italy

B. L. Peierls
Institute of Marine Sciences, University of North Carolina at Chapel Hill
Morehead City, North Carolina

Katja J. M. Philippart
Department of Marine Ecology, Royal Netherlands Institute for Sea Research
Den Burg, the Netherlands

Larissa Podsim
Department of Wildlife and Fisheries Sciences
Texas A&M University
College Station, Texas

Fabio Pranovi
Università di Venezia Ca' Foscari Castello
Venice, Italy

Sasa Raicevich
Istituto Superiore per la Protezione e la Ricerca Ambientale
Chioggia, Italy

S. R. Riggs
Department of Geology, East Carolina University
Greenville, North Carolina

Gregg P. Sakowicz
Rutgers University Marine Field Station
Tuckerton, New Jersey

Ulrich Seeliger
Institute of Oceanography, Federal University of Rio Grande
Rio Grande, Brazil

Adriano Sfriso
Università di Venezia Ca' Foscari Dorsoduro
Venice, Italy

Marco Sigovini
Consiglio Nazionale delle Ricerche, Istituto di Scienze Marine
Venice, Italy

Giorgio Socal
Consiglio Nazionale delle Ricerche, Istituto di Scienze Marine
Venice, Italy

Cosimo Solidoro
Istituto Nazionale di Oceanografia e di Geofisica Sperimentale
Zgonik, Italy

Jennifer W. Stanhope
College of William and Mary, Virginia Institute of Marine Science
Gloucester Point, Virginia

Kenneth W. Staver
Wye Research and Education Center, University of Maryland
Queenstown, Maryland

Dean A. Stockwell
University of Alaska, Institute of Marine Science
Fairbanks, Alaska

Joshua P. Stone
Horn Point Laboratory, Center for Environmental Science, University of Maryland
Cambridge, Maryland

Brian Sturgis
United States Department of the Interior, National Park Service, Assateague Island National Seashore
Berlin, Maryland

Adrienne J. Sutton
Horn Point Laboratory
Center for Environmental Science
University of Maryland
Cambridge, Maryland

Davide Tagliapietra
Consiglio Nazionale delle Ricerche, Istituto di Scienze Marine
Venice, Italy

Mirta Teichberg
Zentrum für Marintropenökologie
Bremen, Germany

Patrizia Torricelli
Università di Venezia Ca' Foscari Castello
Venice, Italy

A. Christy Tyler
Rochester Institute of Technology, Department of Biology
Rochester, New York

Robert E. Ulanowicz
University of Maryland System, Chesapeake Biological Laboratory
Solomons, Maryland
and
Department of Botany and Zoology, University of Florida
Gainesville, Florida

Ivan Valiela
The Ecosystems Center, Marine Biological Laboratory
Woods Hole, Massachusetts

Pierluigi Viaroli
University of Parma, Department of Environmental Sciences, Viale Usberti
Parma, Italy

João P. Vieira
Institute of Oceanography, Federal University of Rio Grande
Rio Grande, Brazil

Christine M. Voss
Department of Biology, Coastal Resources Management Program
East Carolina University
Greenville, North Carolina

Sarah Wallace
The University of Texas at Austin Marine Science Institute
Port Aransas, Texas

Catherine S. Wazniak
Maryland Department of Natural Resources, Tidewater Ecosystem Assessment Division
Annapolis, Maryland

Jie Xu
Division of Environment
Hong Kong University of Science and Technology
Hong Kong SAR, China

Kedong Yin
Australian Rivers Institute, Griffith University (Nathan Campus)
Brisbane, QLD, Australia

1 Coastal Lagoons
Critical Habitats of Environmental Change

Michael J. Kennish and Hans W. Paerl*

CONTENTS

Abstract 1
1.1 Introduction 2
1.2 Classification 2
1.3 Physicochemical Characteristics 3
1.4 Biotic Characteristics 4
1.5 Ecologic and Economic Value 6
1.6 Human Impacts 7
1.7 Natural Stressors 9
1.8 Plan of This Volume 9
Acknowledgments 13
References 14

ABSTRACT

Coastal lagoons rank among the most productive ecosystems on Earth, and they provide a wide range of ecosystem services and resources. Anthropogenic impacts are escalating in many coastal lagoons worldwide because of increasing population growth and associated land-use alteration in adjoining coastal watersheds. The conversion of natural land covers to agricultural, urban, and industrial development has accelerated loading to streams and rivers that discharge into estuaries, leading to cascading water quality and biotic impacts, impairments, and diminishing recreational and commercial uses. Many coastal lagoons, notably those with restricted circulation, freshwater inflow, low flushing rates, and relatively long water residence times, are particularly susceptible to nutrient enrichment from surface runoff, groundwater, and atmospheric inputs. Natural stressors, such as hurricanes and other major storms, as well as impacts of climate change, including rising temperatures, more frequent and extreme floods and droughts can exacerbate these effects. The combined effects of these stressors include accelerated eutrophication, increased frequencies and geographic expansion of harmful algal blooms, and low oxygen bottom waters (hypoxia). Watershed management strategies aimed at controlling these unwanted effects are rightfully focused on reducing nutrient and other contaminant loading to these enclosed systems and include upgrading stormwater controls, adapting low-impact development and best management practices, advancing open space preservation, and encouraging natural treatment of nutrients by establishing riparian buffers and wetlands, and implementing government regulatory measures (e.g., total maximum daily loads [TMDLs] for nutrient limitation). More draconian strategies that have been used in some systems are dredging sediments, diverting rivers, creating inlets, and thereby increasing flushing.

* Corresponding Author. Email: kennish@marine.rutgers.edu

Because of their high biotic production and value to tourism and residential development, coastal lagoons are often overused for fisheries and aquaculture, transportation, energy production, and other human activities. The impacts that result require sound management strategies for long-term environmental and resource sustainability. Human land and water uses pose multiple challenges to managing coastal lagoon ecosystems. Storms, droughts, and other natural stressors complicate these challenges. The co-occurrence of natural and human disturbances causes significant deterioration of coastal lagoon habitats in many regions of the world.

Key Words: coastal lagoons, human impacts, natural stressors, pollution, eutrophication, habitat alteration, biotic responses, climate

1.1 INTRODUCTION

Coastal lagoons are shallow brackish or marine bodies separated from the ocean by a barrier island, spit, reef, or sand bank and connected at least intermittently to the open ocean by one or more restricted tidal inlets (Phleger 1969, 1981; Colombo 1977; Barnes 1980; Kjerfve 1986, 1994; Gonenc and Wolflin 2004). Therefore, coastal lagoons may be partially or wholly enclosed, depending on the extent of the land barrier, which impedes water exchange between the basin and ocean and tends to dampen wave, wind, and current action. According to Bird (1982), the term "coastal lagoon" applies only when the width of the inlets at high tide is less than 20% of the total length of the enclosing barrier.

Coastal lagoons generally occur on low-lying coasts and are usually oriented parallel to shores, often being much longer than they are wide. While common on some seaboards, coastal lagoons occupy only ~13% of the coastal areas worldwide, and they are present on every continent except Antarctica. They are most extensive along the coasts of Africa (17.9% of the coastline) and North America (17.6%), and less so along the coasts of Asia (13.8%), South America (12.2%), Australia (11.4%), and Europe (5.3%) (Barnes 1980). The most extensive stretch of coastal lagoons is along the Atlantic and Gulf coasts of the United States, where they cover ~2800 km of shoreline (Nichols and Boon 1994). However, they are also common along the eastern coasts of South America and India, southern Britain and western France, the western coast of Africa, and southeastern Australia, as well as along the shores of the Baltic, Black, Caspian, and Mediterranean seas (Colombo 1977; Kjervfe 1994). Coastal lagoon systems vary greatly in size from as small as a hectare in area to more than 10,000 km^2 (e.g., Lagoa dos Patos, Brazil) (Bird 1994). They differ considerably in morphological, geological, and hydrological characteristics.

The adjoining coastal watersheds generally cover a smaller area than those bordering river-dominated estuaries. For example, the ratio of watershed area to lagoon area averages 182, which is far less than that of other estuarine types, which average 335 (Bricker et al. 2007). However, because coastal lagoons often have protracted water residence times due to the restrictive enclosure of depositional barriers, the lagoons are susceptible to nutrient and other pollutants, various anthropogenic drivers, and hydrological modifications, as well as the vagaries of climate change (e.g., changes in storm frequencies and intensities, floods, droughts, and warming).

1.2 CLASSIFICATION

Kjerfve (1986, 1994) classified coastal lagoons into three geomorphic types based on water exchange with the coastal ocean: "choked, restricted, and leaky" lagoons. Choked lagoons consist of a series of interconnected elliptical water bodies with a single ocean entrance channel or inlet to the sea. The Lagoa dos Patos in Brazil is an example. Restricted lagoons consist of two or more entrance channels or inlets that connect to a wide, expansive water body usually oriented shore-parallel. The Barnegat Bay–Little Egg Harbor Estuary in New Jersey (USA) is an example. Leaky lagoons have multiple entrance channels leading to an elongated shore-parallel water body. The Wadden Sea, the Netherlands, is an example.

The genesis of coastal lagoons and barrier island systems enclosing them depends on the sea-level history of a region. Rising sea level, such as during the Holocene, promotes the formation of these systems. The coastal extension of spits is an important mechanism for the development of barrier beaches and lagoons during periods of stable or slowly changing sea level. Shoreface dynamics and tidal range play significant roles in the maintenance of these systems (Martin and Dominguez 1994). Sand bars may form across narrow inlet channels either seasonally or over longer periods of time. These ephemeral features impede water exchange between the lagoonal basin and ocean, increasing water residence times and changing biogeochemical and biotic processes in these shallow water bodies. With restricted flushing and protracted water residence times, coastal lagoons are susceptible to nutrient enrichment and toxic pollutant inputs that can culminate in eutrophication and contamination problems. Seasonal closure of coastal lagoons can dramatically alter salinity, with oligohaline to nearly limnetic conditions occurring during periods of high precipitation and riverine inflow and mesohaline to even hypersaline conditions during periods of drought, low riverine inflow, and evaporation. Salinity in coastal lagoons can span an entire spectrum from freshwater to hypersaline levels.

1.3 PHYSICOCHEMICAL CHARACTERISTICS

Coastal lagoons generally average less than 2 m in depth, although deeper waters may be encountered in channels and relict holes. Because they are shallow and well mixed by wave and current actions, only localized stratification may develop in deeper areas. Coastal lagoons are usually characterized by microtidal conditions, with tidal ranges averaging less than 2 m. This is so because advective water movements increase with larger tidal ranges, which accelerates erosion of existing structures and reduces deposition of sediments. As a consequence, gaps between barrier islands enclosing lagoons can expand significantly as stronger ebb and flood currents develop, ultimately leading to open coastal bay systems (Colombo 1977). Where sediment influx from mainland areas is limited, microtidal back-barrier lagoons develop as open water bodies. In these systems, most of the sediment influx accumulates in the lagoonal water body along back-barrier areas via flood tidal delta and washover processes.

The behavior of coastal lagoons can vary considerably despite similar basin morphometry because of human habitat alteration, differences in marine and watershed influences, as well as in-basin hydrological processes. The size and configuration of tidal inlets, expanse and development of bordering watersheds, amount of freshwater input, water depth, and wind conditions greatly influence the physicochemical processes occurring within the lagoonal water bodies (Alongi 1998). Variations in precipitation and evaporation, surface runoff, and groundwater seepage, together with fluxes in wind forcing, account for large differences in advective transport in lagoonal estuaries. Storm and wind surges, overwash events, inlet reconfigurations, land reclamation, construction of dams, dikes, and artificial bars, as well as channel dredging events, are important drivers of hydrological change in these systems. Some catastrophic effects, such as major hurricanes, can inject large pulses of freshwater, reconstruct the barriers enclosing the lagoons, and alter other extensive areas of the system. Seasonal changes in river inflow may also cause marked shifts in salinity structure in coastal lagoons.

Local meteorological conditions strongly influence water temperature and other physicochemical parameters in coastal lagoons. These water bodies are very responsive to meteorological conditions, notably air temperature, because they are very shallow. Coastal lagoons are heavily influenced by waves than by tides, with microtidal conditions predominating in many of these systems (Eisma 1998). Wind direction and speed are key elements of wave genesis in these systems. Wave effectiveness increases as the tidal range decreases (Martin and Dominguez 1994).

A substantial volume of sediment can enter coastal lagoons during storm surge and overwash events. The breaching of barriers and formation of washovers deliver sediments from the coastal ocean and back-barrier areas. Aside from these sources, sediments accumulate in lagoonal basins from rivers draining the mainland, runoff of tidal marsh and other habitats bordering the basin, and internal processes (i.e., organic carbon production, chemical precipitation, as well as erosion and resuspension of older sediments). Nutrients bound in bottom sediments are released to overlying waters by means of microbially mediated regeneration, resuspension of porewater, organic detritus and inorganic particulates, and remobilization primarily associated with geochemical processes in the water column (Thornton et al. 1995). As noted by Nichols and Boon (1994), coastal lagoons are net sediment sinks, but changes in the volume of river inflow, frequency of storms, and in-basin wave and current action all contribute to considerable variation in the rate of biogenic activity and sediment accumulation in these shallow systems.

1.4 BIOTIC CHARACTERISTICS

Coastal lagoons differ from deep estuaries in two fundamental ways: (1) the photic zone extends to most of the seafloor, with the benthos typically accounting for a large fraction of the total primary production of the system; and (2) high rates of metabolism of benthic primary producers mediate nutrient cycling processes and result in strong benthic-pelagic coupling (McGlathery et al. 2007; Anderson et al., Chapter 3, this volume). Coastal lagoon floors generally lie within the photic zone due to the shallowness of the basin, although high turbidity and phytoplankton growth can effectively attenuate light penetration to the estuarine bottom. When sunlight penetrates to the basin floor, prolific growth of benthic algae and seagrasses often occur in these systems. As a result, benthic primary production can exceed phytoplankton production in these systems (Table 1.1). Because the ratio between the lagoonal volume and sediment surface area is low, the significance of biogeochemical cycling and other interactions between bottom sediments and the overlying water column is greater than in deeper, river-dominated estuarine systems. Knoppers (1994) showed that bottom sediments release 5% to 30%, and external inorganic loading 5% to 20%, of the primary production demand in coastal lagoons. In Hog Island Bay, Virginia, Anderson et al. (Chapter 3, this volume) found that, although benthic remineralization was high, net fluxes were low because of uptake by benthic microalgae. In this system, benthic remineralization supplied 55% of the nitrogen required to support primary production plus denitrification.

Furthermore, because many shallow lagoonal systems have relatively long water residence times, nutrient inputs can be recycled many times before they exit to the coastal ocean. This hydrobiogeochemical feature that enables these systems to support relatively high rates of productivity per unit nutrient input is key to why these systems serve as excellent fisheries nurseries. However, this feature also makes lagoonal systems highly sensitive to nutrient over-enrichment and accelerated eutrophication (Paerl et al. 2006a,b).

Annual primary production in coastal lagoons generally ranges from ~50 to >500 g C m^{-2} yr^{-1}. Based on the trophic classification of Nixon (1995), many coastal lagoons fall within the range of eutrophic conditions (300 to 500 g C m^{-2} yr^{-1}). However, others fall within mesotrophic (100 to 300 g C m^{-2} yr^{-1}) and oligotrophic (<100 g C m^{-2} yr^{-1}) conditions. The systems with the highest primary production are those dominated by benthic macroalgae and vascular macrophytes. Phytoplankton-based systems typically have at least seasonal contributions of benthic microalgal and macroalgal populations at shallower depths. However, this may not be true of highly eutrophied systems in which planktonic production can be very high and dominate overall primary production. Seagrasses may occur in high biomasses, most often at depths <1 m. Algal mat systems occur most frequently in hypersaline lagoons and lagoons with intertidal zones subject to protracted periods of exposure during dry seasons (Gamito et al. 2004). Changes in physicochemical conditions foster shifts in the dominant plant communities. For example, altered turbidity, hydraulic conditions, and water residence times play a major role in determining the composition and relative dominance of

TABLE 1.1
Ranking of Principal Anthropogenic Stressors to Estuarine Environments Based on Assessment of Published Literature

Stressor	Principal Impacts
	Tier I Stressors
1. Habitat loss and alteration	Elimination of usable habitat for estuarine biota
2. Eutrophication	Increased primary production, harmful algal blooms, hypoxia and anoxia, increased benthic invertebrate mortality, fish kills, altered community structure, increased turbidity and shading, reduced seagrass biomass, degraded water quality
3. Sewage and organic wastes	Elevated human pathogens, increased nutrient and organic matter loading, increased eutrophication, increased hypoxia, degraded water and sediment quality, reduced biodiversity
4. Fisheries overexploitation	Depletion or collapse of fish and shellfish stocks, altered food webs, changes in the structure, function, and controls of estuarine ecosystems
5. Sea-level rise	Shoreline retreat, loss of wetlands habitat, widening of estuary mouth, altered tidal prism and salinity regime, changes in biotic community structure
6. Storms and hurricanes	Increased nutrient, sediment, organic matter and contaminant loading, increased hypoxia, salinity stress on primary producers and higher trophic levels, altered water residence time with potentially adverse effects on flora and fauna
	Tier II Stressors
7. Chemical contaminants Higher priority synthetic organic compounds Lower priority—oil (PAHs), metals, radionuclides	Adverse effects on estuarine organisms, including tissue inflammation and degeneration, neoplasm formation, genetic derangement, aberrant growth and reproduction, neurological and respiratory dysfunction, digestive disorders and behavioral abnormalities; reduced population abundance; sediment toxicity
8. Freshwater diversions	Altered hydrological, salinity, and temperature regimes; changes in abundance, distribution, and species composition of estuarine organisms
9. Introduced/invasive species	Changes in species composition and distribution, shifts in trophic structure, reduced biodiversity, introduction of detrimental pathogens
10. Subsidence	Modification of shoreline habitat, degraded wetlands, accelerated fringe erosion, expansion of open water habitat
11. Sediment input/ turbidity	Habitat alteration, reduced primary production, shading impacts on benthic organisms
12. Floatables and debris	Increased mortality of seabirds, marine mammals, reptiles, and other animals.

Source: Modified from Kennish et al. (2008).

plant populations (Adolf et al. 2006; Valdes-Weaver et al. 2006). Changes in grazing pressure (top-down effects) can accelerate these biotic responses.

Extensive wetland habitats commonly border coastal lagoons. Salt marshes predominate in temperate lagoonal systems and mangroves in tropical systems, although they co-occur along a latitudinal gradient of ~27° to 38°. Saltmarsh grasses and mangrove forests are highly productive, often contributing 1 to 4 kg dry wt m^{-2} yr^{-1}. Benthic microalgae, macroalgae, phytoplankton, epiphytes, and neuston may account for as much as 50% of the total production in some of these biotopes (Alongi 1998). These vital habitats are now threatened in many regions of the world by climate change, storms, sea-level rise, eutrophication, wastewater discharges, land reclamation, and other human activities (Alongi 1998; Adam et al. 2008; Dodd and Ong 2008).

The high levels of primary production, wide array of habitats, and sheltered back-bays promote the proliferation of biologically productive communities of organisms in coastal lagoons consisting of strictly lagoonal species, together with suites of essentially marine or freshwater forms at least partially dependent on ambient salinity. Thus, coastal lagoons rank among the most biologically productive marine systems. Both the structure (species composition and abundance) and function (productivity, trophic webs, and fluxes) of coastal lagoons vary greatly because of frequent physical and chemical disturbances and perturbations (Gamito et al. 2004). Despite the flux of environmental conditions, coastal lagoons provide ideal nursery and feeding habitats, and many fauna utilize these environments seasonally, entering and leaving on seasonal cues. Alvarez-Borrego (1994) noted that zooplankton productivity is as high as 50% of the benthic macrofaunal productivity, which amounts to about 20 to 200 g ash-free dry weight (AFDW) m^{-2} yr^{-1}. The nekton productivity, in turn, ranges from ~10% to 100% of the zooplankton productivity.

Although coastal lagoons are highly productive ecosystems, relatively few species are generally permanent residents of these systems, and few species numerically dominate the biotic communities (Colombo 1977). Furthermore, the population abundance of individual species varies considerably from year to year and can be extremely high under favorable conditions. Many species migrate into coastal lagoons to feed, reproduce, and use the protected waters as a nursery and refuge. Because of high primary production in the water column and benthos, both pelagic and detrital food chains are quantitatively important (Barnes 1994; Alongi 1998). These food chains support rich recreational and commercial fisheries.

Coastal lagoons are generally more productive than other aquatic ecosystems in terms of fisheries yields; fish productivity averages ~100 kg ha^{-1} yr^{-1} (Macintosh 1994; Pauly and Yáñez-Arancibia 1994). However, they do not have uniformly high yields. A number are extremely productive and others more moderately so. Because of the variability in physicochemical and biotic attributes of coastal lagoons, as well as heavy human use, lagoon fisheries must be carefully and effectively managed for long-term sustainability.

Coastal lagoons provide ideal habitats for various aquaculture operations worldwide (Macintosh 1994; Alongi 1998). The farming of suitable species of fish (e.g., tilapia, milkfish, sea perch, and snappers), shellfish (e.g., shrimp, clams, cockles, mussels, and oysters), and seaweeds (e.g., *Cymodocea, Gracilaria,* and *Laminaria*) have augmented natural fisheries yields in some coastal lagoons. Nations in the Indo-Pacific and Far East are leaders in the development and application of aquaculture in these systems (Alongi 1998). The high natural primary production in coastal lagoon environments is favorable for culturing filter feeding organisms, notably bivalve mollusks. However, these systems are also highly susceptible to the accumulation of pollutants (organochlorine contaminants, polycyclic aromatic hydrocarbon (PAH) compounds, heavy metals, nutrients, toxins, and pathogens) due in large part to their protracted water residence times, which can be detrimental to aquaculture operations because pollutants accumulating in the organisms can pose a serious human health risk (Macintosh 1994; Kennish 1998).

1.5 ECOLOGIC AND ECONOMIC VALUE

Coastal lagoons, as noted above, have exceptional ecologic, recreational, and commercial value. They provide diverse habitats (e.g., open waters, submerged aquatic vegetation, unvegetated bottom sediments, tidal flats and creeks, and fringing wetlands) that serve as nursery, feeding, and refuge areas for numerous estuarine, marine, and terrestrial organisms. Many marine species of recreational and commercial importance spend at least a portion of their life cycles in lagoonal and adjoining coastal wetland habitats. Aside from the value of their fisheries, coastal lagoons are used by humans for aquaculture, electric power generation, biotechnology, transportation, and shipping. Together, these industries inject billions of dollars into the economies of coastal regions worldwide.

Within the land-river-coastal-shelf continuum, coastal lagoons serve a number of vital physical and chemical functions involving the trapping and transformation of nutrients and wastes, filtering of contaminants, and biogeochemical cycling of substances. They therefore can strongly influence the environmental quality of coastal waters. Coastal lagoons also protect coastal watershed areas, buffering the infrastructure from the damaging effects of storms, floods, and erosion.

1.6 HUMAN IMPACTS

Coastal lagoons are highly susceptible to human activities, and many now rank among the most heavily impacted aquatic ecosystems on Earth. They are affected by natural habitat alteration both in adjoining coastal watersheds and in the water bodies themselves. Most of the anthropogenic stressors can be linked to rapid population growth and overdevelopment of the coastal zone (cf. Vitousek et al. 1997). Although anthropogenic stressors on coastal lagoons have received the greatest attention, some natural stochastic events (e.g., major storms, upwelling, severe winds, and coastal flooding) cause environmental perturbations that can have even more profound consequences over both short and longer periods of time (Paerl et al. 2009). However, such natural events often occur less frequently than many anthropogenic stressors which typically result in insidious impacts (Paerl et al. 2006a).

Lagoonal water quality problems, contamination, and habitat alteration generally accompany coastal watershed development. Altered watershed land use/land cover, point and nonpoint source pollution inputs, atmospheric deposition, and groundwater contaminant inflows are the main sources of degraded water quality in coastal lagoons (Stanhope et al. 2009). Oil spills, sanitation tank releases from boats, dredging, and other activities on the water bodies themselves also adversely affect water quality of these systems. For some developing countries, the environmental problems in coastal lagoons can be more overt because of unrestricted disposal of untreated or poorly treated sewage and other wastes. Aquaculture operations have historically degraded estuarine water quality in many regions of the world.

As population growth escalates in the coastal zone, developing infrastructure (e.g., buildings, roadways, bridges, sewer and gas lines, water, and electricity) increases the impervious surface area that facilitates nonpoint source pollution runoff to receiving waters. Expanding human settlements, manifested in particular as suburban sprawl, place greater demands on estuarine and coastal resources. Anthropogenic activities associated with coastal development are conspicuous in the lagoonal basins as well (e.g., harbor and marina development, recreational and commercial fishing, aquaculture and mariculture, dredging and dredged material disposal, and channel and inlet stabilization), along lagoonal shorelines (e.g., docks, boat ramps, bulkheads, revetments, lagoons, and houses), and adjoining watersheds (e.g., domestic and industrial construction, agriculture and silviculture, marsh diking and ditching, dams and reservoirs, channelization and impoundments). Habitat destruction and fragmentation is a primary artifact of coastal development surrounding lagoonal systems.

Kennish (2002) identified 10 principal anthropogenic impacts on estuaries, including coastal lagoons, that pose a threat to their ecological integrity and viability. He later modified this list, dividing it into the more serious Tier I stressors and the less threatening Tier II stressors (Table 1.1; Kennish et al. 2008). The Tier I stressors are most serious because they have the capacity to alter the structure, function, and controls of estuarine systems. A major task, however, is to first accurately document the ecological condition and anthropogenic stresses occurring in a lagoon. Such an effort requires substantial financial and scientific resources and is usually labor intensive. As a result, a paucity of long-term, integrated, ecosystem-level studies exists in most coastal lagoons.

Anthropogenic stressors can be categorized by whether they degrade habitat, compromise water quality, or alter biotic communities. Stressors that degrade habitat are mainly physical factors (e.g., dredging, shoreline modification, and wetland reclamation). Those impacting water quality are primarily chemical and biological in nature (e.g., nutrient enrichment, organic carbon loading, pathogens, heavy metals, and other chemical contaminants). Biotic stressors include significant changes in biological components caused by human activities (e.g., overfishing and introduced/invasive species).

Urbanized coastal lagoons in the northern hemisphere have historically been the most heavily impacted systems. However, coastal lagoons in developing countries of Asia, Africa, and South America may emerge in the twenty-first century as more problematic because these countries either lack government regulatory controls or have less stringent controls than most developed countries. Efforts to mitigate or remediate these impacts will also be hindered by the application of less effective technological innovations, particularly in the poorest, undeveloped tropical countries.

Many coastal lagoons, especially those near urbanized regions, receive a wide array of contaminants from watersheds and airsheds. Particularly noteworthy are particle-reactive contaminants that tend to accumulate in bottom sediments and therefore are likely to have a long residence time. Included here are PAHs, halogenated hydrocarbons, and trace metals. All can degrade sediment quality and, when bioavailable, significantly impact benthic organisms and eventually upper trophic levels. Pollution problems chronicled in coastal lagoons include nutrient over-enrichment, oil spills, sewage and other oxygen-demanding wastes, sedimentation and high turbidity, heavy metals, and pathogens. Radioactive substances, volatile organic compounds, and litter, while potentially problematic, usually are less threatening (Kennish 1998).

The following management strategies are often proposed to improve the environmental state of estuaries and coastal lagoons:

- Increase monitoring and research to identify impacts and to develop remedial actions that restore natural environmental conditions.
- Pursue open space acquisition and smart development in coastal watersheds to protect habitat.
- Implement effective restoration efforts to revitalize altered habitat.
- Formulate tighter regulations to limit nutrient and chemical contaminant inputs to estuaries and lagoons.
- Establish more estuarine reserves to minimize anthropogenic impacts, and to provide protected areas for basic and applied research.
- Improve the management of estuarine fisheries to preclude excessive use or overharvest of commercially and recreationally important finfish and shellfish populations.
- Monitor, assess, and remediate water quality and habitat degradation associated with aquaculture operations.
- Increase the interactions between scientists, resource managers, and policy makers to ensure informed decisions regarding estuaries and coastal lagoons.
- Develop education and outreach programs that inform students and the general public of the importance of maintaining healthy and viable estuarine and lagoonal environments.

Best management practices must be instituted as part of an integrated coastal zone management strategy to protect coastal land and marine resources and mollify anthropogenic impacts threatening the conservation and sustainability of coastal lagoon habitats. It is necessary to formulate coastal management strategies that address human-induced stressors capable of acting synergistically to adversely impact habitats and biotic communities. These strategies must consider and incorporate changing hydrologic conditions and drivers, including flooding and droughts resulting from large storm events, warming, and atmospheric and ocean circulation features. Such drivers are of fundamental importance in transporting, delivering, and cycling of pollutants in coastal ecosystems, and they are changing in frequency and magnitude as our climate undergoes changes (Webster et al. 2005; Emmanuel et al. 2008).

Eutrophication poses perhaps the greatest long-term threat to the ecological integrity of coastal lagoons. Lagoonal systems characterized by restricted water circulation, poor flushing, shallow depths, and heavily populated watersheds are particularly susceptible to nutrient enrichment impacts (Boynton et al. 1996; Kennish 2002). The Barnegat Bay–Little Egg Harbor Estuary and similar embayments such as the Great South Bay (New York), Rehoboth Bay (Delaware), Newport Bay and Sinepuxent Bay (Maryland), and Chincoteague Bay (Maryland and Virginia) provide examples. Even much larger lagoonal systems (e.g., Pamlico Sound) have experienced serious nutrient loading and resultant eutrophication problems (Piehler et al. 2004; Paerl et al. 2005, 2006a,b; Paerl et al., Chapter 2, this volume).

1.7 NATURAL STRESSORS

Coastal lagoons and shallow estuaries, in general, are highly vulnerable to climate change, sea-level rise, droughts, hurricanes and other major storms, and storm surge effects. These factors can significantly increase coastal erosion, flooding, the loss of wetlands, and damage to coastal property and infrastructure. Salt water intrusion into aquifers and potable water supplies are an additional concern. Ecological responses could include acute changes in the abundance, distribution, and diversity of plant and animal populations. Biotic communities in coastal lagoons are faced with adapting to considerable natural stress associated with extreme variability in physicochemical conditions such as daily, seasonal, and annual fluxes in temperature, salinity, organic matter inputs, wave and current action, turbidity, freshwater inflow, winter ice cover, and habitat changes (e.g., loss of seagrass beds). To prepare for these changes, some coastal states in the United States are beginning to incorporate climate change and sea-level rise into strategic land and habitat planning (e.g., Berke and Manta-Conroy 2000; Heinz Center 2002; Brody et al. 2004; Norton 2005).

Since the mid-1990s, the U.S. mid-Atlantic and Gulf Coast regions have witnessed a sudden rise in number of hurricanes and tropical storms. During this time frame, Florida has experienced six major hurricanes, while coastal North Carolina has been impacted by eight hurricanes and six tropical storms; this elevated frequency is forecast to continue for the foreseeable future (Webster et al. 2005; Knutson et al. 2008). These storms have had a range of hydrologic and nutrient loading impacts in the largest lagoonal ecosystem in the United States, the Pamlico Sound (Paerl et al. 2006a,b, 2009; see Paerl et al., Chapter 2, this volume). Different amounts of rainfall from these hurricanes led to variable freshwater, nutrient, and organic matter inputs to Pamlico Sound. This variability differentially affected the lagoon's physicochemical properties (salinity, residence time, transparency, stratification, hypoxia), and phytoplankton primary production and community composition. Floodwaters from the two largest hurricanes, Fran (1996) and Floyd (1999), exerted multi-month to multi-annual effects on hydrology, nutrient loading, productivity, biotic composition, and habitat condition (Paerl et al. 2006b, 2009). In contrast, relatively low rainfall coastal hurricanes, Isabel (2003) and Ophelia (2005) caused strong vertical mixing and storm surges, but relatively minor hydrologic, nutrient, and biotic impacts (Paerl et al. 2006a,b, 2009). Both hydrology and wind forcing are important drivers of change and must be clearly integrated with nutrient, sediment, and other pollutant loadings when assessing, modeling, and managing short- and long-term ecological impacts on coastal lagoons influenced by climate change.

1.8 PLAN OF THIS VOLUME

This volume provides detailed descriptions of the physicochemical and biotic characteristics of major coastal lagoonal systems in different regions of the world. It addresses environmental factors, forcing features, and stressors affecting hydrologic, biogeochemical, and trophic properties of coastal lagoon systems. Lastly, it discusses tools and approaches (i.e., indicators, trend analysis, modeling) for assessing ecological change, evaluating and managing these systems in the context of anthropogenically and climatically induced changes impacting our coastal zones.

Paerl et al. (Chapter 2, Assessing the Response of the Pamlico Sound, North Carolina, USA to Human and Climatic Disturbances: Management Implications) investigate the increasing influence of human activities on the largest coastal lagoon system in the United States. During the past few decades, this system has experienced accelerated nutrient enrichment, accumulation of other pollutants, sedimentation, and habitat disturbance from rapidly expanding agriculture, industry, and urbanization. In addition, Pamlico Sound has, since the mid-1990s, been affected by a dramatic increase in the frequency of tropical storms and hurricanes. Depending on their intensity, direction, and rainfall content (and resultant freshwater runoff), these storms have differential impacts on phytoplankton productivity and community composition, and nutrient and oxygen cycling. Natural and human forcing factors interactively impact water and habitat quality in this ecosystem. Long-term water quality and fisheries management strategies should take these interactions into consideration, as there is a concern that such impacts may escalate in future years.

Anderson et al. (Chapter 3, Sources and Fates of Nitrogen in Virginia Coastal Bays) report on the sources, sinks, and fates of nitrogen in Hog Island Bay, Virginia, a coastal lagoon located along a clearly defined eutrophication gradient on the Delmarva Peninsula (United States). Coastal bays of the Delmarva Peninsula support a wide diversity of benthic autotrophs, which are thought to play a critical role in removal, retention, and transformation of nitrogen. Periodic blooms of ephemeral macroalgae are a symptom of the mesotrophic condition of this system. Benthic and pelagic autotrophy in the Virginia coastal bays is supported primarily by autochthonous nitrogen sources (e.g., benthic remineralization and nitrogen fixation). On an annual basis benthic microalgae accounted for most of the nitrogen demand, with phytoplankton, marsh grass, macroalgae, and denitrification taking up the remainder. Carbon and nitrogen taken up by the benthos were rapidly shuttled between benthic microalgal and bacterial pools, contributing to retention in the sediments; however, macroalgae were shown to interfere with uptake, either by reducing light or by competing with benthic microalgae. The authors hypothesize that increased eutrophication will provide a positive feedback that will reduce the potential of the benthos to take up and retain nitrogen.

Autochthonous nitrogen sources contribute more to total nitrogen input to Hog Island Bay than do allochthonous sources. Benthic nitrogen cycling processes, including nitrogen fixation, nitrogen remineralization, nitrification, denitrification, anammox, and dissimilatory nitrate reduction to ammonium, appear to play an important role in determining nitrogen availability for supporting primary production in both pelagic and benthic habitats and, thus, the potential for eutrophication of the system.

Christian et al. (Chapter 4, Ecosystem Health Indexed through Networks of Nitrogen Cycling) employ ecological network analysis (ENA) to evaluate ecosystem status of nitrogen cycling in two well-studied coastal lagoons, Hog Island Bay (United States) and Sacca di Goro (Italy). The sites represent a large range of nutrient loading and subsequent trophic conditions ranging from mesotrophic to hypereutrophic and dystrophic. Sacca di Goro has much higher nutrient loading and clear signs of cultural eutrophication, while Hog Island Bay is minimally impacted by nutrients. ENA results reflect the trophic status of the two coastal lagoons. The results are consistent with the hypothesis that eutrophication disrupts organization, and that less eutrophic systems may be expected to provide more efficient use of nutrients. ENA is shown to be of value in assessing the status of these lagoonal ecosystems.

Glibert et al. (Chapter 5, Blooms in Lagoons: Different from Those of River-Dominated Estuaries) show that phytoplankton blooms differ in coastal lagoons and river-dominated estuaries. While nutrient loading and availability are necessary for phytoplankton bloom occurrence in coastal lagoons, algal production and accumulation are determined not only by the total quality of nutrient loading, but also by the nutrient composition, the seasonal timing of nutrient inputs, whether nutrient fluxes are episodic or sustained, the physiological status of the primary producers at the time of nutrient delivery, and other physical factors that influence nutrient retention and algal physiology. Lagoonal phytoplankton blooms are often dominated by picoplankton and may be sustained for long periods of time. Blooms in river-dominated estuaries, by contrast, tend to be highly seasonal and dominated by larger-sized phytoplankton (i.e., >10 μm), often diatoms. Furthermore, the total production of coastal lagoons is balanced between that of the water column and that of the benthos. Multiple interactive factors determine the amount and fraction of water column and benthic production.

Montagna et al. (Chapter 6, Relationship between Macroinfaunal Diversity and Community Stability, and a Disturbance Caused by a Persistent Brown Tide Bloom in Laguna Madre, Texas) track persistent brown tide (*Aureoumbra lagunensis*) effects on the macroinfauna in Baffin Bay and Laguna Madre, Texas (United States) over a 7-year period (1990 to 1997). They chronicle the impacts of this long-term hazardous algal bloom (HAB) on the abundance, biomass, and diversity of the benthic macroinfaunal community in Laguna Madre, a coastal lagoon, and Baffin Bay, a secondary bay. Results show that increasing concentrations of brown tide during the study period correlated with decreases in abundance, biomass, and diversity of the macroinfauna.

Fisher et al. (Chapter 7, The Choptank Basin in Transition: Intensifying Agriculture, Slow Urbanization, and Estuarine Eutrophication) summarize their research on nutrient enrichment of the Choptank Estuary, a system that serves as a microcosm of eutrophication processes in Chesapeake Bay. The focus of this work is to delineate how anthropogenic activities on land (primarily fertilizer applications, animal waste production, and human waste disposal) influence nutrient export to the estuary, and how estuarine water quality responds to these nutrient inputs. Although most of the information is taken from the Choptank, the authors also include parallel measurements in other basins of Delmarva for purposes of comparison, notably with the Maryland coastal lagoons.

Kennish et al. (Chapter 8, Seagrass Decline in New Jersey Coastal Lagoons: A Response to Increasing Eutrophication) discuss the effects of increasing nutrient enrichment on seagrass and other biotic indicators (e.g., hazardous algal blooms) in the Barnegat Bay–Little Egg Harbor Estuary (United States). Results of a comprehensive 3-year (2004 to 2006) investigation of seagrass demographics in the Barnegat Bay–Little Egg Harbor Estuary, a highly eutrophic lagoonal system along the central New Jersey coastline, reveal a dramatic decline in plant biomass, density, blade length, and percent cover of bay bottom over the study period. Significant shifts are evident in the ecosystem services provided by the beds associated with increasing eutrophic conditions, such as the loss of critical habitat for shellfish populations and many other estuarine organisms. Declining seagrass abundance is correlated with the occurrence of extensive phytoplankton and benthic macroalgal blooms detrimental to seagrass beds in the estuary. Biotic responses to nutrient enrichment in this system are similar to those observed in other mid-Atlantic estuaries in the United States.

Fox et al. (Chapter 9, Controls Acting on Benthic Macrophyte Communities in a Temperate and a Tropical Estuary) detail the relative importance of bottom-up and top-down controls that act on seagrasses and macroalgae in Waquoit Bay, Massachusetts (United States), and Jobos Bay, Puerto Rico, two shallow enclosed coastal bay systems. In Waquoit Bay, bottom-up controls by nutrients structure the benthic macrophyte communities. Where nutrients stimulate algal growth, even high numbers of grazers cannot control plant biomass. In Jobos Bay, nutrients and grazers are both important in controlling benthic macrophytes, and complex interactions exist between nutrients and grazers. In this system, the relative influence of top-down and bottom-up controls and the response to changes in these controls differ between producers. A complex interplay exists between bottom-up and top-down forces in structuring benthic producer communities in these enclosed water bodies. Nutrients, light, producers, and consumers interact continuously and simultaneously, and these connections result in myriad combinations, many of which can control growth and biomass on different spatial and temporal scales.

Fong and Kennison (Chapter 10, Phase Shifts, Alternative Stable States, and the Status of Southern California Lagoons) provide an overview of the ecological status of southern California coastal lagoons. These lagoonal systems have a long history of physical and hydrological modification that threatens their structure and function. Empirical, experimental, and theoretical evidence indicates that bottom-up forcing has shifted the composition of primary producer communities through at least two states along a nutrient-loading gradient in the southern California lagoons. Results of large-scale experiments demonstrate that, as nitrogen loadings from highly developed watersheds have increased to the lagoons, producer communities dominated by low abundances of phytoplankton and microphytobenthos are being replaced by blooms of opportunistic green macroalgae. Shifts to macroalgal-dominated systems are being documented in coastal lagoons worldwide due to increasing nutrient enrichment.

Oczkowski and Nixon (Chapter 11, Lagoons of the Nile Delta) chronicle changes that have occurred in coastal lagoons of Egypt's Nile Delta (i.e., Burullus, Edku, Manzalah, and Maryut lagoons). An array of human stressors affects these shallow water bodies, including large amounts of agricultural drainage and nutrients, direct sewage inputs, fish pond aquaculture, and urbanization. Despite these factors, the lagoons remain highly productive, and these systems now produce about half of Egypt's national fish catch. Sewage discharges are problematic, particularly in the most altered and industrialized lagoon, Maryut, where the fishery collapsed 30 years ago. Nutrient enrichment is also very high in this lagoon. Nutrient loads from agricultural drains and nearby cities and towns to the delta and coastal lagoons are increasing, and eutrophication may be an escalating problem in future years.

Mutchler et al. (Chapter 12, Origins and Fate of Inorganic Nitrogen from Land to Coastal Ocean on the Yucatan Peninsula, Mexico) assess anthropogenic influences on nutrient loading and dynamics in coastal lagoons and bays on the northeastern Yucatan coast. They note that the nutrient levels in these water bodies are coupled to the amount of freshwater input from the surrounding coastal watershed. Findings from this study suggest that nutrient loading in the region may be increasing in response to accelerating development and tourism. The resulting altered coastal nutrient dynamics pose a potential threat to sensitive seagrass beds and coral reefs.

Herrera-Silveira and Morales-Ojeda (Chapter 13, Subtropical Karstic Coastal Lagoons Assessment: SE Mexico The Yucatan Peninsula Case) evaluate the ecosystem condition of 10 coastal lagoons of the Yucatan Peninsula, analyzing changes in water quality indicators, submerged aquatic vegetation, HABs, and other parameters. Based on this work, the overall ecological health of the coastal lagoons is defined as generally good, although two of the lagoons (i.e., the Chelem and Bojorquez lagoons) appear to be at risk of cultural eutrophication as evidenced by the occurrence of HABs, loss of seagrass cover, and changes in mangrove vegetation condition due to increases in land use alteration and wastewater inputs from the watershed, as well as hydrological modifications. Recommendations are given to preserve ecosystem structure and function (productivity, nutrients dynamics, variability), as well as environmental services (water quality, fisheries, recreation) in these impacted coastal lagoons.

Barbosa (Chapter 14, Seasonal and Interannual Variability of Planktonic Microbes in a Mesotidal Coastal Lagoon [Ria Formosa, SE Portugal]: Impact of Climatic Changes and Local-Human Influences) describes the seasonal and annual variability of planktonic microbes in the Ria Formosa coastal lagoon in southern Portugal in response to human and climatic forcing factors. The effects of increased anthropogenic pressure, both on seasonal and interannual time scales, are evident in areas in close proximity to major urban centers. These effects include decreased oxygen saturation, higher and sustained availability of labile organic substrates during summer, and long-term increases in the concentrations of dissolved inorganic nutrients, particularly phosphate and ammonium, and phytoplankton biomass during summer. Climate variability strongly impacts planktonic microbe flux on seasonal and interannual time scales.

Yin et al. (Chapter 15, Eutrophication Processes in Subtropical Semienclosed Embayments) formulate a comparative analysis of eutrophication processes in contrasting environments in Hong Kong waters. They demonstrate how responses to nutrient enrichment, such as the magnitude of phytoplankton blooms and the formation of hypoxia, differ in three subtropical semienclosed bays (i.e., Deep Bay, Port Shelter, and Tolo Harbour) with different geographical settings and different physical processes, nutrient inputs, and nutrient ratios. In addition to nutrient enrichment, climate change appears to affect the three bays in terms of reduced dissolved oxygen levels in bottom waters.

Philippart and Epping (Chapter 16, The Wadden Sea: A Coastal Ecosystem under Continuous Change) document the time series of abiotic, biotic, and functional changes recorded over many years in the Wadden Sea, a large coastal lagoon bounded by the Dutch, German, and Danish coastlines and the North Sea. Heavy metals, PCBs, pesticides, nutrients (nitrogen and phosphorus), and other pollutants are known to have accumulated in extensive areas of the lagoon during different periods from the 1930s to the 1970s. Nutrient enrichment and eutrophication escalated in the 1970s and 1980s, and the system has yet to recover completely from these factors. Increasing watershed development can be coupled to adverse aquatic impacts recorded in the water body. Climate change and sea-level rise are likely to be major drivers of long-term ecosystem change in the system.

Odebrecht et al. (Chapter 17, The Patos Lagoon Estuary: Biotic Responses to Natural and Anthropogenic Impacts in the Last Decades [1979–2008]) reveal that natural rather than human forcing factors have a greater effect on biotic responses in the Patos Lagoon Estuary (Brazil), a river-dominated system. For example, in the Patos Lagoon the species composition, abundance, and biomass of microalgae, macrobenthic fauna and flora, and fish assemblages respond most profoundly to environmental changes associated with natural events, notably El Niño Southern Oscillation (ENSO), rather than to widespread and long-term local impacts of pollution, dredging, and fisheries in the estuary. It appears that riverine water exchange between the Patos Lagoon system and the Atlantic Ocean has an overriding influence on ecological processes in the estuary, and they reduce or mask the effect of human perturbations.

Eyre and Maher (Chapter 18, Structure and Function of Warm Temperate East Australian Coastal Lagoons: Implications for Natural and Anthropogenic Changes) recount the different stages of maturity (infilling) in three coastal lagoon systems (Wallis Lake, Camden Haven, and Hastings River) in Australia. They also explain that structural changes in these coastal systems through time significantly affect their function as demonstrated by measures of the ratio of benthic to pelagic production and net ecosystem metabolism as a percentage of primary production and wild fisheries production (catch/fisher), which show nonlinear changes with increasing maturity. The higher retention capacity of immature coastal lagoons combined with their larger areas of seagrass also renders them more susceptible to pollutant loads and climate change impacts such as sea-level rise.

Solidoro et al. (Chapter 19, Response of the Venice Lagoon Ecosystem to Natural and Anthropogenic Pressures over the Last 50 Years) review the array of human forcing factors affecting the largest coastal lagoon in Italy. Principal forcing factors acting on the Venice Lagoon include climate change, eustatic sea-level rise, land reclamation, subsidence, channel dredging, sediment erosion, aquaculture activities, fishing, altered freshwater inputs, nutrient loading, and chemical contaminant inputs. The authors analyze the principal drivers of change and the recent biotic and habitat responses observed in this lagoonal system.

Montagna and Li (Chapter 20, Effect of Freshwater Inflow on Nutrient Loading and Macrobenthos Secondary Production in Texas Lagoons) present a bioenergetic model that relates macrobenthic productivity to salinity differences in Texas coastal lagoons. They develop this model for the Lavaca–Colorado, Guadalupe, Nueces, and Laguna Madre systems that lie in a climatic gradient with decreasing rainfall and concordant decreasing freshwater inflow. Their hypothesis is that climatic variability and inflow differences among the systems alter nutrient loading, which in turn affects benthic infaunal communities and maintains secondary production. A long-term dataset on macrobenthic biomass, which serves as an indicator of productivity, is used to calibrate the model.

ACKNOWLEDGMENTS

We would like to thank all of the contributors to this volume on coastal lagoons. This is Contribution Number 2009-47 of the Institute of Marine and Coastal Sciences, Rutgers University.

REFERENCES

Adam, P., M. D. Bertness, A. J. Davy, and J. B. Zedler. 2008. Saltmarsh. Pp. 157–171, In: N. V. C. Polunin (ed.), *Aquatic ecosystems: Trends and global prospects*. Cambridge: Cambridge University Press.

Adolf, J. E., C. L. Yeager, M. E. Mallonee, W. D. Miller, and L. W. Harding, Jr. 2006. Environmental forcing of phytoplankton floral composition, biomass, and primary productivity in Chesapeake Bay, USA. *Estuar. Coastal Shelf Sci.* 67: 108–122.

Alongi, D. M. 1998. *Coastal ecosystem processes*. Boca Raton, Florida: CRC Press.

Alvarez-Borrego, S. 1994. Secondary productivity in coastal lagoons. Pp. 287–309, In: I. E. Gonenc and J. P. Wolflin (eds.), *Coastal lagoons: Ecosystem processes and modeling for sustainable use and development*. Boca Raton, Florida: CRC Press.

Barnes, R. S. K. 1980. *Coastal lagoons*. Cambridge: Cambridge University Press.

Barnes, R. S. K. 1994. Macrofaunal community structure and life histories in coastal lagoons. Pp. 311–362, In: I. E. Gonenc and J. P. Wolflin (eds.), *Coastal lagoons: Ecosystem processes and modeling for sustainable use and development*. Boca Raton, Florida: CRC Press.

Berke, P.R., and M. Manta-Conroy. 2000. Are we planning for sustainable development? An evaluation of 30 comprehensive plans. *J. Am. Plan. Assoc.* 66: 21–33.

Bird, E. C. F. 1982. Changes on barriers and spits enclosing coastal lagoons. *Oceanologica Acta* 45–53.

Bird, E. C. F. 1994. Physical setting and geomorphology of coastal lagoons. Pp. 9–39, In: B. Kjerfve (ed.), *Coastal lagoon processes*. Amsterdam: Elsevier.

Boynton, W. R., J. D. Hagy, L. Murray, C. Stokes, and W. M. Kemp. 1996. A comparative analysis of eutrophication patterns in a temperate coastal lagoon. *Estuaries* 19: 408–421.

Bricker, S. B., B. Longstaff, W. Dennison, A. Jones, K. Boicourt, C. Wicks, and J. Woerner. 2007. Effects of nutrient enrichment in the nation's estuaries: A decade of change. NOAA, National Ocian Service, Special Projects Office and National Centers for Coastal Ocean Science, Silver Spring, Maryland.

Brody, S., W. Highfield, and V. Carrasco. 2004. Measuring the collective planning capabilities of local jurisdictions to manage ecological systems in southern Florida, *Landsc. Urban Plan.* 69: 32–50.

Colombo, G. 1977. Lagoons. Pp. 63–81, In: R. S. K. Barnes (ed.), *The coastline*. London: John Wiley & Sons.

Dodd, R. S. and J. E. Ong. 2008. Future of mangrove systems to 2025. Pp. 172–187, In: N. V. C. Polunin (ed.), *Aquatic ecosystems: Trends and global prospects*. Cambridge: Cambridge University Press.

Eisma, D. 1998. *Intertidal deposits: River mouths, tidal flats, and coastal lagoons*. Boca Raton, Florida: CRC Press.

Emmanuel, K. R., J. Sundararajan, and J. Williams. 2008. Hurricanes and global warming. *Bull. Amer. Meteor. Soc.*, March 2008, pp. 347–367.

Gamito, S., J. Gilabert, C. Marcos Diego, and A. Perez-Ruzafa. 2004. Effects of changing environmental conditions on lagoon ecology. Pp. 193–229, In: I. E. Gonenc and J. P. Wolflin (eds.), *Coastal lagoons: Ecosystem processes and modeling for sustainable use and development*. Boca Raton, Florida: CRC Press.

Gonenc, I. E. and J. P. Wolflin (eds.). 2004. *Coastal lagoons: Ecosystem processes and modeling for sustainable use and development.* Boca Raton, Florida: CRC Press.

Heinz Center, 2002. *Human links to coastal disasters*. Washington, D.C.: The H. John Heinz III Center for Science, Economics and the Environment.

Kennish, M. J. 1998. *Practical handbook of estuarine and marine pollution*. Boca Raton, Florida: CRC Press.

Kennish, M. J. 2002. Environmental threats and environmental future of estuaries. *Environ. Conserv.* 29: 78–107.

Kennish, M. J., R. J. Livingston, D. Raffaelli, and K. Reise. 2008. Environmental future of estuaries. Pp. 188–208, In: Polunin, N. (ed.), *Aquatic ecosystems: Trends and global prospects*. Cambridge: Cambridge University Press.

Kjerfve, B. 1986. Comparative oceanography of coastal lagoons. Pp. 63–81, In: D. A. Wolfe (ed.), *Estuarine variability*. New York: Academic Press.

Kjerfve, B. 1994. Coastal lagoons. Pp. 1–8, In: B. Kjerfve (ed.), *Coastal lagoon processes*. Amsterdam: Elsevier.

Knoppers, B. 1994. Aquatic primary production in coastal lagoons. Pp. 243–286. In: B. Kjerfve (Ed.), *Coastal Lagoon Processes*. Amsterdam: Elsevier.

Knutson, T. R., J. J. Sirutis, S. T. Garner, G. A. Vecchi, and I. M. Held. 2008. Simulated reduction in Atlantic hurricane frequency under twenty-first-century warming conditions. *Nature Geoscience* 1: 359–364, DOI: 1038/nge0202.

Macintosh, D. J. 1994. Aquaculture in coastal lagoons, Pp. 401–442, In: B. Kjerfve (ed.), *Coastal lagoon processes*. Amsterdam: Elsevier.

Martin, L. and J. M. L. Dominguez. 1994. Geological history of coastal lagoons. Pp. 41–68, In: B. Kjerfve (ed.), *Coastal lagoon processes*. Amsterdam: Elsevier.

McGlathery, K. J., K. Sundback, and I. C. Anderson. 2007. Eutrophication in shallow coastal bays and lagoons: The role of plants in the coastal filter. *Mar. Ecol. Prog. Ser.* 348: 1–18.

Nichols, M. M. and J. D. Boon. 1994. Sediment transport processes in coastal lagoons. Pp. 157–219, In: B. Kjerfve (ed.), *Coastal lagoon processes*. Amsterdam: Elsevier.

Nixon, S. W. 1995. Coastal eutrophication: A definition, social causes, and future concerns. *Ophelia* 41: 199–220.

Norton, R. 2005. More and better local planning: State-mandated local planning in coastal North Carolina, *J. Am. Plan. Assoc.* 71: 55–72.

Paerl, H. W., M. F. Piehler, L. M. Valdes et al. 2005. Determining anthropogenic and climatically-induced change in aquatic ecosystems using microbial indicators: An integrative approach. *Verh. Internat. Verin. Limnol.* 29: 89–133.

Paerl, H. W., L. M. Valdes, J. E. Adolf, B. M. Peierls, and L. W. Harding, Jr. 2006. Anthropogenic and climatic influences on the eutrophication of large estuarine ecosystems. *Limnol. Oceanogr.* 51: 448–462.

Paerl, H. W., L. M. Valdes, A. R. Joyner, B. L. Peierls, C. P. Buzzelli, M. F. Piehler, S. R. Riggs, R. R. Christian, J. S. Ramus, E. J. Clesceri, L. A. Eby, L. W. Crowder, and R. A. Luettich. 2006. Ecological response to hurricane events in the Pamlico Sound System, NC and implications for assessment and management in a regime of increased frequency. *Estuaries Coasts* 29: 1033–1045.

Paerl, H. W., K. L. Rossignol, N. S. Hall, B. L. Peierls, and M. S. Wetz. 2009. Phytoplankton community indicators of short- and long-term ecological change in the anthropogenically and climatically impacted Neuse River Estuary, North Carolina, USA. *Estuaries Coasts* DOI: 10.1007/s12237-009-9137-0.

Pauly, D. and A. Yáñez-Arancibia. 1994. Fisheries in coastal lagoons. Pp. 377–399, In: B. Kjerfve (ed.), *Coastal lagoon processes*. Amsterdam: Elsevier.

Phleger, F. B. 1969. Some general features of coastal lagoons. Pp. 5–26, In: A. Ayala-Castaneres (ed.), *Lagunas costeras, un simposio*, Universidad Nacional Autonoma de Mexico, Mexico, DF.

Phleger, F. B. 1981. A review of some general features of coastal lagoons. In: *Coastal lagoon research, past, present, and future*. UNESCO Technical Papers in Marine Science 33: 7–14.

Piehler, M. F., L. J. Twomey, N. S. Hall, and H. W. Paerl. 2004. Impacts of inorganic nutrient enrichment on the phytoplankton community structure and function in Pamlico Sound, NC, USA. *Est. Coastal Shelf Sci.* 61: 197–209.

Stanhope, J. W., I. C. Anderson, and W. G. Reay. 2009. Base flow nutrient discharges from lower Delmarva Peninsula watersheds of Virginia, USA. *J. Environ. Qual.* 38: 2070–2083.

Thornton, J. A., A. J. McComb, and S.-O. Ryding. 1995. The role of sediments. Pp. 205–223, In: A. J. McComb (ed.), *Eutrophic shallow estuaries and lagoons*. Boca Raton, Florida: CRC Press.

Valdes-Weaver, L. M., M. F. Piehler, J. L. Pinkney, K. E. Howe, K. Rosignol, and H. W. Paerl. 2006. Long-term temporal and spatial trends in phytoplankton biomass and class-level taxonomic composition in the hydrological variable Neuse-Pamlico estuarine continuum, NC, USA. *Limnol Oceanogr*. 51: 1410–1420.

Vitousek, P. M., H. A. Mooney, J. Lubchenko, and J. M. Mellilo. 1997. Human domination of Earth's ecosystem. *Science* 277: 494–499.

Webster, P. J., G. J. Holland, J. A. Curry, and H. R. Chang. 2005. Changes in tropical cyclone number, duration, and intensity in a warming environment. *Science* 309: 1844–1846.

2 Assessing the Response of the Pamlico Sound, North Carolina, USA to Human and Climatic Disturbances

Management Implications

Hans W. Paerl, Robert R. Christian, J. D. Bales, B. L. Peierls, N. S. Hall, A. R. Joyner, and S. R. Riggs*

CONTENTS

Abstract ... 17
2.1 Introduction ... 18
2.2 Putting Things in Perspective: Biogeochemical, Geological, and Climatological Time Scales ... 21
2.3 Historic Perspective on Human and Climatic Impacts ... 22
2.3.1 Human Impacts ... 22
2.3.2 Climatic Impacts ... 22
2.3.2.1 Hurricanes ... 23
2.3.2.2 Smaller Storm Events (<74 mph) ... 33
2.4 What Have We Learned? ... 34
2.4.1 Biogeochemical and Trophic Considerations ... 34
2.4.2 Socioeconomic Aspects and Considerations ... 36
2.5 An Integrated Assessment Program in Response to Large-Scale Storm Events ... 36
Acknowledgments ... 38
References ... 38

ABSTRACT

The Pamlico Sound (PS) with its sub-estuaries is the largest lagoonal ecosystem in the United States. It exhibits periodically strong salinity stratification and an average freshwater residence time of 1 year for the sound proper. This relatively long residence time promotes effective use and cycling of nutrients, allowing the system to support high rates of primary and secondary production, and serve as a vitally important fisheries nursery. This hydrologic characteristic also makes the system highly sensitive to nutrient over-enrichment and eutrophication. The PS is experiencing ecological change in response to increasing human activity and climatic perturbations. Human impacts include a rise in nutrient, sediment, and other pollutant loads that accompany urbanization and agricultural and industrial growth in its watersheds and airsheds. Since the mid-1990s, the PS

* Corresponding author. Email: hpaerl@email.unc.edu

has witnessed a sudden rise in tropical storm and hurricane impacts, with eight hurricanes and four tropical storms having made landfall in the PS watershed during the 1996 to 2007 period. Each of these storms had unique hydrologic, nutrient, and other pollutant loading effects. In addition, since the early 2000s, the region has experienced record droughts, which are continuing. Variable freshwater discharges from storms and droughts have caused large oscillations in nutrient enrichment, reflected ultimately in differential phytoplankton production, biomass, and community compositional responses. Floodwaters from the two wettest hurricanes, Fran (1996) and Floyd (1999), and from Tropical Storm Ernesto (2006) exerted long-term (months) effects on hydrology, nutrient loads, and algal production. Windy but relatively dry hurricanes, like Irene (1999) and Isabel (2003), caused strong vertical mixing, storm surges, but relatively minor changes in river flow, flushing, and nutrient loads. These contrasting effects are accompanied by biogeochemical (hypoxia, nutrient cycling) and habitat alterations, and associated food web disturbances. Each storm type influenced algal growth and compositional dynamics; however, their respective ecological impacts differed substantially. Changes in hydrologic and wind forcing resulting from changes in frequency and intensity of storms and droughts strongly influence water and habitat quality. These changes must be integrated with nutrient loading/dilution effects when assessing and predicting ecological responses to nutrient and hydrologic variability on this and other large lagoonal ecosystems.

Key Words: lagoonal estuaries, hurricanes, watersheds, nutrients, water quality management, Pamlico Sound, climate change

2.1 INTRODUCTION

Because lagoons are shallow, semienclosed water bodies, separated from the oceans and seas by sandbars, barrier islands, and reefs, they experience restricted water exchange and relatively long water residence time. North Carolina's Pamlico Sound (PS), the largest lagoonal ecosystem in North America, typifies these conditions. Its five major watersheds (Neuse, Tar-Pamlico, Roanoke-Albemarle, Chowan, and Pasquotank) have a combined area of approximately 80,000 km^2 and discharge ~21 km^3/yr to the PS, which has a surface area of 4350 km^2 (Figure 2.1). Together, these

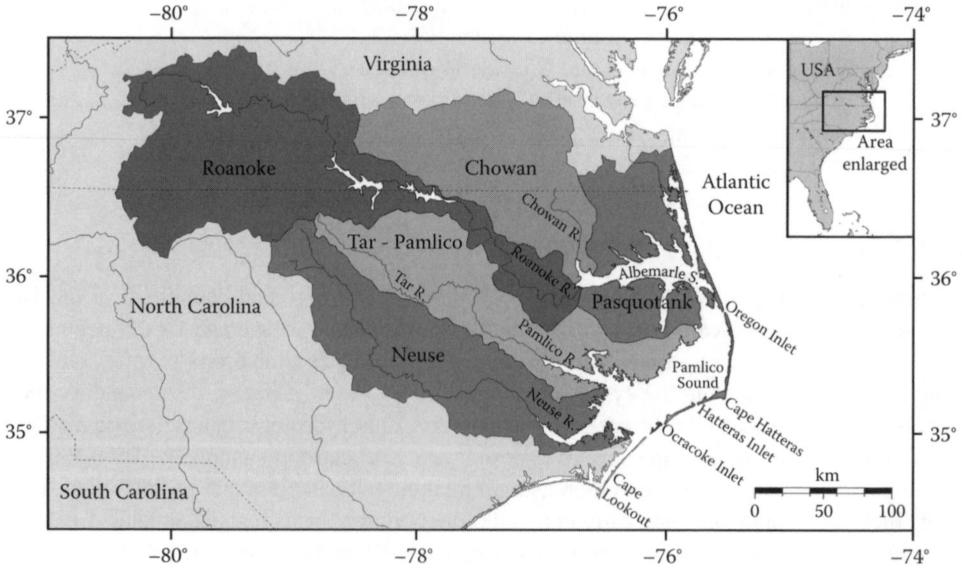

FIGURE 2.1 Map of Pamlico Sound (PS), its major sub-estuaries and their watersheds. Note that PS is the receptacle and processing basin for all discharges from its sub-estuaries.

basins drain about 40% of North Carolina and about 10% of Virginia. Tidal exchange with the coastal Atlantic Ocean and nearby Gulf Stream currently is restricted to three narrow inlets, resulting in an approximately 1-year residence time during average hydrologic years (Pietrafesa et al. 1996). The long residence time provides ample opportunity for resident phytoplankton and vascular plants to assimilate nutrient inputs, resulting in high productivity (per unit nutrient input) and fertility. Moreover, these characteristics allow the PS to serve as a vitally important nursery (Winslow 1889), supporting approximately 80% of mid-Atlantic commercial and recreationally caught finfish and shellfish species (Copeland and Gray 1991).

The PS's efficient use of nutrients and organic matter also makes the system highly sensitive and susceptible to nutrient over-enrichment. The long residence time and low flushing velocities result in the accumulation of allochthonous and autochthonous nutrients and organic matter within the sound. Long residence times and nutrient/organic matter accumulation are conducive to the formation of nuisance algal blooms, hypoxia, anoxia, toxicity, aquatic life diseases, and potentially even mass mortalities of finfish and shellfish, all of which have been previously documented (Paerl et al. 1990, 1995, 2006a; Copeland and Gray 1991; Christian et al. 1991; Rudek et al. 1991; Boyer et al. 1994; Buzzelli et al. 2002).

Like many other estuarine and coastal ecosystems, the PS is under increasing influence of human activities, resulting in nutrient enrichment, sedimentation, accumulation of other pollutants, and habitat disturbance from rapidly expanding agriculture, industry, and urbanization in the watershed and airshed (Stanley 1988; Copeland and Gray 1991; Stow et al. 2001; Burkholder et al. 2006; Paerl et al. 2006a, 2007). Studies of key sub-estuaries going back to the early 1970s have shown a high degree of sensitivity to phosphorus (P) (in freshwater) and nitrogen (N) (in more marine waters) enrichment (Hobbie et al. 1972; Hobbie and Smith 1975; Tedder et al. 1980; Kuenzler et al. 1982; Paerl 1982; Stanley 1983; Christian et al. 1991; Paerl et al. 2006a, 2006b, 2007). The well-documented history of nutrient sensitivity of the PS sub-estuaries has led to the recognition that *both* P and N reductions are necessary to help control eutrophication and its unwanted symptoms (North Carolina Department of Environment and Natural Resources [NC-DENR] 1999, 2001; Paerl 2004, 2006c, 2007).

In addition to being impacted by human activities, the PS is also strongly influenced by climatic perturbations. We have entered a new climatic era in the southeastern and mid-Atlantic coastal region of the United States (Goldenberg et al. 2001; Emanuel 2005; Webster et al. 2005). Coastal North Carolina alone has experienced the effects of at least eight category 2 or higher hurricanes and several tropical storms over the past 13 years, including hurricanes Bertha and Fran in 1996, Hurricane Bonnie in 1998, four visits from three hurricanes (Dennis, Floyd and Irene) in a 6-week period during September and October 1999, Isabel in 2003, Charlie and Alex in 2004, Ophelia in 2005, and Tropical Storm Ernesto in 2006 (Figure 2.2). Two of these hurricanes, Fran and Floyd, led to 100- to 500-year flood events in the PS watershed (Bales et al. 2000; Bales 2003).

The severe effects of the 1996 and 1999 floods on PS and the sub-estuaries resulted from extremely high rainfall in parts of the basin (Bales and Childress 1996; Bales et al. 2000), combined with the results of a long history of human modification and encroachment (e.g., channelization, development of the flood plain) in the riverine systems feeding this lagoon (Riggs 2001). The key riverine features that affect flooding and interaction of the uplands with PS are addressed by Riggs (2001) and summarized here: (1) The morphology and composition of the paleo- and modern channel-flood plain systems including tributary drainages in associated uplands control the rate at which flood waters move through the system; (2) fluctuations in climate and sea level associated with Quaternary glaciation and deglaciation produced complex patterns of erosional incisement and sediment deposition; and (3) the location of a particular river segment along the stream gradient relative to sea level determines the degree of interaction between riverine and oceanic processes at temporal scales ranging from the long term (millennia to decadal) to short term (years to single storm events).

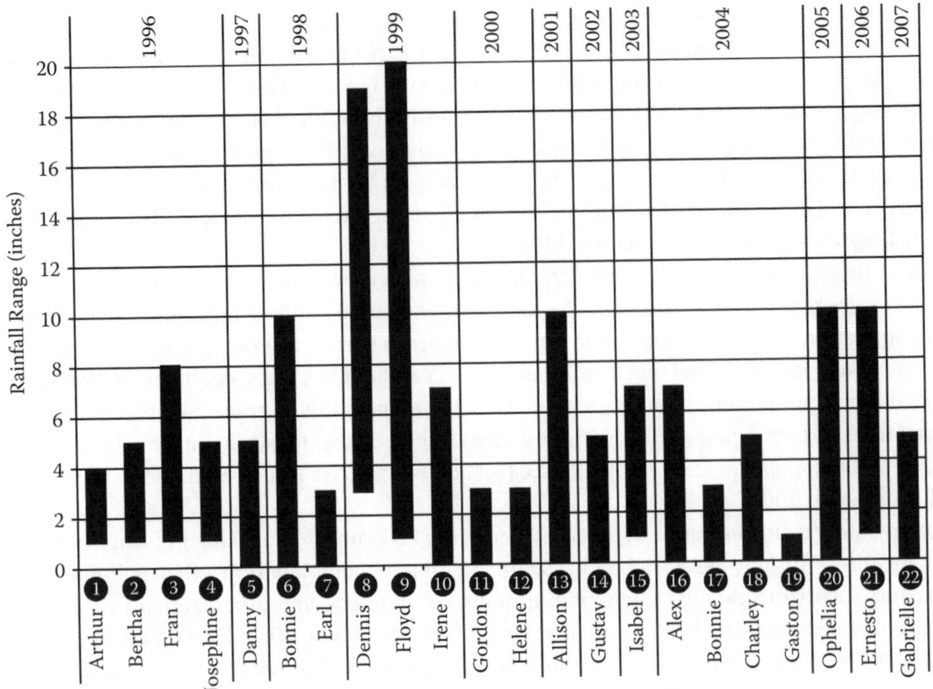

(a) Tropical Cyclone Rainfall Ranges in the Pamlico/Albemarle Sound Watershed (1996–2007)

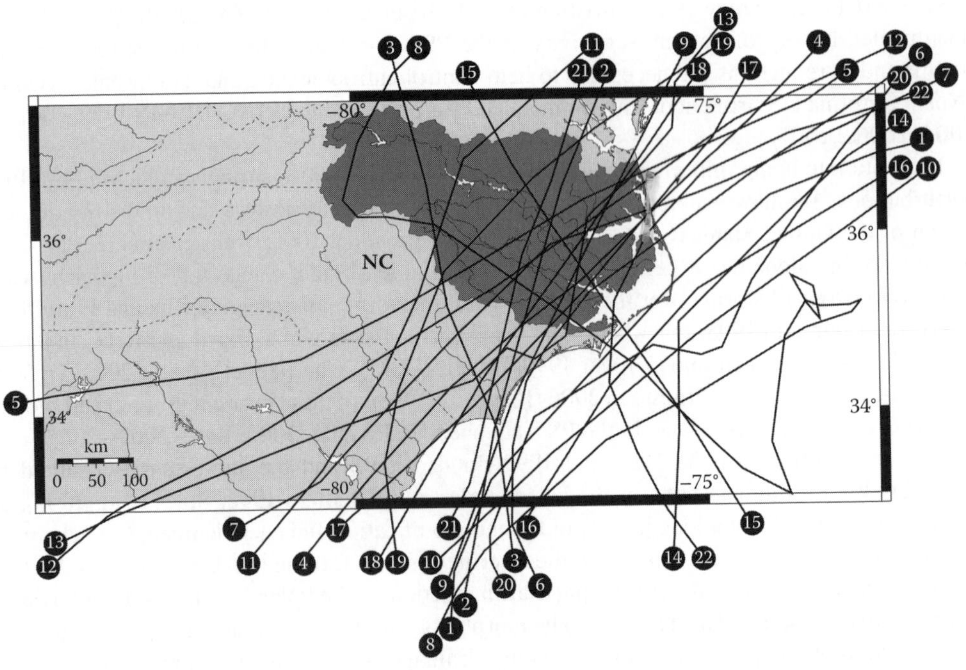

(b) Tropical Cyclone Tracks (1996–2007)

FIGURE 2.2 (a) Rainfall ranges for the various tropical storms and hurricanes that have impacted the PS and its watershed between 1996 and 2007. Each named storm was given a number, which is used to designate storms in part B. (b) Tracks of individual storms across the PS and its watershed. Tropical cyclone data were obtained from the NOAA Coastal Services Center (http://maps.csc.noaa.gov/hurricanes/).

Initial assessments of the effects on water quality, habitat, and fisheries of recent tropical cyclones on the PS system indicate that such impacts vary substantially in magnitude and duration (Paerl et al. 2001; Adams et al. 2003; Peierls et al. 2003), with large storm events leaving multiannual residual impacts (Peierls et al. 2003; Burkholder et al. 2006; Paerl et al. 2006a, 2006b). Climatologists forecast a general increase in storm frequency and intensity of Atlantic tropical cyclones over the next 10 to 40 years (Goldenberg et al. 2001; Webster et al. 2005; Holland and Webster 2007). This poses a repetitive threat to PS and other low-lying Atlantic East Coast and Gulf of Mexico coastal regions, which are characterized by numerous large lagoonal or semi-lagoonal estuaries, bays, and sounds that drain the coastal plains (i.e., Chesapeake Bay, Pamlico Sound, Florida Bay, Lake Ponchartrain, Texas lagoonal estuaries and bays). Our ability to predict the effects of such storms on the function and responses of these coastal ecosystems is limited—we simply have not sufficiently observed the systems under such extremes. Fully evaluating the ecological effects of these hurricanes and other intense storms affords a unique scientific opportunity to understand the linkages between coastal watershed geology, hydrology, biogeochemistry, water quality, and fisheries habitat responses and the resiliency of lagoonal ecosystems.

In this contribution, we will largely focus on the magnitudes, duration, and composition of planktonic primary producer responses to hydrologic and nutrient perturbations that have resulted from these storms as well as droughts. In one key tributary of the PS system, the Neuse River Estuary, phytoplankton accounts for at least 80% of "new" primary production that sustains the planktonic and benthic food webs (Paerl et al. 1998). Hence, gauging the response of the phytoplankton community is of considerable relevance with respect to developing predictive capabilities for productivity-trophic, nutrient cycling, habitat, and ecosystem-level responses. These large-scale events overlay "normal" seasonal and interannual variations in rainfall, including droughts, extratropical storms (local thunderstorms and nor'easters), and anthropogenic perturbations, such as wastewater, sediment, and fertilizer discharges from watershed activities.

Phytoplankton communities were sampled and analyzed prior to and following the storms. Surveys conducted prior to 1999 were mainly confined to the Neuse River and Pamlico River subestuaries of the PS system, but sampling was extended into PS following the 1999 hurricanes (Paerl et al. 2001, 2006b; Peierls et al. 2003). Using these and other studies, we will attempt to reconstruct the ecosystem-level responses to the storms, followed by an assessment of how watershed nutrient, sediment, and water management may need to adapt in response to periods of elevated tropical storm and hurricane activity as well as severe droughts affecting this large lagoonal ecosystem.

2.2 PUTTING THINGS IN PERSPECTIVE: BIOGEOCHEMICAL, GEOLOGICAL, AND CLIMATOLOGICAL TIME SCALES

North Carolina's barrier islands and associated drowned-river lagoonal estuaries are highly dynamic on a geological time scale, and the form of the systems has varied as a function of a changing climate and sea level. Detailed geologic data demonstrate that this evolutionary history contains a record of changing climatic conditions, including periods of both minor and intense storm activity (Riggs 1999). Episodic accretion of barriers, collapse and landward migration of specific barriers, and the dynamics of inlets through the barriers have resulted from periods of intensified storm activity. Local sea level is intimately connected to climatic fluctuations reflecting changing regional oceanographic conditions. These patterns of climatic and sea-level oscillation are superimposed on the ongoing, large-scale Holocene rise in sea level that affects the overall physical-chemical characteristics of the PS.

The associated ecosystem that develops under such conditions should be geologically resilient, characterized by sudden fluctuations followed by rapid recovery. For example, these estuaries and barriers may represent a storm-dominated ecosystem, in which flood events periodically flush the system and reset ecological conditions, comparable to a fire-dominated terrestrial ecosystem (Riggs 2001). Unfortunately, much of the observational data from Atlantic and Gulf Coast systems comes from a period of relatively low Atlantic storm activity (the mid-1960s through mid-1990s). Hence, we have only a rudimentary understanding of system response and recovery to assess either the short- or long-term impacts of high-energy events.

2.3 HISTORIC PERSPECTIVE ON HUMAN AND CLIMATIC IMPACTS

2.3.1 Human Impacts

Over the past 50 years, the PS system has exhibited symptoms of eutrophication associated with expanding use of its watersheds (Harned and Davenport 1990; Copeland and Gray 1991). Much of the blame for eutrophication and its detrimental impacts has been placed on rapidly growing and diversifying nonpoint sources (NPS) of nitrogen entering this largely N-limited system (Tedder et al. 1980; Paerl 1983, 1987, 1997; Christian et al. 1986; Stanley 1988; Rudek et al. 1991; Boyer et al. 1994; Paerl et al. 1995, 2006a, 2006b, 2006c; North Carolina State Senate 1996; Burkholder et al. 2006). In the Neuse River Estuary NPS, dominated by agriculture, contribute ~80% of the external or "new" N loading (NC DENR 1998).

Agricultural expansion during the last 50 years, widespread use of nitrogen fertilizer, proliferating intensive livestock (swine, cattle) and poultry (chicken, turkey) operations, and watershed urbanization have led to unprecedented increases in NPS-N loadings. Loads are estimated to have nearly doubled from the 1960s to the 1990s (Dodd et al. 1993; NC DENR 1998; Stanley 1988; Stow et al. 2001; Paerl et al. 2004; Burkholder et al. 2006). Within the last 20 years, rapidly increasing commercial swine operations increased from <1 million hogs in 1988 to ~10 million in 2008. This has elevated North Carolina to one of the nation's leading pork producers, second only to Iowa (North Carolina Department of Agriculture; http://www.ncagr.com/stats/release/HogRelease12.pdf). Unlike human waste, swine waste is stored in open lagoons and remains largely untreated. Surface and groundwater N releases from expanding animal operations and urbanization have increased since the 1980s (Huffman et al. 1994; Gilliam et al. 1997; Spruill et al. 1998). Additionally, atmospheric deposition of volatilized NH_3 from animal waste contributes substantially to the NPS-N input into the rivers (Paerl and Whitall 1999). For example, volatilization of NH_3 from animal waste and atmospheric emissions from fossil fuel combustion (NO_x) can account for >30% of new N loading to the PS system (Paerl and Fogel 1994). Anthropogenically driven increases in nutrient loading have led to increases in phytoplankton biomass and algal blooms throughout the PS system (Copeland and Gray 1991). This is best documented for the major sub-estuaries of the PS. Nuisance cyanobacterial blooms occurred in the upstream, oligohaline segments of the Chowan and Neuse River estuaries starting in the 1970s (Tedder et al. 1980; Paerl 1982; Craig and Kuenzler 1983; Christian et al. 1986). In addition, increases in chlorophyll *a* concentrations have occurred in oligohaline segments of the Pamlico River Estuary; these frequently exceeded levels that are "acceptable" (typically <40 μg L^{-1}) by the State of North Carolina Department of Environment and Natural Resources (Stanley 1988; NC DENR 2001). Downstream mesohaline estuarine chlorophyll *a* levels also were periodically deemed excessive in the Pamlico and Neuse estuaries, going back as far as the 1970s (Hobbie et al. 1972; Hobbie and Smith 1975). Paleostratigraphic studies based on sediment cores collected in the lower Neuse and Pamlico estuaries have provided additional evidence that these subestuaries of PS have, over the past several decades, experienced accelerating rates of eutrophication and declining habitat conditions (Cooper 2000).

2.3.2 Climatic Impacts

Increases in primary production and algal bloom formation have led to enrichment of organic matter in the sediments (Matson et al. 1983; Matson and Brinson 1990), which appears to have exacerbated the frequencies and spatial extent of bottom water and sediment hypoxia and anoxia in these periodically salinity-stratified sub-estuaries (Stanley and Nixon 1992; Paerl et al. 1998; Buzzelli et al. 2002). The extent to which nutrient-enhanced productivity has led to expanded regions and duration of hypoxia and anoxia, however, is uncertain. This is because bottom water hypoxia likely occurred as a result of salinity stratification and in the absence of human eutrophication. Hypoxic events are highly variable in time and space and are controlled by both chemical and physical processes, such as seasonal and interannual variations in freshwater discharge, wind, and air temperature. The highly variable "drivers" of hypoxia have translated into variable frequencies of finfish and shellfish disease and kills, which have been linked to episodes of bottom-water hypoxia and sulfide emissions (Matson and Brinson 1985; Lenihan and Peterson 1998; Paerl et al. 1998; Eby and Crowder 2002; Baird et al. 2004; Eggleston et al. 2004).

Despite the inherent variability in primary productivity, algal blooms, hypoxia, habitat, and fish kill responses, there are clear increases in trophic state and habitat degradation that have accompanied human nutrient enrichment of this system (Burkholder et al. 2006; Paerl et al. 2006a, 2006b, 2006c, 2007). Nutrient enrichment has also been accompanied by increased sediment loads, which have affected both nutrient status and transparency of the system (Biber et al. 2005). Increased turbidity accompanying enhanced sediment loading has probably not negatively affected phytoplankton primary production in this shallow system. More likely, it has negatively affected benthic micro- and macrophytes, specifically seagrasses, which are known to have relatively high light requirements (>10% of surface irradiance; Biber et al. 2005). Any negative effects on seagrasses may translate into positive effects for phytoplankton, as the latter group can better circumvent light limitation and take advantage of nutrient supplies that would otherwise be used to support seagrass growth.

2.3.2.1 Hurricanes

During the past 12 years, both nutrient and sediment delivery to the PS have been increasingly dominated by episodic hydrologic events, including the recent rise in Atlantic hurricane activity (Goldenberg et al. 2001; Webster et al. 2005; Holland and Webster 2007), as well as intense record periods of drought (http://www.nc-climate.ncsu.edu/climate/drought.php) (Figure 2.2). We appear to have entered a new climatic era, which has impacted delivery, processing, and ecological impacts of nutrients, organic matter, and other pollutants discharged to this large lagoonal ecosystem (Paerl et al. 2007).

Prior to 1996, coastal North Carolina had not been seriously affected by a large hurricane since mid-October 1954, when Hurricane Hazel made landfall in South Carolina. Both in terms of trajectory and meteorology, Category 3 Hazel was remarkably similar to Fran (1996) and Floyd (1999) in that it delivered a massive amount of rainfall to the North Carolina coastal watersheds. This event was followed by a 40-year hiatus in major hurricane impacts on Eastern North Carolina. This lull was suddenly broken in the mid-1990s. Although this lengthy break in hurricane landfalls might seem unusual, history argues that it is not. Analysis of well-kept weather records for the North Carolina coastline indicates that the region has experienced repeated 10- to 40-year periods of elevated Atlantic hurricane activity (NC Climatological Office: http://www.nc-climate.ncsu.edu/climate/hurricane.php). For example, the late 1800s (1880s to 1900) and early to mid-1900s (1930s to 1950s) were particularly active hurricane periods, with hurricane landfall frequencies matching those of the late 1990s.

The most recent period of elevated hurricane activity started in 1996 with the arrival of Bertha (July) and Fran (Sept). Category 2 Bertha raked the North Carolina coastline, making landfall near Jacksonville, North Carolina, and traveling north across Pamlico Sound. Although Bertha's high winds caused significant storm surges, beach erosion, and structural damage, heavy rainfall was confined to the easternmost Outer Banks. Neuse River freshwater discharge records at Kinston (USGS site number 02089500), approximately 60 km upstream from the entrance to the Neuse River Estuary, showed few effects of Bertha on hydrologic loading to the estuary (Figure 2.3). In contrast, Category 3 Hurricane Fran, which struck the North Carolina coast near Wilmington

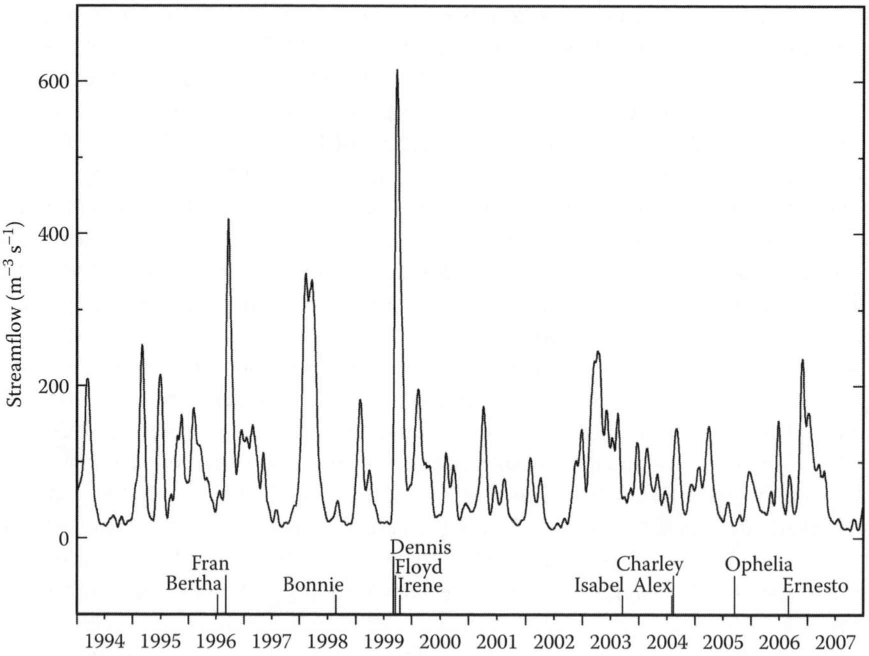

FIGURE 2.3 Summary of daily mean streamflow of the Neuse River (as m^3 s^{-1}) at Kinston, North Carolina (USGS site number 02089500) or approximately 60 km upstream from the entrance to the Neuse River Estuary. (Data courtesy of USGS National Water Information System, waterdata.usgs.gov/nc/nwis/.)

on September 5, 1996, moved inland and stalled over the Piedmont region. During its destructive course, which included the Raleigh-Durham area, this large hurricane delivered as much as 50 cm of rainfall in gauged areas of the PS watershed, causing extensive, long-lasting (4 weeks) flooding in the Neuse River drainage basins (Bales and Childress 1996). In particular, the Neuse River basin was heavily impacted. Hydrograph data at the Kinston gauging station showed a massive amount of freshwater discharge to the Neuse River Estuary (NRE) (Figure 2.3). This freshwater bolus exhibited very low dissolved oxygen (<2 mg O_2 L^{-1}), most likely due to elevated terrestrial organic matter loading. Eventually, the entire NRE turned hypoxic from top to bottom for at least a 2-week period (Paerl et al. 1998; www.marine.unc.edu/neuse/modmon).

Not surprisingly, fish kills were reported throughout the estuary during this period (Figure 2.4) (Paerl et al. 1998). Approximately 2 months after Fran-related flooding, the NRE returned to pre-Fran oxic conditions, although higher than seasonally normal freshwater discharge conditions prevailed well into the following spring months. This was most likely the result of floodwaters continuing to drain from swamps and wetlands, as well as discharge from saturated groundwater sources (Bales 2003). Therefore, it appeared that there was significant residual discharge lasting at least 6 months after Fran.

Large increases in nutrient loading were attributed to Fran (Figure 2.5). It was estimated that nitrogen (N) loading to the NRE associated with Fran's floodwaters approximated the mean annual N load (Paerl et al. 2005), resulting in a doubling of the 1996 annual N load to this estuary (Figure 2.5). Unfortunately, Pamlico Sound proper was not routinely monitored for water quality until late 1999 (following hurricanes Dennis, Floyd, and Irene). Hence, the potential nutrient enrichment effects of Fran's floodwaters on the PS were not documented.

Hurricane Bonnie, in mid-September 1998, was another coastal storm with minimal ecological impact on the sub-estuaries of the PS. Like Bertha, Bonnie was a windy, relatively low rainfall storm, resulting in no significant increase in freshwater runoff or N loading to the NRE or PS

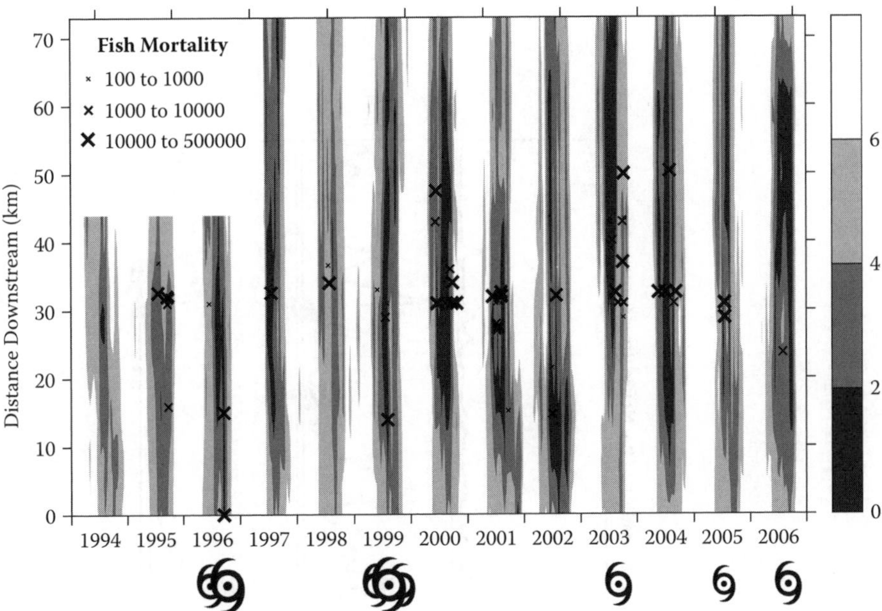

FIGURE 2.4 Bottom water (lowest 0.2 m of water column) values for dissolved oxygen (in mg O_2 L^{-1}) throughout the Neuse River Estuary from 1994 through 2006. Data are from the Neuse River Modeling and Monitoring Program, ModMon (http://www.unc.edu/ims/neuse/modmon/index.htm). The white open area indicates the region of the lower estuary that was not sampled prior to 1997. X marks indicate positions and dates of fish kills reported by the North Carolina Department of Environment and Natural Resources, Division of Water Quality (www.esb.enr.state.nc.us/Fishkill/fishkillmain.htm). Size of X indicates mortality level. Distance downstream refers to the distance from the freshwater end member station near New Bern, North Carolina.

(Figure 2.5b). Hypoxia in the NRE was reduced by strong vertical mixing associated with winds from Bonnie, and no increase in fish kills was reported.

During 1999, following a record summer drought, PS was impacted by three successive hurricanes within a 6-week period in September to October. Hurricanes Dennis, Floyd, and Irene affected the PS in different ways, but the cumulative hydrologic effect was a record 1 m of rainfall over portions of the PS watershed, with peak flows estimated to be greater than the 0.2% exceedance level (500-year flood event) in most tributaries to the PS following Floyd (Paerl et al. 2001, 2006a, 2006b; Bales 2003). Floodwaters inundated the PS sub-estuaries, essentially turning them fresh within a few days after Floyd made landfall in mid-September. Then, over the next 3 to 4 weeks the floodwaters displaced approximately 80% of the volume of PS. Salinity was depressed to near-freshwater condition in the Sound's surface waters, and residence time was reduced from ~1 year to approximately 6 weeks (Table 2.1) (Paerl et al. 2001; Bales 2003). Best estimates based on limited data suggested that PS received approximately its mean annual external N load and somewhat less than its mean annual P load during this 6-week period, meaning that during all of 1999, PS received double the normal nutrient loads (Paerl et al. 2001; Peierls et al. 2003) (Figure 2.5).

Primary production in the PS, assessed as increases in chlorophyll *a* concentration compared to prestorm conditions, showed a sudden, highly significant increase following the influx of nutrient-laden storm runoff. The sub-estuaries, typified by the Neuse, showed no significant increases in chlorophyll *a*, in large part because they served as freshwater "riverine" conduits with residence times of only a few days, compared to the larger PS (Bales 2003), and because the waters were highly turbid. In the Neuse River Estuary, chlorophyll *a* levels initially dropped as the freshwater discharge made its way downstream to PS (Figure 2.6). As a result, much of the nutrient-stimulated

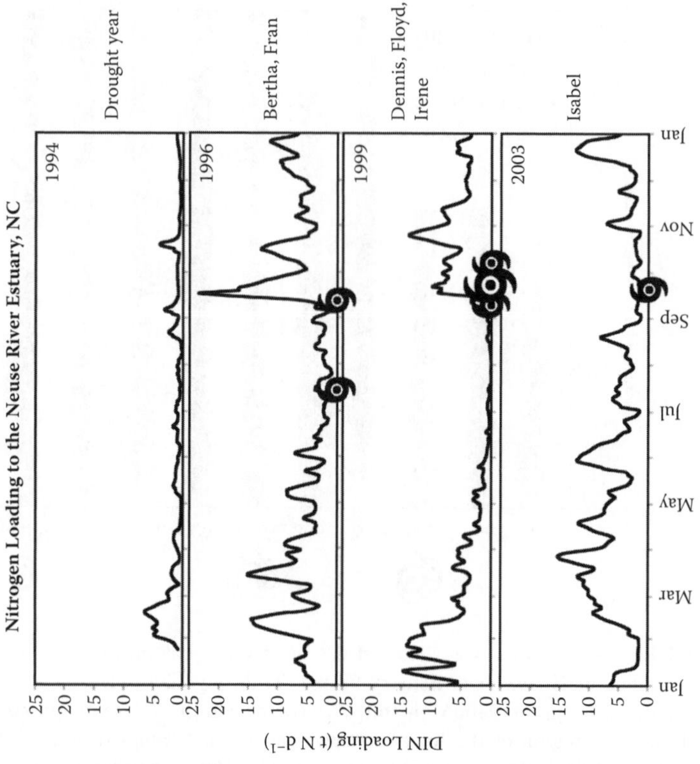
Nitrogen Loading to the Neuse River Estuary, NC
1994
Drought year
1996
Bertha, Fran
1999
Dennis, Floyd, Irene
2003
Isabel
DIN Loading (t N d−1)
0
5
10
15
20
25
Jan
Mar
May
Jul
Sep
Nov
Jan

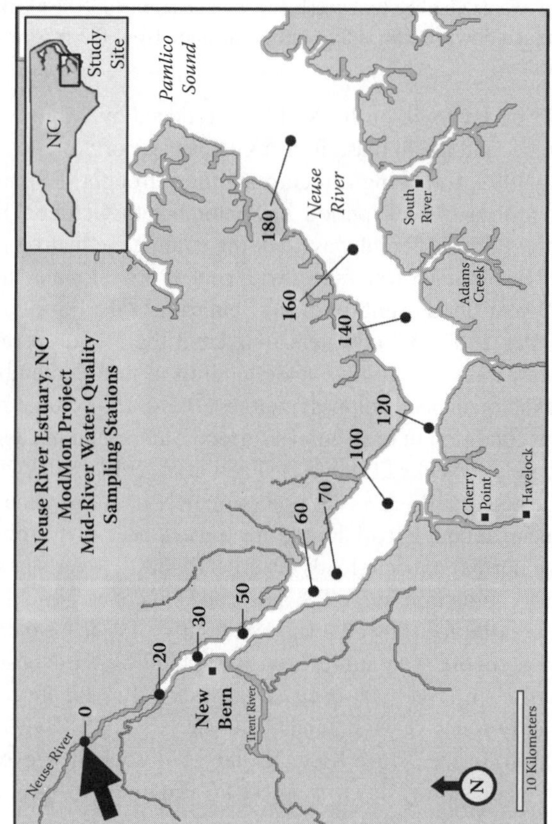
Neuse River Estuary, NC
ModMon Project
Mid-River Water Quality
Sampling Stations
NC
Study Site
Pamlico Sound
Neuse River
Neuse River
New Bern
Trent River
0
20
30
50
60
70
100
120
140
160
180
Cherry Point
Havelock
Adams Creek
South River
N
10 Kilometers

(a)

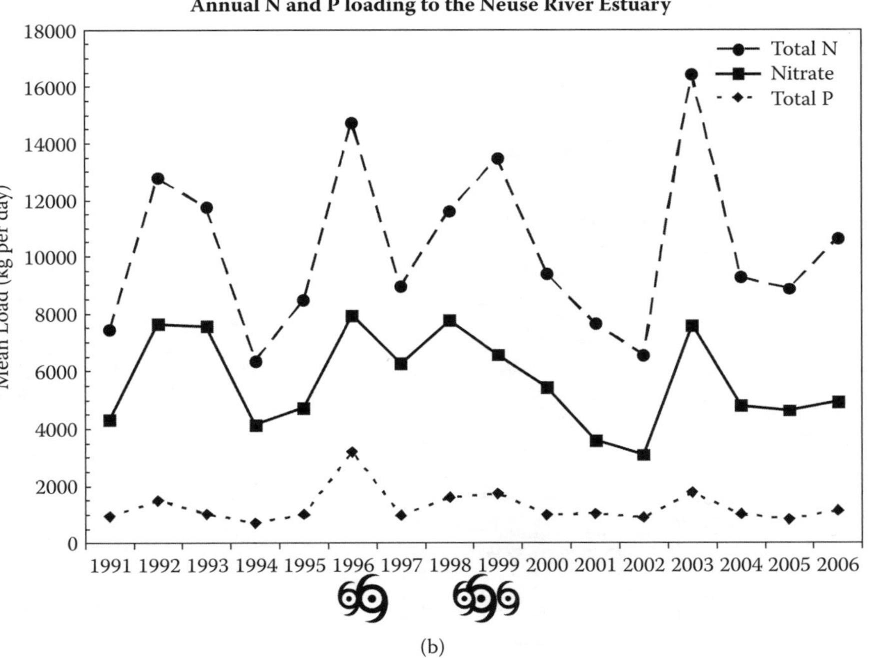

FIGURE 2.5 Watershed-derived "external" nutrient loading to the Neuse River Estuary during 1994–2006. Nutrient loads were determined at the head of the estuary, near New Bern where the Neuse River discharges into the estuary. Nutrient loading was calculated based on in-stream concentrations multiplied by freshwater discharge. Discharge values were derived from the Kinston gauging station (USGS site number 02089500). (a) Annual dissolved inorganic nitrogen (DIN) loading patterns for selected years, including a drought year (1994); a year that included a low rainfall hurricane, Bertha, and a high rainfall hurricane, Fran (1996); 1999, which included the sequential hurricanes Dennis, Floyd, and Irene; and 2003, Isabel, when a powerful, but relatively low rainfall hurricane, made a "direct hit" on the Pamlico Sound. Nutrient data were obtained from the ModMon program. (b) Mean nutrient loads calculated for sequential years 1991–2006. Included were values (kg day^{-1}) for total N, nitrate, and total P. The effects of large hurricanes (shown as symbols) on annual loads are evident. Nutrient data were obtained from the North Carolina Department of Environment and Natural Resources, Division of Water Quality and the ModMon program.

TABLE 2.1
Water Residence Time in Days, Calculated from Monthly Mean Flow, for Two Key Tributaries (Neuse and Pamlico River Estuaries) and the Pamlico Sound Proper

	September		October	
Water Body	**1999**	**Normal**	**1999**	**Normal**
Neuse River Estuary	7	69	11	81
Pamlico River Estuary	7	133	19	175
Pamlico Sound	36	219	79	313

Note: Values are shown for September and October 1999, during the hurricane flood period. These were compared to normal conditions, based on mean flows over the previous 10 years.

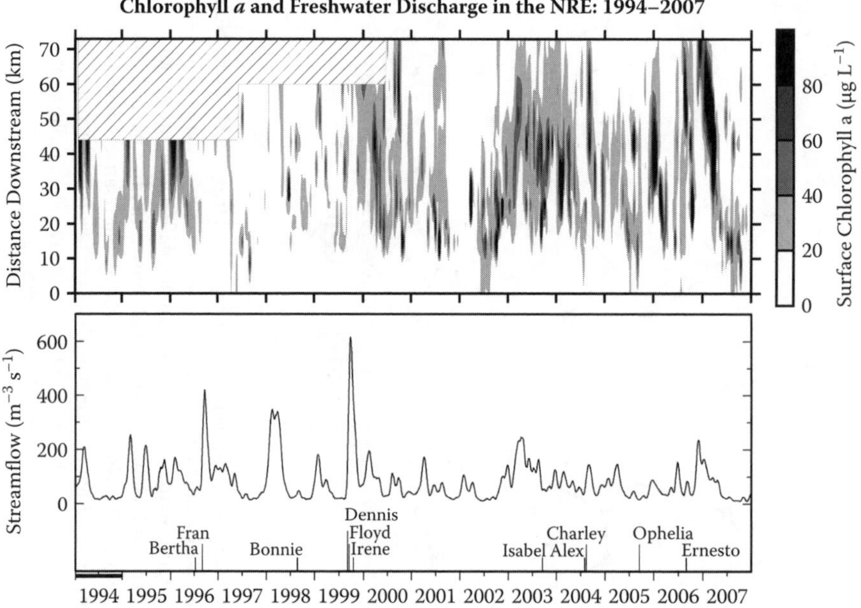

FIGURE 2.6 Surface water chlorophyll *a* concentrations in the Neuse River Estuary (NRE) and freshwater discharge for the Neuse River (Kinston gauging station, USGS site number 02089500) from 1994 to 2007. Also shown are the dates of hurricane and tropical storms that impacted the NRE. Chlorophyll *a* values were derived from 13 ModMon stations that were sampled biweekly on a transect ranging from the head of the estuary to a downstream location near the entrance to Pamlico Sound.

production took place in the PS proper, largely because the sound had a sufficiently long residence time to enable the phytoplankton to assimilate the nutrient load and undergo net growth (Figure 2.7). As the floodwaters receded, 3 weeks after the landfall of Floyd (September 17, 1999), chlorophyll *a* concentrations remained relatively high in the PS, and started to show some increases in the Neuse River Estuary. Due to the long residence time of this system, nutrient-stimulated primary production lasted well into 2000 in the PS (Paerl et al. 2001, 2006a, 2006b; Peierls et al. 2003; Christian et al. 2004) (Figure 2.7).

Dennis and Floyd's floodwaters also delivered large amounts of watershed-derived dissolved and particulate organic carbon DOC/POC (Paerl et al. 2001). Analyses of floodwaters transiting the

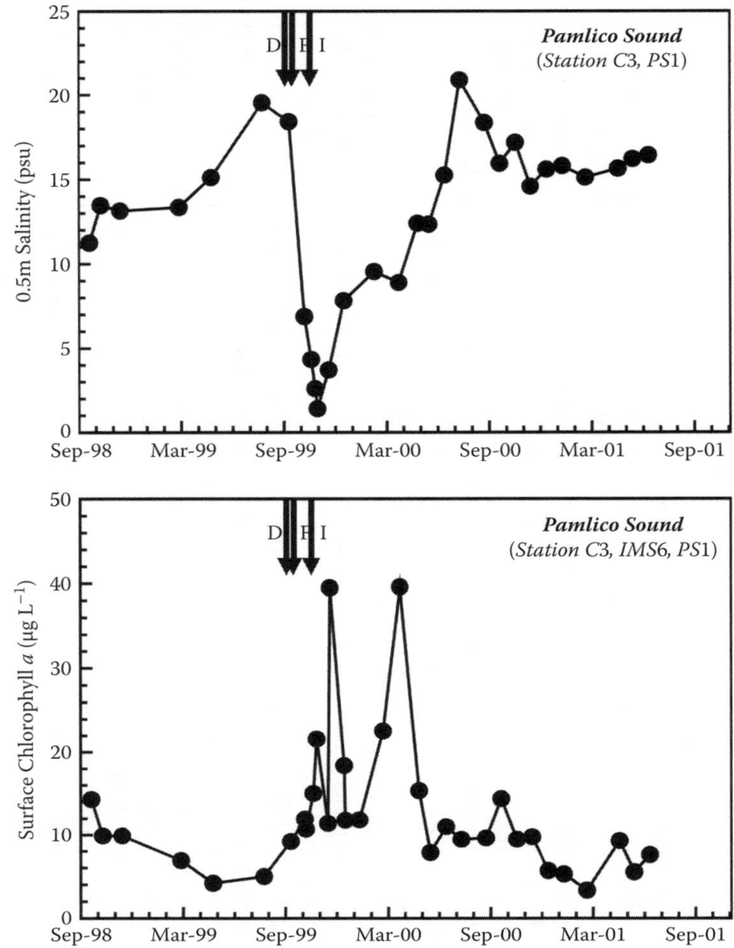

FIGURE 2.7 Effects of the three sequential hurricanes, Dennis (D), Floyd (F), and Irene (I), on salinity and chlorophyll *a* concentrations at a location (X) in the western basin of Pamlico Sound, during fall 1999. The times of landfall of these hurricanes are indicated with arrows. Shown is a SeaWiFs remote sensing image (courtesy of NASA) taken on September 23, showing the discharge of sediment and organic matter–laden freshwater entering the sound. This was one week after landfall of Floyd (September 17, 1999). Due to the huge volume of freshwater runoff discharged to the sound, station X became fresh. It took nearly a year for normal salinity to return to this location. Nutrient-enriched waters led to the stimulation of chlorophyll *a* at this location and throughout the sound; elevated chlorophyll *a* concentrations were observed well into the following spring and summer (2000). (From Paerl, H. W. et al. 2001. *Proc. Natl. Acad. Sci. USA* 98: 5655–5660. With permission.)

Neuse River Estuary and entering PS during the September to November floodwater discharge period indicated that the DOC content was at least twice the normal concentration while POC levels were at least three times the concentrations normally encountered. The PS acted like a large trap for organic matter. During this same period, approximately 2000 metric tons of particulate nitrogen (PN), or 60% of the annual external load, was discharged from the sub-estuaries to PS (Bales 2003).

This large load of organic matter helped fuel extensive bottom-water hypoxia, which was exacerbated by strong vertical stratification due to the fresh floodwaters overlaying the saltier, denser bottom water in the PS. Considering that we were only able to survey about 30% of the PS, the extent of bottom-water hypoxia following the discharge of floodwaters is not accurately known. Extensive bottom-water hypoxia was stressful to both finfish and shellfish, as indicated by an increase in fish sores, diseases, and kills and benthic shellfish and invertebrate mortality following Floyd (Burkholder et al. 2006; Paerl et al. 2006b) (Figure 2.4). Bottom-water hypoxia also forced juvenile fish from their preferred habitats. Hypoxic conditions and decreased surface salinity were alleviated when hurricane Irene totally mixed the PS as it passed by the Outer Banks in mid-October (Paerl et al. 2001). It took several months for nutrient and organic matter levels to return to pre-hurricane conditions after this series of disturbances, and increased incidences of fish disease were evident until well into the fall and winter of 1999 to 2000. Catches of dominant commercial and sport fishes remained markedly low through the following year, 2001 (Crowder and McClellan unpublished data). Low recruitment stocks, leading to poor recruitment during the 2 years following Dennis and Floyd, as well as higher than usual bottom water hypoxic conditions in the sub-estuaries, appear to be leading reasons for this prolonged decline in fisheries resources (Eggleston et al. 2004).

We suspect that the recovery of shellfish and finfish habitats to more desirable conditions would have taken longer if Hurricane Irene had not occurred on the heels of Dennis and Floyd. Irene, which grazed the Outer Banks in mid-October about a month after Floyd, had maximum sustained winds >120 km h^{-1} but brought only a trace of rainfall to the coast. Irene completely mixed the PS, relieving the system of the bottom water hypoxia, which, in retrospect, was of great benefit for finfish and shellfish.

Hurricane Isabel made landfall as a Category 2 storm on September 18, 2003, on the Outer Banks. The storm crossed the PS, and then moved northward through northeastern North Carolina, the Virginia Tidewater, and Chesapeake Bay regions (Figures 2.2 and 2.8). The powerful winds (>160 km h^{-1}), storm surges, and high waves associated with Isabel created a new inlet near Hatteras Village on the Outer Banks. The new inlet had an immediate and measurable effect on salinity in PS, which was documented by the instrumented ferries crossing the PS (see www.ferrymon.org). Salinity in eastern PS was relatively low prior to the inlet opening, in large part because plentiful summer rains delivered high amounts of freshwater to the Sound prior to Isabel (Paerl et al. 2009). The inlet enhanced water exchange between PS and the coastal Atlantic Ocean, increasing the overall salinity of the eastern sound by approximately 3 psu. This elevated salinity condition prevailed until the Army Corps of Engineers filled the inlet 5 weeks later, returning the sound to a lower salinity (Figure 2.8).

FIGURE 2.8 *(see facing page)* Track and hydrologic impacts of Hurricane Isabel, which made landfall on the Outer Banks and crossed Pamlico Sound on September 18, 2003. This hurricane created a new inlet between the Sound and the Atlantic Ocean. (a) The track and the two North Carolina Department of Transportation Ferry Division ferry routes that were instrumented to measure hydrologic conditions (shown as salinity here) before and after the storm struck. (b) The location of the new inlet that was formed along the southern part of Hatteras Island. (c) How the differences in salinity were monitored between the two focal regions of Pamlico Sound that are transected by the two ferries; a western region that is transected by the Cedar Island to Ocracoke ferry and an eastern region that is transected by the Swan Quarter to Ocracoke ferry. The ferries recorded hydrographic data up to 24 h prior to the hurricane and within 36 h after the hurricane struck. (d) The differences in salinity between the west and east focal regions prior to and after the new inlet was formed, and after the inlet was closed by the Army Corps of Engineers 5 weeks after the inlet was formed. Data were obtained by the North Carolina Ferry-Based Water Quality Monitoring Program, FerryMon (www.ferrymon.org).

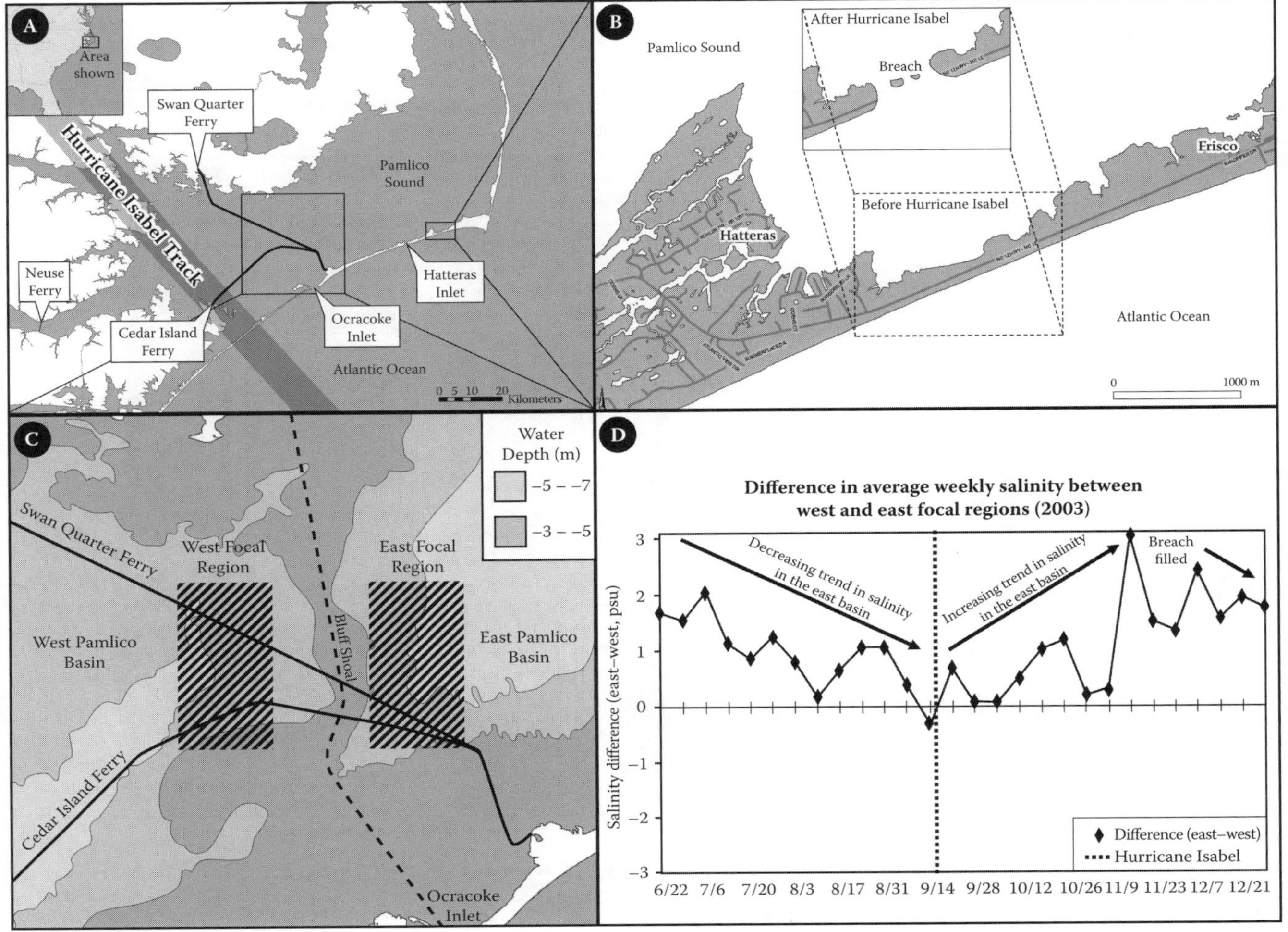
A
Area shown
Swan Quarter Ferry
Hurricane Isabel Track
Pamlico Sound
Neuse Ferry
Hatteras Inlet
Cedar Island Ferry
Ocracoke Inlet
Atlantic Ocean
0 5 10 20 Kilometers
B
Pamlico Sound
After Hurricane Isabel
Breach
Hatteras
Before Hurricane Isabel
Frisco
Atlantic Ocean
0 1000 m
C
Water Depth (m)
−5 – −7
−3 – −5
Swan Quarter Ferry
West Focal Region
East Focal Region
West Pamlico Basin
Bluff Shoal
East Pamlico Basin
Cedar Island Ferry
Ocracoke Inlet
D
Difference in average weekly salinity between west and east focal regions (2003)
Decreasing trend in salinity in the east basin
Increasing trend in salinity in the east basin
Breach filled
Salinity difference (east–west, psu)
3
2
1
0
−1
−2
−3
6/22 7/6 7/20 8/3 8/17 8/31 9/14 9/28 10/12 10/26 11/9 11/23 12/7 12/21
Difference (east–west)
Hurricane Isabel

FIGURE 2.8

Despite high winds, rainfall in North Carolina from Isabel was relatively low (<15 cm in coastal North Carolina) (NOAA National Hurricane Center, www.nhc.noaa.gov/), in part because Isabel was a fast moving storm. Storm surges within the sounds caused locally severe shoreline erosion (up to 28 m), very high sediment loads, and resulting fish kills. Intense vertical mixing within the NRE and PS added to the suspended sediment load for several days after the storm. The sediment resuspension event caused nutrient enrichment in the water column, which led to nutrient-enhanced productivity, accompanied by an increase in phytoplankton biomass (as chlorophyll *a*). This effect was only observed for about a week, however, in contrast to the multi-month chlorophyll *a* stimulation experienced after Dennis and Floyd. No significant hypoxia and/or fish kills were reported as a result of Isabel in either the Neuse River Estuary or PS. Some localized hypoxia, accompanied by fish kills, was reported in the northern Albemarle Sound and Chowan River regions, where shoreline erosion and resultant sediment loadings were greatest. The fish kills likely were due to locally heavy rainfall and washout of organic matter from swamps in the Albemarle region that occurred as the forward speed of Isabel slowed while crossing the northern PS system. This led to an organic matter-enriched freshwater lens overlying saltier bottom water, causing strong vertical stratification and promoting water hypoxia.

In mid-September 2005, Hurricane Ophelia, a Category 1 "dry" coastal storm stalled off the North Carolina coastline for over 12 hours (Figure 2.2). This storm, like Isabel and Irene, produced high winds (Figure 2.9a) but relatively little rainfall. An autonomous vertical profiling system (Reynolds-Fleming et al. 2002) recorded the rapid water quality changes resulting from that storm. The primary initial environmental consequence of Ophelia was intense vertical mixing and sediment

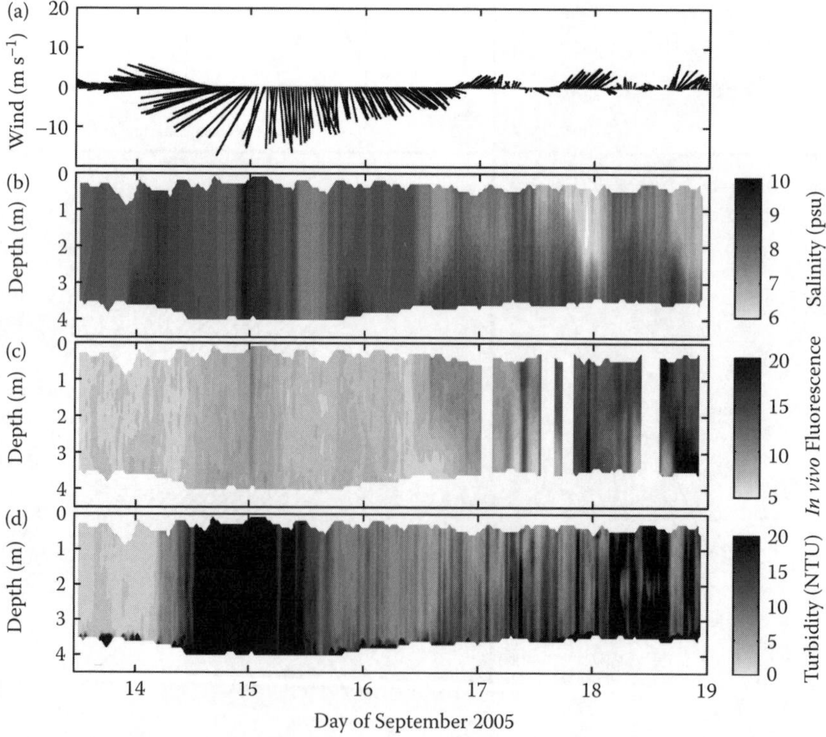

FIGURE 2.9 Impact of Hurricane Ophelia on water quality in the Neuse River Estuary. (a) Wind velocity vectors pointing down wind. (b) Contour plot of vertical salinity profiles. (c) Contour plot of uncalibrated in vivo fluorescence of chlorophyll *a* profiles. (d) Contour plot of vertical turbidity profiles. Profiles were collected every 30 min with 10 cm depth resolution. Contour magnitudes are given by the scale bar to the right of the figure.

resuspension (Figure 2.9b–d) and an abrupt increase in water column turbidity (Figure 2.9d). Within 2 days of the storm's passage, phytoplankton biomass more than doubled (Figure 2.9c), producing a second increase in water column turbidity (Figure 2.9d). We hypothesize that nutrients released from the resuspended sediment were the primary fuel for phytoplankton growth since there was no large-scale riverine discharge event associated with this storm (Figure 2.9b). No accompanying hypoxia and/or fish kills were reported following Ophelia.

2.3.2.2 Smaller Storm Events (<74 mph)

Several tropical storms which delivered high rainfall to significant portions of the PS watershed also have affected water quality and habitat in PS during the past decade. For example, Tropical Storm Helene, which passed over the basin in 2000, was followed by a prolonged (several weeks) increase in phytoplankton biomass (chlorophyll *a*) in the mesohaline section of the NRE (Wetz and Paerl 2008). More recently, Tropical Storm Ernesto, which struck during mid-September 2006, delivered over 30 cm of rainfall within the PS watershed (Figure 2.2) and produced a flood pulse in the NRE that was followed by an eightfold increase in freshwater flow over the prior condition (Figure 2.10a, b) (Hall et al. 2008). Ernesto's discharge brought with it a very large amount of watershed-derived inorganic nitrogen, primarily as nitrate (NO_3^-) (Figure 2.10c) (Hall et al. 2008). Ernesto's freshet caused very

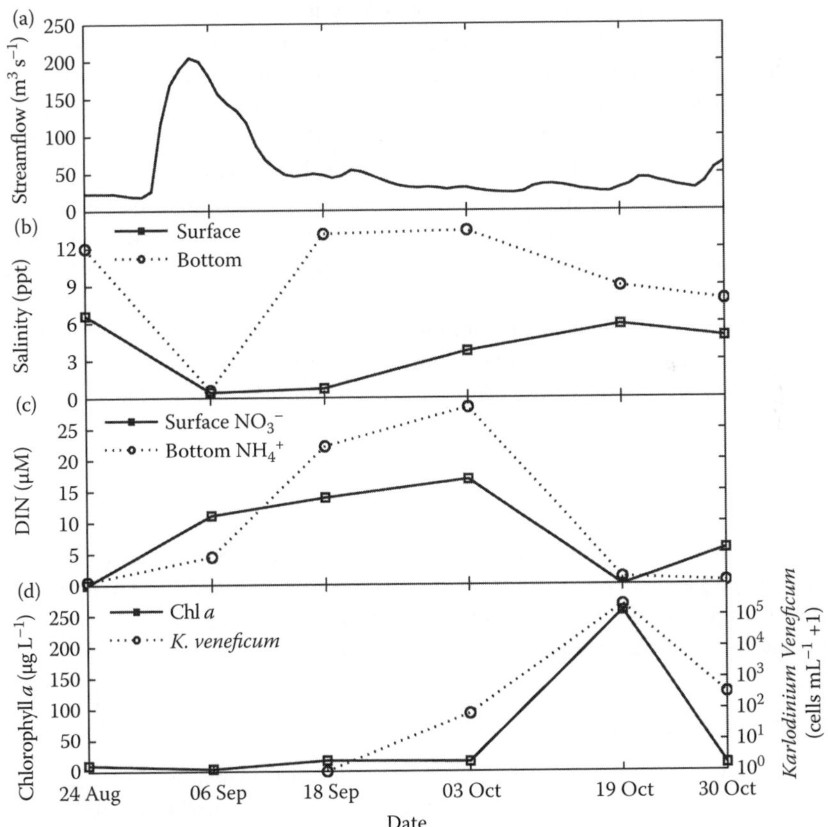

FIGURE 2.10 Environmental conditions leading to the development of a toxic *Karlodinium veneficum* bloom. (a) Neuse River flow at Kinston, NC showing the runoff pulse from Tropical Storm Ernesto during the fall of 2006. (b) Time series of surface (0 m) and bottom (3 m) salinity at station 60 where the peak bloom biomass was observed on October 19, 2006. (c) Time series of the dominant sources of dissolved inorganic nitrogen to the bloom area; riverine NO_3^- input to the surface waters and accumulation of regenerated NH_4^+ in the bottom waters at station 60. (d) Time series of chlorophyll *a* and *K. veneficum* cell abundance at station 60.

strong vertical salinity stratification, which remained in place for several weeks due to ensuing calm weather and a lack of vertical wind mixing (Figure 2.10b). This allowed anomalously high concentrations of regenerated ammonium (NH_4^+) to accumulate in the saline bottom waters (Figure 2.10c). The unusually high nutrient availability fueled phytoplankton growth (Figure 2.10d), and a short-lived frontal circulation pattern additionally concentrated vertically migrating cells. The result was an intense (>200,000 cells mL^{-1}) but highly localized bloom of the toxic dinoflagellate *Karlodinium veneficum* in the mid-estuarine region of the NRE (Figure 2.10d). The collapse of the toxic bloom coincided spatially and temporally with several fish kills (Hall et al. 2008).

In addition to specific tropical storms and hurricane events, abnormally low or high periods of freshwater discharge from local thunderstorms and droughts can also act as important drivers of phytoplankton dynamics in lagoonal estuaries. Hydrologic variability during 2003 provides examples of how short-term and seasonal hydrologic forcing (freshwater discharge) can profoundly affect the magnitude and location of phytoplankton biomass in the NRE. In the NRE phytoplankton biomass, as chlorophyll *a*, accumulates in distinct regions of the estuary. Peaks in chlorophyll *a*, or the "CMAX zone," forms as a result of the combined effects of freshwater discharge, or flow, the nutrient enrichment in this discharge (Figure 2.11), and the residence time of specific segments of the estuary. When freshwater discharge is low, flow and flushing rates are low and water residence times tend to be long. Under these conditions, phytoplankton growth rates (doubling times) are most likely to keep up with relatively low flushing rates, leading to the accumulation of phytoplankton biomass, as CMAX, in upstream segments of the NRE. This can be seen during early February and late June in 2003 (Figure 2.11). Conversely, during elevated rainfall and relatively high freshwater discharge periods, flushing rates are high relative to phytoplankton growth, causing the CMAX to move to the wider downstream segments, where water residence time tends to be longer (Figure 2.11). The composition of phytoplankton communities either in or outside of CMAX can also be changed as a result of variable discharge, flushing, and residence time conditions, since different phytoplankton groups have inherently different maximum growth rates (Pinckney et al. 1998; Valdes-Weaver et al. 2006).

Interestingly, the wet spring and early summer period of 2003 played a much more significant role in shaping phytoplankton community structure than Hurricane Isabel (September 2003), which lashed the NRE with very high winds, caused complete vertical mixing, but delivered relatively little rainfall (Figures 2.3 and 2.11).

2.4 WHAT HAVE WE LEARNED?

The whole-system manipulation that took place in response to the 1999 hurricanes and subsequent hydrologic events provided an opportunity to unravel some of the scientific mysteries of PS and to provide a better framework for management of this and other large lagoonal estuarine and coastal systems that have relatively long water residence times and retention characteristics.

2.4.1 Biogeochemical and Trophic Considerations

Based on the findings from PS, we conclude that an increase in major storm activity affecting lagoonal estuaries will lead to: (1) increased allochthonous and autochthonous nutrient loading; (2) higher frequencies, intensities, and spatial coverage of phytoplankton blooms; (3) expansion of low oxygen conditions; and (4) degraded fisheries with potential reductions in fisheries production. Although it can be argued that primary production and standing stocks of phytoplankton will sporadically increase as a result of the large nutrient loads in accompanying hurricane floodwaters, there was little short-term evidence that the enhanced primary production of PS has translated into increased production at higher trophic levels. If anything, the increased production of phytoplanktonic organic matter added more "fuel" to support bottom-water hypoxia and anoxia in the

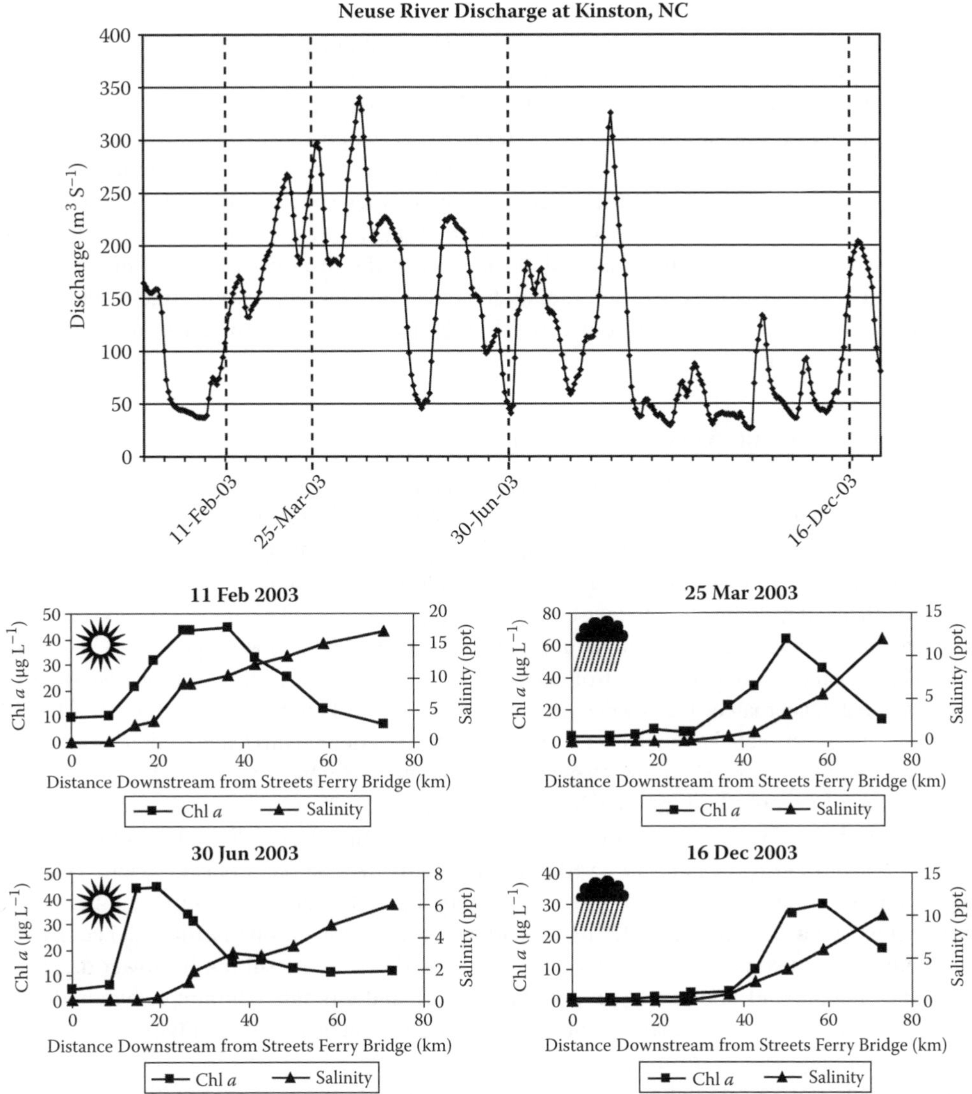

FIGURE 2.11 Influence of freshwater discharge, or flushing, on the zone of maximum phytoplankton biomass production (as chlorophyll *a*), or CMAX, in the Neuse River Estuary during a hydrologically variable year, 2003. Freshwater discharge patterns are shown for 2003 (data from Neuse River gauging station, USGS site number 02089500). When freshwater discharge is low, CMAX tends to form in the upper part of this semi-lagoonal estuary, as illustrated for February and June. Following periods of elevated rainfall and higher freshwater discharge, CMAX is transferred downstream, as shown for March and December.

sound and its sub-estuaries (Paerl et al. 2001). While large amounts of river discharge can enhance shelf fisheries or anadromous fishes in oligotrophic estuaries, high discharge events are more likely to have a negative effect on lagoonal estuaries such as the PS. Flooding not only adds nutrients, organic materials, sediments, and toxic chemicals to the estuary but also leads to strong stratification of the water column, a prerequisite to low oxygen concentrations in the bottom water (Buzzelli et al. 2002).

Although hypoxia reduces habitat availability and can be physiologically challenging or even lethal, oxygenated refuges usually exist in the surface or near-shore waters, allowing mobile organisms to evade areas of low habitat quality. In the wake of the 1999 hurricanes, however, refuge areas were reduced significantly by the rapid and dramatic change in salinity (Eby and Crowder 2002; Eggleston et al. 2004). The combined effects of low dissolved oxygen, low salinity, and rapid flushing as environmental stressors were probably greater than any of them would present alone. In addition, the persistence of these sporadically adverse conditions in the bottom waters of the PS likely exacerbated the fishery effects. Finfish and mobile shellfish responded with changes in abundance, composition, distribution, migration schedules, and resilience, all of which were largely species specific (Eby and Crowder 2002; Eggleston et al. 2004; Paerl et al. 2006b). The changing frequencies of environmental perturbations (storms, floods, droughts) play a defining role in the response and recovery of the ecosystem to these perturbations. Response also depends upon the generation time of the affected organisms, which can be hours to days for phytoplankton, months to years for benthos, and years to decades for fishes.

2.4.2 Socioeconomic Aspects and Considerations

Over the past three centuries, humans have severely modified the vast forested wetlands and associated drainage systems that dominated the North Carolina coastal plain. The large-scale modification and elimination of wetlands was accompanied by ever-increasing human encroachment and utilization of these marginal lands, primarily for agriculture and silviculture. However, the explosive urban growth and development of the past quarter century represented a haphazard approach to land use with almost no zoning, little interaction between various government agencies with different agendas, and disregard for natural hydrologic and geologic processes. The resulting chaotic pattern of development exacerbated the extreme amount of damage and enormously high cost of the 1999 floods, which was estimated to be in excess of $3.5 billion (National Hurricane Center 1999; also see North Carolina Floodplain Mapping Program at http://www.ncfloodmaps.com/default_swf.asp).

Human modifications of the surface drainage systems in the PS watershed fall into three general categories, each with its own suite of problematic ramifications. Riggs (2001) listed the following: (1) ditching and draining of marginal wetlands and channelization of associated tributary streams changed the dynamics of the surface drainage system and led to major land-use changes and loss of wetland habitat and biodiversity; (2) modification of floodplains through filling and construction of road dams across streams severely modified the flow dynamics, resulting in slower drainage of floodwaters, retention of floodwaters in populated areas, and ultimately failure of hundreds of the road dams; and (3) suburban expansion of rural North Carolina has significantly increased the rate and degraded the quality of stormwater runoff.

The geologic framework and evolutionary history of the PS basin determine the terrestrial and hydrologic characteristics of the basin and the resulting flood impacts upon humans and their activities (Riggs 2001). All land is not equal and human land use and subsequent management should be based upon both the geologic and hydrologic dynamics. If future impacts of similar flood events are to be minimized, key geologic and hydrologic data of the riverine systems must form the basis of land-use and management plans for each drainage basin. The floods of 1999 "rezoned" the Coastal Plain for eastern North Carolina at a very high cost.

2.5 AN INTEGRATED ASSESSMENT PROGRAM IN RESPONSE TO LARGE-SCALE STORM EVENTS

The issues discussed for PS and its sub-estuaries are symptomatic of other long-residence-time systems in the United States and worldwide. In the United States, an integrated monitoring, research, and assessment program is needed for East Coast and Gulf of Mexico estuarine systems under the influence of large storm events that cause major hydrologic, nutrient, and trophodynamic alterations. The development of an effective program for lagoonal systems like the PS requires consideration of at least two questions: (1) What type of assessment program is necessary and sufficient to adequately address the range of ecosystem impacts resulting from major human and climatic perturbations? (2) What are the specific issues unique to monitoring and assessment of infrequent but important events, known to be major drivers of ecological change in lagoonal ecosystems?

The monitoring and assessment program must be able to answer the following questions.

- How far away from "normal" conditions is the system changed by these events?
- To what extent do these systems recover; what conditions are required for recovery?
- Are there permanent changes to the ecosystem, and what are they?
- How do human activities affect recovery?
- Are any of the changes (short term or permanent) detrimental to human use and resources?

Important features of a long-term monitoring and assessment program include:

- Sampling to detect (1) trends, (2) changes in state equivalent to a step function, and (3) consequences of infrequent but large-scale events.
- Both a routine monitoring component and an event-response component planned well in advance of the approach of a storm.
- Spatially and temporally extensive monitoring of key environmental variables, using continuous and time-integrative sampling devices to assess water quality, productivity, and turbidity. These would include satellite- and aircraft-flown ocean color and infrared imagery, unattended continuous monitoring from platforms, and ships of opportunity (e.g., ferries) (Harding et al. 1994, 1995; Woodruff 1996; Buzzelli et al. 2003; Ensign and Paerl 2006).
- Measurement of water quality and habitat responses using meaningful, sensitive, and easily deployed indicators of environmental stress. Examples are planktonic community composition and activity indicators (Paerl et al. 2005), and the benthic macrofauna indicators, specifically changes in benthic community structure and composition (i.e., Indices of Biotic Integrity; Miller et al. 1988), plus sediment composition and contaminants (heavy metals, pesticides, PCBs, and PAHs).
- Assessment of sediments. The sediments that have filled estuarine basins contain a wealth of paleo-climate and paleo-sea level information for the past 10,000 years of coastal history (Riggs 1999; Fletcher et al. 2000; Anderson et al. 2001).
- Aggregation of meteorological data on storm paths, winds, rainfall, and flooding to provide a quantitative context for catastrophic storms. In the case of PS, this analysis must consider the fact that, until recently, the most active period of hurricanes in North Carolina (1930 to the 1950s) occurred when the human population was much lower and the amount of landscape modification was significantly less.
- Assessment of effects of hydrologic, chemical, and sediment loading on biotic communities impacting nutrient cycling, finfish and shellfish habitats, including water column, salt marsh, seagrass, and open sediment habitats.
- Fish community behavioral, physiological, and health responses to variation in water depth, salinity, temperature, and dissolved oxygen.
- Analysis of historical and contemporary data on fisheries landings and how landings respond to intense storm activity. Much of our current thinking in estuarine fisheries management developed prior to the period of elevated storm activity. We may need to reevaluate the relationship between human activities in coastal watersheds and their effects on water quality and fisheries under conditions of more severe events.

Modeling of watershed, water quality, habitat, and fisheries effects of intense storms is a key component of an integrated assessment program and is an important tool for synthesizing information and providing options for environmental management. For the PS basin there are models for land-use effects on nutrients (Stanley 1988), transfer of materials through soils (Chescheir et al.

1996), hydrodynamics (Robbins and Bales 1995), water quality and nutrient cycling (Reckhow 1999; Bowen and Hieronymous 2000; Borsuk et al. 2001, 2003; Christian and Thomas 2003; Arhonditsis et al. 2007), and ecosystem response to hypoxia (Baird et al. 2004). These models might be used to address different aspects of storm impacts, but most if not all have a major limitation. The hurricanes of 1999 produced conditions beyond those for which most models have been constructed or validated. Extension of model predictions to cover a wide range of infrequent and extreme conditions and validation of those predictions are both challenges to, and requirements of, an effective assessment program. In lieu of that, models become minimally usable for understanding these macro-events. The challenge to simulating effects of macro storm events is to find ways to quickly integrate the results of various independent models to aid in the scientific and management response to these events.

ACKNOWLEDGMENTS

We appreciate the thorough reviews and helpful comments provided by two anonymous reviewers. This work was supported by the North Carolina Department of Environment and Natural Resources, the Lower Neuse Basin Association, the North Carolina Sea Grant Program, Water Resources Research Institute of the University of North Carolina, the U.S. Department of Agriculture-NRI-CRGP, the U.S. EPA-STAR-EaGLe Program, and the National Science Foundation, Environmental Engineering, Chemical Oceanography, Biological Oceanography and Ecology Programs. Remote sensing imagery was provided by NASA's SeaWiFS Program.

REFERENCES

Adams, S. M., M. S. Greeley, J. M. Law, E. J. Noga, and J. T. Zelikoff. 2003. Application of multiple sublethal stress indicators to assess the health of fish in Pamlico Sound following extensive flooding. *Estuaries* 26: 1365–1382.

Anderson, J. B., A. B. Rodriguez, C. Fletcher, and D. Fitzgerald. 2001. Researchers focus attention on coastal response to climate change. *Eos Trans.* 82 (44): 513–520.

Arhonditsis, G. B., H. W. Paerl, L. M. Valdes-Weaver, C. A. Stow, L. J. Steinberg, and K. H. Reckhow. 2007. Application of Bayesian structural equation modeling for examining phytoplankton dynamics in the Neuse River Estuary (North Carolina, USA). *Estuar. Coastal Shelf Sci.* 72: 63–80.

Baird, D., R. R. Christian, C. H. Peterson, and G. Johnson. 2004. Consequences of hypoxia on estuarine ecosystem function: energy diversion from consumers to microbes. *Ecol. Applic.* 14: 805–822.

Bales, J. D. 2003. Effects of Hurricane Floyd inland flooding, September–October 1999, on tributaries to Pamlico Sound, North Carolina. *Estuaries* 26: 1319–1328.

Bales, J. D. and C. J. O. Childress. 1996. *Aftermath of Hurricane Fran in North Carolina: Preliminary data on flooding and water quality*. U.S. Geological Survey Open-File Report 96-499.

Bales, J. D., C. J. Oblinger, and A. H. Sallenger. 2000. *Two months of flooding in eastern North Carolina, September–October 1999: Hydrologic, water-quality, and geologic effects of Hurricanes Dennis, Floyd, and Irene*. U.S. Geological Survey Water-Resources Investigations Report 00-4093.

Biber, P. D., H. W. Paerl, C. L. Gallegos and W. J. Kenworthy. 2005. Evaluating indicators of seagrass stress to light. Pp. 193–209, In: S. Bartone (ed.), *Proceedings of the Estuarine Indicators Workshop*. Boca Raton, Florida: CRC Press.

Borsuk, M., C. A. Stow, R. A. Luettich, H. Paerl, and J. L. Pinckney. 2001. Probabilistic prediction of hypoxia in the Neuse River Estuary using an empirical model of oxygen dynamics. *Estuar. Coastal Shelf Sci.* 52: 33–49.

Borsuk, M. E., C. A. Stow, and K. H. Reckhow. 2003. Integrated approach to total maximum daily load development for Neuse River Estuary using Bayesian probability network model (Neu-BERN). *J. Water Resour. Plan. Manage.* 129: 271–282.

Bowen, J. D. and J. Hieronymous. 2000. *Neuse River Estuary Modeling and Monitoring Project Stage 1: Predictions and uncertainty analysis of response to nutrient loading using a mechanistic eutrophication model.* Water Resources Research Institute Report No. 325-D, University of North Carolina Water Resources Research Institute, Raleigh, North Carolina.

Boyer, J. N., D. W. Stanley, and R. R. Christian. 1994. Dynamics of NH_4^+ and NO_3^- uptake in the water column of the Neuse River Estuary, North Carolina. *Estuaries* 17: 361–371.

Burkholder, J. M., D. A. Dickey, C. A. Kinder, R. E. Reed, M. A. Mallin, M. R. McIver, L. B. Cahoon, G. Melia, C. Brownie, J. Smith, N. Deamer, J. Springer, H. B. Glasgow, and D. Toms. 2006. Comprehensive trend analysis of nutrients and related variables in a large eutrophic estuary: a decadal study of anthropogenic and climatic influences. *Limnol. Oceanogr*. 51: 463–487.

Buzzelli, C. P., R. A. Luettich, S. P. Powers, C. H. Peterson, J. E. McNinch, J. L. Pinckney, and H. W. Paerl. 2002. Estimating the spatial extent of bottom-water hypoxia and habitat degradation in a shallow estuary. *Mar. Ecol. Prog. Ser.* 230: 103–112.

Buzzelli, C. P., J. R. Ramus, and H. W. Paerl, 2003 Ferry-based monitoring of surface water quality in North Carolina estuaries. *Estuaries* 26:975–984.

Chescheir, G. M., R. W. Skaggs, J. W. Gilliam, M. A. Breve, C. L. Munster, R. B. Leidy, and J. E. Parsons. 1996. *Effects of drainage and water table control on groundwater and surface water quality. Part IIL Experimental results and simulation models.* Water Resources Research Institute Report No. 301. University of North Carolina, Water Resources Research Institute, Raleigh, North Carolina.

Christian, R. R., J. N. Boyer, and D. W. Stanley. 1991. Multi-year distribution patterns of nutrients within the Neuse River Estuary. *Mar. Ecol. Prog. Ser.* 71: 259–274.

Christian, R. R., W. Bryant, and D. W. Stanley. 1986. *The relationship between river flow and* Microcystis aeruginosa *blooms in the Neuse River, NC.* University of North Carolina, Water Resources Research Institute Report No. 223, Raleigh, North Carolina.

Christian, R. R., J. E. O'Neal, B. Peierls, L. M. Valdes, and H. W. Paerl. 2004. *Episodic nutrient loading impacts on eutrophication of the southern Pamlico Sound: The effects of the 1999 hurricanes.* University of North Carolina, Water Resources Research Institute Report No. 349, UNC Water Resources Research Institute, Raleigh, North Carolina.

Christian, R. R. and C. R. Thomas. 2003. Network analysis of nitrogen inputs and cycling in the Neuse River Estuary, North Carolina, USA. *Estuaries* 26: 815–828.

Cooper, S. R. 2000. *The history of water quality in North Carolina estuarine waters as documented in the stratigraphic record.* University of North Carolina, Water Resources Research Institute, Report No. 327. North Carolina State University, Raleigh, North Carolina.

Copeland, B. J. and J. Gray. 1991. *Status and trends report of the Albemarle-Pamlico Estuary*, J. Steel, Ed., Albemarle-Pamlico Estuarine Study Report 90-01. North Carolina Department of Environment, Health and Natural Resources, Raleigh, North Carolina.

Craig, N. M. and E. J. Kuenzler. 1983. *Land use, nutrient yield, and eutrophication in the Chowan River Basin.* University of North Carolina, Water Resources Research Institute Report No. 205. North Carolina State University, Raleigh, North Carolina.

Dodd, R. C., P. A. Cunningham, R. J. Curry, and S. J. Stichter. 1993. *Watershed planning in the Albemarle-Pamlico estuarine system.* Report No. 93-01, Research Triangle Institute, Research Triangle Park, North Carolina Department of Environment, Health and Natural Resources, Raleigh, North Carolina.

Eby, L. A. and L. B. Crowder. 2002. Hypoxia-based habitat compression in the Neuse River Estuary: context-dependent shifts in behavioral avoidance thresholds. *Can. J. Fish. Aquat. Sci.* 59: 952–965.

Eggleston, D. B., J. Hightower, and E. Johnson. 2004. *Population dynamics and stock assessment of the blue crab in North Carolina.* Final Report to the North Carolina Sea Grant/Blue Crab Research Program for Project for Projects 99-FEG-10 and 00-FEG-11.

Emanuel, K. 2005. Increasing destructiveness of tropical cyclones over the past 30 years. *Nature* 436: 686–688.

Ensign, S. H. and H. W. Paerl. 2006. Development of an unattended estuarine monitoring program using ferries as data collecting platforms. *Limnol. Oceanogr. Methods* 4: 399–405.

Fletcher, C., J. Anderson, K. Crook, G. Kaminsky, P. Larcombe, C. V. Murray-Wallace, F. Sansone, D. Scott, S. Riggs, A. Sallenger, I. Shennan, R. Theiler, and J. F. Wehmiller. 2000. Coastal sedimentary research examines critical issues of national and global priority: *EOS* 81: 181–186.

Gilliam, J. W., D. L. Osmond, and R. O. Evans. 1997. *Selected agricultural best management practices to control nitrogen in the Neuse River Basin.* North Carolina Agriculture Research Service Technical Bulletin 311. North Carolina State University, Raleigh, North Carolina.

Goldenberg, S. B., C. W. Landsea, A. M. Mestas-Nuñez, and W. M.Gray. 2001. The recent increase in Atlantic hurricane activity: causes and implications. *Science* 293: 474–478.

Hall, N. S., R. W. Litaker, E. Fensin, J. E. Adolf, A. R. Place, and H. W. Paerl. 2008. Environmental factors contributing to the development and demise of a toxic dinoflagellate (*Karlodinium veneficum*) bloom in a shallow, eutrophic, lagoonal estuary. *Estuaries Coasts* 31: 402–418.

Harding, L. W., E. C. Itsweire, and W. E. Esais. 1994. Estimates of phytoplankton biomass in the Chesapeake Bay from aircraft remote sensing of chlorophyll concentrations, 1989–1992. *Rem. Sens. Environ.* 49: 41–56.

Harding, L. W., E. C. Itsweire, and W. E. Esais. 1995. Algorithm development for recovering chlorophyll concentrations in the Chesapeake Bay using aircraft remote sensing, 1989–91. *Photogramming Eng. Rem. Sens.* 61: 177–185.

Harned, D. A. and M. S. Davenport. 1990. *Water-quality trends and basin activities and characteristics for the Albemarle-Pamlico estuarine system, NC and VA.* Report 90-398, U.S. Geological Survey, Raleigh, North Carolina.

Hobbie, J. E., B. J. Copeland, and W. G. Harrison. 1972. *Nutrients in the Pamlico Estuary, NC, 1969–1971.* Water Resources Research Institute Report No. 76, University of North Carolina Water Resources Research Institute, Raleigh, North Carolina.

Hobbie, J. E. and N. W. Smith. 1975. *Nutrients in the Neuse River Estuary, NC.* Report No. UNC-SG-75-21, UNC Sea Grant Program, North Carolina State University, Raleigh, North Carolina.

Holland, G. J. and P. J. Webster. 2007. Heightened tropical cyclone activity in the North Atlantic: natural variability of climate trend? *Philos. Trans. R. Soc. A.* DOI: 10.1098/rsta.2007.2083.

Huffman, R., P. Westerman, and J. Barker. 1994. *Estimated seepage losses from established swine waste lagoons in the lower coastal plain of North Carolina.* Paper No. BAE-94-13, North Carolina State University, Department of Biological and Agricultural Engineering, Raleigh, North Carolina.

Kuenzler, E. J., K. L. Stone, and D. B. Albert. 1982. *Phytoplankton uptake and sediment release of nitrogen and phosphorus in the Chowan River, NC.* WRRI Report No. 186, University of North Carolina, Water Resources Research Institute, North Carolina State University, Raleigh, North Carolina.

Lenihan, H. and C. H. Peterson. 1998. How habitat degradation through fishery disturbance enhances impacts of hypoxia on oyster reefs. *Ecol. Applic.* 8: 128–140.

Matson, E. A. and M. M. Brinson. 1985. Sulfate enrichments in the estuarine waters of North Carolina. *Estuaries* 8: 279–289.

Matson, E. A. and M. M. Brinson. 1990. Stable carbon isotopes and the C:N ratio in the estuaries of the Pamlico and Neuse Rivers, North Carolina. *Limnol. Oceanogr.* 35: 1290–1300.

Matson, E. A., M. M. Brinson, D. D. Cahoon, and G. J. Davis. 1983. *Biogeochemistry of the sediments of the Pamlico and Neuse River Estuaries, NC.* WRRI Report No. 191, University of North Carolina, Water Resources Research Institute, Raleigh, North Carolina.

Miller, D. L., P. M. Leonard, R. M. Hughes, J. R. Karr, P. B. Moyle, L. H. Schrader, B. A. Thompson, R. A. Daniel, K. D. Fausch, G. A. Fitzhugh, J. R. Gammon, D. B. Halliwell, P. L. Angermeier, and D. J. Orth. 1988. Regional applications of an index of biotic integrity for use in water resource management. *Fisheries* 13: 12–20.

National Hurricane Center. 1999. *Preliminary report: Hurricane Floyd.* NOAA, National Hurricane Center, Miami, Florida.

North Carolina Department of Environment and Natural Resources. 1998. *Neuse River Basin water quality plan.* Available at: http://h2o.enr.state.nc.us/basinwide/Neuse/neuse_wq_management_plan.htm. North Carolina Department of Environment and Natural Resources, Raleigh, North Carolina.

North Carolina Department of Environment and Natural Resources, Division of Water Quality. 2001. *Phase II of the total maximum daily load for total nitrogen to the Neuse River Estuary, NC.* NCDENR-DWQ, Raleigh, North Carolina.

North Carolina State Senate. 1996. *Senate Select Committee on water quality and fish kills, summary report.* State Senate Committee Report, Raleigh, North Carolina.

Paerl, H. W. 1982. *Environmental factors promoting and regulating N_2 fixing blue-green algal blooms in the Chowan River, NC.* WRRI Report No. 176. University of North Carolina, Water Resources Research Institute, Raleigh, North Carolina.

Paerl, H.W. 1983. *Factors regulating nuisance blue-green algal bloom potentials in the lower Neuse River.* WRRI Report No. 177. University of North Carolina, Water Resources Research Institute, Raleigh, NC.

Paerl, H.W. 1987. *Dynamics of blue-green algal (*Microcystis aeruginosa*) blooms in the lower Neuse River, NC: causative factors and potential controls.* WRRI Report No. 229. University of North Carolina, Water Resources Research Institute, Raleigh, North Carolina.

Paerl, H.W. 1997. Coastal eutrophication and harmful algal blooms: importance of atmospheric deposition and groundwater as "new" nitrogen and other nutrient sources. *Limnol. Oceanogr.* 42: 1154–1165.

Paerl, H. W., J. D. Bales, L. W. Ausley, et al. 2001. Ecosystem impacts of 3 sequential hurricanes (Dennis, Floyd and Irene) on the United States' largest lagoonal estuary, Pamlico Sound, NC. *Proc. Natl. Acad. Sci. USA* 98: 5655–5660.

Paerl, H. W. and M. L. Fogel. 1994. Isotopic characterization of atmospheric nitrogen inputs as sources of enhanced primary production in coastal Atlantic Ocean waters. *Mar. Biol.* 119: 635–645.

Paerl, H. W., M. A. Mallin, C. A. Donahue, M. Go, and B. L. Peierls. 1995. *Nitrogen loading sources and eutrophication of the Neuse River Estuary, NC: direct and indirect roles of atmospheric deposition.* WRRI Report No. 291. University of North Carolina, Water Resources Research Institute, Raleigh, North Carolina.

Paerl, H. W., M. A. Mallin, J. Rudek, and P. W. Bates. 1990. *The potential for eutrophication and nuisance algal blooms in the lower Neuse River, N.C.* Albemarle-Pamlico Estuarine Study Report 90-15, North Carolina Department of Natural Resources and Community Development, Raleigh, North Carolina.

Paerl, H. W., M. F. Piehler, L. M. Valdes, et al. 2005. Determining anthropogenic and climatically-induced change in aquatic ecosystems using microbial indicators: an integrative approach. *Verh. Int. Verein. Limnol.* 29: 89–133.

Paerl, H. W., J. L. Pinckney, J. S. Fear, and B. L. Peierls. 1998. Ecosystem response to internal and watershed organic matter loading: consequences for hypoxia in the eutrophying Neuse River Estuary, N.C., USA. *Mar. Ecol. Prog. Ser.* 166: 17–25.

Paerl, H. W., L. M. Valdes, J. E, Adolf, B. M. Peierls, and L. W. Harding Jr. 2006a. Anthropogenic and climatic influences on the eutrophication of large estuarine ecosystems. *Limnol. Oceanogr.* 51: 448–462.

Paerl, H. W., L. M. Valdes, A. R. Joyner, et al. 2006b. Ecological response to hurricane events in the Pamlico Sound System, NC and implications for assessment and management in a regime of increased frequency. *Estuaries Coasts* 29: 1033–1045.

Paerl, H. W., L. M. Valdes, A. R. Joyner, and V. Winkelmann. 2007. Phytoplankton indicators of ecological change in the nutrient and climatically-impacted Neuse River-Pamlico Sound System, NC. *Ecol. Applic.* 17: 88–101.

Paerl, H. W., L. M. Valdes, M. F. Piehler, and M. E. Lebo. 2004. Solving problems resulting from solutions: the evolution of a dual nutrient management strategy for the eutrophying Neuse River Estuary, North Carolina, USA. *Environ. Sci. Tech.* 38: 3068–3073.

Paerl, H. W., L. M. Valdes, M. F. Piehler, and C. A. Stow. 2006c. Assessing the effects of nutrient management in an estuary experiencing climatic change: the Neuse River Estuary, NC, USA. *Environ. Man.* 37: 422–436.

Paerl, H. W. and D. R. Whitall. 1999. Anthropogenically-derived atmospheric nitrogen deposition, marine eutrophication and harmful algal bloom expansion: Is there a link? *Ambio* 28: 307–311.

Peierls, B. L., R. R. Christian, and H. W. Paerl. 2003. Water quality and phytoplankton as indicators of hurricane impacts on large estuarine ecosystems. *Estuaries* 26: 1329–1343.

Pietrafesa, L. J., G. S. Janowitz, T.-Y. Chao, et al. 1996. *The physical oceanography of Pamlico Sound.* University of North Carolina Sea Grant Publication UNC-WP-86-5.

Pinckney, J. L., H. W. Paerl, M. B. Harrington, and K. E. Howe. 1998. Annual cycles of phytoplankton community structure and bloom dynamics in the Neuse River Estuary, NC (USA). *Mar. Biol.* 131: 371–382.

Reckhow, K. H. 1999. Water quality prediction and probability network models. *Can. J. Fish. Aquatic Sci.* 56: 1150–1158.

Reynolds-Fleming, J. V., J. G. Fleming, and R. A. Luettich. 2002. Portable autonomous vertical profiler for estuarine applications. *Estuaries* 25: 142–147.

Riggs, S. R. 1999. Coupled inner shelf-shoreline sediment responses to small-scale Holocene sea-level fluctuations: North Carolina coastal zone. Pp. 165–167, In: C. H. Fletcher and J.V. Matthews (eds.), *The non-steady state of the inner shelf and shoreline: coastal change on the time scale of decades to millennia in the late Quaternary.* Technical Report, University of Hawaii, Honolulu.

Riggs, S. R. 2001. *Anatomy of a flood, in facing our future: Hurricane Floyd and recovery in the coastal plain.* Pp. 29–45, In: J. R. Maiolo, J. C. Whitehead, M. McGee, L. King, J. Johnson and H. Stone (Eds.), *Facing Our Future: Hurricane Floyd and Recovery in the Coastal Plain.* Coastal Carolina Press, Wilmington, North Carolina.

Robbins, J. C. and J. D. Bales. 1995. *Simulation of hydrodynamics and solute transport in the Neuse River Estuary.* USGS Open File Report 94-511. U.S. Geological Survey, Raleigh, North Carolina.

Rudek. J., H. W. Paerl, M. A. Mallin, and P. W. Bates 1991. Seasonal and hydrological control of phytoplankton nutrient limitation in the lower Neuse River Estuary, North Carolina. *Mar. Ecol. Prog. Ser.* 75: 133–142.

Spruill, T. B., D. A. Harned, P. M. Ruhl, J. L. Eimers, G. McMahon, K. E. Smith, D. R. Galeone, and M. D. Woodside. 1998. Water quality in the Albemarle-Pamlico drainage basin: North Carolina and Virginia, 1992–1995. U.S. Geological Survey Circular 1157.

Stanley, D. W. 1983. *Nitrogen cycling and phytoplankton growth in the Neuse River*. North Carolina Water Resources Report No. 204, University of North Carolina, Water Resources Research Institute, Raleigh, North Carolina.

Stanley D. W. 1988. Historical trends in nutrient loading to the Neuse River Estuary, NC. Pp. 155–164, In: W. Lyke and T. Hoban (eds.), *Proceedings of the American Water Resources Association Symposium on Coastal Water Resources*, AWRA Tech. Publ. Ser. TPS-88-1. AWRA, Bethesda, Maryland.

Stanley, D. W. and S. W. Nixon. 1992. Stratification and bottom water hypoxia in the Pamlico River Estuary. *Estuaries* 15: 270–281.

Stow, C. A., M. E. Bursuk, and D. W. Stanley. 2001. Long-term changes in watershed nutrient inputs and riverine exports in the Neuse River, North Carolina. *Water Res.* 35: 1489–1499.

Tedder, S. W., J. Sauber, J. Ausley, and S. Mitchell. 1980. *Working paper: Neuse River investigation 1979.* Division of Environmental Management, North Carolina Department of Natural Resources and Community Development, Raleigh, North Carolina.

Valdes-Weaver, L. M., M. F. Piehler, J. L. Pinckney, K. E. Howe, K. Rosignol, and H. W. Paerl. 2006. Long-term temporal and spatial trends in phytoplankton biomass and class-level taxonomic composition in the hydrologically variable Neuse-Pamlico estuarine continuum, NC, USA. *Limnol. Oceanogr.* 51(3): 1410–1420.

Webster, P. J., G. J. Holland, J. A. Curry, and H. R. Chang. 2005. Changes in tropical cyclone number, duration, and intensity in a warming environment. *Science* 309: 1844–1846.

Wetz, M. S. and H. W. Paerl. 2008. Estuarine phytoplankton responses to hurricanes and tropical storms with different characteristics (trajectory, rainfall, winds). *Estuaries Coasts* 31: 419–429.

Winslow, F. 1889. Report on the sounds and estuaries of North Carolina with reference to oyster culture. *U.S. Coast & Geodetic Survey Bull.* 10: 52–136.

Woodruff, D. L. 1996. *Water clarity of the Neuse River-Pamlico Sound estuary, North Carolina:* In situ *and remote spectral characterization.* UMI Dissertation Services, Ann Arbor, MI.

3 Sources and Fates of Nitrogen in Virginia Coastal Bays

Iris C. Anderson, Jennifer W. Stanhope, Amber K. Hardison, and Karen J. McGlathery*

CONTENTS

Abstract ... 43
3.1 Introduction ... 44
3.2 Sources of Nitrogen to Coastal Bays ... 48
3.2.1 Description of the Virginia Coastal Bays ... 48
3.2.2 Nutrient Gradient across Coastal Bays ... 49
3.2.3 Allochthonous Sources of Nitrogen to Virginia Coastal Bays ... 49
3.2.3.1 Stream Flow ... 50
3.2.3.2 Direct Groundwater Discharge ... 52
3.2.3.3 Atmospheric Deposition ... 54
3.2.4 Autochthonous Sources of Nitrogen to Virginia Coastal Bays ... 56
3.2.4.1 Nitrogen Fixation ... 56
3.2.4.2 Remineralization ... 57
3.3 Sinks of Nitrogen in Coastal Bays ... 58
3.3.1 Abundance of Primary Producers in Coastal Bays: Seasonality ... 58
3.3.2 Seagrasses in HIB ... 58
3.3.3 Nitrogen Assimilation by Primary Producers ... 58
3.3.4 Sources of Nitrogen to Support Benthic Primary Production ... 61
3.4 Fate of Nitrogen Assimilated by Benthic Macro- and Microalgae ... 61
3.5 Discussion: Future Trends and Predictions ... 63
3.5.1 Potential Responses to Seagrass Restoration in the Virginia Coastal Bays ... 65
Acknowledgments ... 66
References ... 67

ABSTRACT

Coastal bays, which are typically shallow, located in the photic zone, and have little freshwater input, respond differently to nutrient enrichment than do deeper estuaries. Because of the diversity of benthic and pelagic autotrophs they support, coastal bays are capable of modulating the effects of nutrient enrichment, derived from both allochthonous and autochthonous sources. We describe a study performed in Hog Island Bay, Virginia, located along a eutrophication gradient on the Delmarva Peninsula, to determine sources, sinks, and fates of nitrogen (N). On an annual basis, allochthonous sources of N, mainly base flow and atmospheric deposition, contributed 13%, whereas organic N remineralization (NRemin) in the sediments, supplied 77% of the total N available to support primary production. On a daily basis, N released by NRemin approximated the demand by benthic microalgae. Annually, N demand by benthic microalgae accounted for 55%,

* Corresponding author. Email: iris@vims.edu

phytoplankton 16%, salt marshes 16%, macroalgae 7%, and denitrification 6% of total N uptake. Maximum daily N assimilation rates for benthic microalgae, macroalgae, and phytoplankton were remarkably similar despite large differences in standing biomass, suggesting that turnover time is a critical parameter regulating availability of N to support continued autotrophic production and retention of N in sediments. Experiments were performed in sediment mesocosms taken from Hog Island Bay to track ^{15}N into macroalgae, benthic microalgae, bulk sediment, and the bacterial-specific biomarker D-alanine (D-Ala). When live, macroalgae competed with benthic microalgae for light and nutrients; when dead, up to half of the incorporated N in macroalgae was transferred to sediments. Within sediments we observed shuttling of N between bacterial and benthic microalgal pools increasing the retention time of N within the system. At nutrient enrichment levels above some "critical threshold," we suggest that pelagic production will become increasingly dominant; benthic autotrophs will no longer function to help retain and recycle nutrients, which will then flux out into the water column to further support eutrophication.

Key Words: coastal bays, benthic microalgae, base flow, nitrogen cycling, macroalgae, nitrogen fixation, nitrification–denitrification, remineralization, d-alanine, compound specific isotopes, biomarkers

3.1 INTRODUCTION

The Delmarva Peninsula, located along the mid-Atlantic coast of the United States and encompassing coastal areas of Delaware, Maryland, and Virginia, is bordered by shallow coastal bays, a common feature of shorelines worldwide. Coastal bays, which occur on all continents except for Antarctica, are typically characterized by shallow depth, extension of the photic zone to the benthic surface, little freshwater input, and diverse benthic primary producers. Because of their location on the land margin and often having a high ratio of watershed area to water surface area, coastal bays are uniquely vulnerable to the nutrient enrichment that results from the activities of a large human population, currently more than half the total U.S. population, who lives along our coasts (NRC 2000; EPA 2001). In many regions, nutrient enrichment by reactive nitrogen of coastal waters is accelerating due to development and intensification of agriculture, which may partly be due to production of biofuels (Galloway et al. 2008; Howarth 2008); in other areas, improved technology has reduced inputs of nutrients from both point and diffuse sources (Nixon 2009).

Nutrient enrichment, especially by nitrogen (N), is thought to be the major cause of eutrophication in coastal systems; however, numerous physicochemical and biotic factors may modulate the eutrophication response, particularly in shallow systems (Kennish and Townsend 2007; Paerl et al. 2007; Nixon 2009). Although the magnitude of N loads to shallow coastal bays is similar to that of estuaries, responses to N enrichment may be very different (Nixon et al. 2001; McGlathery et al. 2007). Based on results of an experimental study performed at the Marine Ecosystems Research Laboratory (MERL), Nixon et al. (2001) concluded that, whereas in deeper estuaries (>5 m) water column chlorophyll and phytoplankton primary production were correlated with N loading, in shallow coastal systems (<2 m), there was no clear correspondence between N inputs and primary production. On the other hand, Boynton et al. (1996) were able to show a strong relationship between N loading and water column chlorophyll *a* for the Maryland coastal bays.

Interactions among a variety of biological and physical factors modulate responses of shallow and deep coastal systems to nutrient enrichment (Cloern 1999, 2001; Kemp et al. 2005). For example, phytoplankton growth in coastal systems varies depending upon water column depth, light availability, benthic microalgal production, and degree of benthic grazing (Lucas et al. 1999a). In San Francisco Bay the presence of large populations of benthic filter feeders, *Potamocorbula amurensis,* was observed to limit phytoplankton abundance despite high nutrient loading. In addition to light, other physical factors, such as tidal energy (Monbet 1992), horizontal transport, water residence time (Lucas et al. 1999b), and stratification (Kemp et al. 2005), may also modulate eutrophication

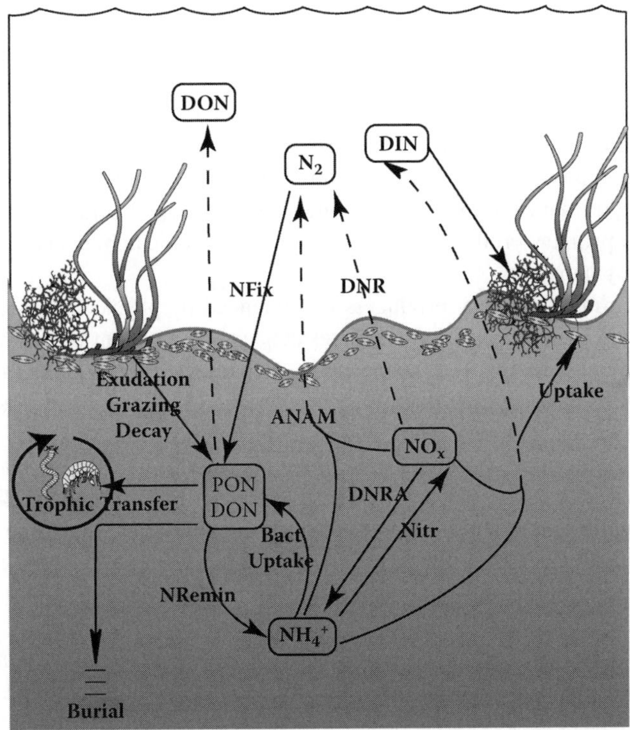

FIGURE 3.1 A conceptual model demonstrating benthic-pelagic coupling in shallow coastal bays, the interactions between benthic autotrophs and microbial nitrogen (N) cycling processes, and the influence of benthic autotrophs on sediment/water fluxes of dissolved inorganic and organic N (DIN, DON). Microbial N processes include nitrogen fixation (NFix), nitrogen remineralization (NRemin), nitrification (Nitr), denitrification (DNR), anammox (Anam), and dissimilatory nitrate reduction to ammonium (DNRA). PON = particulate organic nitrogen. Dashed lines represent fluxes out of the sediment.

responses to nutrient addition. These physical factors also play a role in regulating the relative importance of benthic vs. pelagic primary production and the coupling between benthic and pelagic processes. Whereas phytoplankton may limit light in deep estuarine systems, turbidity caused by wind-driven resuspension may be a more important controlling factor in shallow systems (Lawson et al. 2007). In coastal bays benthic N cycling processes (Figure 3.1), including nitrogen fixation (NFix), nitrogen remineralization (NRemin), nitrification (Nitr), denitrification (Denit), anammox (Anam), and dissimilatory nitrate reduction to ammonium, are thought to play an important role in determining N availability for supporting primary production in both pelagic and benthic habitats, and thus the potential for eutrophication (Sundbäck et al. 2000; Gardner et al. 2006; Fulweiler et al. 2007; Joye and Anderson 2008).

In shallow photic coastal systems, nutrient enrichment may not cause an increase in total system production but instead may cause a shift in dominant species of primary producers with potential feedbacks that may enhance eutrophication (Sand-Jensen and Borum 1991; Taylor et al. 1995; Valiela et al. 1997; McGlathery et al. 2007). Pelagic production increases at the expense of benthic production, due most likely to light limitation (Fear et al. 2004; Kemp et al. 2005). As eutrophication progresses, seagrasses and perennial macroalgae are likely to be replaced by ephemeral macroalgae and epiphytes, and finally by phytoplankton. As this shift occurs among dominant primary producers, we would expect changes in rates of organic matter burial and degradation, shifts in food web structure, and changes in rates of nutrient recycling and retention (McGlathery et al. 2007). The role of benthic microalgae in this cascade remains uncertain, although they clearly influence retention and transformation of nutrients in sediments (Sundbäck et al. 2000, 2006; Eyre and Ferguson 2002, 2005; Anderson et al. 2003; Risgaard-Petersen 2003; Ferguson et al. 2007; Joye and Anderson 2008). Although the structure of the primary producer community is likely to exert a strong influence on biogeochemical process rates, there are currently only limited data available to enable predictions of how succession within autotrophic communities resulting from nutrient enrichment might affect benthic biogeochemistry and its feedbacks (McGlathery et al. 2007; Joye and Anderson 2008). It has been assumed that bottom-up effects are more important than top-down effects in structuring primary producer communities; however, Heck and Valentine (2007) point out that accumulation of primary producer biomass in shallow benthic habitats is more likely controlled by consumers than by nutrients.

Succession of dominant primary producers is influenced by physicochemical drivers and their interactions, including light availability, water residence time, nutrient enrichment, sediment type and biogeochemistry, and grazing. For the most part, N loads to the bays along the northern portions of the Delmarva Peninsula are higher than those in Virginia (Giordano 2009), resulting in a regional gradient of eutrophication, varying from slightly eutrophied at the Virginia end, to moderately eutrophied along the Maryland shoreline, to highly eutrophied in Delaware. The Maryland coastal bays appear to be in a state of transition with available metrics indicating degrading conditions with increasing macroalgal biomass and blooms of harmful algal species, including *Aureococcus anophagefferens* (Wazniak et al. 2007; Bricker et al. 2008). In recent years, there has been a reversal in trend from decreasing to increasing concentrations of total nitrogen (TN), total phosphorus (TP) and chlorophyll *a* (chl *a*), potentially responsible for the leveling off of seagrass abundance, which had shown a 320% increase between 1986 and 2001 (Wazniak et al. 2007). Unlike the Maryland bays, successful seagrass restoration in some of the Virginia coastal bays suggests the potential for improving health (Orth et al. 2006). Because of their location along a regional gradient of eutrophication and the transitional states of some of the systems, the Delmarva coastal bays provide an excellent opportunity to test hypotheses regarding the drivers that regulate dominance by primary producers, trophic status, changes in nutrient cycling and retention that both result from and may feed back to cause changes in autotrophic community composition.

The primary sources of allochthonous nutrients to coastal bays along the mid-Atlantic coast are wastewater treatment plants and nonpoint sources, including fertilizers, septic systems, animal feeding operations, especially poultry production, and combustion of fossil fuels (Giordano 2009; Stanhope et al. 2009). Nutrients are transported from watersheds to the coastal bays in atmospheric deposition, surface water runoff, and groundwater, which provides sustained stream flow, called base flow, during nonprecipitation periods and may also discharge directly to the coastal bays. The relative importance of these sources and modes of transport depends on land use and soil characteristics in the watersheds, rainfall, evapotranspiration, presence of a riparian zone or fringing marshes, and the ratio of watershed area to waterbody surface area of the coastal bay (Peterjohn and Correll 1984; Stanhope et al. 2009). On the Delmarva Peninsula, the shallow unconfined aquifer from which base flow and direct groundwater discharge are derived is contaminated with inorganic N, primarily from agricultural land use although, as shown in Table 3.1, urban development and poultry feeding operations have impacted groundwater in the northern portion of the Delmarva Peninsula (Weil 1990; Boynton et al. 1996; Dillow et al. 2002; Stanhope 2003).

Autochthonous sources of nutrients that may also play an important role in supporting primary production in coastal bays are NFix and NRemin, both occurring principally in the sediments. Since NFix is an anaerobic process, rates are generally higher in the benthos than in the water column (Paerl 2009). Although previous work suggested that NFix is not an important source of N in saline estuaries, recent data suggest that NFix may indeed provide a considerable amount of N in shallow photic coastal systems (Gardner et al. 2006; Fulweiler et al. 2007; Fulweiler and Nixon 2009). NRemin rates are also likely to be high in shallow photic coastal bays, which generally support high rates of primary production. Although some of the organic matter, especially dissolved species, are decomposed in the water column, most NRemin occurs in the benthos producing

TABLE 3.1
Comparison of Hog Island Bay, Virginia vs. Maryland Coastal Bays

	HIB	MD Bays	Sources
Ratio watershed/lagoon area	0.61	1.60	Boynton et al. (1996); Hayden and Porter (2001); Oertel (2001)
% Forested (including wetlands)	51.4	54.2	Porter and Hayden (2001); MD Dept. of Planning (2004)
% Agriculture	45.3	34.5	
% Developed	3.3	9.4	
Total poultry farms[a]	1	143	Google Earth; USDA (2002)
Mean residence/flushing time (days)	16	13	Pritchard (1960); Fugate et al. (2006)
Mean phyto. NPP; summer (gC m^{-2} d^{-1})[b]	1.1	0.6	McGlathery et al. (2001); Giordano (2009)
Mean water column chl *a* (SE) (mg m^{-3})[c]	7.3 (1)	11 (2)	McGlathery et al. (2001); Wazniak et al. (2004); Anderson (unpub); Hardison (unpub)
Max macroalgal biomass (SE) (g dwt m^{-2})[d]	96 (62)	187 (64)	McGlathery et al. (2001); Wazniak et al. (2004); Hardison (unpub)
Direct atmospheric N deposition to bay (kg ha^{-1} yr^{-1})	9.5	7.9–11.6[e]	Meyers et al. (2001); this chapter
Diffuse + point N loadings (kg ha^{-1} yr^{-1})[f]	4.8	35.6	Boynton et al. (1996) and references therein; Stanhope (2003)
Total N loadings (kg ha^{-1} yr^{-1})	14.3	43.4–47.1	

[a] One poultry farm was identified in HIB watershed using aerial maps on Google Earth, 2009 Digital Globe, GeoEye, Commonwealth of Virginia, USDA Farm Service Agency Map data. The number of poultry farms for MD bays is based on Worcester, MD county data.

[b] Net primary production (NPP) is uptake of C. HIB data were collected in August 1998. MD bays data were for Isle of Wight Bay collected in July 2007, July 2008, and August 2008.

[c] HIB data were collected intermittently from 1998 to 2007. MD bays data were based on medians of monthly measurements for each lagoon from 2001 to 2003.

[d] HIB data were collected from 1998 to 2000 and 2006 to 2007. MD bays data were collected from surveys in 2001 and 2003.

[e] Range encompasses estimates for deposition to the water surface and watershed. Meyers et al. (2001) reduced dry deposition by a factor of 4 to estimate atmospheric deposition to water surfaces.

[f] Diffuse loadings include direct groundwater inputs and are normalized by water body surface area; 1 kg ha^{-1} yr^{-1} = 0.1 g m^{-2} yr^{-1}.

dissolved inorganic N (DIN) and dissolved organic N (DON), which can support production of both benthic microalgae (Anderson et al. 2003) and macroalgae (Tyler et al. 2001, 2003). In fact, benthic microalgae are thought to provide an important ecological service in shallow photic systems by capping the sediments and preventing release of ammonium (NH_4^+) to the water column where it could promote eutrophication. However, benthic microalgae only serve to "cap" the sediment nutrient flux if their own biomass or exudates are converted into forms of N and carbon (C) that are either not released to the water column or are released in a less biologically labile form. We will later describe mechanisms responsible for retention of benthic microalgal-derived N and C in sediments.

Coastal bays of the Delmarva Peninsula support a wide diversity of benthic autotrophs, which are thought to play a critical role in removal, retention, and transformation of N in these systems. Seagrasses, mainly *Zostera marina*, were very abundant until the 1930s throughout the Delmarva bays; however, they disappeared from the southern Virginia bays as a result of wasting disease and the hurricane of 1933 (Orth et al. 2006; McGlathery et al. 2007). An effort to restore *Z. marina* to the Virginia bays in recent years has met with a great deal of success (Orth et al. 2006). In the Maryland bays *Z. marina* currently occupies approximately 67% of its potential habitat; although its abundance/coverage has increased by approximately threefold since 1986, that increase has leveled off in the last few years (Wazniak et al. 2007). Other benthic autotrophs common to many of the Delmarva coastal bays include a variety of macroalgae such as *Ulva, Gracilaria, Agardhiella,* and *Chaetomorpha*, benthic microalgae, including diatoms and both filamentous and coccoid cyanobacteria, and seagrasses and their epiphytes (Goshorn et al. 2001; McGlathery et al. 2001, 2004; Glibert et al. 2007).

The presence of benthic autotrophs buffers the effects of nutrient enrichment and reduces the potential for eutrophication. The effectiveness of benthic primary producers in removing and retaining N depends on: (1) the dominant autotrophic species present and the turnover time of the autotroph (McGlathery et al. 2004, 2007); (2) whether the dominant N species is DIN or DON; (3) whether the DON is labile and can be taken up by autotrophs (Tyler et al. 2001, 2003); (4) whether the source of N is the water column or sediment pore water; (5) residence time in the coastal bays (Fugate et al. 2006); (6) resuspension (Lawson et al. 2007) and quality and quantity of light (MacIntyre et al. 1996); and (7) potential limitations by phosphate or silicate. Retention of N by benthic autotrophs is temporary, with retention time in sediments greatest for rooted macrophytes, intermediate for macroalgae, and shortest for benthic microalgae (Duarte and Cebrian 1996). A conceptual model shown in Figure 3.1 describes potential fates of N assimilated by benthic autotrophs, which may be: (1) grazed and transferred to higher trophic levels; (2) exuded by live or grazed autotrophs; (3) decomposed to dissolved organics, which may support benthic and pelagic heterotrophy; (4) immobilized to build microbial biomass; (5) mineralized to DIN and DON to support both benthic and pelagic autotrophy; (6) converted to gaseous forms (N_2, N_2O) by coupled Nitr–Denit or by Anam; (7) exported as dissolved, particulate, or gaseous species; or (8) buried as recalcitrant organic matter (Tyler et al. 2001, 2003; McGlathery et al. 2004; Joye and Anderson 2008). These processes are regulated by complex biological–physical–chemical interactions, including but not confined to resuspension, tidal mixing and residence time, porewater advection, gaseous exchange between water and the atmosphere, and by bioturbation. Given these complexities, it is understandable that much remains to be learned about the fate of N taken up by benthic autotrophs over short (hours) and long (>1 year) time scales.

In this chapter we will focus on research performed in the Virginia coastal bays over approximately a 12-year period, from 1997 to 2009. Most of the work was done in Hog Island Bay (HIB), the largest of the Virginia coastal bays. We will describe the sources, sinks, and potential fates of N and the biological–physical interactions that modulate those fates.

3.2 SOURCES OF NITROGEN TO COASTAL BAYS

3.2.1 Description of the Virginia Coastal Bays

Studies were performed primarily in HIB, located in the southern portion of the Delmarva Peninsula, which forms the eastern shore of the Chesapeake Bay. HIB on Virginia's Eastern Shore is located within the Virginia Coast Reserve (VCR), a National Science Foundation Long Term Ecological Research (LTER) site (Figure 3.2). The lagoon extends over an area of approximately 150 km^2 (Oertel 2001). It is both surrounded and intersected by extensive areas of salt marshes and mudflats, which are drained by a network of intertidal creeks. Relict oyster reefs are common throughout the lagoon. Average water depth within the lagoon is 1 to 2 m at mean low water (MLW); the semidiurnal tidal range is 1.1 m along the barrier islands and 1.5 m along the mainland edge (Santos 1996). Samples were taken at nine stations (three stations per site) along a transect, established across HIB from the mainland border to Hog Island, a barrier island, a distance of approximately 10 km. The "Creeks" sites were located at the mouths of tidal creeks draining salt marshes on the mainland border of the lagoon; the "Shoals" sites were located mid-lagoon adjacent to relict oyster reefs; the "Islands" sites were located adjacent to Hog Island in a back-barrier embayment. The average water depth at the three sites is approximately 0.5 m at MLW.

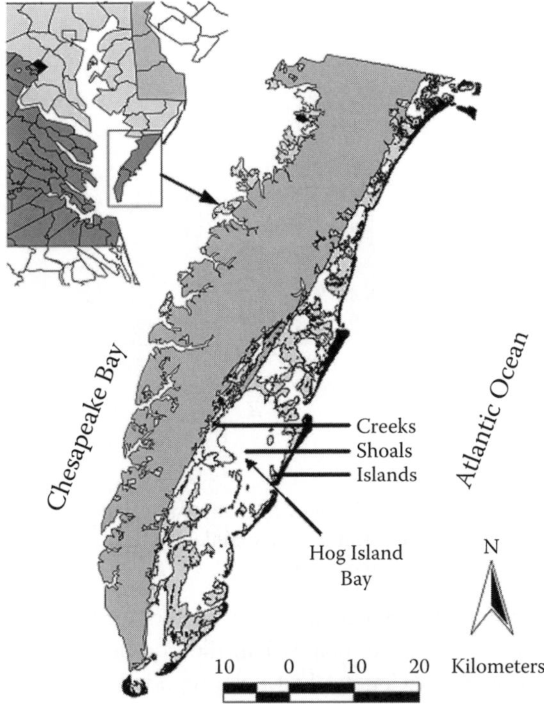

FIGURE 3.2 Hog Island Bay located in the Virginia portion of the Delmarva Peninsula. Sampling sites were located in mainland creeks (Creeks), mid-lagoon shoal areas (Shoals), and adjacent to barrier islands (Islands).

3.2.2 Nutrient Gradient across Coastal Bays

There is typically a nutrient gradient across coastal lagoons, with highest concentrations closest to the mainland, reflecting both watershed nutrient inputs to the lagoon and variations in local water residence times. Figure 3.3 shows data collected along a transect from the mainland to the barrier islands over a 5-year period in HIB. DON concentrations exceeded inorganic N, and represented >75% of total dissolved N. Although DON was the most common form of dissolved N in the lagoon proper, NO_3^- was the dominant N species, comprising 66% to 98% of the TDN pool, in base flow of 14 streams that discharge to the Virginia coastal bays (Stanhope et al. 2009). The high percentage of DON in lagoon waters is likely a product of microbial decomposition and recycling of organic matter in both the water column and sediments (Christian et al., Chapter 4, this volume).

3.2.3 Allochthonous Sources of Nitrogen to Virginia Coastal Bays

Anthropogenic sources of nutrients to surface and groundwater include fossil fuels, fertilizer, symbiotic N fixation from leguminous crops, animal waste, septic systems, and wastewater from treatment plants (Jordan and Weller 1996; Howarth et al. 2002). The dominant nutrient sources to coastal waters vary depending on land-use activities in the watershed. For the watersheds of the Virginia coastal bays, agriculture is the primary land cover (56%), followed by forest (39%), developed land (3%), and wetlands (2%) (Porter and Hayden 2001). Wheat, soybean, corn, and tomatoes are the primary crops grown. Tomato plasticulture, an increasingly popular agricultural practice on Virginia's Eastern Shore, has potentially greater impacts on watershed N sources to the coastal bays than other crops since 1 ha of tomato plasticulture is predicted to leach 433 kg N y^{-1} compared to 35 and 22 kg N y^{-1} for corn and soybeans, respectively (Giordano 2009). Poultry farming is another

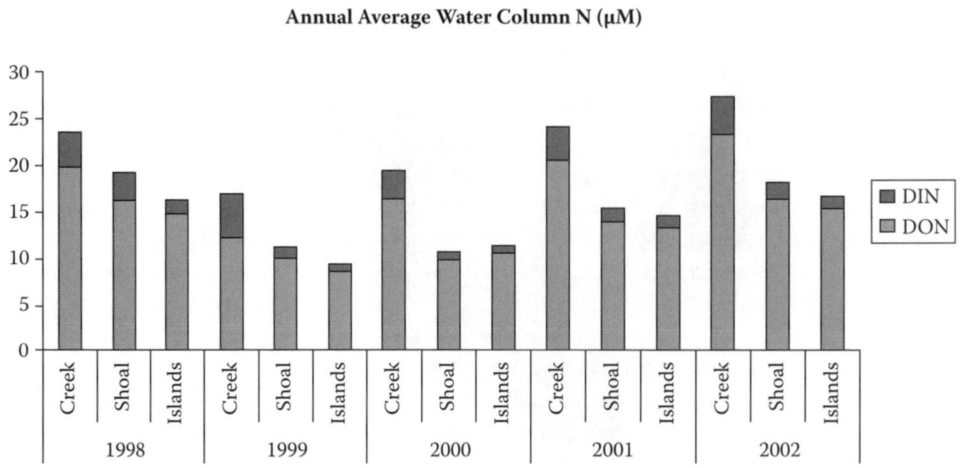

FIGURE 3.3 Nutrient gradient across Hog Island Bay—annual average water column dissolved inorganic and organic nitrogen (DIN, DON) concentrations at mainland creeks (Creeks), mid-lagoon shoal areas (Shoals), and adjacent to barrier islands (Islands) sites.

important activity especially in the northern county of the Virginia Eastern Shore (Accomack County), which had 77 farms and more than 31.7 million broiler chickens raised in 2002, while the southern county (Northampton County) had none reported (USDA 2002). Only two creeks, Parker and Little Mosquito creeks, that drain to Metompkin and Chincoteague bays, respectively, and Chincoteague Channel currently receive permitted wastewater discharges (U.S. Environmental Protection Agency Envirofacts Database, http://www.epa.gov/enviro/#water). As a result, nonpoint sources are the dominant source of N from the rural watersheds of the Virginia coastal bays. These allochthonous sources of nutrients are transported to the coastal bays via base flow, surface water runoff, direct groundwater discharge, direct atmospheric deposition, and exchange with the coastal ocean. This section will discuss each of these pathways, except ocean exchange, and their relative importance in transporting N to the Virginia coastal bays and, in particular, to HIB. Unfortunately, little is currently known about exchanges between the coastal bays and ocean, although modeling activities currently under way to determine residence times will allow future estimates of inputs and exports between bays and the ocean.

3.2.3.1 Stream Flow

Within watersheds, nutrients enter groundwater during precipitation events by recharge of the aquifer and in direct surface water runoff to streams (stormwater runoff). Base flow, groundwater discharge to streams from the saturated zone of shallow aquifers, provides a sustained flow of water in streams when there are no precipitation events. Approximately 56 major streams drain to the Virginia coastal bays (Hayden and Porter 2001). Because of the relatively narrow width of the lower Delmarva Peninsula (5 to 20 km wide), watershed catchments and stream inputs to the Virginia coastal bays are generally smaller than in the upper Delmarva Peninsula and western shore of the Chesapeake Bay. As a result, the contributions of total stream flow (base flow and surface water runoff) to total N loading of the Virginia coastal bays are expected to be less significant in comparison to those lagoons with large watersheds and estuaries with major tributaries that receive greater freshwater input, such as the Chesapeake Bay.

Base flow, in comparison to surface water runoff, can be a major conduit for transporting nutrients and freshwater to coastal waters, especially in regions such as the Delmarva Peninsula where high aquifer permeability and relatively flat topography promote groundwater recharge instead of runoff (Robinson and Reay 2002). In studies conducted in the mid-Atlantic Coastal Plain (Bachman

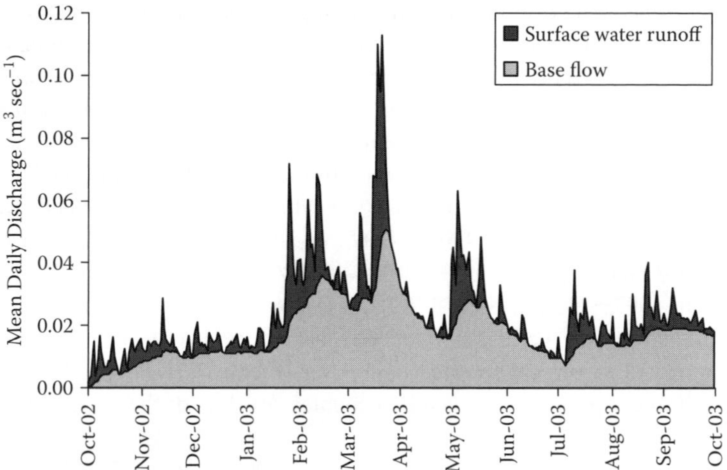

FIGURE 3.4 Total stream flow separated into base flow (light gray filling) and surface water runoff (dark gray filling) components for Nickawampus Creek from October 2002 to October 2003 (Stanhope et al. in prep.).

et al. 1998), and specifically in the upper Delmarva Peninsula of Maryland and Delaware (Bohlke and Denver 1995; Phillips and Bachman 1996; Jordan et al. 1997c), base flow contributed on average between 60% and 80% of total stream flow and represented a significant source of N to low-order streams. Monitoring results for a nontidal stream in a Virginia coastal bay watershed demonstrated that base flow accounted for 72% to 85% of annual total stream flow (Stanhope et al. in prep) (Figure 3.4). Stormwater runoff has also been observed to dilute NO_3^- concentrations in base flow (Bohlke and Denver 1995; Jordan et al. 1997c). Therefore, surface water runoff relative to base flow is likely to contribute a smaller fraction of nutrients to total stream flow in streams of the Piedmont and Coastal Plain regions of the Chesapeake Bay.

Methods for estimating N loadings in base flow and surface water runoff from watersheds include a direct method and use of watershed models. The direct method involves collecting hydrographic data, including stream discharge rates and stage height, developing a ratings curve to estimate continuous stream discharge rates (Rantz 1982), and taking water samples during base and stormwater flow conditions for nutrient analyses (Bachman and Phillips 1996; Jordan et al. 1997b; Volk et al. 2006). Hydrograph separation analysis allows differentiation of base flow from surface water runoff, a process that can be automated using computer programs (Sloto and Crouse 1996; Lim et al. 2005). Statistical and deterministic watershed models, which can be applied to estimate N loadings, vary in complexity and temporal and spatial resolution; examples include the Waquoit Bay Nitrogen Loading model based on a mass balance approach (Valiela et al. 1992) and the Generalized Watershed Loading Function (GWLF) model, which employs a more mechanistic approach (Haith and Shoemaker 1987). Data for these models may include N inputs from various land uses/covers (e.g., agriculture, developed land, septic tanks, Nfix, wastewater discharges), N loss terms and attenuation in the watershed (e.g., Denit, forest uptake, crop harvest), and water balance calculations (e.g., precipitation, aquifer permeability, land-use-specific runoff coefficients, surface elevation, evapotranspiration rates). Alexander et al. (2002) provide a review of a variety of watershed models.

Stanhope et al. (in prep) quantified base flow nutrient yields for one year (May 2001–April 2002) from 14 first-order streams, distributed along a 70 km north–south transect of the Virginia Eastern Shore. During base flow conditions (at least 48-hour antecedent period of no precipitation), stream discharge rates and nutrient concentrations were measured monthly. Relationships between base flow nutrient yields and watershed characteristics (e.g., land cover and soil drainage class) were analyzed and used to develop a multiple regression model to predict base flow N yields. To simplify statistical analyses, nine land cover classes found in the watersheds were grouped into three major land cover categories of agriculture, forest, and developed land, and six soil drainage classes were grouped into three major categories of well drained, poorly drained, and very poorly drained. Base flow total dissolved nitrogen (TDN) yields (normalized to watershed area) from predominantly forested watersheds (67% to 68% forested) ranged from 0.13 to 2.3 kg N ha^{-1} yr^{-1}, and from agricultural watersheds (62% to 75% agricultural), ranged from 1.6 to 9.2 kg N ha^{-1} yr^{-1}. These estimates were generally comparable to total stream flow N yield estimates for watersheds in the Delaware and Maryland portions of the coastal plain of 0.08 to 1.3 kg N ha^{-1} yr^{-1} and 0.94 to 11.2 kg N ha^{-1} yr^{-1}, respectively, for predominantly forested (91% to 93%) and agricultural (58% to 64%) watersheds (Jordan et al. 1997a; Correll et al. 1999). Base flow TDN yields were positively related to percent agricultural land cover, percent developed land cover, and percent very poorly drained soils. It was negatively related to percent forested land cover (Figure 3.5). Forested and developed land covers together with very poorly drained soil accounted for 91% of the variability of TDN yields ($p = 0.0001$).

The base flow study by Stanhope et al. (2009) occurred during a severe drought and likely underestimated TDN loadings to the Virginia coastal bays. To estimate base flow contributions to total stream flow under non-drought conditions, stilling wells were installed at three streams (Nickawampus, Greens, and Gargatha Creeks) following the completion of the base flow study. The drought ended in November 2002, and wet year NO_3^- yields (November 2002 to October 2003) were on average 1.5 times greater than dry year yields (May 2001 to April 2002), largely due to 2.2-fold greater water discharge rates (Stanhope et al. in prep.) (Figure 3.6). NO_3^- concentrations during the wet year were, however, generally lower, suggesting dilution of NO_3^--enriched groundwater by water in soil saturated from rainfall. Because NO_3^- is the dominant N species, representing 66% to 98% of TDN (Stanhope et al. 2009), a 1.5× scaling factor was applied to base flow TDN loads to account for non-drought conditions.

Using the multiple regression model and the non-drought scaling factor, base flow TDN loading to HIB was estimated to contribute 19% of total allochthonous sources (Figure 3.7a), with an average yield from the watershed of 4.5 kg N ha^{-1} yr^{-1}. To estimate surface water runoff loading to HIB, two assumptions were made: (1) surface water runoff represents 30% of the volume of total stream discharge, based on an average measured base flow contribution of 70% to total stream flow (Bohlke and Denver 1995; Phillips and Bachman 1996; Jordan et al. 1997c); (2) TDN concentrations in surface water are similar to those in base flow, which is likely an overestimate since stormwater nutrient concentrations are lower than in base flow (Bohlke and Denver 1995; Jordan et al. 1997c). As a result, TDN loadings from surface water runoff were estimated to be ~40% of base flow loadings and contributed ~8% of total allochothonous sources to HIB (Figure 3.7a).

3.2.3.2 Direct Groundwater Discharge

Direct groundwater discharge (DGWD) to the lagoon is another potentially important N source, but estimates of DGWD to coastal waters are limited due to the difficulty and complexity of measuring flows that vary extensively across spatial and temporal scales (Burnett et al. 2006). In shallow unconfined aquifers, such as the Columbia aquifer in the watersheds of the Virginia and Maryland coastal bays, groundwater pathways and rates vary with topography and sediment grain size and may bypass the stream systems and discharge directly to the coastal bays or ocean (Speiran 1996, Dillow and Greene 1999). Discharge rates to tidal areas are also influenced by hydraulic head gradients in response to water level heights, waves, tides, and currents (Abraham et al. 2003; Burnett et al. 2006). As with base flow, N concentrations in groundwater discharge are affected by land use and sediment biogeochemical processes, such as Denit in organic-rich, fine-grained sediments (Reay et al. 1992; Speiran 1996). Methods for estimating DGWD include use of seepage meters (Reay et al. 1992), radium isotopes and radon-222 tracers (Schwartz 2003), multilevel piezometers (Dillow et al. 2002), hydrological balance calculations (Dillow and Greene 1999; Volk et al. 2006), and numerical models, such as MODFLOW (Robinson and Reay 2002).

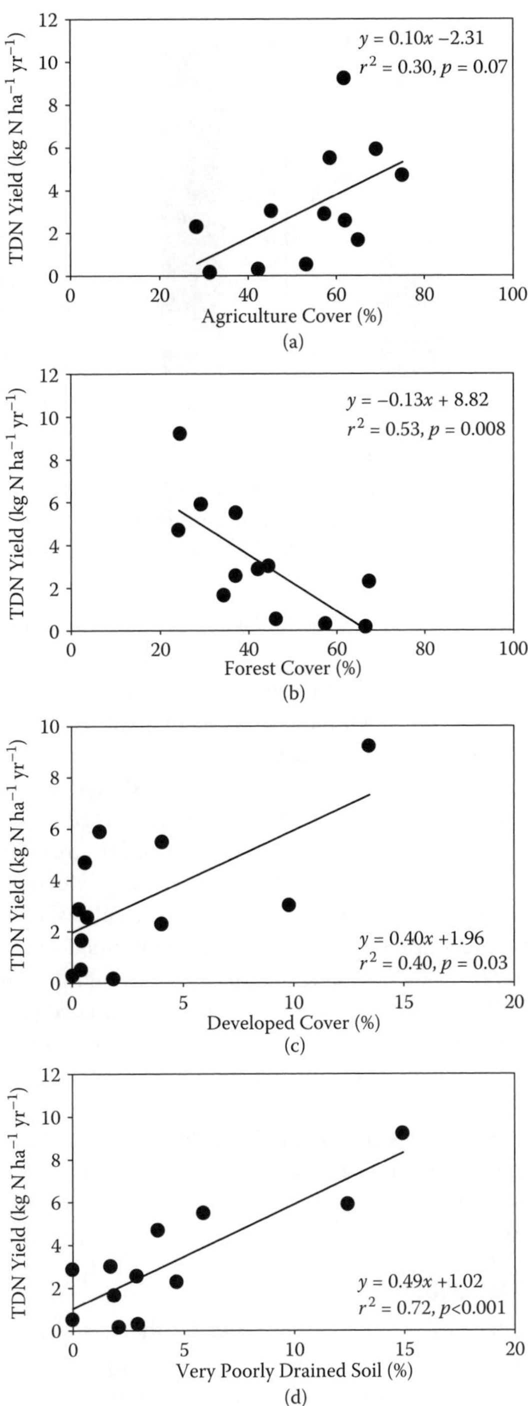

FIGURE 3.5 Linear regression of annual TDN base flow yields (kg N ha^{-1} yr^{-1}) with (a) percent agricultural land cover, (b) percent forested land cover, (c) percent developed land cover, and (d) percent very poorly drained soils for 12 streams on the Virginia Eastern Shore (Stanhope et al. 2009). Two streams in the study were excluded from regression analyses because the first stream had a potential local wastewater source and the second stream was identified as an outlier based on studentized deleted residual analysis. For more information, see Stanhope et al. (2009).

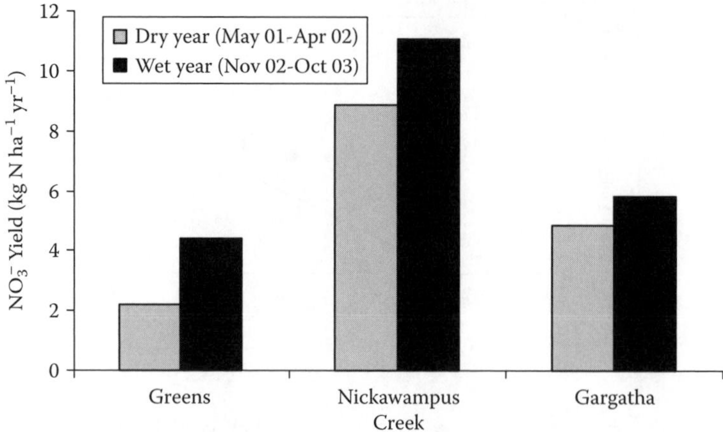

FIGURE 3.6 Comparison of base flow NO_3^- loading rates (kg N ha^{-1} yr^{-1}) for three streams on the Virginia Eastern Shore that discharge to the Virginia coastal bays during dry (84.5 cm precipitation) and wet (125.3 cm) years (Stanhope et al. in prep). The 50-year long-term (1955–2005) precipitation average at Painter, Virginia, is 110.2 cm (NOAA 2005).

Burnett et al. (2006) provide a detailed review of these and other methods, including their strengths and weaknesses.

Efforts in Virginia and Maryland have focused primarily on characterizing the occurrence and biogeochemical properties of DGWD, with few attempts to quantify discharge and N loading rates. For the Maryland coastal bays, Bratton et al. (2004, 2009) have identified fresh groundwater plumes under the bays up to 15 m thick and extending more than 1500 m offshore using offshore sediment coring and piezometers. Although the freshwater plumes contained NO_3^-, their data did not indicate that NO_3^- was discharged to the bays because it was overlain by saline groundwater enriched with NH_4^+, suggesting the occurrence of dissimilatory nitrate reduction to ammonium. Bratton et al. (2004, 2009) also found evidence of Denit. Using electrical resistivity studies, fresh groundwater plumes were located predominantly near shore in fine-grained surficial sediments (Manheim et al. 2004). Based on water mass-balance calculations, groundwater flow analysis, and groundwater nutrient concentrations, Dillow and Greene (1999) indirectly estimated potential NO_3^- loading from DGWD to Maryland coastal bays to be 123,400 kg N yr^{-1} (watershed yield of 2.7 kg N ha^{-1} yr^{-1}), which was one-third of the potential NO_3^- loading from base flow of 391,000 kg N yr^{-1} (watershed yield of 8.7 kg N ha^{-1} yr^{-1}).

To estimate DGWD to HIB, we assumed that the hydrogeology and groundwater systems of the Virginia and Maryland coastal bays are similar and applied the above proportion (33%) to base flow N loadings from HIB watersheds. As a result, DGWD may contribute about 6% of total allochothonous sources (Figure 3.7a).

3.2.3.3 Atmospheric Deposition

Atmospheric deposition includes NH_3 emissions from fertilizer, soil, and animal waste and NO_x emissions from fossil fuel combustion by automobiles and power plants (Russell et al. 1998). Atmospheric deposition to coastal waters can be a major N source, estimated to contribute 20% to 40% of total N inputs to both the watershed and directly to the water surface (Paerl 1997). Direct atmospheric deposition (DAD) can be especially significant for coastal bays that have relatively unimpacted watersheds or have smaller watershed area than water body surface area, such as HIB with a watershed to water body surface area ratio of 0.6.

Meyers et al. (2001) estimated that direct total N (wet + dry; inorganic + organic N) atmospheric deposition to Delaware and Maryland coastal bays and Chesapeake Bay ranged from 7.87 to 9.35 kg

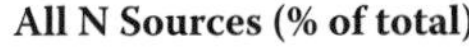

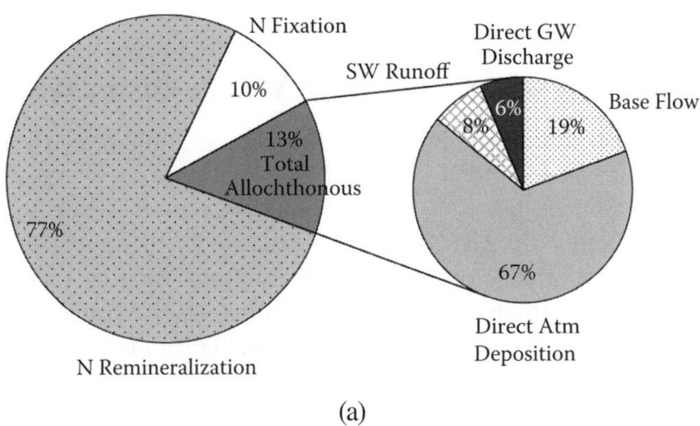

(a)

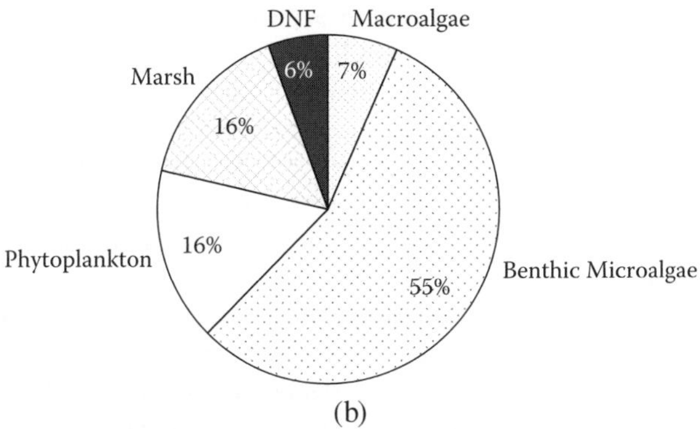

(b)

FIGURE 3.7 (a) The percent contribution of various N sources to Hog Island Bay: both allochthonous and autochthonous N sources (left), including N remineralization and fixation, and the breakout of specific allochthonous sources (right), including surface water (SW) runoff, direct groundwater (GW) discharge, base flow, and direct atmospheric (Atm) deposition to bay surface. (b) The percent contribution of primary producers and denitrification (DNF) to total N sinks in Hog Island Bay.

N ha^{-1} yr^{-1}. Their estimates were based on data from the National Atmospheric Deposition Program (NADP) monitoring of wet deposition to watersheds, the U.S. Environmental Protection Agency's Clean Air Status and Trends Network database of dry deposition to watersheds, and calibrated atmospheric deposition models. Total N wet deposition was measured in precipitation samples; however, it is frequently underestimated by 10% to 20% due to deficient preservation methods and losses of organic N (ON) and NH_4^+ (Keene et al. 2002). Immediate freezing of samples at the time of collection with dry ice and addition of a preservative upon thawing prevented both ON and NH_4^+ losses. Meyers et al. (2001) accounted for underestimation of N in samples analyzed using the NADP's protocol for unfrozen samples by applying a 15% correction factor to wet inorganic N deposition values in the NADP database. Dry N deposition, including particulate inorganic and ON and gaseous NH_3 and HNO_3, contributes about 43% to total atmospheric N deposition (Russell et al. 2003).

We calculated total N atmospheric deposition to HIB by taking the average wet inorganic N deposition measured in samples collected on Hog Island from 1990 to 1999 (Galloway and Keene 1999) and corrected the values for sample preservation error. Dry deposition was estimated as a percentage of the calculated wet deposition and summed with wet deposition to yield total N atmospheric deposition. With a total water surface area of 15,000 ha and a total N deposition rate of 9.49 kg N ha^{-1} yr^{-1}, the total DAD contribution to HIB represents 67% of total allochthonous sources (Figure 3.7a).

3.2.4 Autochthonous Sources of Nitrogen to Virginia Coastal Bays

In Virginia coastal bays and similar shallow bays, which receive little or no riverine input, autochthonous sources contribute more to total N input than allochthonous sources. In HIB 13% of N came from allochthonous sources and 87% from autochthonous sources (Figure 3.7a). The primary autochthonous N sources were NFix and NRemin, accounting for 10% and 77% of total N sources, respectively. The relative importance of these so-called internal loading sources in other coastal lagoons depends largely on autotrophic community structure, which in turn is determined by nutrient loading, light availability, and water residence time (McGlathery et al. 2004, 2007).

3.2.4.1 Nitrogen Fixation

NFix is the biologically mediated transformation of atmospheric nitrogen gas (N_2) to cellular N using the nitrogenase enzyme complex, which is active only under anoxic conditions. In marine systems nitrogen-fixing microbes, including both heterotrophic and autotrophic bacteria (O'Neil and Capone 1989), are active in a wide variety of habitats including the water column, as epiphytes on seagrass leaves or macroalgae, in the rhizosphere of rooted macrophytes, in microbial mats on the benthic surface, and in biofilms (Capone et al. 1992, 1977; Joye and Paerl 1993; McGlathery et al. 1998). Even though autotrophic NFix rates generally exceed heterotrophic rates in coastal systems, heterotrophic NFix is thought to dominate due to light limitation of cyanobacteria (Joye and Anderson 2008). Evidence suggests that heterotrophic NFix by sulfate-reducing bacteria is limited by the availability of labile DOM.

NFix rates are most commonly measured using the acetylene (C_2H_2) reduction technique (Currin et al. 1996; Joye and Anderson 2008), which measures the conversion of C_2H_2 to ethylene by the nitrogenase enzyme complex as a proxy for the reduction of N_2 gas to NH_4^+ + DON. This method has been applied to a variety of micro-habitats including sediment cores, excised roots and rhizomes, and seagrass leaves (Capone and Taylor 1977; Paerl et al. 1994; McGlathery et al. 1998). Most studies apply the theoretical conversion factor of 3 moles of C_2H_2 reduced to 1 mole of N_2 fixed to calculate the NFix rate (Joye and Anderson 2008). Although early attempts to calibrate the acetylene reduction method by comparison with uptake of $^{15}N_2$ suggested that the ratio could vary over an order of magnitude (Seitzinger and Garber 1987), more recent calibrations suggest that a conversion factor of 3 to 4 is suitable for most marine systems (Montoya et al. 1996; Nagel 2004). Nfix rates have been measured in shallow systems worldwide, equaling as much as 2.85 mmol N m^{-2} d^{-1} (see Joye and Anderson 2008, and references therein). Numerous factors limit Nfix, namely the availability of labile organic matter, oxygen, and light. In tropical and temperate systems, Nfix rates are correlated with organic substrate availability and are often associated with photosynthetic or seagrass root exudates (Paerl et al. 1993; McGlathery et al. 1998; Carpenter and Capone 2008). NFix rates have been shown to be higher in seagrass meadows than in unvegetated sediments because of the release of photosynthetic exudates by roots and rhizomes (McGlathery et al. 1998; Welsh 2000). In HIB, we measured NFix rates in surface sediments (0 to 1 cm) collected across the creek–shoal–island transect. Measurements were made seasonally from October 2004 to August 2005 in the light and dark using the C_2H_2 reduction method. Although we did not calibrate C_2H_2 reduction in our study using ^{15}N uptake, we applied a conservative conversion factor of 4:1 moles C_2H_2 reduced:moles N_2 fixed rather than the theoretical ratio of 3:1 (Hardison unpublished data). Our NFix rates likely represent an underestimate of actual rates since all measurements were performed only in surface sediment. Measurements of rates in recently established seagrass meadows in HIB are currently under way.

3.2.4.2 Remineralization

Internal recycling of N through NRemin is the dominant autochthonous source in Virginia's shallow coastal lagoons. Organic matter can be mineralized directly to dissolved ammonium (NH_4^+) or metabolized to dissolved organic N (DON), which can be further mineralized to NH_4^+. Mineralization of organic matter takes place in the water column and sediments of coastal lagoons under both oxic and anoxic conditions. The sediments are thought to be the dominant source of nutrients supporting primary production in coastal lagoons since N concentrations are orders of magnitude higher in sediment porewater than in the water column (Tyler et al. 2003; McGlathery et al. 2004). The dominant organic substrate in coastal bays is plant detritus (Banta et al. 2004). In general, plants lower in structural tissue and higher in nutrient content decay faster; therefore, mineralization rates depend largely on the structure of the autotrophic community. For example, Banta et al. (2004) reviewed numerous studies, which suggest that detritus from micro- and macroalgae have higher decay rates than detritus from seagrasses.

The N produced during NRemin of plant detritus may be immobilized by the heterotrophic bacterial community to build biomass if the elemental composition of the organic matter undergoing remineralization does not satisfy their nutritional requirements. Other fates of NH_4^+ produced in the sediments by NRemin are (1) immobilization by benthic primary producers; (2) flux to the water column; (3) sorption to sediment particles; or (4) transformation to NO_x via Nitr or N_2 gas via Anam or coupled nitrification–dinitrication (Joye and Anderson 2008). Net mineralization, usually estimated as the flux of NH_4^+, NO_2^-, NO_3^-, and N_2, from sediments to the water column in the dark, is generally much less than gross mineralization, which is based on the turnover of the sediment NH_4^+ pool.

Gross mineralization rates are often measured by $^{15}NH_4^+$ isotope pool dilution, following the protocol developed by Blackburn (1979) and modified by Anderson et al. (1997). We measured gross mineralization rates in HIB between 1997 and 2002 (Figure 3.8), and used these values to estimate the contribution of NRemin to total N sources (Figure 3.7a). NRemin is the largest source of N in HIB. Rates measured in HIB were similar to those measured in sandy shallow photic sites across the full salinity range of Chesapeake Bay (CB) (range = 4 to 16 mmol N m^{-2} d^{-1}); however, in

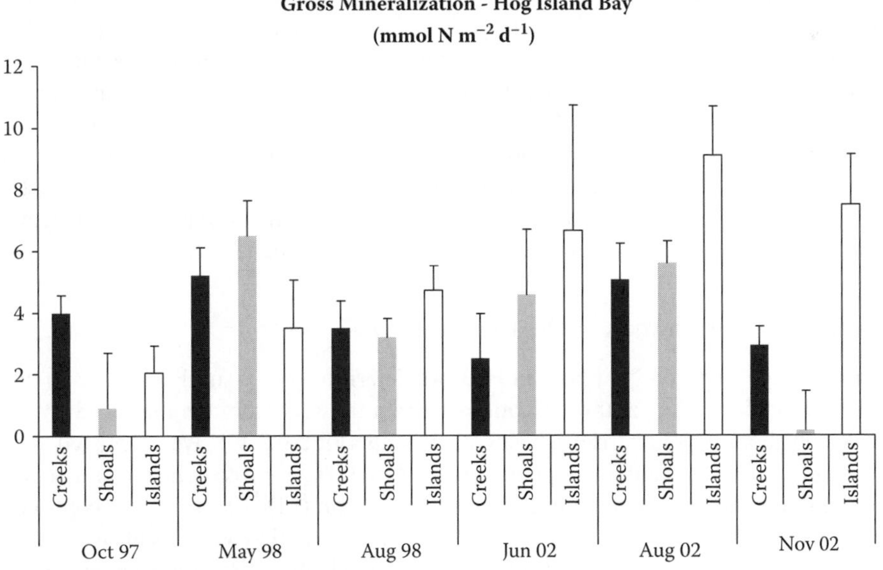

FIGURE 3.8 Gross mineralization measured in sediments sampled across mainland to barrier island transect in HIB. Error bars represent standard errors.

muddy sediments at oligohaline sites in CB, mineralization rates were as high as 86 mmol N m^{-2} d^{-1} (Anderson et al. in prep.). Net mineralization rates measured at the same time in HIB were at least an order of magnitude lower than gross rates (data not shown).

3.3 SINKS OF NITROGEN IN COASTAL BAYS

3.3.1 Abundance of Primary Producers in Coastal Bays: Seasonality

Benthic micro- and macroalgae are both present in the Virginia bays year-round, although biomass varies seasonally as does the relative importance of productivity of the different groups. Benthic microalgal biomass appears to be relatively constant throughout the year, with occasional decreases during summer due to grazing or to shading by overlying macroalgal mats. Macroalgal populations typically go through boom and bust cycles; they reach a maximum biomass in mid-summer (June-July) and typically collapse in late summer, presumably due to high temperatures and self-shading within the macroalgal mats (McGlathery 2001). In late summer when macroalgal biomass is low, McGlathery et al. (2001) found production by benthic microalgae to be the most important contributor to total system production. Phytoplankton populations also increase locally after the macroalgal bloom, probably fueled by the release of nutrients from decomposing macroalgae (McGlathery et al. 2001). Monitoring data of macroalgal biomass and of both benthic and water column chlorophyll *a* from sites along the creek–shoal–island transect in HIB illustrate these patterns (Figure 3.9; Hardison unpublished data).

3.3.2 Seagrasses in HIB

In the 1930s, the Virginia coastal bays underwent a dramatic change in status from a seagrass-vegetated state to one dominated by benthic algae when a large hurricane caused the local extinction of the seagrass (eelgrass, *Zostera marina*) populations, which were already weakened by a pandemic disease (the "wasting disease," *Labarinthula* sp.). Following the die-off of eelgrass, the lagoons went from a clear-water state to one in which sediments were easily suspended in the water column by wind events. An immediate effect of the seagrass loss was the collapse of the scallop (*Argopecten irradians*) fishery and the disappearance of brant geese (*Branta bernicla*), which fed exclusively on seagrass. The lagoon bottoms remained devoid of seagrass until nearly the end of the century, and the waters remained turbid. In the late 1990s, a small patch of seagrass was discovered in the coastal bays, spurring a large-scale restoration effort led by Robert Orth at the Virginia Institute of Marine Science in collaboration with The Nature Conservancy. Since 2001, >23 million seagrass seeds have been introduced into 200 acres of the Virginia coastal bays (including HIB); now some 1400 acres are colonized by seagrass (Orth et al. 2010).

3.3.3 Nitrogen Assimilation by Primary Producers

As a part of a large study on N cycling in HIB, we determined N assimilation rates of autotrophs collected from replicate sites at three locations along a transect from the mainland creeks across the mid-lagoon shoals to the barrier island embayments (Figure 3.10). We calculated N assimilation for benthic microalgae and phytoplankton based on 90% of gross primary production (GPP), measured in sediment cores with an overlying water column (I.D. 10 cm) and in cores containing only water, and a C:N of 9:1 (Sundbäck et al. 1991). Cores were stirred and incubated at *in situ* light and temperature conditions in a Conviron PGR 15 growth chamber. Rates shown may overestimate N assimilation since no correction was made for C fixed that was subsequently exuded as extracellular polymeric substances (EPS) and colloidal C, which may account for as much as 70% of that fixed (Wolfstein et al. 2002). For macroalgae, we calculated N assimilation based on seasonal

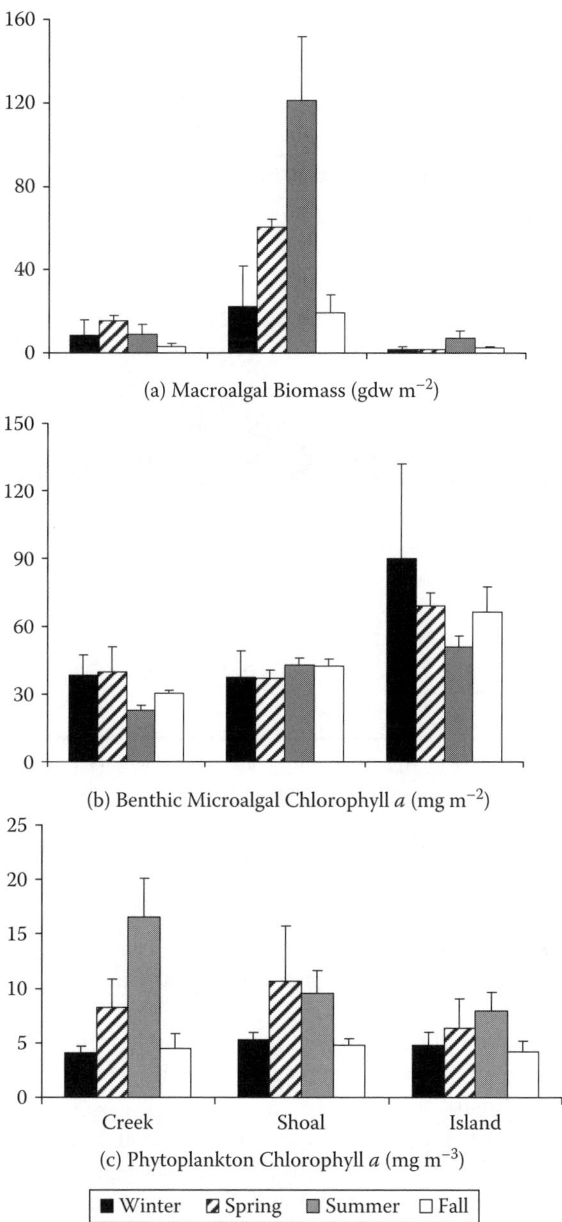

FIGURE 3.9 Seasonal variation in autotrophic biomass at Hog Island Bay (a) creek, (b) shoal, and (c) island sites. Error bars represent standard errors.

measurements of *in situ* biomass, maximum growth rates, and C:N. Aerial growth rates were scaled for self-shading within the macroalgal mat using the model described by McGlathery et al. (2001). Without accounting for this self-shading, N assimilation estimates for macroalgae could be overestimated by as much as 90%.

Marsh N assimilation was estimated based upon results of a flux study performed in Phillips Creek marsh, dominated by short-form *Spartina alterniflora*, and located adjacent to one of the mainland creeks that discharges to HIB. Fluxes of DIN and DON ($n = 5$) were measured bimonthly *in situ* at low tide during daytime (Neikirk 1996; Anderson et al. 1997). We assumed that N uptake

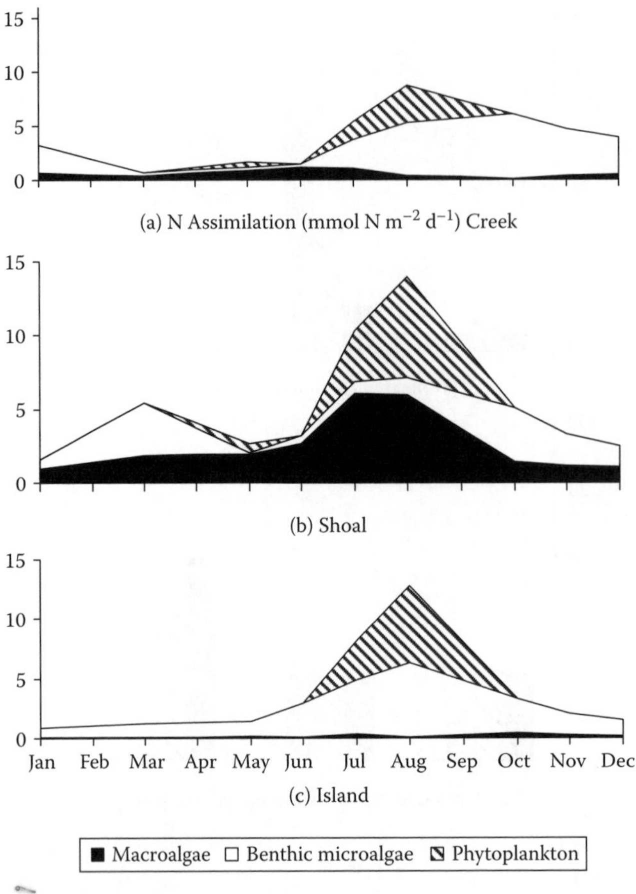

FIGURE 3.10 Total nitrogen assimilation (mmol N m^{-2} d^{-1}) divided into macroalgae (black filling), benthic microalgae (white filling), and phytoplankton (striped filling) components for (a) creek, (b) shoal, and (c) island sites at Hog Island Bay over a year. (Adapted from Sundbäck and McGlathery 2005.)

from overlying water occurred only during daytime immersion. Immersion times were calculated based on hourly tidal heights relative to mean sea level (MSL) for the site and elevations of the marsh, referenced to a benchmark at MSL.

Maximum daily N assimilation rates for the benthic micro- and macroalgae and phytoplankton were remarkably similar despite orders of magnitude differences in standing biomass (Figures 3.9 and 3.10). At the mid-lagoon shoal sites, maximum N assimilation by macroalgae (6.02 mmol N m^{-2} d^{-1}) occurred in mid-summer when macroalgal biomass peaked (average 121 gdw m^{-2}). The benthic microalgal community had highest N assimilation rates during the summer and fall at the mainland creek and barrier island sites where macroalgal biomass was relatively low (0 to 20 gdw m^{-2}). The maximum N assimilation rate by benthic microalgae was 6.2 mmol N m^{-2} d^{-1}. Maximum daily N assimilation by phytoplankton occurred at the shoal sites in August (6.9 mmol N m^{-2} d^{-1}); however, this rate was sustained for only a brief period following the macroalgal die-off (Figure 3.10). The similarity between the short-term N assimilation rates of the different primary producer groups with vastly different standing biomasses implies that the turnover time of nutrients bound in plant tissue is a key factor influencing their role in nutrient retention in these shallow lagoons. More work is needed to determine these turnover rates in order to understand the consequences of the shifting dominance of primary producers as coastal bays become more eutrophied. When scaled to the entire bay and integrated over a year, benthic microalgae accounted for 55%, macroalgae for 7%, phytoplankton for 16%, and marshes for 16% of total TDN uptake (Figure 3.7b).

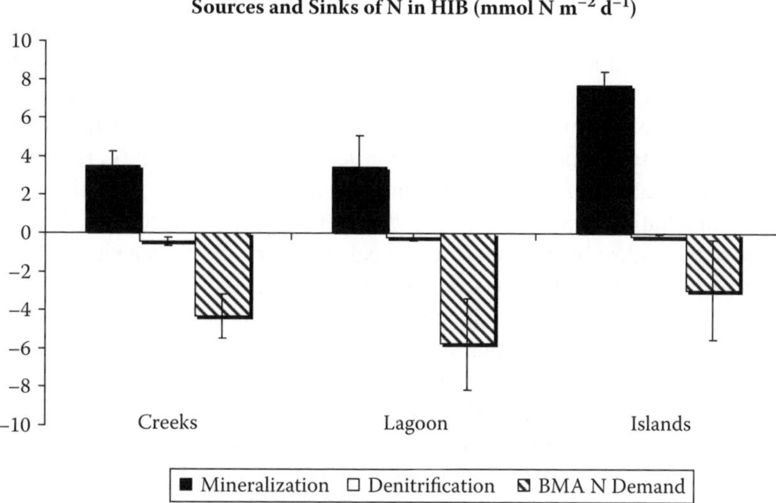

FIGURE 3.11 Sources and sinks of nitrogen in sediments of Hog Island Bay. Bars above the X axis represent sources of N; bars below the X axis represent sinks of N. Error bars represent standard errors.

3.3.4 Sources of Nitrogen to Support Benthic Primary Production

Figure 3.7a graphically demonstrates that allochthonous sources provide only ~13% of the N required to support the diverse primary producers in HIB. The remainder is supplied primarily by NRemin of organic N in the sediments. In Figure 3.11, we compare N supplied by NRemin to that removed by Denit and by benthic microalgal N demand. For calculation of benthic microalgal N-demand in this figure and Figure 3.7b, we conservatively assumed that 50% of the C fixed was exuded as EPS or colloidal C. The N necessary to support biomass production was based on a C:N of 9 (Sundbäck et al. 1991). The N supplied by benthic NRemin approximated that required for benthic microalgal growth. Denit, measured here in the dark by membrane inlet mass spectrometry (MIMS), accounted for only 3% to 10% of the N taken up by benthic microalgae on a daily basis and 5% of the total DIN taken up within the bay on an annual basis (Figure 3.7b). Others have similarly observed that, in shallow photic systems when N is limiting, benthic autotrophs outcompete denitrifiers for substrate (Sundbäck and Miles 2000; Risgaard-Petersen 2003; Joye and Anderson 2008). We noted above that assimilation rates of benthic microalgae were similar to those for macroalgae, albeit in different seasons. Turnover times for these benthic autotrophs are on time scales of days to weeks. Thus, in these photic coastal bays, retention of N in the sediments depends on continuous recycling and uptake of N by autotrophs on short time scales.

3.4 FATE OF NITROGEN ASSIMILATED BY BENTHIC MACRO- AND MICROALGAE

Attempts to trace the fate of C and N assimilated by benthic microalgae or macroalgae in sediments have been fraught with methodological difficulties. It is thought that the benthic microalgal and sediment bacterial communities are tightly coupled, which would imply longer N retention within the sediments compared to that in macroalgal tissue. Recently, we conducted a series of stable isotope tracer experiments to better constrain N uptake and retention by the sediment microbial community in HIB. We also considered the influence of living and decaying macroalgal biomass on microbial N cycling.

Stable isotope tracers (^{15}N, ^{13}C) are particularly useful for tracking C and N through the sediment microbial pool because microbial biomass is difficult to separate from the sediment matrix. In recent studies stable isotopes incorporated into microbial biomarkers have been used to track and distinguish nutrient uptake into benthic microalgal and bacterial biomass pools (Boschker et al. 1998; Boschker and Middelburg 2002; Tobias et al. 2003). Compound specific isotope analysis (CSIA) is perhaps one of the most powerful geochemical tools available to unambiguously trace C and N through a system. CSIA is commonly used in microbial ecology because it provides the best tool for tracing C into microbial fatty acid biomarkers (Bouillon and Boschker 2006). However, most lipids do not contain N, thus making them unsuitable for ^{15}N studies. This is problematic because often the ability to trace N in addition to, or instead of, C is preferred. Amino acids (AA) have an advantage over lipids because they allow both C and N to be analyzed using CSIA. Specifically, D-amino acids (D-AA) from peptidoglycan, a cell wall component found uniquely in almost all species of bacteria, have been used as bacterial biomarkers since bacteria are the only organisms to incorporate D-AA into their biomass (Pelz et al. 1998; Tobias et al. 2003; Veuger et al. 2005); all other organisms utilize only L-amino acids (L-AA). Recently, Veuger and colleagues were the first to combine CSIA analysis of bacterial-specific D-alanine (D-Ala) and L-AA with ^{15}N and ^{13}C labeling to quantify ^{15}N and ^{13}C incorporation by sediment microbes (Veuger et al. 2005, 2006). Using bulk sediments and AA biomarkers, we conducted two ^{15}N isotope tracer experiments in HIB to study N cycling within the sediment microbial community. In the first experiment we tracked the uptake and retention of ^{15}N-labeled DIN ($DI^{15}N$) by benthic microalgae and bacteria in the presence or absence of bloom-forming nuisance macroalgae (*Gracilaria*) to assess the direct and indirect effects of macroalgae on bacteria–benthic microalgae coupling (Hardison et al. in prep.-b). In the second experiment we tracked the fate of macroalgal-bound ^{15}N after a simulated macroalgal die-off using sediments and macroalgae from HIB and Isle of Wight Bay, Maryland to evaluate the role of the sediment microbial community in retaining N following a macroalgal die-off (Hardison et al. in prep.-a).

The results from both experiments support a tight coupling between benthic microalgae and bacteria in shallow illuminated systems. In the first experiment, we observed $DI^{15}N$ uptake into the bulk sediment pool (Figure 3.12a), which was dominated by benthic microalgae, the major source of total hydrolysable amino acids (THAA) in this experiment (Figure 3.12b). Labeling of the bacterial biomarker D-Ala indicated that bacteria were also actively incorporating the ^{15}N labels (Figure 3.12c), most likely through direct $DI^{15}N$ uptake along with organic exudates (EPS) from benthic microalgae. Macroalgae, however, significantly decreased ^{15}N uptake by benthic microalgae and bacteria. We suggest that macroalgae were competing directly with benthic microalgae for light and/or nutrients, thereby decreasing benthic microalgal activity. This resulted in an indirect macroalgal effect on sediment bacteria. If bacteria utilized benthic microalgal exudates, then diminished benthic microalgal activity likely decreased EPS exudation, which decreased EPS and $DI^{15}N$ uptake by bacteria. Our second experiment demonstrated incorporation of up to 30% of macroalgal N into the sediments following a simulated macroalgal die-off. Immediately following the addition of dead, isotopically labeled macroalgae, bulk sediments were enriched in ^{15}N derived from macroalgae (Figure 3.13a). These enrichments increased throughout the experiment, peaking 1 to 2 weeks following the die-off. Thus, some of the N temporarily retained within macroalgal biomass was deposited in sediments and retained for >2 weeks following the macroalgal die-off. Further evidence suggests that this retention was due to incorporation of N by the microbial community. The ratio of ^{15}N label in the bacterial biomarker, D-Ala, vs. that in L-Ala provided an indication of the relative importance of bacterial uptake to total microbial uptake (Figure 3.13b). The observed ratio suggested that bacteria were responsible for the initial isotopic uptake; however, as bacteria remineralized macroalgal biomass, benthic microalgae incorporated relatively more ^{15}N (Figure 3.13b).

We conducted our second experiment using sediments and macroalgae from HIB as well as Isle of Wight Bay, Maryland, and found similar results for both systems. This suggested that our results may be applicable to other shallow coastal lagoons along the Delmarva Peninsula. Overall, in our isotopic tracer experiments, we observed evidence for shuttling of N between the benthic microalgal and bacterial pools. In the presence and absence of living macroalgae, bacterial activity was

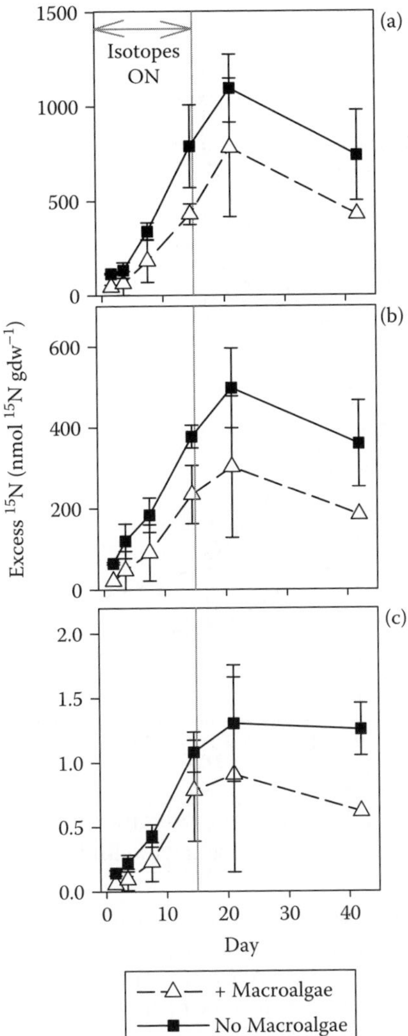

FIGURE 3.12 Excess ^{15}N in (a) bulk sediments, (b) total hydrolyzable amino acids, and (c) D-alanine over the course of the experiment for treatments with macroalgae (triangles, dashed lines) and without macroalgae (squares, solid lines). The vertical gray line at day 15 indicates the end of the isotope addition period. Values represent mean ± 1 SD, $n = 2$. (Adapted from Hardison et al. in prep.-b.)

dependent on benthic microalgal activity, suggesting that bacteria used benthic microalgal-derived metabolites to support some portion of their production. But for a brief time period following the macroalgal die-off, benthic microalgae depended on bacteria to remineralize organic matter into inorganic substrates.

3.5 DISCUSSION: FUTURE TRENDS AND PREDICTIONS

In relatively pristine, temperate coastal bays such as HIB, allochthonous sources of N, delivered via base flow, stormwater flow, direct groundwater discharge, and atmospheric deposition, are insufficient to support the N demand for observed primary production by macroalgae and benthic microalgae, the dominant autotrophs in the system. Pedersen et al. (2004) in a review of nutrient uptake by autotrophic communities in shallow coastal marine bays indicated that nutrient assimilation by

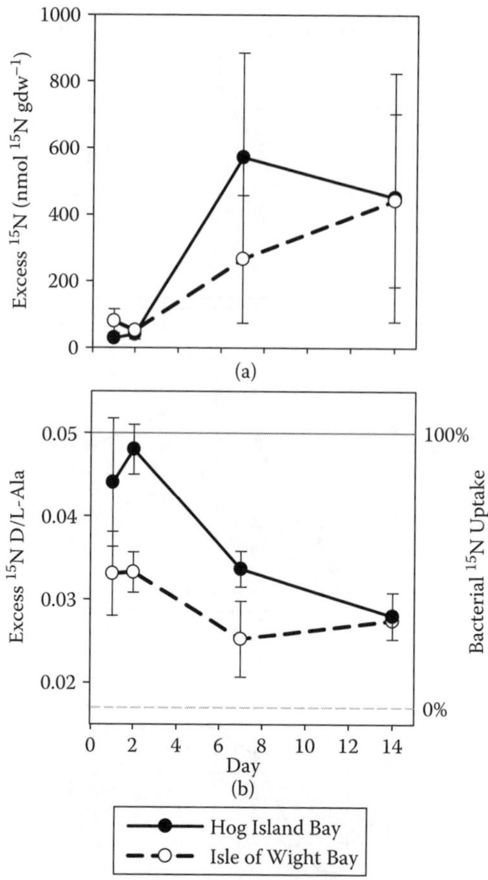

FIGURE 3.13 Excess ^{15}N for (a) bulk sediments, (b) in D-alanine vs. L-alanine over the course of the experiment for treatments from Hog Island Bay (filled symbols, solid lines) and Isle of Wight Bay, MD (open symbols, dotted lines). The ratio of excess ^{15}N in D-Ala vs. L-Ala represents the percent contribution of bacteria vs. BMA to total microbial ^{15}N uptake, denoted on the right axis in panel b. Figure adapted from Hardison et al. (in prep.-a). Values represent mean ± 1 SE, $n = 3$.

autotrophs exceeds inputs when nutrient loads are <50 g N m^{-2} y^{-1}. In HIB, the N load from all allochthonous sources was estimated to be 1.4 g N m^{-2} y^{-1} (normalized by water body surface area) (Stanhope 2003); thus, the deficit of N required to support the N demand for autotrophic primary production, estimated to be 8.8 g N m^{-2} y^{-1}, must be supplied by remineralization of organic matter derived from sinking phytoplankton, macroalgae, and benthic microalgae. The ephemeral macroalgae found in HIB undergo a "boom and bust" cycle, with maximum biomass in June to early July, and decay during July to early August. Tyler et al. (2001) observed almost complete decomposition of accumulated macroalgal biomass within a 2-week period following the crash of a macroalgal bloom in August. Hardison et al. (in prep.-a) observed that up to half of N in dead macroalgae may be incorporated into sediments, first into bacterial biomass with transfer later into benthic microalgal biomass. Although retention of N in autotrophic biomass and sediment is temporary, its retention may be sustained for longer periods of time by the shuttling of N between benthic microalgal and bacterial pools (Hardison et al., in prep.-a, in prep.-b) and by transfer to higher trophic levels (van Oevelen et al. 2006). Some of the N derived during decomposition of macroalgal or benthic microalgal biomass may be recalcitrant and thereby preserved either in sediments (Veuger et al. 2006) or released as unavailable DON to the water column, where it may accumulate and, thus, not support phytoplankton growth (Nagata et al. 2003). Components of peptidoglycans have been shown to have turnover times of 20 to 67 days in sediments (Veuger et al. 2006) and 10 to 167 days in seawater (Nagata et al. 2003). Based on the high D/L amino acid ratio observed in ultra-filtered dissolved organic matter (UDOM) from seawater, McCarthy et al. (1998) suggested that a substantial fraction of seawater DON is derived from bacteria. Given the high percentage of DON as TDN in coastal bay waters, accumulation of recalcitrant forms of DON is a likely fate of N in coastal bays, an area in much need of additional study.

It is expected that the mass of N exported to the coastal ocean or to recalcitrant, buried forms in sediments balances allochthonous sources, and is small relative to the mass of N that turns over and is available to support primary production in HIB. It has been demonstrated that shallow, photic bays do not behave in a predictable manner to increasing N loads. For example, primary production does not appear to increase as nutrient loads increase (Nixon et al. 2001). At moderate nutrient loads, uptake, retention, and transformation of N by sediments and benthic primary producers may be sufficient to modulate a eutrophic response (McGlathery et al. 2007). As nutrient loads increase, the autotrophic community will respond with shifts in the dominance of primary producers and a trend toward decreasing benthic/pelagic GPP; however, total primary production may remain constant due to interactions between regulators including light, grazers, oxygen, and physical factors (Sand-Jensen and Borum 1991; McGlathery et al. 2007; Nixon 2009). However, there is likely to be a threshold above which the benthos loses its ability to modulate responses to nutrient enrichment. Once past this threshold, it is difficult to restore ecosystem function; studies of eutrophied systems undergoing oligotrophication suggest that the reversion of systems to their original state is difficult if not impossible to accomplish (Knowlton 2004; Duarte et al. 2009).

Conceptual models portraying ecosystem responses to nutrient enrichments below and above the critical threshold for sustaining benthic primary productivity are shown in Figures 3.14a and b. At nutrient loads below some "critical threshold," benthic production, which dominates over pelagic production, serves as a sink for N. Benthic autotrophs and heterotrophs cap the sediments, capturing N produced by NRemin and buffering the system from eutrophication. At nutrient loads above some "critical threshold," pelagic production is dominant as phytoplankton limit light available to support benthic autotrophy. The sediments are net heterotrophic, releasing remineralized DIN and DON into the water column and supporting eutrophication. These models are based on the assumption that sustained benthic autotrophy and shuttling of N among benthic autotrophs, heterotrophs, and higher trophic levels will sequester N for extended periods of time, and that release will be out of phase with the optimum time for pelagic production. However, that may not be the case in other systems. In a field and modeling study by Cerco and Seitzinger (1997) in the Delaware Inland Bays, annual phytoplankton production was greater in the presence of benthic microalgal activity than in its absence, due to benthic microalgae uptake of N in late winter–early spring and release during summer supporting phytoplankton production. The influence of other autotrophs such as macroalgae in the Delaware Bays study was not addressed. Much additional work is needed to better understand the role of benthic microorganisms in modulating nutrient enrichment in shallow coastal systems.

3.5.1 Potential Responses to Seagrass Restoration in the Virginia Coastal Bays

The success of recent seagrass restoration efforts in the Virginia coastal bays offers the opportunity to pose hypotheses related to the trajectory of change if seagrasses become the dominant benthic autotrophs in these systems. For example, will seagrasses provide long-term retention of N such that we will no longer see blooms of ephemeral macroalgae every June? Will DON concentrations in the lagoons decrease as a result? How will establishment of seagrasses affect sediment biogeochemistry, and how will the benthic food web respond? Just as the Virginia coastal bays are in a state of transition with a trend toward increasing health, the Maryland coastal bays may also be in a state

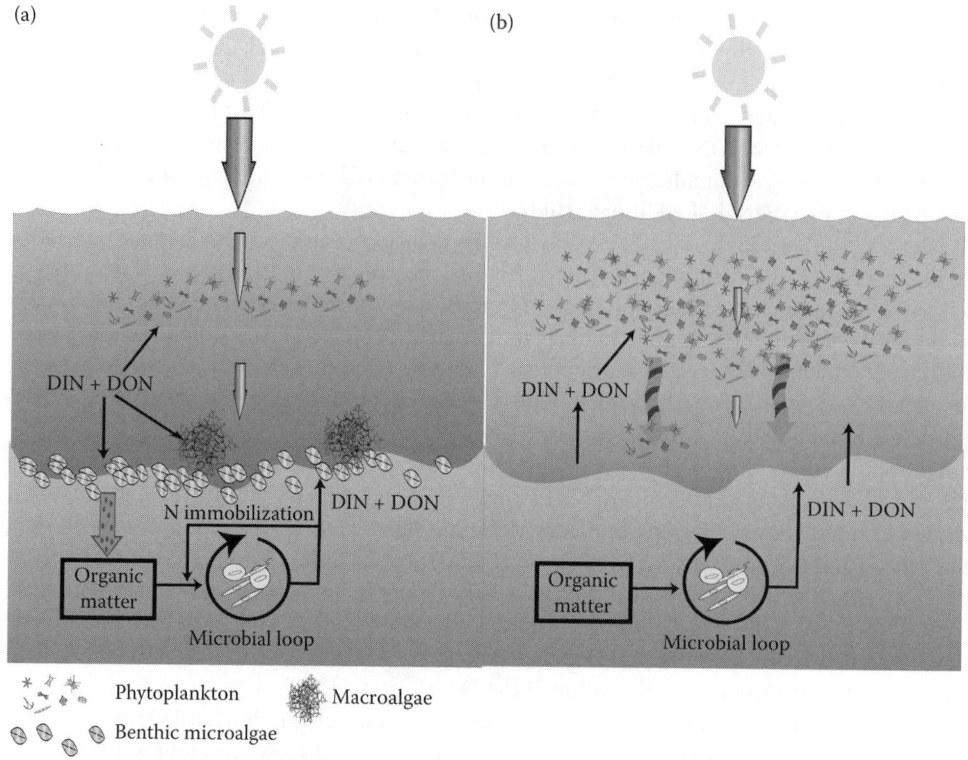

FIGURE 3.14 (a) Conceptual model of a shallow photic coastal bay at nutrient loads below the "critical threshold." Light is sufficient to support benthic primary production, which serves as a sink for DIN and DON from both the water column and sediment pore water. Organic matter derived from benthic autotrophs and sedimenting phytoplankton is remineralized and the N released is taken up by benthic autotrophs and bacteria. (b) When nutrient loads are above the "critical threshold," available light limits benthic autotrophy. Pelagic GPP is dominant relative to benthic GPP. Organic matter from sedimenting phytoplankton and other water column detritus is remineralized in the sediments. With no benthic autotrophs to take up N, the DIN and DON are released to the water column to support eutrophication.

of transition but with a trajectory opposite to that observed in Virginia. Available indicators suggest degrading conditions with increased nutrient loads and macroalgal blooms and a leveling off of seagrass abundance. Have the Maryland coastal bays reached a "tipping point"? Can their health be restored? How? Much remains to be done to answer these questions. The Delmarva Peninsula offers just such an opportunity.

ACKNOWLEDGMENTS

This research was supported by the National Science Foundation (DEB Ecosystems 0542645) and through the Virginia Coast Reserve LTER project (DEB 0080381 and DEB 0621014). Additional support was provided by the U.S. Department of Agriculture-NRICGP (2001-35101-09873). Symbols used in Figures 3.1 and 3.14 are courtesy of the Integration and Application Network, University of Maryland Center for Environmental Science. We thank our many colleagues who contributed substantially to this work, including A. C. Tyler (RIT); C. R. Tobias (UNCW); E. Canuel (VIMS); B. Veuger and J. Middelburg (NIOO); M. Luckenbach, S. Fate, and R. Bonniwell (VIMS Eastern Shore Laboratory); B. Neikirk, H. Walker, C. Funkey, D. Maxey, S. Salisbury, and E. Lerberg (VIMS); as well as J. Burton, M. Thomsen, and S. Lawson (UVA). This is VIMS Contribution Number 3033.

REFERENCES

Abraham, D. M., M. A. Charette, M. C. Allen, A. Rago, and K. D. Kroeger. 2003. Radiochemical estimates of submarine groundwater discharge to Waquoit Bay, Massachusetts. *Biol. Bull.* 205: 246–247.

Alexander, R. B., P. J. Johnes, E. W. Boyer, and R. A. Smith. 2002. A comparison of models for estimating the riverine export of nitrogen from large watersheds. *Biogeochemistry* 57–58: 295–339.

Anderson, I.C., K. J. McGlathery, and A. C. Tyler. 2003. Microbial mediation of "reactive" nitrogen transformations in a temperate lagoon. *Mar. Ecol. Prog. Ser.* 246: 73–84.

Anderson, I. C., L. Schaffner, and J. W. Stanhope. In preparation. Drivers influencing metabolic and nitrogen cycling rates in shallow sites along the salinity gradient of Chesapeake Bay. School of Marine Science, College of William and Mary, Virginia Institute of Marine Science, Department of Biological Studies, Gloucester Point, Virginia.

Anderson, I. C., C. R. Tobias, B. B. Neikirk, and R. L. Wetzel. 1997. Development of a process-based nitrogen mass balance model for a Virginia (USA) *Spartina alterniflora* salt marsh: Implications for net DIN flux. *Mar. Ecol. Prog. Ser.* 159: 13–27.

Bachman, L. J., V. Lindsay, J. W. Brakebill, and D. S. Powars. 1998. *Groundwater discharge and base-flow nitrate loading rates of nontidal streams, and their relation to a hydrogeomorphic classification of the Chesapeake Bay Watershed, Middle Atlantic Coast.* USGS Water Resources Investigations Report 98-4059, Baltimore, Maryland.

Bachman, L. J. and P. J. Phillips. 1996. Hydrologic landscapes on the Delmarva Peninsula. II. Estimates of base-flow nitrogen load to Chesapeake Bay. *Water Resour. Bull.* 32: 779–791.

Banta, G. T., M. F. Pedersen, and S. L. Nielsen. 2004. Decomposition of marine primary producers: consequences for nutrient recycling and retention in coastal ecosystems. Pp. 187–216, In: S. L. Nielsen, G. T. Banta, and M. E. Pedersen (eds.), *Estuarine nutrient cycling: the influence of primary producers.* Dordrecht, the Netherlands: Kluwer.

Blackburn, T. H. 1979. Method for measuring rates of Nh4+ turnover in anoxic marine-sediments using a N-15-Nh4+ dilution technique. *Appl. Environ. Microbiol.* 37: 760–765.

Bohlke, J. K. and J. M. Denver. 1995. Combined use of groundwater dating, chemical, and isotopic analyses to resolve the history and fate of nitrate contamination in two agricultural watersheds, Atlantic Coastal Plain, Maryland. *Water Resour. Res.* 31: 2319–2339.

Boschker, H. T. S. and J. J. Middelburg. 2002. Stable isotopes and biomarkers in microbial ecology. *FEMS Microbiol. Ecol.* 40: 85–95.

Boschker, H. T. S., S. C. Nold, P. Wellsbury, D. Bos, W. De Graaf, R. Pel, R. J. Parkes, and T. E. Cappenberg. 1998. Direct linking of microbial populations to specific biogeochemical processes by ^{13}C-labelling of biomarkers. *Nature* 392: 801–805.

Bouillon, S. and H. T. S. Boschker. 2006. Bacterial carbon sources in coastal sediments: a cross-system analysis based on stable isotope data of biomarkers. *Biogeosciences* 3: 175–185.

Boynton, W. R., J. D. Hagy, L. Murray, C. Stokes, and W. M. Kemp. 1996. A comparative analysis of eutrophication patterns in a temperate coastal lagoon. *Estuaries* 19: 408–421.

Bratton, J. F., J. K. Böhlke, D. E. Krantz, and C. R. Tobias. 2009. Flow and geochemistry of groundwater beneath a back-barrier lagoon: the subterranean estuary at Chincoteague Bay, Maryland, USA. *Mar. Chem.* 113: 78–92.

Bratton, J. F., J. K. Böhlke, F. T. Manheim, and D. E. Krantz. 2004. Groundwater beneath coastal bays of the Delmarva Peninsula: ages and nutrients. *Groundwater* 42: 1021–1034.

Bricker, S. B., B. Longstaff, W. C. Dennison, A. Jones, K. Boicourt, C. Wicks, and J. Woerner. 2008. Effects of nutrient enrichment in the nation's estuaries: a decade of change. *Harmful Algae* 8: 21–32.

Burnett, W. C., P. K. Aggarwal, A. Aureli, H. Bokuniewicz, J. E. Cable, M. A. Charette, E. Kontar, S. Krupa, K. M. Kulkarni, A. Loveless, W. S. Moore, J. A. Oberdorfer J. Oliveira, N. Ozyurt, P. Povinec, A. M. G. Privitera, R. Rajar, R. T. Ramessur, J. Scholten, T. Stieglitz, M. Taniguchi, and J. V. Turner. 2006. Quantifying submarine groundwater discharge in the coastal zone via multiple methods. *Sci. Total Environ.* 367: 498–543.

Capone, D. G., S. E. Dunham, S. G. Horrigan, and L. E. Duguay. 1992. Microbial nitrogen transformations in unconsolidated coral-reef sediments. *Mar. Ecol. Prog. Ser.* 80: 75–88.

Capone, D. G. and B. F. Taylor. 1977. Nitrogen fixation (acetylene reduction) in phyllosphere of *Thalassia testudinum*. *Mar. Biol.* 40: 19–28.

Capone, D. G., D. L. Taylor, and B. F. Taylor. 1977. Nitrogen fixation (acetylene reduction) associated with macroalgae in a coral-reef community in Bahamas. *Mar. Biol.*, 40: 29–32.

Carpenter, E. J. and D. Capone. 2008. Nitrogen fixation in the marine environment. Pp. 141–198, In: D. Capone, D. Bronk, M. R. Mulholland, and E. J. Carpenter (eds.), *Nitrogen in the marine environment*, 2nd ed. New York: Elsevier.

Cerco, C. F. and S. P. Seitzinger. 1997. Measured and modeled effects of benthic algae on eutrophication in Indian River-Rehoboth Bay, Delaware. *Estuaries* 20: 231–248.

Cloern, J. E. 1999. The relative importance of light and nutrient limitation of phytoplankton growth: a simple index of coastal ecosystem sensitivity to nutrient enrichment. *Aquat. Ecol.* 33: 3–16.

Cloern, J. E. 2001. Our evolving conceptual model of the coastal eutrophication problem. *Mar. Ecol. Prog. Ser.* 210: 223–253.

Correll, D. L., T. E. Jordan, and D. E. Weller. 1999. Effects of precipitation and air temperature on nitrogen discharges from Rhode River watersheds. *Water Air Soil Pollut.* 115: 547–575.

Currin, C. A., S. B. Joye, and H. W. Paerl. 1996. Diel rates of N-2-fixation and denitrification in a transplanted *Spartina alterniflora* marsh: implications for N-flux dynamics. *Estuar. Coastal Shelf Sci.* 42: 597–616.

Dillow, J. J. A. and E. A. Greene. 1999. *Groundwater discharge and nitrate loading to the coastal bays of Maryland.* USGS Water-Resources Investigations Report 99-4167, Baltimore, Maryland.

Dillow, J. A. A., W. S. L. Banks, and M. J. Smigaj. 2002. *Groundwater quality and discharge to Chincoteague and Sinepuxent Bays adjacent to Assateague Island National Seashore, Maryland.* USGS Water-Resources Investigations Report 02-4029, Baltimore, Maryland.

Duarte, C. M. and J. Cebrian. 1996. The fate of marine autotrophic production. *Limnol. Oceanogr.* 41: 1758–1766.

Duarte, C. M., D. J. Conley, J. Carstensen, and M. Sanchez-Camacho. 2009. Return to Neverland: shifting baselines affect eutrophication restoration targets. *Estuaries Coasts* 32: 29–36.

EPA. 2001. *National Coastal Condition Report*. EPA-620/R-01/005, Office of Research and Development and Office of Water, U.S. Environmental Protection Agency, Washington, D.C.

Eyre, B. D. and A. J. P. Ferguson. 2002. Comparison of carbon production and decomposition, benthic nutrient fluxes and denitrification in seagrass, phytoplankton, benthic microalgae- and macroalgae-dominated warm-temperate Australian lagoons. *Mar. Ecol. Prog. Ser.* 229: 43–59.

Eyre, B. D. and A. J. P. Ferguson. 2005. Benthic metabolism and nitrogen cycling in a subtropical east Australian estuary (Brunswick): temporal variability and controlling factors. *Limnol. Oceanogr.* 50: 81–96.

Fear, J., T. Gallo, N. Hall, J. Loftin, and H. W. Paerl. 2004. Predicting benthic microalgal oxygen and nutrient flux responses to a nutrient reduction management strategy for the eutrophic Neuse River Estuary, North Carolina, USA. *Estuar. Coastal Shelf Sci.* 61: 491–506.

Ferguson, A., B. Eyre, J. Gay, N. Emtage, and L. Brooks. 2007. Benthic metabolism and nitrogen cycling in a sub-tropical coastal embayment: spatial and seasonal variation and controlling factors. *Aquat. Microb. Ecol.* 48: 175–195.

Fugate, D. C., C. T. Friedrichs, and A. Bilgili. 2006. Estimation of residence time in a shallow back-barrier lagoon, Hog Island Bay, Virginia, USA. Pp. 319–337, In: M. L. Spaulding (ed.), *Estuarine and coastal modeling 2005.* Reston, Virginia. American Society of Civil Engineers.

Fulweiler, R. W. and S. W. Nixon. 2009. Responses of benthic-pelagic coupling to climate change in a temperate estuary. *Hydrobiologia* 629: 147–156.

Fulweiler, R. W., S. W. Nixon, B. A. Buckley, and S. L. Granger. 2007. Reversal of the net dinitrogen gas flux in coastal marine sediments. *Nature* 448:180–182.

Galloway, J. N. and W. C. Keene. 1999. Chemical composition of precipitation at the Virginia Coast. VCR-LTER dataset VCR02082.

Galloway, J. N., A. R. Townsend, J. W. Erisman, M. Bekunda, Z. Cai, J. R. Freney, L. A. Martinelli, S. P. Seitzinger, and M. A. Sutton. 2008. Transformation of the nitrogen cycle: recent trends, questions, and potential solutions. *Science* 320: 889–892.

Gardner, W. S., M. J. McCarthy, S. An, and D. Sobolev. 2006. Nitrogen fixation and dissimilatory nitrate reduction to ammonium (DNRA) support nitrogen dynamics in Texas estuaries. *Limnol. Oceanogr.* 51: 558–568.

Giordano, J. P. 2009. Nutrient loading and system response in the coastal lagoons of the Delmarva Peninsula. M.S. Thesis, College of William and Mary, Virginia Institute of Marine Science, Gloucester Point, Virginia.

Glibert, P. M., C. E. Wazniak, M. R. Hall, and B. Sturgis. 2007. Seasonal and interannual trends in nitrogen and brown tide in Maryland's coastal bays. *Ecol. Applic.* 17: S79–S87.

Goshorn, D., M. McGinty, C. Kennedy, C. Jordan, C. Wazniak, K. Schwenke, and K. Coyne. 2001. *An examination of benthic macroalgae communities as indicators of nutrients in middle Atlantic coastal estuariesL Maryland component.* Final Report 1998–1999, Maryland Department of Natural Resources, Annapolis, MD.

Haith, D. A. and L. L. Shoemaker. 1987. Generalized watershed loading functions for stream-flow nutrients. *Water Resour. Bull.* 23: 471–478.

Hardison, A. K., E. A. Canuel, I. C. Anderson, and B. Veuger. In preparation-a. *Fate of macroalgal biomass following a die-off event: uptake by the benthic microbial community.* Virginia Institute of Marine Science, Department of Biological Studies, Gloucester Point, Virginia.

Hardison, A. K., E. A. Canuel, B. Veuger, C. Tobias, and I. C. Anderson. In preparation-b. *Benthic macroalgae influence carbon and nitrogen cycling within the sediment microbial community.* Virginia Institute of Marine Science, Department of Biological Studies, Gloucester Point, Virginia.

Hayden, B. P. and J. H. Porter. 2001. VCR-LTER Dataset VCR02076: Terrestrial-marine watershed boundaries on the Delmarva Peninsula of Virginia. Virginia Coast Reserve LTR dataset, Charlottesville, Virginia.

Heck, K. L., Jr. and J. F. Valentine. 2007. The primacy of top-down effects in shallow benthic ecosystems. *Estuaries Coasts* 30: 371–381.

Howarth, R. W. 2008. Coastal nitrogen pollution: a review of sources and trends globally and regionally. *Harmful Algae* 8:14–20.

Howarth, R. W., A. Sharpley, and D. Walker. 2002. Sources of nutrient pollution to coastal waters in the United States: Implications for achieving coastal water quality goals. *Estuaries* 25: 656–676.

Jordan, T. E., D. L. Correll, and D. E. Weller. 1997a. Effects of agriculture on discharges of nutrients from coastal plain watersheds of Chesapeake Bay. *J. Environ. Quality* 26: 836–848.

Jordan, T. E., D. L. Correll, and D. E. Weller. 1997b. Nonpoint source discharges of nutrients from Piedmont watersheds of Chesapeake Bay. *J. Am. Water Resour. Assoc.* 33: 631–645.

Jordan, T. E., D. L. Correll, and D. E. Weller. 1997c. Relating nutrient discharges from watersheds to land use and streamflow variability. *Water Resour. Res.* 33: 2579–2590.

Jordan, T. E. and D. E. Weller. 1996. Human contributions to terrestrial nitrogen flux. *Bioscience* 46: 655–664.

Joye, S. B. and I. C. Anderson. 2008. Nitrogen cycling in coastal sediments. Pp. 867–915, In: D. G. Capone, D. A. Bronk, M. R. Mulholland, and E. J. Carpenter (eds.), *Nitrogen in the marine environment.* New York: Academic Press.

Joye, S. B. and H. W. Paerl. 1993. Contemporaneous nitrogen-fixation and denitrification in intertidal microbial mats: rapid response to runoff events. *Mar. Ecol. Prog. Ser.* 94: 267–274.

Keene, W. C., J. A. Montag, J. R. Maben, M. Southwell, J. Leonard, T. M. Church, J. L. Moody, and J. N. Galloway. 2002. Organic nitrogen in precipitation over eastern North America. *Atmos. Environ.* 36: 4529–4540.

Kemp, W. M., W. R. Boynton, J. E. Adolf, D. F. Boesch, W. C. Boicourt, G. Brush, J. C. Cornwell, T. R. Fisher, P. M. Glibert, J. D. Hagy, L. W. Harding, E. D. Houde, D. G. Kimmel, W. D. Miller, R. E. E. Newell, M. R. Roman, E. M. Smith, and J. C. Stevenson. 2005. Eutrophication of Chesapeake Bay: Historical trends and ecological interactions. *Mar. Ecol. Prog. Ser.* 303: 1–29.

Kennish, M. J. and A. R. Townsend. 2007. Nutrient enrichment and estuarine eutrophication. *Ecol. Applic.* 17: S1–S2.

Knowlton, N. 2004. Multiple "stable" states and the conservation of marine ecosystems. *Prog. Oceanogr.* 60: 387–396.

Lawson, S. E., P. L. Wiberg, K. J. McGlathery, and D. C. Fugate. 2007. Wind-driven sediment suspension controls light availability in a shallow coastal lagoon. *Estuaries Coasts* 30: 102–112.

Lim, K. J., B. A. Engel, Z. Tang, J. Choi, K. Kim, S. Muthukrishnan, and D. Tripathy. 2005. Automated web GIS based hydrograph analysis tool, WHAT. *J. Am. Water Resour. Assoc.* 41: 1407–1416.

Lucas, L. V., J. R. Koseff, J. E. Cloern, S. G. Monismith, and J. K. Thompson. 1999a. Processes governing phytoplankton blooms in estuaries. I. The local production-loss balance. *Mar. Ecol. Prog. Ser.* 187: 1–15.

Lucas, L. V., J. R. Koseff, S. G. Monismith, J. E. Cloern, and J. K. Thompson. 1999b. Processes governing phytoplankton blooms in estuaries. II. The role of horizontal transport. *Mar. Ecol. Prog. Ser.* 187: 17–30.

MacIntyre, H. L., R. J. Geider, and D. C. Miller. 1996. Microphytobenthos: The ecological role of the "secret garden" of unvegetated, shallow-water marine habitats. I. Distribution, abundance and primary production. *Estuaries* 19: 186–201.

Manheim, F. T., D. E. Krantz, and J. F. Bratton. 2004. Studying ground water under Delmarva coastal bays using electrical resistivity. *Ground Water* 42: 1052–1068.

Maryland Department of Planning. 2004. Land use data in Maryland. Surf your watershed: Watershed profiles for MD coastal bays. http://mddnr.chesapeakebay.net/wsprofiles/surf/prof/prof.html.

McCarthy, M. D., J. I. Hedges, and R. Benner. 1998. Major bacterial contribution to marine dissolved organic nitrogen. *Science* 281: 231–234.

McGlathery, K. J. 2001. Macroalgal blooms contribute to the decline of seagrass in nutrient-enriched coastal waters. *J. Phycol.* 37: 453–456.

McGlathery, K. J., I. C. Anderson, and A. C. Tyler. 2001. Magnitude and variability of benthic and pelagic metabolism in a temperate coastal lagoon. *Mar. Ecol. Prog. Ser.* 216: 1–15.

McGlathery, K. J., N. Risgaard-Petersen, and P. B. Christensen. 1998. Temporal and spatial variation in nitrogen fixation activity in the eelgrass *Zostera marina* rhizosphere. *Mar. Ecol. Prog. Ser.* 168: 245–258.

McGlathery, K., K. Sundbäck, and I. Anderson. 2004. The importance of primary producers for benthic nitrogen and phosphorus cycling. Pp. 231–261, In: S. L. Nielsen, G. T. Banta, and M. Pedersen (eds.), *Estuarine nutrient cycling: the influence of primary producers*. Dordrecht, the Netherlands: Kluwer Academic Publishers.

McGlathery, K. J., K. Sundbäck, and I. C. Anderson. 2007. Eutrophication in shallow coastal bays and lagoons: the role of plants in the coastal filter. *Mar. Ecol. Prog. Ser.* 348: 1–18.

Meyers, T. P., J. Sickles, R. Dennis, K. M. Russell, J. N. Galloway, and T. Church. 2001. Atmospheric nitrogen deposition to coastal estuaries and their watersheds. In: R. A. Valigura, R. B. Alexander, M. S. Castro, T. P. Meyers, H. W. Paerl, P E. Stacey, and R. E. Turner (eds.), *Nitrogen loading in coastal water bodies: an atmospheric perspective.* American Geophysical Union, Washington, D.C.

Monbet, Y. 1992. Control of phytoplankton biomass in estuaries: a comparative analysis of microtidal and macrotidal estuaries. *Estuaries* 15: 563–571.

Montoya, J. P., M. Voss, P. Kahler, and D. G. Capone. 1996. A simple, high-precision, high sensitivity tracer assay for N_2 fixation. *Appl. Environ. Microbiol.* 62: 986–993.

Nagata, T., B. Meon, and D. L. Kirchman. 2003. Microbial degradation of peptidoglycan in seawater. *Limnol. Oceanogr.* 48: 745–754.

Nagel, E. D. 2004. Nitrogen fixation in benthic microalgal mats: an important, internal source of "new" nitrogen to benthic communities in Florida Bay. M.S. Thesis, University of Maryland, College Park.

Neikirk, B. B. 1996. Exchanges of dissolved inorganic nitrogen and dissolved organic carbon between salt marsh sediments and overlying tidal water. M.S. Thesis, College of William and Mary, Virginia Institute of Marine Science, Gloucester Point, Virginia.

Nixon, S. W. 2009. Eutrophication and the macroscope. *Hydrobiologia* 629: 5–19.

Nixon, S., B. Buckley, S. Granger, and J. Bintz. 2001. Responses of very shallow marine ecosystems to nutrient enrichment. *Human Ecol. Risk Assess.* 7: 1457–1481.

NOAA. 2005. National Climatic Data Center, Weather Station "Painter 2 W," located in Painter, VA. Daily precipitation data available at http://www.ncdc.noaa.gov/oa/climate/stationlocator.html.

NRC. 2000. Clean coastal waters: understanding and reducing the effects of nutrient pollution. National Research Council, National Academy Press, Washington, D.C.

O'Neil, J. M. and D. G. Capone. 1989. Nitrogenase activity in tropical carbonate marine-sediments. *Mar. Ecol. Prog. Ser.* 56: 145–156.

Oertel, G. F. 2001. Hypsographic, hydro-hypsographic and hydrological analysis of coastal bay environments, Great Machipongo Bay, Virginia. *J. Coastal Res.* 17: 775–783.

Orth, R. J., M. L. Luckenbach, S. R. Marion, K. A. Moore, and D. J. Wilcox. 2006. Seagrass recovery in the Delmarva coastal bays, USA. *Aquat. Bot.* 84: 26–36.

Orth, R. J., S. R. Marion, K A. Moore, and D. J. Wilcox. 2010. Eelgrass (*Zostera marina* L.) in the Chesapeake Bay region of the mid-Atlantic coast of the USA: Challenges in conservation and restoration. *Estuaries Coasts* 33: 139–150.

Paerl, H. W. 1997. Coastal eutrophication and harmful algal blooms: importance of atmospheric deposition and groundwater as "'new"' nitrogen and other nutrient sources. *Limnol. Oceanogr.* 42: 1154–1165.

Paerl, H. W. 2009. Controlling eutrophication along the freshwater–marine continuum: dual nutrient (N and P) reductions are essential. *Estuaries Coasts* 32: 593–601.

Paerl, H. W., S. B. Joye, and M. Fitzpatrick. 1993. Evaluation of nutrient limitation of CO_2 and N_2 fixation in marine microbial mats. *Mar. Ecol. Prog. Ser.* 101: 297–306.

Paerl, H. W., L. E. Prufertbebout, and C. Z. Guo. 1994. Iron-stimulated N-2 fixation and growth in natural and cultured populations of the planktonic marine cyanobacteria *Trichodesmium* spp. *Appl. Environ. Microbiol.* 60: 1044–1047.

Paerl, H. W., L. M. Valdes-Weaver, A. R. Joyner, and V. Winkelmann. 2007. Phytoplankton indicators of ecological change in the eutrophying Pamlico Sound system, North Carolina. *Ecol. Applic.* 17: 88–101.

Pedersen, M. F., S. L. Nielsen, and G. T. Banta. 2004. Interactions between vegetation and nutrient dynamics in coastal marine ecosystems: an introduction. Pp. 1–15, In: S. L. Nielsen, G. T. Banta, and M. F. Pedersen (eds), *Estuarine nutrient cycling: The influence of primary producers.* Dordrecht: Kluwer Academic Publishers.

Pelz, O., L. A. Cifuentes, B. T. Hammer, C. A. Kelley, and R. B. Coffin. 1998. Tracing the assimilation of organic compounds using delta C-13 analysis of unique amino acids in the bacterial peptidoglycan cell wall. *FEMS Microbiol. Ecol.* 25: 229–240.

Peterjohn, W. T. and D. L. Correll.1984. Nutrient dynamics in an agricultural watershed. *Ecology* 65: 1466–1475.

Phillips, P. J. and L. J. Bachman. 1996. Hydrologic landscapes on the Delmarva Peninsula. I. Drainage basin type and base-flow chemistry. *Water Resour. Bull.* 32: 767–778.

Porter, J. H. and B. P. Hayden. 2001. VCR-LTER Dataset VCR00075: Land Cover for VCR-LTER Watersheds. Charlotttesville, Virginia.

Pritchard, D. W. 1960. Salt balance and exchange rate for Chincoteague Bay. *Chesapeake Science* 1: 48–57.

Rantz, S. E. 1982. Measurement and computation of streamflow, Volume 1 and 2. USGS Water Supply Paper 2175, Washington, D.C.

Reay, W. G., D. L. Gallagher, and G. M. Simmons. 1992. Groundwater discharge and its impact on surface-water quality in a Chesapeake Bay Inlet. *Water Resour. Bull.* 28: 1121–1134.

Risgaard-Petersen, N. 2003. Coupled nitrification-denitrification in autotrophic and heterotrophic estuarine sediments: on the influence of benthic microalgae. *Limnol. Oceanogr.* 48: 93–105.

Robinson, M. A. and W. G. Reay. 2002. Groundwater flow analysis of a mid-Atlantic outer coastal plain watershed, Virginia, USA. *Ground Water* 40: 123–131.

Russell, K. M., J. N. Galloway, S. A. Macko, J. L. Moody, and J. R. Scudlark. 1998. Sources of nitrogen in wet deposition to the Chesapeake Bay region. *Atmos. Environ.* 32: 2453–2465.

Russell, K. M., W. C. Keene, J. R. Maben, J. N. Galloway, and J. L. Moody. 2003. Phase partitioning and dry deposition of atmospheric nitrogen at the mid-Atlantic U.S. coast. *J. Geophys. Res.-Atmos.* 108(D21), 4656. DOI: 10.1029/2003JD003736, 2003.

Sand-Jensen, K. and J. Borum. 1991. Interactions among phytoplankton, periphyton, and macrophytes in temperate freshwaters and estuaries. *Aquat. Bot.* 41: 137–175.

Santos, M. 1996. Intertidal salinity buildup and salt flat development in temperate salt marshes: A case study of the salt flats at the Virginia barrier islands. Ph.D. Thesis, University of Virginia, Charlottesville.

Schwartz, M. C. 2003. Significant groundwater input to a coastal plain estuary: assessment from excess radon. *Estuar. Coastal Shelf Sci.* 56: 31–42.

Seitzinger, S. P. and J. H. Garber. 1987. Nitrogen fixation and 15N2 calibration of the acetylene reduction assay in coastal marine sediments. *Mar. Ecol. Prog. Ser.* 37: 65–73.

Sloto, R. A. and M. Y. Crouse. 1996. HYSEP: a computer program for streamflow hydrograph separation and analysis. U.S. Geological Survey Water-Resources Investigations Report 96-4040, Reston, VA.

Speiran, G. K. 1996. *Geohydrology and geochemistry near coastal groundwater-discharge areas of the Eastern Shore, Virginia*. USGS Water Supply Paper 2479, Washington, D.C.

Stanhope, J. W. 2003. Relationships between watershed characteristics and base flow nutrient discharges to Eastern Shore coastal lagoons, Virginia. M.S. Thesis, College of William and Mary, Virginia Institute of Marine Science, Gloucester Point, Virginia.

Stanhope, J. W., I. C. Anderson, and K. J. McGlathery. In preparation. Effects of high vs. low rainfall conditions on base flow nutrient discharges to VA Coastal Lagoons. Virginia Institute of Marine Science, Department of Biological Studies, Gloucester Point, Virginia.

Stanhope, J.W., I. C. Anderson, and W. G. Reay. 2009. Base flow nutrient discharges from lower Delmarva Peninsula watersheds of Virginia. *J. Environ. Quality* 38: 2070–2083.

Sundbäck, K., V. Enoksson, W. Graneli, and K. Pettersson.1991. Influence of sublittoral microphytobenthos on the oxygen and nutrient flux between sediment and water: a laboratory continuous-flow study. *Mar. Ecol. Prog. Ser.* 74: 263–279.

Sundbäck, K. and K. McGlathery. 2005. Interactions between benthic macroalgal and microalgal mats. Pp. 7–29, In: E. Kristensen, R. R. Haese, and J. E. Kostka (eds.), *Interactions between macro- and microorganisms in marine sediments*. AGU Series: Coastal and Estuarine Studies, Volume 60. Washington, D.C.: American Geophysical Union.

Sundbäck, K. and A. Miles. 2000. Balance between denitrification and microalgal incorporation of nitrogen in microtidal sediments, NE Kattegat. *Aquat. Microb. Ecol.* 22: 291–300.

Sundbäck, K., A. Miles, and E. Goransson. 2000. Nitrogen fluxes, denitrification and the role of microphytobenthos in microtidal shallow-water sediments: an annual study. *Mar. Ecol. Prog. Ser.* 200: 59–76.

Sundbäck, K., A. Miles, and F. Linares. 2006. Nitrogen dynamics in nontidal littoral sediments: role of microphytobenthos and denitrification. *Estuaries Coasts* 29: 1196–1211.

Taylor, D. I., S. W. Nixon, S. L. Granger, B. A. Buckley, J. P. Mcmahon, and H. J. Lin. 1995. Responses of coastal lagoon plant communities to different forms of nutrient enrichment: A mesocosm experiment. *Aquat. Bot.* 52: 19–34.

Tobias, C. R., M. Cieri, B. J. Peterson, L. A. Deegan, J. Vallino, and J. Hughes. 2003. Processing watershed-derived nitrogen in a well-flushed New England estuary. *Limnol. Oceanogr.* 48: 1766–1778.

Tyler, A. C., K. J. McGlathery, and I. C. Anderson. 2001. Macroalgae mediation of dissolved organic nitrogen fluxes in a temperate coastal lagoon. *Estuar. Coastal Shelf Sci.* 53: 155–168.

Tyler, A. C., K. J. McGlathery, and I. C. Anderson. 2003. Benthic algae control sediment-water column fluxes of organic and inorganic nitrogen compounds in a temperate lagoon. *Limnol. Oceanogr.* 48: 2125–2137.

USDA. 2002. U.S. Census of Agriculture Accomack County, VA, Northampton County, VA, and Worchester County, MD Data. National Agricultural Statistic Service, Washington, D.C.

Valiela, I., K. Foreman, M. Lamontagne, D. Hersh, J. Costa, P. Peckol, B. Demeoandreson, C. Davanzo, M. Babione, C. H. Sham, J. Brawley, and K. Lajtha. 1992. Couplings of watersheds and coastal waters: sources and consequences of nutrient enrichment in Waquoit Bay, Massachusetts. *Estuaries* 15: 443–457.

Valiela, I., J. McClelland, J. Hauxwell, P. J. Behr, D. Hersh, and K. Foreman. 1997. Macroalgal blooms in shallow estuaries: Controls and ecophysiological and ecosystem consequences. *Limnol. Oceanogr.* 42:1105–1118.

van Oevelen, D., L. Moodley, K. Soetaert, and J. J. Middelburg. 2006. The trophic significance of bacterial carbon in a marine intertidal sediment: Results of an in situ stable isotope labeling study. *Limnol. Oceanogr.* 51: 2349–2359.

Veuger, B., J. J. Middelburg, H. T. S. Boschker, and M. Houtekamer. 2005. Analysis of N-15 incorporation into D-alanine: A new method for tracing nitrogen uptake by bacteria. *Limnol. Oceanogr. Methods* 3: 230–240.

Veuger, B., D. van Oevelen, H. T. S. Boschker, and J. J. Middelburg. 2006. Fate of peptidoglycan in an intertidal sediment: An in situ C-13 labeling study. *Limnol. Oceanogr.* 51: 1572–1580.

Volk, J. A., K. B. Savidge, J. R. Scudlark, A. S. Andrews, and W. J. Ullman. 2006. Nitrogen loads through baseflow, stormflow, and underflow to Rehoboth Bay, Delaware. *J. Environ. Quality*. 35: 1742–1755.

Wazniak, C., M. Hall, C. Cain, D. Wilson, R. Jesien, J. Thomas, T. Carruthers, and W. Dennison. 2004. *State of the Maryland coastal bays*, Maryland Department of Natural Resources, Maryland Coastal Bays Program, and University of Maryland Center for Environmental Science, College Park, Maryland.

Wazniak, C. E., M. R. Hall, T. J. B. Carruthers, B. Sturgis, W. C. Dennison, and R. J. Orth. 2007. Linking water quality to living resources in a mid-Atlantic lagoon system, USA. *Ecol. Applic.* 17 :S64–S78.

Weil, J. C. 1990. A diagnosis of the asymmetry in top-down and bottom-up diffusion using a Lagrangian stochastic model. *J. Atmos. Sci.* 47: 501–515.

Welsh, D. T. 2000. Nitrogen fixation in seagrass meadows: regulation, plant-bacteria interactions and significance to primary productivity. *Ecol. Lett.* 3: 58–71.

Wolfstein, K., J. F. C. de Brouwer, and L. J. Stal. 2002. Biochemical partitioning of photosynthetically fixed carbon by benthic diatoms during short-term incubations at different irradiances. *Mar. Ecol. Prog. Ser.* 245: 21–31.

4 Ecosystem Health Indexed through Networks of Nitrogen Cycling

Robert R. Christian, Christine M. Voss, Cristina Bondavalli, Pierluigi Viaroli, Mariachiara Naldi, A. Christy Tyler, Iris C. Anderson, Karen J. McGlathery, Robert E. Ulanowicz, and Victor Camacho-Ibar*

CONTENTS

Abstract 73
4.1 Introduction 74
4.2 Methods 76
4.2.1 Study Sites 76
4.2.2 Network Construction 77
4.2.3 Network Analysis and Sensitivity to Uncertainty 82
4.3 Results and Discussion 83
4.3.1 Trophic Status 83
4.3.2 Ecosystem Health Framework 84
4.4 Conclusions 86
Acknowledgments 88
References 88

ABSTRACT

Assessing ecosystem-level status is a challenge, and several indices from information theory are used to assess the status of ecosystems. Food web networks are generally used in these assessments, but we have applied them to simpler networks of nitrogen cycling. Specifically, we indexed status in the framework of ecosystem health and three of its attributes: organization, vigor, and resilience, respectively indexed by ascendency (A), total system throughput (TST), and overhead (O). These indices were derived from seasonal networks of nitrogen cycles in two coastal lagoons: the Hog Island Bay, Virginia, and the Sacca di Goro, Emilia Romagna, Italy. The sites represent a large range of nutrient loading and subsequent trophic conditions from mesotrophic to hypereutrophic and dystrophic. As the degree of eutrophication (indexed through TN loading) increased, the cycling, efficiency of use of nitrogen, and degree of organization decreased. Indeterminacy of flows, as redundancy, increased. A problem of spurious correlations arose in the empirical analysis of ascendency vs. overhead because they share the same TST as a component. We propose that log transformations of TST can lessen the problem and might be used more generally in the empirical applications of information indices. Our results

* Corresponding author. Email: christianr@ecu.edu

are encouraging but illustrate the point that, while ecosystem-level assessments with theoretical foundations are increasingly possible, there remains considerable need for validation with empirical information.

Key Words: eutrophication, coastal lagoons, Sacca di Goro, Hog Island Bay, macroalgae, ascendency, spurious correlation

4.1 INTRODUCTION

Ecosystem-based management is challenged by the need for assessments of system-level status (Dame and Christian 2006). Traditional assessments of coastal aquatic ecosystems focused on population dynamics or community structure. These have been used in numerous situations in estuaries, coastal lagoons, and coastal waters among other systems (Niemi and McDonald 2004; Jørgensen et al. 2005). Assessments of ecosystem function or structure have had a shorter and less pervasive history. This is not to say that coastal aquatic ecosystems have not been assessed. Bricker et al. (1999, 2007) provided national assessments of the impacts of eutrophication on coastal aquatic ecosystems, and the Millennium Ecosystem Assessment (2005) included coastal systems from a global perspective. These are excellent frameworks for large-scale policy making, but may lack detailed observations necessary for management of individual systems.

Ecological network analysis (ENA) has been used to assess ecosystem status of individual (Baird and Ulanowicz 1989) and multiple aquatic ecosystems (Christensen 1995). ENA begins with the formal description of ecosystem components and the interactions between them as networks. The interactions are mostly associated with feeding and trophic structure, but biogeochemical cycles (Christian et al. 1996, 1998; Bondavalli 2003) have also been studied. Analyses result in indicators of: (1) strength of direct and indirect interactions; (2) relative importance of components to overall activity; (3) amount of flow involved in cycling vs. passing through the system; and (4) overall status of development, structure, and activity. Agencies focused on fisheries management use ENA of trophic structure to characterize aquatic ecosystems and to aid decision making (Christensen and Pauly 1993; Dame and Christian 2006).

Here we explore the use of ENA of nitrogen cycles in coastal lagoons to assess ecosystem status as "ecosystem health" defined by Costanza (1992) and interpreted for ENA by Ulanowicz (1997). Healthy ecosystems function well and are structured to do so, whereas less healthy ecosystems are not performing these functions as well (Costanza 1992). The image of health for ecosystems is less clear than for individual organisms with better defined boundaries. Furthermore, the term conveys a bias that promotes normative science (Lackey 2001). However, thoughtful implications and activities have arisen from the concept (Jørgensen et al. 2005). Ecosystem health is also linked to other aspects of status. Westra (1995) used ecosystem "integrity" as a generalization and extension of "health." The chief difference is that the notion of health is good performance (at processing materials), whereas ecosystem integrity references the "wild" condition. Integrity is perceived in the broader context of human cultural, political, and socioeconomic activities.

Synthesizing a longer list, Costanza (1992) defined ecosystem health through three attributes: organization, resilience, and vigor. He gave examples for each attribute, but his examples were not meant to be all-inclusive. Organization could be represented as biodiversity or connectivity. Resilience could be represented as response to perturbation to either resist change or return to some nominal condition. In addition, vigor could be represented as productivity or rate of metabolism. His "health index" was the product of indices of these three attributes. He recognized the value of ecological network analysis in developing the indices but did not fully develop specific ENA indices. He did, however, suggest that eutrophication would increase metabolism (vigor), while decreasing organization and resilience. Ulanowicz (1997) interpreted the three aspects of ecosystem health in the context of his information theoretic ENA indices (Ulanowicz 1986, 1992; Ulanowicz and Norden 1990), including flow diversity (H), constrained flow structure or average mutual information (AMI), and unconstrained flow structure (S). The total sum of flows (i.e., total system throughput, TST) is multiplied by each index, yielding developmental capacity, ascendency, and overhead, respectively.

Information theoretic indices reflect Costanza's three components of ecosystem health quite appropriately (Ulanowicz 1992). Mutual information reflects the degree to which two variables are dependent on each other. Organization appears to be characterized by the average mutual information (AMI), as a system measure of the direct mutual influence of different compartments directly on one another. Resilience can be represented by the unconstrained flows through S. S increases with evenness of flows for both donor and recipient compartments and number of flows. These flows include imports, exports, dissipation or respiration, and interactions within the system. A major contribution to S consists of the number of parallel pathways among these flows, where resiliency and S increase, as more options exist for a unit of material or energy to pass through the system (functional redundancies). In other words, removal or reduction of a small number of links has less impact on overall flow as resiliency increases. Finally, vigor can be represented by TST, a measure of the activity of the system. Vigorous systems are expected to have higher TST. Because ascendency is the product of AMI and TST and overhead is the product of S and TST, Ulanowicz (1997) proposed a graph of ascendency vs. overhead to represent a phase plane of ecosystem health (Figure 4.1).

Coastal aquatic ecosystems are subject to numerous human impacts and stressors (Kennish 1999; Peterson et al. 2008). These include enrichment of nutrients and organic matter, alterations to sediment delivery, habitat disruption, contaminant pollution, overharvesting of resources, and now the consequences of climate change. These alter ecosystem functions and health. Of particular interest are the impacts of eutrophication. Viaroli et al. (2008) reviewed how eutrophication of shallow, coastal lagoons alters community structure and biogeochemistry. Dominant primary producer communities often shift away from rooted aquatic vegetation to macroalgae and blooming phytoplankton as nutrient loading increases. These changes alter biogeochemical cycling. To assess these changes, we construct and analyze nitrogen networks of different habitats, containing different distributions of growth forms of primary producers, of two coastal lagoons with different levels of eutrophication. We focused on nitrogen, as it is often the primary controlling nutrient to primary production and thereby the health of coastal ecosystems (Kennish 1999; Bricker et al. 1999, 2007; Viaroli et al. 2008).

Ecological network analyses of nitrogen cycles have not been conducted as extensively as those of trophic structure. Christian et al. (1996) compared the nitrogen networks of three coastal lagoons, an estuary, and a rice field. These systems represented a range of eutrophication, primary production, hydrogeomorpology, and distribution of primary production associated with phytoplankton, macroalgae, and phanerogams. Residence time of water within the systems promoted recycling, where the

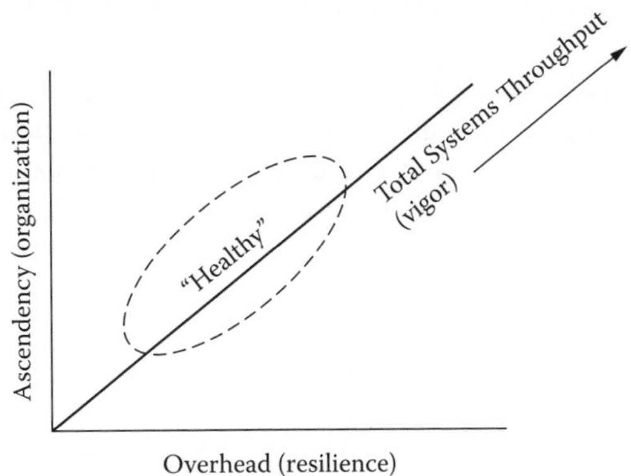

FIGURE 4.1 The three components of ecosystem health (organization, resilience, and vigor) proposed by Costanza (1992) as diagrammed, using network indices (ascendency, overhead, and total system throughput). (From Mageau, M.T., R. Costanza, and R.E. Ulanowicz, *Ecosys. Health* 1: 201–213, 1995. With permission.)

degree of recycling was also positively correlated to the proportion of primary productivity associated with phytoplankton. This was also found in a more developed analysis of 16 seasons of nitrogen cycling within the Neuse River Estuary (Christian and Thomas 2003).

Information indices were not used to evaluate these networks of nitrogen cycling. The networks of nitrogen cycling contained only five to seven compartments with highly aggregated consumer compartments, in contrast to networks focused on trophic structure with perhaps tens of compartments. Christian et al. (1996) expressed concern that small networks may not provide the flexibility to use information theory. We reevaluate this contention using networks of nitrogen cycling for two coastal lagoons that represent mesotrophic and hypereutrophic conditions. Specifically, we evaluate the ability of information indices to quantify ecosystem status within the framework of Ulanowicz (1997) based on Costanza (1992). Further, we assess other ENA indices and compare them to the information indices. This represents the first such empirical test of this ecosystem health framework.

4.2 METHODS

4.2.1 Study Sites

Two well-studied coastal lagoons with very different trophic conditions were assessed for network construction and analysis: Hog Island Bay (HIB), Virginia, USA (37° 26′ N, 75° 45′ W) and Sacca di Goro (GOR), Italy (44° 48′ N, 12° 18′ E). Hog Island Bay is a large, shallow coastal lagoon (area ≈100 km^2, depth <1–2 m at mean low water with the exception of a deep main channel) on the eastern shore of the Delmarva Peninsula and is part of the Virginia Coast Reserve (VCR) Long Term Ecological Research (LTER) Site (http://www.vcrlter.virginia.edu/). The Sacca di Goro (area ≈26 km^2, mean depth ≈1.5 m) is at the mouth of the Po River, receiving water from nutrient-rich distributaries, the Po di Volano and Po di Goro.

Large differences in nutrient loading are represented with numerous low and high rates to Hog Island Bay and the Sacca di Goro, respectively. Since no large rivers feed Hog Island Bay, nutrient inputs from the watersheds are via groundwater and atmospheric deposition (Anderson et al. unpublished data). The shallow aquifer is enriched in nitrogen primarily from agricultural activities which represent 55% of the Hog Island Bay watershed (442 km^2) (Bohlke and Denver 1995; Hamilton and Helsel 1995; Gu et al. 2008). Baseflow nutrient loading rates to Hog Island Bay are low compared to other lagoons (Chauhan and Mills 2002; Stanhope 2003; Stanhope et al. 2009), in part due to a high removal of NO_3^- from groundwater discharging to the streams by denitrification in a narrow band of sediments and bank materials (Flewelling personal communication). The Po di Volano is the major source of nutrients into the Sacca di Goro with a total loading of dissolved nitrogen from all sources exceeding demand most of the time (Christian et al. 1998; Viaroli et al. 2005; Giordani et al. 2008).

Both systems have macroalgae, microphytobenthos, and phytoplankton but their importance within the systems and among habitats differs. Hog Island Bay was formerly vegetated with seagrasses, which disappeared in the 1930s as a result of the storm disturbance of disease-weakened populations (Orth et al. 2006). Microphytobenthos and macroalgae are currently the dominant primary producers in the lagoon; however, both natural recolonization of seagrass and recent restoration efforts (since 2006) have increased seagrass areal coverage in localized areas. Benthos is generally net autotrophic except in localized, mid-lagoon areas where macroalgae accumulate and decay in mid-summer (McGlathery et al. 2001). The water column is typically net heterotrophic throughout the lagoon except following the mid-summer macroalgal collapse.

Sacca di Goro locations exhibit large amounts of primary production and standing stocks by macroalgae with mid- to late summer dystrophy (Viaroli et al. 2001; Giordani et al. 2008). *Gracilaria* and *Ulva* spp. dominate different regions of the lagoon, with the dynamics of *Ulva* spp. most significant to dystrophy (Viaroli et al. 1992). Little microphytobenthic or phytoplankton production has been noted (Christian et al. 1998; Viaroli and Christian 2003).

We consider Hog Island Bay and the Sacca di Goro to be mesotrophic and hypereutrophic lagoons, respectively, largely on the basis of nitrogen loading and general characteristics of primary production. External nutrient loading and variations in water residence time from the ocean inlet to the mainland creeks result in gradients of nutrient inputs and sediment organic matter across Hog Island Bay, with the highest concentrations of dissolved nitrogen and sediment organic matter found closest to the mainland (McGlathery et al. 2001). We captured that gradient in our sampling sites as Creek (Crk) near the mainland, Shoal (Shl), and near Hog Island (HI) (Figure 4.2a). High production rates are supported mostly by remineralization in the sediments and efficient internal nutrient cycling. Microphytobenthos control benthic–pelagic nutrient coupling (Anderson et al. 2003); in particular, nitrogen uptake by benthic algae suppresses denitrification to negligible rates and prevents the efflux of mineralized nitrogen from the sediment to the water column (Havens et al. 2001; Tyler et al. 2001, 2003; Anderson et al. 2003).

Two habitats [Gorino (Gor) and Giralda (Gir)] were assessed for the Sacca di Goro (Figure 4.2b). Dystrophy is most common in the very shallow and hydrologically isolated area of Gorino with intense growth of *Ulva* spp. (Viaroli et al. 1992). In contrast, Gir is largely devoid of macroalgae (Bartoli et al. 2001a), and microphytobenthos contribute significantly to overall primary production. One important aspect of the ecology of the Sacca di Goro is the farming of the Philippine clam, *Tapes philippinarum*. This invasive but farmed species dominates much of the benthos and can control much of the flux of oxygen and nutrients between the sediments and water column (Bartoli et al. 2001b; Nizzoli et al. 2007).

4.2.2 Network Construction

We constructed our networks of nitrogen cycling during two periods at each habitat (three in Hog Island Bay (HIB) and two in the Sacca di Goro (GOR)). A presumably more productive period in spring and early summer (Sp; March to June) was compared to a presumably less productive period in late summer and early autumn (Su; August to October).

The generalized network (Figure 4.3), and specific modifications for period and habitat were similar to those described by Christian et al. (1996). Dimensional units of standing stock were mmol N m^{-2}, and flows were mmol N m^{-2} d^{-1}. Three compartments were used for primary producers: phytoplankton (PHY), macroalgae (MAC), and microphytobenthos (MPB). Rooted phanerogams were not present in any of the habitats. Dissolved nitrogen was divided into separate inorganic (DIN) and organic (DON) compartments. The trypton (TRP) compartment contained non-phytoplankton N in seston including any consumer organisms. Estimates of the importance of size of aquatic consumers to nitrogen cycling indicate that planktonic-sized consumers, included in seston, dominate fluxes (Christian et al. 1992). The sediment (SED) compartment contained all forms of nitrogen within sediments, including nitrogen within and outside of organisms, except MPB. Standing stocks and flows were derived from a number of sources.

We quantified both flows between compartments as well as flows connecting internal components to the outside of the system. The latter consisted of imports and two forms of outputs: exports, and dissipations. Compartments PHY, TRP, SED, DIN, and DON received nitrogen from outside the location. All compartments, except MPB, have the potential to export nitrogen from the system. Growth of MAC was represented as either export (Hog Island Bay) or dissipation (Sacca di Goro) to allow mass balancing of networks. Inadvertently each group chose a different route for growth as biomass removal, but the route did not affect our interpretations. Dissipation for both locations involved the loss of nitrogen due to denitrification from the sediment. Compartments were internally connected via feeding, biogeochemical, and detrital pathways. When a flow was considered possible but not detectable or could not be inferred from available data, it was given a minimal value of 0.001 mmol N m^{-2} d^{-1}.

Table 4.1 provides a summary of imports, exports, dissipations, and intercompartmental flows initially used for the networks. Comments on the sources, logic, or assumptions associated with

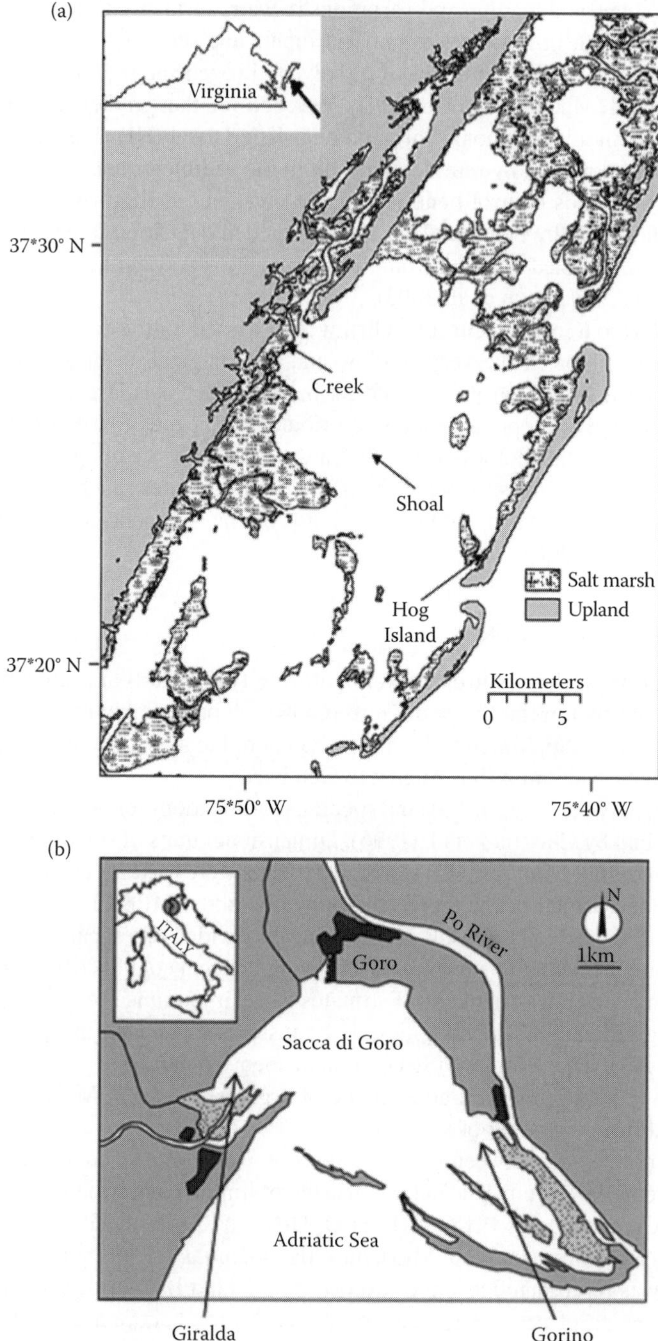

FIGURE 4.2 Maps of (a) Hog Island Bay, Virginia, and (b) Sacca di Goro, Italy. Maps show the three sampling sites in Hog Island Bay (Creek, Shoal, and Hog Island) and the two sampling sites in Sacca di Goro (Giralda and Gorino).

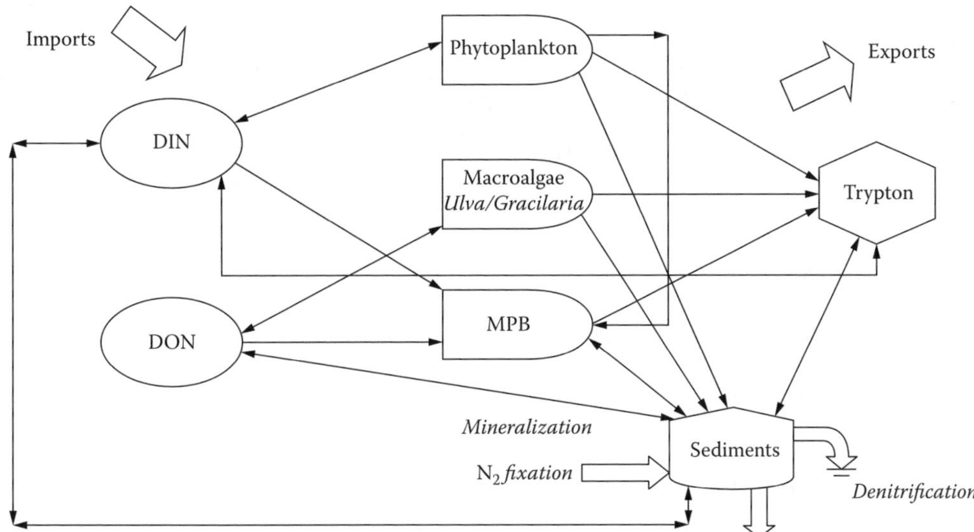

FIGURE 4.3 Generalized network of N cycling within Hog Island Bay and Sacca di Goro. Compartments include N in three primary producers: phytoplankton, macroalgae, and microphytobenthos (MPB); N in trypton (non-phytoplankton seston including any consumer organisms); dissolved organic (DON) and dissolved inorganic (DIN) nitrogen; and all forms of N within sediments (including organisms except MPB).

each flow are included. The specific reference is not given for each flow or standing stock in the interests of brevity. However, many of the flows derive from the references given in "Study Sites." Much of the information for Hog Island Bay comes from both published and unpublished research of McGlathery, Anderson, Tyler, and their colleagues (e.g., McGlathery et al. 2001; Havens et al. 2001; Tyler et al. 2001, 2003; Anderson et al. 2003). Most of this work was done in the late 1990s and early 2000s. Information for the Sacca di Goro largely came from 1997 because the available dataset was most complete from investigations under the NICE (Nitrogen Cycling in Estuaries) project (Bartoli et al. 2001a; Viaroli et al. 2001, 2006). These data were integrated with additional data from different field campaigns, laboratory studies, and literature (e.g., Naldi and Viaroli 2002).

Network analysis is more easily interpreted when flows of matter entering and leaving a given compartment, and the ecosystem as a whole, are steady state (Allesina and Bondavalli 2003). The construction of nitrogen cycle networks in coastal ecosystems required quantification of flows between different compartments measured directly (e.g., sediment–water N exchange) or indirectly (e.g., N burial, phytoplankton imports and exports) (Table 4.1). The natural variability of coastal ecosystems, the application of different flow-measuring approaches, and errors associated with measurements, among other factors, resulted in unsteady flow networks. Therefore, to construct robust N networks, rules and procedures for converting field data to flows and standing stocks were established (see details in Christian et al. 1992; Christian and Thomas 2003). Once the flow networks were constructed using initial values, flows for each compartment were balanced.

In this study, balancing was achieved by a three-step process. The first step, based on expert judgment, changed flows in such a way that: (1) it made ecological sense; (2) it went in the direction of balancing; (3) it tried to minimize the change in flows perceived to be most reliable; and (4) it promoted final closure on imports or exports. In this step and based on the degree of uncertainty in the measurements, the original flows were categorized as highly reliable (e.g., standing stocks of all compartments but trypton, denitrification, all flows from sediments), intermediately reliable (e.g., trypton standing stocks, DIN and DON consumption by macroalgae and phytoplankton), and least reliable (e.g., imports of all primary producers, DIN and DON consumption by trypton). In each case, the maximum adjustment to the flow was allowed to increase with decreased reliability.

TABLE 4.1
Compartments, Flows, and Processes Considered in Network Analysis of Nitrogen Cycling in Sacca de Goro (GOR) and Hog Island Bay (HIB)

From	To	Flow ID	Process Represented	Flow Estimation Sources
Input to	PHY	F_{01}	Loading	Chlorophyll *a* direct measurement multiplied by inflow
Input to	TRP	F_{04}	Loading	Particulate direct measurement multiplied by inflow for both and atmospheric inputs at HIB
Input to	DIN	F_{06}	Loading	DIN direct measurement multiplied by inflow for both and atmospheric inputs at HIB
Input to	DON	F_{07}	Loading	DON direct measurement multiplied by inflow for both and atmospheric inputs at HIB
Export from	PHY	F_{10}	Water flow from the lagoon to the sea	GOR = tide measurements multiplied by the average concentration determined in the lagoon HIB = by balancing and modeled residence time
Export from	MAC	F_{20}	Flow from the lagoon to the sea	GOR = quantified as 10% of the compartment total uptake HIB = growth of MAC
Export from	TRP	F_{40}	Water flow from the lagoon to the sea	GOR = tide measurements multiplied by the average concentration determined in the lagoon HIB = by balancing and modeled residence time
Export from	SED	F_{50}	Sediment burial	GOR = sedimentation rate times direct determination of N content in sediment HIB = considered insignificant
Export from	DIN	F_{60}	Water flow from the lagoon to the sea	GOR = tide measurements multiplied by the average concentration determined in the lagoon HIB = by balancing and modeled residence time
Export from	DON	F_{70}	Water flow from the lagoon to the sea	GOR = tide measurements multiplied by the average concentration determined in the lagoon HIB = by balancing and modeled residence time
Dissipation from	MAC	R_{20}	Macroalgae biomass increase	GOR = growth as 60% of the compartment total uptake HIB = considered as export
Dissipation from	SED	R_{50}	Denitrification	From experimental measurements and *in situ* estimations
PHY	MPB	F_{13}	Phytoplankton sedimentation rate	GOR = 1 turnover of stock per day HIB = 50% stock per day
PHY	TRP	F_{14}	Phytoplankton degradation and grazing	GOR = PHY mass balance difference from other processes HIB = 12.5% of donor's uptake
PHY	SED	F_{15}	Phytoplankton grazing by benthos	GOR = measured clam grazing rates HIB = 12.5% of donor's uptake
PHY	DIN	F_{16}	Phytoplankton release	GOR = considered insignificant HIB = 5% of donor's uptake
PHY	DON	F_{17}	Phytoplankton release	35% of donor's uptake for both lagoons
MAC	TRP	F_{24}	*Ulva* degradation	GOR = 10% of donor's uptake HIB = 10%–25% of donor's uptake
MAC	SED	F_{25}	*Ulva* sedimentation rate	GOR = 10% of donor's uptake HIB = 4%–7% of donor's uptake
MAC	DIN	F_{26}	Macroalgal release	GOR = considered insignificant HIB = measured directly
MAC	DON	F_{27}	Macroalgal release	GOR = 10% of donor's uptake HIB = measured directly

TABLE 4.1 (continued)
Compartments, Flows, and Processes Considered in Network Analysis of Nitrogen Cycling in Sacca de Goro (GOR) and Hog Island Bay (HIB)

From	To	Flow ID	Process Represented	Flow Estimation Sources
MPB	PHY	F_{31}	Resuspension (alive MPB)	Minimum flow (0.001 mmol N/m^2/d)
MPB	TRP	F_{34}	Resuspension (dead MPB)	GOR = 10% of donor's uptake
MPB	SED	F_{35}	MPB mortality and decomposition	GOR = MPB compartment mass balance HIB = 90% of donor's uptake
MPB	DON	F_{37}	MPB release	GOR = 35% of the compartment total uptake HIB = 0 because assumed all goes to sediment
TRP	SED	F_{45}	Trypton sedimentation rate	GOR =1 times the standing stock per day HIB = 35% of donor's uptake as sedimented fecal pellets
TRP	DIN	F_{46}	Trypton mineralization	GOR = DIN compartment mass balance HIB = 15% of donor's uptake
TRP	DON	F_{47}	Trypton excretion	GOR = considered insignificant HIB = 50% of donor's uptake
SED	MPB	F_{53}	MPB uptake	Experimental measurements and *in situ* estimations HIB = 80% uptake from sediments
SED	TRP	F_{54}	Sediment resuspension	Experimental measurements and *in situ* estimations HIB = 0%
SED	DIN	F_{56}	Sediment release to the water column	Experimental measurements and *in situ* estimations
SED	DON	F_{57}	Sediment release to the water column	Experimental measurements and *in situ* estimations
DIN	PHY	F_{61}	Phytoplankton uptake	Experimental measurements and *in situ* estimations
DIN	MAC	F_{62}	Macroalgal uptake	Experimental measurements and *in situ* estimations
DIN	MPB	F_{63}	MPB uptake	GOR = minimum flow (0.001 mmol N/m^2/d) HIB = 20% of uptake from water column
DIN	SED	F_{65}	Sediment uptake	Experimental measurements and *in situ* estimations
DON	PHY	F_{71}	Phytoplankton uptake	Experimental measurements and *in situ* estimations
DON	MAC	F_{72}	Macroalgal uptake	Experimental measurements and *in situ* estimations
DON	MPB	F_{73}	MPB uptake from water column	GOR = considered insignificant HIB = 20% of uptake from water column
DON	TRP	F_{74}	Net trypton consumption	DON compartment mass balance
DON	SED	F_{75}	Uptake by sediments	GOR = uptake fluxes undetectable HIB = *in situ* estimations
DON	DIN	F_{76}	Mineralization	GOR = considered insignificant HIB = measured experimentally

Note: PHY = phytoplankton (compartment 1), MAC = macroalgae (compartment 2), MPB = microphytobenthos (compartment 3), TRP = trypton (compartment 4), SED = sediments (compartment 5), DIN = dissolved inorganic nitrogen (compartment 6), DON = dissolved organic nitrogen (compartment 7). Much of the information for Hog Island Bay comes from both published and unpublished research of McGlathery, Anderson, Tyler and their colleagues (e.g., McGlathery et al. 2001; Havens et al. 2001; Tyler et al. 2001, 2003; Anderson et al. 2003). Most of this work was done in the late 1990s and early 2000s. Information for the Sacca di Goro largely derived from 1997 because the available dataset was most complete from investigations under the NICE project (Bartoli et al. 2001; Viaroli et al. 2001a; Viaroli et al. 2006). These data were integrated with additional data from different field campaigns, laboratory studies, and literature (e.g., Naldi and Viaroli 2002). Values for which there were no measurements on site or literature values were estimated by best professional judgment or by difference in mass balance calculations. More detail is given in the text.

Balancing proceeded until the balance between imports and exports and dissipations was within 10%, or when further flow modifications were not ecologically meaningful. In step two, we applied the Output-Input and/or the Average algorithms for balancing ecosystem networks using the WAND Balance package (see Allesina and Bondavalli 2003). In the case of Hog Island Bay, the Average balance routine resulted in a substantial distortion of the most dominant flows (i.e., from microphytobenthos to sediments and vice versa) than with the Output-Input balance. As flow distortion between these compartments was high even after balancing with the Output-Input algorithm, a third step in the balancing procedure was included for Hog Island Bay. In this step, the MPB to sediment and the sediment to MPB flows in the matrix obtained from the Output-Input balance were substituted by their original values, and an Average balance was performed on the modified matrix. This was not necessary for networks from the Sacca di Goro, where only the Average balance algorithms were used. ENA was applied on 10 balanced matrices using WAND (see Allesina and Bondavalli 2004). We have posted each of the balanced networks for input to and the results of all as separate analyses within WAND as Excel files on www.verlter.virginia.edu/cgi-bin/w3-msg12/data/query/datasets/show_data.html?Qdata_ID=VCR10171

4.2.3 Network Analysis and Sensitivity to Uncertainty

For this chapter, we focus on information indices that could be used to indicate ecosystem health and variables that might correlate to the degree of health. We considered average mutual information (AMI), unconstrained flow diversity (S), ascendency (A), developmental capacity (C), and overhead (O), along with total system throughput (TST) (Ulanowicz 1986, 1997). Correlative variables include TST, the sum of imports (total N loading), summed N uptake by primary producers, and average path length (APL, a measure of degree of cycling) (Finn 1976).

Flow measurement uncertainty resulting from natural and/or experimental variability is a concern during the ENA estimation of indices. A simple relative sensitivity analysis was conducted using primary production variability in the network of spring Hog Island Bay Shoals as a case study to determine the robustness of selected information indices. Primary production (i.e., the flows of N from DIN, DON, and sediments to phytoplankton, macroalgae, and MPB) for each producer was varied by ±10%, ±25%, and ±40% in the previously balanced flow matrix, and the resulting matrix was balanced with the Average algorithm before running the network analysis.

At the largest variation in primary production (±40%), the resulting information indices generally showed variation of less than 10% (Table 4.2). This result is partially from the fact that the

TABLE 4.2
Sensitivity Analysis Results Showing Information Theoretic Index Values before and after an Increase by 40% in Microphytobenthos (MPB) Primary Production at Shoals Site in Hog Island Bay during Spring

Producer	% Change	O	A	C	AMI	S	H	TN Loading	APL	TST	PP
MPB	0	46.1	29.1	75.2	1.46	2.31	3.76	0.96	19.83	20.0	8.02
MPB	+40	47.9	31.8	79.6	1.45	2.05	3.50	0.96	21.81	21.9	8.92
% change in index values		3.9	9.3	5.9	–0.7	–11.3	–6.9	0.0	10.0	9.5	11.2

Note: O = overhead; A = ascendency; C = capacity; AMI= average mutual information; S = unconstrained flow structure; H = flow diversity; TN loading = total nitrogen loading; APL = average path length; TST = total system throughput; PP = primary production. % change in values = [(value with +40% in MPB – original value)/(original value)] × 100 information indices are log-transformed values and partially from the dampening effect that the balancing step had on the modified matrix. If the antilogarithms were used, then AMI, S, and H would have higher percentage variations; however, our focus is on the more commonly used log transformed indices. We explain the dampening effect of balancing as follows. We increased the primary production of MPB in the "original" (balanced) matrix from 4.7 to 6.6 mmol N m^{-2} d^{-1} (i.e., +40%). MPB primary production was one of the highest flows in the spring Hog Island Bay Shoals network. Subsequent balancing of this new matrix with the average algorithm decreased the MPB primary production to 5.7 mmol N m^{-2} d^{-1}, a value 21.2% above the original. Balancing reduced the influence of large uncertainty in the data. Furthermore, no output variable approached this percentage (Table 4.2). Therefore, integrating and log transformed indices estimated through ENA appear modulated relative to flow measurement uncertainty. This limited analysis supports similar assessments (Baird et al. 1998; Christian and Thomas 2003).

4.3 RESULTS AND DISCUSSION

4.3.1 Trophic Status

Hog Island Bay (HIB) is characterized as biologically active but with relatively low nutrient loading (McGlathery et al. 2001; Havens et al. 2001; Tyler et al. 2001, 2003; Anderson et al. 2003; Stanhope 2003), while the Sacca di Goro (GOR) has larger nutrient loading and is eutrophic or hypereutrophic with dystrophic crises (Viaroli et al. 1992, 2005; Christian et al. 1998; Giordani et al. 2008). We categorize Hog Island Bay as mesotrophic to highlight the differences in trophic status as summarized in Table 4.3. Six networks from Hog Island Bay represented three habitats, and four networks of two habitats represented the Sacca di Goro, with each habitat being modeled for two seasonal periods. We used three measures of trophic status: TN loading, N uptake for primary production (PP), and total system throughput (TST) as the sum of all flows. TN loading rates ranged from 0.52 to 1.03 mmol N m^{-2} d^{-1} for the mesotrophic Hog Island Bay, which were one to two orders of magnitude less than the 16.1 to 25.3 mmol N m^{-2} d^{-1} for the hypereutrophic Sacca di Goro. Primary production was less

TABLE 4.3
Ecosystem-Level Status Indicators Estimated through Nitrogen Network Analyses for Hog Island Bay (HIB) and Sacca di Goro (GOR) during Spring (Sp) and Summer (Su)

Ecosystem	Model	O	A	C	AMI	S	H	TN Loading	APL	TST	PP
HIB	Crk-Sp	32.5	21.7	54.2	1.36	2.03	3.39	0.99	15.16	16.0	6.23
HIB	Crk-Su	43.9	31.6	75.5	1.40	1.94	3.34	0.52	42.46	22.6	8.98
HIB	Shl-Sp	46.1	29.1	75.2	1.46	2.31	3.76	0.96	19.83	20.0	8.02
HIB	Shl-Su	77.7	51.3	129	1.44	2.18	3.62	1.03	33.56	35.6	14.8
HIB	HI-Sp	17.8	13.2	31.0	1.43	1.92	3.35	0.37	24.00	9.25	3.64
HIB	HI-Su	37.2	21.0	58.2	1.48	2.62	4.10	0.97	13.64	14.2	5.10
GOR	Gor-Sp	375	188	563	1.41	2.82	4.23	25.3	4.26	133	27.1
GOR	Gor-Su	279	125	404	1.33	2.98	4.31	19.6	3.78	93.7	14.3
GOR	Gir-Sp	249	111	360	1.28	2.88	4.16	22.1	2.92	86.6	8.39
GOR	Gir-Su	182	80.6	263	1.24	2.81	4.05	16.1	3.04	65.0	5.58

Note: HIB sites are Creek (Crk), Shoal (Shl), and Hog Island (HI) and GOR sites are Gorino (Gor) and Giralda (Gir). See site locations in Figure 4.2. O = overhead; A = ascendency; C = capacity; AMI= average mutual information; S = unconstrained flow structure; H = flow diversity; TN loading = total nitrogen loading; APL = average path length; TST = total system throughput; PP = primary production. TN loading, TST and PP are indicators of trophic status.

than 10 mmol N m^{-2} d^{-1} in five of six networks for Hog Island Bay (3.64 to 8.98 mmol N m^{-2} d^{-1}). Primary production for Giralda fell within this range. Large macroalgal standing stocks and growth in Gorino during both periods and in the Shoal site of Hog Island Bay contributed the only primary productivities greater than 10 mmol N m^{-2} d^{-1}. The high primary productivities in Gorino fostered the two highest TST values. The next most active site was Giralda. While the most active site in Hog Island Bay was where macroalgae contributed to high primary productivity (Shoal in summer), all Hog Island Bay sites had TST values less than 55% of the lowest site in the Sacca di Goro. Spring in the Sacca di Goro had higher values of PP than in summer, whereas the opposite occurred in Hog Island Bay, where summer PP had higher values than spring.

Average path length (APL) is a measure of the degree to which material remains in the system and thus may be linked to the degree of recycling. It was higher within Hog Island Bay, ranging from 13 to 42 path lengths, than in the Sacca di Goro, ranging from 3 to 4 path lengths (Table 4.3). The equation for APL calculates higher APL with higher TST and lower TN loading. As TST is lower for Hog Island Bay than for the Sacca, the longer path lengths reflected the low rates of loading in the mesotrophic lagoon.

Information indices capture both the amount of activity and degree of organization by containing the product of TST and some algorithm of probabilistic interrelationships. Not surprisingly, given the higher TST in the Sacca di Goro, its networks' ascendency (A), capacity (C), and overhead (O) are the highest (Table 4.3). The organizational components remove the direct influence of TST, but still show differences between networks of the two lagoons. H is flow diversity, reflecting the structure of both constrained and unconstrained flows. H of Hog Island Bay networks ranged from 3.34 to 4.10 bits and were generally less than those of the Sacca di Goro (4.05 to 4.31 bits). AMI, the indicator of constrained flow structure, ranged from 1.24 to 1.48 bits. The lowest values were for Giralda, while the highest generally occurred within HIB. The spring Gorino network at 1.41 bits had a value within the range of HIB networks. S indicates unconstrained flow structure and is the difference between H and AMI. S demonstrates considerably greater unconstrained flow structure in the Sacca di Goro networks (2.81 to 2.98 bits) than for Hog Island Bay networks (1.92 to 2.32 bits). Thus, the higher H values for the Sacca are a result of higher degrees of unconstrained flows or evenness of parallel flows.

Clearly, the organizational structure of nitrogen networks from the Sacca di Goro is different from that of Hog Island Bay, and the direction of differences is consistent with expectations of trophic status. However, are differences associated with geographical location within the systems also consistent with expectations? We used pair-wise correlation analysis between measures of trophic status (i.e., TN loading, primary production, and TST) with each of the information indices (i.e., AMI, S, and H). Although significant Pearson correlations ($p < 0.01$) were found for either TST or TN loading with either S ($r = 0.793$ for TST and 0.857 for TN loading) or H ($r = 0.767$ for TST and 0.804 for TN loading), graphic representation failed to show any patterns across locations within lagoons. Furthermore, when correlations of these results from locations within each lagoon were determined, no statistically significant correlation was found. Therefore, the information indices did not track trophic status within a lagoon, although they did across the two very different lagoons. This is likely because H, AMI, and S are logarithmic terms and variations are thereby muted. In addition, the simple biogeochemical networks of few compartments may limit sensitivity by limiting opportunities for differences in structure (Christian et al. 1996). Unfortunately, the statistics of these information indices remain unstudied.

4.3.2 Ecosystem Health Framework

Costanza's (1992) identification of ecosystem health as organization, resilience, and vigor is represented through network analysis as average mutual information (AMI), unconstrained flow structure (S), and total system throughput (TST), respectively (Ulanowicz 1997). Ulanowicz

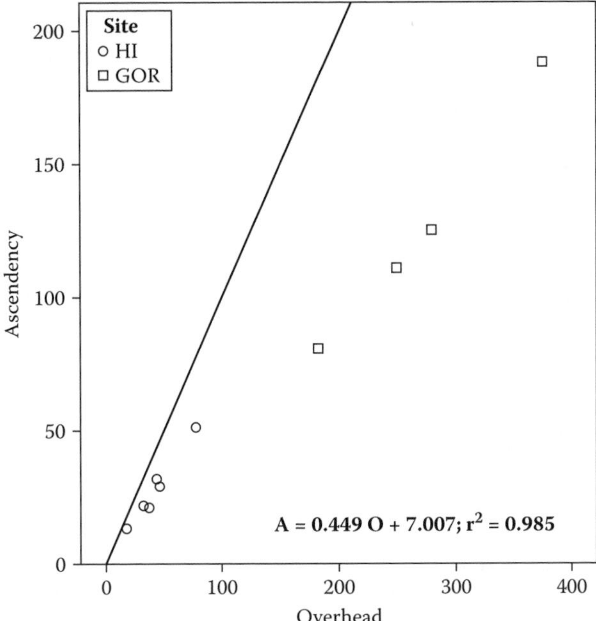

FIGURE 4.4 Position of nitrogen networks, within the framework for ecosystem health proposed by Ulanowicz (1997), for three sites in Hog Island Bay (HI) and two sites in Sacca di Goro (GOR) during two sampling periods (see Section 4.2 for details). Units for Ascendency and Overhead are mmol N m^{-2} d^{-1} bits. The line represents the 1:1 relationship and not the linear regression equation.

(1997) proposed a plot of ascendency (A) vs. overhead (O) that represents a phase plane of ecosystem health. A contains AMI, and O contains S. Both increase as TST increases, as TST is multiplied by AMI and S for A and O, respectively. The A vs. O of nitrogen networks provides the linear relationship suggested by Ulanowicz (1997) (Figure 4.4). The linear regression equation is A = 0.449 O + 7.007 with an $r^2 = 0.985$. This relationship demonstrates the relative loss of organization (lower A relative to O) with increased eutrophication. Ulanowicz (1997) previously postulated that eutrophication can be defined as increased TST with a decrease in organization (AMI or A). Our findings support this hypothesis at least in a relative sense. AMI was generally lower in the Sacca di Goro networks, although A was higher in the eutrophic system. Given that TST values in the Sacca were over 10 times those of the bay, it is not surprising that A values for the Sacca di Goro, which are the product of TST and AMI, were higher than for Hog Island Bay. Perhaps more significant, the S values in the Sacca are higher, reflecting greater disorganization or indeterminacy.

Empirical evaluation of Ulanowicz's framework (Figure 4.4) has a statistical complication—the potential for spurious correlation (Brett 2004). Spurious correlations can arise from shared variables embedded in the correlated variables. In this case, TST is shared as A = TST × AMI and O = TST × S. Such a relationship increases the value of r^2 needed to reject the null hypothesis (Brett 2004). The amount of r^2 for the null hypothesis of zero correlation increases as the coefficient of variation (CV) of TST increases above that for either AMI or S. As these latter two variables are based on logarithms and TST is not, the differences are large. The coefficients of variation of AMI, S, and TST are 5.6, 17.0, and 85.9%, respectively. The ratio of CVs of TST:AMI was therefore 15, and of TST:S, as 5. Using Monte Carlo simulations, Brett (2004) found random nonshared variables of X and Y in X × Z vs. Y × Z gave coefficients of determination over 0.9 when the CV ratios of shared to unshared variables were only 4. Thus, a high correlation and well-fit regression would be expected for the relationship of A vs. O simply through spurious correlation.

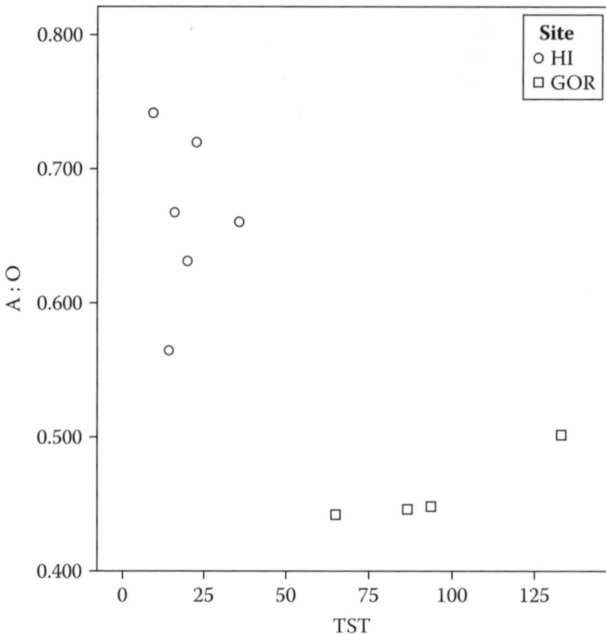

FIGURE 4.5 The ratio of Ascendency (A) to Overhead (O), representing the relationship of organization over resilience, plotted against TST, representing vigor, for three sites in Hog Island Bay (HI) and two sites in Sacca di Goro (GOR) during two sampling periods (see Section 4.2 for details). A:O is nondimensional, and TST units are mmol N m^{-2} d^{-1}.

This complication led to an evaluation of the framework in two alternate ways more appropriate for evaluating empirical information. First, A/O is equal to AMI/S and represents the relative amounts of organization to disorganization. This relative index was related to TST (Figure 4.5). The A/O values for the Sacca di Goro, where TST is higher, were lower than for Hog Island Bay, supporting our previous interpretations. Also, no trends were obvious within each lagoon. Second, we log-transformed TST to reduce its CV, making it more similar to those of AMI and S and evaluated the transformed A vs. O framework (i.e., [(log TST) × AMI] vs. [(log TST) × S]). The CV of log TST reduced to 27.0%, but still gave ratios of shared to unshared CVs of 1.6 to 4.8. The resultant r^2 of the transformed A vs. O lowered to 0.875 (Figure 4.6), which remains near values for random relationships by Brett (2004). While this did not fully equalize coefficients of variation in our study, it did reduce the differences significantly. We propose that log transformations of TST might be used more generally in the empirical applications of information indices to reduce this statistical problem.

While our recommendation provides some correction to the issue of spurious correlation, the problem of shared variables among information indices is much more complicated. H, AMI, and S are constructed on the same flows and network structure and represent different probability conditions of those flows. The flows themselves are interdependent and constrained by mass balance. Thus, shared variables occurred across hierarchical levels of calculation, and statistical inference of the interactions of ascendency, capacity, and overhead is a challenge for which we have only provided a beginning.

4.4 CONCLUSIONS

Measures and indices of the ecosystem status at the system level are necessary if ecosystem-based management is to be a reality. The Millennium Ecosystem Assessment (2005) provides the most ambitious attempt, but is focused on global and regional scale conditions. Bricker et al. (1999,

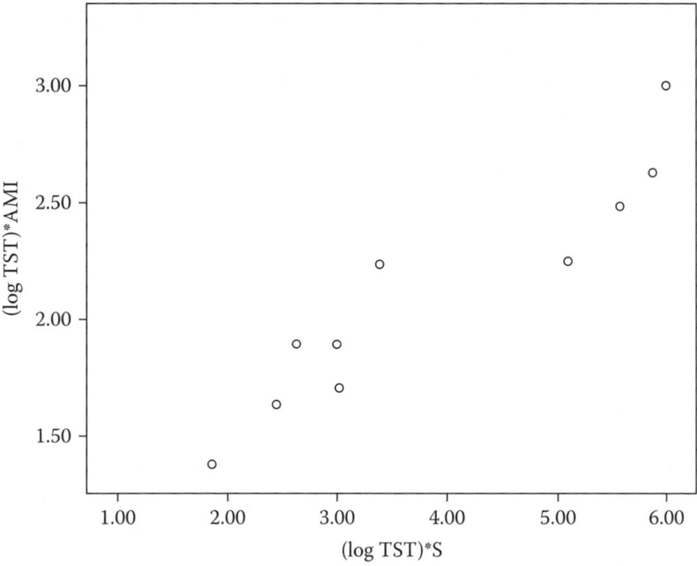

FIGURE 4.6 Position of nitrogen networks within framework for ecosystem health proposed by Ulanowicz (1997), using versions of A and O in which TST is log transformed and multiplied by AMI or S, respectively. Networks are for three sites in Hog Island Bay (HI) and two sites in Sacca di Goro (GOR) during two sampling periods (see Section 4.2 for details).

2007) developed an assessment of eutrophication of coastal systems within the United States with ecosystem-specific information. However, neither approach uses the in-depth observational and experimental information that exists for some of the better studied coastal ecosystems. ENA provides this opportunity with several system-level indices (Jørgensen et al. 2005). Our networks have tracked nitrogen as an important controlling element to lagoonal dynamics. We have not directly addressed the human causes for the differences in nitrogen loading or the consequences to humans, as in the broader assessments. Our efforts only address the consequences to the lagoon's trophic status. One might infer that lagoons with less healthy ecosystems are less valuable to humans, but this is untested. In fact, the Sacca di Goro is a highly valued ecosystem for its economically important seafood production (Bartoli et al. 2001b; Viaroli et al. 2006). Ecological network analysis can be extended to include other currencies and human dimensions directly. This is one of the next major steps for the methodology.

While most ENA studies have focused on trophic structure, we evaluated ecosystem status using nitrogen cycling of two well-studied coastal lagoons: Hog Island Bay, USA, and Sacca di Goro, Italy. The latter has much higher nutrient loading with clear signs of cultural eutrophication, while the former is minimally impacted by nutrients. The less impacted system, Hog Island Bay, demonstrated greater cycling of nitrogen (as measured by average path length) with more efficient use of the lower amount of loading compared to the highly eutrophic Sacca di Goro. This confirms the findings of Anderson et al. (2003). Previously, Christian et al. (1996) found cycling (as measured by the Finn Cycling Index) in coastal aquatic ecosystems to be largely associated with primary producer growth form, increasing with greater relative importance of phytoplankton to primary productivity. However, among the systems with rooted macrophytes and macroalgae, systems with higher loading had reduced cycling. Thus, as eutrophication alters the sources of primary production (Viaroli et al. 2008); it may also significantly retard nitrogen cycling and reduce efficiency of the element's use.

We applied information theoretical metrics to index "ecosystem health" properties (*sensu* Costanza 1992) of organization, resilience, and vigor (Ulanowicz 1997). We found ecological network analysis results reflected the trophic status of the two coastal lagoons. Ulanowicz (1997) proposed that eutrophication promoted the increase in total system throughput (TST) while decreasing average mutual information within ascendency. In a sense, eutrophication sets back ecosystem development. We found this generally to be the case in the comparisons of the two coastal lagoons. Thus, "organization" decreased with eutrophication, while TST rose. Additionally, the "resilience" component or unconstrained flow structure (S), representing indeterminacy and redundancy, increased with eutrophication. Thus, ecosystem health as portrayed in this manner does not simply improve as all of the three components increase. Rather, the ecosystem status reflects the balance of the three. Our results are consistent with the hypothesis that eutrophication disrupts organization, and less eutrophic systems may provide more determinant and efficient use of nutrients. While this hypothesis appears reasonable, extended testing and generalization remain.

ACKNOWLEDGMENTS

Funding for this study was granted by the National Science Foundation through the VCR-LTER and grants DEB-0080381and DEB-0621014 including supplements for international exchanges between the VCR-LTER and University of Parma. V. Camacho Ibar's contributions were made during his sabbatical at East Carolina University. We thank the numerous investigators of the Sacca di Goro and Hog Island Bay who have contributed to our ability to construct the networks. We thank the reviewers for their excellent suggestions to improve an earlier manuscript and give special thanks to Ann Krause for her insights into the complexity of information indices.

REFERENCES

Allesina, S. and C. Bondavalli. 2003. Steady state ecosystem flow networks: A comparison between balancing procedures. *Ecol. Model.* 165: 221–229.

Allesina, S. and C. Bondavalli. 2004. WAND: An ecological network analysis user-friendly tool. *Environ. Model. Software* 19: 337–340.

Anderson, I. C., K. J. McGlathery, and A. C. Tyler. 2003. Microbial mediation of "reactive" nitrogen transformations in a temperate lagoon. *Mar. Ecol. Prog. Ser.* 246: 73–84.

Baird, D. and R. E. Ulanowicz. 1989. The seasonal dynamics of the Chesapeake Bay. *Ecol. Monogr.* 59: 329–364.

Baird, D., J. Luczkovich, and R. R. Christian. 1998. Assessment of spatial and temporal variability in ecosystem attributes of St. Marks National Wildlife Refuge, Apalachee Bay, Florida. *Estuar. Coastal Shelf Sci.* 47: 329–349.

Bartoli, M., G. Castaldelli, D. Nizzoli, L. G. Gatti, and P. Viaroli. 2001a. Benthic fluxes of oxygen, ammonium and nitrate and coupled-uncoupled denitrification rates within communities of three different primary producer growth forms. Pp. 225–233, In: F. M. Faranda, L. Guglielmo, and G. Spezie (eds.), *Mediterranean ecosystems: Structure and processes*. Milan: Springer-Verlag.

Bartoli, M., D. Nizzoli, P. Viaroli, E. Turolla, G. Castaldelli, E. A. Fano, and R. Rossi. 2001b. Impact of *Tapes philippinarum* farming on nutrient dynamics and benthic respiration in the Sacca di Goro. *Hydrobiologia* 455: 202–212.

Bohlke, J. K. and J. M. Denver. 1995. Combined use of groundwater dating, chemical, and isotopic analyses to resolve the history and fate of nitrate contamination in two agricultural watersheds, Atlantic coastal plain, Maryland. *Water Resour. Res.* 31: 2319–2339.

Bondavalli, C. 2003. Effect of eutrophication upon radionuclide dynamics in the Sacca di Goro lagoon (Po river delta, Italy): A combined field, experimental and modeling study. *Environ. Pollut.* 125: 433–446.

Brett, M. T. 2004. When is a correlation between non-independent variables "spurious"? *Oikos* 105: 647–656.

Bricker, S. B., C. G. Clement, D. E. Pirhalla, S. P. Orlando, and D. R. G. Farrow. 1999. *National Estuarine Eutrophication Assessment: Effects of nutrient enrichment in the nation's estuaries*. NOAA, National Ocean Service, Special Projects Office and the National Centers for Coastal Ocean Science, Silver Spring, Maryland.

Bricker, S., B. Longstaff, W. Dennison, A. Jones, K. Boicourt, C. Wicks, and J. Woerner. 2007. *Effects of nutrient enrichment in the nation's estuaries: A decade of change*. NOAA Coastal Ocean Program Decision Analysis Series No. 26, National Centers for Coastal Ocean Science, Silver Spring, Maryland.

Chauhan, M. J. and A. L. Mills. 2002. Modeling baseflow nitrate loading on the eastern shore of Virginia. Presented at the Atlantic Estuarine Research Federation Spring Meeting, Lewes, Delaware.

Christensen, V. 1995. Ecosystem maturity: Towards quantification. *Ecol. Model.* 77: 3–32.

Christensen, V. and D. Pauly (eds.). 1993. *Trophic models of aquatic ecosystems*. ICLARM Conference Proceedings 26, Manila, the Philippines.

Christian, R. R., J. N. Boyer, D. W. Stanley, and W. M. Rizzo. 1992. *Network analysis of nitrogen cycling in an estuary*. Pp. 217–247, In: C. Hurst (ed.), Modeling the metabolic and physiologic activities of microorganisms. New York: John Wiley & Sons.

Christian, R. R., E. Forès, F. Comín, P. Viaroli, M. Naldi, and I. Ferrari. 1996. Nitrogen cycling networks of coastal ecosystems: Influence of trophic status and primary producer form. *Ecol. Model.* 87: 111–129.

Christian, R. R., M. Naldi, and P. Viaroli. 1998. Construction and analysis of static, structured models of nitrogen cycling in coastal ecosystems. Pp. 162–195. In: A. L. Koch, J. A. Robinson, and G. A. Milliken (eds.), *Mathematical modeling in microbial ecology*. New York: Chapman & Hall.

Christian, R. R. and C. R. Thomas. 2003. Network analysis of nitrogen inputs and cycling in the Neuse River Estuary, North Carolina, USA. *Estuaries* 26: 815–828.

Costanza, R. 1992. Toward an operational definition of ecosystem health. Pp. 239–256, In: R. Costanza, B. G. Norton, and B. D. Haskell (eds.), *Ecosystem health: New goals for environmental management.* Washington, D.C.: Island Press.

Dame, J. K. and R. R. Christian. 2006. Uncertainty and the use of network analysis for ecosystem-based fishery management. *Fisheries* 31: 331–341.

Finn, J. T. 1976. Measures of ecosystem structure and function derived from analysis of flows. *J. Theor. Biol.* 56: 363–380.

Giordani G., M. Austoni, J. M. Zaldívar, D. P. Swaney, and P. Viaroli. 2008. Modelling ecosystem functions and properties at different time and spatial scales in shallow coastal lagoons: An application of the LOICZ biogeochemical model. *Estuar. Coastal Shelf Sci.* 77: 264–277.

Gu, C. H., G. M. Hornberger, J. S. Herman, and A. L. Mills. 2008. Effect of freshets on the flux of groundwater nitrate through streambed sediments. *Water Resour. Res.* 44, W05415. DOI: 10.1029/2007WR006488.

Hamilton P. A. and D. R. Helsel. 1995. Effects of agriculture on ground-water quality in five regions of the United States. *Ground Water* 33: 217–226.

Hassan, R., R. Scholes, and N. Ash (eds.). Millennium Ecosystem Assessment. 2005. *Ecosystems and human well-being: Current state and trends*, Vol. 1. Washington, D.C.: Island Press.

Havens, K. E., J. Hauxwell, A. C. Tyler, S. Thomas, K. J. McGlathery, J. Cebrian, I. Valiela, A. D. Steinman, and S. J. Hwang. 2001. Complex interactions between autotrophs in shallow marine and freshwater ecosystems: Implications for community responses to nutrient stress. *Environ. Pollut.* 113: 95–107.

Jørgensen, S. E., F.-L. Xu, F. Sala, and J. C. Marques. 2005. Application of indicators for the assessment of ecosystem health. Pp. 5–66, In: S. E. Jørgensen, R. Costanza, and F.-L. Xu (eds.), *Handbook of ecological indicators for the assessment of ecosystem health*. Boca Raton, Florida: CRC Press.

Kennish, M. J. 1999. *Estuary restoration and maintenance: The National Estuary Program*. Boca Raton, Florida: CRC Press.

Lackey, R. T. 2001. Values, policy, and ecosystem health. *BioScience* 51: 437–443.

Mageau, M.T., R. Costanza, and R.E. Ulanowicz. 1995. The development, testing and application of a quantitative assessment of ecosystem health. *Ecosys. Health* 1: 201–213.

McGlathery, K. J., I. C. Anderson, and A. C. Tyler. 2001. Magnitude and variability of benthic and pelagic metabolism in a temperate coastal lagoon. *Mar. Ecol. Prog. Ser.* 216: 1–15.

Naldi, M. and P. Viaroli. 2002. Nitrate uptake and storage in the seaweed *Ulva rigida* C. Agardh in relation to nitrate availability and thallus nitrate content in a eutrophic coastal lagoon (Po River Delta, Italy). *J. Exp. Mar. Biol. Ecol.* 269: 65–83.

Niemi, G. J. and M. E. McDonald. 2004. Applications of ecological indicators. *Annu. Rev. Ecol. Evol. Syst.* 35: 89–111.

Nizzoli, D., M. Bartoli, and P. Viaroli. 2007. Nitrogen and phosphorus budgets during a farming cycle of the bivalve *Ruditapes philippinarum*. *Hydrobiologia* 587: 25–36.

Orth, R. J., M. L. Luckenbach, S. R. Marion, K. A. Moore, and D. J. Wilcox. 2006. Seagrass recovery in the Delmarva coastal bays, USA. *Aquat. Bot.* 84: 26–36.

Peterson, C. H., R. T. Barber, K. L. Cottingham, H. K. Lotze, C. A. Simenstad, R. R. Christian, M. F. Peihler, and J. Wilson. 2008. National Estuaries. Pp. 7-1–7-108, In: S. H. Julius and J. M. West (eds.), [J. S. Baron, L. A. Joyce, P. Kareiva, B. D. Keller, M. A. Palmer, C. H. Peterson, and J. M. Scott (authors)]. CCSP. 2008: *Preliminary review of adaptation options for climate-sensitive ecosystems and resources*. A Report by the U.S. Climate Change Science Program and the Subcommittee on Global Change Research. Washington, D.C.: U.S. Environmental Protection Agency.

Stanhope J. W., I. Anderson, and W. G. Reay. 2009. Base flow nutrient discharges from lower Delmarva Peninsula watersheds of Virginia, USA. *J. Environ. Qual.* 38: 2070–2083.

Stanhope, J. W. 2003. Relationships between watershed characteristics and base flow nutrient discharges to Eastern Shore coastal lagoons, Virginia. M.S. Thesis, College of William and Mary, Gloucester Point, VA.

Tyler, A. C., K. J. McGlathery, and I. C. Anderson. 2001. Macroalgae mediation of dissolved organic nitrogen fluxes in a temperate coastal lagoon. *Estuar. Coastal Shelf Sci.* 53: 155–168.

Tyler, A. C., K. J. McGlathery, and I. C. Anderson. 2003. Benthic algae control sediment-water column fluxes of organic and inorganic nitrogen compounds in a temperate lagoon. *Limnol. Oceanogr.* 48: 2125–2137.

Ulanowicz, R. E. 1986. *Growth and development: Ecosystem phenomenology*. New York: Springer-Verlag.

Ulanowicz, R. E. 1992. Ecosystem health and trophic flow networks. Pp. 190–206, In: R. Constanza, B. G. Norton, and B. D. Haskell (eds.), *Ecosystem health: New goals for environmental management*. Washington, D.C.: Island Press.

Ulanowicz, R.E. 1997. *Ecology: The ascendent perspective.* New York: Columbia University Press.

Ulanowicz, R. E. and J. S. Norden, 1990. Symmetrical overhead in flow networks. *Int. J. Systems Sci.* 21: 429–437.

Viaroli, P., R. Azzoni, M. Bartoli, G. Giordani, and L. Tajè. 2001. Evolution of the trophic conditions and dystrophic outbreaks in the Sacca di Goro lagoon (Northern Adriatic Sea). Pp. 467–475, In: F. M. Faranda, L. Guglielmo, and G. Spezie (eds.), *Mediterranean ecosystems: Structure and processes*. Milan: Springer-Verlag.

Viaroli, P., M. Bartoli, R. Azzoni, G. Giordani, C. Mucchino, M. Naldi, D. Nizzoli, and L. Tajè. 2005. Nutrient and iron limitation to *Ulva* blooms in an eutrophic coastal lagoon (Sacca di Goro, Italy). *Hydrobiologia* 550: 57–71.

Viaroli, P., M. Bartoli, G. Giordani, M. Naldi, S. Orfanidis, and J. M. Zaldivar. 2008. Community shifts, alternative stable states, biogeochemical controls and feedbacks in eutrophic coastal lagoons: A brief overview. *Aquat. Conserv.: Mar. Freshw. Ecosyst.* 18(1):S105–S117.

Viaroli, P. and R. R. Christian. 2003. Description of trophic status, hyperautotrophy and dystrophy of a coastal lagoon through a potential oxygen production and consumption index. *Ecological Indicators* 3: 237–250.

Viaroli, P., G. Giordani, M. Bartoli, M. Naldi, R. Azzoni, D. Nizzoli, I. Ferrari, J. M. Zaldivar Comenges, S. Bencivelli, G. Castaldelli, and E. A. Fano. 2006. The Sacca di Goro Lagoon and an arm of the Po River. Pp. 197–232, In: P. J. Wangersky (ed.), *The handbook of environmental chemistry, Estuaries*, Volume 5, Part H. Berlin: Springer-Verlag.

Viaroli, P., A. Pugnetti, and I. Ferrari. 1992. *Ulva rigida* growth and decomposition processes and related effects on nitrogen and phosphorus cycles in a coastal lagoon (Sacca di Goro, Po River Delta). Pp. 77–84, In: G. Colombo, I. Ferrari, V. U. Ceccherelli, and R. Rossi (eds.), *Marine eutrophication and population dynamics*. Fredensborg, Denmark: Olsen & Olsen.

Westra, L. 1995. Ecosystem integrity and sustainability: The foundational value of the wild. Pp. 12–33, In: L. Westra and J. Lemons (eds.), *Perspectives on ecological integrity*. Dordrecht, the Netherlands: Kluwer.

5 Blooms in Lagoons: *Different from Those of River-Dominated Estuaries*

Patricia M. Glibert, Joseph N. Boyer, Cynthia A. Heil, Christopher J. Madden, Brian Sturgis, and Catherine S. Wazniak*

CONTENTS

Abstract 91
5.1 Introduction 92
5.2 Site Descriptions and Data Resources 94
5.3 Seasonality and Intensity of Algal Blooms in Lagoons 97
5.4 Dominant Algal Species Composition and Algal Size Spectrum 99
5.5 Nutrient Composition and Sources in Lagoons 100
5.6 Differential Use of Nutrient Forms by Algal Groups: Algal Physiology 102
5.7 Poised on the Edge 103
5.8 Conclusions 106
Acknowledgments 108
References 108

ABSTRACT

Coastal lagoons differ substantially from river-dominated estuaries in the type of high-biomass phytoplankton blooms that generally proliferate, the seasonal timing of the blooms, the forms of nutrients that support these blooms, and the influence of hydrology on blooms and their nutrient dynamics. Coastal lagoons often support blooms of picoplankton (<3 μm in size) that can be sustained for months to years, whereas riverine estuaries often support high-biomass spring diatom blooms. These picoplankton blooms are also more likely to be sustained on regenerated forms of nitrogen, such as ammonium, urea, or dissolved organic substrates, compared to riverine spring blooms, which are generally supported by seasonal nitrate inputs. Regenerated forms of nutrients can be sustained in lagoons due to the higher surface:volume ratios and long residence times of these systems, both of which lead to a strong coupling between benthic nutrient fluxes and the plankton. The picoplankton genera that dominate have physiological characteristics that make them well suited to compete effectively for these substrates, but they must have sustained nutrient sources and limited grazing losses to maintain bloom biomass for prolonged periods. Using long-term monitoring data as well as experimental results, these characteristics are illustrated for Florida Bay and the Maryland–Virginia Chincoteague Bay.

Key Words*:* coastal lagoons, picoplankton, cyanobacteria, brown tide, diatoms, Florida Bay, Coastal Bays, Chincoteague Bay, nitrate, ammonium, urea, organic nutrients, HABs, EDABs

* Corresponding author. Email: glibert@hpl.umces.edu

5.1 INTRODUCTION

It has long been recognized, perhaps simplistically so, that increased nutrient inputs to an estuary should lead to higher algal biomass. While sufficient nutrient availability is a prerequisite for blooms, the concept of nutrient limitation of algal production is complex (e.g., Nixon 1995; Cloern 2001). The relationships between nutrient availability, nutrient loading, and algal production and accumulation are now recognized to be determined not only by the total quantity of nutrient loading, but also by the nutrient composition, the seasonal timing of nutrient inputs, whether nutrient fluxes are episodic or sustained, the physiological status of the primary producers at the time of nutrient delivery, and other physical factors that influence nutrient retention and algal physiology (Glibert and Burkholder 2006; Heisler et al. 2008). In attempting to understand the effects of nutrient availability on an ecosystem it is important to make the distinction between effects on total biomass and effects on nutrient assimilation or production processes. As initially developed by Caperon et al. (1971), and applied more recently to classical river-dominated estuaries, Chesapeake Bay, USA, and Moreton Bay, Australia (Malone et al. 1996; Glibert et al. 2006a), primary production can be viewed in a manner analogous to a saturation curve in response to nutrient enrichment (Figure 5.1). Primary production, the product of biomass and growth rate, may fall in either the maximum response or the minimum response region of the curve. The maximum physiological response is the initial slope of the saturation curve, where growth rates, but not necessarily biomass, are high. Under maximum response mode, the responses to changes in nutrient availability are rapid, usually driven by physiological adaptation, but may not necessarily result in large biomass changes. Alternatively, the minimum physiological response region is that in which maximum biomass is reached. In the minimum response mode, incremental increases in nutrient availability do not result in significant changes in growth, as nutrient uptake processes or productivity may be operating at maximal levels. Thus, different nutrients (or other abiotic factors such as light or temperature) may limit biomass and production.

Coastal lagoons represent a class of estuary characterized as shallow, highly enclosed, and typically quiescent in terms of wind, current, and wave energy relative to their deeper, more dynamic counterparts, river-dominated estuaries (Madden et al. 2010). Coastal lagoons also tend to have

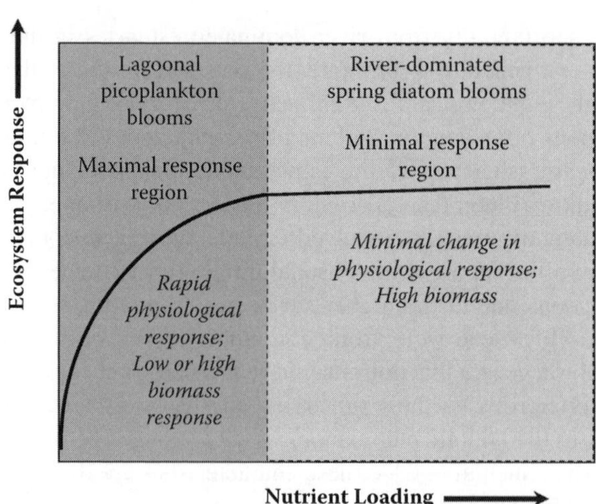

FIGURE 5.1 Generalized relationship between nutrient loading and ecosystem productivity response. The "maximum response" region of the curve indicates the region in which physiological rate changes such as nutrient uptake or growth rate would be rapid upon changes in nutrient loading, but biomass changes could range from small to large. Alternatively, the "minimum response" region reflects the region in which physiological processes may be near maximum levels, thus would show little change upon additional nutrient loading, but high biomass may be attained.

high surface to volume ratios and can have high wetland to water ratios as well (Bricker et al. 2007; Madden et al. 2010). These conditions collectively tend to promote a different dynamic in terms of algal proliferation than do river-dominated estuaries. Differences in the quality of the nutrient pool, the seasonal timing of delivery, and the source of nutrients, as well as in the resident phytoplankton community can lead to fundamentally different types of algal blooms in coastal lagoons than in river-dominated estuaries. Here, using data from two well-studied lagoons in the United States, Florida Bay and the Chincoteague Bay, part of the Maryland–Virginia Coastal Bays, the nutrient and algal bloom dynamics and characteristics of coastal lagoons are described, with an aim of highlighting those features that are generally unique to, or are typically significantly different from, those of river-dominated estuaries.

Numerous features differentiate lagoonal blooms from classical estuarine blooms. Lagoonal blooms are often dominated by picoplankton and may be sustained for long periods of time. Blooms in river-dominated estuaries, by contrast, tend to be highly seasonal and dominated by larger-sized phytoplankton (i.e., >10 μm), typically diatoms. Furthermore, blooms in lagoons tend to be supported by regenerated nutrients that may be lower in concentration, but which are made available through higher regeneration rates from both water column and benthic processes. (Note that the term "bloom" will hereafter refer to significant increases in algal biomass, as opposed to those events in which toxic or harmful algal species develop in relatively low numbers in proportion to the total phytoplankton community but are nevertheless often referred to as "blooms.") Highest phytoplankton biomass in riverine systems generally occurs in the spring, following maximum runoff (e.g., Pennock 1985; Malone et al. 1996). Coastal lagoons, in contrast, may lack the characteristic spring blooms of river-dominated systems. Only in summer do temperate river-dominated systems resemble those of coastal lagoons when temperatures are warm, flow is reduced, and the spring diatom bloom begins to be succeeded by flagellates, cyanobacteria, and other picoplankton that thrive on the nutrients regenerated from the spring bloom (Malone et al. 1996). Episodic blooms of large dinoflagellates occur in both types of systems, but are generally not long-lived. Furthermore, the total production of coastal lagoons is balanced between that of the water column and that of the benthos. River-dominated systems, due to their deeper comparative water columns (and reduced light penetration), tend to have proportionately more production in the water column, although in both cases many factors determine the amount and fraction of the water column and benthic production.

It has long been hypothesized that shifts in nitrogen (N) form from NO_3^- to NH_4^+ lead to community shifts away from plankton communities dominated by diatoms to those dominated by flagellates, cyanobacteria, and bacteria, in turn, resulting in a shift in composition of higher food webs (e.g., Legendre and Rassoulzadegan 1995; Glibert 1998). Food web models based on diatoms as the dominant algal producer are likely to consist of zooplankton grazers and to ultimately support a larger biomass of secondary producers. Higher export production may result in the development of sustained hypoxia or anoxia as this biomass decays (Figure 5.2a). In contrast, where the algal community is dominated by flagellates, cyanobacteria, or other picoplankton, the system generally sustains a proportionately greater flow through the microbial loop (Figure 5.2b; Azam et al. 1983; Legendre and LeFevre 1995; Legendre and Rassoulzadegan 1995). As shown herein, coastal lagoons are more likely to have blooms of the latter category. It is more likely for such blooms to be sustained via nutrient regeneration from the benthos, which is proportionately higher than in riverine systems due to the high surface to volume ratio of these systems, and the long water residence times (Table 5.1), both of which contribute to nutrient retention and nutrient regeneration. Thus, using the classical oceanographic terminology (*sensu* Dugdale and Goering 1967), coastal lagoons tend to be supported by *regenerated* nutrients, whereas river-dominated systems are generally supported by *new* nutrients. Coastal lagoons thus can support sustained blooms that conceptually fall in the maximal response region of the nutrient response curve, while river-dominated systems tend to support spring blooms that fall in the minimal response region of the response curve (Figure 5.1).

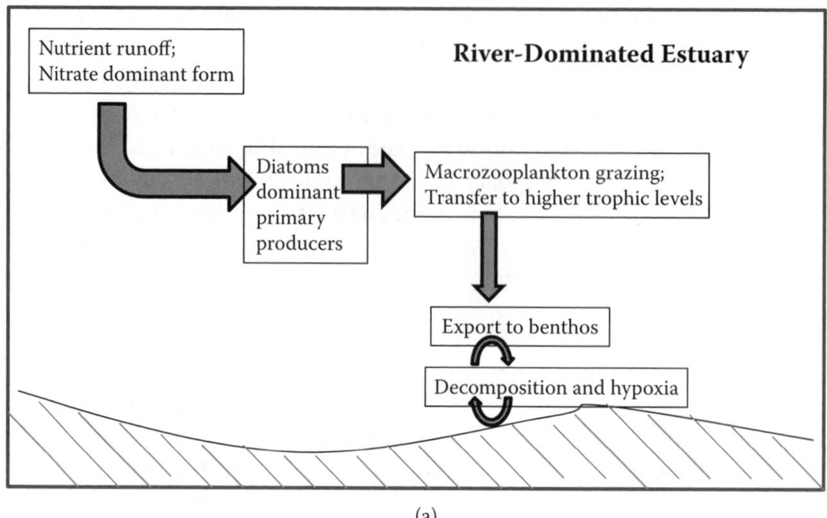

(a)

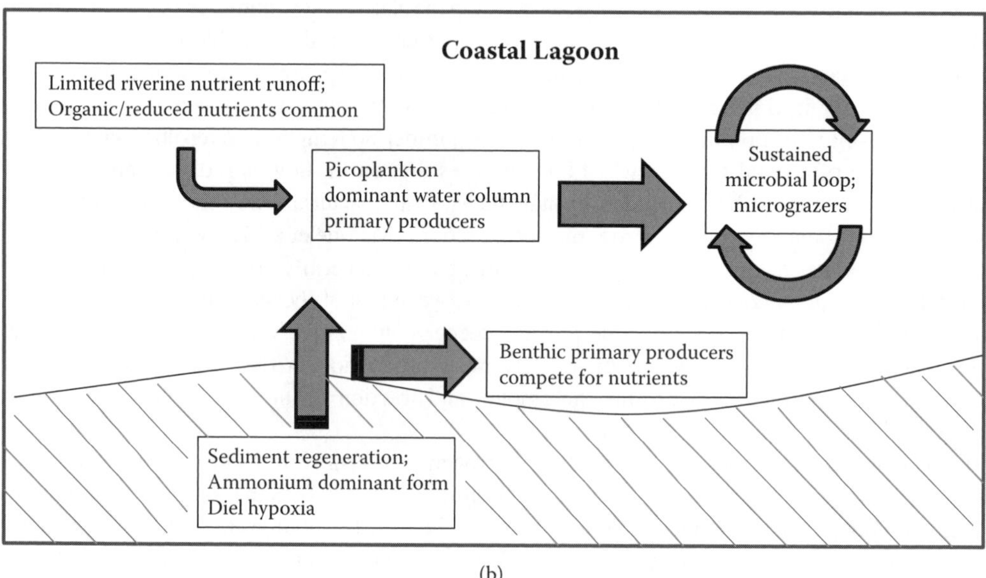

(b)

FIGURE 5.2 Conceptual sketch of the fate of primary production based on oxidized vs. reduced forms of nitrogen. (a) River-dominated estuary; (b) coastal lagoon.

5.2 SITE DESCRIPTIONS AND DATA RESOURCES

Florida Bay is a shallow (1 to 2 m), wedge-shaped subtropical lagoon, approximately 2200 km^2 in area, located at the southern end of the Florida peninsula. It is bounded by the Atlantic Ocean and the Gulf of Mexico (Figure 5.3a). The northern boundary of the bay is formed by the Everglades wetland system on the Florida mainland, and the eastern and southern boundaries are formed by the arc of the Florida Keys and the associated reef track (McIvor et al. 1994). Limited exchange with the Atlantic Ocean occurs through the tidal passes between the Keys. On its western border, the bay exchanges freely with the Gulf of Mexico, where a small diurnal tide (amplitude of ~0.5 m) and westerly currents circulate Gulf water in the bay (Smith 1998). The subtropical ecosystem has an average temperature of ~25°C and two distinct meteorological seasons: a November to April

TABLE 5.1
Representative Surface Areas, Total Volumes, and Surface:Volume Ratios for Some Coastal Lagoons and River-Dominated Estuaries of the U.S. East Coast

System Type	Estuary	Surface Area (km^2)	Total Volume ($1000 \times m^3$)	Surface:Volume (m^{-1})
Lagoon	Florida Bay	1663	1,031,060	0.0016
	N. Coastal Bay	54	103,680	0.0005
	S. Coastal Bay	335	649,900	0.0005
	New Jersey bays	273	308,580	0.0009
	Great South Bay	383	421,300	0.0009
Riverine	Chesapeake Bay mainstem	6974	51,119,420	0.0001
	Delaware Bay	2070	12,668,400	0.0002
	Narragansett Bay	416	3,456,960	0.0001
	Neuse River	456	1,304,160	0.0003

Note: Values from Bricker et al. (2007).

dry season, and a May to October rainy season, during which 75% of the average 152 cm annual precipitation occurs (Duever et al. 1994).

Hydrological influences dominate the ecological dynamics in Florida Bay and its Everglades watershed. During the rainy season, Florida Bay receives proportionately more freshwater flow from the Everglades. However, during the dry season the bay receives very little freshwater input and evaporation processes drive salinity. The hydrology of Florida Bay is also complex because the water budget and flow have been significantly altered by the filling of Atlantic tidal passes and by engineered channelized water flow at the bay's northern fringes (Madden 2010). Florida Bay's unique geomorphology includes a system of banks and shoals that create barriers to hydrologic circulation (Nuttle et al. 2003). Thus, almost all the tide and hydrologic circulation in eastern and central Florida Bay is wind driven, with limited oceanic exchange. The region sporadically experiences climatic extremes, including occasional frost, drought, and intense windstorms. Numerous tropical storms and hurricanes have impacted the system, particularly through the extirpation of benthic macrophytes and mats, resuspension of sediments and interstitial nutrients into the water column, and redistribution of sediments and muds (Nuttle et al. 2003; Davis et al. 2004).

Florida Bay has witnessed many ecological changes in the past decades. Since the onset of industrialization in the 1880s, the health of the Florida Bay ecosystem has been negatively impacted on both decadal (e.g., increasing eutrophication) and centurial (e.g. changes in land use and water management practices within southern Florida) time scales (Fourqurean and Robblee 1999). These anthropogenic changes have led to a significant alteration in freshwater flow patterns within the Everglades, causing declines in submerged aquatic vegetation (SAV) distribution and abundance (Zieman et al. 1989; Robblee et al. 1991; Fourqurean et al. 1993), increases in pelagic algae blooms (Butler et al. 1995; Phlips et al. 1995; Phlips and Badylak 1996; Glibert et al. 2004, 2009a), decreases in coral health (Chiappone and Sullivan 1994; Szmant and Forrester 1994), and economically important fisheries (e.g., Tortugas shrimp; Nance 1994; Costello and Allen 1996), sport fisheries (Tilmant 1989), and manatees (McIvor et al. 1994) within the bay. Although Florida Bay receives its freshwater flow from the Everglades, this flow results from managed discharge rather than from natural hydrological conditions (Rudnick et al. 1999). Agricultural and human development have been intensive in recent decades in southern Florida and the Florida Keys and are thought to have changed nutrient inputs to Florida Bay (Lapointe and Clark 1992; Rudnick et al. 2005).

The Coastal Bays are a network of shallow (0.7 to 1.2 m) lagoons located behind the barrier island of Assateague Island (Figure 5.3b). These embayments are connected to the Atlantic Ocean

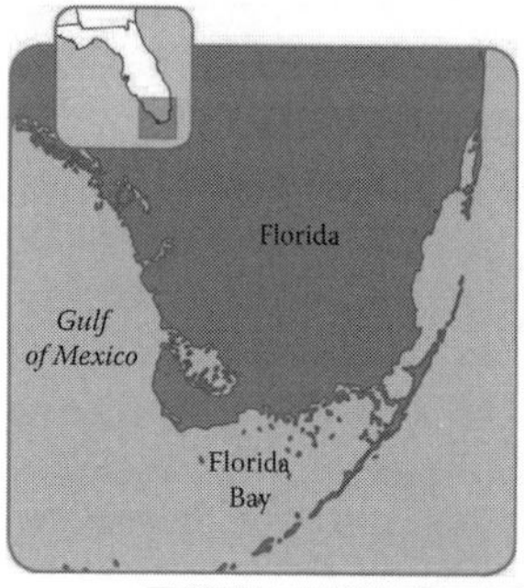

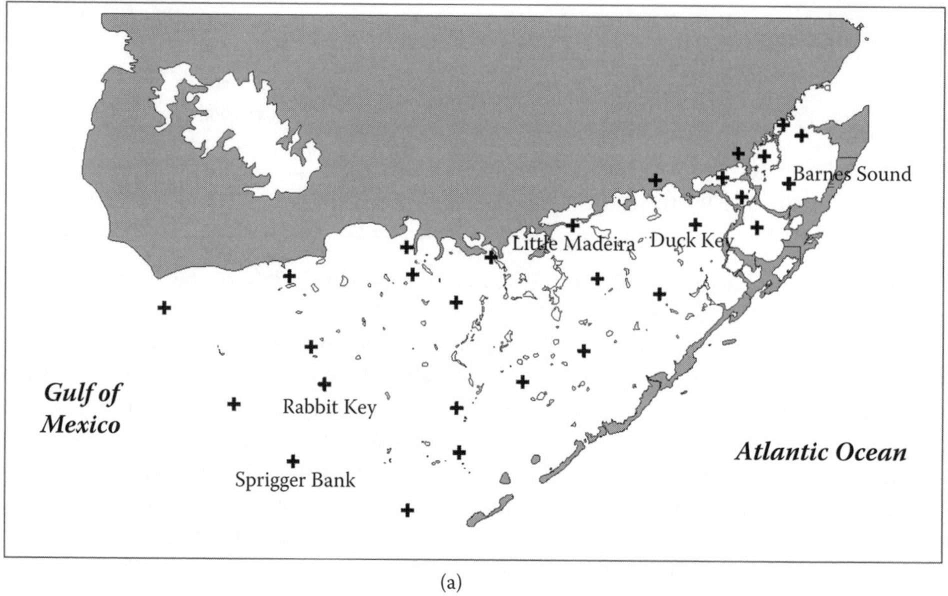

(a)

FIGURE 5.3 Maps of (a) Florida Bay and (b) mid-Atlantic coastal bays, indicating the spatial distribution of monitoring stations in Chincoteague Bay.

by two inlets at the northern and southern ends of Assateague Island. Like Florida Bay, the Coastal Bays are poorly flushed, nonstratified, and have long residence times (Boynton et al. 1996). In fact, flushing rates have been estimated to be on the order of 7% day^{-1} (Pritchard 1969), which approximate 10 to 20 days in the northern segments and >60 days in the largest of the sub-bays, Chincoteague Bay (Pritchard 1969; Lung 1994). Collectively the Coastal Bays are about 330 km^2 in area, about one-seventh the size of Florida Bay, but have a watershed that is about half the size of that of Florida Bay (Bricker et al. 2007; Glibert et al. 2007). The watershed of the Coastal Bays has traditionally been dominated by farming and forestry, but rapidly increasing residential development is occurring (Wazniak et al. 2007). A significant amount of wetland loss has also occurred in recent years through the construction of canals and bulkheads (Wazniak et al. 2007). In fact, this region is the fastest growing in the state of Maryland and one of the fastest growing regions in the United States (Crosset et al. 2004).

Both systems have had extensive, ongoing monitoring programs for many years. Florida Bay has been monitored extensively on at least a monthly basis for nearly two decades for a range of chemical, physical, and biological parameters (Boyer and Briceño 2008), and intensive process-oriented studies have been carried out several times a year for the past several years to assess nutrient–algal relationships. All monitoring data are available at http://serc.fiu.edu/wqmnetwork. Intensive monitoring in the Coastal Bays has been ongoing for nearly two decades as well, with more focused

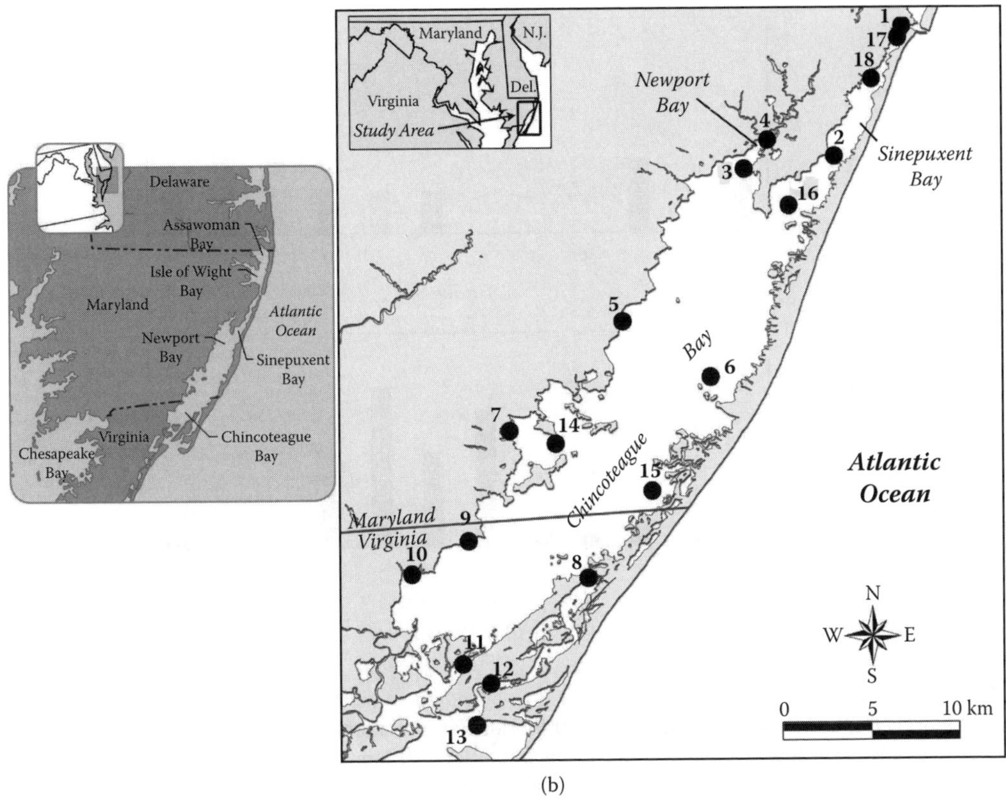

FIGURE 5.3 (continued).

studies taking place occasionally as blooms or other conditions warrant. All monitoring data are available from the Maryland Department of Natural Resources (http:www.eyesonthebay.net) and the National Park Service.

Thus, both Florida Bay and the Coastal Bays, especially the largest sub-bay, Chincoteague Bay, represent coastal lagoons in regions that are experiencing land-use changes with resulting alterations in land-derived nutrient loading. Both have similar dominant physical forces, very low riverine inputs, and are increasingly impacted by multiple stressors leading to increasing frequency of algal blooms.

5.3 SEASONALITY AND INTENSITY OF ALGAL BLOOMS IN LAGOONS

Florida Bay and Chincoteague Bay experience their annual phytoplankton biomass maximum in summer, fall, or early winter, not in spring. Annual data from 10 sites in Florida Bay and 18 regularly monitored sites in Chincoteague Bay for the 5-year period of 1999 to 2003 inclusive show that the peak algal biomass was reached ~45% of the time in October and November in Florida Bay, whereas ~80% of the time peak biomass was reached in Chincoteague Bay during the months of June to September (Figure 5.4). In Florida Bay most of the blooms tend to occur in the dry season, with no instances observed in this 5-year period of maximal algal biomass during the wet season of April through June (Figure 5.4a).

Florida Bay has historically been a more oligotrophic system than Chincoteague Bay, with generally low average water column chlorophyll *a* (Chl *a*) concentrations, <1 to 2 μg L^{-1} (Boyer and Keller 2007). Blooms, when they do occur, can be significant, with chlorophyll *a* levels exceeding as much as 30 μg L^{-1}, depending on the subregion of the bay where they occur (Table 5.2). The average concentration of chlorophyll *a* in the Coastal Bays, by contrast, is ~4 μg L^{-1}, but it can reach levels

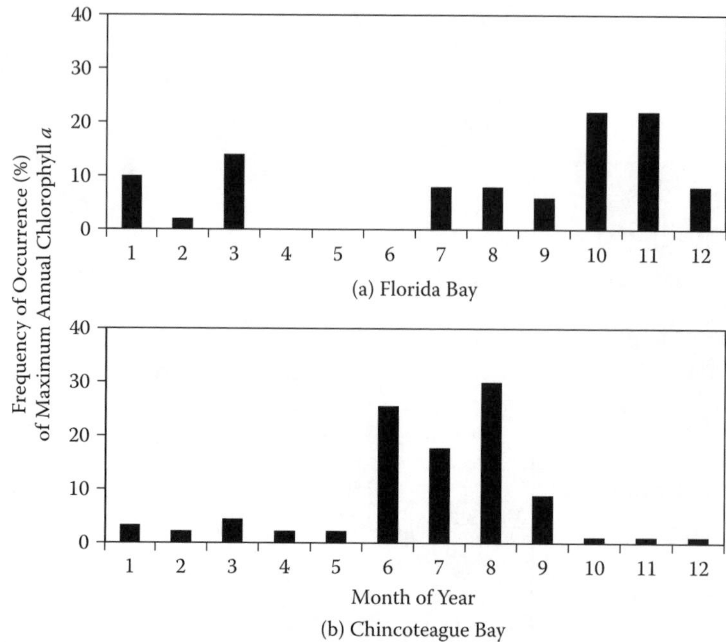

FIGURE 5.4 Frequency diagrams of the distribution, by month, of the annual peak chlorophyll *a* maximum for (A) Florida Bay and (B) Chincoteague Bay. For Florida Bay, the sampling period encompasses 1999–2003, and 10 sites that span the longitudal gradient, for a total of 120 months surveyed. For Chincoteague Bay the sampling period encompasses 1999 to 2003, and 18 sites that span the latitudinal gradient, for a total of 216 months surveyed.

TABLE 5.2
Long-Term Averages and Ranges of Chlorophyll *a* (μg L^{-1}) in Florida Bay and Chincoteague Bay by Subregion

System	Sub-Region	Median	Minimum	Maximum	*n*
Florida Bay	All	0.84	<0.03	35.61	3612
	Central bay	1.79	0.11	35.61	542
	Eastern bay	0.55	<0.03	11.35	2284
	Western bay	1.55	0.14	22.08	786
Chincoteague Bay	All	4.11	0.20	78.05	1044
	Northern bay	5.86	0.26	78.05	426
	Mid-bay	4.77	0.48	49.80	290
	South bay	3.06	0.20	25.7	348

Note: Values based on data from the long-term monitoring programs ongoing in each system. For Florida Bay, the sampling period encompasses 1989–2003; in Chincoteague Bay the sampling period encompasses 1999–2003. The northern bay included stations 1, 2, 3, 4, 16, 17, and 18; the mid-bay included stations 5, 6, 7, 14, and 15; and the southern bay included sites 8, 9, 10, 11, 12, and 13 (see Figure 5.3).

twice those of Florida Bay, occasionally exceeding 60 μg L^{-1} during blooms (Table 5.2). In both cases the blooms represent at least an order of magnitude increase in chlorophyll *a* over non-bloom conditions.

5.4 DOMINANT ALGAL SPECIES COMPOSITION AND ALGAL SIZE SPECTRUM

Blooms in lagoons are often dominated by phytoplankton that are considerably smaller in cell size than those that dominate riverine estuaries. Dominant algal species, although there are certainly exceptions, are typically those that are <3 μm in size. In Florida Bay, the dominant bloom-forming algal genus is the cyanobacterium *Synechococcus*, whereas in the Chincoteague Bay the dominant species of early summer is the pelagophyte *Aureococcus anophagefferens*. High-biomass picoplankton blooms in lagoons may reach near mono-specific proportions. During blooms in Florida Bay in 2002 and 2005, for example, *Synechocococcus* spp. comprised >99% of the phytoplankton assemblage and reached 10^8 cells L^{-1} (Glibert et al. 2004; Hitchcock et al. 2007; Heil et al. unpublished data) and were dominated by three organisms from the same clade V of MC-A cluster: *Synechococcus* sp. WH 8101 (Woods Hole), *Synechococcus* sp. CB 0201 (Chesapeake Bay), and *Synechococcus* sp. RS 9708 (Gulf of Aqaba). The planktonic cyanobacterial community was completely distinct from the benthic community (Boyer unpublished data). Similarly, in the Chincoteague Bay, *A. anophagefferens* has been found to comprise from 85% to 95% of the phytoplankton community during bloom periods (e.g., Wazniak and Glibert 2004).

These picoplankton are far more prevalent than diatoms, the typical dominant phytoplankton group in spring blooms of riverine estuaries. Diatoms may be found in coastal lagoons as one of many algal groups in the ambient flora during periods when blooms are not prevalent, for example, on the western fringe of Florida Bay and in the northern sub-lagoons of both Florida and Chincoteague Bays (Glibert et al. 2004; Tango et al. 2004). The most common diatoms found in the western edges of Florida Bay are *Rhizosolenia* spp. and *Chaetoceros* spp. (Hitchcock et al. 2007). During winter diatoms may comprise up to 40% of the phytoplankton community in Chincoteague Bay (Tango et al. 2004). However, the classic long-chain diatom species, such as *Skeletonema costatum*, that often dominate in spring blooms of river-dominated systems are generally not found in high abundance. Where significant abundances of dinoflagellates are observed in these coastal lagoons, they are either associated with benthos, as in *Gambierdiscus* in Florida Bay (e.g., Babinchak et al. 1986), localized to specific subsegments of the estuary, as in the case of *Pfiesteria* sp. or *Prorocentrum minimum* in Chincoteague Bay (Glibert et al. 2001; Tango et al. 2004) or transported into specific regions of the bay, as in *Karenia brevis* blooms in the western region of Florida Bay.

Once picoplankton blooms occur, they tend to be sustained for periods from weeks to months, and in a few cases, years. Sustained blooms of picoplankton such as these have been termed "ecosystem disruptive algal blooms, EDABs" (Sunda et al. 2006). Blooms of *Synechococcus* spp. in Florida Bay are not recent phenomena, although major blooms have occurred in recent years. In Florida Bay, one such bloom was observed in the early 1990s in the central and western regions of the bay (Boyer et al. 1999; Stumpf et al. 1999). These blooms continued through the 1990s and into the early 2000s (Richardson and Zimba 2002; Glibert et al. 2004, 2009a). The *Synechococcus* spp. bloom which began in late 2005 was unprecedented because it occurred in the eastern and northeastern region of the bay, which had previously been unaffected by such blooms, and it reached chlorophyll *a* concentrations of ~30 μg L^{-1}, levels not previously observed (Rudnick et al. 2006; Gilbert et al. 2009a). Ecosystem impacts of these blooms have been severe in Florida Bay. The SAV community declined significantly in areal coverage (Robblee et al. 1991; Hall et al. 1999; Durako et al. 2002). A 100% mortality of sponges also resulted (Fourqurean and Robblee 1999). Landings of spiny lobster and pink shrimp also plunged during the period of the blooms in the early 1990s (Madden 2010).

In Chincoteague Bay, *A. anophagefferens*, commonly known as "brown tide," has bloomed annually for more than a decade, the period over which such data are available (Trice et al. 2004; Glibert et al. 2007). In fact, there is considerable evidence that the intensity of these blooms increased in the decade of the 1990s (Trice et al. 2004; Glibert et al. 2007). Blooms tend to last in this system for a period of several weeks, although some brown tide blooms have been known to last months to years in other coastal lagoonal systems (e.g., the Texas brown tide of 1989 to 1997; Buskey et al. 1997, 2001). Annual blooms are now sufficiently dense to cause severe impacts or mortality on shellfish and also result in losses of SAV due to reduced light penetration (Gastrich and Wazniak 2002; Wazniak and Glibert 2004; Glibert et al. 2007).

5.5 NUTRIENT COMPOSITION AND SOURCES IN LAGOONS

Like all algal blooms regardless of location, lagoonal algal blooms must have nutrient sources to sustain the biomass that is produced, and the relative composition of the nutrient pool, not just the total nutrient quantity, influences which species are likely to proliferate. Nutrient composition within coastal lagoons often tends to differ significantly from that of river-dominated systems. Lagoonal systems do not generally have elevated concentrations of NO_3^- that are frequently observed in river-dominated estuaries, particularly in late winter–spring. Whereas concentrations of NO_3^- in many riverine systems can exceed many tens of μM (e.g., Chesapeake Bay, Glibert et al. 1995; Kemp et al. 2005; Neuse River Estuary, Christian et al. 1991; Burkholder et al. 2006), lagoonal systems may develop higher relative concentrations of the N from NH_4^+ than NO_3^- (Burkholder et al. 2006). In fact, some lagoons can accumulate very significant levels of NH_4^+, >30 μM (Boyer et al. 1999; Burkholder et al. 2006; Glibert et al. 2007; Sturgis 2007). In addition, organic forms of N and P tend to dominate the nutrient pool in lagoons compared to their respective inorganic nutrient forms (Boyer et al. 1999, 2006; Glibert et al. 2007).

In Florida Bay, an examination of nutrient availability at five stations, spanning the entire bay, from the period of 1987 through 2006, shows that concentrations of NH_4^+ frequently exceed 1 μM, and that concentrations of NH_4^+ at some stations exceed 5 μM up to 20% of the time (Figure 5.5a). In contrast, these same stations rarely have concentrations of NO_3^- that exceed 5 μM (Figure 5.5b). Likewise, an examination of four stations spanning the length of the Chincoteague Bay, for the time period of 1999 to 2004, shows that concentrations of NH_4^+ often exceed 5 μM, but rarely do those of NO_3^- (Figure 5.5c and d). The accumulation of N in these estuaries is not solely a function of limitation by another nutrient. While eastern Florida Bay has historically been considered P-limited, and ratios of total N:P clearly demonstrate a proportionate lack of P, such is not the case in Chincoteague Bay, where P remains in relatively high concentrations throughout the year, and the total N:P ratio is significantly lower than that of Florida Bay (Table 5.3). Both systems also have concentrations of silicate that range from several to tens of μM; this nutrient is neither limiting nor generally required to sustain the common bloom-forming species of these lagoons. The highly divergent nutrient ratios between Florida Bay and Coastal Bays are driven to at least some extent by the differing sediment types. Florida Bay is dominated by carbonate sediments, particularly in the eastern region, whereas Chinocoteague Bay is dominated by silt and clay. Strong carbonate binding of P is one reason for P limitation in Florida Bay (Orem et al. 1999).

Lagoonal estuaries have several sources of NH_4^+. Benthic fluxes are an important source (Yarbro and Carlson 2008), and these lagoons have significant faunal communities. Sponges, for example, in Florida Bay may contribute significantly to N (Corredor et al. 1988; Southwell et al. 2008). The shallow nature of these systems also means that nutrient regeneration from benthic fauna would be more important to the overlying algal populations than would be the case for deeper estuaries. Microbially mediated dissimilatory NO_3^- reduction to NH_4^+ (DNRA), a process by which NO_3^- is converted to NH_4^+ in the presence of organic matter, has also been shown to be significant in Florida Bay where measured rates have ranged from 11 to 170 μM m^{-2} h^{-1} (Gardner and McCarthy 2005). While no direct measurements of this process are available for Chincoteague Bay, there is no reason to assume that it would not be occurring: DNRA is thought to be more common in aquatic systems

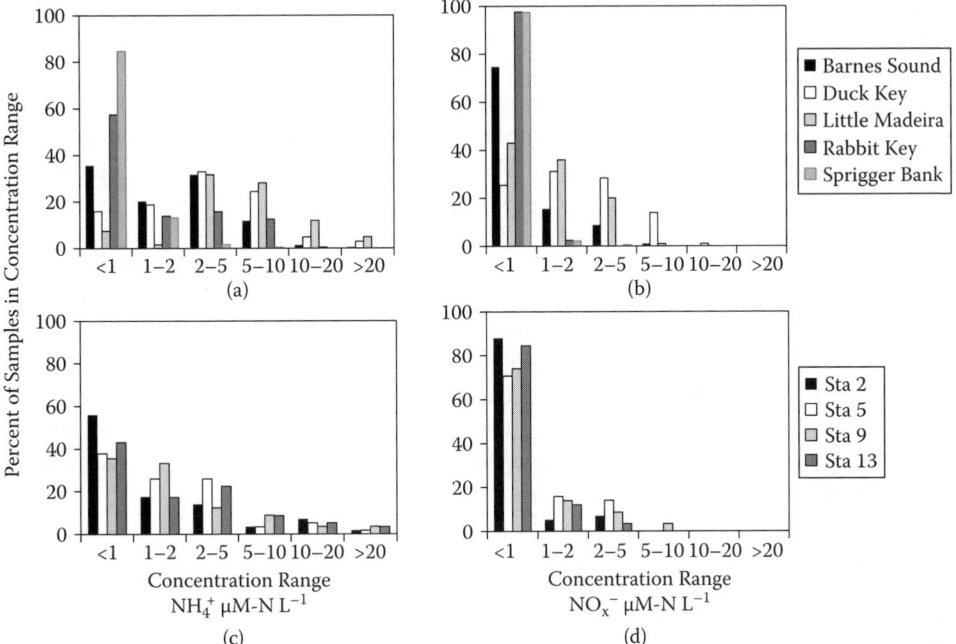

FIGURE 5.5 Frequency diagrams of the concentrations of NH_4^+ (a, c) and NOx^- (b, d) for Florida Bay (a, b) and Chincoteague Bay (c, d) for representative stations spanning the length of each estuary. Data are based on monthly monitoring in both estuaries. For Florida Bay, the sampling period encompasses 1989 to 2003, representing a total of 206 measurements for each station, while in Chincoteague Bay, the sampling period encompasses 1999 to 2003, representing a total of 58 measurements for each station.

TABLE 5.3
Summary of Average Nutrient Ratios Indicated for Selected Stations from the Long-Term Monitoring Data from Florida Bay and Chincoteague Bay

System	Station	NH_4^+:NO_3^-	TN:TP	*n*
Florida Bay	Barnes Sound	13.29	181	208
	Duck Key	2.92	222	209
	Little Madeira	11.69	213	202
	Rabbit Key	71.58	124	211
	Sprigger Bank	13.51	64.6	189
Chincoteague Bay	Station 2	16.2	20.9	56
	Station 5	7.6	47.8	57
	Station 9	10.3	23.7	56
	Station 13	13.2	14.4	58

Note: For Florida Bay, the sampling period encompasses 1989–2003; for Chincoteague Bay the sampling period encompasses 1999–2003.

where the water column is rich in labile organic carbon and low in NO_3^- (Tiedje 1988; Burgin and Hamilton 2007). Benthic resuspension events are also more prevalent in shallow lagoons and may be a significant source of nutrients to the water column (Lawrence et al. 2004).

While sustained hypoxia or anoxia is not common in coastal lagoons due to their shallow and well-mixed nature, localized hypoxia can occur, especially on a diel basis, and may be related to changes in benthic metabolism due to algal blooms. SAV or benthic macroalgae may die off as light penetration is reduced from algal blooms (e.g., Borum et al. 2005). This, too, would lead to increased flux of reduced forms of N from the sediment to the water column.

In Florida Bay, it has also been shown that the bacteria in the water column process nutrients quite differently from active algal (cyanobacterial) bloom communities (Glibert et al. 2004). In the eastern bay region, which has been shown to be P-limited for the phytoplankton, the bacteria are more responsive to organic N than P (Glibert et al. 2004). Recently described differences in the biochemical demand for P by cyanobacteria and heterotrophic bacteria may explain these different responses to N and P (Van Mooey et al. 2009). This suggests that bacteria in the water column could also be significant contributors to the regeneration of NH_4^+, although direct assays would be required to confirm this.

5.6 DIFFERENTIAL USE OF NUTRIENT FORMS BY ALGAL GROUPS: ALGAL PHYSIOLOGY

Phytoplankton groups differ in their requirements for, and in their ability to utilize, both inorganic and organic forms of N. Use of varying substrates depends on physiological ability of the cells to use specific substrates, and on the physiological state (nutrient status, growth rate, etc.) of the cells at the time of nutrient supply (Glibert and Burkholder 2006). While diatoms are generally NO_3^- opportunists, autotrophic and heterotophic picoplankton compete for reduced forms of N. Thus, the occurrence of many fast-growing diatoms has been found to be highly correlated with the large and/or frequent additions of NO_3^- found in river-dominated estuaries (e.g., Goldman 1993; Lomas and Glibert 1999). In contrast, recent studies in both enriched coastal areas and oligotrophic oceanic systems have shown that while productivity may increase quantitatively with overall N availability, the organic N component may contribute disproportionately to the uptake by dinoflagellates, pelagophytes, and cyanobacteria (Paerl 1988; Berg et al. 1997; LaRoche et al. 1997; Lomas et al. 2001; Glibert et al. 2001, 2006b; Zubkov et al. 2003).

In fact, urea has been previously implicated as an important N source for both *A. anophagefferens* (e.g., Berg et al. 1997; Lomas et al. 2001; Glibert et al. 2001, 2007) and *Synechococcus* spp. in both fresh and marine systems (Berman and Chava 1999; Collier et al. 1999; Sakamoto and Bryant 2001). The assimilation of urea by algae requires the enzyme urease. A survey of 13 algal species from multiple algal genera revealed that the highest urease activities per cell were found for *A. anophagefferens* and *Synechococcus* (Solomon et al. 2010). Berg et al. (1997) also found that *A. anophagefferens* had a higher potential for uptake of urea than for uptake of other N substrates, including inorganic and organic amino acids. Several urease genes have been well characterized in cyanobacteria, including *Synechococcus* (clone WH 7805; Collier et al. 1999), and higher rates of urease activity have been found for those cells grown on urea than for cells grown on NH_4^+ (Collier et al. 1999). In laboratory experiments, it has also been observed that at low light, growth of *A. anophagefferens* was higher when urea was provided as a growth N source than when growth N did not include this form (Pustizzi et al. 2004).

The potential for different fractions of the phytoplankton community to respond to inorganic and organic substrates was examined in Florida Bay. Using stable isotopic tracer techniques to track the relative use of different forms of N in both bloom and non-bloom conditions, it was found that the zeaxanthin:chlorophyll *a* ratio (an indicator of the relative contribution of cyanobacteria to phytoplankton biomass) was positively correlated with the rate of uptake of urea, and negatively

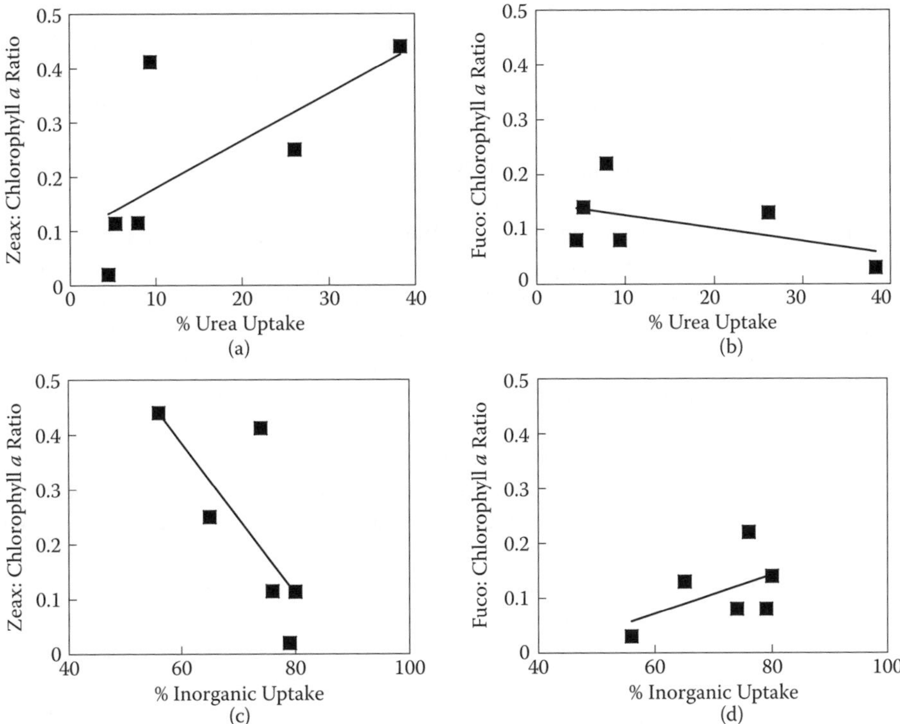

FIGURE 5.6 Relationships between the percent contribution of different N substrates to total phytoplankton N uptake and the composition of the phytoplankton community in Florida Bay in November 2002 as indicated by pigment ratios. (a) Gives the relationships between the fraction of urea uptake relative to total N uptake and the zeaxanthin: chlorophyll *a* ratio (indicative of cyanobacteria); (b) shows the fraction of urea uptake relative to the fucoxanthin: chlorophyll *a* ratio (indicative of diatoms and chrysophytes). (c and d) Show the same relationships except relative to the fraction of inorganic nitrogen uptake as NO_3^-. The fractional nitrogen uptake was determined using ^{15}N tracer techniques as described. (From Gilbert, P.M. et al. *Mar. Ecol. Prog. Ser.*, 280: 73–83, 2004. With permission.)

correlated with the rate of uptake of NO_3^- (Figure 5.6). The opposite pattern was observed for the fucoxanthin:chlorophyll *a* ratio (indicative of relative diatom and chrysophyte biomass) and the zeaxanthin:chlorophyll *a* ratio (indicative of cyanobacterial biomass; Figure 5.6), suggesting that different algal groups were using different N substrates (Glibert et al. 2004). Similar results have previously been observed in mesocosm experiments which varied in their surface to volume ratio. In that case, as in Florida Bay, there was a good agreement between the percentage of NO_3^- uptake and the percentage of diatoms in the phytoplankton community. Conversely, where cyanobacteria dominated the assemblages, the percentage of reduced forms of N uptake varied in proportion to the percent cyanobacteria (Glibert and Berg 2009). Moreover, in the mesocosm experiments, diatoms tended to dominate in the mesocosms with a smaller surface to volume ratio, while cyanobacteria dominated in those with a larger surface to volume ratio. Patterns in N utilization in the larger and deeper mesocosms were thus comparable to those observed in river-dominated estuaries in spring, whereas patterns of N utilization in the smaller and shallower mesocosms were more comparable to those observed in coastal lagoons in summer (Glibert and Berg 2009).

5.7 POISED ON THE EDGE

Blooms in lagoons are "poised on the edge" of success in several ways. First, they compete with other primary producers in the system, mainly microphytobenthos, macroalgae, and SAV. Also, due to the small size of the cells, they do not have the physiological capability to store large pools of nutrients internally, so nutrient limitation is a continual challenge (e.g., Sunda et al. 2006). In the context of the Caperon et al. (1971) nutrient response curve (Figure 5.1), production and biomass are generally limited by the same nutrient factor. River-dominated estuaries, by contrast, may have biomass limited by one factor, such as total N availability (as in the cases of Chesapeake Bay and Moreton Bay; Malone et al. 1996; Glibert et al. 2006a), while production may be limited by another factor, such as light.

Owing to the generally shallow nature of coastal lagoons, benthic primary producers are in competition with water column primary producers for nutrients. Coverage of SAV can be substantial, and is >60% for both Florida Bay and the Coastal Bays, although in both systems SAV has been under stress and has had recent die-backs (Durako et al. 2002; Wazniak and Hall 2004). In lagoonal systems, benthic chlorophyll *a* can be a significant contributor to production, and the presence of a significant population of benthic producers can help sustain dissolved oxygen levels in the water column. In some subregions of Chincoteague Bay, for example, there is more chlorophyll *a* in microphytobenthos than in the phytoplankton. Macroalgae, also, can be abundant (e.g., McGinty et al. 2004). The SAV of Florida Bay has been particularly well studied, and their role in physico-chemical, primary, and secondary trophic functions of the bay are well established (e.g., Matheson et al. 1999; Stumpf et al. 1999; Fourqurean et al. 2002). While these roles are not detailed here, the point is that with the relatively high biomass and high production capacity of SAV and benthic microalgae, the competition for nutrients with water column phytoplankton is greater than that in river-dominated systems. Such competition would also favor those phytoplankton with a high uptake capacity for regenerated nutrients (Glibert 1998). A conceptual model relating the trade-offs between benthic primary producers and the production of brown tide has been proposed for the coastal lagoons of Long Island (MacIntyre et al. 2004). In this model, conditions leading to a benthic-dominated state, including high input of NO_3^- from groundwater, will lead to low brown tide biomass, while conditions leading to a pelagic nutrient environment, including low NO_3^- input from groundwater and high organic input, will be more suitable to sustain brown tide. The relationship between high brown tide years and low flow (and consequently low NO_3^- input) documented for Peconic Bay, New York, by LaRoche et al. (1997) substantiates this model and further underscores the dependence of brown tide on organic or regenerated sources of nutrients.

Picoplankton such as *Synechococcus* are especially well adapted to low inorganic nutrient conditions because their small size reduces diffusion limitations in nutrient uptake. However, unlike diatoms which have large internal nutrient storage pools (e.g., Dortch et al. 1984; Lomas and Glibert 1999), picoplankton, because of their small size, are limited in their ability to store nutrients. Laboratory experiments on various clones of *Synechocococcus* have shown that these cells may either arrest cell division immediately upon N starvation, or may divide slowly, but only for approximately one division when they become N-limited (Glibert et al. 1986; Glibert and Ray 1990). A demonstrated response of cyanobacteria is the mobilization of phycobilisome-bound phycobiliproteins during periods of N deprivation (Yamanaka and Glazer 1980). Laboratory experiments in which *Synechococcus* was grown over a gradient of light intensities show that cells grown in low light had about 20% or more of their N in phycoerythrin, but cells grown in high light had <3% of their cell N bound in phycoerythrin (Kana and Glibert 1987). From these experiments it was concluded that rapidly grown cells sequester little N in phycoerythrin, and that all phycoerythrin is regulated photosynthetically; during N depletion, depending on the clone and its growth conditions, the ability to grow beyond one doubling is apparently very limited (Glibert et al. 1986; Kana and Glibert 1987). Although exactly comparable experiments have not been conducted for *A. anophagefferens*, the small size of this cell as well also would limit the extent to which N from pigment pools could be mobilized to support growth upon N depletion (Pustizzi et al. 2004). Thus, in order for blooms of either species, *Synechococcus* or *A. anophagefferens*, to be sustained, nutrients must be supplied or regenerated on a continual basis. Pulsed delivery of nutrients will not be able to be acquired and stored by the cells to sustain growth.

Grazing must also be reduced for lagoonal blooms to be sustained. Sunda et al. (2006) note that these EDAB species are often unpalatable or toxic, and thereby reduce grazing. The positive feedbacks of reduced grazing and/or bottom shading contribute to the availability of nutrients for these blooms. While no direct data on microbial grazing are available from the Coastal Bays, previous blooms of *A. anophagefferens* in Narragansett Bay, Rhode Island, Great South Bay, New York, and of *A. lagunensis* in Laguna Madre, Texas, provide some insight into grazer dynamics. In Narragansett Bay, during a bloom in 1989, the "normal" community of ciliates and heterotrophic flagellates was unusually sparse, suggesting reduced grazing (Smayda and Villareal 1989; Smayda 2008). Depression of grazing of brown tide blooms in New York embayments has also been shown (Gobler et al. 2002, 2005; Caron et al. 2004). In Laguna Madre the density of protozoan grazers was found to be greatly reduced during blooms, and it was suggested that a thick polysaccharide layer around the cells may make it difficult for the protozoa to feed (Buskey and Stockwell 1993; Buskey et al. 2001). Allelopathic chemicals may play an important role in maintaining EDAB species also (Sunda et al. 2006; Graneli et al. 2008). It has also been suggested that elevation of pH during the bloom may have some inhibitory effect on grazers (Buskey 2008).

While sponges have been shown to graze on blooms in Florida Bay (e.g., Peterson et al. 2006), the common microbial grazers of *Synechococcus* are heterotrophic protozoa and ciliates (Campbell and Carpenter 1986; Caron et al. 1991; Strom 1991; Jochem 2003). Recently, some red-tide dinoflagellates, including autotrophic dinoflagellates, have been found to consume *Synechococcus* (Legrand et al. 1998; Jeong et al. 2005; Glibert et al. 2009b). A modified dilution experiment from Florida Bay conducted with a Florida Bay *Synechococcus* spp. bloom population shows that grazing by a mixed flagellate community can be significant (Figure 5.7). However, natural blooms in Florida Bay are found to contain >90% *Synechocococcus* spp., and thus grazing impacts from heterotrophic flagellates and ciliates are localized to small subembayments where these microbes are able to develop. Microzooplankton grazing also provides an important source of regenerated nutrients (Glibert 1998).

The net effects of grazer and nutrient limitation can be seen from nutrient-enrichment bioassay experiments conducted similarly in Florida Bay and Chincoteague Bay (Figure 5.8). In the examples shown, typical of a larger set of data (Glibert and Heil unpublished), the control treatment showed a 20% to 40% decline in chlorophyll *a* in 24 h. A similar decline in chlorophyll *a* was also observed in the P-enrichment treatments in the Coastal Bays, underscoring the lack of limitation by this element in that system (see also Table 5.3). Thus, microzooplankton grazing indeed appeared to be regulating biomass to some extent. The largest response in chlorophyll *a* in the bioassay experiments, in both systems, however, was typically observed when both organic N and organic P were

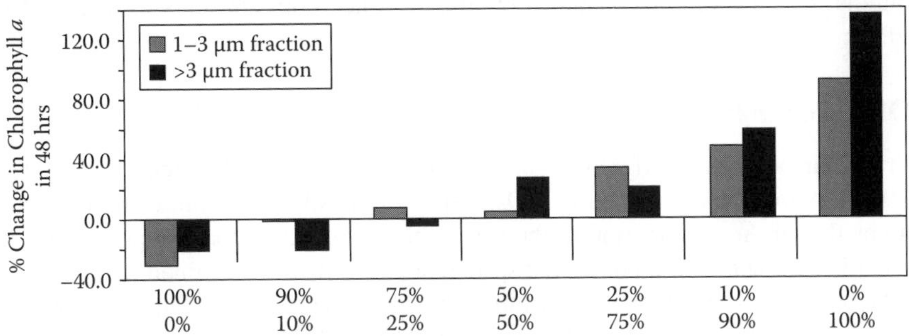

FIGURE 5.7 Representative result for one modified dilution experiment in which sample waters from two sites in Florida Bay were mixed in the proportions indicated. The top line on the x axis indicates the percentage of water from Blackwater Sound that was dominated by flagellated grazers, whereas the second line indicates the percentage of water from Barnes Sound that was dominated by *Synechococcus* spp. The change in size-fractionated chlorophyll *a* was determined after 48 h of incubation.

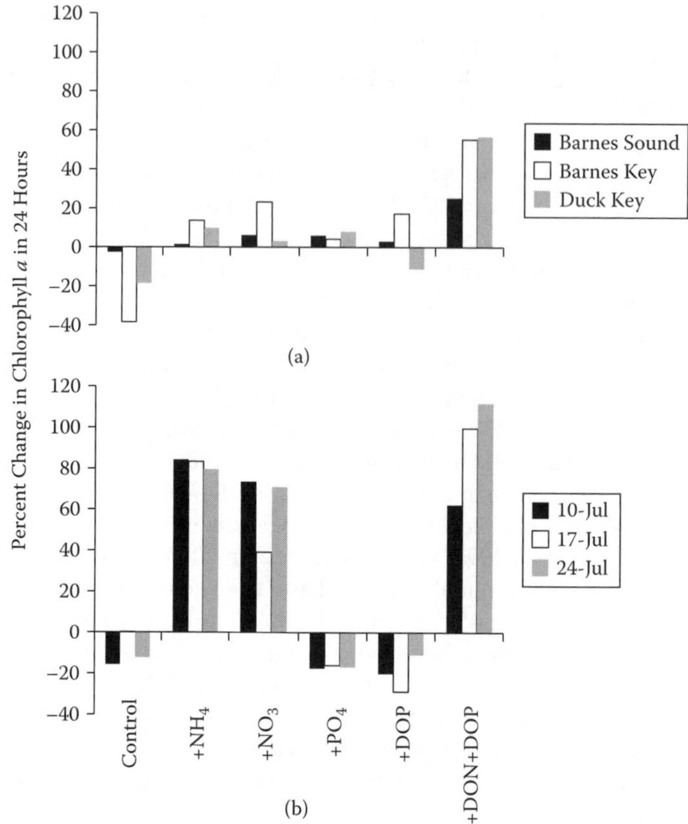

FIGURE 5.8 Bioassay results for representative stations or periods of sampling from (a) Florida Bay and (b) Chincoteague Bay. In each case, samples were collected from a site that was dominated by its respective bloom species. The percent change in chlorophyll *a* was determined after *in situ* incubation for 24 h. The enrichment concentrations were 10 μM N for NO_3^- and NH_4^+ and DON (as either all urea-N in Chincoteague Bay, or 50% urea, 25% arginine, and 25% glutamine in Florida Bay), and 2.5 μM P in all treatments. The DOP was glycerophosphate.

provided in combination (Figure 5.8). Although the bloom populations experienced some limited grazing control, as well as modest to severe nutrient limitation, they were able to grow at a level higher than that which could be consumed by grazing when both nutrients were provided in their preferred form.

5.8 CONCLUSIONS

Blooms in lagoons are often different from those in river-dominated systems for several reasons (Figure 5.9). They may be dominated by EDAB species, mostly picoplankton, compared to large-sized diatoms that are most common in the spring blooms of river-dominated systems. They frequently occur in summer or fall, not during spring, but may also be sustained for very long periods of time in some systems. They are not sustained by river-derived NO_3^-, but rather by regenerated sources of nutrients. They compete well for organic forms of N and P. Owing to their physiology, they must have sustained nutrient sources in order for bloom biomass to be sustained; they do not have large internal nutrient storage pools. Grazing losses (during blooms) are greatly reduced, and when grazing does occur, microzooplankton rather than macrozooplankton are the dominant water column grazers. The shallow, retentive nature of lagoons and the high degree of coupling between the benthos and the pelagic environment in terms of fluxes of nutrients are conditions that can

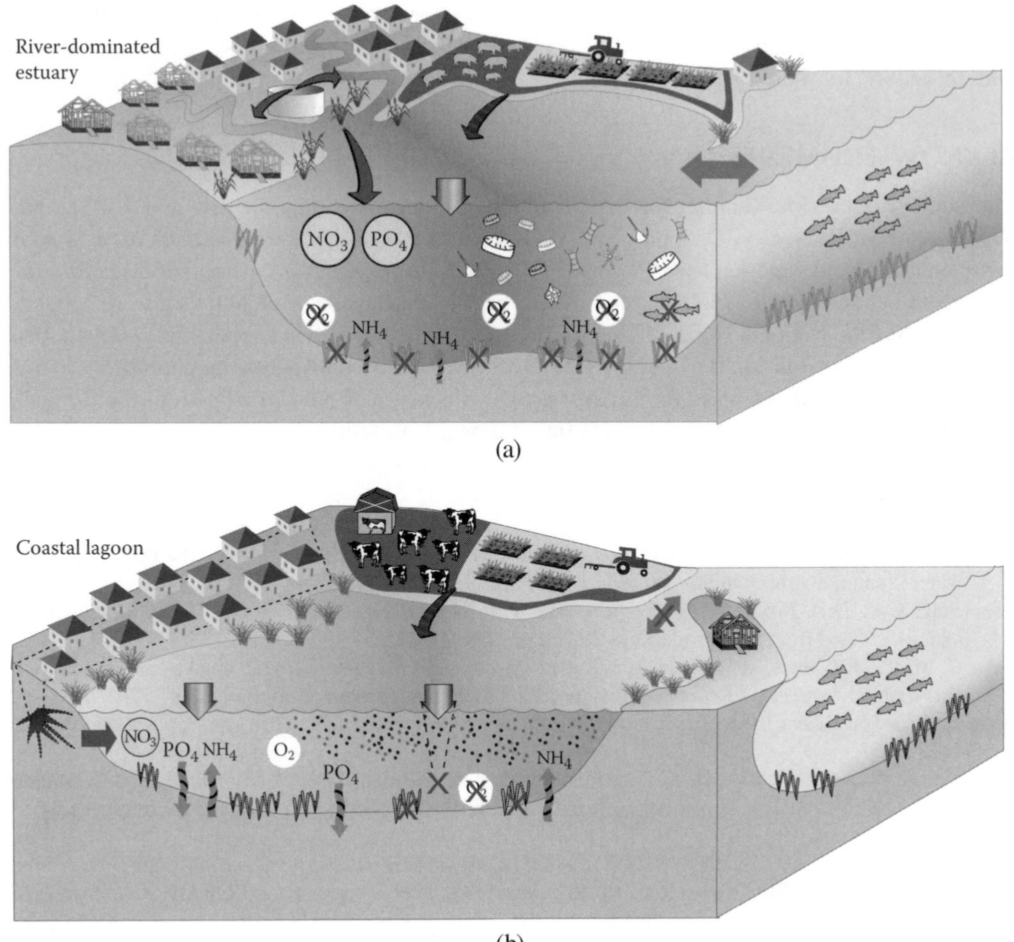

FIGURE 5.9 Synthesis schematics contrasting riverine and coastal lagoons in terms of dominant nutrient sources, fluxes, and algal blooms. In the river-dominated systems (a), the major source of nutrients is from riverine input. Most of the N from this source is in the form of NO_3^-. The major sources of nutrients are sewage from the human population as well as runoff and groundwater input from agricultural and/or animal operations in the watershed. Diatoms are common in the resulting phytoplankton community, and occur often as spring blooms. Nutrient regeneration in the benthos can be significant especially when hypoxia or anoxia develop. In the coastal lagoons (b), the major nutrient sources are from nonpoint sources. These may include sewage from human population, often septic systems rather than treated sewage, as well as runoff and groundwater input from the agricultural and/or animal operations in the watershed. Although variable, the dominant nitrogen form is usually in reduced or organic form, namely NH_4^+, urea, or DON. Phytoplankton blooms are dominated by cells that are small in size, such as cyanobacteria or brown tide, referred to as ecosystem disruptive blooms, EDABs (*sensu* Sunda et al. 2006). Nutrient fluxes from the benthos are important in regulating the nutrient availability in these systems.

lead to prolonged blooms of these picoplankton, with significant ecosystem consequences. Thus, these systems diverge along the classic lines delineated for herbivorous food web vs. microbial food web systems (e.g., Legendre and Rassouzadegan 1995). In general, river-dominated systems, which have proportionately greater inputs of oxidized forms of N, appear to have greater abundance of large phytoplankton, and the episodic nature of those inputs favors diatoms (Goldman 1993). The dominance of regenerated forms of N, and high organic loads, derived from both microzooplankton grazing and/or benthic fluxes, favors small-sized phytoplankton with high growth rates (Glibert

1998). These fundamental differences in blooms between lagoons and river-dominated estuaries have important implications for nutrient management as well as for the development of estuarine nutrient criteria for these systems.

ACKNOWLEDGMENTS

The authors thank M. Kennish for the invitation to contribute this chapter, and J. Alexander, S. Murasko, M. B. Neeley, D. Johns, and M. Kimmel for assistance with various aspects of the experimental results reported here. PMG, CAH, and CM thank the NOAA-South Florida Program, Grant Number NA06NOS4780075, from which the Florida Bay experimental data were collected, and the U.S. EPA National Nutrient Criteria Program, which provided support for various aspects of the syntheses for Florida Bay and the Coastal Bays. This is contribution number 4293 from the University of Maryland Center for Environmental Science and number 468 from the Southeast Environmental Research Center at Florida International University.

REFERENCES

Azam, F., T. Fenchel, J. G. Field, J. S. Gray, L. A. Meyer-Reil, and F. Thingstad. 1983. The ecological role of water-column microbes in the sea. *Mar. Ecol. Prog. Ser.* 10: 257–263.

Babinchak, J. A., D. J. Jollow, M. S. Voegtline, and T. B. Higert. 1986. Toxin production by *Gambierdiscus toxicus* isolated from the Florida Keys. *Mar. Fish. Rev.* 48(4): 53–56.

Berg, G. M., P. M. Glibert, M. W. Lomas, and M. Burford. 1997. Organic nitrogen uptake and growth by the chrysophyte *Aureococcus anophagefferens* during a brown tide event. *Mar. Biol.* 129: 377–387.

Berman, T. and S. Chava. 1999. Algal growth on organic compounds as nitrogen sources. *J. Plank Res.* 21: 1423–1437.

Borum, J., O. Pedersen, T. M. Greve, T. A. Frankovich, J. C. Zieman, J. W. Fourqurean, and C. J. Madden. 2005. The potential role of plant oxygen and sulphide dynamics in die-off events of the tropical seagrass, *Thalassia testudinum. J. Ecol.* 93: 148–158.

Boyer, J. N. and H. O. Briceño. 2008. *FY2007 Annual report of the South Florida Coastal Water Quality Monitoring Network.* SERC Tech. Rep. #T-407. http://serc.fiu.edu/wqmnetwork/Report%20Archive/2007_CWQMN.pdf.

Boyer, J. N., S. K. Dailey, P. J. Gibson, M. T. Rogers, and D. Mir-Gonzalez. 2006. The role of dissolved organic matter bioavailability in promoting phytoplankton blooms in Florida Bay. *Hydrobiologia* 561: 71–85.

Boyer, J. N., J. W. Fourqurean, and R. D. Jones. 1999. Seasonal and long-term trends in the water quality of Florida Bay (1989–1997). *Estuaries* 22(2B): 417–430.

Boyer, J. N. and B. Keller. 2007. Nutrient dynamics. Pp. 55–76, In: J. H. Hunt and W. Nuttle (eds.), *Florida Bay Science Program: A synthesis of research on Florida Bay.* Fish and Wildlife Research Institute Technical Report TR-11. Florida Fish and Wildlife Research Institute, St. Petersburg, Florida.

Boynton, W. R., J. D. Hagy, L. Murray, C. Stokes, and W. M. Kemp. 1996. A comparative analysis of eutrophication patterns in a temperate coastal lagoon. *Estuaries* 19: 408–421.

Bricker, S. B., B. Longstaff, W. Dennison, A. Jones, K. Boicourt, C. Wicks, and J. Woerner. 2007. *Effects of nutrient enrichment in the nation's estuaries: A decade of change.* NOAA Coastal Ocean Program Decision Analysis Series No. 26. National Center for Coastal Ocean Science, Silver Spring, Maryland.

Burgin, A. and S. K. Hamilton. 2007. Have we overemphasized the role of denitrification in aquatic ecosystems? A review of nitrate removal pathways. *Frontiers Ecol.* 5: 89–96.

Burkholder, J. M., D. A. Dickey, C. A. Kinder, R. E. Reed, M. A. Mallin, M. R. McIver, L. B. Cahoon, G. Melia, C. Brownie, J. Smith, N. Deamer, J. Springer, H. B. Glasgow, and D. Toms. 2006. Comprehensive trend analysis of nutrients and related variables in a large eutrophic estuary: A decadal study of anthropogenic and climatic influences. *Limnol. Oceanogr.* 51: 463–487.

Buskey, E. J. 2008. How does eutrophication affect the role of grazers in harmful algal bloom dynamics? *Harmful Algae* 8: 152–157.

Buskey, E. J., H. Liu, C. Collumb, and J. G. F. Bersano. 2001. The decline and recovery of a persistent Texas brown tide algal bloom in the Laguna Madre (Texas, USA). *Estuaries* 24: 337–346.

Buskey, E. J., P. A. Montagna, A. F. Amos, and T. E. Whitledge. 1997. Disruption of grazer populations as a contributing factor to the initiation of the Texas brown tide algal bloom. *Limnol. Oceanogr.* 42: 1515–1222.

Buskey, E. J. and D. A. Stockwell. 1993. Effects of a persistent "brown tide" on zooplankton populations in the Laguna Madre of South Texas. Pp. 659–666, In: T. J. Smayda and Y. Shimizu (eds.), *Toxic phytoplankton blooms in the sea.* Proceedings 5th International Conference on Toxic Marine Phytoplankton. Amsterdam: Elsevier.

Butler, M. H., J. H. Hunt, W. F. Herrkind, M. H. Childress, R. Bertlesen, W. Sharp, T. Matthews, J. M. Field, and H. G. Marshall. 1995. Cascading disturbances in Florida Bay, USA: Cyanobacterial blooms, sponge mortality, and implications for juvenile spiny lobsters *Panulirus argus*. *Mar. Ecol. Prog. Ser.* 129: 119–125.

Campbell, L. and E. J. Carpenter 1986. Estimating the grazing pressure of heterotrophic nanoplankton on *Synechococcus* spp. using the sea water dilution and selective inhibitor techniques. *Mar. Ecol. Prog. Ser.* 132: 121–129.

Caperon, J., S. A. Cattell, and G. Krasnick. 1971. Phytoplankton kinetics in a subtropical estuary: Eutrophication. *Limnol. Oceanogr*. 4: 599–607.

Caron, D. A., C. J. Gobler, D. J. Lonsdale, R. M. Cerrato, R. A. Schaffner, J. M. Rose, N. J. Buck, G. Taylor, K. R. Boissonneault, and R. Mehran. 2004. Microbial herbivory on the brown tide alga, *Aureococcus anophagefferens*: Results from natural ecosystems, mesocosms and laboratory results. *Harmful Algae* 3: 439–457.

Caron, D. A., E. L. Lim, G. Miceli, J. B. Waterbury, and F. W. Valois. 1991. Grazing and utilization of chroococcoid cyanobacteria and heterotrophic bacteria by protozoa in laboratory cultures and a coastal plankton community. *Mar. Ecol. Prog. Ser*. 76: 205–217.

Chiappone, M. and K. M. Sullivan. 1994. Ecological structure and dynamics of nearshore hard-bottom communities in the Florida Keys. *Bull. Mar. Sci.* 54: 747–756.

Christian, R. R., J. N. Boyer, and D. W. Stanley. 1991. Multi-year distribution patterns of nutrients within the Neuse River Estuary, North Carolina. *Mar. Ecol. Prog. Ser.* 71: 259–274.

Cloern, J. 2001. Our evolving conceptual model of the coastal eutrophication problem. *Mar. Ecol. Prog. Ser.* 210: 223–253.

Collier, J. L., B. Brahamsha, and B. Palenik. 1999. The marine cyanobacterium *Synechococcus* sp. WH7805 requires urease (urea amidohydrolase, EC 3.5.1.5) to utilize urea as a nitrogen source: Molecular-genetic and biochemical analysis of the enzyme. *Microbiol.* 145: 447–459.

Corredor, J. E., C. R. Wilkinson, V. P. Vicente, J. M. Morell, and E. Otero. 1988. Nitrate release by Caribbean reef sponges. *Limnol. Oceanogr*. 33: 114–120.

Costello, T. J. and D. M. Allen. 1996. Migrations and geographic distributions of pink shrimp, *Penaeus duorarum*, of the Tortugas and Sanibel ground, Florida. *Fish. Bull. Fish. Wildl. Serv. U.S.* 65: 449–459.

Crosset, K. M., T. J. Culliton, P. C. Wiley, and T. R. Goodspeed. 2004. *Population trends along the coastal United States: 1980–2004.* National Oceanic and Atmospheric Administration (NOAA). Management and Budget Office, Special Projects.

Davis, S., J. E. Cable, D. L. Childers, C. Coronado-Molina, J. W. Day, Jr., C. D. Hittle, C. J. Madden, E. Reyes, D. Rudnick, and F. Sklar. 2004. Importance of storm events in controlling ecosystem structure and function in a Florida Gulf coast estuary. *J. Coastal Res.* 20(4): 1198–1208.

Dortch, Q., J. R. Clayton, Jr., S. S. Thoresen, and S. I. Ahmed. 1984. Species differences in accumulation of nitrogen pools in phytoplankton. *Mar. Biol.* 81: 237–250.

Duever, M. J., J. F. Meeder, L. C. Meeder, and J. M. McCollom. 1994. The climate of south Florida and its role in shaping the Everglades ecosystem. Pp. 225–248, In: S. M. Davis and J. C. Ogden (eds.), *Everglades: The ecosystem and its restoration.* Delray Beach, Florida: St. Lucie Press.

Dugdale, R. C. and J. J. Goering. 1967. Uptake of new and regenerated forms of nitrogen in primary productivity. *Limnol. Oceanogr.* 12: 196–206.

Durako, M. J., M. O. Hall, and M. Merello. 2002. Patterns of change in the seagrass dominated Florida Bay hydroscape. Pp. 523–537, In: J. W. Porter and K. G. Porter (eds.), *The Everglades, Florida Bay and coral reefs of the Florida Keys: An ecosystem sourcebook.* Boca Raton, Florida: CRC Press.

Fourqurean, J. W., M. J. Durako, M. O. Hall, and L. N. Hefty. 2002. Seagrass distribution in south Florida: A multiagency coordinated monitoring program. Pp. 497–522, In: J. W. Porter and K. G. Porter (eds.), *The Everglades, Florida Bay and coral reefs of the Florida Keys: An ecosystem sourcebook.* Boca Raton, Florida: CRC Press.

Fourqurean, J. W., R. D. Jones, and J. C. Zieman. 1993. Processes influencing water column nutrient characteristics and phosphorus limitation of phytoplankton biomass in Florida Bay, FL, USA: Inferences from spatial distributions. *Estuar. Coastal Shelf Sci.* 36: 295–314.

Fourqurean, J. W. and M. B. Robblee. 1999. Florida Bay: A history of recent ecological changes. *Estuaries* 22(2B): 345–357.

Gardner, W. S. and M. J. McCarthy. 2005. Denitrification versus dissimilatory nitrate reduction to ammonium (DNRA) or nitrite (DNRN) in hypersaline Florida Bay sediments in August 2004 and January 2005. Pp. 85–86, In: 2005 Florida Bay and Adjacent Marine Systems Science Conference, Duck Key, Florida. Abstract.

Gastrich, M. D. and C. E. Wazniak. 2002. A brown tide bloom index based on the potential harmful effects of the brown tide alga, *Aureococcus anophagefferens*. *Aquat. Ecosyst. Health Manag.* 5: 1–7.

Glibert, P. M. 1998. Interactions of top-down and bottom-up control in planktonic nitrogen cycling. *Hydrobiologia* 363: 1–12.

Glibert, P. M. and G. M. Berg. 2009. Nitrogen form, fate and phytoplankton composition. In: V. S. Kennedy, W. M. Kemp, J. E. Peterson, and W. C. Dennison (eds.), *Experimental ecosystems and scale: Tools for understanding and managing coastal ecosystems*. New York: Springer.

Glibert, P. M. and J. M. Burkholder. 2006. The complex relationships between increasing fertilization of the earth, coastal eutrophication and proliferation of harmful algal blooms. Pp. 341–354, In: E. Granéli and J. Turner (eds.), *Ecology of harmful algae*. New York: Springer.

Glibert, P. M., J. M. Burkholder, T. M. Kana, J. A. Alexander, C. Schiller, and H. Skelton. 2009b. Grazing by *Karenia brevis* on *Synechococcus* enhances their growth rate and may help to sustain blooms. *Aquat. Microb. Ecol.* 55: 17–30.

Glibert, P. M., J. Harrison, C. A. Heil, and S. Seitzinger. 2006b. Escalating worldwide use of urea—a global change contributing to coastal eutrophication. *Biogeochemistry* 77: 441–463.

Glibert, P. M., D. J. Conley, T. R. Fisher, L. W. Harding, Jr., and T. C. Malone. 1995. Dynamics of the 1990 winter/spring bloom in Chesapeake Bay. *Mar. Ecol. Prog. Ser.* 122: 22–43.

Glibert, P. M., C. A. Heil, D. Hollander, M. Revilla, A. Hoare, J. Alexander, and S. Murasko. 2004. Evidence for dissolved organic nitrogen and phosphorus uptake during a cyanobacterial bloom in Florida Bay. *Mar. Ecol. Prog. Ser.* 280: 73–83.

Glibert, P. M., C. A. Heil, J. M. O'Neil, W. C. Dennison, and M. J. H. O'Donohue. 2006a. Nitrogen, phosphorus, silica and carbon in Moreton Bay, Queensland, Australia: Differential limitation of phytoplankton biomass and production. *Estuaries Coasts* 29: 107–119.

Glibert, P. M., C. A. Heil, D. T. Rudnick, C. J. Madden, J. N. Boyer, and S. P. Kelly. 2009a. Florida Bay: Water quality status and trends, historic and emerging algal bloom problems. *Contrib. Mar. Schi.* 38: 5–17.

Glibert, P. M., T. M. Kana, R. J. Olson, D. L. Kirchman, and R. S. Alberte. 1986. Clonal comparison of growth and photosynthetic responses to nitrogen availability in marine *Synechococcus* spp. *J. Exp. Mar. Biol. Ecol.* 101: 199–208.

Glibert, P. M., R. Magnien, M. W. Lomas, J. Alexander, C. Fan, E. Haramoto, M. Trice, and T.M. Kana. 2001. Harmful algal blooms in the Chesapeake and Coastal Bays of Maryland, USA: Comparisons of 1997, 1998, and 1999 events. *Estuaries* 24: 875–883.

Glibert, P. M. and R. T. Ray. 1990. Different patterns of growth and nitrogen uptake in two clones of marine *Synechococcus* spp. *Mar. Biol.* 107: 273–280.

Glibert, P. M., C. E. Wazniak, M. Hall, and B. Sturgis. 2007. Seasonal and interannual trends in nitrogen in Maryland's Coastal Bays and relationships with brown tide. *Ecol. Appl.* 17(5): S79–S87.

Gobler, C. J., D. J. Lonsdale, and G. L. Boyer. 2005. A synthesis and review of causes and impact of harmful brown tide blooms caused by the alga, *Aureococcus anophagefferens*. *Estuaries* 28: 726–749.

Gobler, C. J., M. J. Renaghan, and N. J. Buck. 2002. Impacts of nutrients and grazing mortality on the abundance of *Aureococcus anophagefferens* during a New York brown tide bloom. *Limnol. Oceanogr.* 47: 129–141.

Goldman, J. C. 1993. Potential role of large oceanic diatoms in new primary production. *Deep Sea Res.* 40: 159–168.

Granéli, E., M. Weberg, and P. Salomon. 2008. Harmful algal blooms of allelopathic species: The role of eutrophication. *Harmful Algae* 8: 94–102.

Hall, M. O., M. J. Durako, J. W. Fourqurean, and J. C. Zieman. 1999. Decadal changes in seagrass distribution and abundance in Florida Bay. *Estuaries* 22(2B): 445–459.

Heisler, J., P. Glibert, J. Burkholder, D. Anderson, W. Cochlan, W. Dennison, Q. Dortch, C. Gobler, C. Heil, E. Humphries, A. Lewitus, R. Magnien, H. Marshall, K. Sellner, D. Stockwell, D. Stoecker, and M. Suddleson. 2008. Eutrophication and harmful algal blooms: A scientific consensus. *Harmful Algae* 8: 3–13.

Hitchcock, G., E. Phlips, L. Brand, and D. Morrison. 2007. Plankton blooms. In: J. Hunt and W. Nuttle (eds.), *Florida Bay Science Program: A synthesis of research on Florida Bay*. Florida Fish and Wildlife Conservation Commission Technical Report TR-11.

Jeong, H. J., J. Y. Park, J. H. Nho, M. O. Park, J. H. Ha, K. A. Seong, C. Jeng, C. N. Seong, K. Y. Lee, and W. H. Yih. 2005. Feeding by red-tide dinoflagellates on the cyanobacterium *Synechococcus*. *Aquat. Microb. Ecol.* 41: 131–143.

Jochem, F. J. 2003. Photo- and heterotrophic pico- and nanoplankton in the Mississippi River plume: Distribution and grazing activity. *J. Plank. Res.* 25: 1201–1214.

Kana, T. M. and P. M. Glibert. 1987. Effect of irradiances up to 2000 μE m^{-2} sec^{-1} on marine *Synechococcus* WH7803: I. Growth, pigmentation, and cell composition. *Deep-Sea Res.* 34: 479–495.

Kemp, W. M., W. R. Boynton, J. E. Adolf, D. F. Boesch, W. C. Boicourt, G. Brush, J. C. Cornwell, T. R. Fisher, P. M. Glibert, J. D. Hagy, L. W. Harding, E. D. Houde, D. G. Kimmel, W. D. Miller, R. I. E. Newell, M. R. Roman, E. M. Smith, and J. C. Stevenson. 2005. Eutrophication in Chesapeake Bay: Historical trends and ecological interactions. *Mar. Ecol. Prog. Ser.* 303: 1–29.

Lapointe, B. and M. W. Clark. 1992. Nutrient inputs from the watershed and coastal eutrophication in the Florida Keys. *Estuaries* 15: 465–476.

LaRoche, J., R. Nuzzi, R. Waters, K. Wyman, P. G. Falkowski, and D. W. R. Wallace. 1997. Brown tide blooms in Long Island's coastal waters linked to interannual variability in groundwater flow. *Global Change Biol.* 3: 397–410.

Lawrence, D., M. J. Dagg, H. Liu, S. R. Cummings, P. B. Ortner, and C. Keble. 2004. Wind events and benthic–pelagic coupling in a shallow subtropical bay in Florida. *Mar. Ecol. Prog. Ser.* 266: 1–13.

Legrand, C., E. Graneli, and P. Carlsson. 1998. Induced phagotrophy in the photosynthetic dinoflagellate *Heterocapsa triquetra*. *Aquat. Microb. Ecol.* 15: 65–75.

Legendre, L. and J. Lefevre 1995. Microbial food webs and the export of biogenic carbon in oceans. *Aquat. Microb. Ecol.* 9: 69–77.

Legendre, L. and F. Rassoulzadegan. 1995. Plankton and nutrient dynamics in marine waters. *Ophelia* 14: 153–172.

Lomas, M. W. and P. M. Glibert. 1999. Temperature regulation of nitrate uptake: A novel hypothesis about nitrate uptake and reduction in cool-water diatoms. *Limnol. Oceanogr*. 44: 556–572.

Lomas, M. W., P. M. Glibert, D. A. Clougherty, D. E. Huber, J. Jones, J. Alexander, and E. Haramoto. 2001. Elevated organic nutrient ratios associated with brown tide blooms of *Aureococcus anophagefferens* (Pelagophyceae). *J. Plank. Res.* 23: 1339–1344.

Lung, W. S. 1994. *Water quality modeling of the St. Martin River, Assawoman, and Isle of Wight Bays*. Annapolis, Maryland: Maryland Department of the Environment.

MacIntyre, H. L., M. W. Lomas, J. Cornwell, D. J. Suggett, C. J. Gobler, E. W. Koch, and T. M. Kana. 2004. Mediation of benthic-pelagic coupling by microphytobenthos: An energy- and material-based model for initiation of blooms of *Aureococcus anophagefferens*. *Harmful Algae* 3: 403–437.

Madden, C. In press. Florida Bay. In: P. Glibert, C. Madden, W. Boynton, D. Flemer, C. Heil, and J. Sharp (eds.), *Estuarine nutrient criteria development: State of the science*. Washington, D.C.: EPA Office of Water.

Madden, C., R. Smith, E. Dettmann, J. Kurtz, W. Nelson, N. Detenback, J. Latimer, and S. Bricker. In press. Estuarine typology development and application. In: P. Glibert, C. Madden, W. Boynton, D. Flemer, C. Heil, and J. Sharp (eds.), *Estuarine nutrient criteria development: State of the science.* Washington, D.C.: EPA Office of Water.

Malone, T. C., D. J. Conley, P. M. Glibert, L. W. Harding, Jr., and K. Sellner. 1996. Scales of nutrient limited phytoplankton productivity: The Chesapeake Bay example. *Estuaries* 19: 371–385.

Matheson, R. E., Jr., .K. Camp, S. M. Sogard, and K. Bjorgo. 1999. Changes in seagrass-associated fish and crustacean communities on Florida Bay mud banks: The effects of recent ecosystem changes? *Estuaries* 22(2B): 534–551.

McGinty, M., C. Wazniak, and M. Hall. 2004. Results of recent macroalgae surveys in the Maryland coastal bays. Pp. 6-23–6-29, In: C. Wazniak and M. Hall (eds.), *Maryland's coastal bays ecosystem health assessment 2004.* DNR-12-1202-0009. Maryland Department of Natural Resources, Tidewater Ecosystem Assessment, Annapolis, MD. http://dnr.maryland.gov/coastalbays/publications/Chapter6.3.pdf.

McIvor, C. C., J. A. Ley, and R. D. Bjork. 1994. Changes in freshwater inflow from the Everglades to Florida Bay including effects on biota and biotic processes. Pp. 117–146, In: S. M. Davis and J. C. Ogden (eds.), *Everglades: The ecosystem and its restoration.* Delray Beach, Florida: St. Lucie Press.

Nance, J. M. 1994. *A biological review of the Tortugas pink shrimp fisheries through December 1993.* Galveston Laboratory, Southeast Fisheries Science Center, National Marine Fisheries Service, Galveston, Texas.

Nixon, S.W. 1995. Coastal marine eutrophication: A definition, social causes and future concerns. *Ophelia* 41: 199–219.

Nuttle, W., J. Hunt, and M. Robblee. 2003. *Florida Bay Science Program: A synthesis of research on Florida Bay.* Gainesville, Florida: Florida Caribbean Science Center.

Orem, W. H., C. W. Holmes, C. Kendall, H. E. Lerch, A. L. Bates, S. R. Silva, A. Boylan, M. Corum, M. Marot, and C. Hedgman. 1999. Geochemistry of Florida Bay sediments: Nutrient history at five sites in eastern and central Florida Bay. *J. Coastal Res.* 15: 1055–1071.

Paerl, H. W. 1988. Nuisance phytoplankton blooms in coastal, estuarine, and inland waters. *Limnol. Oceanogr.* 33: 823–847.

Pennock, J. R. 1985. Chlorophyll distributions in the Delaware Estuary: Regulation by light limitation. *Estuar. Coastal Shelf Sci.* 21: 711–725.

Peterson, B. J., C. M. Chester, F. J. Jochem, and J. W. Fourqurean. 2006. Potential role of the sponge community in controlling phytoplankton blooms in Florida Bay. *Mar. Ecol. Prog. Ser.* 328: 93–103.

Phlips, E. J. and S. Badylak. 1996. Spatial variability in phytoplankton standing crop and composition in a shallow inner-shelf lagoon, Florida Bay, Florida. *Bull. Mar. Sci.* 58: 203–216.

Phlips, E. J., T. C. Lynch, and S. Badylak. 1995. Chlorophyll *a*, tripton, color, and light availability in a shallow tropical inner shelf lagoon, Florida Bay, USA. *Mar. Ecol. Prog. Ser.* 127: 223–234.

Pritchard, D. W. 1969. Salt balance and exchange rate for Chincoteague Bay. *Chesapeake Sci.* 1: 48–57.

Pustizzi, F., H. MacIntyre, M. E. Warner, and D. A. Hutchins. 2004. Interaction of nitrogen source and light intensity on the growth and physiology of the brown tide algae *Aureococcus anophagefferens. Harmful Algae* 3: 343–360.

Richardson, L. L. and P. V. Zimba. 2002. Spatial and temporal patterns of phytoplankton in Florida Bay: Utility of algal accessory pigments and remote sensing to assess bloom dynamics. Pp. 461–478, In: J. W. Porter and K. G. Porter (eds.), *The Everglades, Florida Bay and coral reefs of the Florida Keys: An ecosystem sourcebook*. Boca Raton, Florida: CRC Press.

Robblee, M. B., T. R. Barber, P. R. Carlson, M. J. Durako, J. W. Fourqurean, L. K. Muehlstein, D. Porter, L. A. Yarbro, R. T. Zieman, and J. C. Zieman. 1991. Mass mortality of the tropical seagrass *Thalassia testudinum* in Florida Bay (USA). *Mar. Ecol. Progr. Ser.* 71: 297–299.

Rudnick, D., Z. Chen, D. Childers, J. Boyer, and T. Fontaine. 1999. Phosphorus and nitrogen inputs to Florida Bay: The importance of the Everglades watershed. *Estuaries* 22: 398–416.

Rudnick, D., C. Madden, S. Kelly, R. Bennett, and K. Cunniff. 2006. Algae blooms in eastern Florida Bay and southern Biscayne Bay. Report to the South Florida Water Management District.

Rudnick, D., P. B. Ortner, J. A. Browder, and S. M. Davis. 2005. A conceptual ecological model of Florida Bay. *Wetlands* 25: 870–883.

Sakamoto, T. and D. A. Bryant. 2001. Requirement of nickel as an essential micronutrient for the utilization of urea in the marine cyanobacterium *Synechococcus* sp. PCC 7002. *Microb. Environ.* 16:177–184.

Smayda, T. J. 2008. Complexity on the eutrophication-harmful algal bloom relationship, with comment on the importance of grazing. *Harmful Algae* 8: 140–151.

Smayda, T .J. and T. Villareal. 1989. The 1985 "brown-tide" and the open phytoplankton niche in Narragansett Bay during summer. Pp. 159–187, In: E. M. Cosper, E. J. Carpenter, and V. M. Bricelj (eds.), *Novel phytoplankton blooms: Causes and impacts of recurrent brown tides and other unusual blooms*. Coastal and Estuarine Studies No. 35. Berlin: Springer-Verlag.

Smith, N. P. 1998. Tidal and long-term exchanges through channels in the middle and upper Florida Keys. *Bull. Mar. Sci.* 62: 199–211.

Solomon, C. M., J. L. Collier, G. M. Berg, and P. M. Glibert. 2010. Role of urea in microbe metabolism in aquatic systems: A biochemical and molecular review. *Aquat. Microb. Ecol.* doi: 10.3354/ame01390.

Southwell, M. W., B. N. Popp, and C. S. Martens. 2008. Nitrification controls on fluxes and isotopic composition of nitrate from Florida Keys sponges. *Mar. Chem.* 108: 96–108.

Strom, S. 1991. Growth and grazing rates of the herbivorous dinoflagellate *Gymnodinium* sp. from the open subarctic Pacific Ocean. *Mar. Ecol. Prog. Ser.* 176: 103–113.

Stumpf, R. P., M. L. Frayer, M. J. Durako, and J. C. Brock. 1999. Variations in water clarity and bottom albedo in Florida Bay from 1985 to 1997. *Estuaries* 22: 431–444.

Sturgis, R. B. 2007. *Annual report for Assateague Island National Seashore Water Quality Monitoring Program chemical and physical properties*. Assateague Island National Seashore, Berlin, Maryland.

Sunda, W. G., E. Granéli, and C. J. Gobler. 2006. Positive feedback and the development and persistence of ecosystem disruptive algal blooms. *J. Phycol.* 42: 963–974.

Szmant, A. M. and A. Forrester. 1994. Sediment and water column nitrogen and phosphorus distribution patterns in the Florida Keys: SEAKEYS. *Bull. Mar. Sci.* 54:1085–1086.

Tango, P., W. Butler, and C. Wazniak. 2004. Assessment of harmful algae bloom species in the Maryland coastal bays. In: C. Wazniak and M. Hall (eds.), *Maryland's Coastal Bays Ecosystem health assessment* 2004. Annapolis, Maryland: Maryland Department of Natural Resources.

Tiedje, J. M. 1988. Ecology of denitrification and dissimilatory nitrate reduction to ammonium. In: A. J. B. Zehnder (ed.), *Biology of anaerobic microorganisms*. New York: John Wiley & Sons.

Tilmant, J. T. 1989. A history and an overview of recent trends in the fisheries of Florida Bay. *Bull. Mar. Sci.* 44:3–22.

Trice, T. M., P. M. Glibert, C. Lea, and L. Van Heukelem. 2004. HPLC pigment records provide evidence of past blooms of *Aureococcus anophagefferens* in the Coastal Bays of Maryland and Virginia, USA. *Harmful Algae* 3: 295–304.

Van Mooey, B. A. S., H. F. Fredricks, B. E. Pedler, S. T. Dyhrman, D. M. Karl, M. Koblizek, M. W. Lomas, T. J. Mincer, L. R. Moore, T. Moution, M. S. Rappe, and E. A. Webb. 2009. Phytoplankton in the ocean use non-phosphorus lipids in response to phosphorus scarcity. *Nature* 458: 69–72.

Wazniak, C. E. and P. M. Glibert. 2004. Potential impacts of brown tide, *Aureococcus anophagefferens*, on juvenile hard clams, *Mercenaria mercenaria*, in the Coastal Bays of Maryland, USA. *Harmful Algae* 3: 321–329.

Wazniak, C. and M. Hall (eds.). 2004. *Maryland's Coastal Bays ecosystem health assessment 2004*. Annapolis, Maryland: Maryland Department of Natural Resources.

Wazniak, C. E., M. R. Hall, T. Carruthers, and R. Sturgis. 2007. Linking water quality to living resources in a mid-Atlantic lagoon system, USA. *Ecol. Appl.* 17(5): S64–S78.

Yamanaka, G. and A. N. Glazer. 1980. Dynamic aspects of phycobilisome structure: Phycobilisome turnover during nitrogen starvation in *Synechococcus* spp. *Arch. Microbiol.* 124: 39–47.

Yarbro, L. A. and P. R. Carlson, Jr. 2008. Community oxygen and nutrient fluxes in seagrass beds of Florida Bay, USA. *Estuaries Coasts* 31: 877–897.

Zieman, J. C., J. W. Fourqurean, and R. L. Iverson. 1989. Distribution, abundance and productivity of seagrasses and macroalgae in Florida Bay. *Bull. Mar. Sci.* 44: 292–311.

Zubkov, M. V., B. M. Fuchs, G. A. Tarran, P. H. Burkill, and R. Amann. 2003. High rate of uptake of organic nitrogen compounds by *Prochlorococcus* cyanobacteria as a key to their dominance in oligotrophic oceanic waters. *Appl. Environ. Microbiol.* 69: 2991–3304.

6 Relationship between Macroinfaunal Diversity and Community Stability, and a Disturbance Caused by a Persistent Brown Tide Bloom in Laguna Madre, Texas

Paul A. Montagna, Mary F. Conley, Richard D. Kalke, and Dean A. Stockwell*

CONTENTS

Abstract ... 115
6.1 Introduction ... 116
6.2 Materials and Methods ... 117
6.3 Results ... 119
  6.3.1 Water Column ... 119
  6.3.2 Macrofauna ... 121
    6.3.2.1 Abundance ... 121
    6.3.2.2 Biomass ... 122
    6.3.2.3 Diversity ... 123
  6.3.3 Community Structure ... 124
6.4 Discussion ... 128
Acknowledgments ... 132
References ... 132

ABSTRACT

A brown tide, *Aureoumbra lagunensis*, bloom persisted throughout the 1990s in Baffin Bay and upper Laguna Madre, Texas, which resulted in increased phytoplankton biomass and decreased bottom light levels. Coincidently, changes in benthic macroinfauna abundance, biomass, and diversity were studied from 1988 to 2000. The combination of the bloom and benthic study allows us to perform a post hoc analysis to determine if the brown tide affected macroinfauna. Three hypotheses were tested: (1) initial disturbance leads to a diversity loss, (2) more diverse systems resist change, and (3) more diverse systems recover more rapidly. Four stations were sampled quarterly, two in Baffin Bay in an open bay, muddy bottom habitat, and two in Laguna Madre in seagrass and an

* Corresponding author. Email: paul.montagna@tamucc.edu

adjacent unvegetated habitat. Approximately 75% of invertebrate biomass was lost in both Baffin Bay and upper Laguna Madre prior to the brown tide. This loss is linked to the high salinities in the bays from 1988 to 1990. In addition, freezes in December 1989 to January 1990 caused two large fish kills and probable invertebrate kills. These die-offs resulted in increased ammonium in the system, which fueled the brown tide bloom. The loss of macroinfauna decreased the grazing potential of benthos on the brown tide organism. Both systems revealed decreases in abundance, biomass, and diversity concurrent with the increased concentrations of brown tide. The upper Laguna Madre seagrass stations, with the highest diversity, recovered in abundance and diversity during later dates, while the muddy bottom habitats, with low diversity, did not. On the other hand, biomass recovered at the muddy Baffin Bay stations, but not in the seagrass bed. Thus, the effect of brown tide on the macroinfauna is dependent on the habitat and diversity. Community structure changed with the onset of brown tide. Species diversity decreased and disturbance tolerant species, such as *Streblospio benedicti* and *Prionospio heterobranchia*, increased dominance in the system. There appear to be two different equilibrium states in this ecosystem: one prior to brown tide and one after onset of brown tide. The new equilibrium state is less diverse and lower in abundance and biomass. The effect of brown tide on the macroinfauna is important because the benthos, in Baffin Bay and Laguna Madre, constitute an important food source for commercial and recreational fish and play a role in the regeneration of nutrients to the shallow water ecosystem.

Key Words: benthos, infauna, harmful algal bloom, pelagophyte, Baffin Bay

6.1 INTRODUCTION

A monospecific algal bloom began in upper Baffin Bay and Laguna Madre, Texas, in January 1990 and persisted until October 1997 (Buskey et al. 2001). The species, *Aureoumbra lagunensis* (DeYoe et al. 1997), is small, 4 to 5 µm in diameter. It is in the same class, Pelagophyceae, as the brown tide alga *Aureococcus anophagefferens* found in the northeastern United States. Densities of this species as high as 1×10^7 cells mL^{-1} have been recorded. Brown tide is ecologically significant because it effectively out-competes other phytoplankton (Stockwell et al. 1993). The origin of the Texas brown tide is linked to drought conditions in south Texas during 1988 and 1989, and two freezes in the winter of 1989–1990, because these events killed fish and invertebrates and their decomposition led to increased ammonium levels in the water (Whitledge 1993; Buskey et al. 1998). The persistence of the brown tide bloom may be due to a long, one year, water residence time in this ecosystem (Shormann 1992), and the ability of the brown tide species to survive high salinities (Buskey et al. 1996; Buskey et al. 2001) and utilize ammonium (DeYoe and Suttle 1994). Previous studies have measured the impact of brown tide on zooplankton (Buskey and Stockwell 1993; Buskey and Hyatt 1995), light attenuation (Dunton 1994), nutrients (Whitledge 1993), and bivalve filter feeding (Montagna et al. 1993). Disruption of grazer control has contributed to persistence of the bloom (Buskey et al. 1997).

The relationship between brown tide and macroinfauna is less well known and is the subject of the current study. From other studies, it is known that brown tide can directly affect the benthic mollusks by interfering with filter feeding (Tracey 1988; Bricelj and Kuenstner 1989), or altering larval populations (Ward et al. 2000). Trophic structure could change if the carbon associated with brown tide is utilized by the benthic food web, which would result in phytoplankton playing a greater role as a primary food source than seagrass (Street et al. 1997). Shading caused by brown tide reduced seagrass growth (Dunton 1994), and this led to an 18% to 27% loss of seagrass habitat (Onuf 1996, 2000). Reduced seagrass area can have a serious consequence for macrofauna because vegetated areas have 3 to 50 times more benthos than bare areas (Orth et al., 1984). A diverse benthic community utilizes seagrass habitats for protection from predators, larval recruitment, and as a detrital food source (Orth et al. 1984). Understanding the effect of brown tide on the benthic community is important, because in shallow coastal environments the benthos serve as a food source for higher trophic levels and play a role in nutrient regeneration. In particular, upper Laguna Madre supports a large commercial black drum fishery (90% of Texas landings) and is a premier destination for sport fishing (especially trophy-sized seatrout). With the known links between brown tide and benthos, and the importance of benthos in the ecosystem functioning, it is necessary to know what effect brown tide has on macrobenthos.

The current study is one of opportunity, not design. So it is useful to frame a context in which to perform post hoc analyses. The persistent brown tide was a large-scale disturbance of the ecosystem. Numerous studies have shown that disturbance leads to loss in diversity (Valiela 1995). Elton (1958) suggested that a large number of species provides a stabilizing effect in communities. The persistent brown tide provides an opportunity to perform a post hoc examination of the relationship between disturbance, stability, and diversity. Three hypotheses are tested: (1) that initial disturbance leads to diversity loss, (2) more diverse systems resist change, and (3) more diverse systems recover more rapidly. The benthic macrofauna community was studied for 12 years between 1988 and 2000 (2 pre-bloom and 10 during-bloom years) in two different habitat types (open bay bottoms and seagrass beds). Abundance and biomass were measured, and diversity was calculated. Both univariate and multivariate statistical approaches were used to test for community change across temporal treatments (i.e., pre- vs. during-bloom sampling events) and spatial treatments (i.e., Laguna Madre vs. Baffin Bay).

6.2 MATERIALS AND METHODS

The study area is Baffin Bay and upper Laguna Madre (Figure 6.1). Laguna Madre is a shallow lagoon separated from the Gulf of Mexico by barrier islands. It is one of only four hypersaline estuaries in the world (Javor 1989). Upper Laguna Madre is analogous to a primary bay because it is connected to the Gulf of Mexico, but this system does not have a direct connection to the Gulf of Mexico because exchange occurs from connections in the south (Mansfield Pass in Lower Laguna Madre) and from the north (through Aransas Pass in Corpus Christi Bay). In general, water flow in Laguna Madre is from south to north, and driven by the dominant southeasterly winds, and increasing in salinity as it moves north (Breuer 1962). Northern cold fronts, mainly in winter, reverse the wind-driven circulation allowing some water movement into the northern end of Laguna Madre from Corpus Christi Bay, lowering salinity (Copeland et al. 1968). Water movement is primarily through the deepened Gulf Intracoastal Waterway. The dredging has reduced salinities in upper Laguna Madre over the long term and this has caused increases in seagrass cover (Quammen and Onuf 1993). Baffin Bay is a secondary bay that receives freshwater pulses from the watershed via three small tertiary bays. Precipitation is low (74 cm year^{-1}) and freshwater inflow is restricted to periodic flow in ephemeral creeks that empty into tertiary bays, which then connect to Baffin Bay (Orlando et al. 1991). One exception is Petronila Creek, which has a baseflow provided by wastewater treatment plants. On average, 65% of the total inflow to Baffin Bay is direct precipitation onto the bay surface (http://midgewater.twdb.state.tx.us/bays_estuaries/hydrologypage.html). Overall, the combination of low freshwater inflow and high evaporation rates results in a seasonal hypersaline environment. Turbidity is high in Baffin Bay because of its shallow depth (3 m), high temperatures, and high average wind speeds.

Two sampling stations each are located in open bay and seagrass habitats (Figure 6.1). The Baffin Bay stations are located in the open bay near navigation channel Markers 6 and 24, and are thus labeled stations 6 and 24. Both sites are bare, muddy bottom habitats at depths near 3 m. The upper Laguna Madre stations are located in a seagrass bed near navigation channel Marker 189 at a water depth of about 1 m. Station 189G is located in a vegetated portion of the seagrass bed while station 189S is located in an adjacent unvegetated sand patch. Sampling, at all stations, was performed quarterly (January, April, July, and October) beginning in April 1988 and ending in October 2000.

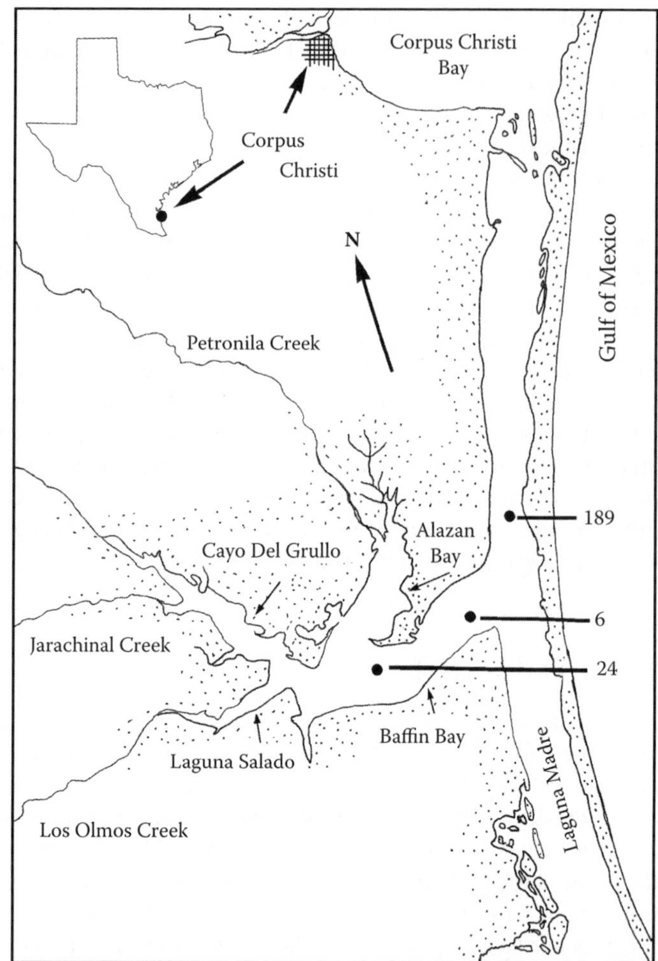

FIGURE 6.1 Map of Baffin Bay and upper Laguna Madre with sampling locations. Locations in degrees: stations 189 (189G and 189S) = 27.34990 N, 97.39238 W; station 6 = 27.27697 N, 97.42690 W; and station 24 = 27.26388 W, 97.55142 W.

Thus, the treatments are arranged in a two-way, partially hierarchical analysis of variance experimental design, where the two main effects are: (1) before and during the brown tide event, with unique sampling dates nested within the two periods; and (2) bay systems, with the stations unique to Baffin Bay and Laguna Madre nested within bay systems.

At each station a multiparameter sonde was used to measure salinity, temperature, dissolved oxygen, and pH just below the surface and at the bottom. Surface and bottom water samples were collected for chlorophyll and nutrient analyses. Bottom samples were collected using a Van Dorn bottle and kept cold until analysis. Nutrient concentrations were analyzed using standard automated nutrient seawater analyses on a Technicon AutoAnalyzer (Whitledge et al. 1981; Whitledge 1993). Chlorophyll measurements were made using a fluorometric method (Holm-Hansen et al. 1965) on samples extracted using 60% acetone and 40% DMSO (Dagg and Whitledge 1991).

At each station, three replicate sediment samples were taken using a 6.72 cm-diameter plastic tube to a depth of 10 cm, preserved with 10% formalin solution, and sorted using a 0.5-mm sieve (Montagna and Kalke 1992). The retained organisms were identified to the lowest possible taxa (generally species), and counted. Biomass was measured by combining the organisms into larger taxonomic groups (Crustacea, Mollusca, Polychaeta, etc.), which were dried at 55°C for 24 hours and weighed to the nearest 0.01 mg. All carbonate shells from the mollusks were removed with 1 M hydrochloric acid before weighing.

Species diversity was calculated by pooling all three replicate cores for each date-station combination. Diversity indices were then calculated using Hill's number 1 (N1) (Ludwig and Reynolds 1988). Hill's N1 is calculated by: $N1 = e^{H'}$, where H′ is the Shannon–Weaver index (Hutcheson 1970). Hill's N1 represents the number of abundant species (Ludwig and Reynolds 1988), and thus is easier to interpret than most diversity indices.

All statistical analyses were performed using SAS software (SAS Institute Inc. 1991). A two-way factorial ANOVA was used to test for differences in macrofaunal abundance, biomass, and diversity among stations, dates, and the interaction. Total core abundance and biomass were tested against the hydrologic data (salinity, temperature, and dissolved oxygen) as well as bottom water column nutrient concentrations ($NO_3^- + NO_2^-$, NH_4^+, PO_4^{3-}, SiO_4) using a principal components analysis (PCA).

A multivariate analysis was used to test for community structure change between before and after brown tide dates. June 1990 was used as the brown tide origination date because high brown tide cell counts were first recognized in the bays at this time (Buskey et al. 1996). Species were analyzed by nonmetric multidimensional scaling (MDS). All multivariate statistical analyses were performed using PRIMER software (Clarke and Warwick 2001; Clarke and Gorley 2006). The MDS procedure was used to compare average abundances of individuals of each species for each station-date combination. The MDS analysis was completed using a Bray–Curtis similarity matrix on log-transformed (ln + 1) data. Differences and similarities among communities were highlighted based on cluster analysis calculated from the similarity matrix. A subset of species that represented the spatial pattern in an MDS plot was determined using the BVSTEP procedure. The BVSTEP procedure employs a step-wise approach to determine the minimum subset of species that can yield the same pattern of community structure obtained from the entire dataset (Clarke and Warwick 1998). Cluster analysis was performed on samples and significant differences among clusters were computed using the SIMPROF procedure. Rank-dominance plots were created to identify diversity trends before and during the brown tide event by pooling samples in Baffin Bay and Laguna Madre.

6.3 RESULTS

6.3.1 Water Column

There is a close correspondence between the salinities of Baffin Bay and Laguna Madre, and the overall average salinity was 37.7 practical salinity units (psu) (Figure 6.2). From 1988 to 1990, there was below-average rainfall in southeast Texas (Buckner et al. 1989, 1990; Buckner and Shelby 1990), which caused increased salinities in Baffin Bay and Laguna Madre. Salinities remained above 30 psu to 1992. In 1992 and 1993, there was above-average rainfall associated with an El Niño event, and salinities in the system dropped, ranging between 10 and 30 psu. Salinity increased again from 1994 to 1996. Precipitation controls salinity concentrations in Laguna Madre and Baffin Bay due to low freshwater runoff and high evaporation.

Bottom water dissolved oxygen (DO) was generally higher in Laguna Madre than Baffin Bay (Figure 6.2). The average bottom DO in Laguna Madre was 8.1 mg L^{-1}, and in Baffin Bay, it was 6.7 mg L^{-1}. The higher DO in Laguna Madre is due to photosynthesis by seagrass, which supersaturates the water with DO during the day. Concordantly, pH is higher (8.4 on average) in Laguna Madre than in Baffin Bay (8.1 on average).

The quantity of brown tide in the system was estimated as chlorophyll *a* concentration because the water column has contained almost an exclusively monospecific bloom (Buskey and Stockwell 1993). Chlorophyll *a* concentrations correspond with brown tide cell counts, which first appeared in high concentrations in June of 1990 (Buskey et al. 1996). Chlorophyll concentrations throughout

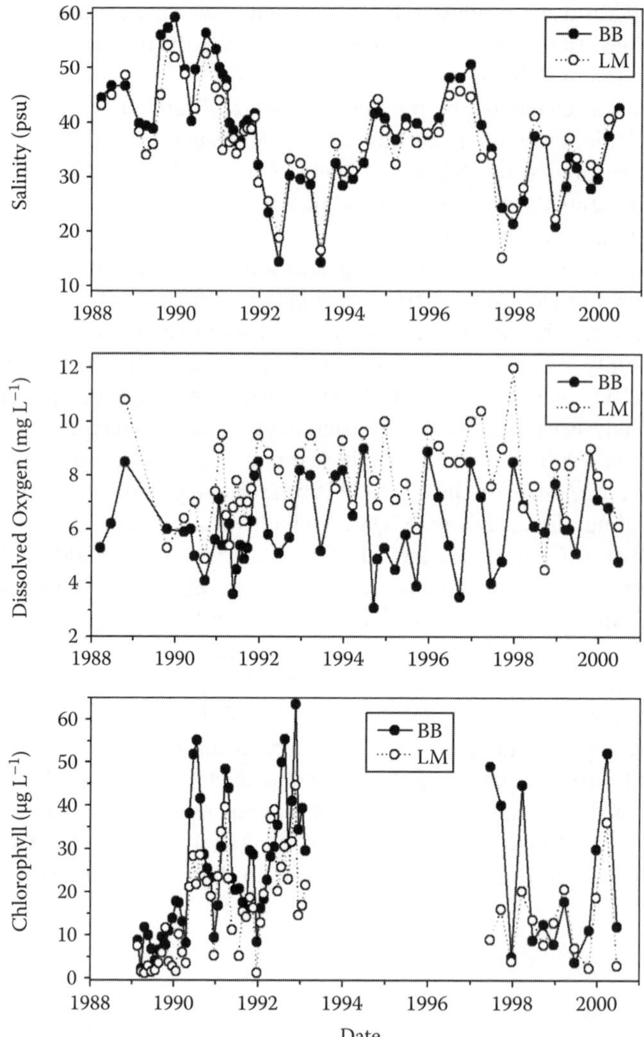

FIGURE 6.2 Environmental conditions in Baffin Bay and upper Laguna Madre. Water column salinity (psu), bottom dissolved oxygen (mg L^{-1}), and water column chlorophyll *a* (µg L^{-1}) over time. Chlorophyll data for July 1997 through January 1998 are from Buskey et al. (2001).

the sampling period were similar at all stations and averaged 18.9 µg L^{-1} for the overall samples (Figure 6.2). Laguna Madre was usually slightly lower, with an average of 14.5 µg L^{-1} compared with Baffin Bay, which had an average chlorophyll *a* of 23.0 µg L^{-1}. Chlorophyll *a* levels increased in June 1990 from concentrations below 20 mg L^{-1} to 50 mg L^{-1}. There were lower chlorophyll *a* concentrations in winter months. Although there are data gaps between 1993 and 1997, concentrations in spring and summer remained high until the bloom receded in January 1998 and 1999 (Buskey et al. 2001).

A multivariate analysis revealed that dissolved inorganic nitrogen (NO_x and NH_4^+) was typically higher when salinity was high, and chlorophyll *a* increased when PO_4^{3-} increased (Figure 6.3). The nutrient regime was slightly different in Baffin Bay where it was typically at higher concentrations (1.8 µM for NO_x, 5.4 µM for NH_4^+, 1.1 µM for PO_4^{3-}, and 92 µM for SiO_4) than in Laguna Madre (1.3 µM for NO_x, 2.1 µM for NH_4^+, 0.9 µM for PO_4^{3-}, and 78 µM for SiO_4). The higher nutrient concentrations in Baffin Bay are correlated with the higher chlorophyll values found there.

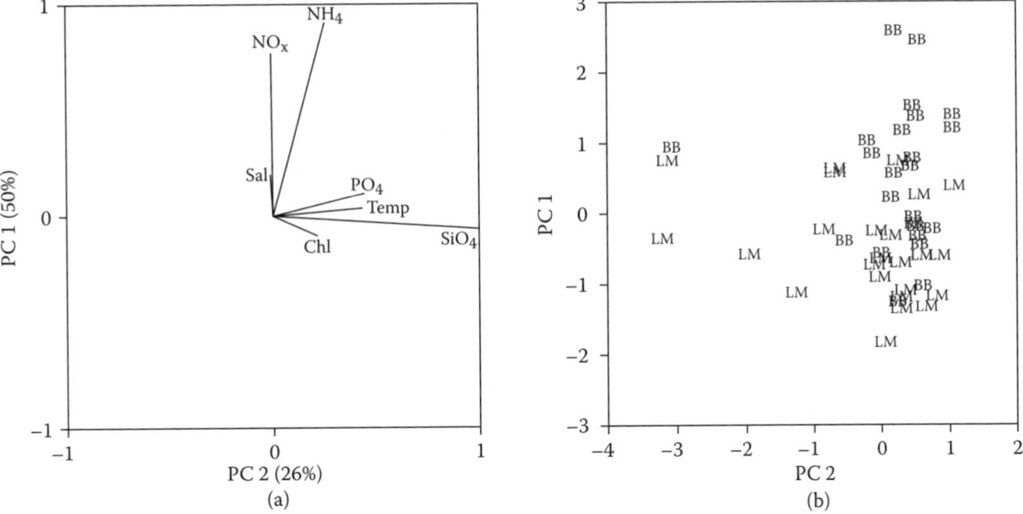

FIGURE 6.3 Principal components (PC) analysis of water column variables. (a) Vector variable loads for ammonium (NH_4^+), nitrate+nitrite (NOx), salinity (Sal), phosphate (PO_4^{3-}), temperature (Temp), silicate (SiO_4), and chlorophyll *a* (Chl). (b) Sample score plots using bay systems as the symbol where BB = Baffin Bay and LM = upper Laguna Madre.

6.3.2 Macrofauna

There was no significant difference between the two bay systems (Baffin Bay and Laguna Madre) in biomass or abundance ($P = 0.0511$, 0.2124, respectively), but there was a significant difference for diversity ($P = 0.0254$) (Table 6.1). There was a significant difference in biomass, abundance, and diversity between the two sampling periods, before and during brown tide ($P \leq 0.0001$ for all). There was a significant interaction between bay and sampling period for biomass, abundance, and diversity (Figure 6.4).

6.3.2.1 Abundance

The overall average abundance for macrofauna was 20,179 m^{-2}. Although the average abundance was twice as high in Laguna Madre (27,045 m^{-2}) than in Baffin Bay (13,062 m^{-2}), it was not significant. There was a peak abundance between July 1989 and January 1990 (Figure 6.4). High abundances were the result of large pulses of a few species. The Baffin Bay maxima corresponded to a large increase in the population of *Streblospio benedicti*, which is both a deposit- and suspension-feeding spionid polychaete. Abundance increases in Laguna Madre were caused by large populations of the spionid polychaete *Prionospio heterobranchia* and members of the family Syllidae. By July 1990, both systems had low abundance, less than 20,000 individuals m^{-2}. Station 189G, which is located in the seagrass bed, maintained higher abundance throughout the sampling period, rarely dropping below 40,000 m^{-2}. In Laguna Madre, there was resurgence in abundance starting in the winter of 1993 and continuing through 2000, except for short periods in 1995 and 1996. Seasonal maxima in Baffin Bay occurred in the winter and early spring. Laguna Madre had a seasonal maximum only in spring. There was a significant difference in abundance between the two sampling periods. Before brown tide, the average abundance was 35,786 m^{-2}, and during the brown tide, the average abundance was 17,741 m^{-2}. The difference between the two periods (before and after June 1990) was more pronounced in Baffin Bay, where the abundance decreased 3.7 times, from 35,159 to 9,536 m^{-2}. In contrast, the average abundance in Laguna Madre dropped only 1.4 times, from 36,413 to 25,611 m^{-2}.

TABLE 6.1A
Analysis of Macrofaunal Characteristics: Analysis of Variance

Source	F-Test	DF	Abundance $P(n\ m^{-2})$	Biomass $P(g\ m^{-2})$	Diversity P(N1)
BT	BT/Date (BT)	1	0.0081	<.0001	0.0056
Date (BT)	BT/MSE	57	<.0001	<.0001	0.0103
Bay	Bay/Station (Bay)	1	0.2142	0.0511	0.0254
Station (Bay)	Station (Bay)/MSE	2	<.0001	<.0001	<.0001
BT*Bay	BT* Bay/MSE	1	0.0032	0.0408	0.0005
MS (Error)		603			

Note: ANOVA table with two main effects: before and during brown tide (BT), and bay system (Bay), i.e., Baffin Bay (BB) or Laguna Madre (LM).

TABLE 6.1B
Analysis of Macrofaunal Characteristics: Summary of Means

BT	Bay	Sample Number	Abundance $(n\ m^{-2})$	Biomass $(g\ m^{-2})$	Diversity (N1)
		666	20,179	5.24	6.1
	BB	327	13,062	1.80	2.4
	LM	339	27,045	8.56	9.7
B		90	35,786	14.19	7.4
D		576	17,741	3.84	5.8
B	BB	45	35,159	3.62	2.1
B	LM	45	36,413	24.75	12.7
D	BB	282	9,536	1.51	2.4
D	LM	294	25,611	6.09	9.1

Note: Summary of (main effects) means for macrofaunal abundance ($n\ m^{-2}$), biomass ($g\ m^{-2}$), and diversity (N1 index, number of dominant species 0.8 m^{-1}).

6.3.2.2 Biomass

The overall average biomass was 5.24 g m^{-2}. Biomass was four times lower in Baffin (1.80 g m^{-2}) than in Laguna Madre (8.56 g m^{-2}). There was a decrease in biomass in both bay systems between April and July 1990 (Figure 6.4). In Baffin Bay, the biomass remained close to zero from April 1990 through 1992 and between 1996 and 1999. Laguna Madre also had lower biomass in 1991 and 1992, dropping from more than 30 g m^{-2} to 10 g m^{-2} or less. The biomass lost before the onset of brown tide represented ~75% of the average biomass prior to June 1990. The difference between the two periods (before and after June 1990) was more pronounced in Laguna Madre, where the biomass decreased 4.1 times, from 24.8 to 6.1 g m^{-2}. In contrast, the average abundance in Baffin Bay dropped only 2.4 times, from 3.6 to 1.5 g m^{-2}.

The biomass maxima corresponded with larger quantities of bivalves, in particular *Mulinia lateralis*, and large, deeper-dwelling polychaetes, including *Branchioasychis americana*, *Scolopolos rubra*, and *Magelona pettiboneae*. Unlike abundance, biomass never recovered to the levels that existed prior to 1990.

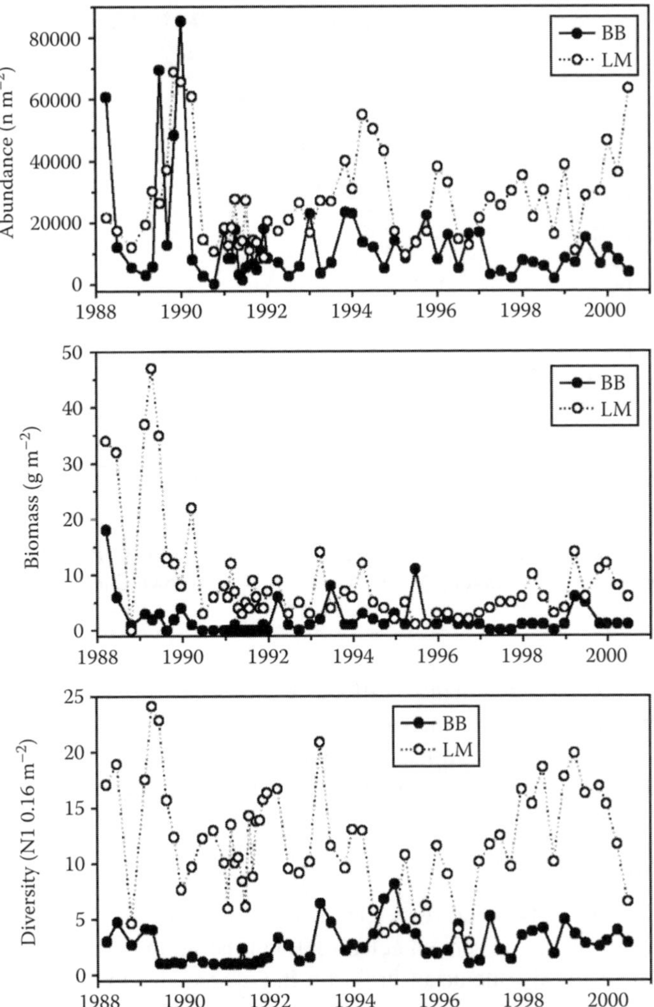

FIGURE 6.4 Macrofaunal change over time in Baffin Bay and upper Laguna Madre. Abundance (n m^{-2}), dry weight biomass (g m^{-2}), and diversity (Hill's N1 index).

6.3.2.3 Diversity

Laguna Madre had higher diversities (Hill's N1 index average of 9.7) than Baffin Bay stations (N1 average of 2.4). There was some seasonality in diversity indices (Figure 6.4). Most often the diversity peak fell in the spring and early summer (i.e., April and July). Exceptions occurred at station Laguna Madre in 1993 and 1994, when a diversity maximum occurred in January.

In Baffin Bay, N1 diversity was around three to five species prior to brown tide, and then it decreased to near one after the onset of brown tide and remained low until 1993. Often from 1989 through 1993 only one species, *Streblospio benedicti*, was present in samples.

In Laguna Madre, the vegetated station 189G had the highest maximum and average diversity of all stations (23 species and 12.07 species, respectively). There was a large variation in diversity throughout the study, but there was no direct increase or decrease with time. Diversity was higher than average until January 1990. It then remained at or below average until April 1993. April usually had the highest diversity throughout the sampling period. Diversity in Laguna Madre recovered to pre-brown tide levels in 1998.

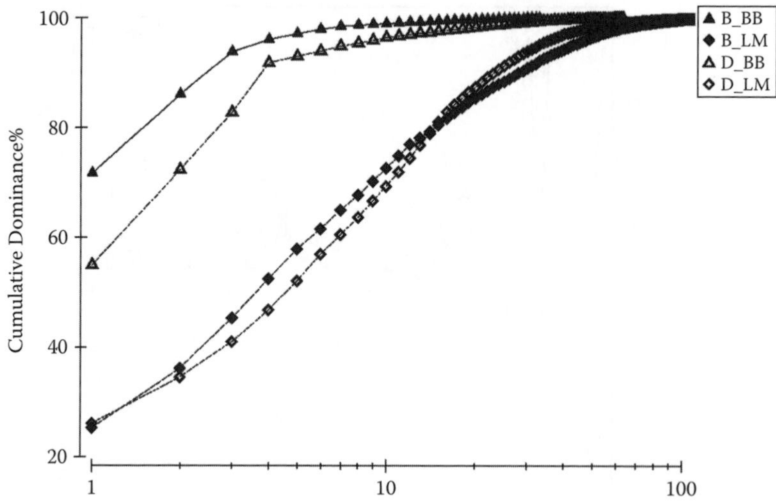

FIGURE 6.5 Species rank-dominance curves. Treatment abbreviations: B_BB = before brown tide, Baffin Bay; B_LM = before brown tide, Laguna Madre; D_BB = during brown tide, Baffin Bay; D_LM = during brown tide, Laguna Madre.

There was distinct rank-dominance difference between Baffin Bay and Laguna Madre (Figure 6.5). The relative differences between the period prior to and during the brown tide events were small in the effect on rank-dominance in both systems. However, the systems behaved very differently, with Baffin Bay having much higher dominance than Laguna Madre.

6.3.3 Community Structure

There was a distinct break between recurrent species in the Baffin Bay and Laguna Madre. A recurrent species is defined as present on at least 50% of the sampling dates, and is independent of abundance. Exceptions to this rule occur in Baffin Bay where, due to the low number of different species, percentages as low as 30 were counted. Baffin Bay had fewer recurrent species than Laguna Madre. *Streblospio benedicti*, *Mulinia lateralis*, and *Ampelisca abdita* were the recurrent species in Baffin Bay. Laguna Madre stations 189G and 189S had 29 and 11 recurrent species, respectively. The same species were found in both habitats, making 189S a subset of 189G. Found on at least 90% of the sampling dates at both stations were oligochaetes, *Capitella capitata*, *Syllis cornuta*, and *Prionospio heterobranchia*. Only *Streblospio benedicti* was recurrent across bays.

A total of 171 species were found over the entire study period; however, 15 dominant species represented a total of 84% of all organisms found (Table 6.2). Of these, the polychaete *Streblospio benedicti* represented 30% over all, but it represented 78% of all organisms found in Baffin Bay. A total of 9 of the 15 dominant species were polychaetes, save the bivalve *Mulinia lateralis*, the gastropod *Caecum pulchellum,* the amphipods *Ampelisca abdita* and *Grandidierella bonnieroides*, oligochaetes, and nemerteans.

The multivariate (MDS) analysis of the community indicated that there were large differences between Baffin Bay and Laguna Madre (ANOSIM Global R = 0.300, $p \leq 0.001$) because they separate completely from right to left (Figure 6.6). There was also a difference in community structure before and after the brown tide in Laguna Madre (ANOSIM Global R =: 0.865, $p \leq 0.001$) because the before samples separated from the after samples from top to bottom on the right side of Figure 6.6. Overlaying a cluster analysis on top of the MDS indicates that there was seriation in the temporal change of community structure over time (SIMPROF $p = 0.05$ at the 40% similarity level).

TABLE 6.2
Average Abundance (n m^{-2}) of Dominant Species of Benthic Macroinfauna in Baffin Bay (Mean of Stations 6 and 24) and Laguna Madre (Mean of Stations 189S and 189G) before and during the Brown Tide Event

			Baffin Bay		Laguna Madre			
Rank	**Taxa**	**Species Name**	**Before**	**During**	**Before**	**During**	**Mean**	**Percent**
1	P	*Streblospio benedicti*	22,630	5,233	1,324	1,177	7,591	29.9%
2	P	*Prionospio heterobranchia*	–	14	9,062	1,594	2,667	10.5%
3	O	Oligochaeta (unidentified)	5	1	3,337	6,398	2,435	9.6%
4	P	*Mediomastus ambiseta*	4,519	871	761	648	1,700	6.7%
5	P	*Syllis cornuta*	–	2	3,905	2,074	1,495	5.9%
6	A	*Ampelisca abdita*	2,435	1,657	208	98	1,100	4.3%
7	P	*Exogone* sp.	–	17	1,967	1,443	857	3.4%
8	G	*Caecum pulchellum*	–	–	2,524	569	773	3.0%
9	P	*Brania furcelligera*	–	–	1,243	774	504	2.0%
10	B	*Mulinia lateralis*	742	987	5	171	476	1.9%
11	A	*Grandidierella bonnieroides*	38	88	969	541	409	1.6%
12	P	*Capitella capitata*	–	43	269	1,306	405	1.6%
13	P	*Heteromastus filiformis*	5	–	927	300	308	1.2%
14	N	Nemertea (unidentified)	175	12	378	616	295	1.2%
15	P	*Branchioasychis americana*	5	7	837	230	269	1.1%
		156 other species	997	591	8,235	6,668	4,123	16.2%
		Total 171 species	31,550	9,523	35,951	24,608	25,408	100.0%

Note: Species averaging <1% of the total abundance are combined into the row of other species. Taxa abbreviations: A = Amphipoda, B = Bivalvia, G = Gastropoda, O = Oligochaeta, N = Nemertea, P = Polychaeta.

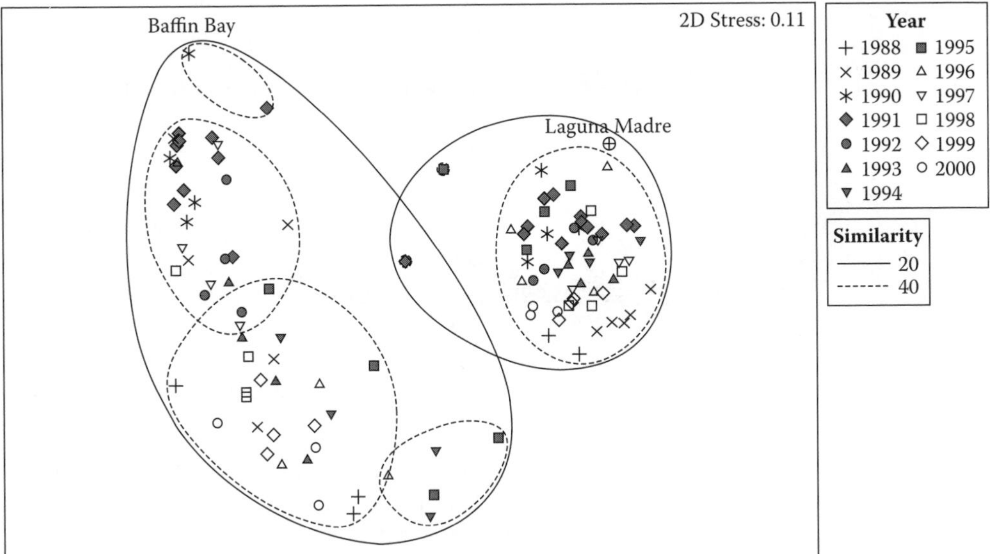

FIGURE 6.6 Nonmetric multidimensional scaling (MDS) analysis on species composition by station–date combination. Baffin Bay on the right, and Laguna Madre on the left, with symbols representing the year when the sample was taken.

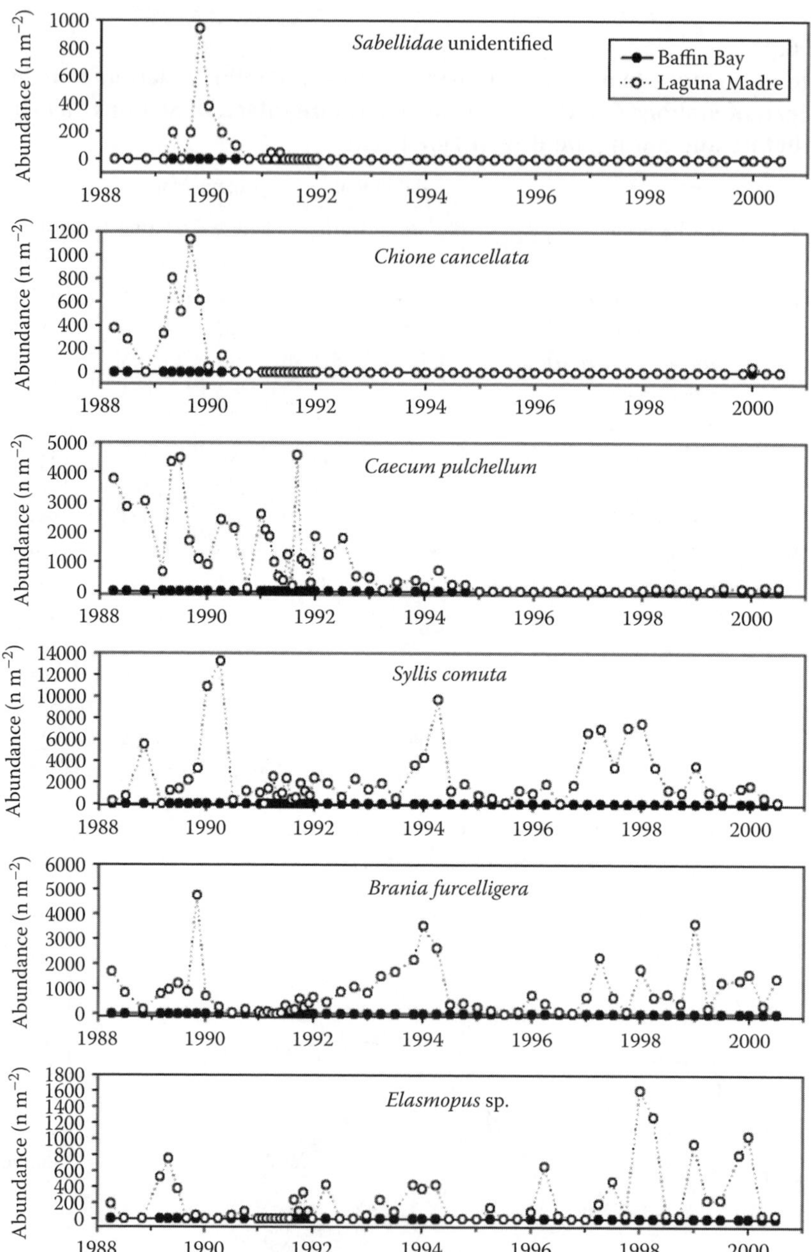

FIGURE 6.7 Abundance over time for minimum set of six species that can replicate the MDS plot (Figure 6.6).

For example, there is a transition in the pattern of points from years 1988 and 1989 at the bottom to 1991–1993 at the top, and a return to the bottom in later years past 1998.

A total of six species were able to duplicate the same MDS pattern as in Figure 6.7 (BVSTEP procedure, Rho = 0.96). These species were *Brania furcelligera, Caecum pulchellum, Chione cancellata, Elasmopus* sp., *Sabellidae* unidentified, and *Syllis cornuta.* An examination of these six species over time can help explain the effect of brown tide on the benthic community structure pattern. The *Sabellidae* and *Chione cancellata* were present prior to brown tide, and then virtually disappeared. *Caecum pulchellum* had high abundance through 1992 and then disappeared. *Syllis*

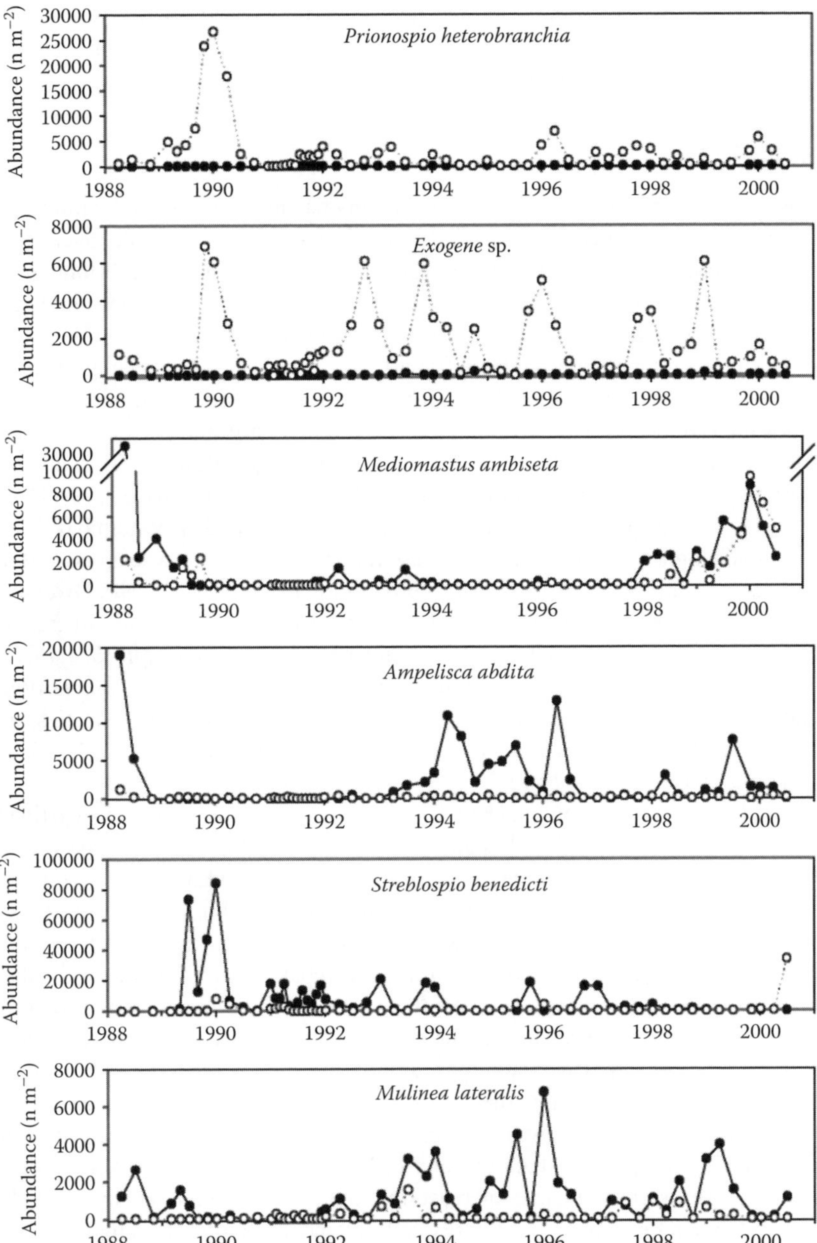

FIGURE 6.8 Abundance over time of six dominant species not included in Figure 6.7.

cornuta and *Brania furcelligera* crashed after the brown tide appeared, but then recovered in 1994, crashed again, and recovered again in 1997. *Elasmopus* sp. was more abundant after 1998 than earlier in the record.

It is also useful to examine the six dominant species that were not found in the BEST analysis (Figure 6.8). *Exogone* sp. and *Prionospio heterobranchia* were dominant in Laguna Madre, but rarely found in Baffin Bay. *Mediomastus ambiseta* and *Ampelisca abdita* were dominant in Baffin Bay, and occurred in high abundances prior to the onset of brown tide but did not occur after onset. *Streblospio benedicti* was dominant throughout, but in higher abundances prior to the brown tide.

Mulinia lateralis was present prior to brown tide, but disappeared for a 2-year period after the brown tide, and recovered through 2000. Three species: *Exogone* sp., Oligochaeta, and *Polydora ligni,* occurred primarily after brown tide onset.

6.4 DISCUSSION

The relationship between brown tide and macroinfauna has to be interpreted within the context of other environmental factors that drive benthic community dynamics. Benthic macrofauna in Texas bays react to variation in salinity through changes in community structure (Montagna and Kalke 1992, 1995). There are no major sources of freshwater discharge into Baffin Bay and Laguna Madre, and direct precipitation to the water contributes ~65% of the total freshwater inflow into the estuary (Orlando et al. 1991). From 1988 through 1990 southeast Texas had below-average rainfall (Buckner et al. 1989, 1990; Buckner and Shelby 1990). The increased salinities that resulted were reflected by the benthos through decreases in overall biomass and diversity, and peaks in abundance of species that are pioneers or tolerate either a wide range or very high salinities. Diversity at all stations decreased prior to brown tide. In Baffin Bay, this decline occurred in July of 1989. Overall, Baffin Bay diversity was low due to the small number of species in the muddy-bottom bay. Laguna Madre stations had a much higher diversity index, because they were located in a seagrass bed. Diversity at both Laguna Madre stations decreased prior to brown tide in January of 1990. Diversity decreases corresponded with peaks in abundance, which were primarily monospecific. In Baffin Bay, high abundance samples were dominated by the spionid polychaete *Streblospio benedicti*, while the Laguna Madre samples were dominated by the spionid polychaete *Prionospio heterobranchia* and the syllid polychaetes *Syllis cornuta* and *Exogone* sp. The high abundance species, in particular *Streblospio benedicti*, are opportunistic and tolerant of stressed environments (Grassle and Grassle 1974, 1984; Dauer et al. 1981; Dauer 1993).

Approximately 75% of macroinfaunal biomass was lost prior to the onset of brown tide (Table 6.1). Biomass loss was correlated with high salinities in 1988 and 1989 and freezes in the winter of 1989–1990. In shallow water environments, such as Baffin Bay and Laguna Madre, the benthos are strongly affected by the water column (Montagna and Kalke 1995). During coastal wind events, bottom sediments are disturbed and mixed with the water column. The decomposition of the dead macrofauna provided a source of carbon and nitrogen, in particular ammonium, for the water column. Detrital material, including the dead benthic macrofauna, is broken down by bacteria in the sediment to form ammonia (NH_3), which in the presence of hydrogen (H^+) forms ammonium (NH_4^+) (Libes 1992). Under most estuarine conditions, ammonium flux is from the sediments to the water column. After the benthic invertebrate die-off there was an increase in particulate organic nitrogen in the sediments. More ammonium was probably fixed in the sediments by bacteria, increasing the flux into the water column. In the shallow water of Baffin Bay and Laguna Madre, disturbance of the sediments by storms could further increase the amount of ammonium flux. Benthic biomass loss, along with the fish kill during the freezes of 1989 and 1990, produced high concentrations of ammonium that could have facilitated the brown tide bloom.

Brown tide was first recognized at the stations in the spring of 1990 when chlorophyll concentrations were markedly higher (Buskey and Stockwell 1993). The plankton shifted from a mixed to a heavily monospecific community. Dominating water column primary production, the brown tide alga could affect the macrobenthos in many ways. The large concentration of the brown tide alga represented a new carbon source, which might affect trophic structure if incorporated into the benthic food web (Street et al. 1997). Brown tide could directly damage the macroinfauna through toxicity or interference with filter-feeding structures (Montagna et al. 1993). Brown tide could damage the larval stages of benthic animals while they inhabit the water column (Buskey et al. 1997). Finally, brown tide could alter the benthic habitat.

The bloom of brown tide alga brought a new, large carbon source into Baffin Bay and Laguna Madre. As it was incorporated into the food web, there was likely a change in the trophic structure of the benthic community (Street et al. 1997). If animals were able to incorporate brown tide, then they could replace those that cannot. Phytoplankton, such as brown tide, has a more negative stable carbon isotope value than do seagrasses. Filter-feeding polychaetes from Alazan Bay, a tertiary bay of Baffin Bay, had a more negative carbon isotope signature than filter-feeders in Laguna Madre. The negative carbon isotope value in Alazan Bay indicated that brown tide was being incorporated into the food web. The brown tide signature was not recognized in Laguna Madre, where seagrass detritus remained dominant (Street et al. 1997). The lower regions of Alazan Bay are comparable to Baffin Bay in salinity, sediment type, and lack of macrophytes. The presence of the brown tide signature in Alazan Bay indicated that brown tide could be incorporated into the Baffin Bay food web and might affect the benthic community structure.

It is possible that the brown tide alga directly affects the benthic macrofauna through toxicity or interference with filter-feeding structures. *Aureococcus anophagefferens*, the brown tide alga on the northeast coast, was shown to block the filter-feeding structures of scallops and other bivalves (Tracey 1988; Bricelj and Kuenstner 1989). In this manner, it was directly killing local populations of shellfish. The dwarf surf clam, *Mulinia lateralis*, is the dominant bivalve in Baffin Bay and the Laguna Madre. In laboratory studies, *M. lateralis* utilized brown tide as a primary food source (Montagna et al. 1993).

Little is known about the affect of the brown tide alga on benthic macroinfauna larvae. In field studies, the increase in brown tide cells coincided with decreases in both microzooplankton and mesozooplankton populations and brown tide could be a poor food source or could secrete a substance that interferes with feeding (Buskey and Hyatt 1995). In the laboratory, some zooplankton species had negative growth rates and increased mortality when fed brown tide as a primary food source (Buskey and Hyatt 1995). The brown tide may affect the meroplankton, including benthic larval stages, in ways similar to the studied zooplankton. Harm to larval stages could explain the continuation of decreased macrobenthos populations, and why some species were more affected, if survival was dependent on the length of stay in the water column and food sources. *Streblospio benedicti* larval growth rates and swimming speeds, but not mortality rates, were reduced in cultures with brown tide relative to the alga *Isochrysis galbana*, which is about the same size as brown tide (Ward et al. 2000). Thus, the brown tide does have harmful sublethal effects for one dominant species of meroplanktonic larvae, which could help explain reduced adult population size.

The brown tide alga may alter habitats within Baffin Bay and Laguna Madre, and in turn may influence the macroinfauna. Two distinct habitats were studied: (1) the bare, muddy bottom of Baffin Bay; and (2) the seagrass habitat of Laguna Madre. There were some similarities in temperature and salinity between stations in Baffin Bay and Laguna Madre, but small differences also existed, such as dissolved oxygen levels, which were slightly higher in Laguna Madre; nutrient concentrations, which were slightly higher in Baffin Bay; and chlorophyll *a* concentrations, which were slightly lower in Laguna Madre.

Depth was greater at the Baffin Bay stations (3 m) than at the Laguna Madre stations (1 m). Disturbance of the sediments by wind-induced waves occurs at both depths. The other effect of depth is light penetration, which was reduced in both bays by the brown tide. There were some distinct differences between Baffin Bay and Laguna Madre. Both bays supported high abundances, but Laguna Madre stations had higher macrofaunal biomass and diversity throughout the study. The benthic fauna within the two ecosystems responded differently over time because the temporal dynamics in Baffin Bay were always dampened relative to those in Laguna Madre. There were more distinct differences between the ecosystems than within the bays.

Baffin Bay stations represented bare, muddy-bottom habitats. The sediment is nearly 75% clay (Montagna unpublished data). Macrofaunal abundance was generally below 40,000 m^{-2}, with occasional maximum values of 100,000 m^{-2}. These peaks were usually monospecific, corresponding to increases in opportunistic species, including *Streblospio benedicti* and *Ampelisca abdita*. Biomass in Baffin Bay was low, near 2 g m^{-2}. High biomass generally indicated the presence of the dwarf surf clam, *Mulinia lateralis*. *Mulinia lateralis* abundance was unusually low during the first 2 years of the brown tide bloom, 1990 to 1992, as was evident in the low biomass numbers. However, *M. lateralis* was able to successfully feed on brown tide in the laboratory (Montagna et al. 1993). The increase in macrofaunal biomass later in the sampling period, 1993, corresponded with the return of *M. lateralis. Mulinia lateralis* is considered a hardy species covering a large range of salinities and temperatures (Parker 1975; Montagna and Kalke 1995). It is also an opportunist, being able to colonize quickly after a disturbance (Flint and Yuonk 1983; Montagna et al. 1993). Opportunistic behavior may explain the relatively quick recovery 2 years after the onset of brown tide. Diversity in Baffin Bay was also low. Hill's N1 Index fell below four species most of the time. From July 1989 to January 1992, diversity remained near one, indicating the samples were almost monospecific, being heavily dominated by the opportunistic polychaete, *Streblospio benedicti*.

Seagrass habitats have high abundance and diversity of biota, including macroinfauna (Orth et al. 1984). Large, diverse populations exist because seagrasses provide protection from predators, food abundance, habitat complexity, and sediment stability (Orth et al. 1984). However, a seagrass bed is not a uniform habitat, and not all seagrass species provide the same benefits. Seagrass species differ in blade shape, canopy complexity, and rhizome biomass. The seagrasses at Laguna Madre station 189 were primarily *Halodule wrightii* (Dunton 1994), a species with narrow leaves and small, shallow, finely branched roots and rhizomes (Orth et al. 1984). It has a large surface area to biomass ratio, which supports high abundance (Stoner 1980). The seagrass bed is a heterogeneous environment with bare sand patches interdispersed within the seagrasses. These patches do not provide the aboveground cover or the belowground habitat complexity of the seagrasses. Larval fish use the seagrasses for protection from predators, while feeding in unvegetated patches where there is better visibility (Holt et al. 1983). Therefore, sand patches provide less protection and are typified by greater predation than nearby seagrass habitat. The habitat difference between vegetated and unvegetated patches and the grasses explains the lower diversity and abundance of macroinfauna in the unvegetated samples. Previous studies showed that abundance and diversity of macrofauna were lower within unvegetated patches (Orth 1971; Santos and Simon 1974; Stoner 1980). Macroinfaunal abundances in both Laguna Madre habitats were high relative to most previous reports.

Although Laguna Madre stations were different at different times due to habitat variation, the two Baffin Bay stations were similar all the time. While the hydrologic data for the stations were the same, macroinfaunal abundance varied between the habitats. In the seagrass patch, macroinfaunal abundance averaged 38,592 m^{-2}, more than double the macroinfaunal abundance in the sand patches, which averaged 16,094 m^{-2}. However, the temporal abundance trend was similar for both locations. Abundance was low in the early brown tide years, 1990 to 1992, but increased from 1993 to the end of the study. Based on community structure analysis, the animals present before the brown tide were different from those during the bloom.

Laguna Madre macrofaunal biomass was consistently low from 1991 to the end of the study, indicating that small organisms, including oligochaetes and syllid polychaetes, dominate the system. Peaks in biomass were related to a combination of molluscs and large, deep-dwelling polychaetes. High biomass values were evenly distributed between polychaetes and mollusks at the seagrass station. At the unvegetated station, large polychaetes occupying the deeper 3- to 10-cm section of the core accounted for much of the biomass (unpublished data). Large polychaetes included *Branchioasychis americana*, *Scolopolos rubra*, and *Magelona pettiboneae*. Large polychaetes burrowed more effectively at the Laguna Madre stations due to the larger grain size of the sediments.

Diversity at both Laguna Madre stations was higher than that at the Baffin Bay stations. Diversity was higher within the seagrass bed than in the sand patches. Seagrass provides protection as well as habitat differentiation to the macroinfauna. At the seagrass station (189G), diversity changes corresponded with changes in abundance, indicating that the ratio between the total number of individuals and number of abundant species was fairly consistent. Diversity remained at or below average from January 1990 to October 1992. Seasonal winter peaks found in 1989 returned in 1993 and 1994. Sand station (189S) diversity did not decline with the onset of brown tide. Diversity in the sand patch decreased later in the study, beginning in July 1992 and continuing through 1994. Diversity in Laguna Madre was dependent on habitat, unlike biomass and abundance, which were temporally similar at both Laguna Madre stations.

Macroinfaunal biomass, abundance, and diversity decreased from before (April 1988 to May 1990) the onset of brown tide to after (June 1990 to December 1995) at all the stations. Stations 6 and 24 were combined since there was no significant difference between the two stations. The aforementioned decreases were greater during the 2 years immediately following the onset of the brown tide. Since 1993, there has been some recovery in macrofaunal biomass, abundance, and diversity, but values have not returned to the maximal levels found prior to the brown tide. Increases in biomass, abundance, and diversity at later dates were linked to a new macrofaunal community, and this is indicated by seriation in the multivariate analysis.

Laguna Madre and Baffin Bay are estuaries and therefore undergo large changes in hydrologic variables with time. Within coastal environments, there are many natural variations within a 7-year period that could account for changes in the benthic community. Salinity is one such variable. In Baffin Bay and Laguna Madre, salinity varied with precipitation. Brown tide originated in hypersaline waters following a drought. Laboratory studies have shown that the brown tide species grows more rapidly in high salinities (40 psu), but the species is tolerant of low salinities, having been found in salinities down to 5 psu (Buskey et al. 1996). Primary production measurements had no direct relationship to a variety of salinities (Stockwell et al. 1993). Salinity was high from 1989 to 1991, and lower in 1992 and 1993 during an El Niño. It increased from 1994 to 1996, but then decreased during an El Niño in 1997. The macrofaunal abundance and biomass decreased prior to these salinity variations. Salinity and macroinfaunal variation did not correspond well throughout the sampling period. Other underlying changes during a 7-year study, such as nutrients, storms, and temperature, may also explain the variation (Smayda 1984).

There was change in macroinfaunal communities at all four stations beginning at the onset of the brown tide. Changes in abundance, biomass, and diversity occurred by January 1990, before brown tide was dominant at the studied stations in Baffin Bay and Laguna Madre. Brown tide dominated the phytoplankton beginning in June 1990. The loss of biomass, abundance, and diversity was related to the high salinities in the area caused by drought conditions. However, throughout the study, community structure did not follow very closely the changes in salinity. Other factors, such as brown tide, were also affecting the temporal changes.

Baffin Bay and Laguna Madre stations, although similar hydrographically, differ from one another due to habitat variation. Isotopic studies show that, while the brown tide is being incorporated into a food web similar to muddy-bottom Baffin Bay, it is not replacing seagrass as the primary food source in Laguna Madre (Street et al. 1997). Likewise, macrofaunal abundance recovered in Laguna Madre but not in Baffin Bay. The two bays reacted differently, implying that habitat is influencing the macroinfaunal reaction to brown tide.

Considerable research has been performed on how diversity affects stability of an ecosystem. It is thought that diversity confers stability. Average diversity over all samples is higher in Laguna Madre (N1 9.7) than in Baffin Bay (N1 2.4); therefore, it is reasonable to hypothesize that Laguna Madre, the more diverse system, will be more resistant and/or resilient in response to the brown tide disturbance. Indeed, this was the case. During the 2 years after the initial onset of brown tide, Baffin Bay N1 diversity decreased by 110% from 2.5 to 1.2, and most of the time there was only one species, *Streblospio benedicti*, which is well know as a disturbance indicator species. In contrast, N1 diversity decreased in Laguna Madre by 40% from 15 to 11. The smaller decrease in Laguna Madre can be interpreted as higher resistance to change. The brown tide abated during 1992 and 1993, and N1 diversity increased to 14 in Laguna Madre and 2.6 in Baffin Bay, indicating both systems recovered equally and were resilient. The only indicator that long-term resilience differed in the two systems was the complete recovery of abundance in Laguna Madre from 1997 through 2000. In contrast, abundance in Baffin Bay never recovered to pre–brown tide levels.

Brown tide affected Laguna Madre and Baffin Bay by lowering macroinfaunal biomass, abundance, and diversity. The timing of the decreases, as well as the lack of correlation between brown tide and other hydrographic parameters, established this relationship. Habitat altered the way brown tide and the macroinfauna interacted during the study period. There were two equilibrium states, one before and one after the brown tide bloom. The system prior to the brown tide bloom had a mixed plankton community that supported a more diverse community, higher in abundance and biomass. The 2 years directly after the onset of the brown tide were indicative of a disturbed environment where diversity, abundance, and biomass decreased at all stations. Through time the systems reached a second equilibrium dependent on the habitat. Baffin Bay stations that incorporated brown tide into the food web did not recover in abundance, but increased in biomass and diversity. The seagrass stations returned to high abundance, but retained a lower biomass. Diversity in the vegetated site returned to pre-bloom levels, while the unvegetated Laguna Madre site did not.

The variation in macrofaunal reaction was significant, because as the brown tide continued more seagrasses were lost due to low light levels (Onuf 1996). The loss of seagrasses equaled loss of the macrofaunal diversity and abundance. The loss of macroinfaunal diversity and abundance could have had additional negative impacts on the ecosystem, because the benthos were important in the food web and in nutrient recycling. Therefore, understanding the complex relationship between brown tide and the macroinfauna may help explain responses at all trophic levels to disturbance in general and, specifically, harmful plankton blooms.

ACKNOWLEDGMENTS

This research was partially supported by funding provided by the Texas Higher Education Coordinating Board, Advanced Research Program, under Grant Numbers 4541 and 3658-426. The authors especially thank Ms. Carol Simanek for her vital role in data management.

REFERENCES

Breuer, J. P. 1962. An ecological survey of the lower Laguna Madre of Texas, 1953–1959. *Publ. Univ. Tex. Inst. Mar. Sci.* 8: 153–185.

Bricelj, V. M. and S. H. Kuenstner. 1989. Effects of the brown tide on the feeding, physiology and growth of bay scallops and mussels. Pp. 491–509, In: E. M. Cosper, V. M. Bricelj, and E. J. Carpenter (eds.), *Novel phytoplankton blooms*. Coastal and Estuarine Studies no. 35. Berlin: Springer-Verlag.

Buckner, H. D., E. R. Carrille, H. J. Davidson, and W. J. Shelby. 1989. *Water resource data for Texas, water year 1988*. U.S. Geological Survey water-data report TX-88-3, Austin, Texas.

Buckner, H. D. and W. J. Shelby. 1990. *Water resource data for Texas, water year 1990*. U.S. Geological Survey water-data report TX-88-3, Austin, Texas.

Buckner, H. D., W. J. Shelby, and H. J. Davidson. 1990. *Water resource data for Texas, water year 1989*. U.S. Geological Survey water-data report TX-88-3, Austin, Texas.

Buskey, E. J. and C. J. Hyatt. 1995. Effects of the Texas (USA) "brown tide" alga on planktonic grazers. *Mar. Ecol. Prog. Ser.* 126: 285–292.

Buskey, E. J., H. Liu, C. Collumb, and J. G. F. Bersano. 2001. The decline and recovery of a persistent Texas brown tide algal bloom in the Laguna Madre (Texas, USA). *Estuaries* 24: 337–346.

Buskey, E. J., P. A. Montagna, A. F. Amos, and T. E. Whitledge. 1997. Disruption of grazer populations as a control factor to the initiation of the Texas brown tide algal bloom. *Limnol. Oceanogr*. 42: 1515–1222.

Buskey, E. J., S. Stewart, J. Peterson, and C. Collumb. 1996. *Current status and historical trends of brown tide and red tide phytoplankton blooms in the Corpus Christi Bay National Estuary Program study area.* CCNEP Report No. 07.

Buskey, E. J. and D. A. Stockwell. 1993. Effects of a persistent brown tide on zooplankton populations in the Laguna Madre of south Texas. Pp. 659–666, In: T. J. Smayda and Y. Shimizu (eds.), *Toxic phytoplankton blooms in the sea*. New York: Elsevier Science Publications.

Buskey, E. J., B. Wysor, and C. J. Hyatt. 1998. The role of hypersalinity in the persistence of the Texas "brown tide" in the Laguna Madre. *J. Plankton Res.* 20: 1553–1565.

Clarke, K. R. and R. N. Gorley. 2006. *PRIMER v6: User manual tutorial*. Plymouth, U.K.: PRIMER-E.

Clarke, K. R. and R. M. Warwick. 1998. Quantifying structural redundancy in ecological communities. *Oecologia* 113: 278–289.

Clarke, K. R. and R. M Warwick. 2001. *Change in marine communities: An approach to statistical analysis and interpretation*. 2nd ed. Plymouth, U.K.: PRIMER-E.

Copeland, B. J., J. H. Thompson, and W. B. Ogletree. 1968. Effects of wind on water levels in the Texas Laguna Madre. *Tex. J. Sci.* 20: 168–169.

Dagg, M. J. and T. E. Whitledge. 1991. Concentrations of copepod nauplii associated with the nutrient-rich plume of the Mississippi River. *Cont. Shelf Res.* 11: 1409–1423.

Dauer, D. M. 1993. Biological criteria, environmental health and estuarine macrobenthic community structure. *Mar. Poll. Bull.* 26: 249–257.

Dauer, D. M., C. A. Maybury, and R. M. Ewing. 1981. Feeding behavior and general ecology of several spionid polychaetes from the Chesapeake Bay. *J. Exp. Mar. Biol. Ecol.* 54: 21–38.

DeYoe, H. R., D. A. Stockwell, R. R. Bidgare, M. Latasa, P. W. Johnson, P. E. Hargraves, and C. A. Suttle. 1997. Description and characterization of the algal species *Aureoumbra lagunensis* gen. et sp. nov. and referral of *Aureoumbra* and *Aureococcus* to Pelagophyceae. *J. Phycol.* 33: 1042–1048.

DeYoe, H. R. and C. A. Suttle. 1994. The inability of the Texas "brown tide" alga to use nitrate and the role of nitrogen in the initiation of a persistent bloom of the organism. *J. Phycol.* 30: 800–806.

Dunton, K. H. 1994. Seasonal growth and biomass of the subtropical seagrass *Halodule wrightii* in reference to continuous measurements of underwater irradiance. *Mar. Biol.* 120: 479–489.

Elton, C. S. 1958. *The ecology of evasions by plants and animals*. London: Methuen and Company.

Flint, R. W. and J. A. Younk. 1983. Estuarine benthos: Long-term variations, Corpus Christi, Texas. *Estuaries* 6: 126–141.

Grassle, J. P. and J. F. Grassle. 1974. Opportunistic life histories and genetic systems in marine benthic polychaetes. *J. Mar. Res.* 32: 253–284.

Grassle, J. P. and J. F. Grassle. 1984. The utility of studying the effects of pollutants on single species populations in benthos of mesocosms and coastal ecosystems. Pp. 621–642, In: *Concepts in marine pollution measurements*. College Park, Maryland: Maryland Sea Grant Program.

Holm-Hansen, O., C. J. Lorenzen, R. Holmes, and J. D. H. Strickland. 1965. Fluorometric determination for chlorophyll. *J. Cons. Perm. Int. Explor. Mer* 30: 3–15.

Holt, S. A., C. L. Kitting, and C. R. Arnold. 1983. Distribution of young red drums among different sea-grass meadows. *Trans. Amer. Fisheries Soc.* 112: 267–271.

Hutcheson, K. 1970. A test for comparing diversities based on the Shannon formula. *J. Theor. Biol.* 29: 151–154.

Javor, B. 1989. *Hypersaline environments*. New York: Springer-Verlag.

Libes, S. M. 1992. *An introduction to marine biogeochemistry*. New York: John Wiley & Sons, Inc.

Ludwig, J. A. and J. F. Reynolds. 1988. *Statistical ecology*. New York: John Wiley & Sons.

Montagna, P. A. and R. D. Kalke. 1992. The effect of freshwater inflow on meiofaunal and macrofaunal populations in the Guadalupe and Nueces estuaries, Texas. *Estuaries* 15: 307–326.

Montagna, P. A. and R. D. Kalke. 1995. Ecology of infaunal Mollusca in south Texas estuaries. *Amer. Malacol. Bull.* 11: 163–175.

Montagna, P. A., D. A. Stockwell, and R. D. Kalke. 1993. Dwarf surfclam *Mulinia lateralis* (Say, 1822) populations and feeding during the Texas brown tide event. *J. Shellfish Res.* 12: 433–442.

Onuf, C. P. 1996. Seagrass responses to long-term light reduction by brown tide in the upper Laguna Madre, Texas: Distribution and biomass patterns. *Mar. Ecol. Prog. Ser.* 138: 219–231.

Onuf, C. P. 2000. Seagrass responses to and recovery from seven years of brown tide. *Pacific Conserv. Biol.* 5: 306–313.

Orlando, S. P., Jr., L. P. Rozas, G. H. Ward, and C. J. Klein. 1991. *Analysis of salinity structure and stability for Texas estuaries*. Rockville, Maryland: National Ocean Service, National Oceanic and Atmospheric Administration, Strategic Assessment Branch.

Orth, R. J. 1971. *The effect of turtle grass,* Thalassia testudinum, *on the benthic infauna community structure in Bermuda*. St. Georges West, Bermuda, Bermuda Biological Station Special Report No. 9: 18–37.

Orth, R. J., K. L. Heck, and J. van Montfrans. 1984. Faunal communities in seagrass beds: A review of the influence of plant structure and prey characteristics on predator-prey relationships. *Estuaries* 7: 339–350.

Parker, R. H. 1975. *The study of benthic communities: A model and review*. New York: Elsevier.

Quammen, M. L. and C. P. Onuf. 1993. Laguna Madre: Seagrass changes continue decades after salinity reduction. *Estuaries* 16: 303–311.

SAS Institute Inc. 1991. *SS/STAT guide for personal computers*. Cary, North Carolina: SAS Institute Inc.

Santos, S. L. and J. L. Simon. 1974. Distribution and abundance of the polychaetous annelids in a south Florida estuary. *Bull. Mar. Sci.* 24: 669–689.

Shormann, D. E. 1992. The effects of freshwater inflow and hydrography on the distribution of brown tide in south Texas bays. M.A. Thesis, University of Texas at Austin, Austin, Texas.

Smayda, T. W. 1984. Variations and long-term changes in Narragansett Bay, a phytoplankton-based coastal marine ecosystem: relevance to field monitoring for pollution assessment. In *Concepts in marine pollution measurements*. Maryland Sea Grant Program, College Park, Maryland. Pp. 663–679.

Stockwell, D. A., E. J. Buskey, and T. E. Whitledge. 1993. Studies on the conditions conducive to the development and maintenance of a persistent "brown tide" in Laguna Madre, Texas. Pp. 693–698, In: T. J. Smayda and Y. Shimizu (eds.), *Toxic phytoplankton blooms in the sea*. Amsterdam: Elsevier.

Stoner, A. W. 1980. Abundance, reproductive seasonality and habitat preferences of amphipod crustaceans in seagrass meadows of Apalachee Bay, Florida. *Contrib. Mar. Sci.* 23: 63–77.

Street, G. T., P. A. Montagna, and P. L. Parker. 1997. Incorporation of brown tide into an estuarine food web. *Mar. Ecol. Prog. Ser.* 152: 67–78.

Tracey, G. A. 1988. Feeding reduction, reproductive failure, and mortality in *Mytilus edulis* during the 1985 "brown tide" in Narragansett Bay, Rhode Island. *Mar. Ecol. Prog. Ser.* 50: 73–81.

Valiela, I. 1995. *Marine ecological processes*. 2nd ed. New York: Springer-Verlag.

Ward, L. A., P. A. Montagna, R. D. Kalke, and E. J. Buskey. 2000. Sublethal effects of Texas brown tide on *Streblospio bendicti* (Polychaeta) larvae. *J. Exp. Mar. Biol. Ecol.* 248: 121–129.

Whitledge, T. E. 1993. The nutrient and hydrographic conditions prevailing in Laguna Madre, Texas before and during a brown tide bloom. Pp. 711–716, In: T. J. Smayda and Y. Shimizu (eds.), *Toxic phytoplankton blooms in the sea*. Amsterdam: Elsevier.

Whitledge, T. E., S. C. Malloy, C. J. Patton, and C. D. Wirick. 1981. *Automated nutrient analysis in seawater.* Brookhaven National Laboratory Formal Report 51938, Brookhaven, New York.

7 The Choptank Basin in Transition

Intensifying Agriculture, Slow Urbanization, and Estuarine Eutrophication

Thomas R. Fisher, Thomas E. Jordan, Kenneth W. Staver, Anne B. Gustafson, Antti I. Koskelo, Rebecca J. Fox, Adrienne J. Sutton, Todd Kana, Kristen A. Beckert, Joshua P. Stone, Gregory McCarty, and Megan Lang*

CONTENTS

Abstract 136
7.1 Introduction 136
7.2 Study Site 137
7.3 Methods 139
7.3.1 Hydrology 139
7.3.2 Surface Water Chemistry 139
7.3.3 Drainage Control Structures 141
7.3.4 Groundwater Piezometers 141
7.3.5 Excess N_2, O_2, N_2O, and CH_4 142
7.3.6 Estuarine Water Quality Data 143
7.4 Results 145
7.4.1 Regional Hydrology 145
7.4.2 Water Quality Trends: Nontidal Monitoring Stations 147
7.4.3 Water Quality Drivers of Change 148
7.4.4 Agricultural N and P Reduction 151
7.4.5 Estuarine Water Quality Conditions 157
7.5 Discussion 158
7.5.1 Water Quality Status 158
7.5.2 Water Quality Improvement 159
Acknowledgments 162
References 162

* Corresponding author. Email: fisher@hpl.umces.edu

ABSTRACT

The Choptank Basin and Estuary are located on the Delmarva Peninsula in the mid-Atlantic coastal plain. The regional hydrology is characterized by nearly uniform seasonal rainfall but large seasonal variations in temperature, evapotranspiration, groundwater levels, and stream discharge. Water quality in nontidal streams is largely determined by agricultural land use and animal feeding operations, and nitrogen (N) and phosphorus (P) concentrations have been increasing for decades. Inputs from nontidal streams, together with increasing human populations and wastewater discharges, have resulted in degraded estuarine water quality, including increases in chlorophyll *a* in surface waters and declining oxygen in bottom waters. Attempts to reduce losses of N and P from nontidal streams in agricultural areas have met with limited success. One targeted watershed in the Choptank Basin showed stabilized concentrations of base flow N, along with small decreases in base flow P, a decade after extensive application of some best management practices (BMPs) in contrast to the nearby Greensboro watershed which was not targeted for BMPs and exhibited increases in base flow N and P. An attempt to improve water quality using increased stream buffers has yet to be successful, probably because new stream buffers represented only an 11% increase over existing ones.

Based on our observations, we suggest policies to improve water quality in the Choptank basin and the mid-Atlantic region in general. We recommend application of water quality standards at the watershed scale, reduced caps for wastewater discharges, lower fertilizer applications on agricultural areas, mandatory stream buffers and winter cover crops on farms, and banning of lawn fertilizers. Anthropogenically impacted systems, such as the Choptank and Delmarva coastal bays, require a more regulated approach at the watershed scale, with long-term monitoring to improve water quality.

Key Words: Chesapeake Bay, Choptank River, Delmarva coastal bays, eutrophication, hydrology, agriculture, wastewater

7.1 INTRODUCTION

The Choptank Basin and Estuary are components of a coastal plain tributary of Chesapeake Bay on the Delmarva Peninsula. The eutrophication of the Choptank Estuary (Fisher et al. 2006b) can be viewed as a microcosm of the eutrophication processes in Chesapeake Bay as a whole. The disturbance and degradation of both the Chesapeake Bay and the Choptank Estuary over the last few centuries have been well documented (e.g., Cooper and Brush 1993; Staver et al. 1996; Benitez 2002; Benitez and Fisher 2004; Kemp et al. 2005), and the major drivers of N and P export from the terrestrial basins to the estuaries leading to eutrophication are intensive agriculture and urbanization (Lee et al. 2001; Koroncai et al. 2003). In the mainstem Chesapeake, algal blooms are common in surface waters (Glibert et al. 1995; Harding and Perry 1997; Sellner and Fonda-Umani 1999), and oxygen is depleted in bottom waters in summer (Officer et al. 1984; Hagy et al. 2004). The Choptank Estuary is currently undergoing the same degradation (Fisher et al. 2006b) that is now routinely observed in Chesapeake Bay.

An important difference between the Choptank and the main stem of the Chesapeake Bay is the depth and stratification where initial consumption of terrestrial nutrients occurs. Nutrients (N and P) enter estuaries in rivers with considerable amounts of turbidity, and the estuarine circulation further concentrates particles near the head of an estuary in a region known as the turbidity maximum (Meade 1968). In this region, little primary production and nutrient consumption occur due to the limited availability of light in the water column (Harding et al. 1986); however, as the water clears farther downstream, phytoplankton production and nutrient consumption increase, resulting in a chlorophyll *a* maximum (Fisher et al. 1988). The chlorophyll *a* maximum in the Chesapeake Bay develops just south of the Bay Bridge at Annapolis, where depths are 10 to 50 m, and stratification is strong, resulting in vertical isolation of bottom waters and oxygen depletion (Officer et al. 1984;

Harding et al. 1986; Fisher et al. 1988). In contrast, the chlorophyll *a* maximum in the Choptank occurs in shallower water (5 to 15 m), with weaker stratification and less isolation of bottom waters (Berndt 1999). As a result, there is less oxygen depletion in the Choptank despite higher annual average chlorophyll *a* concentrations of 15 to 20 μg L^{-1} at the chlorophyll *a* maximum in the water body (Fisher et al. 2006b), compared to 10–15 μg L^{-1} in the upper Chesapeake (Harding and Perry 1997). The dinoflagellate blooms known as mahogany tides that commonly occur in the Chesapeake Bay mainstem (Glibert et al. 2001; Tango et al. 2002; Marshall et al. 2004) are still only an occasional occurrence in the Choptank Estuary. Despite the somewhat higher average chlorophyll in the Choptank, the shallower bathymetry and weaker stratification has limited hypoxia in bottom waters, although it is increasing (Fisher et al. 2006b). Oxygen conditions in the Choptank, therefore, are somewhat similar to those in Chesapeake Bay in an earlier era, and an understanding of the forcing of eutrophication in the Choptank may provide useful information on the Chesapeake and other systems undergoing eutrophication, such as the Delmarva coastal bays (Wazniak et al. 2004).

The primary sources of terrestrial nutrients in the Chesapeake Bay and the Choptank Estuary are similar. Intensive agriculture and human waste disposal provide the largest sources of N and P (Staver et al. 1996; Lee et al. 2001; Koroncai et al. 2003). Agriculture and forests are the dominant land uses in the basins of both the Choptank and mainstem bay, and population density in the Choptank (59 km^{-2} in 2000; Fisher et al. 2006b) is comparable to the other Chesapeake basins (30 to 70 km^{-2}; Carlozo et al. 2008). However, urbanization and wastewater discharges are increasing in both areas (Fisher et al. 2006a; Williams et al. 2006), and in the Choptank the largest wastewater discharges are concentrated in density-stratified areas of the estuary (Lee et al. 2001), which are susceptible to algal blooms.

In the Choptank Basin (Figure 7.1) we have been attempting to quantify the effects of agriculture, wetlands, and hydric soils on export of N and P from the land to the estuary. There are long-term USGS measurements of hydrology (since 1948) and water chemistry (since 1964) at the gauging station near Greensboro, Maryland, and in 2004, we established 17 gauged and monitored watersheds (1 to 50 km^2) within or near the Choptank Basin with varying proportions of forest (10% to 100%), agriculture (0 to 84%), and hydric soils (15% to 97%) for the study of surface waters. We have also installed ~80 piezometers (wells sampling from a limited depth stratum) to record groundwater temperature and depth and to measure the process of denitrification of agricultural nitrate (NO_3^-) as accumulation of excess N_2 and N_2O in groundwaters. The long-term goal of these projects is to evaluate the effects on water quality of agricultural Best Management Practices (BMPs) such as stream buffers and water management of agricultural ditches.

In addition to providing an example of eutrophication in the Chesapeake region, our research in the Choptank also provides useful comparisons with the lagoonal systems that are the main subject of this volume. The Delmarva coastal bays are nearby (<80 km) on the eastern margin of the Delmarva Peninsula (Figure 7.1), and there are many useful parallels between the Choptank and the coastal bays. Here we summarize our research on the Choptank and attempt to relate our results to parallel observations in the basins and lagoonal systems of the Delmarva Peninsula.

The main focus of this chapter is an analysis of how anthropogenic activities on land (primarily fertilizer applications, animal waste production, and human waste disposal) influence nutrient export from land to the estuary, and how estuarine water quality responds to these nutrient inputs. We also explore attempts to reduce nutrient losses from land to estuary, and how these might be improved in the future. Most of the information is taken from the Choptank Basin, but we also compare these results with related measurements from other Delmarva basins, including the coastal bays.

7.2 STUDY SITE

The Choptank Basin (1756 km^2) and Estuary (280 km^2) are located on the eastern shore of Chesapeake Bay on the Delmarva Peninsula, a region of the mid-Atlantic coastal plain of North

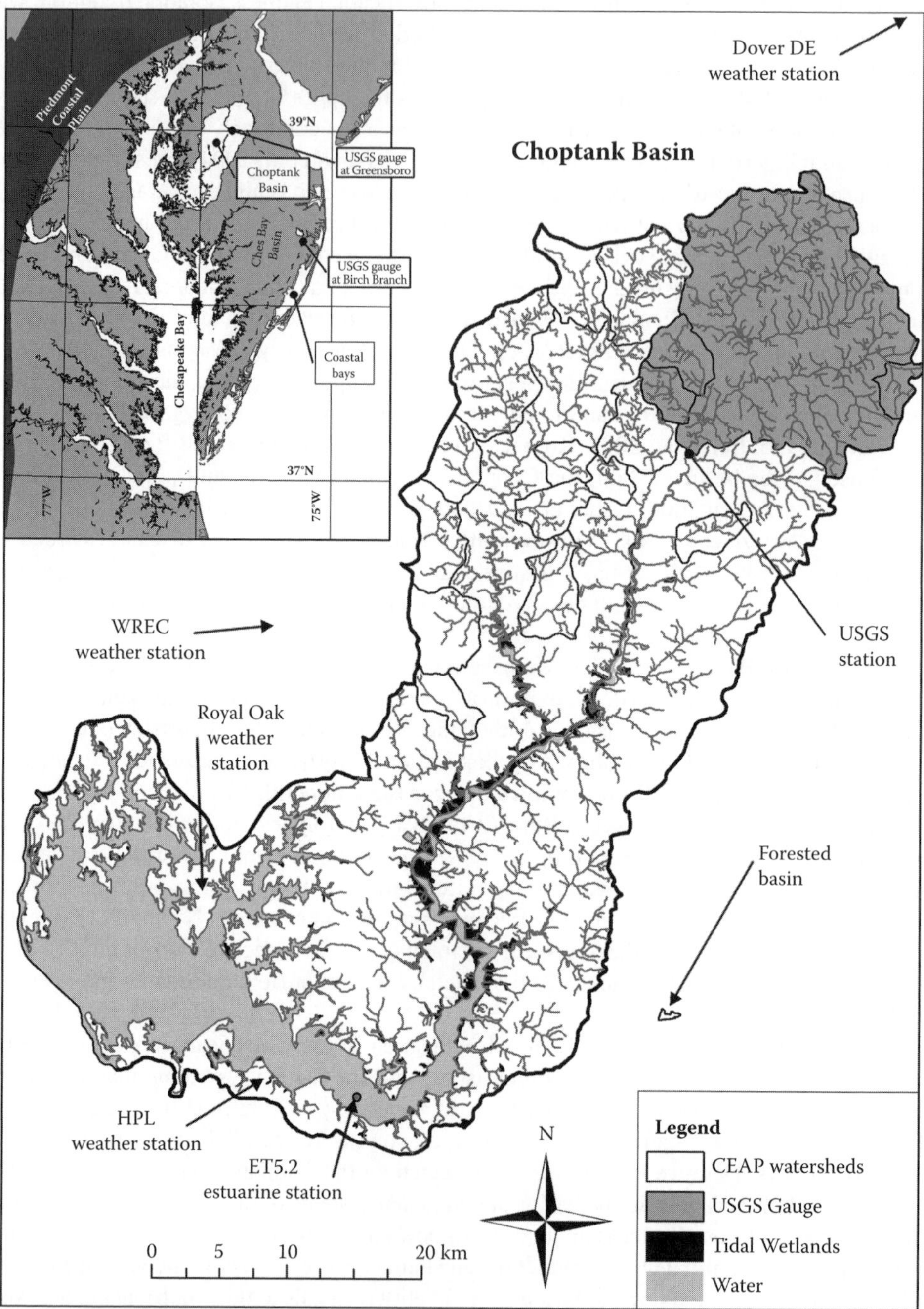

FIGURE 7.1 Inset: The location of the Choptank Basin and coastal bays on the Delmarva Peninsula in the mid-Atlantic region of North America. Basins of two USGS gauging stations are also shown. Main figure: Choptank Basin and Estuary showing locations of weather stations, the gauged basins in and near the Choptank, and the EPA Bay Program Monitoring station ET5.2. The forested basin lies just outside the Choptank in the adjacent Nanticoke Basin.

America (Figure 7.1). The Delmarva Peninsula has low topographic relief (<30 m elevation) and incised nontidal stream valleys with poorly drained, hydric soils. The stream valleys are typically forested in second- and higher-order streams, and ditched in first-order streams (Norton and Fisher 2000). The land use of the adjacent uplands is usually dominated by row crops, but increasingly row crops are being converted to low-density urban areas (Lee et al. 2001; Benitez and Fisher 2004, in review; Fisher et al. 2006a, 2006b for more details). Soils range from well-drained, oxic sandy loams to poorly drained clays that are usually hydric and hypoxic (Norton and Fisher 2000). Several hydrogeomorphic provinces have been described on Delmarva (Hamilton et al. 1993), consisting of the poorly drained uplands in the center of Delmarva, the well-drained lowlands, and the poorly drained lowlands close to Chesapeake Bay. Hydric soils and wetlands are more commonly found in the poorly drained uplands and lowlands than in the well-drained lowlands.

The Choptank Estuary is a broad, shallow, flooded coastal plain valley with salinities of 0 to 15. The tidal region of the former river valley is ~100 km in length, and the maximum depth in a horizontally restricted portion of the estuary is ~30 m. The former river thalweg forms a narrow channel of 0.5 to 2 km in width, with typical channel depths of 5 to 10 m. Broad shallows of 1 to 5 km in width flank both sides of the channel, with depths of 0 to 3 m; numerous shallow tidal creeks penetrate far inland, particularly in the lower Choptank (Figure 7.1). The upper half of the estuary is bordered by tidal wetlands of 10 to 500 m in width, and dense stands of emergent macrophytes grow annually to high densities (Traband 2003). Sediments in the estuary are largely soft, organic-rich muds, much of which was formerly dominated by oyster bars (Newell et al. 2004).

7.3 METHODS

7.3.1 Hydrology

Data on surface water discharge were obtained from two sources. Daily discharge data were downloaded from the water website of the USGS (www.water.usgs.gov) for the gauging station near Greensboro, Maryland (01491000) in the Choptank Basin (Lee et al. 2001; Fisher et al. 2006b). The gauged watershed is 293 km^2, lies within the poorly drained uplands of Delmarva, and has been gauged since 1948 by the USGS. The Greensboro watershed contains less agriculture, more forest, and more hydric soils than the Choptank Basin as a whole (Fisher et al. 1998). In 2003, the USDA Conservation Effects Assessment Project (CEAP) selected 17 smaller watersheds (1 to 50 km^2 in area) for study of the effects of agricultural management practices on water quality. All of these watersheds lie within the Choptank Basin (Figure 7.1) except the 100% forested reference watershed, which is in the adjacent Nanticoke Basin. Two watersheds are dominated by forests, and in the remainder agriculture is the dominant land use (Table 7.1). Water discharge from the CEAP watersheds was calculated from stage measurements collected at 30-min intervals using Solinst Leveloggers (model 3001 LT F15/M5) installed in anchored cinderblocks. The raw stage data were corrected for variations in barometric pressure using a separate Solinst Barologger (model 3001 F5/M1.5). Exponential rating curves relating stage (cm) to discharge (m^3 s^{-1}) were developed from direct discharge measurements using a Gurley pygmy meter at low flows and a StreamPro ADCP during storm flows ($r^2 = 0.92$ to 0.99; Koskelo 2008; Fisher et al. unpublished).

7.3.2 Surface Water Chemistry

Surface water chemistry was sampled as both base flow and storm flow. Base flow was collected monthly as single surface grab samples during 1986 to 1987 (Norton and Fisher 2000) and during 2003 to 2008 (Sutton 2006; Sutton et al. 2009b) when there had been no rain for 3 days. Storm flow was sampled selectively during seven to eight storms in six basins during 2006 to 2007 (Koskelo 2008); samples were collected hourly using ISCO model 6700 samplers (two samples composited

TABLE 7.1
Summary of Physical Characteristics in the 17 Gauged Watersheds of the Choptank Basin

	Area,	Land Use, % of Subbasin Area					Hydrologic Soil Class, % of Subbasin Area				%	# CAFOs	
Subbasin	km^2	Agriculture	Developed	Feedlots	Forest	CREP	A	B	C	D	Hydric	1990	2006
USGS (Greensboro)	293.3	48.9	4.6	0.4	45.7	NA	15.4	12.4	13.0	59.2	63.2	NA	NA
Kittys Corner	13.5	64.3	2.0	0.9	32.1	0.7	46.2	22.1	23.3	2.5	26.0	2	NA
Cordova	26.5	75.1	4.0	1.3	18.4	1.1	53.8	22.4	16.9	1.2	14.6	5	NA
Norwich Creek	24.5	69.6	1.8	0.4	23.1	5.2	11.7	35.7	21.7	26.5	32.6	1	NA
Blockston Branch	17.0	63.3	0.0	0.3	28.3	8.1	2.3	39.5	20.1	38.0	34.3	2	NA
Piney Branch	14.7	78.0	3.6	1.6	16.2	0.6	57.9	13.6	23.3	0.8	24.1	3	NA
Oakland	10.0	83.8	4.4	1.3	10.3	0.2	74.2	5.9	15.9	0.9	16.8	3	1
German Branch	51.4	67.8	0.2	0.9	26.8	4.2	0.7	36.6	11.7	51.0	45.2	9	NA
Beaverdam Ditch	23.3	62.3	0.8	0.0	32.2	4.6	0.7	27.9	5.3	66.1	64.1	0	NA
Long Marsh Ditch	40.5	54.1	0.4	0.5	40.8	4.2	13.7	22.3	9.5	54.5	63.7	2	NA
Broadway Branch	16.2	61.5	2.3	0.7	35.1	0.4	29.0	12.5	16.9	41.6	58.4	3	NA
Oldtown Branch	11.6	54.3	8.4	1.2	32.3	3.7	29.5	9.6	27.5	32.3	59.9	3	NA
Spring Branch	12.2	74.3	0.3	0.3	21.6	3.5	59.4	8.3	26.8	5.2	32.0	1	NA
North Forge Branch	25.0	59.6	2.1	0.2	30.7	7.4	31.0	16.8	29.8	21.0	51.2	0	NA
South Forge Branch	8.5	62.9	5.3	1.4	28.2	2.2	45.7	11.7	34.3	3.9	38.2	1	NA
Downes	23.4	76.8	5.1	1.7	15.6	0.8	66.6	11.9	19.0	0.4	19.4	6	NA
Willow Grove	8.59	24.0	0.6	0.0	75.4	0.0	0.0	0.5	2.3	97.2	97.2	NA	NA
Marshy Hope Forested	0.7	0.0	0.0	0.0	100.00	0.0	50.72	49.28	0.0	0.0	54.4	0	0

per bottle) over 48 hours. Sampling tubes for the ISCOs ran to a perforated tube mounted on top of the anchored cinderblocks housing the data loggers in mid-stream.

Samples were transported to the laboratory on the day of collection of base flow samples and within 24 h after the end of a storm event for storm samples. Particulates were filtered at low vacuum on tared Whatman GF/F filters, dried at 50°C, and reweighed to 0.01 mg for computation of total suspended solids (TSS, mg L^{-1}). Controls to correct for filter weight loss during handling (three per storm event) were performed using filtrate. The filtrate was also analyzed for NH_4^+, $NO_2^- + NO_3^-$, and PO_4^{3-} using automated colorimetric methods in the HPL Analytical Services Laboratory and in the USDA Agricultural Research Service Analytical Laboratory. Colorimetric nutrient analyses in both labs have been successfully compared using split samples and standards (McConnell and McCarty 2005). Unfiltered samples were autoclaved with persulfate oxidizing reagents to convert all forms of N and P to NO_3^- and PO_4^{3-} (Valderrama 1981). Following digestion, autoclaved samples were neutralized to pH 7 and analyzed for NO_3^- and PO_4^{3-} as described above.

7.3.3 Drainage Control Structures

Many areas of Delmarva have been ditched during the last 200 years to improve the drainage for agricultural or residential purposes (Bell 2000). Many of these ditched areas were formerly wooded wetlands on hydric soils, which were likely to have been very strong landscape sinks for anthropogenic nutrients (see below). These ditched areas now under agricultural land use are large sources of nutrients, often discharging water with N concentrations of 700 to 1000 μM (10 to 15 mg N L^{-1}) during base flow, primarily as NO_3^-. Concentrations of P during storms are 15 to 30 μM (0.5 to 1.0 mg P L^{-1}), primarily in the form of particulate P and PO_4^{3-} (Hively et al. in review).

In an attempt to restore some of the former wetland function in these drained wetlands, the USDA and Maryland Department of Agriculture are installing experimental drainage control structures that enable the regulation of water level within the drainage ditches via drop-in boards. We monitored both surface waters and adjacent groundwaters (see below) of a 1-km long ditch on a farm in Caroline County, Maryland, that had a drainage control structure installed in the middle of the ditch. Groundwater levels were ~2 m below the soil surface in the original unflooded section of the ditch, but damming of surface water by the drainage control raised groundwater to ~1 m below the soil surface in soils adjacent to the flooded section of the ditch. The goals of raising the water levels in the flooded section of the ditch were to improve water availability to crops during the summer months as well as to increase the rate of denitrification of agricultural nitrate in groundwater by increasing the exposure of nitrate-rich groundwater to C-rich surface soils (for details, see Hively et al. in review). We monitored the surface chemistry of this ditch at both the end of the flooded and unflooded sections at monthly intervals to assess the impact of the drainage control structure on nitrate in ditch waters. Chemistry methods were the same as those described above for the watershed sampling.

7.3.4 Groundwater Piezometers

Groundwaters were monitored by installation of piezometers which sampled defined depth strata in ditch and wetland studies. We augured from the soil surface to depths of 1 to 4 m, and in some locations, piezometers were installed at multiple depths within a 2- to 3-m radius in order to sample vertical differences in groundwater chemistry in the top of the unconfined aquifer. We pumped piezometers dry 24 h prior to sampling to ensure fresh groundwater, and we isolated the inflowing groundwaters from exposure to the atmosphere using a sphere with a diameter of ~5 cm floating within the piezometer. After removing the float, a 5-cm Teflon bailer was lowered as deeply as possible to obtain groundwater at the bottom of the piezometer with the least exposure to the atmosphere. A Teflon stopcock and tubing were inserted into the bottom of the bailer to transfer groundwater samples from the bailer to 15-mL ground glass stoppered tubes. The Teflon tubing reached the bottom of the glass tube, and the tube was filled with one volume of overflow to reduce air exposure and eliminate bubbles. After stoppering, the tubes were stored in ice until analyses the next day to prevent bubble formation. An additional sample was transferred to a plastic bottle for colorimetric chemistry (NH_4^+, NO_3^-, PO_4^{3-}) or for electrode measurements (pH, conductivity).

Most piezometers were equipped with the same model Solinst data loggers used for stream stage. Groundwater depth and temperature in the piezometers were recorded at 30-min intervals, and loggers were downloaded at 3- to 6-month intervals. The pressure data were corrected for variations in atmospheric pressure, as described above for the stream loggers. Groundwater depth below the soil surface was computed from variations in water depth within the piezometers and the fixed depth from the soil surface to the bottom of the piezometer.

7.3.5 Excess N_2, O_2, N_2O, and CH_4

Denitrification is an important process in groundwater which occurs under low or zero oxygen conditions, resulting in the conversion of NO_3^- into N_2 and N_2O gases. Quadruplicate ground glass stoppered tubes were used for analysis of dissolved gases (N_2, O_2, and Ar) using a Pfeiffer Vacuum model QMG422 quadrupole mass spectrometer fitted with a membrane inlet (MIMS; Kana et al. 1994, 1998; Kana and Weiss 2002). Concentrations of Ar, N_2, and O_2 in the samples (μM) were computed using sample signals (μamps) and air calibration factors (μamps/μM). Ar concentrations were used as a tracer of exchange with the atmosphere prior to and during infiltration of rainwater to groundwater, and an inverted solubility curve was used to estimate the effective recharge temperature (Figure 7.2a). We have compared these effective groundwater recharge temperatures to *in situ* temperatures observed at the time of collection (Figure 7.2b), and the cooler recharge temperatures indicate that most recharge occurs in fall, winter, and spring at relatively low temperatures (Figure 7.3), which agrees with direct observations of infiltration (e.g., Staver and Brinsfield 1998; Figure 7.4).

Equilibrium O_2 and N_2 concentrations were computed from their respective solubility curves in freshwater (Colt 1984) and the Ar-based recharge temperature. The equilibrium concentrations were combined with the observed N_2 and O_2 concentrations measured in the MIMS to compute excess N_2-N (μM) and % saturation of O_2 as follows:

$$\text{excess } N_2\text{-N} = 2 * (\text{observed } [N_2] - \text{equilibrium } [N_2]) \tag{7.1}$$

$$\%\text{ saturated } O_2 = 100 * (\text{observed } [O_2]/\text{equilibrium } [O_2]) \tag{7.2}$$

Excess N_2 was expressed per unit N for comparison with NO_3-N and usually ranges from near-zero to ~500 μM N_2-N (in excess of the equilibrium N_2-N of 1100 to 1500 μM, depending on recharge temperature). Excess N_2-N represents the amount of NO_3^- that has been converted to N_2 gas that is retained in the groundwater. In contrast to supersaturated N_2 in groundwater, O_2 is usually undersaturated in groundwater and varies from near-zero to 90% saturation.

The respiratory gases N_2O and CH_4 were also measured using the same groundwater sampling protocol described above. These are end-products of soil metabolism (nitrification, denitrification, and methanogenesis), and we use their concentrations to infer processes in the soil upslope of the piezometer sampling location. Aliquots of the groundwater samples were removed by syringe from the ground glass stoppered tubes and injected into 12-mL borosilicate Labco Exetainer® vials previously purged with N_2 gas. The vials were manually shaken for 2 min to ensure full equilibration of the N_2 headspace and water. A sample of each respective headspace was injected into both a Shimadzu GC-14B equipped with an electron capture detector (ECD) with a HayeSep Q column for N_2O analysis and a separate Shimadzu GC-8A equipped with a flame ionization detector (FID) with a HayeSep A column for CH_4 analysis. The concentration of each gas was injected at a volume

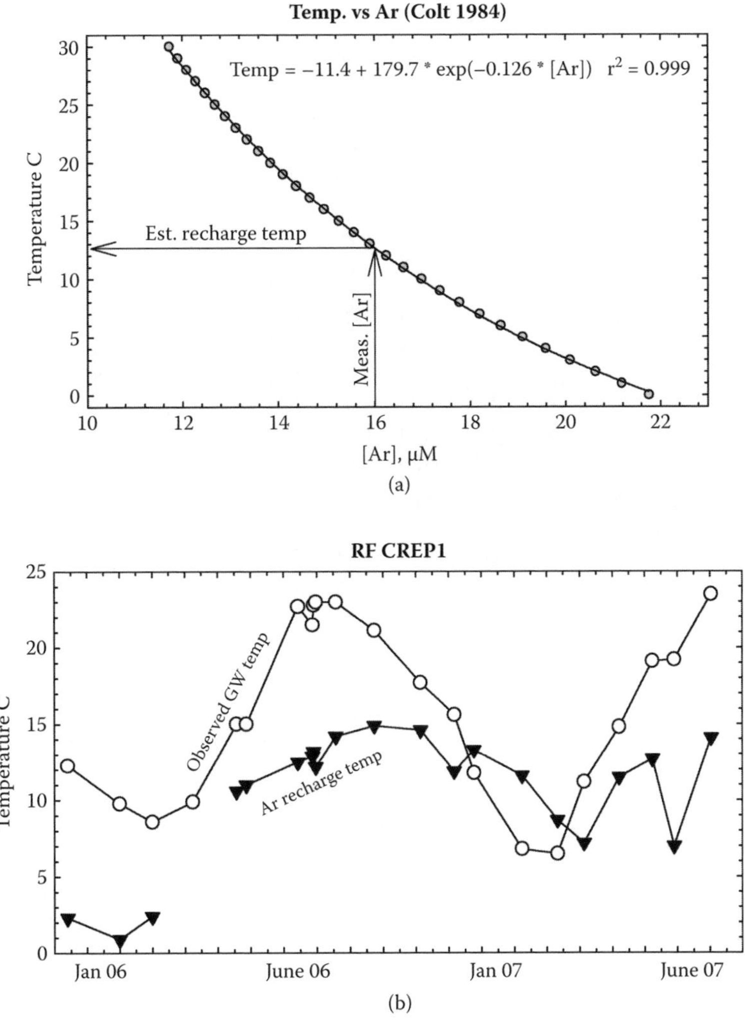

FIGURE 7.2 (a) Inverted Ar solubility curve used to estimate groundwater recharge temperatures from observed Ar concentrations. (b) Comparison of Ar recharge temperatures and observed groundwater temperatures in the top of the unconfined aquifer sampled by a piezometer within a forested buffer between an agricultural field and tidal waters. Groundwater temperature cycles annually due to heat fluxes through the soil, but the Ar recharge temperatures exhibit a damped seasonal cycle primarily reflecting the recharge history integrated over several months during cooler seasons.

that lies within the linear range of each instrument, and sample concentrations were corrected for the influences of pressure, temperature, and solubility.

7.3.6 Estuarine Water Quality Data

Data for Choptank station ET5.2 (Figure 7.1) were downloaded from the Chesapeake Information Management System (CIMS, www.chesapeakebay.net/data/index.htm). The data consist of monthly vertical profiles of temperature (°C), salinity, oxygen (mg O_2 L^{-1}), chlorophyll *a* (chl*a*, μg L^{-1}), and nutrients (NH_4^+, NO_3^-, TN, PO_4^{3-}, TP) collected by Maryland Department of Natural Resources as part of the EPA Chesapeake Bay Water Quality Monitoring Program. Here we have used the data on

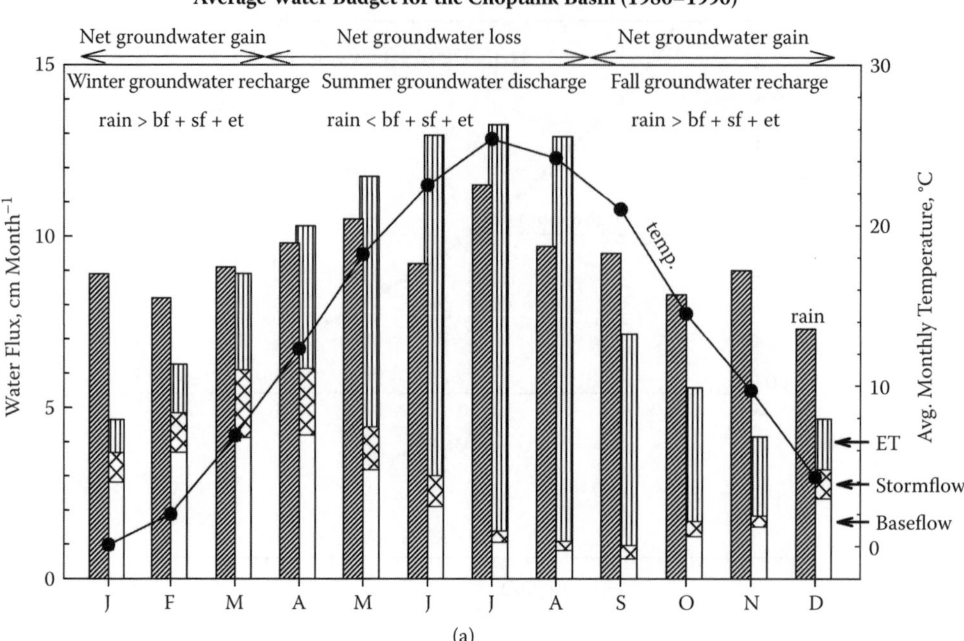

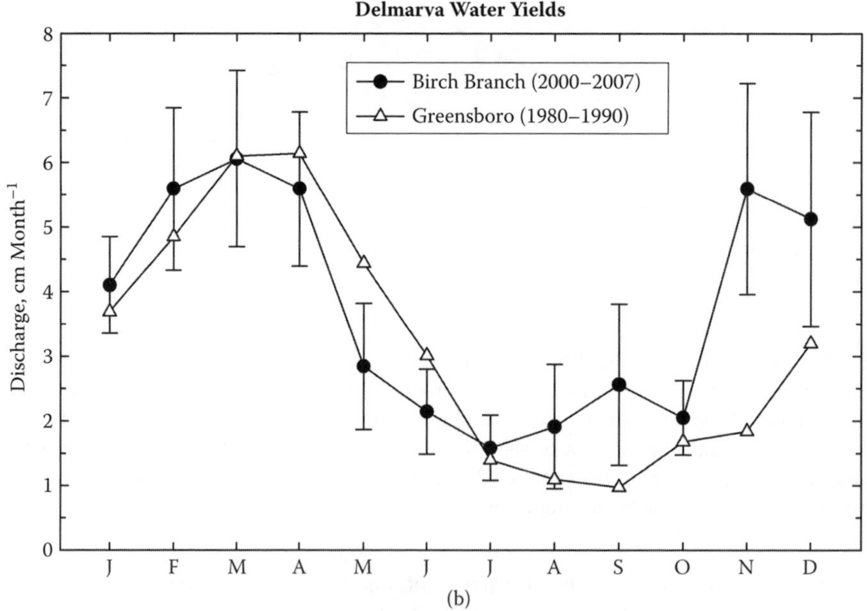

FIGURE 7.3 (a) Regional, long-term precipitation and discharge records (cm month^{-1}) for the Delmarva Peninsula based on daily data collected at NADP site MD13 at WREC, the NWS observer station at Royal Oak, Maryland, the automated NWS station at Dover, Delaware, and the Horn Point Laboratory weather station (see Figure 7.1). Base flow, storm flow, and evapotranspiration for the Choptank River at Greensboro, Maryland (USGS sta. 01491000) are from Lee et al. (2001). (b) Comparison of long-term monthly water yields (base + storm flow) at Greensboro in the Choptank River and at Birch Branch in the St. Martin Basin of the Delmarva coastal bays (USGS sta. 0148471320). Water yields are discharge (m^3 month^{-1}) normalized to basin area (m^2).

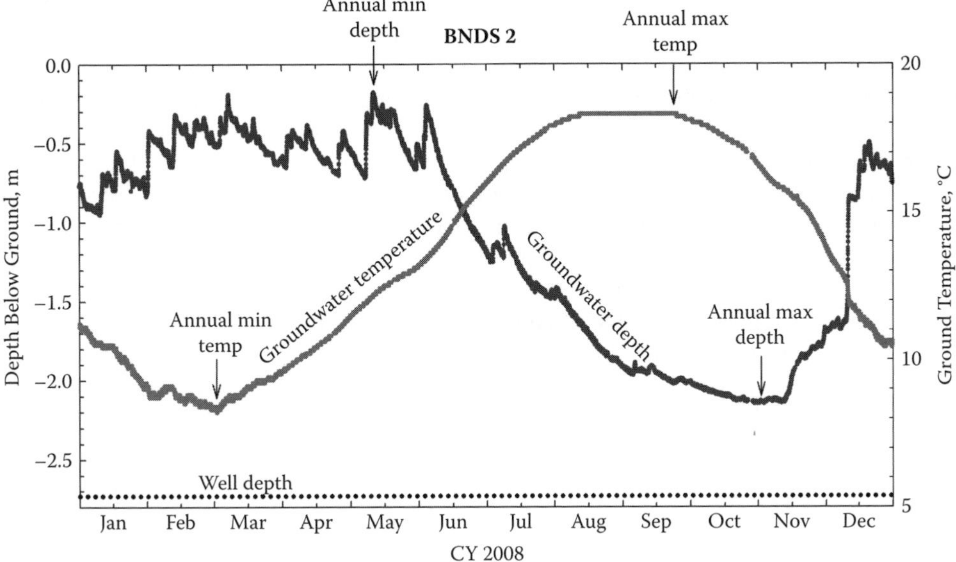

FIGURE 7.4 Annual summary of groundwater temperature and water depth below ground surface (water table depth) in a piezometer sampling the top of the surficial aquifer at 2.8 m depth below ground at a location within the basin of the USGS gauge on the Choptank River at Greensboro (see Figure 7.1). Multiple recharge events occurred in cooler months (January to May and November to December in 2008), with fewer and smaller recharge events in summer (Figure 7.3a). The overall pattern was driven by infiltration events in cooler periods and an excess of ET over recharge in summer (Figure 7.3a). Groundwater temperature exhibited a similar but inverted cycle primarily due to heat conduction through the ground from the overlying air.

annual average surface chl*a* and summer (June to August) bottom dissolved oxygen (deepest depth reported) to quantify interannual trends at this station.

7.4 RESULTS

7.4.1 Regional Hydrology

The hydrology of the region is controlled by rainfall, temperature, evapotranspiration, topography, and soil drainage properties. Rainfall in the mid-Atlantic region is relatively uniform throughout the year, and Figure 7.3a shows a climatic regional summary for 1980 to 1990. Despite the relatively uniform precipitation throughout the year (diagonally striped bars, 8 to 13 cm mo^{-1}) with a slight maximum in July, a large seasonal variation in discharge occurs (1 to 6 cm mo^{-1}) with a minimum in September due to a large seasonal variation in air temperature (0°C to 26°C); these, in turn, drive a parallel pattern of evapotranspiration (ET, 1 to 11 cm mo^{-1}) with a maximum in July, diverting rainfall back to the atmosphere as water vapor rather than discharging it as stream water (base flow + storm flow). In the warmer months with highest insolation (April to August), ET plus stream discharge exceeds rainfall, resulting in a net water loss and falling groundwater levels (Figure 7.4). As a result, base flow is diminished (Figure 7.3a, open bars). Because soils are very dry in summer and absorb much of the precipitation, storm flows are also reduced compared to cooler months with less insolation, plant activity, and ET (Figure 7.3a, hatched bars).

Long-term mean stream discharge for the Choptank River and Birch Branch in the drainage basin of the coastal lagoons (Figure 7.1) exhibits a similar seasonal pattern. Each has a maximum in spring and a minimum in late summer or early fall (Figure 7.3b), which is consistent with the groundwater levels observed regionally (Figure 7.4). However, the Choptank has later spring discharge and recharges more slowly in the fall (Figure 7.3b), an effect of the more poorly drained soils in the gauged basin of the Choptank, which lies largely in the poorly drained uplands of the Delmarva Peninsula (Lee et al. 2001). Under very dry conditions (e.g., summer of 2007), surface discharges from the coastal lagoon drainages such as Birch Branch virtually dry up due to the small watershed area and moderately well-drained soils (Beckert et al. in review). The stream beds become a series of interconnected ponds with little flow between them, although it is likely that groundwater flow continues below the surface.

At shorter time scales, moderately sized rain storms induce discharge events of 2- to 4-day durations (Figure 7.5a). Koskelo (2008) separated base flow (groundwater-based discharge) from storm flow (event-based discharge) in six Choptank watersheds using a new approach incorporating discharge patterns and rainfall (Koskelo et al. in review). Using this approach, base flow was found to be significantly and inversely related to hydric soils in long-term data due to surface ponding and root zone retention of rainfall and subsequent ET (Koskelo 2008). Storm flow was not related to

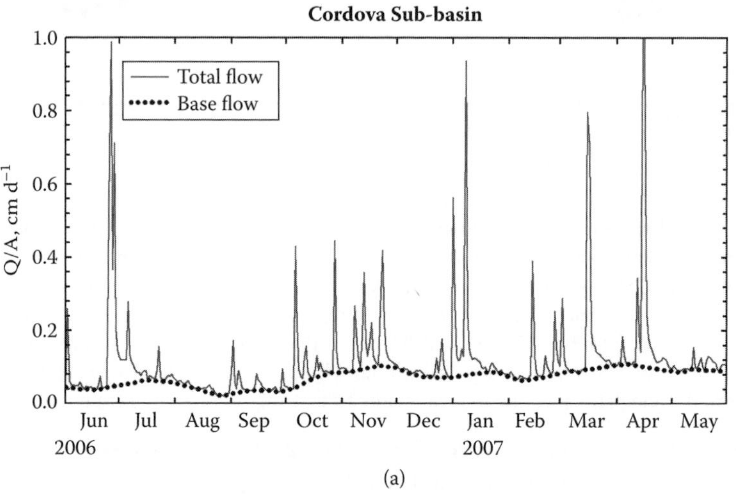

(a)

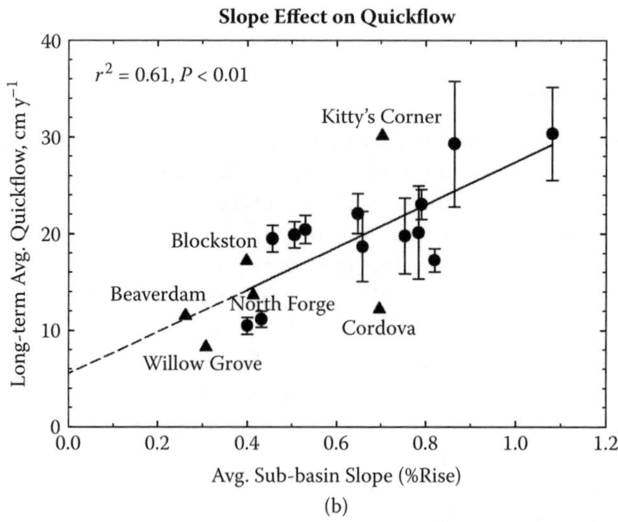

(b)

FIGURE 7.5 (a) Separation of quick flow and base flow in the Cordova watershed. Q = total discharge (m^3 d^{-1}); A = watershed area (m^2). (b) The effect of basin slope on the annual average volume of quickflow from storm events. Steeper average slopes in a watershed generate greater quickflow. (Data from Koskelo, A.I., Ms. Thesis, University of Maryland, College Park, 2008.)

soil drainage properties, but increased significantly with average surface topographic slope of the watershed (Figure 7.5b).

7.4.2 Water Quality Trends: Nontidal Monitoring Stations

Water quality at the USGS gauging station at Greensboro, Maryland, has been declining for many decades (Figure 7.1). In primarily base flow sampling, NO_3^- has been increasing since the 1960s when water quality monitoring began, and total N (TN) has also increased since 1975 (Figure 7.6a). NO_3^- in base flow represents about 70% of the TN and is primarily derived from application of fertilizers on agricultural fields, which passes to streams via groundwater (Hamilton et al. 1993). As a

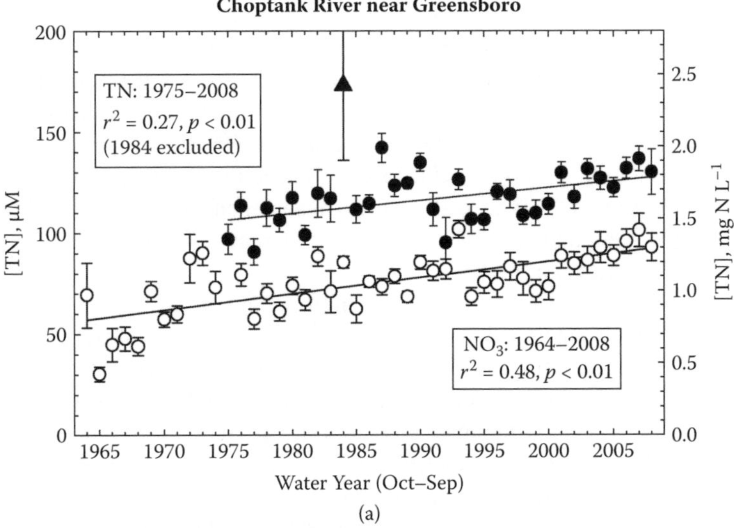

(a)

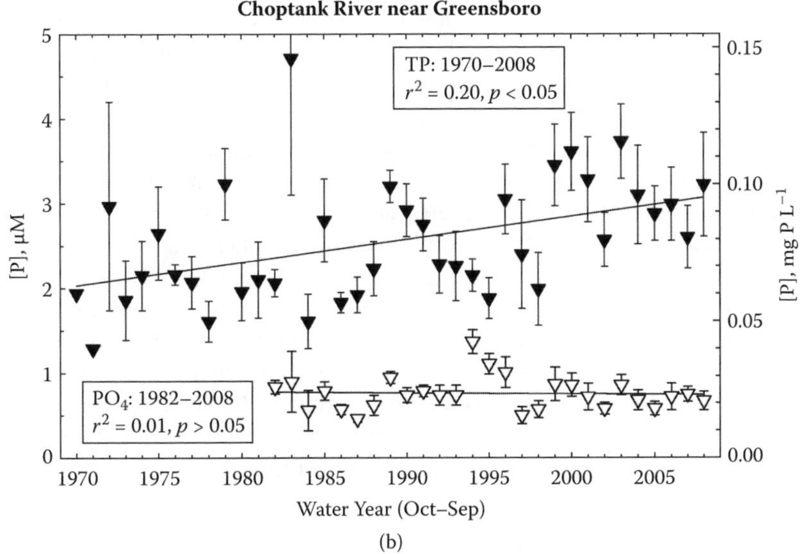

(b)

FIGURE 7.6 Changes in annual average concentrations of (a) total N (TN) and nitrate (NO_3^-) and (b) total P (TP) and phosphate (PO_4^{3-}) over the last 45 years at the USGS gauging station on the Choptank River near Greensboro, Maryland (01491000). (Data from USGS.)

result, both annual average TN and NO_3^- in Choptank streams are strongly correlated with percent agriculture in their watersheds (Fisher et al. 2006b; Figure 7.8).

Annual average total P (TP) concentrations (primarily base flow) have also significantly increased since monitoring began in 1970 (Figure 7.6b). Whereas NO_3^- is the major component of TN, phosphate (PO_4^{3-}) is a smaller component of TP and has not increased significantly over time (Figure 7.6b). In both panels of Figure 7.6, which illustrates trends in N and P, the data for 1984 were excluded from the regressions because of a small number of samples in that water year which included a single storm sample with very high concentrations. Overall, the data in this figure indicate that base flow N and P concentrations in the gauged portion of the Choptank Basin (Figure 7.1) have been increasing for several decades. There is significant interannual variation in water discharge, but no significant trend; therefore, trends in N and P fluxes (concentrations × water discharge) are primarily determined by changes in concentrations.

Similar trends have been observed for base flow in other watersheds within the Choptank Basin. Sutton et al. (2009b) compared the concentrations at 15 Choptank CEAP watersheds dominated by agriculture (Figure 7.1). These watersheds were sampled in 1985 to 1986 and in 2003 to 2006 (Figure 7.7). While lacking the detailed time series available for Greensboro (Figure 7.6), the data for these watersheds (Figure 7.7) show no evidence of systematic decreases in base flow N or P concentrations between 1986 and 2003 to 2006 due to management actions within these watersheds. In fact, one watershed (Oakland) exhibited significant increases in TN and NO_3^- concentrations, similar to Greensboro (Figure 7.6).

One of the agriculturally dominated watersheds in the Choptank Basin shown in Figure 7.7 is the German (or Jarmin) Branch watershed, the site of the first targeted watershed program in Maryland during 1990 to 1996 (Primrose et al. 1997). Within this watershed there was nearly 100% application of the agricultural BMPs of soil conservation and water quality plans, nutrient management plans, and conservation tillage, with small increases in riparian buffers and winter cover crops. Continuous quantitative data on the management practices are not available except for 1990 to 1995, and fertilizer and manure applications and other agricultural practices probably varied during 1996 to 2006. With this caveat, 10 years after the targeted watershed project (more than the median groundwater residence time of 8 years in this basin), there were no significant changes in base flow N during this 10-year interval (Sutton et al. 2009a). However, base flow N concentrations at German Branch did not increase, unlike the Choptank River at Greensboro (Figure 7.6a). Furthermore, TP in base flow at German Branch decreased significantly (~50%), again in contrast with increasing P at Greensboro (Figure 7.6b). P is difficult to sample quantitatively because it increases dramatically during short-term storm events with large discharges (Fisher et al. 2006b). More than 80% of the annual P export from Choptank watersheds occurs during brief storm events (Koskelo 2008), making base flow measurements of P unrepresentative of annual export. Furthermore, storm flow is rarely sampled adequately in monitoring programs, and the data in Figures 7.6 and 7.7 are essentially base flow data. Nevertheless, the stable N concentrations and decreasing P concentrations in base flow at the German Branch watershed compared to the continuous increases at Greensboro (Figure 7.6) indicate that intensive management in a targeted watershed program can have a detectable effect. Since changes in base flow chemistry were relatively small, insufficient applications of effective practices (winter cover crops, stream buffers) or relatively ineffective management actions (nutrient management plans, which are not monitored, nor have they ever been tested) resulted in small water quality effects detectable only after a decade of time.

7.4.3 Water Quality Drivers of Change

There are two major drivers of the poor water quality trends reported above (i.e., agriculture and sewage disposal). Agricultural lands are the primary driver of water quality in nontidal streams of the Choptank Basin and the Delmarva coastal bays due to application of fertilizers and distribution of manure from animal feeding operations (Figure 7.8). As a result, nontidal streams in coastal

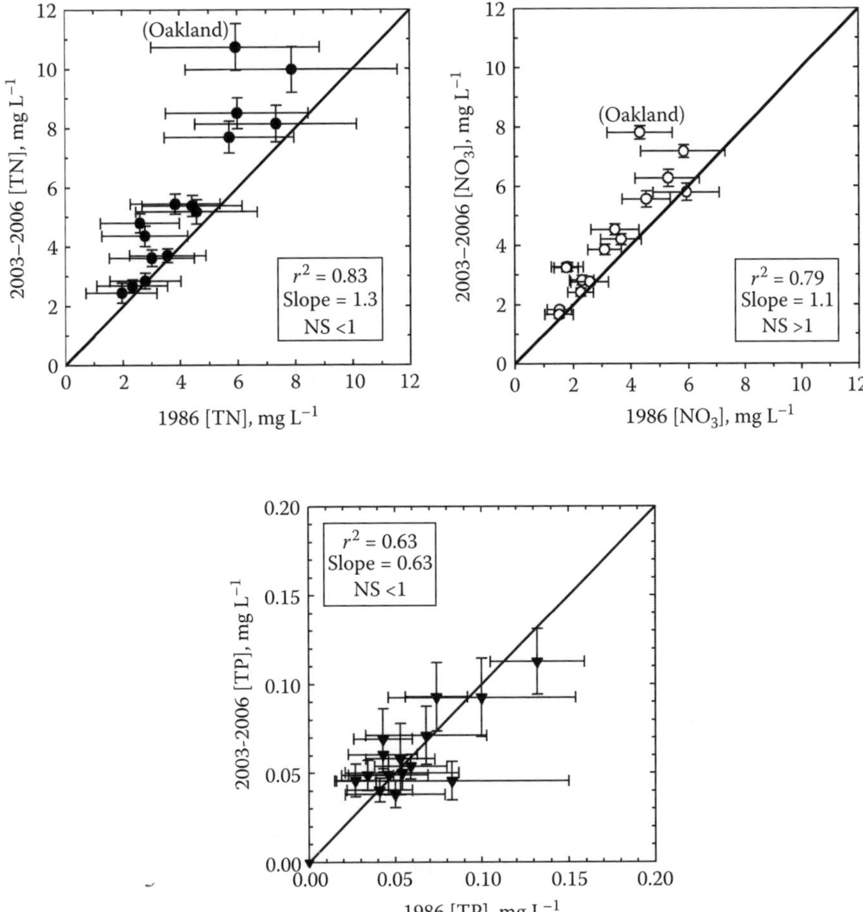

FIGURE 7.7 Comparison of multi-year averages of N and P concentrations in stream waters (base flow) of 15 agriculturally dominated watersheds in the Choptank Basin. Overall, there were no significant changes between the two time periods (all slopes not significantly different from 1). However, total N (TN) and nitrate (NO_3^-) increased significantly in base flow of the Oakland watershed (see Table 7.1). (Data from Sutton et al., 2009a.)

plain areas, including Delmarva, carry large quantities of agricultural NO_3^- (Figure 7.8a), and the relationship between NO_3^- concentration and percent agriculture is nonlinear due to the substitution of croplands for landscape sinks for NO_3^- such as riparian forests and wetlands as the percent agricultural land use increases beyond 50%. Loss of the remaining wetland and riparian sites results in both new agricultural source areas for NO_3^- and losses of denitrifying areas, accelerating the increase of NO_3^- with additional agricultural land use. In the Maryland coastal bays (St. Martin Basin, Figure 7.8b), the percent land use of animal feeding operations, the area covered by the footprint of chicken houses within each watershed, was the most important determinant of TN and NO_3 concentrations in nontidal streams. Despite the small area of the actual structure for the feeding operation, the distribution of manure from the houses to nearby fields has a significant impact on N concentrations in streams. A similar significant relationship between base flow PO_4 levels and density of chicken houses (number per km^{-2}) was also found among the Choptank CEAP watersheds ($r^2 = 0.47$, $p < 0.01$; Koskelo 2008).

Land use effects on P concentrations in streams are more difficult to quantify. As described above, most P is mobilized from watersheds during storms, which are usually vastly undersampled

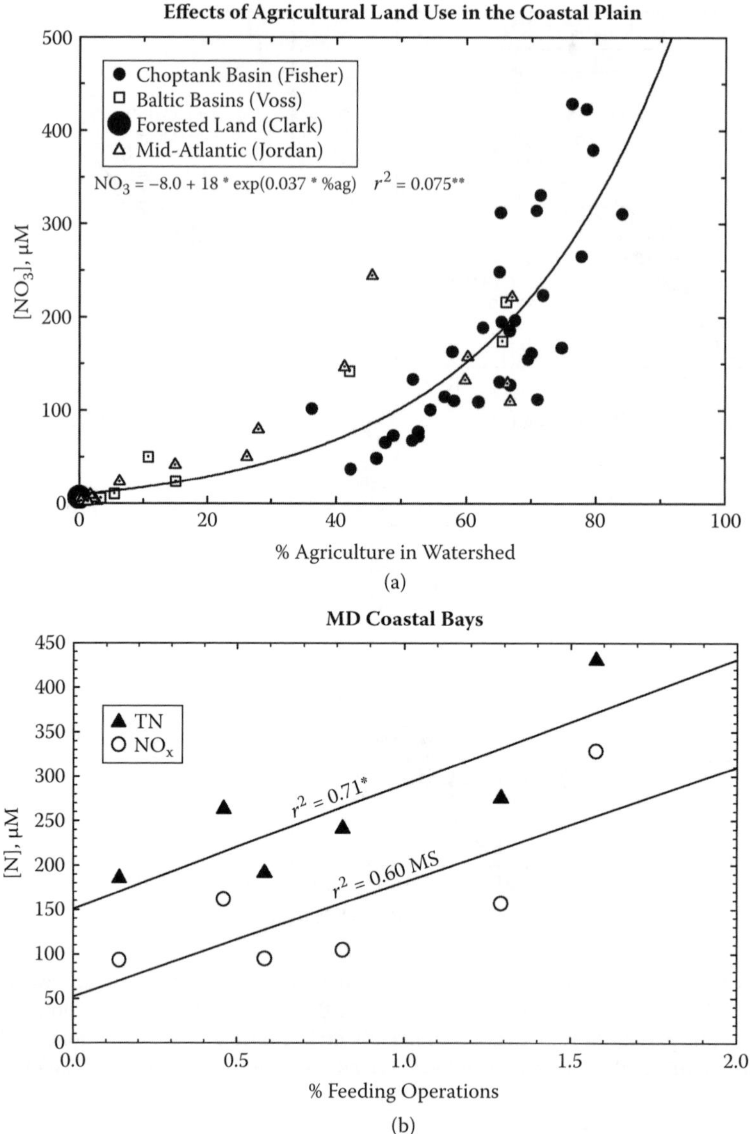

FIGURE 7.8 (a) Effects of percent agriculture (cropland) on nitrate concentrations [NO_3^-] in coastal plain watersheds. The exponential curve was forced through the extensive summary by Clark et al. (2000) on forested lands (= 0% agriculture). (b) Effects of animal feeding operations on average stream N in the St. Martin Basin in the Maryland coastal bays (Beckert 2008). The correlation with TN is significant, whereas the nitrate correlation is marginally significant ($p = 0.07$ and not significant for ammonium ($p > 0.10$).

in watershed studies (Koskelo 2008). However, Stone et al. (in review) has documented land use effects on N and P in both storm and base flows in the Little Blackwater watershed (Figure 7.9), which lies just south of the Choptank Basin (Figure 7.1). In this example from 2005 (Figure 7.9), PO_4^{3-} concentrations in storm flows were up to four times higher than in base flows from areas dominated by agriculture; in contrast, areas dominated by forested wetlands showed no such enhancement of PO_4 in storm flows. Fisher et al. (2006b), Sutton (2006), and Koskelo (2008) also showed P enhancement in storm flows >10 times higher than base flow concentrations in agriculturally dominated watersheds of the upper Choptank Basin.

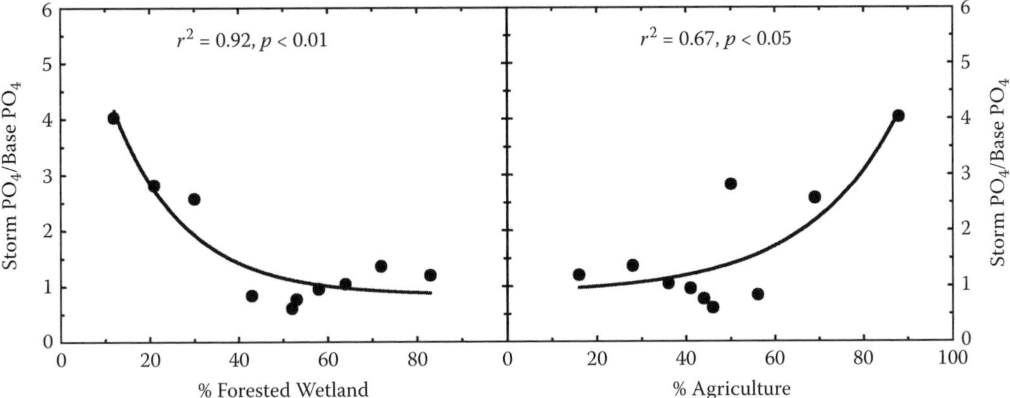

FIGURE 7.9 The ratio of storm phosphate (PO_4^{3-}) and base flow PO_4^{3-} in streams of the Little Blackwater Basin, just south of the Choptank Basin (Figure 7.1). In watersheds with >20% forested wetlands, there was no change in PO_4 between base flow and storm flow (storm PO_4^{3-}/base PO_4^{3-} = 1); however, as the forested wetlands decreased to <20% and percent agriculture increased beyond 50% land use, the amount of PO_4^{3-} in storm flow rose to four times the base flow value. (Data from Stone et al. (in review).)

Wastewater discharge is the second major driver of poor water quality. There are 10 wastewater treatment plants in the Choptank Basin, which discharge an average of 4.0 kg N and 1.2 kg P person^{-1} yr^{-1} primarily to tidal, estuarine waters (Figure 7.10). While the inputs of wastewater N are a small fraction (4%) of the Choptank nutrient budget due to the importance of agricultural nitrate (Figure 7.8), wastewater P is a large fraction (49%) of the P budget (Lee et al. 2001). The Choptank Basin has been slowly urbanizing (Benitez 2002), which has resulted in increases in the size of the small towns, the service populations, and wastewater N and P inputs to the estuary. Additions of tertiary treatment at several of the larger plants since 1990 have decreased N and P concentrations in some discharges, but larger wastewater volumes from increasing service populations have offset the effect on fluxes (Fisher unpublished).

7.4.4 Agricultural N and P Reduction

Due to the importance of agriculture in determining stream N concentrations (Figure 7.8), many BMPs have been implemented in the Choptank Basin to reduce N and P losses from croplands. The BMPs include: (1) nutrient management plans; (2) conservation tillage; (3) drainage control structures; (4) grassed or forested stream buffers (e.g., Conservation Reserve Enhancement Program or CREP); and (5) winter cover crops (Staver and Brinsfield 1998). Nutrient management plans are documents submitted by farmers to Maryland Department of Agriculture providing details on how their farming operations will avoid nutrient losses to surface water and groundwater. Conservation tillage is the use of herbicides to control weeds in place of plowing to avoid soil disturbance and reduce soil erosion. Drainage control structures are weir-like devices used to control water levels in cropland drainage ditches originally dug to improve the drainage of wet areas. Stream buffers are vegetated areas that border flowing surface waters (seasonally mowed grasses or managed forests) in which no fertilization takes place, but denitrification, plant uptake of N and P, and particle trapping do occur. Winter cover crops are cold-season grasses, such as rye or winter wheat, which stabilize the soil, reduce erosion, and provide N and P uptake during the fall through spring seasons. The function of these BMPs is (1) to physically trap soil and P during overland flow events (conservation tillage, cover crops, buffers); (2) to induce cool season plant N uptake to reduce nitrate leaching to groundwater during infiltration events (cover crops); and (3) to encourage denitrification, primarily

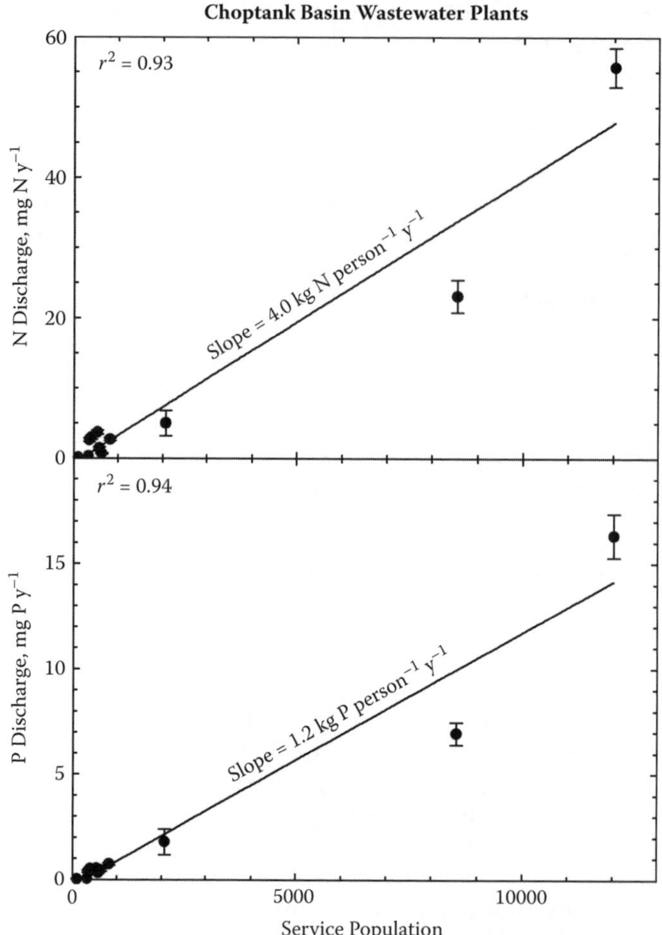

FIGURE 7.10 Total N and Total P discharge by the 10 NPDES wastewater treatment plants in the Choptank Basin. As the population serviced by the plants increases, discharges of N and P increase by approximately 4.0 kg N and 1.2 kg P person^{-1} yr^{-1}. (Data from Lee et al., 2001.)

in groundwater as it flows through a C-rich and O_2-poor environment (drainage structures, buffers). Lowrance et al. (1997) has reviewed riparian buffers in the Chesapeake region.

The concept behind nutrient management plans is to consider the nonpoint source nutrient contributions of individual farms. These plans have six components involving disposal and storage of manure, field management to reduce nutrient losses, and farm management documentation. In the plan, farmers are asked to demonstrate attempts to minimize environmental effects of livestock and row crop agriculture, and annual plans are now mandatory in the state of Maryland. However, the effectiveness of these plans has never been demonstrated at any scale, and there is no monitoring to validate whether plans are being followed.

Conservation tillage has two components ("low till" and "no-till"). Both substitute varying amounts of herbicide control of weeds for mechanical tillage, and the major goals are to reduce energy costs, stabilize soils, and reduce soil erosion. In a paired watershed study, Staver et al. (1994) showed that soil erosion was significantly less in the watershed with conservation tillage compared to the watershed with conventional tillage. However, despite the reduced soil erosion in the watershed with conservation tillage, leaching and transport of crop P increased during overland flows because of the P-rich crop litter concentrated at the soil surface. Therefore, conservation tillage appears to reduce soil erosion while increasing P losses from crop fields.

Drainage control structures are being experimentally applied to farm ditches originally cut through wetlands to enable adequate drainage for conversion to agriculture. The soils are usually hydric (poorly drained, with high water content during the cool season), and the drainage ditches often export water with millimolar concentrations of NO_3^- (700 to 1000 μM or 10 to 15 mg NO_3^--N L^{-1}) from the adjacent agricultural fields. This value is consistent with the extrapolated right y intercept (720 μM or 10.1 mg NO_3^--N L^{-1}) for the regression line at 100% agriculture in Figure 7.8A. Drainage control structures are used to raise the water levels close to the root zone both to store water in groundwater for summer crop use as well as to induce denitrification in C-rich upper soil horizons.

We have strong evidence of denitrification in groundwater entering a drainage controlled ditch (Figure 7.11). In this example from the flooded section of the ditch above the drainage control structure, excess N_2-N accumulated in groundwater near the ditch at concentrations of 200 to 400 μM above that of atmospheric N_2-N equilibrium (~1200 μM). NO_3^- shows strong seasonal variations, varying from 0 to 700 μM with groundwater levels. During winter and spring, when groundwater levels are highest, NO_3^- is typically 400 to 700 μM; during late summer and fall, when groundwater gradients relax (Figure 7.4), NO_3^- is much lower (0 to 200 μM). A particularly dry summer in 2007 caused low groundwater levels and low concentrations of NO_3^-, which did not recover until March 2008. Note that less excess N_2-N accumulated in groundwater (~350 μM) than the net disappearance of NO_3^- (~700 μM). This appears to result from diffusive loss of excess N_2-N from the supersaturated groundwater into the vadose zone above and ultimately to the atmosphere (Fox et al. 2009). In contrast with the flooded section of the ditch upstream of the drainage control structure, the unflooded control section of the ditch downstream of the drainage control structure had groundwater NO_3^- levels continuously >1000 μM, and excess N_2-N was <100 μM, with little seasonal variation (data not shown).

Other gases also accumulate in groundwater associated with the flooded section of the ditch. Nitrous oxide-N (N_2O-N) is a by-product of both denitrification and nitrification (Wrage et al. 2001), and in the example in Figure 7.11, N_2O accumulates up to 6 μM, considerably higher than the atmospheric background of 0.02 nM. Both here and in other datasets (R. Fox, unpublished), N_2O-N typically accumulates to 0.5% to 5% of the excess N_2-N. Methane (CH_4) also accumulates in groundwater when NO_3^- decreases to <100 μM (Figure 7.11). Very high concentrations of CH_4 typically occur at the end of summer when water levels (Figure 7.4), hydrostatic head, and NO_3^- are low. These high accumulations of CH_4 may lead to brief ebullition (degassing) events in which gases dissolved in groundwater are partially removed by CH_4 bubble formation (August 2007 in Figure 7.11).

The drainage control structure improved water quality in surface waters (Figure 7.12). When we first began sampling the ditch waters in 2006 (about 1 year after the installation of the drainage control structure), there was a large contrast between surface water in the upper, flooded section of the ditch (NO_3^- = 0 to 200 μM) and the lower unflooded section of the ditch (NO_3^- = 500 to 1200 μM). However, it is clear that over time, NO_3^- in the unflooded section of the ditch has decreased, either as a result of the very dry year 2007 or as a result of NO_3^--depleted groundwater from the upper perched section bypassing the drainage control structure. Regardless of the cause of the NO_3^- decrease in the unflooded section downstream of the control structure, it is clear that the drainage control structure has a large impact on agricultural NO_3^- discharged in upstream surface waters, allowing significant NO_3^- to pass only during the cool season (Figure 7.12) when temperatures are lower and hydrologic gradients are greater (Figure 7.4). However, this BMP is still experimental and has not yet been widely applied in the Choptank.

Strong land cover effects on the concentrations of dissolved gases are observed in groundwater (Figure 7.13). CH_4 concentrations are highest (average = 23 ± 13 μM) at wetland sites, followed

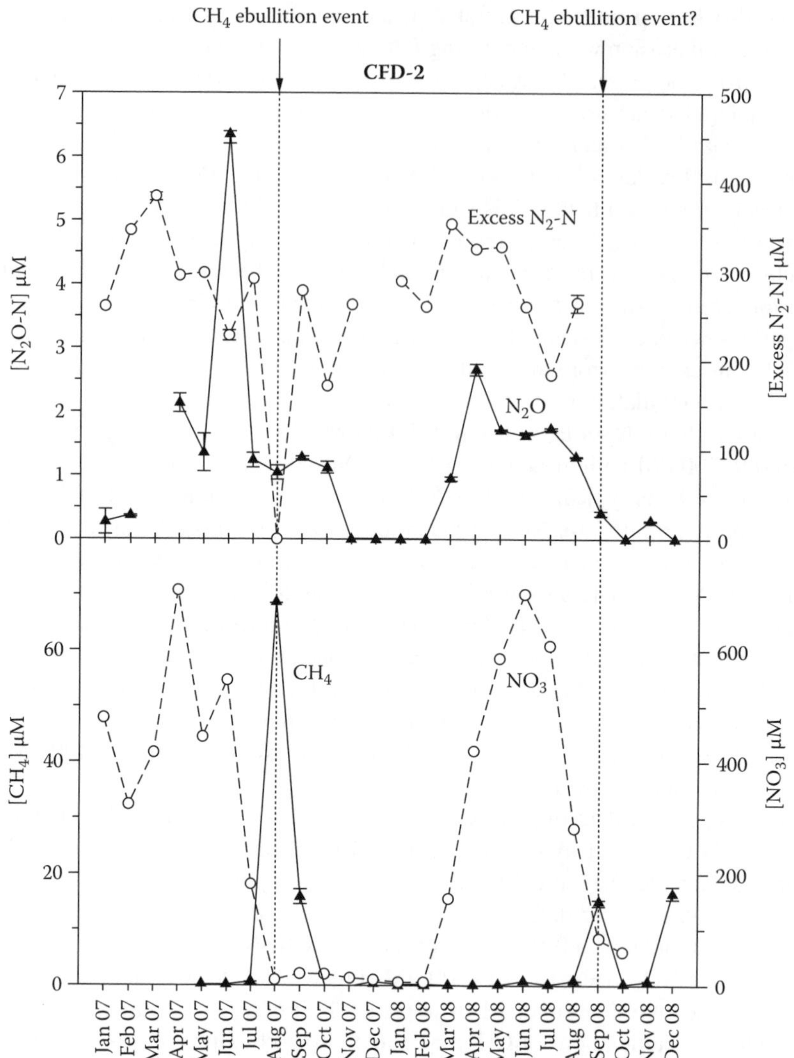

FIGURE 7.11 An example of a time series of concentrations of nitrate (NO_3^-) and the metabolic dissolved gases nitrous oxide (N_2O), methane (CH_4), and excess N_2-N dissolved in shallow groundwaters in the Choptank Basin. Excess N_2-N routinely accumulated in groundwater up to 25% to 50% of atmospheric N_2-N (~1200 μM at 12°C) as NO_3^- was denitrified as an alternate electron acceptor within the soil matrix under suboxic conditions (8% to 15% saturation, not shown). N_2O-N also accumulated to 1–10 μM (~20,000 times the atmospheric background of ~0.02 nM) and was typically 0.5% to 5% of the excess N_2-N. CH_4 accumulated when NO_3^- was depleted and CO_2 became the terminal, alternative electron acceptor. Accumulations of CH_4 > ~50 μM (~20,000 times the atmospheric background of 2 nM) resulted in ebullition events (bubble formation) which stripped out other dissolved gases (August 2007).

by wet areas close to wetlands (average = 4 ± 2 μM) due to low or depleted O_2 and NO_3^- as electron acceptors for respiration. CH_4 is lowest under farm fields, grassed CREP buffers, and forests (<1 μM) due to the presence of significant amounts of O_2 and/or NO_3^-. Similarly, N_2O is highest in wetlands (average = 0.8 ± 0.4 μM) and CREP buffers (average = 0.6 ± 0.4 μM) due to denitrification of NO_3^- from adjacent agricultural fields, and N_2O is lowest in wet areas, farm fields, and forests (<0.4 μM) due to apparent reduction (denitrification) of N_2O (wet areas) or the presence of O_2 (farm fields and forests).

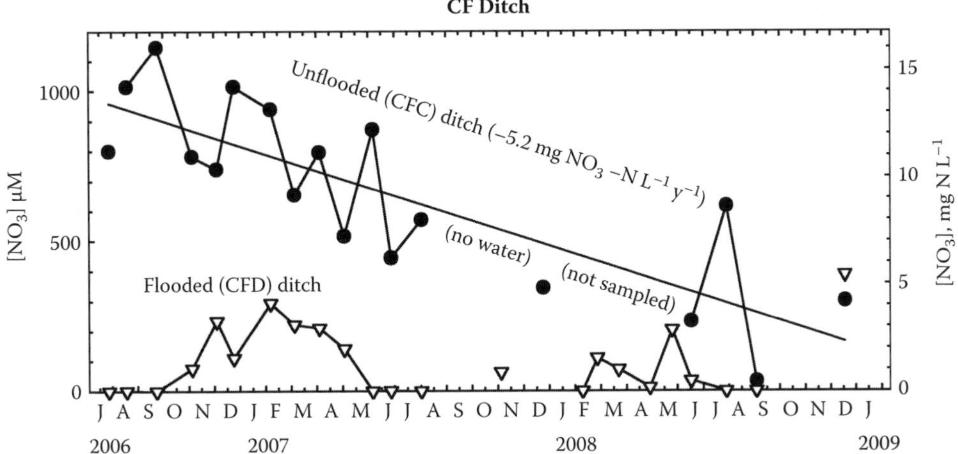

FIGURE 7.12 Chronology of NO_3^- concentrations in surface water of a 1 km-long farm ditch divided in half by a drainage control structure. The upper, flooded section has raised groundwater levels to within ~1 m of the soil surface, whereas in the lower, unflooded section the groundwater levels are ~2 m below the soil surface.

Another BMP implemented in the Choptank Basin to reduce losses of agricultural N and P is restoration of stream buffers. These areas were originally cleared for agriculture where slopes were low or where topography was altered, primarily on first- or second-order, nontidal streams on farms. The CREP program of USDA now makes funds available to farmers to reforest or plant grasses on these streamside areas to reduce transport of agricultural N and P from crop fields to streams.

Restoration of forested stream buffers has been quantified for the agriculturally dominated Choptank watersheds listed in Table 7.1 by Sutton et al. (2009b). During 1998 to 2005, 11% of streamside vegetation (range 1% to 30% within each watershed) was replanted under USDA's CREP Program, increasing the preexisting forested buffers (average = 33%, range = 10% to 48%). Streamside lengths were defined as 2 × stream lengths within each watershed. In 2005 total buffered streamsides (preexisting + CREP) averaged 44% (range = 12% to 61%) within the watersheds. Multi-year water quality data were used to examine the effect of the 11% average increase in streamside vegetation resulting from the CREP program, but no significant effects were detectable. Concentrations of N and P in streams of these watersheds have not changed or have slightly increased in some streams over the previous 20 years (Figure 7.7). Sutton et al. (2009b) attributed the lack of a statistically detectable effect to (1) the young age of the buffers resulting in low plant uptake and denitrification; (2) possible increases in agricultural N and P loading (e.g., increased fertilizers or manure applications, double cropping, etc.); and/or (3) insufficient implementation of buffers (33% preexisting to 44% preexisting + CREP). Bypassing of buffers by deeper groundwater flow paths (see Lowrance et al. 1997) is another possible explanation for the lack of a significant effect.

The use of winter cover crops is the last BMP implemented in the Choptank Basin to reduce losses of agricultural N. Planting of these winter annual grasses as soon as possible after crop harvest has been shown to stabilize soil (reducing erosion and P losses) and create demand for N in soils after N mineralization and oxidation produce highly soluble NO_3^- (Staver and Brinsfield 1998; Staver 2001a). Under low evapotranspiration rates in winter, rainfall produces large infiltration events which replenish groundwater levels (Figure 7.4) and transports soil NO_3^- to the groundwater (Staver and Brinsfield 1998; Staver 2001a). Larger rainstorms also generate overland flow which transports soil P rapidly to streams (e.g., Figure 7.9; Fisher et al. 2006b). The presence of cover crops acts to immobilize N and P during the winter in organic forms, making these available to crops during the following growing season. Field-scale tests of winter cover crops have been shown to reduce fluxes of both N and P (Staver and Brinsfield 1998), but the aggregate effects of winter cover crops

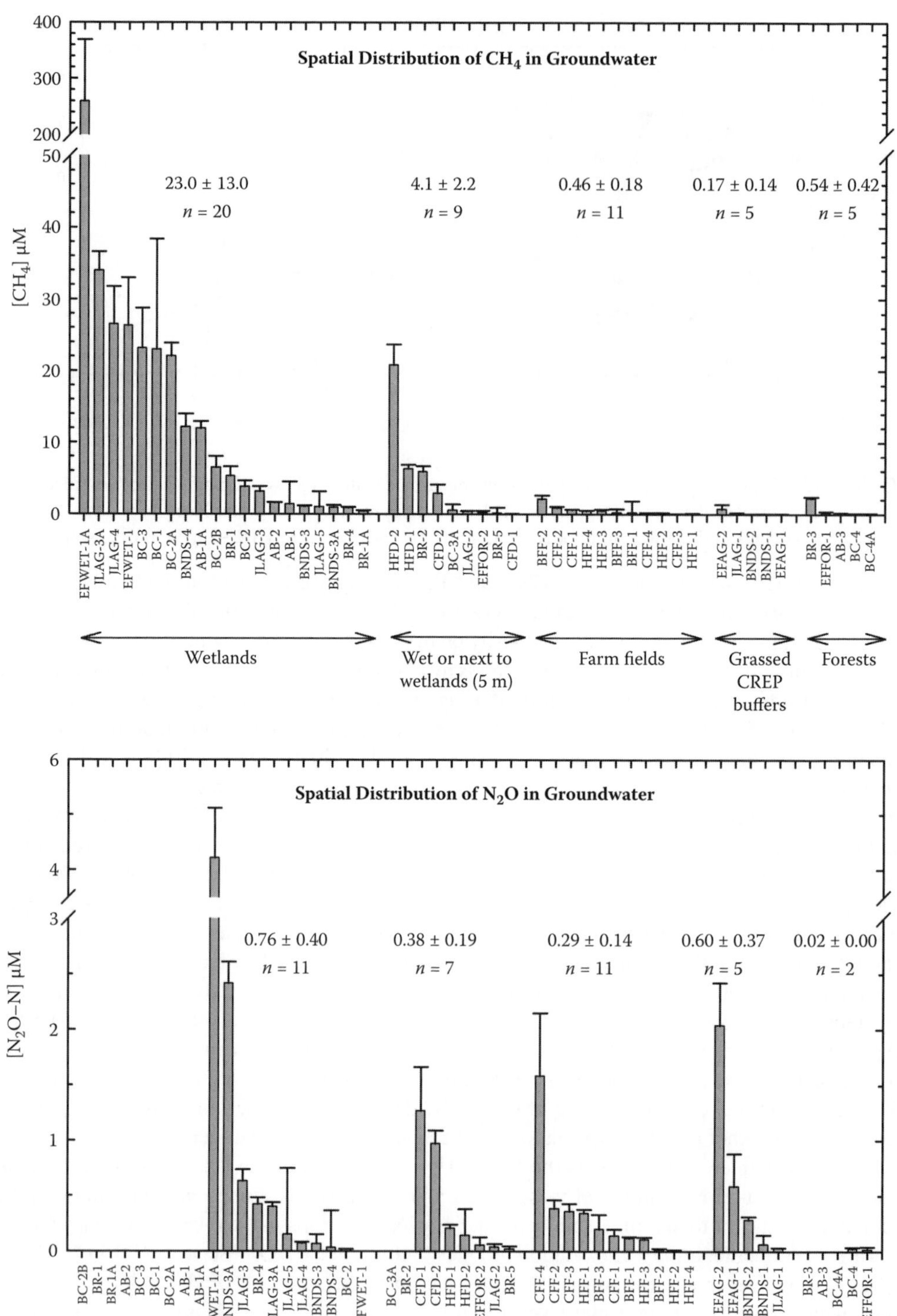

FIGURE 7.13 Spatial distributions of methane (CH_4) and nitrous oxide (N_2O-N) dissolved in groundwater in the Choptank Basin. Groundwater piezometers are grouped in classes (wetlands, wet areas, crop fields, buffers, forests) and sorted by the average (±SE) of 2008 data (12 monthly observations). Wetlands and wet areas have the highest CH_4 due to a lack of or depleted NO_3^-, whereas all classes except forest have high N_2O due to production during both nitrification and denitrification.

has not been adequately tested at the watershed scale. In the German Branch watershed (described above), Sutton et al. (2009a) found no significant changes in base flow N and decreases in base flow P after widespread application of BMPs, including winter cover crops. However, winter cover crops were one of several BMPs, and only 0% to 4% of crop fields in the German Branch watershed had winter cover crops.

7.4.5 Estuarine Water Quality Conditions

During 23 years of water quality monitoring at station ET5.2 in the Choptank Estuary (Figure 7.1), there has been no discernable improvements in water quality (Figure 7.14). Annual average values of chlorophyll *a* have increased, with a relatively large interannual variability driven primarily by variations in annual river discharge (positive relationship, $r^2 = 0.57$, $p < 0.01$). Likewise, there has been a decrease in summer bottom dissolved oxygen concentrations; however, these are not significantly related to river discharge but are negatively correlated with annual average chlorophyll *a* in surface waters ($r^2 = 0.47$, $p < 0.01$), the source of the sedimenting organic matter that results in strong biological oxygen demand in bottom waters.

Projections for water quality over the next decade are worrisome (Figure 7.14). Linear extrapolation of the trend in the upper panel to 2015 predicts annual average chlorophyll *a* >20 μg L^{-1}, a threshold often associated with increased bloom frequency (U.S. Environmental Protection Agency 2007). Likewise, linear extrapolation of the trend in the lower panel indicates that the average summer bottom dissolved oxygen will be approaching the EPA Bay Program's 30-d water quality

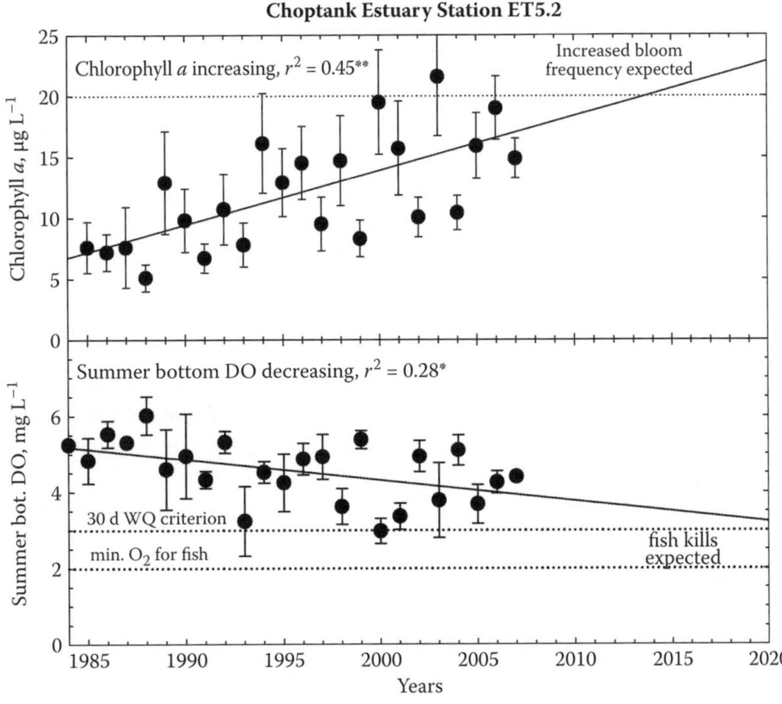

FIGURE 7.14 Trends in annual average values of chlorophyll *a* (chl*a*) concentrations in surface waters and summer (June to August) dissolved O_2 in bottom waters of the Choptank Estuary at EPA Bay Program station ET5.2 (see Figure 7.1 for location). There have been significant changes in both of these water quality parameters, and projections of the trends into the next decade suggest increased algal blooms in surface waters as annual average chl*a* increases beyond 20 μg L^{-1} and fish kills in bottom waters as the normal range of dissolved O_2 results in summers with <2 mg O_2 L^{-1}.

criterion (3 mg L^{-1}). Negative excursions in wetter years with higher chlorophyll *a* are likely to result in average summer dissolved oxygen less than the 2 mg L^{-1}, causing severe biological impacts. These conditions are not limited to the Choptank Estuary or the Chesapeake Bay, and Wazniak et al. (2004) and Beckert (2008) have reported similar conditions in the Maryland coastal bays, particularly in summer tidal waters at the northerly, more populated end of these systems.

7.5 DISCUSSION

Most agricultural and urban BMPs applied nationally have not been tested quantitatively for effectiveness at the watershed scale (Bernhardt et al. 2005). Despite national programs encouraging various economic and agricultural policies (e.g., nutrient management plans, CREP, winter cover crops, etc.), little is known about time scales of watershed responses or incremental reductions in export of N or P expected for application of BMPs at the watershed scale. The examples of our research given above strongly support the need for quantitative evaluation of BMPs to establish time scales of response as well as the expected decrease in N or P concentrations per unit BMP applied at the watershed scale. Examples of adequate tests are longitudinal studies in one watershed (e.g., Sutton et al. 2009a, for the German Branch watershed) or parallel studies in multiple watersheds with similar amounts of agriculture but varying amounts of a single BMP (e.g., Sutton et al. 2009b). Our strongest research recommendation is for more testing of BMPs at the watershed scale to provide quantitative information on response time and reductions of N and P per unit BMP applied.

7.5.1 Water Quality Status

It is clear that water quality in the Choptank Estuary is approaching a threshold. Driven by nutrient inputs from the terrestrial basin, projections of current trends in the estuary suggest algal blooms in surface waters and losses of benthos and fish from bottom waters in the coming decade. Most indicators of watershed nutrient inputs from the two largest sources, agriculture (Figures 7.6 to 7.8) and sewage (Figure 7.10) indicate that no significant reductions and some increases have occurred in the last 40 years (Figures 7.6 and 7.7). Given the relative stability of agricultural land use (Benitez and Fisher in review) and increases in human populations (Fisher et al. 2006a), there is little hope for improved water quality in the estuary under current conditions and policies.

There are four primary reasons for the lack of progress in water quality in the Choptank Estuary and elsewhere: (1) the voluntary approach taken by the EPA Bay Program to improve water quality; (2) insufficient implementation and watershed-scale testing of BMPs; (3) time lags for water quality improvements to appear; and (4) strong economic incentives to continue business as usual without regard to environmental consequences. Each of these reasons is discussed below, and in the next section, we present a summary of suggested changes to counteract these reasons for the lack of progress in improvements in water quality.

The single largest problem in the Chesapeake Bay watershed is excess N and P entering the bay from land (U.S. Environmental Protection Agency 2007). However, despite this, the EPA Bay Program has taken a largely voluntary approach, encouraging farmers and homeowners to reduce their use of fertilizers, with no oversight or enforcement. However, there are relatively few incentives to reduce nutrient losses. Farmers have strong economic incentives to continue to maximize crop yields to get the largest net income by applying fertilizers or manures, and Figure 7.8 provides strong evidence that excess agricultural fertilizers and manure are the primary drivers of N concentrations in Delmarva streams. Vitousek et al. (2009) show an average of +10 kg ha^{-1} yr^{-1} surplus of agricultural inputs of N (155 kg ha^{-1} yr^{-1}) over outputs (145 kg ha^{-1} yr^{-1}) in the U.S. Midwest, although it may be the outliers at the high end of input distributions embedded within a watershed that drive patterns such as those shown in Figure 7.8. Similarly, homeowners are given advice by the Cooperative Extension Service and other sources to apply lawn fertilizers at a rate of 87 to 174 lbs N $acre^{-1}$ yr^{-1} (equivalent to 99–195 kg N ha^{-1} yr^{-1}, see http://agbiopubs.sdstate.edu/articles/

ExEx1016.pdf, accessed August 25, 2009). This rate is equivalent to N applications for corn (100 to 150 lbs N acre^{-1} yr^{-1} or 110 to 170 kg N ha^{-1} yr^{-1}), one of the most heavily fertilized crops in the Choptank Basin. In a study of turf in New England, Guillard and Kopp (2004) used four fertilizer types applied at a rate of 147 kg ha^{-1} yr^{-1}, in the range given above, and found volume-weighted mean NO_3^- concentrations in leachate from inorganic fertilizers of 5 mg N L^{-1}, in the mid-range of average stream NO_3^- in agriculturally dominated basins in Figure 7.7. Fertilized lawns can potentially make a large, distributed, and heterogeneous contribution to high groundwater nitrate which becomes base flow nitrate in streams. However, the distribution of fertilized lawns is poorly known, making a quantitative assessment problematic. Nonetheless, in a nutrient overloaded system, reducing fertilizers applied for aesthetic reasons should take place before fertilizer reductions on crop fields used for food production.

The second reason for lack of progress in water quality in the Choptank Basin is insufficient implementation and watershed-scale testing of BMPs. In the targeted watershed study in German Branch referred to above (Primrose et al. 1997), nearly 100% compliance with some BMPs was achieved (e.g., nutrient management plans, conservation tillage); however, these have not, to our knowledge, ever been tested individually at the watershed scale to quantify their effectiveness. Sutton et al. (2009a) evaluated the collective effectiveness of all applied BMPs; however, the water quality effect was small, and it was impossible to attribute the observed effects to individual BMPs. Furthermore, newer and more promising BMPs such as winter cover crops, were only lightly used (0% to 4%), and even the CREP program funded by USDA to restore streamside vegetation increased the average stream buffer coverage only from 33% to 44% in 15 agriculturally dominated watersheds in the Choptank Basin (Sutton et al. 2009b). On average, more than half of all first- and second-order streams in the Choptank Basin had no buffers in 2005. More extensive implementation would provide a stronger water quality signal to detect, and more rigorous testing of individual BMPs at the watershed scale would provide a rational and quantitative basis for expectations of water quality improvements following implementation.

The third reason for a lack of progress toward improved stream water quality is time lags associated with groundwater emergence. Groundwater residence times in the unconfined surface aquifer are typically years to decades (Winter 1983; Staver 2001b), and some water quality improvements in groundwater may have already been achieved; however, there is less systematic monitoring of groundwater compared to streams, and groundwater time lags may obscure the effects of BMPs influencing groundwater chemistry. Even in our study of the controlled drainage structure described above (Figure 7.12), several years were required until the effects of the raised water table on water quality in a first-order stream (the ditch) were observed. We may yet detect more effects of some current BMPs in surface waters over the coming decade as younger, potentially cleaner groundwaters reach surface waters, but continued monitoring will be required to quantify these effects.

The final reason for the lack of progress in nontidal water quality is the lack of economic incentives for improved water quality. Currently, economic incentives favor a "business-as-usual" approach in the absence of regulations or enforcement, and high export rates of N and P are expected from land to water under these conditions. Neither farmers nor land owners are penalized for over-applying fertilizers (other than the cost of the fertilizer), and there are no economic incentives for improved environmental conditions or disincentives for environmental degradation. Only point sources have been systematically regulated in response to increases in human populations that provide wastewater. In the long term, economic incentives due to avoidance of areas with poor water quality may provide some pressure to solve these problems, but current prospects are dim, except for pressure from a small fraction of the population active in local water quality issues.

7.5.2 Water Quality Improvement

We have used the data presented here to develop a series of policy recommendations which could potentially provide a rational basis for improvements in water quality. These include (1) applications of water quality standards in nontidal watersheds; (2) lower caps on wastewater discharge volume and concentrations; (3) lower fertilizer application rates on farms (subsidized); (4) buffers and cover crops on crop fields; and (5) limited applications of lawn fertilizers. All of these recommendations are described below, and we emphasize that they are based on our own observations. More detailed policy considerations by government agencies (e.g., U.S. EPA, USDA NRCS, MD DNR, MDE, etc.) will be required to place them into practice.

The first recommendation for improving water quality on Delmarva is the implementation of numeric water quality standards for nutrients in nontidal waters at the watershed scale. The U.S. EPA (2000) has proposed water quality standards for nutrients in nontidal waters, but to our knowledge these standards have neither been adopted nor applied in the mid-Atlantic region or elsewhere. U.S. EPA (2000) recommended 0.71 mg N L^{-1} (51 μM) and 0.031 mg P L^{-1} (1 μM) for the nutrient ecoregion XIV, which includes Delmarva. These water quality standards are equivalent to ~5 times the concentrations found in local nontidal streams draining forested areas and are much lower than observed concentrations in most agriculturally dominated basins in the Choptank Basin (e.g., Figure 7.8) and elsewhere (Jordan et al. 1997; Beckert et al. in review). This indicates widespread regional violation of the recommended, but unenforced, water quality standards.

The original goal of the EPA Bay Program was a 40% reduction of the 1985 N and P loads (e.g., Belval and Sprague 1999). From the monitoring data presented above, it is clear that this goal has not been achieved in the Choptank Basin, despite optimistic model predictions, and the trends are not even in the right direction. A further complication is the fact that the Greensboro watershed gauged by USGS (Figure 7.1, 18% of the basin) is not representative of the remaining ungauged portion of the Choptank Basin (Lee et al. 2001). The Greensboro watershed has significantly less agriculture, more forest, and more hydric soils than most of the basin, which results in relatively low N and P concentrations (Lee et al. 2001). Jurisdictionally, half of the Greensboro watershed also lies in the state of Delaware, which does not participate in the Maryland tributary strategy.

Attainment of the 40% reductions of the 1985 N and P loads would undoubtedly have large impacts on water quality in the Choptank nontidal streams and estuary. In about 1985, N concentrations ranged from 110 μM (1.3 mg L^{-1}) at the Greensboro watershed to 430 μM (6.0 mg L^{-1}) at the Oakland watershed (Figures 7.6 and 7.7). If the efforts of the EPA Bay Program had achieved 40% reductions in the Choptank, N concentrations would now range over 67 to 260 μM (0.9 to 3.6 mg L^{-1}), considerably lower than those currently observed for total N (Figures 7.6 and 7.7). Parallel computations for 40% P reductions would result in current P concentrations of 0.5 to 2.5 μM (0.02 to 0.08 mg L^{-1}), also lower than those currently observed for total P (Figures 7.6 and 7.7). We strongly support the original goals of the EPA Bay Program's 40% reductions, and the numbers estimated above could be used as local water quality standards for the Choptank Basin. Attainment of the stricter EPA water quality standards should remain a long-term but much more difficult goal to achieve in this nutrient-rich basin. Because water yields are relatively uniform regionally at monthly to annual time scales (Sutton et al. 2009b), concentrations could be substituted for nutrient loads in water quality standards to simplify management protocols.

Evaluation of violation of a water quality standard would require extensive water quality sampling. Quarterly base flow sampling of all HUC 12 or 14 nontidal watersheds for a year could be undertaken by an independent agent (e.g., USGS, a local university, or a private consulting company) to determine attainment or violation of N and P standards in base flow. Validation, certification, and standardization of sampling protocols and chemical analyses would be essential. N is easier to sample than P, but base flow P can provide an initial value for P status in nontidal streams. Incorporating storm flow sampling is desirable, especially for P, but probably too expensive for the widespread application suggested here. Following a year of monitoring, watersheds in violation of water quality standards should be required to adopt an appropriate mix of the BMPs described below, with quarterly monitoring at 5-year intervals to assess progress toward water quality standards. Selected research watersheds such as those in Table 7.1 or the NAWQA watersheds of USGS could be used for more detailed studies. To provide an economic incentive, failure to attain a water quality standard after BMP implementation within a watershed after a decade (close to median groundwater residence times) should result in economic consequences for those living and using land within the watershed. This approach is similar to the TMDL process of EPA, but would be focused on nontidal streams, with fixed deadlines for compliance with concentration standards developed locally.

The second recommendation for improving water quality on Delmarva is to lower caps on wastewater volumes and concentrations. To our knowledge, there are currently only loosely enforced caps for wastewater volumes and nutrient mass discharge from wastewater plants, even in areas such as the Choptank with degrading estuarine waters (Figure 7.14). The current strategy of many municipalities is to add tertiary treatment to reduce the N and P concentrations in order to accommodate urban growth and increased wastewater volumes. A better approach would be to add wastewater volume caps and to lower N and P mass caps for a tributary or region. This would force wastewater plants to find alternative uses for human wastewater, a valuable product that can be better used for production of compost (e.g., Milorganite), methane, irrigation, and fertilizer products, rather than discharging wastewater into our swimming and fishing waters. Similarly, denitrifying septic systems should be mandatory on all new construction, and existing septic systems should be retrofit within a 5- to 10-year time period.

The third recommendation for improving water quality on Delmarva is to lower fertilizer application rates on farms. Fertilizer applications are currently targeted to maximize crop yields under the best weather conditions, with little consideration for the environmental consequences of unused N and P when crop uptake of N and P is limited by droughts or floods. We suggest that farmers apply fertilizers at lower rates to match crop yields. However, N budgets of crop fields are poorly known, especially under varying weather conditions (Vitousek et al. 2009). The data of Figure 7.8 indicate that agricultural land is a significant source of N, but the magnitude and spatial extent of excess N applications on crop fields is not clear. We recommend the development of national GIS databases on watershed N budgets (HUC 12 or 14), including application rates and crop yields to provide a rational basis for better N and P use efficiency. Much of this information is already compiled, but not systematically at appropriate spatial scales in a watershed context, and usually without spatial reference except at the county level. This recommendation simply endorses better use of information already being collected by different agencies.

In a recently funded pilot project, some of the authors are testing an alternative approach to reduced fertilizer applications. Economic incentives are being offered to a small group of farmers in the Choptank Basin to maintain their soil P and NO_3^- concentrations low in the fall at the end of the growing season to reduce cold season losses of P in overland flow and NO_3^- in infiltrating rain water during fall and winter storms. Achieving low soil P and NO_3^- in the fall will require repeated applications of fertilizers and manures at the rate that they can be utilized by the crops rather than a small number of larger applications. The result should be reduced losses of N and P from better managed crop fields, an outcome that will be monitored.

The focus of most agricultural policies has been on the production of inexpensive food with maximum profits by the producer, usually from increased production in decreasing areas (Benitez 2002; Fisher et al. 2006a). However, we suggest that an additional consideration must be the reduction of the environmental impact of high-intensity agriculture (Figures 7.6 through 7.8). We support the continued economic viability of agriculture on Delmarva, but we are attempting to provide recommendations for policies that reduce the impact of agriculture on water quality (Figure 7.8). We also suggest that the deleterious water quality effects of corn-based ethanol production should be added to its other problems, including the greater C storage of unfertilized grasslands harvested for biomass (Tilman et al. 2006), greater energy production per unit area for cellulosic biofuels (Cambell et al. 2009), and competition with crop production for arable land (Tilman et al. 2009).

The fourth recommendation for improving water quality on Delmarva is mandatory stream buffers, drainage control structures, and winter cover crops on at least 50% of agricultural fields. Priorities for the 50% could be set by crop type (e.g., corn) or by fertilizer application type or rate (poultry or sludge) with the highest N and P loss rates. Buffers will take land out of agricultural production, as many first-order or zero-order streams are found within farm fields or are farm ditches. Buffering these areas will segment many fields and remove land that is currently farmed. An alternative to buffers for ditches could be the drainage control structures described above. Given the apparent, initial success of drainage control structures in reducing NO_3^- losses from a farm ditch in a former wetland converted to cropland (Figure 7.12), mandatory water controls on similar ditches might also be a good recommendation after testing at the watershed scale. Likewise, mandating winter cover crops (or unfertilized commodity crops) will have economic costs for which farmers should continue to be compensated. Prior to implementation of this recommendation, testing of the assumptions and the previous field-scale measurements should be done at the watershed scale.

The last recommendation for improving water quality on Delmarva is to substantially limit the use of lawn fertilizers. Given that excess N and P is the single largest problem in Chesapeake Bay and most coastal waters, the high rates of fertilizer applications to lawns (see above) seems to have no justification. If we are going to ask farmers to sacrifice their highest yields in perfect weather in order to reduce the losses of N and P from their fields under less perfect weather conditions, then we should also be willing to have a less green turf in our yards for a common cause. However, we acknowledge that there has been little quantitative study of the effects of nutrient losses from lawns at the watershed scale. There are clearly large losses of N from fertilized turf (e.g., Guillard and Kopp 2004), but the areal extent of fertilized lawns and the quantitative significance has yet to be rigorously determined at the watershed scale.

Many of these recommendations may seem draconian or invasive. Enforcement of water quality standards and BMPs will conflict with individual property rights and established agricultural and municipal practices. However, our current high-intensity society generates large amounts of N and P in surface waters draining the land that we inhabit (Figures 7.6 through 7.8), with significant impacts on coastal and estuarine waters (Figure 7.14). If we want to have clean waters in which to swim and fish, we have to take at least some of the steps recommended above. The fluxes of N and P from land to water are well known and defined, and the production of our food and disposal of our wastes are the largest sources. We have a choice of leaving cleaner waters to future generations if we take the steps listed above, or we can leave a legacy of green waters with lifeless bottom conditions.

ACKNOWLEDGMENTS

The research on surface hydrology and chemistry has been supported by subcontracts to TRF and TEJ from funding provided by the USDA's CEAP program to Laura McConnell and Greg McCarty at the USDA ARS lab in Beltsville, Maryland. Funding on groundwater and denitrification was provided to TRF and TEJ by the USDA CSREES program. Partial funding for land use and GIS support was also provided to TRF by NASA's Land Cover Land Use Change Program (NEWS04-2-0000-0420). RJF and KAB were supported by graduate fellowships and teaching assistantships from the Horn Point Laboratory (UMCES). We thank Jim Hagy and an anonymous reviewer for substantial improvements in the original manuscript.

REFERENCES

Beckert, K. A. 2008. Watershed land use and nutrient dynamics in MD coastal bays USA. M.S. Thesis, University of Maryland, College Park.

Beckert, K. A., T. R. Fisher, J. M. O'Neil, R. V. Jesien. In review. Characterization and comparison of stream nutrients, land use, and loading patterns in Maryland coastal bay watersheds. *Estuar. Coastal Shelf Sci.*

Bell, W. H. 2000. Moving water. Report to the Chesapeake Bay Cabinet by Public Drainage Task Force, Annapolis, Maryland.

Belval, D. L. and L. A. Sprague. 1999. Monitoring nutrients in the major rivers draining to Chesapeake Bay. USGS Water-Resources Investigations Report 99-4238, Baltimore, Maryland.

Benitez, J. A. 2002. Historical land cover changes (1665–2000) and impact on N and P export from the Choptank watershed. Ph.D. Thesis, University of Maryland, College Park.

Benitez, J. A. and T. R. Fisher. 2004. Historical land cover conversion (1665–1850) in the Choptank watershed, eastern USA. *Ecosystems* 7: 219–232.

Benitez, J. A. and T. R. Fisher. In review. Land cover history (1665–2000), landscape patterns (1850–2000), and their effect on nutrient yields in the Choptank watershed, eastern USA. *Landscape Ecology*.

Berndt, H. 1999. Effects of nutrients and turbidity from point and non-point source inputs on phytoplankton in the Choptank estuary. M.S. Thesis, University of Maryland, College Park.

Bernhardt, E. S., M. A. Palmer, J. D. Allan, G. Alexander, K. Barnas, S. Brooks, J. Carr, S. Clayton, C. Dahm, J. Follstad-Shah, D. Galat, S. Gloss, P. Goodwin, D. Hart, B. Hassett,, R. Jenkinson, S. Katz, G. M. Kondolf, P. S. Lake, R. Lave, J. L. Meyer, T. K. O'Donnell, L. Pagano, B. Powell, and E. Sudduth. 2005. Synthesizing U.S. river restoration efforts. *Science* 308: 636–637.

Cambell, J. E., D. B. Lobell, and C. B. Field. 2009. Greater transportation energy and GHG offsets from bioelectricity than ethanol. *Science* 324: 1055–1057.

Carlozo, N., G. Radcliffe, and T. R. Fisher. 2008. Trends in water quality in response to human populations and land use in the Delaware, Hudson, and Chesapeake Basins. Abstract, American Society of Limnology and Oceanography Conference, Orlando, Florida.

Clark, G. M., D. K. Mueller, and M. A. Mast. 2000. Nutrient concentrations and yields in undeveloped stream basins of the United States. *J. Am. Water Res. Assoc.* 36: 849–860.

Colt, J. 1984. Computation of dissolved gas concentrations in water as functions of temperature, salinity, and pressure. *Am. Fish. Soc. Spec. Pub.* 14, pp. 1–154.

Cooper, S. R. and G. S. Brush. 1993. A 2,500 year history of anoxia and eutrophication in Chesapeake Bay. *Estuaries* 16: 617–626.

Fisher, T. R., L. W. Harding, Jr., D. W. Stanley, and L. G. Ward. 1988. Phytoplankton, nutrients, and turbidity in the Chesapeake, Delaware, and Hudson estuaries. *Estuar. Coastal Shelf Sci.* 27: 61–93.

Fisher, T. R., K.-Y. Lee, H. Berndt, J. A. Benitez, and M. M. Norton. 1998. Hydrology and chemistry of the Choptank River Basin in the Chesapeake Bay drainage. *Water Air Soil Pollut.* 105: 387–397.

Fisher, T. R., J. A. Benitez, K.-Y. Lee, and A. J. Sutton. 2006a. History of land cover change and biogeochemical impacts in the Choptank River Basin in the mid-Atlantic region of the US. *Int. J. Remote Sens.* 27: 3683–3703.

Fisher, T. R., J. D. Hagy III, W. R. Boynton, and M. R. Williams. 2006b. Cultural eutrophication in the Choptank and Patuxent Estuaries of Chesapeake Bay. *Limnol. Oceanogr.* 51: 435–447.

Fox, R. J., T. R. Fisher, T. M. Kana, and A. B. Gustafson. 2009. Concentration gradients of excess N_2 and N_2O in the vadose zone and groundwater of the mid-Atlantic coastal plain, U.S.A. Abstract, EGU, Vienna, Austria.

Glibert, P. M., D. J. Conley, T. R. Fisher, L. W. Harding, Jr., and T. C. Malone. 1995. Dynamics of the 1990 winter/spring bloom in Chesapeake Bay. *Mar. Ecol. Prog. Ser.* 122: 27–43.

Glibert, P. M., R. Magnien, M. W. Lomas, J. Alexander, C. Fan, E. Haramoto, M. Trice, and T. M. Kana. 2001. Harmful algal blooms in the Chesapeake and coastal bays of Maryland, USA: Comparisons of 1997, 1998, and 1999 events. *Estuaries* 24: 875–883.

Guillard, K. and K. L. Kopp. 2004. Nitrogen fertilizer form and associated nitrate leaching from cool-season lawn turf. *J. Environ. Qual.* 33: 1822–1827.

Hagy, J. D., W. R. Boynton, C. W. Keefe, and K. V. Wood. 2004. Hypoxia in Chesapeake Bay, 1950–2001: Long-term change in relation to nutrient loading and river flow. *Estuaries* 27: 634–658.

Hamilton, P. A., J. M. Denver, P. J. Phillips, and R. J. Shedlock. 1993. Water-quality assessment of the Delmarva Peninsula, Delaware, Maryland, and Virginia: Effects of agricultural activities on, and distribution of, nitrate and other inorganic constituents in the surficial aquifer. US Geol. Surv., Open-File Rept. 93-40.

Harding, L. W., B. W. Meeson, and T. R. Fisher. 1986. Phytoplankton production in two east coast estuaries: Photosynthesis-light functions and patterns of carbon assimilation in Chesapeake and Delaware Bays. *Estuar. Coastal Shelf Sci.* 23: 773–806.

Harding, L. W., Jr. and E. S. Perry. 1997. Long-term increase of phytoplankton biomass in Chesapeake Bay. 1950–1994. *Mar. Ecol. Prog. Ser.* 157: 39–52.

Hively, W. D., C. J. Hapeman, L. L. McConnell, T. R. Fisher, C. P. Rice, G. W. McCarty, P. M. Downey, G. T. Nino de Gusman, K. Bialek, M. W. Lang, A. M. Sadeghi, D. R. Whitall, A. B. Gustafson, A. J. Sutton, K. A. Sefton, J. A. H. Fetcho. In review. Relating nutrient and herbicide fate with landscape features in 15 sub-basins of the Choptank River watershed: Implications for management. *Sci. Total Environ.*

Jordan, T. E., D. L. Correll, and D. E. Weller. 1997. Effects of agriculture on discharges of nutrients from coastal plain watersheds of Chesapeake Bay. *J. Environ. Qual.* 26: 836–848.

Kana, T. M., C. Darkangelo, M. D. Hunt, J. B. Oldham, G. E. Bennett, and J. C. Cornwell. 1994. Membrane inlet mass spectrometer for rapid high-precision determination of N_2, O_2, and Ar in environmental water samples. *Anal. Chem.* 66: 4166–4170.

Kana, T. M., M. B. Sullivan, J. C. Cornwell, and K. M. Groszkowski. 1998. Denitrification in estuarine sediments determined by membrane inlet mass spectrometry. *Limnol. Oceanogr.* 43: 334–339.

Kana, T. M. and D. L. Weiss. 2002. Comment on "Comparison of isotope pairing and N2:Ar methods for measuring sediment denitrification" by B. D. Eyre, S. Rysgaard, T. Dalsgaard, and P. B. Christianson. *Estuaries* 25: 1077–1087.

Kemp, W. M., W. R. Boynton, J. E. Adolf, D. F. Boesch, W. C. Boicourt, G. Brush, J. C. Cornwell, T. R. Fisher, P. M. Glibert, J. D. Hagy, L. W. Harding, E. D. Houde, D. G. Kimmel, W. D. Miller, R. I. E. Newell, M. R. Roman, E. M. Smith, and J. C. Stevenson. 2005. Eutrophication of Chesapeake Bay: Historical trends and ecological interactions. *Mar. Ecol. Prog. Ser.* 303: 1–29.

Koroncai, R., L. Linker, J. Sweeney, and R. Batiuk. 2003. Setting and allocating the Chesapeake Bay Basin and sediment loads: The collaborative process, technical tools, and innovative approaches. US EPA Region III, Chesapeake Bay Program Office, Annapolis, Maryland.

Koskelo, A. I. 2008. Hydrologic and biogeochemical storm response in Choptank Basin headwaters. M.S. Thesis, University of Maryland, College Park.

Koskelo, A. I., T. Jordan, and T. R. Fisher. In review. A new, precipitation-based method of base flow separation. *J. Hydrol.*

Lee, K.-Y., T. R. Fisher, and E. Rochelle-Newall. 2001. Modeling the hydrochemistry of the Choptank River basin using GWLF and Arc/Info: 2. Model Application. *Biogeochemistry* 56: 311–348.

Lowrance, R. L., S. Altier, J. D. Newbold, R. R. Schnabel, P. M. Groffman, J. M. Denver, D. L. Correll, J. W. Gilliam, J. L. Robinson, R. B. Brinsfield, K. W. Staver, W. Lucas, and A. H. Todd. 1997. Water quality functions of riparian forest buffer systems in Chesapeake Bay watersheds. *Environ. Manage.* 21: 687–712.

Marshall, H.G., T. Egerton, T. Stern, J. Hicks, and M. Kokocinski. 2004. Extended bloom concentrations of the toxic dinoflagellate *Dinophysis acuminata* in Virginia estuaries during late winter through early spring, 2002. Pp. 364–366, In: K. A. Steidinger, J. H. Landsberg, C. R. Tomas, and G. A. Vargo (eds.), *Harmful algae 2002*. Florida Fish & Wildlife Conservation Commission, Florida Institute of Oceanography, IOC of UNESCO, St. Petersburg, Florida.

McConnell, L. L. and G. W. McCarty. 2005. Annual report. CEAP Special Emphasis—Choptank River Watershed. 10/1/2004–9/31/2005, U.S. Department of Agriculture, ARS, Towson, Maryland.

Meade, R. H. 1968. Relations between suspended matter and salinity in estuaries of the Atlantic seaboard USA. *Internat. Assoc. Science Hydrol. General Assembly Bern* 4: 96–109.

Newell, R. I. E., T. R. Fisher, R. R. Holyoke, and J. C. Cornwell. 2004. Influence of eastern oysters on N and P regeneration in Chesapeake Bay, USA. In: R. Dame and S. Olenin (eds.), *The comparative roles of suspension feeders in ecosystems*. NATO Science Series: IV – Earth and Environmental Sciences. Dordrecht, the Netherlands: Kluwer.

Norton, M. G. M. and T. R. Fisher. 2000. The effects of forest on stream water quality in two coastal plain watersheds of the Chesapeake Bay. *Ecol. Eng.* 14: 337–362.

Officer, C. B., R. B. Biggs, J. L. Taft, L. E. Cronin, M. A. Tyler, and W. R. Boynton. 1984. Chesapeake Bay anoxia: Origin, development, and significance. *Science* 223: 22–27.

Primrose, N. L., J. L. Millard, P. E. McCoy, M. G. Sturm, S. E. Bowen, and R. J. Windschitl. 1997. German Branch Targeted Watershed Project: Report on 5 years of biotic and water quality monitoring 1990 through 1995. Chesapeake and Coastal Watershed Service, Watershed Restoration Division, Maryland Department of Natural Resources, Annapolis, Maryland.

Sellner, K. G. and S. Fonda-Umani. 1999. Dinoflagellate blooms and mucilage production. Pp. 173–206, In: T. C. Malone, A. Malej, L. W. Harding, Jr., N. Smodlaka, and R. E. Turner (eds.), *Ecosystems at the land-sea margin: Drainage basin to coastal sea*. Washington, D.C.: American Geophysical Union.

Staver, K. W. 2001a. *Increasing N retention in coastal plain agricultural watersheds*. Kirkkonummi, Finland: The Scientific World.

Staver, K. W. 2001b. The effect of agricultural best management practices on subsurface nitrogen transport in the German Branch watershed. Final Report, Maryland Department of Natural Resources, Project 14-198-34CZM025, Annapolis, Maryland.

Staver, K. W. and R. B. Brinsfield. 1998. Using cereal grain winter cover crops to reduce groundwater nitrate contamination in the mid-Atlantic coastal plain. *J. Soil Water Conserv.* 53: 230–240.

Staver, K. W., R. B. Brinsfield, and W. L. Magette. 1994. Tillage effects on phosphorus transport from Atlantic coastal plain watersheds. Pp. 215–222, In: P. Hill and S. Nelson (eds.), *Toward a sustainable coastal watershed: the Chesapeake experiment*. Proceedings of the 1994 Chesapeake Research Consortium Conference, Edgewater, Maryland.

Staver, L. W., K. W. Staver, and J. C. Stevenson. 1996. Nutrient inputs to the Choptank River Estuary: Implications for watershed management. *Estuaries* 19: 342–358.

Stone, J. P., T. R. Fisher, and R. Stone. In review. Water quality of non-tidal streams in Dorchester County, Maryland. *Water Air Soil. Pollut.*

Sutton, A. J. 2006. Evaluation of agricultural nutrient reductions in restored riparian buffers. Ph.D. Thesis, University of Maryland, College Park.

Sutton, A. J., T. R. Fisher, and A. B. Gustafson. 2009a. Historical changes in water quality at German Branch in the Choptank River Basin. *Water Air Soil Pollut.* 199: 353–369.

Sutton, A. J., T. R. Fisher, and A. B. Gustafson. 2009b. Effects of restored stream buffers (CREP) on water quality in non-tidal streams in the Choptank River Basin. *Water Air Soil Pollut.* DOI 10.1007/s11270-009-0152-3.

Tango, P., W. Butler, R. Lacouture, D. Goshorn, R. Magnien, B. Michael, S. Hall, K. Brohawn, R. Wittman, and W. Beatty. 2002. An unprecedented bloom of *Dinophysis acuminate* in Chesapeake Bay. Pp. 358–360, In: K. A. Steidinger, J. H. Landsberg, C. R. Tomas, and G. A. Vargo (eds.), *Harmful algae 2002*. Florida Fish & Wildlife Conservation Commission, Florida Institute of Oceanography, IOC UNESCO, St. Petersburg, Florida.

Tilman, D., J. Hill, and C. Lehman. 2006. Carbon-negative biofuels from low-input high diversity grassland biomass. *Science* 314: 1598–1600.

Tilman, D., R. Socolow, J. A. Foley, J. Hill, E. Larson. L. Lynd, S. Pacala, J. Reilly. T. Searchinger, C. Somerville, and R. Williams. 2009. Beneficial biofuels: The food, energy, and environment trilemma. *Science* 325: 270–271.

Traband, J. J. 2003. Removal of wastewater nitrogen and phosphorus by an oligohaline marsh. M.S. Thesis, University of Maryland, College Park.

U.S. Environmental Protection Agency. 2000. Information supporting the development of state and tribal nutrient criteria for rivers and streams in nutrient ecoregion XIV. EPA 822-B-00-022. Washington, D.C.: US EPA.

U.S. Environmental Protection Agency. 2007. Ambient water quality criteria for dissolved oxygen, water clarity and chlorophyll *a* for the Chesapeake Bay and its tidal tributaries—Chlorophyll *a* Addendum. October 2007. EPA 903-R-07-005. Region III Chesapeake Bay Program Office, Annapolis, Maryland.

Valderrama, J.C. 1981. The simultaneous analysis of total nitrogen and total phosphorus in natural waters. *Mar. Chem.* 10: 109–122.

Vitousek, P. M., R. Naylor, T. Crews, M. B. David, L. E. Drinkwater, E. Holland, P. J. Johnes, J. Katzenberger, L. A. Martinelli, P. A. Matson, G. Nziguheba, D. Ojima, C. A. Palm, G. P. Robertson, P. A. Sanchez, A. R. Townsend, and F. S. Zhang. 2009. Nutrient imbalances in agricultural development. *Science* 324: 1519–1520.

Voss, M., B. Deutsch, R. Elmgren, C. Humborg, P. Kuuppo, M. Pastuszak, C. Rolff, and U. Schulte. 2006. Source Identification of nitrate by means of isotopic tracers in the Baltic Sea catchments. *Biogeosciences* 3: 663–676.

Wazniak, C. E., M. R. Hall, C. Cain, D. Wilson, R. Jesien, J. Thomas, T. Carruthers, and W. C. Dennison. 2004. State of the Maryland Coastal Bays Report, Annapolis, Maryland.

Williams, M. R., T. R. Fisher, K. N. Eshleman, W. C. Boynton, W. M. Kemp, C. F. Cerco, S.-C. Kim, R. R. Hood, D. A. Fiscus, and G. Radcliffe. 2006. An integrated modeling system for management of the Patuxent River Estuary and Basin, Maryland, USA. *Int. J. Remote Sens.* 27: 3705–3726.

Winter, T. C. 1983. The interaction of lakes with variably saturated porous media. *Water Resourc. Res.* 19: 1203–1218.

Wrage, N., G. L. Velthof, M. L. Beusichem, and O. Oenema. 2001. Role of nitrified denitrification in the production of nitrous oxide. *Soil. Biol. Biochem.* 33: 1723–1732.

8 Seagrass Decline in New Jersey Coastal Lagoons

A Response to Increasing Eutrophication

Michael J. Kennish, Scott M. Haag, and Gregg P. Sakowicz*

CONTENTS

Abstract ... 168
8.1 Introduction ... 168
8.2 Study Area ... 170
8.3 Materials and Methods ... 170
8.3.1 Sampling Design ... 170
8.3.2 Sampling Methods ... 172
8.3.2.1 Quadrat Sampling ... 172
8.3.2.2 Macroalgae Sampling ... 173
8.3.2.3 Core Sampling ... 173
8.3.2.4 Sediment Sampling ... 173
8.3.2.5 Water Quality Sampling ... 173
8.3.2.6 Videographic Imaging ... 173
8.4 Results: Physicochemical Conditions ... 174
8.5 Biotic Indicators ... 175
8.5.1 Seagrass Distribution ... 175
8.5.2 Aboveground Biomass ... 176
8.5.3 Belowground Biomass ... 179
8.5.4 Seagrass Density ... 184
8.5.5 Seagrass and Macroalgae Cover ... 184
8.5.6 Seagrass Blade Length ... 186
8.5.7 Macroalgae Composition ... 187
8.5.8 Brown Tide ... 190
8.6 Discussion ... 190
8.7 Summary and Conclusions ... 196
Acknowledgments ... 198
References ... 198

* Corresponding author. Email: kennish@marine.rutgers.edu

ABSTRACT

Results of a comprehensive 3-year (2004 to 2006) investigation of the seagrass (*Zostera marina* L.) demographics in the Barnegat Bay–Little Egg Harbor Estuary, a lagoonal system located along the central New Jersey coastline, reveal a dramatic decline in plant biomass (g dry wt m^{-2}), density (shoots m^{-2}), blade length, and percent cover of bay bottom associated with increasing eutrophic conditions. Quadrat, core, and hand sampling, as well as digital camera imaging, at 120 transect stations in four disjunct seagrass beds of the estuary during the June to November period in 2004, 2005, and 2006 indicate distinct changes in demographic patterns leading to significant shifts in ecosystem services provided by the beds, such as the loss of habitat for bay scallops (*Argopecten irradians*), blue mussels (*Mytilus edulis*), blue crabs (*Callinectes sapidus*), and seatrout (*Cynoscion nebulosus*). Of all the metrics analyzed, seagrass biomass exhibited the most significant decrease over the study period, which was evident estuary-wide. For example, the mean aboveground biomass and belowground biomass of *Z. marina* decreased by 50.0% to 87.7% between 2004 and 2006, with the greatest decrease occurring in Little Egg Harbor. Shoot density of *Z. marina* also decreased from a mean of 292 shoots m^{-2} in 2005 to 241 shoots m^{-2} in 2006. The mean blade length of *Z. marina* was relatively consistent during 2004 (31.83 to 34.02 cm) and 2005 (25.89 to 32.71 cm), but decreased greatly during 2006 (18.61 to 19.37 cm), thereby reducing overall plant biomass. A progressive seasonal reduction in the percent cover of seagrass in the estuary was apparent during each year of the study and correlated well with diminishing eelgrass biomass. The percent cover of seagrass in the study area over the June to November period dropped from 45% to 21% in 2004, 43% to 16% in 2005, and 32% to 19% in 2006. The reduced growth and spatial cover of *Z. marina* during 2006 correlated with lower shoot density and biomass measurements. Declining seagrass abundance and biomass in the coastal lagoons of New Jersey signal an ongoing negative response to nitrogen over-enrichment, which has also been correlated with the occurrence of phytoplankton and benthic macroalgal blooms which are detrimental to seagrass habitat. Macroalgae completely cover extensive areas of the bay bottom during blooms, particularly when comprised of sheet-like forms such as *Ulva lactuca*. These blooms appear to cause significant dieback of seagrass in some areas of the estuary. Recovery of seagrass beds in the estuary will depend on management intervention, including an accelerated remediation program involving a reduction of nitrogen loading from coastal watershed areas.

Key Words: Barnegat Bay–Little Egg Harbor Estuary, seagrass, *Zostera marina*, nitrogen loading, eutrophication

8.1 INTRODUCTION

Eutrophication poses a serious threat to the long-term health of estuaries throughout the world (Kennish 2002a; Orth et al. 2006a). On a global basis, the two dominant sources of nitrogen to estuaries and coastal ecosystems are inorganic fertilizers and fossil fuels. Inputs of nitrogen from both of these sources are intensifying (Howarth et al. 2002), and forecasts indicate significant continued degradation of estuarine habitats (Kennish 2002a; Bricker et al. 2007). Nitrogen over-enrichment adversely affects water quality and the benthic habitat regime by promoting phytoplankton and macroalgal blooms, epiphytic growth on seagrass blades and shoots, and anoxic conditions in bottom sediments (Moore and Wetzel 2000; Kennish 2001; Brush and Nixon 2002; Campbell et al. 2003; Irlandi et al. 2004; Lamote and Dunton 2006; Kennish et al. 2007; McGlathery et al. 2007). In addition, high concentrations of nitrate in the water column can be toxic to seagrasses (Burkholder et al. 1992, 1994). Anoxia and associated sulfide production have been linked to the overgrowth of thick macroalgal mats on seagrass bed surfaces (Carlson et al. 2004), and sulfide buildup in bottom sediments is particularly detrimental to benthic organisms in these habitats. For example, Pregnall et al. (1984) noted that nutrient uptake and the energy status of eelgrass (*Zostera marina*) decreased as sulfide accumulated in the rhizosphere. Much (>70%) of the recent loss of seagrass beds in many estuaries has been attributed to eutrophication (Plus et al. 2003; Lamote and Dunton 2006).

High nutrient loading and water column turbidity typically elicit aberrant seagrass responses that compromise the ecosystem services the beds provide, notably organic carbon production, enhanced biodiversity, and habitat stabilization (Larkum et al. 2006). Seagrasses exhibit both physiological and morphological responses to nitrogen enrichment (Lee et al. 2004). Altered seagrass habitat commonly impacts resource species, such as bay scallops (*Argopecten irradians*), hard clams (*Mercenaria mercenaria*), blue mussels (*Mytilus edulis*), blue crabs (*Callinectes sapidus*), and fish (*Cynoscion regalis*) (Bologna 2006; Orth et al. 2006b). In severe cases, the food web structure changes significantly (Rabalais 2002). Disturbance of seagrass beds due to nutrient over-enrichment can be as damaging as major storms, oil spills, dredging events, wasting disease, or other stressors (Plus et al. 2003). Because of the devastating ecosystem effects of nutrient over-enrichment, early detection and intervention are necessary for effective management actions (Lee et al. 2004).

Seagrasses are important bioindicators of water and sediment quality, and thus have been used in ecological studies as biological sentinels (Hemminga 1998; Duarte, 1999; Orth et al. 2006a). As demonstrated by Longstaff and Dennison (1999) and Carruthers et al. (2002), seagrasses integrate environmental impacts over considerable timescales. Therefore, they are extremely valuable for assessing the environmental condition of water bodies. Historically, the declines of seagrass beds have been most often ascribed to anthropogenic factors, especially excess nutrients and turbidity that adversely affect the structure and function of shallow estuarine systems (Lee et al. 2004; Orth et al. 2006a). Shallow coastal lagoons are particularly susceptible to nutrient loading because they typically have relatively long water residence times, low freshwater inflow and dilution, and they tend to retain nutrients in plant biomass and sediments. Although the consequences of eutrophication in these bays have been well chronicled, predictive models have been more elusive because of insufficient data on the mechanisms responsible for the cascading impacts that typically arise. Current conceptual models of coastal bay eutrophication reveal little change in the total system metabolism but significant change in biotic structure (McGlathery et al. 2007). Seagrasses have high light requirements (Dennison et al. 1993), and light attenuation due to elevated turbidity generally leads to degrading habitat conditions.

Most of the seagrass beds in New Jersey (~75%) occur in the Barnegat Bay–Little Egg Harbor Estuary. While more than 6000 ha of seagrass habitat have been reported in this system (Lathrop et al. 2006), some studies indicate that significant losses of seagrass have taken place during the past 30 years, possibly reducing the beds by 30% to 60% (Bologna et al. 2000). A GIS spatial comparison analysis of submerged aquatic vegetation (SAV) surveys by Lathrop et al. (2001) suggests that a contraction of the seagrass beds to shallow subtidal areas (<2 m) may have occurred during this period due to a decrease in available light in response to phytoplankton and macroalgal blooms as well as epiphytic attenuation. Eutrophy has progressively increased in the estuary. Designated as moderately eutrophic in the early 1990s, the estuary was later classified as highly eutrophic by application of NOAA's National Ecosystem Assessment Model (Bricker et al. 2007; Kennish et al. 2007).

Increased nutrient loading to the Barnegat Bay–Little Egg Harbor Estuary has raised concern over potential impacts on seagrass, as well as biotic communities and other habitats in the system (Hunchak-Kariouk and Nicholson 2001). Nutrient enrichment and associated organic carbon loading in this shallow coastal bay system have been linked to an array of cascading environmental problems such as increased micro- and macroalgal growth, harmful algal blooms (HABs), high turbidity, altered benthic invertebrate communities, impacted harvestable fisheries, and loss of essential habitat (e.g., seagrass and shellfish beds) (Kennish 2001; Kennish et al. 2007). The net insidious effect of progressive eutrophication is the potential for the permanent alteration of biotic communities and debilitating ecosystem-level impacts. Because the estuary is shallow, poorly flushed with low freshwater inflow, and bordered by highly developed watershed areas, it is particularly susceptible to the effects of nutrient loading and other anthropogenic stressors. Consequently, research and monitoring programs are targeting priority indicators of eutrophication and associated water quality changes in both Barnegat Bay and Little Egg Harbor.

In this study, we examine the demographics of *Z. marina* in the estuary over the 2004 to 2006 study period to determine its status and trends. We also discuss the environmental factors responsible for changes in the structure and function of the seagrass beds. Assessment of the distribution, abundance, and biomass of seagrasses is important for tracking escalating eutrophic problems. To this end, comprehensive *in situ* sampling of seagrass beds was conducted in the estuary during the spring to fall period in 2004, 2005, and 2006. This *in situ* work was preceded by aerial mapping of the beds in 2003, which generated a seagrass-bottom-type map for the estuary (Lathrop et al. 2006).

A major scientific question addressed by this study is whether the biotic responses to nitrogen enrichment of the estuary represent a stable, continuous gradient or exhibit significant seasonal and interannual variability. The principal objectives therefore are to determine: (1) the spatial habitat changes of seagrass in the estuary over annual growing periods; (2) the species composition, relative abundance, spatial distribution, and potential impacts of macroalgae on the seagrass beds; and (3) the occurrence and impacts of nuisance and toxic phytoplankton blooms, notably brown tide (*Aureococcus anophagefferans*). The overall hypothesis, in turn, is that nutrient enrichment of the estuary correlates with increasing negative biotic responses.

8.2 STUDY AREA

Barnegat Bay–Little Egg Harbor is a lagoonal estuary located along the central New Jersey coastline (Figure 8.1). The irregular tidal basin is about 70 km long, 2 to 6 km wide, and 1.5 m deep, with a wet surface area of about 280 km^2 and a volume of 3.54×10 m^8 (Kennish 2001). The barrier island complex (Island Beach and Long Beach Island) restricts exchange of estuarine water with the coastal ocean and contributes to a protracted flushing time of about 74 days in summer. Exchange of bay and ocean water occurs through Barnegat Inlet, Little Egg Inlet, and the Point Pleasant Canal.

The adjoining Barnegat Bay watershed covers an area of 1730 km^2, and the watershed:estuary areal ratio is 6.5:1. A total of 562,493 people live in the surrounding watershed year round, but the population approaches 1,500,000 people during the summer tourist season. A north-to-south gradient of decreasing population density occurs in the watershed. As a result, nutrient loading is highest in the northern segment of the estuary (Seitzinger et al. 2001), where low dissolved oxygen levels have also been recorded.

8.3 MATERIALS AND METHODS

8.3.1 Sampling Design

We established a series of 12 transects within four distinct and representative seagrass beds located in Little Egg Harbor (~1700 ha) and Barnegat Bay (~1550 ha) (Figure 8.2). The seagrass beds occur on the eastern side of the estuary in well-sorted, fine to medium sands. We conducted water quality, quadrat, core, and hand sampling at multiple sampling stations along the transects over the spring to fall period (June to November in 2004, 2005, 2006). In 2004 and 2005, we sampled at 10 equally spaced stations along each of the 12 east–west trending transects (transects 1 to 6 in Little Egg Harbor in 2004 and transects 7 to 12 in Barnegat Bay in 2005), and each transect was sampled three times over a 6-month period (June to November). A total of 180 seagrass samples were collected at 60 transect stations each year, along with more than 1000 biotic and abiotic measurements. In 2006, seagrass samples were also collected three times over the June to November sampling period for 80 of the 120 sampling stations, and underwater videographic imaging was conducted at all 120 sampling stations to characterize the habitat and determine the percent cover of seagrass and macroalgae. We also recorded more than 1000 biotic and abiotic measurements during the 2006 sampling period.

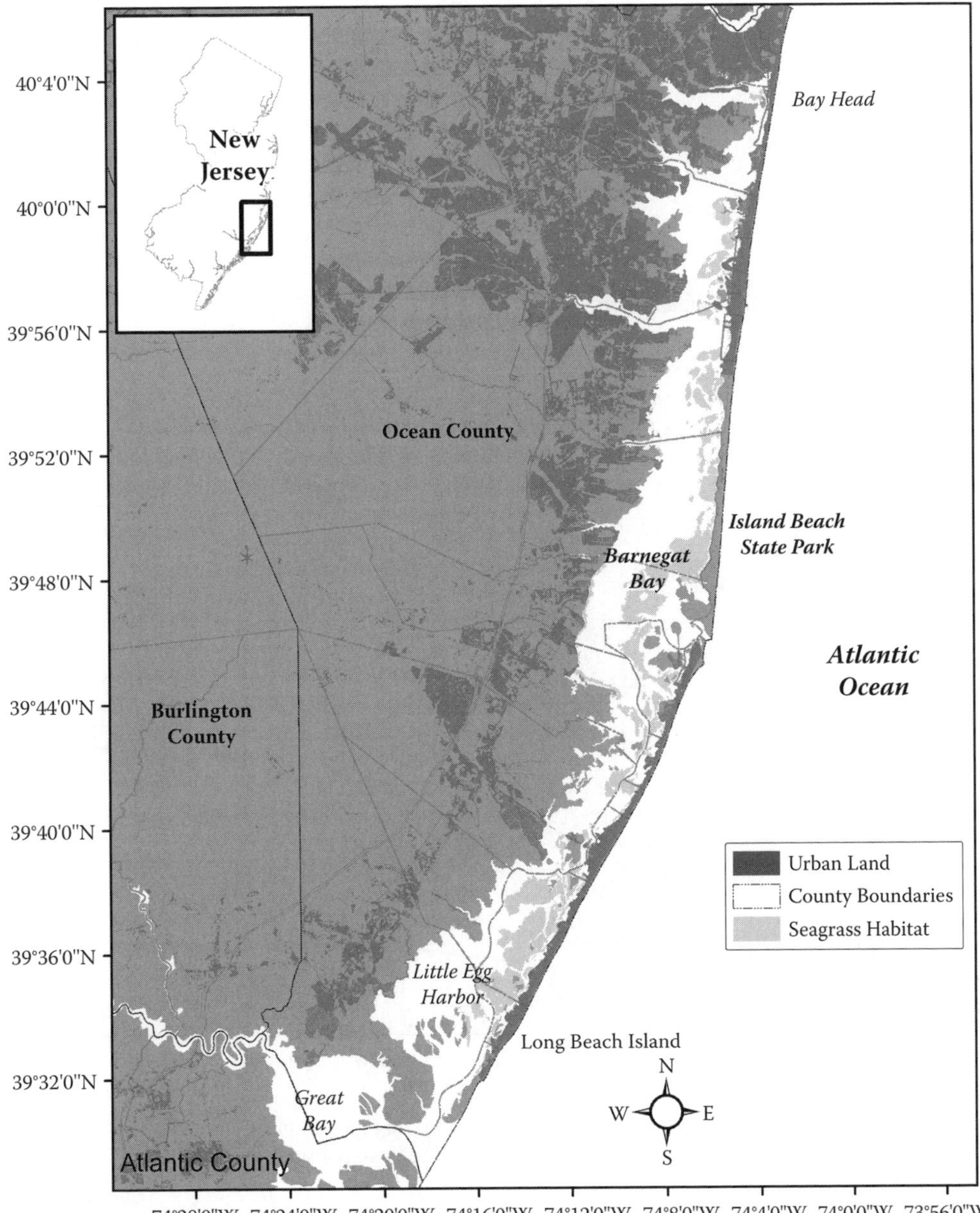

FIGURE 8.1 Barnegat Bay–Little Egg Harbor Estuary. Inset shows the location of the estuary with respect to the state of New Jersey.

Sampling stations along each transect were permanently located with a Differential Global Positioning System (Trimble®GeoXT™ handheld unit). Our three field sampling periods commenced in June, August, and October and continued until all targeted stations were sampled. No samples were collected after November in any year. The *in situ* sampling methods used in this study followed the standard protocols of Short et al. (2002).

The following demographic data were recorded on all sampling dates: (1) presence/absence (+/–) and spatial cover (% cover) of seagrass and macroalgae; (2) seagrass biomass (aboveground and belowground); (3) seagrass shoot density (in 2005 and 2006); and (4) seagrass blade length. Physicochemical data (temperature, salinity, pH, dissolved oxygen, turbidity, and depth) were also

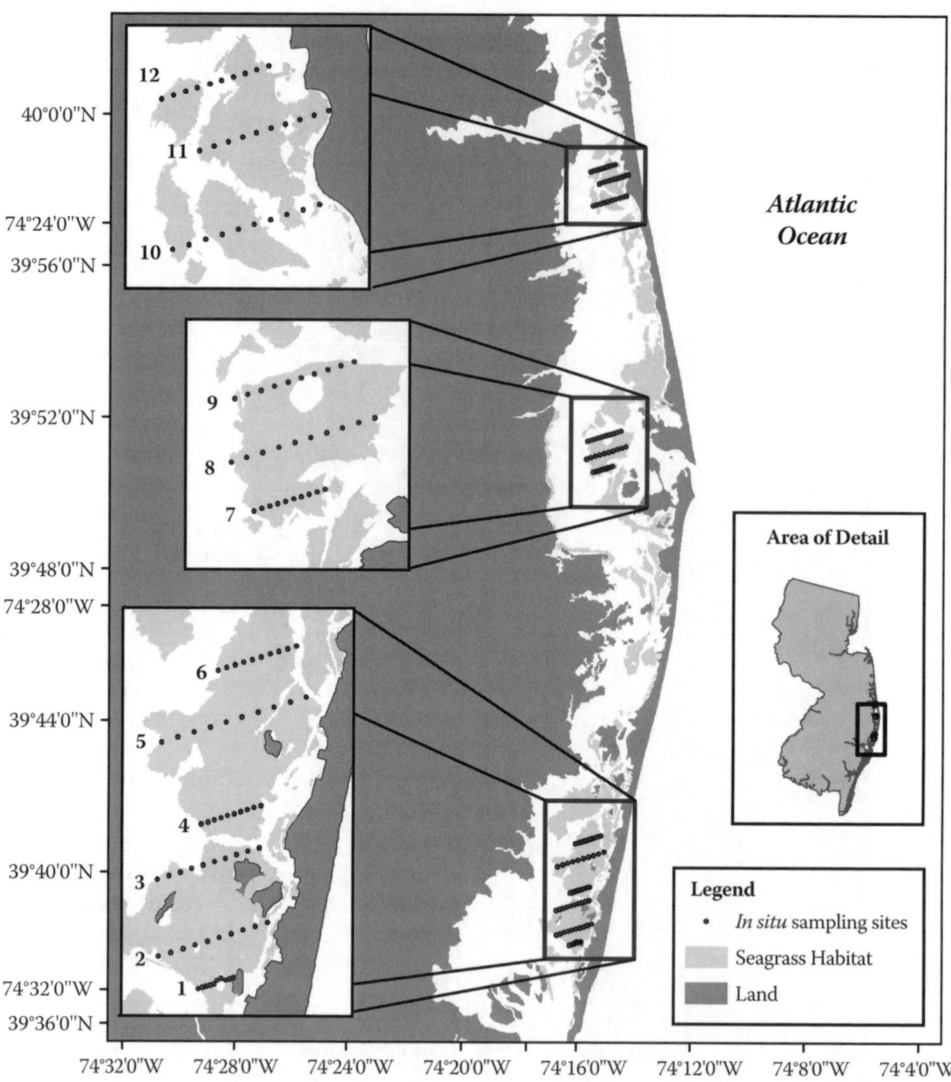

FIGURE 8.2 Barnegat Bay–Little Egg Harbor Estuary illustrating the sampling stations and transects used in this study. Inset shows the location of the study area with respect to the state of New Jersey.

collected using either a handheld YSI 600 XL data sonde coupled with a handheld YSI 650 MDS display unit, an automated YSI 6600 unit (equipped with a turbidity probe), or a YSI 600 XLM automated datalogger. However, in 2006 depth measurements were taken using a depth gauge. Secchi disk measurements were likewise collected in the survey area together with measurements of several nutrient parameters (i.e., nitrate plus nitrite, ammonium, total dissolved nitrogen, phosphate, and silica). Sediment composition (percent sand, silt, and clay) was determined at all sampling stations.

8.3.2 Sampling Methods

8.3.2.1 Quadrat Sampling

Field sampling was conducted within the shallow seagrass beds by divers following the field methods of Short et al. (2002). At each sampling station, a 0.25-m^2 metal quadrat was haphazardly tossed at the sampling stations to obtain measurements of seagrass and macroalgae presence or absence and areal coverage. The percent cover of seagrass and macroalgae was estimated within each quadrat on a scale of 0 to 100% with 5% increments. Lengths were measured with a meter-stick for a subset of the seagrass blades, and the mean values were recorded. The diver then visually inspected the seagrass bed within the quadrat for evidence of grazing, boat scarring, macroalgae, epiphytic loading, and eelgrass wasting disease.

8.3.2.2 Macroalgae Sampling

A diver collected macroalgae samples at each of the sampling stations during 2004 and 2005. These samples were removed from the seagrass bed and placed in 1-L nalgene bottles containing formalin adjusted to approximate ambient salinities. The macroalgae samples were subsequently transported to the Rutgers University Marine Field Station (RUMFS) and later examined for taxonomic identification.

8.3.2.3 Core Sampling

We used a 10-cm (0.00785-m^2)-diameter PVC coring device to collect the seagrass samples, with care taken not to cut or damage the aboveground plant tissues, following the methods of Short et al. (2002). The diver-deployed corer extended deep enough into the soft sediment to extract all belowground fractions (roots and rhizomes). Each core was placed into a mesh bag (3 × 5 mm mesh) and rinsed in the field to separate plant material from the sediment. After removing the seagrass sample from the mesh bag, the sample was placed in a labeled bag and stored on ice in a closed container prior to transport back to RUMFS. The seagrass samples were carefully sorted in the laboratory and separated into aboveground (shoots) and belowground (roots and rhizomes) components. The aboveground and belowground fractions were then oven dried at 50°C to 60°C for a minimum of 48 hours. The dry weight biomass (g dry wt m^{-2}) of each fraction was then measured to the third decimal place.

8.3.2.4 Sediment Sampling

Sediment cores were collected at each sampling site during 2004 and 2005 to a depth of ~15 cm using a 10-cm-diameter coring device. These samples were analyzed in the laboratory for the percent composition of sand and silt (dry sieving) as well as clay (wet sieving through a 63-μm sieve). The sand component was further analyzed for five component size classes.

8.3.2.5 Water Quality Sampling

Water quality parameters (temperature, salinity, dissolved oxygen, pH, and turbidity) were measured at all sampling stations using the protocols noted above. The data were obtained prior to biotic sampling at each sampling site. Water quality data were collected at a uniform depth (~10 cm) above the sediment–water interface.

Water samples were also collected and analyzed for nutrient concentrations during each sampling period. Nitrate plus nitrite, ammonium, total dissolved nitrogen, phosphate, and silica concentrations were measured at sampling stations within the seagrass beds. Laboratory analysis of the nutrients followed standard methods.

8.3.2.6 Videographic Imaging

Underwater video images of bottom habitats were obtained using a high-resolution digital video camera and recording unit as described by Kennish et al. (2006) and Haag et al. (2008). The camera (i.e., Seaviewer Sea Drop Camera) is a color unit attached to the video in channel on the recording device. The recorder is a SONY mini-DVD camera system (Model# DCR-DVD301). It incorporates the audio signal from the Red Hen Systems Video Mapping System (VMS 300) and the video signal from the underwater camera. The camera records data in a DVD format, which is later downloaded to a personal computer and into a Geographic Information System for analysis. Numerous images of the seagrass beds were collected with the Sea Drop Camera along the sampling transects.

8.4 RESULTS: PHYSICOCHEMICAL CONDITIONS

Table 8.1 summarizes the physicochemical data collected in this study. Mean water temperature over the three sampling periods each year (i.e., June–July, August–September, and October–November) ranged from 11.29°C to 24.03°C, with highest temperatures recorded during the August–September period. Salinities were less variable, ranging between 22.65‰ and 30.42‰. The highest salinities occurred in 2004 (29.59‰ to 30.42‰), and the lowest salinities in 2005 (22.65‰ to 25.07‰). Greater freshwater inflow in spring and fall produced lower salinities than in mid-summer. The pH values were consistent across the sampling stations (7.63 to 8.17); lowest mean pH (7.63 to 7.70) occurred in 2004. No anoxic or hypoxic events were observed, as dissolved oxygen measurements ranged from 6.78 to 9.83 mg/L. The estuary is shallow, well mixed, and often saturated or supersaturated with oxygen. Turbidity varied considerably from a mean value of 1.09 to 6.41 NTU. Although Secchi depth was commonly less than 1 m during the summer (Lathrop et al. 2006), it averaged ~2 to 2.5 m. Unlimited visibility was documented most frequently in spring and fall at times of low phytoplankton abundance.

Bottom sediments in the seagrass beds consisted mainly of silt and sand. At nearly all transect stations, for example, the total amount of sand and silt exceeded 80%. The highest concentration

TABLE 8.1
Mean Values of Key Physicochemical Parameters Recorded in the Barnegat Bay–Little Egg Harbor Estuary during Three Sampling Periods in 2004, 2005, and 2006

	Parameter				
Sample Period	Temp (°C)	Salinity (‰)	DO (mg/L)	pH	Turbidity (NTU)
2004-1	20.97	30.01	7.91	7.64	1.88
	(2.87)	(0.38)	(1.53)	(0.15)	(1.50)
2004-2	24.79	30.42	8.53	7.63	2.23
	(1.11)	(1.12)	(1.42)	(0.21)	(1.99)
2004-3	11.29	29.59	10.49	7.70	1.09
	(1.52)	(1.27)	(1.48)	(0.13)	(0.64)
2005-1	22.23	22.65	8.69	8.06	1.85
	(1.99)	(4.04)	(1.55)	(0.15)	(3.19)
2005-2	24.63	25.07	6.78	7.97	3.40
	(1.29)	(2.99)	(1.01)	(0.10)	(2.43)
2005-3	12.26	23.52	8.94	8.10	6.41
	(3.67)	(4.51)	(0.32)	(0.18)	(3.20)
2006-1	21.94	26.67	8.44	8.17	—
	(2.90)	(4.21)	(0.78)	(0.12)	
2006-2	24.03	25.67	7.37	8.13	—
	(7.14)	(4.40)	(1.25)	(0.22)	
2006-3	14.37	24.97	9.83	7.91	—
	(3.31)	(3.65)	(0.94)	(0.28)	

Note: Sample Period 1 = June–July; Sample Period 2 = August–September; Sample Period 3 = October–November; Standard Deviation Values in Parentheses

TABLE 8.2
Range of Nutrient Values (μM) Recorded in the Barnegat Bay–Little Egg Harbor during the Three Sampling Periods in 2004, 2005, and 2006

NO_3^- plus NO_2^-	NH_4^+	TDN	PO_4	Si
		2004		
0.1–0.3[a]	0.0–2.1	8.2–15.3	0.03–1.21	9.8–26.4
0.0–0.8[b]	0.0–1.5	0.0–24.2	0.67–0.89	0.0–18.7
		2005		
0.00–0.58[a]	0.60–8.30	8.40–14	0.00–0.24	10.67–37.01
0.00–0.32[b]	0.02–4.50	0.10–28	0.00–1.00	0.00–44.8
		2006		
0.00–1.20[a]	0.00–2.60	1.20–16.6	0.10–2.60	19.00–57.00
0.00–2.70[b]	0.10–34.00	1.10–33.60	0.00–24.70	0.00–107.3
0.00–2.90[c]	0.00–5.70	0.00–18.70	0.00–1.10	0.00–50.3

[a] Sample Period 1 = June–July
[b] Sample Period 2 = August–September
[c] Sample Period 2 = October–November

of sand was recorded at the northernmost transect stations in the estuary (transects 11 and 12; see Figure 8.2).

The concentrations of dissolved inorganic nutrients are low during the warmer months, primarily due to uptake by seagrasses, macroalgae, and phytoplankton (Table 8.2), a finding also reported by Seitzinger et al. (2001). For example, nitrate plus nitrite levels ranged from 0 to 2.9 μM during the June–September sampling period each year, and the highest levels occurred in 2006. A wider range of ammonium values was recorded (0 to 8.30 μM), with highest measurements in 2005. Total dissolved nitrogen concentrations were similar from year to year, ranging from 0 to 24.2 μM in 2004, 0.10 to 28.0 μM in 2005, and 10.2 to 33.6 μM in 2006. Phosphate levels (0 to 1.21 μM) were lower than those of nitrate and ammonium, typically less than 1 μM. Silica concentrations (0 to 107.3 μM) were usually much higher than nitrate or phosphate, peaking in 2006.

8.5 BIOTIC INDICATORS

8.5.1 Seagrass Distribution

The demographic analysis focused on *Z. marina*, the dominant seagrass species in the estuary, because few *Ruppia maritima* samples were collected in the current study or the recent past. For example, Bologna et al. (2001) found that eelgrass covered an area of 1299 ha in Little Egg Harbor in 1998, while the aerial extent of widgeon grass was only 6.8 ha. Similarly, Lathrop et al. (2001) reported even greater areal coverage of *Z. marina* in the estuary, amounting to ~2000 ha vs. only 5 ha for *R. maritima*.

In the present study, we observed eelgrass along all transects sampled at depths ranging from <1 to 2 m. However, the biomass and areal coverage of *Z. marina* varied considerably both in space and time. Conspicuous declining trends were evident in 2004 and 2005, with the highest biomass and percent cover of eelgrass occurring during the June–July period and gradually decreasing values observed into the fall season.

8.5.2 Aboveground Biomass

Biomass of *Z. marina* was measured at 60 stations in Little Egg Harbor in 2004, at 60 stations in Barnegat Bay in 2005, and at 80 stations in both Little Egg Harbor and Barnegat Bay in 2006. The sampling stations covered a broad area in the estuary and a range of habitat conditions. Seagrass samples were collected during three sampling periods each year: (1) June–July; (2) August–September; and (3) October–November.

In 2004, aboveground biomass peaked during the June–July sampling period and then declined through November (Figure 8.3). This pattern held for all transects except transect 2 (Figure 8.4). For example, the mean aboveground biomass of the seagrass samples collected during the June–July sampling period (106.05 g dry wt m^{-2}) was nearly twice that during the August–September (54.61 g dry wt m^{-2}) period and more than five times that during the October–November (18.22 g dry wt m^{-2}) period. The highest mean aboveground biomass (87.20 g dry wt m^{-2}) was recorded at the northernmost transect (6) in Little Egg Harbor, and the lowest mean aboveground biomass (38.76 g dry wt m^{-2}) was observed at transect 1 (Table 8.3). An analysis of variance (ANOVA) test showed statistically significant differences ($F = 45.02$; $P < 0.0001$) in the aboveground biomass values between the three sampling periods. A Tukey HSD test applied to the data revealed that all mean aboveground biomass values per sampling period were significantly different.

A similar temporal pattern of decreasing aboveground biomass of *Z. marina* was found in Barnegat Bay during 2005 (Figure 8.5). This pattern was clearly evident at most of the transects (Figure 8.6). For example, the mean aboveground biomass of the *Z. marina* samples collected during the June–July sampling period in 2005 (51.69 g dry wt m^{-2}) was nearly twice that during the August–September (28.79 g dry wt m^{-2}) period and more than three times that during the October–November (15.66 g dry wt m^{-2}) period. During the 2005 study period, mean aboveground biomass of *Z. marina* was highest at transect 9 (80.23 g dry wt m^{-2}). The lowest mean aboveground biomass value (5.65 g dry wt m^{-2}) was recorded at transect 12 (Table 8.4). An ANOVA test showed statistically significant differences in the aboveground biomass values of *Z. marina* among the three sampling periods in 2005

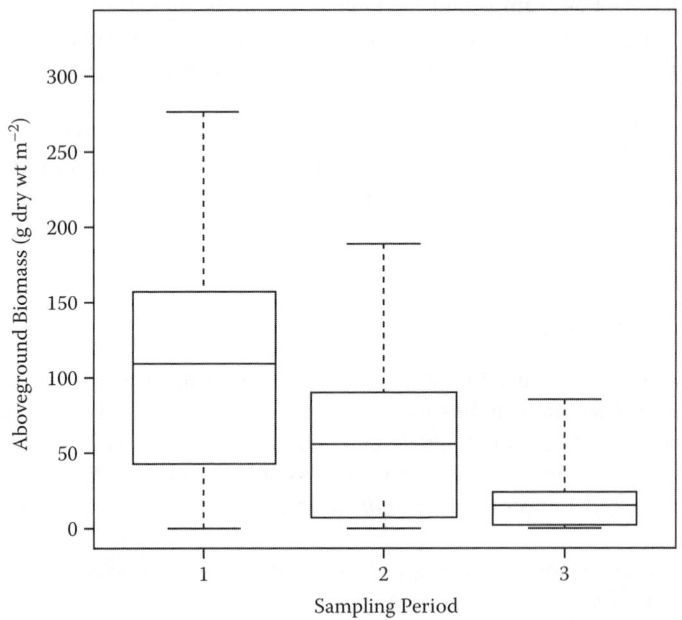

FIGURE 8.3 Mean aboveground biomass of all *Z. marina* samples collected in Little Egg Harbor during three sampling periods in 2004. Sampling period 1, June–July; sampling period 2, August–September; and sampling period 3, October–November.

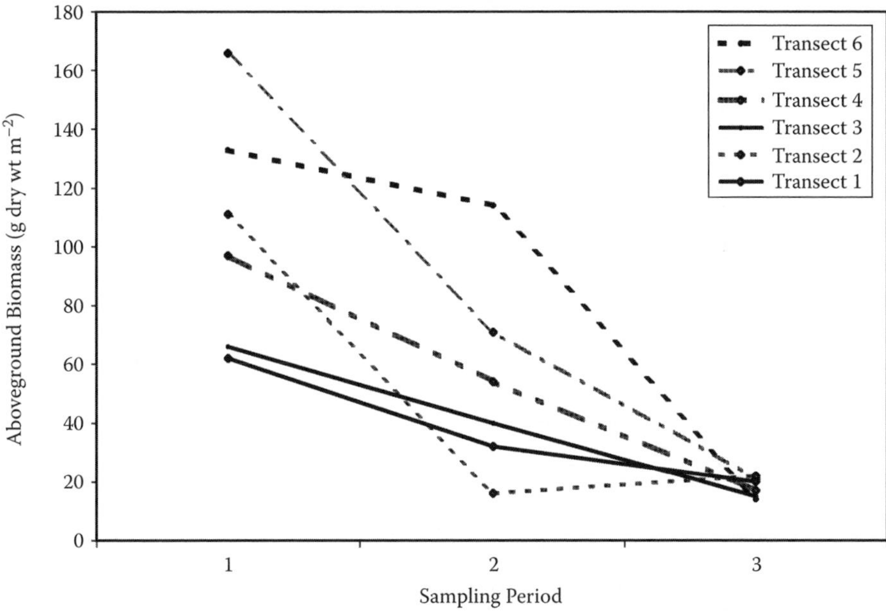

FIGURE 8.4 Mean aboveground *Z. marina* biomass at six transects (1 to 6) in Little Egg Harbor during three sampling periods in 2004. Sampling period 1, June–July; sampling period 2, August–September; and sampling period 3, October–November.

TABLE 8.3
Mean Aboveground and Belowground Biomass of *Zostera marina* along Six Sampling Transects in Little Egg Harbor during the June–November Period in 2004

Transect	Mean Aboveground Biomass	Mean Belowground Biomass	Aboveground/Belowground Biomass Ratio
6	87.20	131.01	0.67
5	86.24	104.48	0.82
4	55.69	52.41	1.06
3	40.13	44.80	0.90
2	49.72	72.24	0.69
1	38.76	48.67	0.78

Note: Values in g dry wt m^{-2}.

($F = 7.349$; $P = 0.0008$). A Tukey HSD test applied to the data revealed that the mean aboveground biomass values were significantly different between sampling periods 1 and 2 and sampling periods 1 and 3.

The mean aboveground biomass of the *Z. marina* samples collected during the June–July sampling period in 2006 (11.35 g dry wt m^{-2}) was nearly equal to that for the August–September period (13.77 g dry wt m^{-2}) and the October–November (12.74 g dry wt m^{-2}) period (Figure 8.7). An ANOVA test showed that the differences in aboveground biomass values of *Z. marina* were not statistically significant between the three sampling periods in 2006 ($F = 0.18$; $P = 0.8314$). A paired *t*-test was used to compare aboveground biomass between 2004 and 2006 and between 2005 and

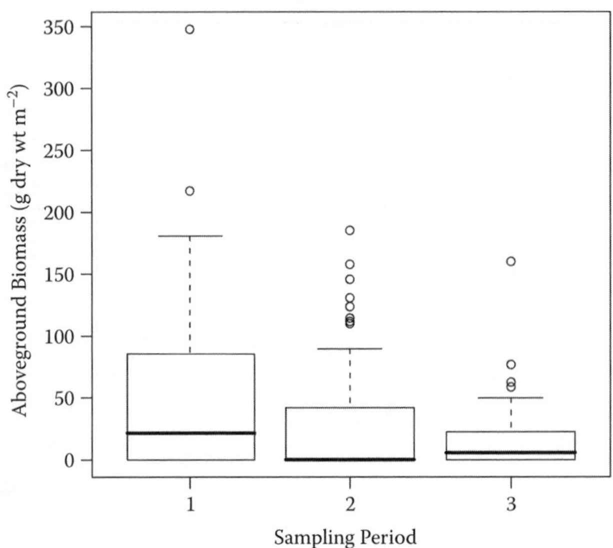

FIGURE 8.5 Mean aboveground biomass of all *Z. marina* samples collected in Barnegat Bay during three sampling periods in 2005. Sampling period 1, June–July; sampling period 2, August–September; and sampling period 3, October–November.

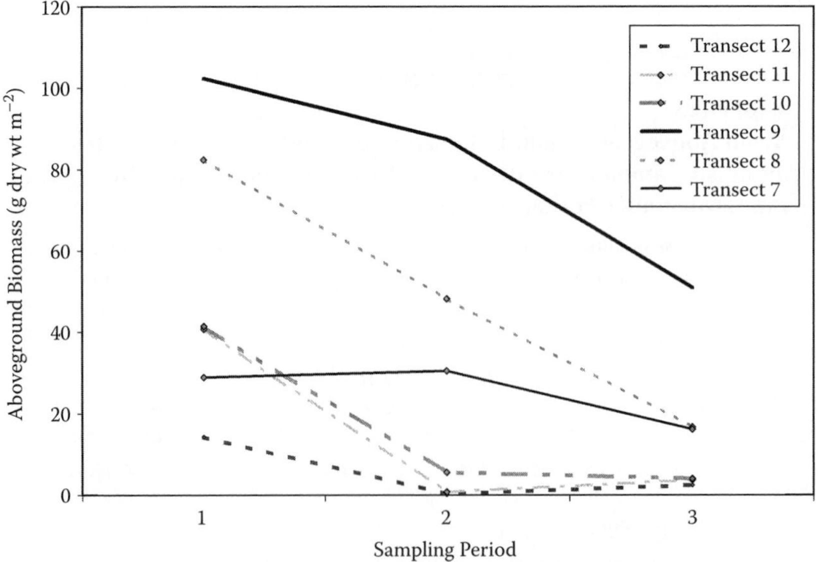

FIGURE 8.6 Mean aboveground biomass of *Z. marina* at six transects (7 to 12) in Barnegat Bay during three sampling periods in 2005. Sampling period 1, June–July; sampling period 2, August–September; and sampling period 3, October–November.

2006. Both tests showed significant differences at the 95% CI, with the 2004–2006 test p-value = 8.21×10^{-15} and the 2005–2006 p-value = 2.7×10^{-4}.

In 2006, the highest aboveground biomass values of *Z. marina* were found along transects 7 to 9 (Figure 8.8). The highest mean aboveground biomass value (41.60 g dry wt m^{-2}) was recorded at transect 9, and the lowest mean aboveground biomass value (0.79 g dry wt m^{-2}), at transect 12 (Table 8.5). The biomass measurements were lowest for the northernmost and southernmost

TABLE 8.4
Mean Aboveground and Belowground Biomass of *Zostera marina* along Six Sampling Transects in Barnegat Bay during the June–November Period in 2005

Transect	Mean Aboveground Biomass	Mean Belowground Biomass	Aboveground/Belowground Biomass Ratio
12	5.65	41.47	0.14
11	15.04	85.00	0.18
10	16.99	53.37	0.32
9	80.23	140.23	0.57
8	49.13	121.17	0.41
7	25.23	66.30	0.38

Note: Values in g dry wt m^{-2}.

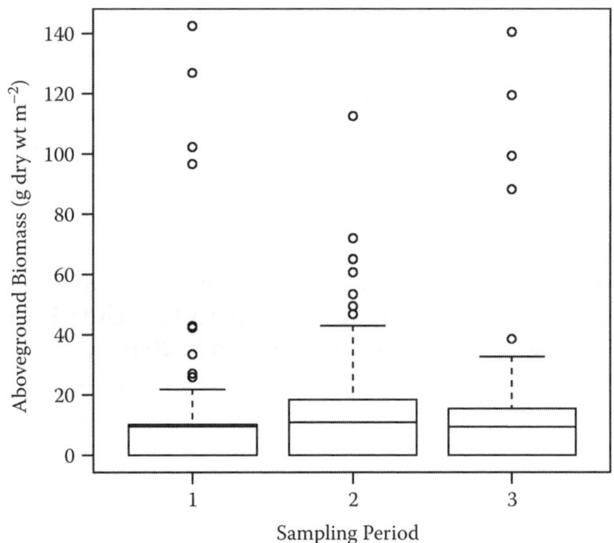

FIGURE 8.7 Mean aboveground biomass of all *Z. marina* samples collected in the Barnegat Bay–Little Egg Harbor Estuary during three sampling periods in 2006. Sampling period 1, June–July; sampling period 2, August–September; and sampling period 3, October–November.

transects. The aboveground biomass measurements for Barnegat Bay (transects 7 to 12) in 2006 were all less than the biomass values for these transects in 2005.

The mean aboveground biomass values for the interior sampling stations (3, 4, 5, 6, 7, and 8) exceeded those for the exterior sampling stations (1, 2, 9, and 10) during all 3 years of sampling. Environmental conditions were generally more favorable for seagrass growth in the interior areas of each bed than near the edges. At the deeper marginal sampling stations (9 and 10), light attenuation played a more important role in limiting seagrass biomass than it did in shallower interior stations.

8.5.3 Belowground Biomass

A distinct trend of decreasing belowground biomass of *Z. marina* was also observed each year, consistent with that of declining aboveground biomass. For example, the highest mean belowground

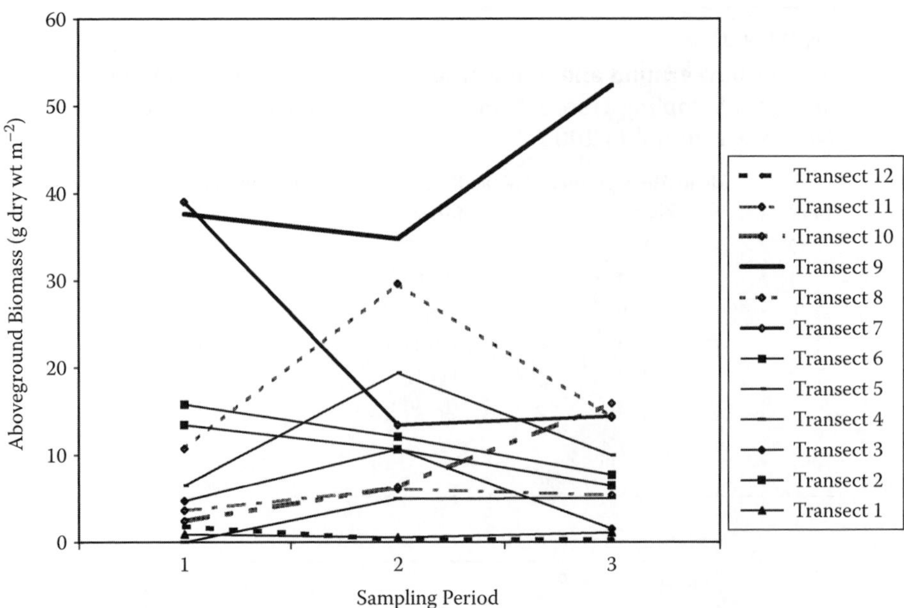

FIGURE 8.8 Mean aboveground biomass of *Z. marina* at 12 transects (1 to 12) in the Barnegat Bay–Little Egg Harbor Estuary during three sampling periods in 2006. Sampling period 1, June–July; sampling period 2, August–September; and sampling period 3, October–November.

TABLE 8.5
Mean Aboveground and Belowground Biomass of *Zostera marina* along 12 Sampling Transects in the Barnegat Bay–Little Egg Harbor Estuary during the June–November Period in 2006

Transect	Mean Aboveground Biomass	Mean Belowground Biomass	Aboveground/Belowground Biomass Ratio
12	0.79	11.23	0.07
11	5.04	22.62	0.22
10	8.21	18.74	0.44
9	41.60	81.73	0.51
8	18.24	59.76	0.31
7	22.32	47.45	0.47
6	11.88	56.97	0.21
5	11.95	61.30	0.19
4	3.33	9.93	0.34
3	5.64	36.18	0.16
2	10.18	17.89	0.57
1	0.85	1.88	0.45

Note: Values in g dry wt m^{-2}.

biomass of seagrass collected in 2004 was recorded during the June–July sampling period (107.64 g dry wt m^{-2}), and the lowest mean belowground biomass was observed during the October–November sampling period (50.48 dry wt m^{-2}). An intermediate mean belowground biomass value was obtained during the August–September sampling period (68.69 g dry wt m^{-2}) (Figure 8.9). Although the mean belowground biomass measurements for the June–July and August–September sampling

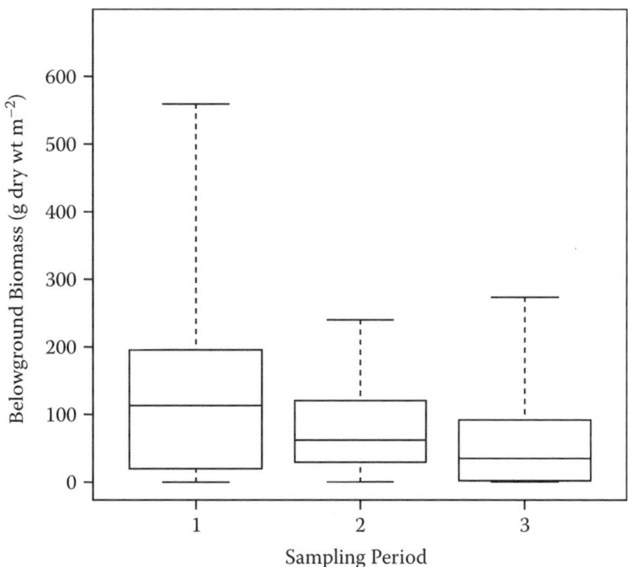

FIGURE 8.9 Mean belowground biomass of all *Z. marina* samples collected in Little Egg Harbor during three sampling periods in 2004. Sampling period 1, June–July; sampling period 2, August–September; and sampling period 3, October–November.

periods followed those of the mean aboveground biomass measurements for these periods, the mean belowground biomass for the October–November sampling period was nearly three times higher than that of the aboveground biomass. An ANOVA test used to compare belowground biomass values between the three sampling periods in 2004 showed statistically significant differences (F = 6.89; P = 0.0013). A Tukey HSD test applied to the data revealed that the mean belowground biomass values in Little Egg Harbor were significantly different between sampling periods 1 and 2 and sampling periods 1 and 3.

The highest mean belowground biomass (131.01 g dry wt m^{-2}) was recorded at the northernmost transect (6), and the lowest mean belowground biomass (44.80 g dry wt m^{-2}) was documented at transect 3 (Table 8.3). The biomass values for both aboveground and belowground samples followed similar spatial distribution patterns; however, temporal differences were evident. Belowground biomass values exhibited more variable temporal trends compared to aboveground biomass values, most notable between sampling periods 1 (June–July) and 2 (August–September), when the mean biomass increased along two transects (3 and 4) (Figure 8.10).

A distinct trend of decreasing belowground biomass was evident for *Z. marina* over the entire 6-month sampling period in 2005, consistent with that of declining aboveground biomass (Figure 8.11). For example, the highest mean belowground biomass of *Z. marina* samples was recorded during the June–July sampling period (141.95 g dry wt m^{-2}), and the lowest mean belowground biomass was registered during the October–November sampling period (42.78 dry wt m^{-2}). An intermediate mean belowground biomass value was obtained during the August–September sampling period (69.03 g dry wt m^{-2}). The mean belowground biomass measurements of *Z. marina* were substantially higher than the aboveground biomass measurements during all three sampling periods in 2005. An ANOVA test showed statistically significant differences in the belowground biomass values of *Z. marina* between the three sampling periods (F = 8.88; P = 0.0002). A Tukey HSD test applied to the data revealed that the mean belowground biomass values of *Z. marina* were significantly different between sampling periods 1 and 2 and sampling periods 1 and 3.

The highest mean belowground biomass during 2005 (140.23 g dry wt m^{-2}) was recorded for transect 9, and the lowest mean belowground biomass (41.47 g dry wt m^{-2}) was observed for

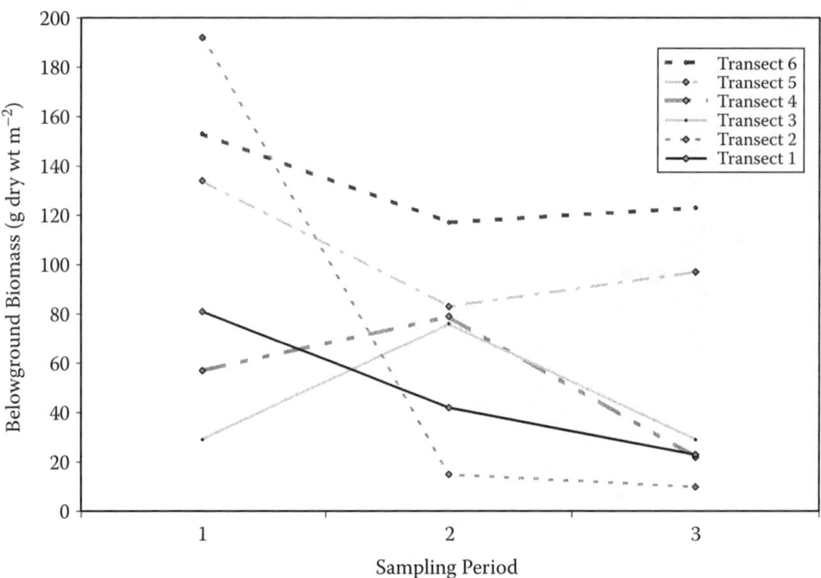

FIGURE 8.10 Mean belowground biomass of *Z. marina* at six transects (1 to 6) in Little Egg Harbor during three sampling periods in 2004. Sampling period 1, June–July; sampling period 2, August–September; and sampling period 3, October–November.

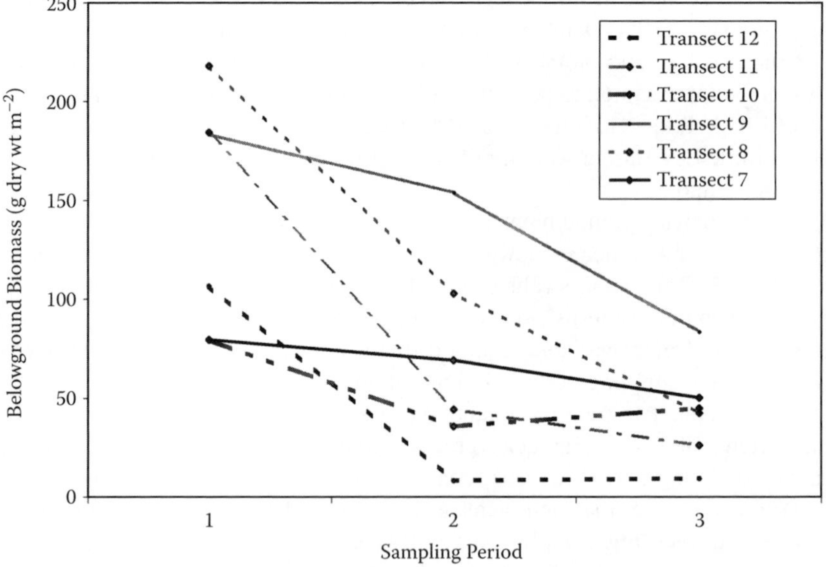

FIGURE 8.11 Mean belowground biomass of *Z. marina* at six transects (7 to 12) in Barnegat Bay during three sampling periods in 2005. Sampling period 1, June–July; sampling period 2, August–September; and sampling period 3, October–November.

transect 12 (Table 8.4). Belowground biomass followed similar spatial patterns along each transect. For example, interior sampling stations (3 to 8) along transects had significantly higher mean belowground biomass values during the 2005 sampling period than did exterior sampling stations (1, 2, 9, and 10), a condition also observed in 2004 and 2006.

The mean belowground biomass of *Zostera marina* was two to nearly five times greater than the mean aboveground biomass during 2006. A trend of decreasing belowground biomass of *Z. marina*

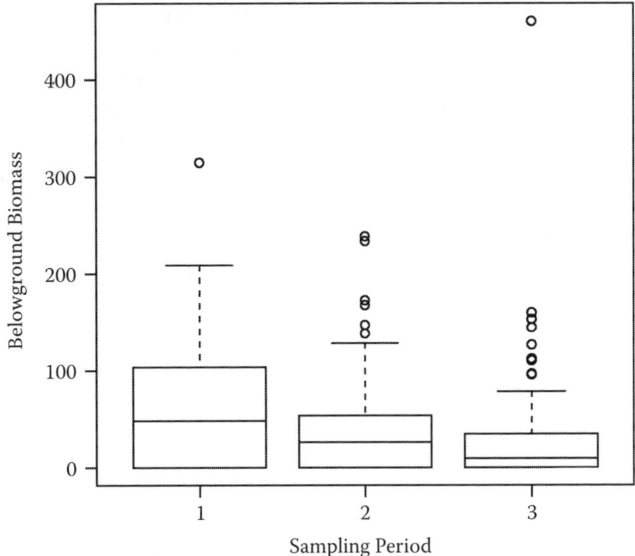

FIGURE 8.12 Mean belowground biomass of all *Z. marina* samples collected in the Barnegat Bay–Little Egg Harbor Estuary during three sampling periods in 2006. Sampling period 1, June–July; sampling period 2, August–September; and sampling period 3, October–November.

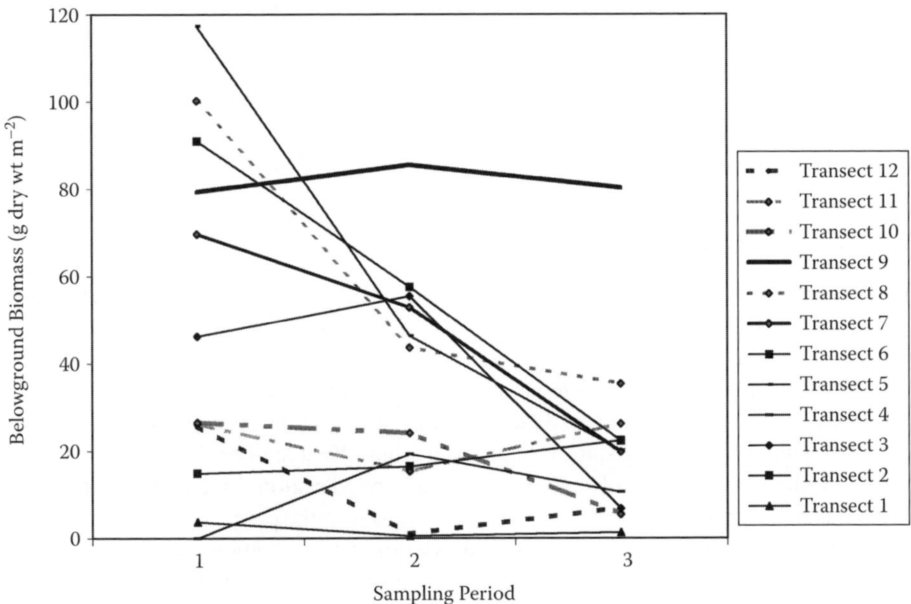

FIGURE 8.13 Mean belowground biomass of *Z. marina* at 12 transects (1 to 12) in the Barnegat Bay–Little Egg Harbor Estuary during three sampling periods in 2006. Sampling period 1, June–July; sampling period 2, August–September; and sampling period 3, October–November.

was evident over the 6-month sampling period (Figure 8.12), but the decline was not consistent from transect to transect (Figure 8.13). The highest mean belowground biomass during 2006 (81.73 g dry wt m^{-2}) was recorded for transect 9, and the lowest mean belowground biomass (1.88 g dry wt m^{-2}) was registered for transect 1 (Table 8.5). The mean belowground biomass of the *Z. marina* samples collected during the June–July sampling period in 2006 (51.54 g dry wt m^{-2}) exceeded that during the August–September period (36.08 g dry wt m^{-2}) and the October–November period (24.23 g dry wt m^{-2}). An ANOVA test used to compare belowground biomass values of *Z. marina* between the three sampling periods in 2006 showed that the differences were statistically significant ($F = 4.60$; $P = 0.0110$). A Tukey HSD test applied to the data revealed that the mean belowground biomass in 2006 was significantly different between sampling periods 1 (June–July) and 3 (October–November) but not significantly different between periods 1 (June–July) and 2 (August–September) and periods 2 (August–September) and 3 (October–November).

A paired *t*-test used to compare belowground biomass between 2004 and 2006 and between 2005 and 2006 showed a significant difference at the 95% CI, with the 2004 to 2006 test *p*-value = 1.4 10^{-4} and the 2005 to 2006 *p*-value = 1.8 10^{-4}. Belowground biomass usually exceeded aboveground biomass at the sampling stations during all 3 years of study. As a result, aboveground to belowground biomass ratios were less than 1.0 for nearly all transects in all years. The ratios were lowest in 2006, when both aboveground and belowground biomass values were significantly reduced relative to 2004 and 2005.

8.5.4 Seagrass Density

Counts were made of *Z. marina* density (shoots m^{-2}) over the June–November sampling period in 2005 and 2006. The highest density measurements in 2005 were recorded during June–July (mean = 479 shoots m^{-2}). Significantly lower densities of *Z. marina* were observed during August–September (mean = 163 shoots m^{-2}) and October–November (mean = 233 shoots m^{-2}). The decrease in *Z. marina* density later in the season followed the decline in eelgrass biomass shown previously.

Zostera marina density peaked in 2006 during the June–July period when the mean value was 378 shoots m^{-2}. Subsequently, density measurements decreased markedly, with a mean value of 171 shoots m^{-2} during the August–September period and 174 shoots m^{-2} during the October–November period. The seasonal trend of *Z. marina* density was consistent with that observed in 2005. A paired *t*-test applied to the data showed a significant difference in seagrass density between 2005 and 2006 at the 95% CI, $p = 0.0065$.

8.5.5 Seagrass and Macroalgae Cover

The mean percent cover of the estuarine floor occupied by seagrass was 45% during June–July (sample period 1), 38% in August–September (sample period 2), and 21% in October–November (sample period 3) in 2004 (Figure 8.14). Spatial cover of seagrass was considerably higher at the interior sampling stations (3 to 8) in comparison with the exterior sampling stations (1, 2, 9, and 10). In contrast, the spatial cover by macroalgae during these sampling periods was substantially less, averaging 13% (June–July), 21% (August–September), and 14% (October–November). As illustrated in Figure 8.15, the percent cover of macroalgae (trend and variation) by transect differed substantially from that of seagrass. Similar to seagrass cover, however, the percent cover of macroalgae was distinctly higher at interior transect stations than exterior transect stations during the study period. Both seagrass growth and areal coverage decreased within the deeper waters.

By comparison, seagrass exhibited a mean spatial cover of 37% during June–July, 43% during August–September, and 16% during October–November in 2005 (Figure 8.16). As in 2004, the percent cover of eelgrass was higher at interior sampling stations than exterior sampling stations. In contrast, the percent cover by macroalgae during these periods was substantially less, averaging 14% (June–July), 7% (August–September), and 2% (October–November) in 2005. As illustrated in Figure 8.17, the percent cover of macroalgae (trend and variation) by transect differed considerably from that of seagrass.

In 2006, the mean percent cover by seagrass during periods 1 (June–July), 2 (August–September), and 3 (October–November) was 32%, 23%, and 19%, respectively (Figure 8.18). By comparison, the percent cover by macroalgae during these periods was substantially less, averaging 2% (June–July),

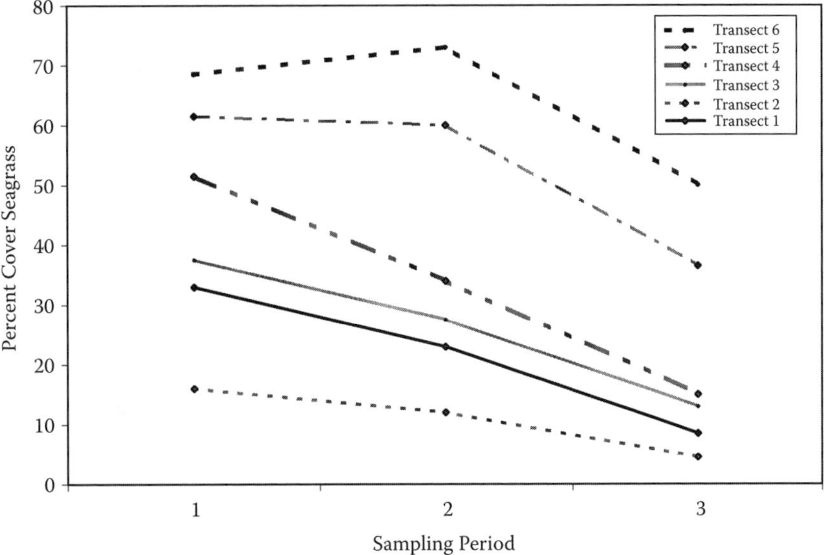

FIGURE 8.14 Mean percent cover of *Z. marina* at six transects (1 to 6) in Little Egg Harbor during three sampling periods in 2004. Sampling period 1, June–July; sampling period 2, August–September; and sampling period 3, October–November.

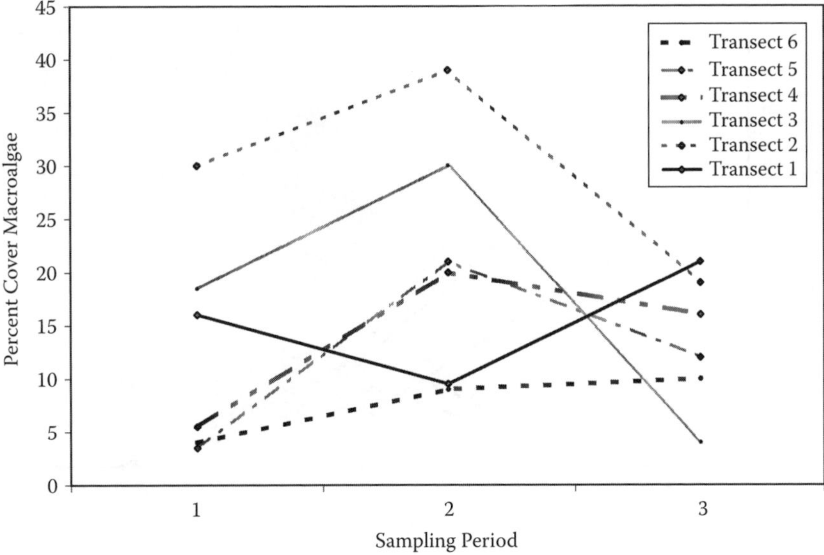

FIGURE 8.15 Mean percent cover of macroalgae at six transects (1 to 6) in Little Egg Harbor during three sampling periods in 2004. Sampling period 1, June–July; sampling period 2, August–September; and sampling period 3, October–November.

7% (August–September), and 7% (October–November) over the sampling period. As in prior years, the percent cover of macroalgae by transect differed greatly from that of seagrass. Similar to seagrass cover, however, the percent cover of macroalgae was generally higher at interior transect stations than exterior transect stations during the study period. The percent cover of seagrass and percent cover of macroalgae was significantly different between 2004 and 2006 as shown by application of a paired *t*-test (seagrass 2004 to 2006, 3.7×10^{-7}; macroalgae 2004 to 2006, 1.5×10^{-9}).

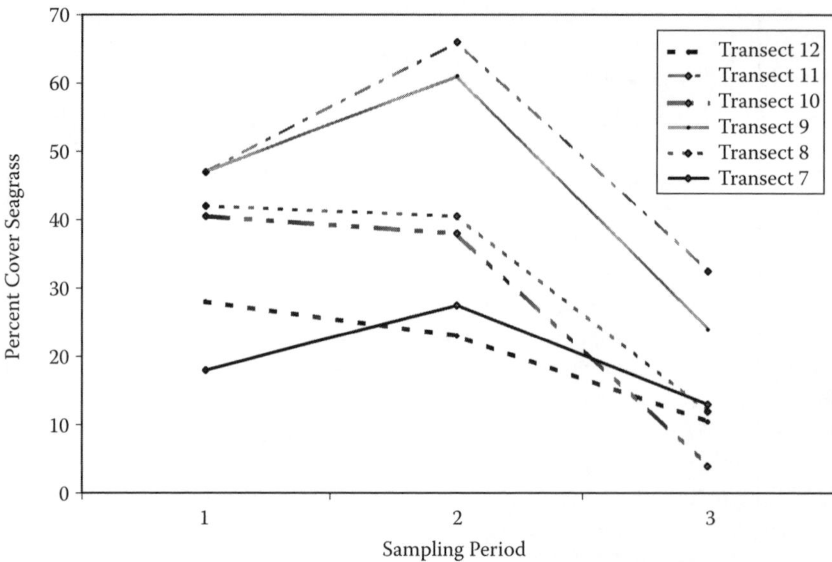

FIGURE 8.16 Mean percent cover of *Z. marina* at six transects (7 to 12) in Barnegat Bay during three sampling periods in 2005. Sampling period 1, June–July; sampling period 2, August–September; and sampling period 3, October–November.

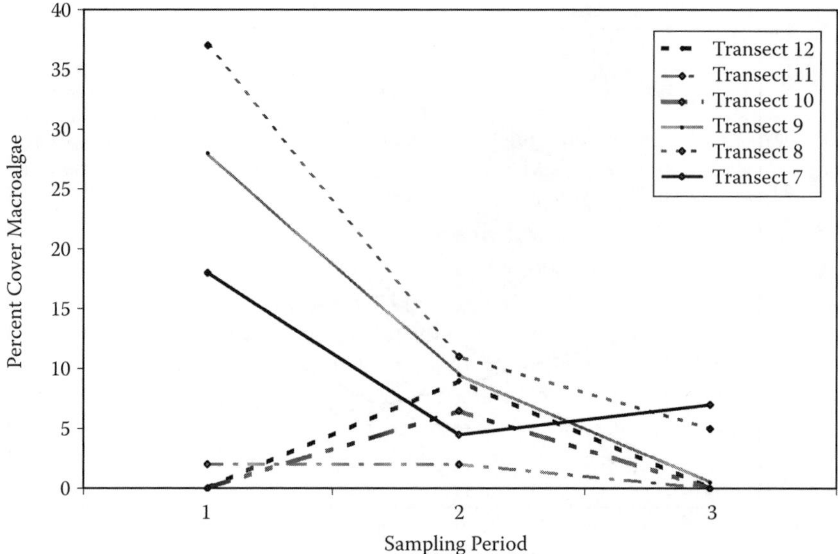

FIGURE 8.17 Mean percent cover of macroalgae at six transects (7 to 12) in Barnegat Bay during three sampling periods in 2005. Sampling period 1, June–July; sampling period 2, August–September; and sampling period 3, October–November.

8.5.6 Seagrass Blade Length

In 2004, the highest mean length of *Z. marina* blades (34.02 cm) was observed during sampling period 1 (June–July). Subsequently, the mean length of the blades decreased to 32.21 cm during sampling period 2 (August–September) and 31.83 cm during sampling period 3 (October–November). In 2005, the highest mean length of *Z. marina* blades (32.71 cm) was also observed during June–July. Subsequently, the mean length of the blades decreased to 25.89 cm during August–September, but

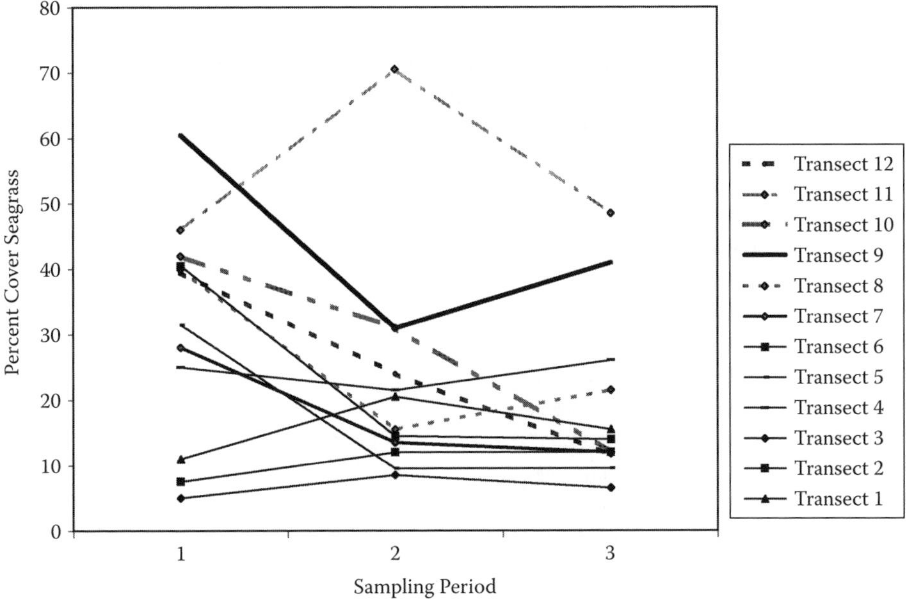

FIGURE 8.18 Mean percent cover of *Z. marina* on 12 transects (1 to 12) in the Barnegat Bay–Little Egg Harbor Estuary during three sampling periods in 2006. Sampling period 1, June–July; sampling period 2, August–September; and sampling period 3, October–November.

then increased again to 28.47 cm during October–November. In 2006, the blade lengths were substantially shorter. For example, the highest mean blade length of *Z. marina* (19.37 cm) was observed during June–July in 2006. Subsequently, the mean length of the blades decreased to 18.65 cm during August–September and 18.61 cm during October–November. An ANOVA test used to compare blade length values between the three sampling periods in 2004, 2005, and 2006 showed no statistically significant differences ($F = 0.90$; $P = 0.4078$), ($F = 2.04$; $P = 0.134$), and ($F = 0.06$; $P = 0.9427$), respectively. These data indicate that the blades of *Z. marina* grew to consistent lengths from spring to fall each year.

8.5.7 Macroalgae Composition

Thirty-two species of benthic macroalgae were recorded during the 2004 sampling period, and the majority were red algae (19; Table 8.6). Far fewer species of green algae (11) and brown algae (2) were collected. The most common species was the green seaweed, *Ulva lactuca*, which occurred in 59% of the samples, followed by three red seaweeds, *Spyridia filamentosa* (55%), *Gracilaria tikvahiae* (30%), and *Champia parvula* (23%). Several other species appeared in at least 10% of the samples: two greens, *Ulothrix flacca* and *Enteromorpha intestinalis,* and four reds, *Ceramium deslongchampsii, C. cimbricum, C. strictum*, and *Neosiphonia harveyi*. Only two species of brown algae were observed, and they occurred very infrequently. There were several species of two red algae genera, *Polysiphonia* (4) and *Ceramium* (4); in both cases some samples contained fragments that could not be identified to species. The average number of species per sample was 3.1, but there was a high degree of variability (SD = 1.6).

A seasonal change in the number of macroalgal species was evident, with maximum numbers observed in June and October and minimum numbers in September. However, these changes should be viewed with caution because they were associated with disparate sample sizes among the months. Despite the differences in sample sizes per month, some seasonal relationships were apparent. With the exception of an isolated sample in October, *Enteromorpha* spp. were only present in samples

TABLE 8.6
Percentage Occurrence of Macroalgae in Bottom Samples Collected from the Seagrass Survey Area during the June–November Period in 2004 and 2005

	Occurrence	
Species	**2004**	**2005**
Chlorophyta		
Ulothrix flacca	10.4	
Ulva lactuca	58.9	25.9
Enteromorpha intestinalis	9.8	3.7
Enteromorpha clathrata	3.7	
Enteromorpha compressa	0	1.9
Enteromorpha prolifera	5.5	
Uropsora penicilliformis	0.6	
Percusaria percursa	0.6	
Chaetomorpha linum	3.1	
Cladophora spp.	0.6	
Cladophora servica	1.8	
Codium fragile	2.5	
Bryopis plumose	0	1.9
Rhodophyta		
Audouinella spp.	3.1	3.7
Gracilaria tikvahiae	30.1	70.4
Agardhiella subulata	3.1	
Scinaia interupta	2.5	
Champia parvula	23.3	18.5
Lomentaria baileyana	9.2	
Ceramium deslongchampsii	12.3	11.1
Ceramium cimbricum	10.4	1.9
Ceramium strictum	14.1	9.3
Ceramium diaphanum	9.2	5.6
Ceramium rubrum	0	3.7
Ceramium sp.	3.7	18.5
Spyridia filamentosa	54.6	46.3
Bostrychia radicans	0.6	
Neosiphonia harveyi	11.0	1.9
Polysiphonia fucoids	0.6	
Polysiphonia harveyi	1.2	
Polysiphonia stricta	6.1	7.4
Polysiphonia subtilissima	5.5	3.7
Polysiphonia spp.	4.9	1.9
Bonnemaisonia hamifera	0	55.6
Gellidum pusillum	0	7.4
Euthora cristata	0	1.9
Hypnea musciformis	0	5.6
Phaeophyta		
Sphacelaria cirrhosa	1.8	
Ectocarpus siliculosus	1.2	1.9

FIGURE 8.19 Bloom of *Ulva lactuca* in the seagrass study area in Little Egg Harbor during summer 2004. Rapid growth of *U. lactuca* covers extensive area of the estuarine floor during bloom periods.

taken in June. The only green alga found in September was *Ulva lactuca*, which occurred in 29% of the samples, compared with more than 45% during the other months. The increase in ephemeral green algae might indicate the onset of summer nutrient (nitrogen) limitation. These species typically form blooms during periods of high nutrient availability, providing other factors (e.g., solar radiation) are also favorable. In 2004, the blooms were corroborated by field observations which showed *Ulva lactuca* completely blanketing extensive areas of the estuarine floor during the summer (Figure 8.19). There were no obvious patterns in the abundance of the red or brown algae.

No major seasonal differences were noted in the number of macroalgal species per sample (mean and median both ~3). On average, samples contained about two species of red algae and one species of green algae. A few samples (3% of the total) contained no algae.

In 2005, nearly a third fewer macroalgal species were found with the majority being red algae (Table 8.6). In contrast to 2004 when the most common species was the green seaweed *Ulva lactuca*, the most abundant species in 2005 were the red seaweeds *Gracilaria tikvahiae* (present in 70% of samples), *Bonnemaisonia hamifera* (56%), *Spyridia filamentosa* (46%), and *Champia parvula* (19%). *Ulva lactuca* was still abundant and occurred in 26% of all samples. Two other red algae appeared in at least 10% of the samples: *Ceramium deslongchampsii* (11%) and *Ceramium* sp. unidentifiable to species (18.5%). The average number of species per sample was similar to that for 2004 (3.1), amounting to 3.1, 2.5, and 3.5 in August, October, and November, respectively. As in 2004, red algae dominated the samples, but there were fewer species of green algae, and these forms were primarily collected in August, with one species (*Ulva lactuca*) recorded in November.

Because the macroalgal samples were analyzed in both years on the basis of presence/absence, no data were recorded on biomass or relative amounts; a small fragment of *Ceramium* was equivalent to a large specimen of *Ulva lactuca* with a biomass of 100 times or more. The vast majority of the species found were either ephemeral (e.g., most of the green algae) or small epiphytic forms (e.g., the majority of the red algae). There were some relatively large species such as *Codium fragile*, *Gracilaria tikvahiae,* and *Lomentaria baileyana.* Although the total number of species found per sample was similar to 2004, green algae were fewer, and red algae more frequent, especially in samples taken in October and November.

We observed several differences in the species composition of macroalgal communities between 2004 and 2005. For example, 15 species were absent in 2005 that were present in 2004, including several species that were very common in the previous year, such as *Ulothrix flacca* and *Lomentaria baileyana*. It is important to note that the sampling effort completed in 2004 not only covered a greater part of the year, but also represented a much greater number of samples. These factors may contribute to the absence of some of the macroalgal species, which were highly seasonal or relatively rare in 2005, but probably do not explain the absence of common species such as *Ulothrix flacca* and *Lomentaria baileyana*. We observed seven new species in 2005, including the red alga *Bonnemaisonia hamifera*, that were recorded in more than half of the samples.

8.5.8 Brown Tide

Brown tide blooms have commonly occurred in the Barnegat Bay–Little Egg Harbor Estuary since 1995. These blooms are problematic because they increase the attenuation of light in the water column and adversely affect seagrass, other shallow water habitats, and negatively impact the growth of shellfish (e.g., *Mercenaria mercenaria*) and the survival of bay scallops (*Argopecten irradians*) (Bricelj and Lonsdale 1997; Kennish et al. 2008). Brown tide blooms are caused by a minute alga, *Aureococcus anophagefferans* (Pelagophyceae), which consists of spherical cells about 2 to 4 μm in diameter. During past years, peak numbers of *A. anophagefferans* have commonly occurred in June, often discoloring bay waters brown (Olsen and Mahoney 2001). Maximum abundances of *A. anophagefferans* documented by the New Jersey Department of Environmental Protection during 2000, 2001, and 2002 exceeded 10^6 cells mL^{-1} each year. Gastrich et al. (2002, 2004) noted that the magnitude of brown tide blooms in the Barnegat Bay–Little Egg Harbor Estuary was elevated relative to that in other estuaries where impacts on resources had been recorded, with the worst conditions observed in Little Egg Harbor. The peak numbers of *A. anophagefferans* declined each year from 2000 (2.155×10^6 cells mL^{-1}) and 2001 (1.883×10^6 cells mL^{-1}) to 2002 (1.561×10^6 cells mL^{-1}). The years of significant brown tide blooms in the estuary were characterized by the occurrence of extended drought conditions, corresponding low freshwater inputs, and elevated bay salinity. The maximum abundance of *A. anophagefferans* declined greatly in 2003 (5.4×10^4 cells mL^{-1}) and 2004 (4.9×10^4 cells mL^{-1}), with no blooms reported during either year. Prior to 2000, brown tide blooms were observed in Little Egg Harbor during 1995, 1997, and 1999 (Olsen and Mahoney 2001).

In 2004, the numbers of *A. anophagefferans* were well below the level necessary to create bloom conditions. This observation was consistent for stations located throughout the estuary, including three stations in the seagrass study area. Thus, the probability of phytoplankton shading impacts on the seagrass beds was reduced in 2004 relative to 2000–2002.

The abundance of *A. anophagefferans* appears to have been much lower in the spring and summer of 2005 than in the peak abundance years of 2000, 2001, and 2002, when cell counts exceeded 10^6 cells mL^{-1} each year. Concentrations of *A. anophagefferans* ranged from $<5,000$ to 47,000 cells mL^{-1} in Little Egg Harbor during the May to August sampling period in 2005; highest cell counts were found in June. Brown tide was not monitored in the estuary during 2006.

8.6 DISCUSSION

Seagrasses provide vital ecosystem services in estuaries as major sources of primary production, habitat for benthic communities, nurseries for finfish and shellfish, and a food source for waterfowl, turtles, and other organisms. Seagrass beds also serve as Essential Fish Habitat, and they support many commercially and recreationally important species. Numerous studies in tropical, subtropical, and temperate estuarine waters have clearly demonstrated the ecological and functional significance of seagrasses (e.g., Kemp 1983, 2000; Dennison et al. 1993; Heck et al. 1995; Hauxwell et al.

2003; Larkum et al. 2006). These studies have also shown conclusively that seagrasses are sensitive indicators of estuarine water quality and long-term ecosystem health.

Light conditions (i.e., quality and quantity of photosynthetically active radiation; 400 to 700 nm) and water clarity are important factors controlling the spatial distribution and productivity of seagrasses. As light attenuation increases in the water column, seagrasses will be gradually displaced to shallower waters from deeper subtidal areas (Moore et al. 1996; Moore 2004). Because these vascular plants are broadly distributed in shallow estuarine waters, they are also exposed to a wide array of other natural and anthropogenic stressors that can be detrimental (Short and Wyllie-Echeverria 1996; Bortone 2000; Kenworthy et al. 2001; Kennish 2002a; Kennish et al. 2007, 2008). For example, nutrient over-enrichment, prop scarring by boats, hurricanes and other major storms, temperature increases, sediment influx from coastal watersheds, in-basin dredging, infestation by *Labyrinthula zosterae* (i.e., wasting disease), invasive species, as well as fish farming and other aquaculture practices have all contributed to seagrass decline in many regions of the world (den Hartog 1987; Kennish 2002b; Livingston 2002; Hauxwell et al. 2003; Moore 2004; Orth et al. 2006a; Kennish et al. 2007). Escalating coastal development and associated human activities have led to increasing impacts on seagrass meadows worldwide, resulting in their retreat to shallower depths (Dixon 2000). In some estuaries, human impacts have completely eliminated seagrass beds (e.g., Waquoit Bay, Massachusetts) (Short and Burdick 1996; Kennish 2004). Over the past four decades, seagrass loss has increased nearly 10-fold in temperate and tropical regions (Orth et al. 2006a).

The decrease of seagrass beds in Barnegat Bay–Little Egg Harbor during the study period is not unprecedented for estuarine systems in the mid-Atlantic region. Orth and Moore (1983, 1984) and Moore (2004) reported a dramatic decline in *Zostera marina* beds in Chesapeake Bay and its tributaries (e.g., York River Estuary). They also chronicled considerable variations in the growth of eelgrass in the bay, a finding corroborated by similar studies of other estuaries (Orth and Moore 1986; Bortone 2000; Larkum et al. 2006; Orth et al. 2006a). Surveys of seagrass areal coverage have revealed broad-scale changes in these subsystems (Robbins 1997; Lathrop et al. 2001). The Maryland and Virginia coastal bays are highly susceptible to nutrient overenrichment, and escalating nutrient levels have led to greater water quality degradation and living resource impacts (Wazniak et al. 2007). Increasing seagrass abundance observed in these coastal bays between 1986 and 2001 has leveled off in recent years; as stated by Wazniak et al. (2007, p. S76): "Large areas of the bays exhibited nutrient enrichment above threshold levels needed to maintain biotic communities." Bricker et al. (2007) reported that 75% of the coastal lagoonal estuaries in the United States are highly impacted by eutrophication, which exceeds that of other estuarine types.

Eutrophication poses a serious threat to the long-term health of seagrass beds in U.S. estuaries (Nixon 1995; Howarth et al. 2000; Rabalais 2002; Kennish et al. 2007). Nutrient loading, particularly nitrogen, has been responsible for a greater incidence of nuisance and toxic algal blooms and epiphytic growth, which has caused shading stress on seagrass beds via light attenuation, especially in shallow lagoonal estuaries. Macroalgal overburden can likewise impact seagrasses by directly smothering the beds or altering the sediment geochemistry. Sheet-like masses of drifting macroalgae (e.g., *Ulva lactuca* and *Enteromorpha* spp.) are especially problematic because they grow rapidly under favorable light and nutrient conditions, often forming a dense canopy up to 0.5 m thick over the seagrass. High macroalgal biomass can totally block sunlight transmission to the seagrass beds, lead to sulfide buildup in the rhizophere, and cause damage to seagrass habitat and associated benthic faunal communities. For example, Bologna et al. (2001) reported significant losses of *Zostera marina* habitat in Little Egg Harbor during 1998 as a consequence of macroalgal (e.g., *Ulva*, *Codium*, and *Gracilaria*) loading effects. They showed that large increases in algal-detrital biomass during the July to October period, which exceeded 400 g AFDW m^{-2}, resulted in complete elimination of the aboveground biomass of *Z. marina* in affected areas of Little Egg Harbor by October 1998. A reduction of bay scallop (*Argopecten irradians*) population density during 1999 in these impacted areas may have been caused by the loss of eelgrass in 1998. Additional losses of bay scallops were reported in 2006. As noted previously, the Barnegat Bay–Little Egg Harbor Estuary has been classified as a highly eutrophic system (Kennish et al. 2007, 2008).

Other symptoms of eutrophication problems that signal ecosystem stress have surfaced in the Barnegat Bay–Little Egg Harbor Estuary. For instance, recurring phytoplankton blooms have been documented, including serious brown tides (*Aureococcus anophagefferans*) that occurred in 1995, 1997, and 1999 to 2002 (Olsen and Mahoney 2001; Gastrich et al. 2004; Lathrop et al. 2006). Brown tides not only cause shading problems for seagrass beds (Dennison et al. 1989), but also adversely affect shellfish populations such as hard clams (*Mercenaria mercenaria*) and bay scallops (*Argopecten irradians*) (Bricelj and Lonsdale 1997; Bologna et al. 2001). Epiphytic overgrowth on seagrass blades significantly reduces photosynthetic oxygen production (McGlathery et al. 2006). In severe cases, eutrophication can lead to permanent loss of essential habitat, diminished life support, and a sharp decrease in human use of impacted systems.

External nitrogen inputs to the Barnegat Bay–Little Egg Harbor Estuary amounting to ~6.5 × 10 kg^5 N/yr are largely delivered by surface runoff (66%) and atmospheric deposition (22%), and secondarily by groundwater (12%) (Wieben and Baker 2009). The mean total nitrogen yield estimate for the Barnegat Bay watershed is 5.4 kg N ha^{-1} yr^{-1} (Kennish et al. 2007). Seitzinger et al. (2001) found that nutrient levels were highest in the northern part of the estuary due to the effects of heavy coastal watershed development in the northern sector. They reported that mean concentrations of nitrate plus nitrite were less than 4 μM. Because of biotic uptake, nitrate plus nitrite levels were lowest in the summer. Highest levels were recorded in the winter when autotrophic production was at a minimum. The New Jersey Department of Environmental Protection (1996) has chronicled more variable nitrate plus nitrite values in the estuary. Mean ammonium concentrations compiled by Seitzinger et al. (2001) were less than 2.5 μM. Total nitrogen concentrations ranged from ~20 to 80 μM. Most nitrogen in the estuary (87% to 90%) occurred in organic form. Phosphate concentrations were less than those of nitrate and ammonium, typically less than 1 μM.

The escalating eutrophication problems in the estuary due to nitrogen loading have raised concern regarding the long-term condition of seagrass habitat in the system. Bologna et al. (2000, 2001) documented dramatic losses of seagrass cover in Little Egg Harbor during the summer of 1998, reporting a 62% reduction in coverage of seagrass there between 1975 and 1999. However, color imagery obtained later in the spring (May) of 2003 and complemented with boat-based surveys by Rutgers University throughout the estuary revealed that seagrass distribution in the system remained reasonably stable between 1998 and 2003 (Lathrop et al. 2006). The apparent decline of seagrass beds reported during the 1970s, 1980s, 1990s has not been verified because of the application of different mapping techniques. Habitat loss of wetland and estuarine habitat is an ongoing major ecological concern in this system (Lathrop and Bognar 2001).

The Barnegat Bay–Little Egg Harbor Estuary is susceptible to eutrophication because of its low freshwater inflow, long water residence time in summer (74 days), limited water exchange through narrow tidal inlets, restricted circulation, and escalating population growth and development in the Barnegat Bay watershed. Recently (2008), the northern part of Barnegat Bay has experienced low dissolved oxygen levels characteristic of estuarine waters subjected to progressive eutrophication. Other mid-Atlantic coastal lagoons, for example, Great South Bay (New York), Rehoboth Bay (Delaware), Newport Bay (Maryland), and Chincoteague Bay (Virginia), have similar physical characteristics and comparable eutrophic conditions.

When nutrient enrichment persists in estuaries, there are often shifts from large to small phytoplankton species and from diatoms to dinoflagellates that can impact benthic fauna. Additional impacts include a shift from filter-feeding to deposit-feeding benthos, and a progressive change from larger, long-lived benthic forms to smaller, rapidly growing but shorter-lived species (Rabalais 2002). The net effect is the potential for insidious and persistent alteration of biotic communities in the system associated with food web alteration.

Schramm (1999) and Rabalais (2002) described a predictable series of changes in autotrophic components of estuarine and shallow marine ecosystems in response to progressive eutrophication.

For those systems that are uneutrophied, the predominant benthic macrophytes inhabiting soft bottoms typically include perennial seagrasses and other phanerogams, with long-lived seaweeds occupying hard substrates. As slight to moderate eutrophic conditions arise, bloom-forming phytoplankton species and fast-growing, ephemeral epiphytic macroalgae gradually replace the longer-lived macrophytes; hence, perennial macroalgal communities decline. With greater eutrophy, dense phytoplankton blooms occur along with drifting macroalgal species (e.g., *Ulva* and *Enteromorpha*), ultimately eliminating the perennial and slow-growing benthic macrophytes, a situation that may be taking place in some areas of the Barnegat Bay–Little Egg Harbor Estuary (Kennish 2001; Kennish et al. 2007). With hypereutrophic conditions, benthic macrophytes become locally extinct, and phytoplankton dominate the autotrophic communities (Wazniak et al. 2007). Excessive nitrification also affects secondary production through altered food web interactions (Livingston 2002).

In the present study, seagrass beds were intensely sampled in the Barnegat Bay–Little Egg Harbor Estuary during the 2004 to 2006 growing seasons. Metric measurements of the seagrass beds in Little Egg Harbor during 2004 showed that the maximum mean aboveground biomass (106.05 g dry wt m^{-2}) and belowground biomass (107.64 g dry wt m^{-2}) of *Z. marina* occurred during the June–July period. This is the time when the photoperiod peaks and nutrient levels are favorable for plant growth in the system. Eelgrass biomass declined significantly during the succeeding sampling periods (August–September and October–November), and by the end of the survey the mean aboveground biomass had decreased by 82.8% and the belowground biomass by 53.1%.

The biomass measurements were not only higher during the June–September period in 2004, but also more variable than those obtained during the October–November period, when the biomass was consistently low at all transects. This pattern indicates that the seagrass responds uniformly across the study area to decreasing photoperiod, light intensity, and temperature late in the growing season. These major controlling factors are more favorable for seagrass growth between June and September; however, other factors such as macroalgal abundance, nutrient concentrations, and turbidity may be more locally variable at this time, accounting for the greater range of biomass values found along the transects.

A similar trend of *Z. marina* biomass was evident in Barnegat Bay during 2005. Highest mean aboveground and belowground biomass values of *Z. marina* were recorded during the June–July period, which amounted to 51.69 g dry wt m^{-2} and 141.95 g dry wt m^{-2}, respectively. The highest mean density of *Z marina* also occurred at the time (479 shoots m^{-2}). The mean aboveground and belowground biomass also declined after the June–July period. The mean aboveground biomass had dropped by 69.7%, and the belowground biomass, by 69.9% at the end of the survey period.

During 2006 sampling, the peak density of *Z. marina* occurred in June–July (378 shoots m^{-2}), with lower density observed in August–September (171 shoots m^{-2}) and October–November (174 shoots m^{-2}). The highest mean aboveground biomass of *Z. marina* (13.77 g dry wt m^{-2}) was recorded during the August–September period. In contrast, the highest mean belowground biomass of *Z. marina* (51.54 g dry wt m^{-2}) was registered during the June–July period. While the belowground biomass declined appreciably after the June–July period, the aboveground biomass of *Z. marina* did not exhibit a significant seasonal reduction in 2006, although it was very low from the inception of the survey period. The substantial estuary-wide decline in seagrass biomass in 2006 correlated with elevated turbidity levels, suggesting that water column shading effects may have been responsible, possibly due to brown tide or other phytoplankton blooms.

One major concern is the marked decrease in *Z. marina* biomass observed over the course of this 3-year investigation. For example, between 2004 and 2006, the mean aboveground biomass of eelgrass at transects 1 to 6 declined in Little Egg Harbor by 87.7% from 59.62 g dry wt m^{-2} to 7.31 g dry wt m^{-2}, and the mean belowground biomass decreased by 59.4% from 75.60 g dry wt m^{-2} to 30.69 g dry wt m^{-2}. Similarly, between 2005 and 2006, the mean aboveground biomass of eelgrass at transects 7 to 12 in Barnegat Bay decreased by 50% from 32.04 g dry wt m^{-2} to 16.03 g dry wt m^{-2}, and the mean belowground biomass, by 52.4% from 84.59 g dry wt m^{-2} to 40.25 g dry wt m^{-2}. Aboveground and belowground *Z. marina* biomass varied considerably among the sampling transects. Beem and Short (2009) observed a similar reduction of seagrass beds in the Great Bay Estuary, New Hampshire, where the seagrass decline was recorded over a longer time frame. Additional seagrass biomonitoring is therefore strongly recommended in this estuarine system to track potential ongoing degrading conditions.

In an earlier study, Ohori (1982) reported mean aboveground biomass values of *Z. marina* at a limited number of sampling stations in Barnegat Bay amounting to 23.6 g dry wt m^{-2} in 1979 and 42.3 g dry wt m^{-2} in 1980. These values, despite being low, are higher than the mean aboveground biomass measurements of *Z. marina* found in the estuary during 2006. Wootton and Zimmerman (1999), working in the same region of the bay as Ohori (1982), documented decreasing seagrass biomass over the July to November sampling period, similar to that reported here in 2004 to 2006. At the Sedge Islands near Barnegat Inlet, their most productive seagrass sampling location, aboveground biomass decreased from a mean of 141.23 g AFDW m^{-2} in July, to 25.05 g AFDW m^{-2} in August, and 0 g AFDW m^{-2} in November. Belowground biomass, which was higher, differed somewhat in trend from that of the aboveground biomass, with a mean value of 221.0 g AFDW m^{-2} in July and 270.69 g AFDW m^{-2} in August dropping to 152.69 g AFDW m^{-2} in November. No clear trend in the relationship between aboveground and belowground biomass was observed during the June to November sampling period.

In a study of seagrass demographics in Little Egg Harbor during 1999, Bologna et al. (2000) found gradually increasing monthly biomass from May to August, followed by progressively decreasing biomass through October. The maximum biomass of *Z. marina* (230.0 g AFDW m^{-2}) was ~1.5 times greater than that recorded by Vaughn (1982) (149.0 g AFDW m^{-2}). This seasonal (temporal) sequence of seagrass biomass, with peak values in August 1999, differed from that reported here and by Ohori (1982), with highest biomass values recorded in June–July and subsequent declining values into the fall. Vaughn (1982) also ascertained an early (May–June) peak biomass of eelgrass in Little Egg Harbor in 1980 to 1981, and he reported a primary production value of 466 g dry wt m^{-2} yr^{-2}. Hence, there may be considerable seasonal variation in seagrass biomass from year to year in the estuary based on other studies.

Bologna (2006) also delineated significant differences in habitat complexity and seagrass characteristics between the edge and interior portions of *Z. marina* beds in Little Egg Harbor. Although the conditions in the interior zone of *Z. marina* beds were clearly better for plant growth than those at the edge, the differences in habitat structural complexity did not result in greater faunal abundance in the interior zone as reflected by the higher faunal density and secondary production observed at the edge of the beds. In addition, noteworthy differences in faunal species composition were documented between the two habitat zones due to differences in plant structural features.

The decreasing biomass of seagrass observed in the study area over the June–November period during the survey years also corresponded with a marked decrease in the percent cover of seagrass from 45% to 21% in 2004, 43% to 16% in 2005, and 32% to 19% in 2006. The change in percent cover of macroalgae was less consistent. In 2004, the percent cover of macroalgae in Little Egg Harbor increased from 13% to 21% from June to September and then decreased to 14% in November. The increase in areal coverage of macroalgae from June to September was likely a significant stressor on seagrass at this time and possibly a key factor in its decline. The decrease of macroalgal areal coverage was more consistent in Barnegat Bay during 2005, from 14% in June–July to 7% in August–September, and 2% in October–November. Reduced macroalgal areal coverage continued during 2006 with a range from 2% during the June–July period to 7% during the October–November period. Low macroalgal and seagrass cover in 2006 suggest that water column light attenuation may have played a factor.

The frequency and magnitude of blooms of ephemeral green macroalgae, notably *Ulva lactuca*, appear to be a primary indicator of seagrass success or failure in the estuary. When both the frequency and magnitude of these blooms are high, the reduction in seagrass biomass and areal coverage can exceed 50% in local areas and in extreme cases can approach 100%. Seagrass dieback due to high nutrient availability and associated macroalgal blooms can significantly impact the estuarine ecosystem structure and function.

A decrease in the abundance of bay scallops (*A. irradians*) in the estuary has been ascribed to the loss of seagrass habitat (Bologna et al. 2001). It is likely that other benthic faunal populations in the study area, such as the hard clam (*M. mercenaria*), have also been adversely affected by these acute events, but more investigations must be conducted on the benthic faunal communities to determine the overall effect. If these communities in the heavily impacted areas have also been severely downgraded by the loss of critical habitat, then it would be useful to assess potential food chain effects on upper-trophic-level organisms. The loss of bay scallops by previous macroalgal blooms shows that resource species can be highly susceptible to seagrass decline in the estuary. Bologna et al. (2005) reported that blue mussels (*Mytilus edulis*) settled in Barnegat Bay eelgrass beds in late spring at densities greater than 170,000 m^{-2}. Therefore, diminishing seagrass coverage could have a dramatic effect on the abundance of this important bivalve as well. Other resource species that use seagrass beds extensively, such as the blue crab (*Callinectes sapidus*), may also be affected.

Several natural and anthropogenic factors create stressful conditions for seagrasses in Little Egg Harbor. For example, high water temperatures (>28°C) in summer concomitant with increasing light attenuation caused by phytoplankton (e.g., brown tide) and macroalgal blooms, as well as epiphytic overgrowth, can significantly reduce seagrass growth and survival. Infestation of wasting disease can exacerbate these effects. Bologna et al. (2000) determined that less than 10% of the seagrass samples collected in their surveys were infected by *Labyrinthula zosterae*, although the infection rate varied greatly (0 to 50%). Early summer appeared to be the time of greatest impact of wasting disease on the seagrass. They also recorded serious brown tide blooms in the estuary during 1999. Only a small fraction of the eelgrass samples collected in our study exhibited wasting disease. In addition, no brown tide blooms were recorded in the estuary during 2004 and 2005. The occurrence of brown tide and macroalgal blooms, turbidity, severity of wasting disease, and magnitude of summer temperatures all are likely factors influencing the density, biomass, and spatial coverage of seagrass in the estuary.

Moore (2004) examined the relationships between seagrass bed development and water quality in the lower region of the York River, Virginia, by sampling along transect stations across vegetated and formerly vegetated areas. Based on the analysis of water quality and macrophyte samples collected in this region of the lower Chesapeake Bay, Moore (2004) concluded that the influence of seagrass beds on water quality in shallow waters of this system varied seasonally, reflecting the capacity of the seagrass to act as sources or sinks for suspended particulates and nutrients. The success of seagrasses in the study area may be dependent on the plants' capacity to regulate high levels of suspended particulate concentrations during spring.

A major concern of progressive eutrophication of the Barnegat Bay–Little Egg Harbor Estuary is the loss of ecosystem structure and function. Seagrass meadows located in nutrient-enriched areas typically experience increasing fragmentation in their spatial configuration. The loss of biotope function is manifested first by lost physical structure of the beds followed by food web alteration (Deegan 2002). Increased fragmentation of seagrass beds creates greater edge effects that often lead to increased mortality and reduced growth of fish, shellfish, and other species inhabiting the beds (Debinski and Holt 2002). Through time, the patchiness of the seagrass will also increase, further impacting benthic infaunal communities. In severe cases, seagrass meadows can be completely decimated, potentially impairing ecosystem function, production, and resource protection.

Declining seagrass biotopes are important indicators of degrading water quality conditions in lagoonal estuaries often linked to excessive loading of nitrogen from adjoining coastal watersheds and overlying airsheds (Kennish 2002a; Orth et al. 2006a; Bricker et al., 2007). As coastal population growth and development continue unabated in many regions of the United States and other developed countries in North America, Europe, Asia, and other continents, nutrient management strategies and goals are being formulated to remediate impacted estuarine ecosystems. Management plans commonly entail the application of nutrient load reductions from surrounding watershed areas (e.g., Total Daily Maximum Load in the United States), restoration of seagrass and other in-basin habitat, and long-term environmental monitoring of impaired systems. In the United States, there are currently no nationwide numerical criteria in place to limit nitrogen pollution, although the U.S. Environmental Protection Agency (2001) published a document that outlines the preferred regional approach for developing nutrient criteria and standards for estuarine waterbodies, and a National Estuarine Experts Workgroup convened in 2006 has provided guidance on it (Hameedi et al. 2007).

Effective management programs can reverse the effects of nutrient over-enrichment of estuaries. For example, remedial measures implemented in Tampa Bay and surrounding watershed areas resulted in the areal expansion of seagrass cover in the bay by more than 30% between 1982 and 1997. The actions taken to mitigate nitrogen loading problems consisted of wastewater treatment upgrades, improved stormwater systems, and changes in fertilizer manufacturing processes and agricultural practices. The net effect of this nutrient management strategy was reduced eutrophy of the bay manifested by increased water clarity and restoration of seagrass habitat (Greening and Janicki 2006). It is clear that collaborative efforts by government agencies, industry, citizens groups, and academic institutions can be effective in mitigating nitrogen loading and eutrophication problems in impacted estuarine systems.

8.7 SUMMARY AND CONCLUSIONS

The Barnegat Bay–Little Egg Harbor Estuary, similar to other coastal lagoon systems in the mid-Atlantic region, is subject to an array of natural and anthropogenic stressors that pose a potential threat to the structure of seagrass habitat and the function of the estuarine ecosystem, including nutrient over-enrichment, phytoplankton and macroalgal blooms, high turbidity, prop scarring, and other factors. With continued population growth and development in the coastal watershed surrounding this shallow estuary, future impacts on seagrass and other vital habitats are likely to escalate. Nutrient enrichment and excessive algal growth resulting from anthropogenic activities are ongoing problems in this system, and they must be effectively addressed to mollify future impacts. However, they require comprehensive research, monitoring, and remediation programs to meet the ecosystem-level challenges of environmental problems that have, to this point in time, hampered various management intervention strategies.

This investigation of the Barnegat Bay–Little Egg Harbor Estuary yielded a number of important findings. For example, the biomass of eelgrass beds in the Barnegat Bay–Little Egg Harbor during the 3-year study period (2004 to 2006) exhibited important temporal and spatial reduction patterns. The density as well as the aboveground and belowground biomass of seagrass varied considerably during the spring to fall period, but generally declined from June to November. This temporal pattern is attributed to more favorable light conditions during the late spring and summer. Aboveground and belowground biomass also varied spatially due to a wide range of physicochemical conditions over small spatial scales, including marked differences in shading, light availability, macroalgae cover, and other factors. Of most concern is the low aboveground and belowground biomass of *Z. marina* recorded along transects during 2006 compared to those in 2004 and 2005, indicating a 50% to 87.7% decline over the 3-year period. The mean density of shoots also decreased from 292 shoots m^{-2} in 2005 to 241 shoots m^{-2} in 2006. Diminishing seagrass biomass, density, and percent cover of *Z. marina* in 2006 appear to signal an estuary-wide problem, likely coupled to ongoing nutrient enrichment.

Although considerable temporal and spatial variation of eelgrass biomass was observed, eelgrass blade length was very consistent each year across sampling stations and sampling periods. For example, in 2004 there was only a slight decrease in mean eelgrass blade length in Little Egg Harbor from June to July (34.02 cm), August to September (32.21 cm), and October to November (31.83 cm) despite the gradually declining photoperiod and variable water temperature over the 6-month study period. The maximum blade length did not vary substantially between eelgrass beds. In 2005, the mean eelgrass blade length in Barnegat Bay was more variable, with the highest measurement (32.71 cm) obtained for the June–July period, the lowest measurement (25.89 cm) for the

August–September period, and an intermediate measurement (28.47 cm) for the October–November period. In 2006, the mean eelgrass blade length was substantially lower, amounting to 19.37 cm in June–July, 18.65 cm in August–September, and 18.61 cm in October–November. The reduced eelgrass blade length also correlated with reduced aboveground biomass values.

The percent cover of seagrass decreased from 2004 to 2006 in concert with the decline of biomass, density, and eelgrass blade length. In 2004, there was decreasing cover of seagrass from spring to fall in Little Egg Harbor. The highest mean percent cover of seagrass in June–July (45%) was significantly greater than that in August–September (38%) and October–November (21%). In contrast, the percent cover of macroalgae was lower and more seasonally variable than the percent cover of seagrass. For example, the mean percent cover of macroalgae increased from 13% in June–July to 21% in August–September and then declined to 14% in October–November. The highest percent cover of macroalgae in August–September probably reflects the greater growth and abundance of different algal species at this time.

In 2005, the percent cover of seagrass during June–July, August–September, and October–November sampling periods in Barnegat Bay amounted to 37%, 43%, and 16%, respectively. The percent cover by macroalgae during these periods was 14% (June–July), 7% (August–September), and 2% (October–November). Once again, the percent cover of both seagrass and macroalgae declined rapidly from summer into the fall.

The percent cover of seagrass was significantly reduced estuary-wide in 2006, concomitant with declining biomass measurements. It amounted to 32% in June–July, 23% in August–September, and 19% in October–November. The percent cover of macroalgae was similarly reduced in 2006, being 2% in June–July, 7% in August–September, and 7% in October–November. The percent cover of both seagrass and macroalgae in 2006, as well as in 2004 and 2005, was generally highest in interior areas of the seagrass beds than in marginal areas.

Most of the macroalgal species in the Barnegat Bay–Little Egg Harbor Estuary belong to a drift community. However, macroalgal blooms and patches that blanket the estuarine floor can be particularly detrimental to seagrass beds and associated benthic fauna. They hinder seagrass growth by shading or blocking sunlight and can render the estuarine floor unsuitable for regrowth of seagrass for extended periods. Hence, excessive growth of macroalgae in the estuary can be extremely damaging to seagrass habitat, a finding corroborated by studies conducted in other coastal bays in the mid-Atlantic region and elsewhere.

In 2004, 32 macroalgal species were documented in the Little Egg Harbor survey area. Red algae (n = 19) accounted for 59% of the species collected, with green algae (n = 11) comprising 34% and brown algae only 6%. *Ulva lactuca* was the most common algal species, being found in 59% of the samples. Sheet-like species, such as *U. lactuca*, posed the most serious threat to seagrass beds because they formed extensive patches that blanketed and damaged the seagrass. In 2005, nearly a third fewer macroalgal species were recorded in Barnegat Bay with most species once again being red algae. *Gracilaria tikvahiae* (present in 70% of samples), *Bonnemaisonia hamifera* (56%), *Spyridia filamentosa* (46%), and *Champia parvula* (19%) were the most abundant forms.

While brown tide (*Aureococcus anophagefferans*) blooms may be equally detrimental to seagrass beds due to their shading effects, no blooms were observed during the 2004 and 2005 sampling periods, but high turbidity measurements in 2006 suggest that brown tide or other major phytoplankton blooms may have occurred. The maximum cell counts of *A. anophagefferans* reported in the estuary during 2004 and 2005 amounted to 4.9×10^4 cells mL^{-1} and 4.7×10^4 cells mL^{-1}, respectively. These numbers are far less than those recorded during the bloom years of 2000 to 2002 (>1×10 cells mL^{-1}). Thus, it is very unlikely that *A. anophagefferans* had any adverse impact on the eelgrass beds in the estuary during these 2 survey years.

Annual aerial and *in situ* surveys of seagrass beds are strongly recommended along with extensive water quality measurements to assess ongoing eutrophic conditions in the estuary. In addition, other biotic responses to nitrogen loading must be investigated in greater detail, such as phytoplankton and macroalgal blooms and their effects on seagrass habitat. Finally, remedial management actions to reduce nitrogen loading from the surrounding watershed should be implemented as soon as possible to mitigate impacts on essential habitat in the system.

ACKNOWLEDGMENTS

This is Contribution Number 2009-48 of the Institute of Marine and Coastal Sciences, Rutgers University. Funding for this work was provided by research grants to the senior author from the Jacques Cousteau National Estuarine Research Reserve, National Oceanic and Atmospheric Administration (Estuarine Reserves Division, Office of Ocean and Coastal Resource Management, National Ocean Service), U.S. Environmental Protection Agency, and New Jersey Department of Environmental Protection.

REFERENCES

Beem, N. and F. T. Short. 2009. Subtidal eelgrass declines in the Great Bay Estuary, NH-ME. *Estuaries Coasts.* 32: 202–205.

Biber, P. D. and E. A. Irlandi. 2006. Temporal and spatial dynamics of macroalgal communities along an anthropogenic salinity gradient in Biscayne Bay (Florida, USA). *Aquat. Bot.* 85: 65–77.

Bologna, P. A. X. 2006. Assessing within habitat variability in plant demography, faunal density, and secondary production in an eelgrass (*Zostera marina* L.) bed. *J. Exp. Mar. Biol. Ecol.* 329: 122–134.

Bologna, P. A. X., M. L. Fetzer, S. McDonnell, and E. M. Moody. 2005. Assessing the potential benthic-pelagic coupling in episodic blue mussel (*Mytilus edulis*) settlement events within eelgrass (*Zostera marina*) communities. *J. Exp. Mar. Biol. Ecol.* 316: 117–131.

Bologna, P. A. X., R. Lathrop, P. D. Bowers, and K. W. Able. 2000. *Assessment of the health and distribution of submerged aquatic vegetation from Little Egg Harbor, New Jersey.* Technical Report, Contribution #2000-11, Institute of Marine and Coastal Sciences, Rutgers University, New Brunswick, New Jersey.

Bologna, P., A. Wilbur, and K. Able. 2001. Reproduction, population structure, and recruitment limitation in a bay scallop (*Argopecten irradians* Lamarck) population from New Jersey, USA. *J. Shellfish Res.* 20: 89–96.

Bortone, S. A. (Ed.). 2000. *Seagrasses: monitoring, ecology, physiology, and management.* Boca Raton, Florida: CRC Press.

Bricelj, M. and D. Lonsdale. 1997. *Aureococcus anophagefferens*: Causes and ecological consequences of brown tides in U.S. mid-Atlantic coastal waters. *Limnol. Oceanogr.* 42: 1023–1038.

Bricker, S. B., B. Longstaff, W. Dennison, A. Jones, K. Boicourt, C. Wicks, and J. Woerner. 2007. *Effects of nutrient enrichment in the nation's estuaries: A decade of change.* NOAA, Coastal Ocean Program Decision Analysis Series No. 26, National Centers for Coastal Ocean Science, Silver Spring, Maryland.

Brush, M. J. and S. Nixon. 2002. Direct measurements of light attenuation by epiphytes on eelgrass *Zostera marina*. *Mar. Ecol Prog. Ser.* 238: 73–79.

Burkholder, J. M., H. B. Glasgow, and J. E. Cooke. 1994. Comparative effects of water column nitrate enrichment on eelgrass *Zostera marina*, shoalgrass *Halodule wrightii*, and widgeongrass *Ruppia maritima*. *Mar. Ecol Prog. Ser.* 105: 121–138.

Burkholder, J. M., K. M. Mason, and H. B. Glasgow. 1992. Water column nitrate enrichment promotes decline of eelgrass *Zostera marina*: Evidence from seasonal mesocosm experiments. *Mar. Ecol Prog. Ser.* 81: 163–178.

Campbell, S., C. Miller, A. Steven, and A. Stephens. 2003. Photosynthetic responses of two temperate seagrasses across a water quality gradient using chlorophyll fluorescence. *J. Exp. Mar. Biol. Ecol.* 291: 57–78.

Carlson, P. R., Jr., L. A. Yarbro, and T. R. Barber. 2004. Relationships of sediment profile to mortality of *Thalassia testudinum* in Florida. *Bull. Mar. Sci.* 54: 733–746.

Carruthers, T. J. B., W. C. Dennison, B. J. Longstaff, M. Waycott, E. G. Abal, L. J. McKenzie, and W. J. L. Long. 2002. Seagrass habitats in Northeast Australia: Models of key processes and controls. *Bull. Mar. Sci.* 71: 1153–1169.

Debinski, D. M. and R. D. Holt. 2002. A survey and overview of habitat fragmentation experiments. *Conserv. Biol.* 14: 342–355.

Deegan, L.A. 2002. Lessons learned: The effects of nutrient enrichment on the support of nekton by seagrass and salt marsh ecosystems. *Estuaries* 25: 727–742.

den Hartog, C. 1987. Wasting disease and other dynamic phenomena in *Zostera* beds. *Aquat. Bot*. 27: 3–14.

Dennison, W., G. Marshall, and C. Wigand. 1989. Effect of brown-tide shading on eelgrass (*Zostera marina* L.) distributions. Pp. 675–695, In: E. Cosper, V. Bricelj, and E. Carpenter (eds.), *Novel phytoplankton blooms*. New York: Springer-Verlag.

Dennison, W. C., R. J. Orth, K. A. Moore, J. C. Stevenson, V. Carter, S. Kollar, P. W. Bergstrom, and R. A. Batiuk. 1993. Assessing water quality with submersed aquatic vegetation: Habitat requirements as barometers of Chesapeake Bay health. *Bioscience* 43: 86–94.

Dixon, L. K. 2000. Establishing light requirements for the seagrass *Thalassia testudinum*: An example from Tampa Bay, Florida. Pp. 9–31, In: S. A. Bortone (ed.), *Seagrasses: Monitoring, ecology, physiology, and management*. Boca Raton, Florida: CRC Press.

Duarte, C. M. 1999. Seagrass ecology at the turn of the millennium: Challenges for the new century. *Aquat. Bot*. 65: 7–20.

Garono, R. J., C. A. Simenstad, R. Robinson, and H. Ripley. 2004. Using high spatial resolution hyperspectral imagery to map intertidal habitat structure in Hood Canal, Washington, U.S.A. *Can. J. Remot. Sens*. 30: 54–63.

Gastrich, M. D., O. R. Anderson, and E. M. Cosper. 2002. Viral-like particles (VLPs) in the alga, *Aureococcus anophagefferens* (Pelagophyceae), during 1999–2000 brown tide blooms in Little Egg Harbor, New Jersey. *Estuaries* 25: 938–943.

Gastrich, M. D., R. Lathrop, S. Haag, M. P. Weinstein, M. Danko, D. A. Caron, and R. Schaffner. 2004. Assessment of brown tide blooms, caused by *Aureococcus anophagefferans*, and contributing factors in New Jersey coastal bays: 2000–2002. *Harmful Algae* 3: 305–320.

Gastrich, M. D., J. A. Leigh-Bell, C. J. Gobler, O. R. Anderson, S. W. Wilhelm, and M. Bryan. 2004. Viruses as potential regulators of regional brown tide blooms caused by the alga, *Aureococcus anophagefferens*. *Estuaries* 27: 112–119.

Greening, H. and A. Janicki. 2006. Toward reversal of eutrophic conditions in a subtropical estuary: Water quality and seagrass response to nitrogen loading reduction in Tampa Bay, Florida, USA. *J. Environ. Manage*. 38: 163–178.

Haag, S. M., M. J. Kennish, and G. P. Sakowicz. 2008. Seagrass habitat characterization in estuarine waters of the Jacques Cousteau National Estuarine Research Reserve using underwater videographic imaging techniques. *J. Coastal Res*., SI 55: 171–185.

Hameedi, J., H. Paerl, M. Kennish, and D. Whitall. 2007. Nitrogen deposition in U.S. coastal bays and estuaries. *EM Magazine*, December, pp. 19–26.

Hauxwell, J., J. Cebrian, and I. Valiela. 2003. Eelgrass *Zostera marina* loss in temperate estuaries: Relationships to land-derived nitrogen loads and effect of light limitation imposed by algae. *Mar. Ecol. Prog. Ser*. 247: 59–73.

Heck, K. L., K. W. Able, C. T. Roman, and M. Fahay. 1995. Composition, abundance, biomass, and production of macrofauna in a New England estuary: Comparison among eelgrass meadows and other nursery habitats. *Estuaries* 18: 379–389.

Hemminga, M. A. 1998. The root/rhizome system of seagrasses: An asset and a burden. *J. Sea Res*. 39: 183–196.

Howarth, R. W., D. M. Anderson, T. M. Church, H. Greening, C. S. Hopkinson, W. C. Huber, N. Marcus, R. J. Nainman, K. Segerson, A. N. Sharpley, and W. J. Wiseman. 2000. *Clean coastal waters: Understanding and reducing the effects of nutrient pollution*. Ocean Studies Board and Water Science and Technology Board, National Academy Press, Washington, D.C.

Howarth, R. W., A. Sharpley, and D. Walker. 2002. Sources of nutrient pollution to coastal waters in the United States: Implications for achieving coastal water quality goals. *Estuaries* 25: 656–676.

Hunchak-Kariouk, K. and R. S. Nicholson. 2001. Watershed contributions of nutrients and other nonpoint source contaminants to the Barnegat Bay–Little Egg Harbor Estuary. *J. Coastal Res*., SI 32: 28–81.

Irlandi, E. A., B. A. Orlando, and P. D. Biber. 2004. Drift algae-epiphyte seagrass interactions in a tropical *Thalassia testudinum* meadow. *Mar. Ecol. Prog. Ser*. 279: 81–91.

Kemp, W. M. 1983. Seagrass communities as a coastal resource. *Mar. Technol. Soc. J*. 17: 3–5.

Kemp, W. M. 2000. Seagrass ecology and management: An introduction. Pp. 1–6, In: S. A. Bortone (ed.), *Seagrasses: Monitoring, ecology, physiology, and management*. Boca Raton, Florida: CRC Press.

Kennish, M. J. (ed.). 2001. Barnegat Bay–Little Egg Harbor, New Jersey: Estuary and watershed assessment. *J. Coastal Res*., Special Issue 32.

Kennish, M. J. 2002a. Environmental threats and environmental future of estuaries. *Environ. Conserv*. 29: 78–107.

Kennish, M. J. (ed.). 2002b. Impacts of motorized watercraft on shallow estuarine and coastal marine environments. *J. Coastal Res.*, Special Issue 37.

Kennish, M. J. (ed.). 2004. *Estuarine research, monitoring, and resource protection*. Boca Raton, Florida: CRC Press.

Kennish, M. J., S. B. Bricker, W. C. Dennison, P. M. Glibert, R. J. Livingston, K. A. Moore, R. T. Noble, H. W. Paerl, J. M. Ramstack, S. Seitzinger, D. A. Tomasko, and I. Valiela. 2007. Barnegat Bay–Little Egg Harbor Estuary: Case study of a highly eutrophic coastal bay system. *Ecol. Applic.* 17(5): S3–S16.

Kennish, M. J., S. M. Haag, and G. P. Sakowicz. 2005. *Submersed aquatic vegetation: Tier 2 monitoring in the Jacques Cousteau National Estuarine Research Reserve.* Technical Report, National Estuarine Research Reserve Program, National Oceanic and Atmospheric Administration, Silver Spring, Maryland.

Kennish, M. J., S. M. Haag, and G. P. Sakowicz. 2006. Application of underwater videography to characterize seagrass habitats in Little Egg Harbor, New Jersey. *Bull. N.J. Acad. Sci.* 51: 1–6.

Kennish, M. J., S. M. Haag, and G. P. Sakowicz. 2008. Seagrass demographic and spatial habitat characterization in Little Egg Harbor, New Jersey, using fixed transects. *J. Coastal Res.*, SI 55: 148–170.

Kenworthy, W. J., M. S. Fonseca, P. E. Whitfield, and K. K. Hammerstrom. 2001. Analysis of seagrass recovery in experimental excavations and propeller-scar disturbances in the Florida Keys National Marine Sanctuary. *J. Coastal Res.*, SI 37: 75–85.

Lamote, M. and K. H. Dunton. 2006. Effects of drift macroalgae and light attenuation in chlorophyll fluorescence and sediment sulfides in the seagrass *Thalassia testudinum*. *J. Exp. Mar. Biol. Ecol.* 334: 174–186.

Larkum, W. D., R. J. Orth, and C. M. Duarte (eds.). 2006. *Seagrasses: Biology, ecology, and conservation.* Dordrecht, the Netherlands: Springer.

Lathrop, R. G. and J. Bognar. 2001. Habitat loss and alteration in the Barnegat Bay region. *J. Coastal Res.*, SI 32: 212–228.

Lathrop, R. G., Jr., J. A. Bognar, A. C. Henrickson, and P. D. Bowers. 1999. *Data synthesis effort for the Barnegat Bay Estuary Program: Habitat loss and alteration in the Barnegat Bay region.* Technical Report, Center for Remote Sensing and Spatial Analysis, Rutgers University, New Brunswick, New Jersey.

Lathrop, R. G., P. Montesano, and S. Haag. 2006. A multi-scale segmentation approach to mapping seagrass habitats using airborne digital camera imagery. *Photogr. Eng. Remote Sens.* 72: 665–675.

Lathrop, R. G., R. Styles, S. Seitzinger, and J. Bognar. 2001. Use of GIS mapping and modeling approaches to examine the spatial distribution of seagrasses in Barnegat Bay, New Jersey. *Estuaries* 24: 904–916.

Lee, K.-S., F. T. Short, and D. M. Burdick. 2004. Development of a nutrient pollution indicator using the seagrass, *Zostera marina*, along nutrient gradients in three New England estuaries. *Aquat. Bot.* 78: 197–216.

Livingston, R. J. 2002. *Trophic organization in coastal systems*. Boca Raton, Florida: CRC Press.

Longstaff, B. J. and W. C. Dennison. 1999. Seagrass survival during pulsed turbidity events: The effects of light deprivation on the seagrasses *Halodule pinifolai* and *Halophila ovalis*. *Aquat. Bot.* 65: 105–121.

McGlathery, K. J., K. Sundback, and I. C. Anderson. 2007. Eutrophication in shallow coastal bays and lagoons: The role of plants in the coastal filter. *Mar. Ecol. Prog. Ser.* 348: 1–18.

Moore, K. A. 2004. Influence of seagrasses on water quality in shallow regions of the lower Chesapeake Bay. *J. Coastal Res.* SI 45: 162–178.

Moore, K. A., H. A. Neckles, and R. J. Orth. 1996. *Zostera marina* (eelgrass) growth and survival along a gradient of nutrients and turbidity in the lower Chesapeake Bay. *Mar. Ecol. Prog. Ser.* 142: 247–259.

Moore, K. A. and R. L. Wetzel. 2000. Seasonal variations in eelgrass (*Zostera marina* L.) responses to nutrient enrichment and reduced light availability in experimental ecosystems. *J. Exp. Mar. Biol. Ecol.* 244: 1–28.

Moore, K. A., D. J. Wilcox, and R. J. Orth. 2000. Analysis of abundance of submersed aquatic vegetation communities in the Chesapeake Bay. *Estuaries* 23: 115–127.

Morris, L. J., R. W. Virnstein, J. D. Miller, and L. M. Hall. 2000. Monitoring seagrasses changes in Indian River Lagoon, Florida, using fixed transects. Pp. 167–176, In: S.A. Bortone (ed.), *Seagrasses: Monitoring, ecology, physiology, and management*. Boca Raton, Florida: CRC Press.

New Jersey Department of Environmental Protection. 1996. *Shellfish growing Area 6: Manasquan River 1990–1995*. Water Monitoring Project Report, New Jersey Department of Environmental Protection, Trenton, New Jersey.

Nixon, S. W. 1995. Coastal eutrophication: A definition, social causes, and future concerns. *Ophelia* 41: 199–220.

Norris, J. G., S. Wyllie-Echeverria, T. Mumford, A. Bailey, and T. Turner. 1997. Estimating basal area coverage of subtidal seagrass beds using underwater videography. *Aquat. Bot.* 58: 269–287.

Ohori, V. L. 1982. The distribution, abundance, and importance of the seagrass, *Zostera marina* L., in the vicinity of Oyster Creek Nuclear Generating Station, Barnegat Bay, New Jersey, 1979–1980. M.S. Thesis, Rutgers University, New Brunswick, New Jersey.

Olsen, P. S. and J. B. Mahoney. 2001. Phytoplankton in the Barnegat Bay–Little Egg Harbor estuarine system: Species composition and picoplankton bloom development. *J. Coastal Res.*, SI 32: 115–143.

Orth, R .J., T. J. B. Carruthers, W. C. Dennison, C. M. Duarte, J. W. Fourqurean, K. L. Heck, Jr., A. R. Hughes, G. A. Kendrick, W. J. Kenworthy, S. Olyarnik, F. T. Short, M. Waycott, and S. L. Williams. 2006a. A global crisis for seagrass ecosystems. *Bioscience* 56: 987–996.

Orth, R. J., M. L. Luckenbach, S. R. Marion, K. A. Moore, and D. J. Wilcox. 2006b. Seagrass recovery in the Delmarva coastal bays, USA. *Aquat. Bot.* 84: 26–36.

Orth, R. J. and K. A. Moore. 1983. An unprecedented decline in submerged aquatic vegetation. *Science* 222: 51–53.

Orth, R. J. and K. A. Moore. 1984. Distribution and abundance of submerged aquatic vegetation in Chesapeake Bay: An historical perspective. *Estuaries* 7: 531–540.

Orth, R. J. and K. A. Moore. 1986. Seasonal and year-to-year variations in the growth of *Zostera marina* L. (eelgrass) in lower Chesapeake Bay. *Aquat. Bot.* 24: 335–341.

Plus, M., J.-M. Deslous-Paoli, and F. Dagault. 2003. Seagrass (*Zostera marina* L.) bed recolonization after anoxia-induced full mortality. *Aquat. Bot.* 77: 121–134.

Pregnall, A. M., R. D. Smith, T. A. Kursar, and R. S. Alberte. 1984. Metabolic adaptations of *Zostera marina* (eelgrass) to diurnal periods of root anoxia. *Mar. Biol.* 83: 141–147.

Provancha, J. A. and D. M. Scheidt. 2000. Long-term trends in seagrass beds in the Mosquito Lagoon and Northern Banana River, Florida. Pp. 177–193, In: S. A. Bortone (ed.), *Seagrasses: Monitoring, ecology, physiology, and management*. Boca Raton, Florida: CRC Press.

Rabalais, N. N. 2002. Nitrogen in aquatic ecosystems. *Ambio* 21: 102–112.

Robbins, B. 1997. Quantifying temporal change in seagrass areal coverage: The use of GIS and low resolution aerial photography. *Aquat. Bot.* 58: 259–267.

Rumrill, S. S. and V. K. Poulton. 2004. *Ecological role and potential impacts of molluscan shellfish culture in the estuarine environment of Humboldt Bay, CA*. Final Report, Western Regional Aquaculture Center, University of Washington, Seattle.

Schramm, W. 1999. Factors influencing seaweed responses to eutrophication: Some results from EU-project EUMAC. *J. Appl. Phycol.* 11: 69–78.

Seitzinger, S. P., R. M. Styles, and I. E. Pilling. 2001. Benthic microalgal and phytoplankton production in Barnegat Bay, New Jersey (USA): Microcosm experiments and data synthesis. *J. Coastal Res.*, SI 32: 144–162.

Short, F. T. and D. M. Burdick. 1996. Quantifying eelgrass habitat loss in relation to housing development and nitrogen loading in Waquoit Bay, Massachusetts. *Estuaries* 19: 730–739.

Short, F. T., L. J. McKenzie, R. G. Coles, and K. P. Vidler. 2002. SeagrassNet manual for scientific monitoring of seagrass habitat. (QDPI, QFS, Cairns.)

Short, F. T. and S. Wyllie-Echeverria. 1996. Natural and human-induced disturbance of seagrasses. *Environ. Conserv.* 23: 17–27.

U.S. Environmental Protection Agency. 2001. Nutrient criteria technical guidance manual: Estuarine and coastal marine waters. EPA-822-B-01-003, U.S. Environmental Protection Agency, Washington, D.C.

Vaughn, D. 1982. Production ecology of eelgrass (*Zostera marina* L.) and its epiphytes in Little Egg Harbor, New Jersey. Ph.D. Thesis, Rutgers University, New Brunswick, New Jersey.

Ward, D. H., C .J. Markon, and D. C. Douglas. 1997. Distribution and stability of eelgrass beds at Izembek Lagoon, Alaska. *Aquat. Bot.* 58: 229–240.

Wazniak, C. E., M. R. Hall, T. J. B. Carruthers, B. Sturgis, W. C. Dennison, and R. J. Orth. 2007. Linking water quality to living resources in a mid-Atlantic lagoon system, USA. *Ecol. Applic.* 17(5): S64–S78.

Wieben, C.M. and R. J. Baker. 2009. Contributions of nitrogen to the Barnegat Bay–Little Egg Harbor Estuary: Updated loading estimates. Technical Report, U.S. Geological Survey, West Trenton, New Jersey.

Wootton, L. and R. W. Zimmerman. 1999. Water quality and biomass of *Zostera marina* (eelgrass) beds in Barnegat Bay. Technical Report, Georgian Court College, Lakewood, New Jersey.

9 Controls Acting on Benthic Macrophyte Communities in a Temperate and a Tropical Estuary

Sophia E. Fox, Ylra S. Olsen, Mirta Teichberg, and Ivan Valiela*

CONTENTS

Abstract 203
9.1 Introduction 204
9.2 Benthic Macrophyte Community: Structure and Function 204
9.3 Bottom-Up and Top-Down Controls 204
9.4 Case Studies 207
    9.4.1 Waquoit Bay, Massachusetts: Bottom-Up and Top-Down Controls 207
    9.4.2 Jobos Bay, Puerto Rico: Bottom-Up and Top-Down Controls 215
9.5 Conclusions 219
Acknowledgments 221
References 221

ABSTRACT

Seagrass meadows and macroalgal canopies are critical coastal habitats that are widely recognized for their vital ecological functions and services. Both seagrass and macroalgal growth and community structure can be controlled by nutrients and herbivory. Although these benthic macrophyte groups have been studied extensively worldwide, there is still no consensus as to the relative roles of the different control mechanisms on their structure and function. To examine the relative influence of nutrients and herbivory on growth of seagrasses and macroalgae, we conducted a set of field experiments, manipulating the nutrient supply, grazing pressure, or both in a temperate and a tropical estuary. In these estuarine systems, seagrass and macroalgae were controlled by numerous and dynamic interactions between nutrients and herbivory, whereby nutrient supply considerably altered the potential for herbivore control of producer biomass. These results highlight the major role that human-driven nutrient loading from land plays in adjacent coastal habitats, and underscore the need for greater understanding of the benthic floral community dynamics in shallow estuarine environments.

Key Words: nutrients, herbivory, macroalgae, seagrass, estuaries, eutrophication

* Corresponding author. Email: sophia_fox@nps.gov

9.1 INTRODUCTION

Shallow estuaries are important coastal ecosystems situated at the interface of land and sea, and, therefore, are heavily influenced by human activities and natural processes taking place on land, along fringing wetland habitats, and in adjacent bays and oceans. They are shallow in depth, usually less than a few meters, with light often penetrating to the bottom. The flora and fauna vary considerably in these coastal systems, but the benthic producer communities generally include microalgae, macroalgae, and seagrasses.

Human impacts in the coastal zone have led to remarkable restructuring of estuarine benthic macrophyte communities (Hauxwell et al. 2001; McGlathery 2001; Fox et al. 2008). Where the intensity of human activity on adjacent coastal watersheds is low, seagrass meadows dominate the benthic producer communities in estuaries, and diverse macroalgal and microalgal communities comprise a smaller proportion of the benthic producer biomass and production. As human activities increase in these watersheds and the nearby coastal ocean, estuarine producer communities tend to shift from seagrass-dominated ecosystems to diatom- and macroalgae-dominated systems. Shifts in producer community structure at the base of the food web have major implications for ecosystem function and higher trophic level production. It is important to understand which mechanisms control shifts in the abundance and species composition of benthic producers in a changing environment. In this chapter, we explore the role of different control mechanisms in structuring benthic seagrass and macroalgal communities in shallow estuaries.

9.2 BENTHIC MACROPHYTE COMMUNITY: STRUCTURE AND FUNCTION

Seagrasses and macroalgal canopies are important habitats commonly found in estuaries worldwide. These benthic producers contribute much of the primary production at the base of most estuarine food webs, both temperate and tropical. Macroalgae are an important nutritional food source for numerous invertebrates and fish. Seagrasses tend to be less palatable, and enter the benthic food web mainly as detritus in temperate estuaries (Cebrián 1999; Cebrián and Duarte 2001). In contrast, in tropical systems seagrasses are consumed directly by fish, invertebrates, and mammals, and can constitute a significant component of consumer diets (Valentine and Heck 1999; Kirsch et al. 2002). In addition, seagrasses typically support a diverse algal epiphyte community that is heavily grazed by small invertebrates in both temperate and tropical estuaries (Orth and van Montfrans 1984; Neckles et al. 1993; Jernakoff et al. 1996; Frankovich and Zieman 2005; Peterson et al. 2007; Jephson et al. 2008). Seagrass and macroalgal canopies also provide structurally complex environments in which prey can escape their predators (Graham et al. 1998; Bologna and Heck 1999). Seagrass ecosystems are widely recognized as critical coastal habitats that provide numerous ecological functions and services. Characterized by high biomass and productivity, seagrass beds provide an important source of organic carbon to coastal ecosystems (Duarte and Chiscano 1999; Green and Short 2003). They also provide food and shelter for fish and invertebrates (Lubbers et al. 1990; Heck and Valentine 1995), stabilize sediments (Harlin et al. 1982), alter biogeochemical processes (Blaabjerg and Finster 1998; Risgaard-Petersen et al. 1998), and support near-shore fisheries by serving as nursery grounds for juvenile fish (Nagelkerken et al. 2002; Heck et al. 2003; Dorenbosch et al. 2004).

9.3 BOTTOM-UP AND TOP-DOWN CONTROLS

Both seagrasses and macroalgae are subject to control by nutrients from the bottom up and by herbivory and predation from the top down (Valiela et al. 1997; Geertz-Hansen et al. 1993; A. R. Hughes et al. 2004; Burkepile and Hay 2006; Heck and Valentine 2007). Nutrients can stimulate growth and increase producer biomass, and herbivores can consume producer tissue and reduce biomass.

Predators may also exert control from the top of the food web down by consuming herbivores, thereby reducing grazing pressure and increasing producer biomass. Although the taxa involved may differ, interacting bottom-up and top-down control mechanisms come into play as controls for both seagrasses and macroalgae.

Nutrients have been shown to control primary productivity in many environments. Nutrient enrichment has led to increased leaf growth, production, canopy height, and biomass of seagrasses in temperate (Orth 1977; Murray et al. 1992; van Lent et al. 1995) and tropical sites (Powell et al. 1989; Short et al. 1990; Udy et al. 1999; Ferdie and Fourqurean 2004; Armitage et al. 2005). Some studies have also demonstrated a lack of seagrass response to nutrients (Dennison et al. 1987; Erftmeijer et al. 1994; Lee and Dunton 2000), perhaps because their nutrient requirements had been met, or possibly light may have limited growth (Ibarra-Obando et al. 2004; Ralph et al. 2007). High macroalgal productivity in coastal waters is partly driven by increased nutrient inputs (Lapointe and O'Connell 1989; Sfriso and Marcomini 1997; Valiela et al. 1997; Fox et al. 2008). Opportunistic macroalgae can readily take up available nutrients from the water column to grow rapidly (Pedersen and Borum 1996; Teichberg et al. 2007, 2008).

Top-down control by consumers can also structure producer communities (Valentine and Duffy 2006; Heck et al. 2006; Heck and Valentine 2007). Today, relatively few vertebrate herbivores (fishes, sirenians, turtles, and waterfowl) and sea urchins graze directly on seagrasses (Valentine and Heck 1999). Historically, large, specialized seagrass grazers, such as fishes, manatees, and turtles, were common, and likely had a large impact on seagrass standing crop (Domning 2001; Heck and Valentine 2007). Some fishes, however, still exert a major grazing pressure on seagrasses and macroalgae in the tropics (Mumby et al. 2006; Valentine et al. 2007). The more common, smaller meso-grazers (crustaceans, gastropods, urchins, and other invertebrates) usually feed on epiphytic algae growing on seagrass blades (Orth and van Montfrans 1984; Jernakoff et al. 1996; Frankovich and Zieman 2005; Moksnes et al. 2008). Seagrasses are, thus, subject to lower grazing pressure, so that most seagrass populations might lose <10% of aboveground production to direct herbivory (Mateo et al. 2006). Herbivore controls on seagrasses may have either a stimulatory or a negative effect on seagrass production (Thayer et al. 1984; Zieman et al. 1984). Moderate grazing can stimulate plant growth and increase shoot density (A. R. Hughes et al. 2004), while intense grazing can decrease biomass (Heck and Valentine 1995; McGlathery 1995; Cebrián and Duarte 1998; Alcoverro and Mariani 2002).

Grazers have also been shown to control macroalgal biomass (Lewis 1986; Geertz-Hansen et al. 1993; Menge et al. 1997; T. Hughes et al. 1999; Lotze and Worm 2000; Duffy and Harvilicz 2001; Tewfik et al. 2005). Mesograzers, such as amphipods, isopods, and small gastropods, tend to dominate the grazer assemblages in temperate estuaries (Geertz-Hansen et al. 1993; Duffy and Hay 1994; Hauxwell et al. 1998; Fox et al. 2009), while large macroherbivores, mainly large gastropods, urchins, and fishes, tend to dominate in tropical estuaries, seagrass meadows, and coral reefs. Differences in the behavior, grazing rates, territory size, and abiotic and biotic controls of mesograzers and macroherbivores suggest substantial differences in the relative abilities of temperate and tropical grazer assemblages to control producer growth. Meta-analyses of nutrient and herbivore control of seagrasses and macroalgae have shown that the relative influence of control mechanisms in marine environments is highly variable (Burkepile and Hay 2006; Gruner et al. 2008), the more so in soft bottom environments, such as estuaries, and for macroalgae (Gruner et al. 2008).

Herbivory effects may also interact with bottom-up effects. Such interactions may take place when increased nutrients alter producer and consumer assemblages (Worm et al. 2000; Oesterling and Pihl 2001; A. R. Hughes et al. 2004; Heck and Valentine 2007; Fox et al. 2009), and producer nutrient content (Heck et al. 2000, 2006; Hemmi and Jormalainen 2002; Boyer et al. 2004). To date, there is still no clear understanding of how these processes interact, despite experimental studies in seagrass meadows (McGlathery 1995; Heck et al. 2000, 2006; Ruiz et al. 2001; Heck and Valentine 2007) and in macroalgal canopies (Geertz-Hansen et al. 1993; Hauxwell et al. 1998; Lotze et al.

2001; Burkepile and Hay 2006; Worm and Lotze 2006; Fox 2008; Fox et al. submitted). There is some support, however, for the notion that a nutrient "threshold" exists, at which nutrient effects may overwhelm control by herbivory. For macroalgae, the effects of nutrient enrichment, herbivory, and their interactions are more significant and larger in magnitude for tropical species than for temperate species (Burkepile and Hay 2006). The reasoning behind these differences has not been well explained, but there is an indication that in low-productivity compared to high-productivity environments, there is more likely to be an effect of herbivory on macroalgae. Meta-analyses have shown that the experimental results, however, were highly variable and depended on algal type and how producer responses were measured (Burkepile and Hay 2006; Gruner et al. 2008).

Discerning the relative impacts of top-down and bottom-up mechanisms as ecosystem controls is particularly relevant, because we are carrying out a global-scale, human-driven experiment in bottom-up and top-down control (Jackson et al. 2001; Valiela 2006; Bricker et al. 2007). Human development of the coastal zone has led to increased nutrient inputs to waters, mainly through terrestrial runoff and groundwater flows. Coastline development has also modified coastal areas by decreasing natural vegetation and destroying coastal wetlands. Wetlands are a critical habitat, since they serve to protect adjacent water quality, by acting as coastal filters that intercept land-derived particulates and nutrients (Valiela and Cole 2002). Coastal wetlands, however, are being destroyed at alarming rates worldwide, with incredible losses of salt marsh in temperate zones (Valiela and Cole 2002) and mangrove areas in tropical regions (Valiela et al. 2001).

The losses of wetland area have been linked to decreases in seagrass habitat extent (Valiela et al. 1997; Valiela and Cole 2002), as a result of increased throughput of nutrients and particulates, and subsequent eutrophication and habitat degradation. In a study of 12 estuaries and coastal seas, 67% of wetlands and 65% of seagrasses have been lost since the time of human settlement (Lotze et al. 2006). These losses trump the widely publicized estimated losses of the world's rainforests, which some environmental groups argue to be up to 50%, and losses of coral reefs, estimated at about 20% (IUCN 2008). Seagrass losses in both temperate and tropical regions have been increasing rapidly over the last 40 years, suggesting high rates of seagrass decline worldwide (Orth et al. 2006).

Coastal seas and estuaries with the longest histories of human impacts and highest human populations are the most degraded (Lotze et al. 2006). Receiving coastal waters and estuaries are among the most nutrient-enriched environments on Earth (Nixon et al. 1986; Valiela 2006). Increased nutrient enrichment of estuarine waters leads to a cascade of events in benthic habitats, where macroalgal and epiphytic algal biomass increase and macroalgal blooms form, leading to significant losses of seagrass habitats and seagrass-associated taxa. Increased nutrient inputs interact with top-down controls whereby in eutrophied estuaries, and macroalgal blooms lead to frequent hypoxic events, during which oxygen concentrations in estuarine bottom waters plunge below levels sufficient to support invertebrates and fishes. As a result of the hypoxia in eutrophied estuaries, there are fewer invertebrates (Oesterling and Pihl 2001; Fox et al. 2009), and fish (Baden et al. 1990; Graham et al. 1998). There are further consequences of humans on controls of benthic producers, since herbivorous fishes have been subject to intense harvesting pressure (Mumby et al. 2006). The lower abundances of herbivorous invertebrates and fishes lead to reduced grazing on benthic macrophytes (Valentine and Heck 1999; Domning 2001; Mumby et al. 2006; Heck and Valentine 2007; Fox 2008; Fox et al. submitted).

It seems, therefore, imperative to understand the relative and potentially interactive impacts of bottom-up and top-down mechanisms in different environments, to develop strategies to mitigate the myriad ways in which we are altering coastal marine ecosystems. Below, we present results from experimental manipulations designed to examine the relative influence of bottom-up and top-down controls acting on benthic seagrasses and macroalgae in a temperate estuary, Waquoit Bay, and a tropical estuary, Jobos Bay.

9.4 CASE STUDIES

9.4.1 Waquoit Bay, Massachusetts: Bottom-Up and Top-Down Controls

The Waquoit Bay estuarine system, located on Cape Cod, Massachusetts, is representative of temperate, shallow, groundwater-fed estuarine embayments in the northeastern United States. The subestuaries of Waquoit Bay are shallow, from 1 to 3 m in depth, and are similar in many other physical aspects, including water residence time, sediment character (Carmichael and Valiela 2005), and light penetration to 1.5 m depth (M. Teichberg, unpublished). The watersheds of its subestuaries, however, differ in their land covers, and therefore deliver different nitrogen loads to receiving estuaries (Table 9.1), mainly through freshwater discharges to estuaries from groundwater inputs (Bowen and Valiela 2001). The major forcing mechanism of change on the Waquoit Bay estuarine ecosystem is development of the watershed, namely the increase in number of houses and decrease in fringing salt marsh, which has increased delivery of nitrogen to estuarine waters (Valiela et al. 1992, 1997).

The Waquoit Bay estuarine system is an excellent location to investigate alterations of estuarine communities in response to changes in nitrogen inputs to coastal waters, since its subestuaries are subject to a range of nitrogen loads (Bowen and Valiela 2001). Childs River has the highest nitrogen load owing to extensive suburbanization of the watershed and high nitrate inputs from individual septic systems, while Timms Pond has the lowest nitrogen load and inputs from a predominantly forested watershed, a result of its location within a state park (Bowen and Valiela 2001) (Table 9.1, Figure 9.1). The Quashnet River watershed has a mix of land covers, and delivers an intermediate load to the estuary.

For decades, losses of seagrass area have been increasing in Waquoit Bay (Figure 9.2a), and the decrease in seagrass cover is related to nitrogen loading from adjacent watersheds (Figure 9.2b). Concurrently, macroalgal biomass has been increasing in Waquoit Bay (Valiela et al. 1997; Fox 2008; Fox et al. 2008), and macroalgal biomass is approximately four times higher in estuaries

TABLE 9.1
Land-Derived Nitrogen (N) Loads and Eutrophication Status for the Estuaries of Waquoit Bay and Jobos Bay

Estuary	N load (kg N ha^{-1} y^{-1})	Eutrophication Status
Waquoit Bay, MA, USA		
Childs River (CR)[d,e,g,h]	586[a]	Eutrophic[b]
Quashnet River (QR)[d,e,g,h]	436[a]	Eutrophic[b]
Jehu Pond (JP)[d,e]	34[a]	Oligotrophic[b]
Sage Lot Pond (SLP)[d,e,f,g,h]	10[a]	Oligotrophic[b]
Timms Pond (TP)[d,e]	5[a]	Oligotrophic[b]
Jobos Bay, Puerto Rico[i,j]	—	Oligotrophic[c]

[a] E. Kinney, unpublished data.
[b] Fox et al. 2008.
[c] Bowen and Valiela 2008.
[d] Sites included in Waquoit Bay seagrass and macroalgae surveys (Figure 9.2).
[e] Sites included in Waquoit Bay fauna survey (Figure 9.5).
[f] Sites included in Waquoit Bay seagrass experiment (Figure 9.6).
[g] Sites included in Waquoit Bay macroalgae enrichment experiment (Figure 9.7).
[h] Sites included in Waquoit Bay macroalgae grazing experiment (Figure 9.8).
[i] Sites included in Jobos Bay seagrass experiment (Figures 9.10 and 9.11).
[j] Sites included in Jobos Bay macroalgae experiment (Figures 9.12 and 9.13).

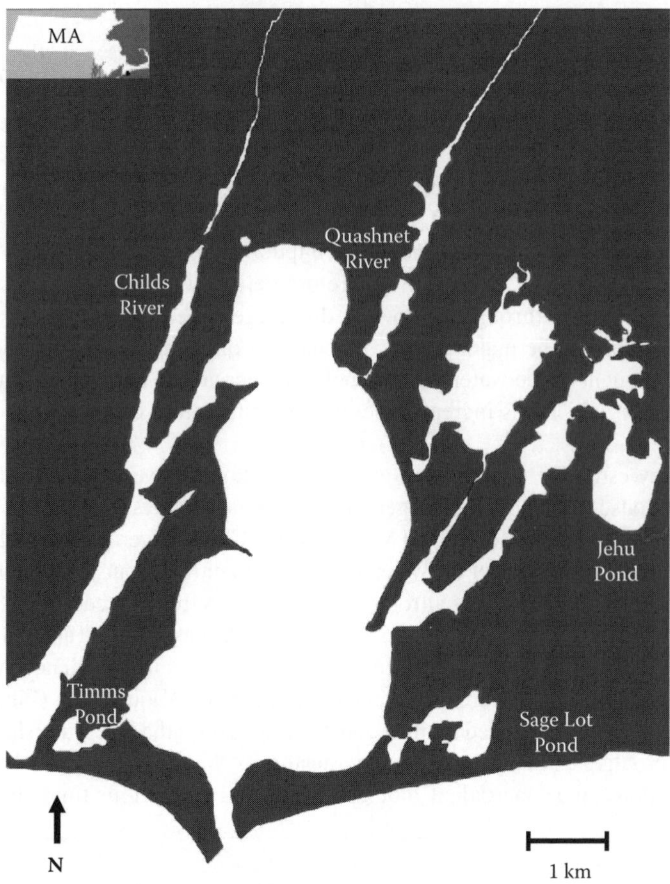

FIGURE 9.1 Map of Waquoit Bay. Inset shows the location of the bay with respect to the state of Massachusetts.

with high nitrogen loads compared to estuaries with low nitrogen loads (Figure 9.2b). Eutrophied estuaries of Waquoit Bay have high macroalgal biomass and low seagrass biomass, while in the non-eutrophied estuaries the reverse is true (Figure 9.2c). The decline of seagrass biomass and the increase in macroalgal biomass have been related to both nitrogen loading rates (Figure 9.2b) and area of salt marsh fringing the estuary (Figure 9.3). Where there was a large area of fringing salt marsh, seagrass biomass was high and macroalgal biomass was low. Where there was less salt marsh area, macroalgal biomass was high, and seagrass was absent (Figure 9.3).

In the higher nitrogen loaded estuaries with smaller area of fringing salt marsh, several symptoms of eutrophication have been observed, including the disappearance of seagrass meadows and the prevalence of macroalgal blooms (Hauxwell et al. 2001; Fox et al. 2008). In the eutrophied estuaries, the high macroalgal biomass controls the oxygen regime by releasing and consuming oxygen during photosynthesis and respiration (D'Avanzo and Kremer 1994). During summer, when respiration and growth rates are highest, the bottom waters of the eutrophied estuary become hypoxic on a daily basis (Figure 9.4). In the oligotrophic, low nitrogen loaded estuaries that are dominated by high seagrass and low macroalgal biomass, estuarine waters are rarely hypoxic. Higher nitrogen loads, eutrophication, and the associated hypoxia have effects on the benthic consumer communities (Fox et al. 2009). Results of benthic invertebrate surveys in seagrass and macroalgal canopies indicate significantly fewer consumers in estuaries with higher nitrogen loads compared to estuaries with lower loads (Figure 9.5), most likely the result of more frequent hypoxic conditions in the

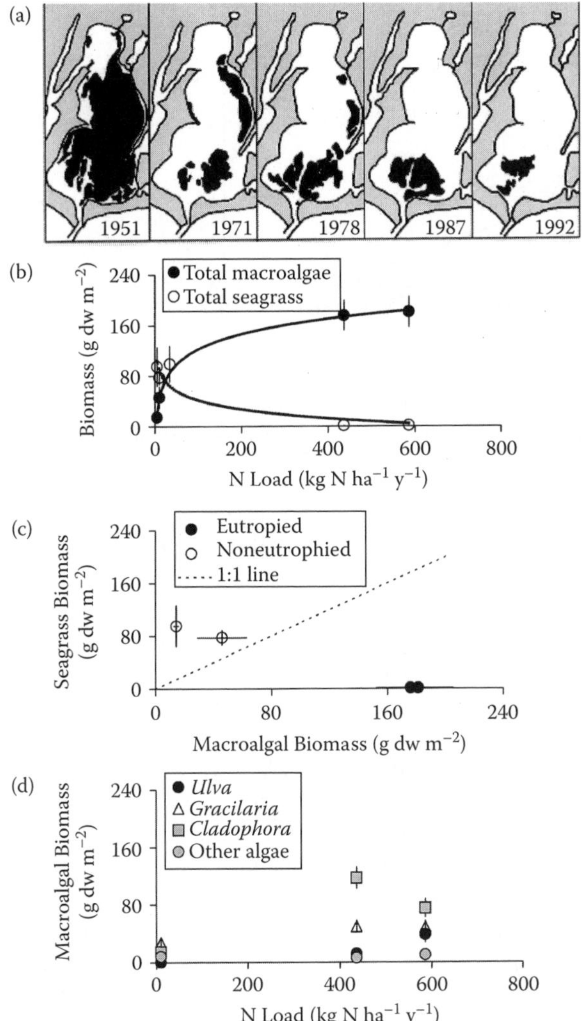

FIGURE 9.2 Results from surveys of seagrass and macroalgae in Waquoit Bay. (a) Map showing seagrass cover (in black) over time in Waquoit Bay. (b) Biomass of seagrass (*Zostera marina*) and macroalgae in estuaries of Waquoit Bay with different land-derived nitrogen loads. Lines represent significant regressions, for seagrass $F = 39.3$, $p = 0.008$, and for macroalgae $F = 35.9$, $p = 0.027$. (c) Relationship between seagrass and macroalgal biomass in eutrophied and noneutrophied estuaries of Waquoit Bay. (d) Biomass of different macroalgal taxa in estuaries of Waquoit Bay with different land-derived nitrogen loads.

eutrophied estuaries (Fox et al. 2009). The abundance of grazing epifauna declined two orders of magnitude in estuaries with high relative to low nitrogen loads (Figure 9.5), which is an important interaction between bottom-up controls and herbivory in coastal estuaries.

The conditions observed in the estuaries of Waquoit Bay are an example of changes taking place in macrophyte and consumer community structure related to bottom-up forcings owing to nutrient loads and eutrophication status, as well as to top-down effects related to differences in the assemblage of consumers. In this framework, we carried out *in situ* experimental studies examining the relative influence of control mechanisms acting on seagrasses and macroalgae.

To assess the response of the eelgrass, *Zostera marina*, to experimental nutrient enrichment, we fertilized sediments and removed macroalgal biomass in a set of seagrass plots during the summer months in a Waquoit Bay subestuary with a low nitrogen load, Sage Lot Pond (Table 9.1). The

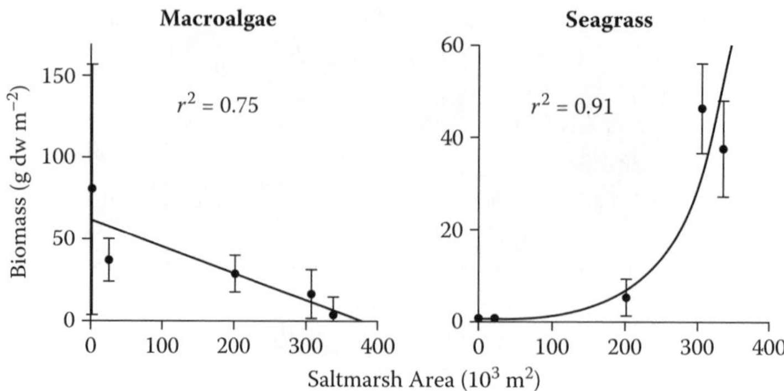

FIGURE 9.3 Biomass of macroalgae and seagrass in relation to saltmarsh area. (From Valiela, I. et al. 1997. *Limnol. Oceanogr.* 42: 1105–1118 © (1997) by the American Society of Limnology and Oceanography Inc. With permission.)

experimental design consisted of control (C) and fertilized (F) plots, in which we did (–A) or did not (+A) remove ambient macroalgal mats (for a total of 12 plots with three replicates of each of the four treatments, C + A, C – A, F + A, and F – A). After 12 weeks, *Z. marina* biomass response showed an interactive effect of nutrients and macroalgae. There was no effect of sediment nutrient enrichment on growth of eelgrass. Aboveground biomass and shoot density of the eelgrass in fertilized plots were not significantly different from those in control plots (Figure 9.6). In plots where the unattached macroalgae were removed, the biomass and shoot density tended to be higher than in plots with macroalgae, although not significantly ($F = 4.6$, $p = 0.063$). Further, the mean biomass and shoot density of eelgrass in the fertilized plots where algae were removed were higher than any other treatment; however, these differences were not statistically significant.

These results suggest that macroalgal mats may impose a primary limitation to seagrass growth in Waquoit Bay, probably owing to shading and reduction of light to the seagrasses. Light limitation has been demonstrated as the primary cause of seagrass decline in many estuaries (den Hartog 1994; Short et al. 1995; Hauxwell et al. 2001; McGlathery 2001; Ibarra-Obando et al. 2004), although sulfide invasion of seagrasses exposed to anoxic macroalgal mats may also contribute to lower seagrass growth (Holmer and Nielsen 2007). Sulfide invasion was not likely a mechanism causing lower seagrass growth in our study, since the macroalgal mats were not dense enough to create anoxic conditions and increase sulfide concentrations in sediments. Top-down impacts of grazing were not explicitly tested in this manipulation, but grazer abundance in the estuary was high (Figure 9.5), and grazing pressure was likely high in all treatments. Grazers of eelgrass in these estuaries are relatively rare, and the abundant mesograzers would likely be feeding on epiphytic algae growing on the eelgrass blades and unattached macroalgae in the plots (Orth and van Montfrans 1984; Neckles et al. 1993; Jernakoff et al. 1996; Moksnes et al. 2008). Grazing in the plots might, therefore, stimulate eelgrass growth by removing algal competitors for light and nutrients (Neckles et al. 1993).

The limitation of seagrass growth by algal shading is corroborated by previous work by Hauxwell et al. (2001, 2003), and by the low seagrass biomass in estuaries with high macroalgal biomass (Figures 9.2b and 9.2c). Macroalgal biomass is higher in estuaries with higher nitrogen loads than those with lower nitrogen loads (Figure 9.2b). In high nutrient environments, interactions among macroalgal species lead to domination by a few highly competitive taxa, which are able to proliferate to form macroalgal blooms. This notion is supported by the results from macrophyte surveys in Waquoit Bay that reveal the dominance of the macroalgae, *Cladophora*, *Gracilaria*, and *Ulva* spp., and the absence of the seagrass, *Z. marina*, in the eutrophied estuaries (Figure 9.2d) (Fox 2008; Fox et al. 2008). These taxa are common macroalgae that have been shown to bloom in response to increased anthropogenic nutrient inputs to coastal waters worldwide (Thorne-Miller et al. 1983; Lapointe and O'Connell 1989; Sfriso et al. 1989; Lavery et al. 1991; Thybo-Christensen et al. 1993).

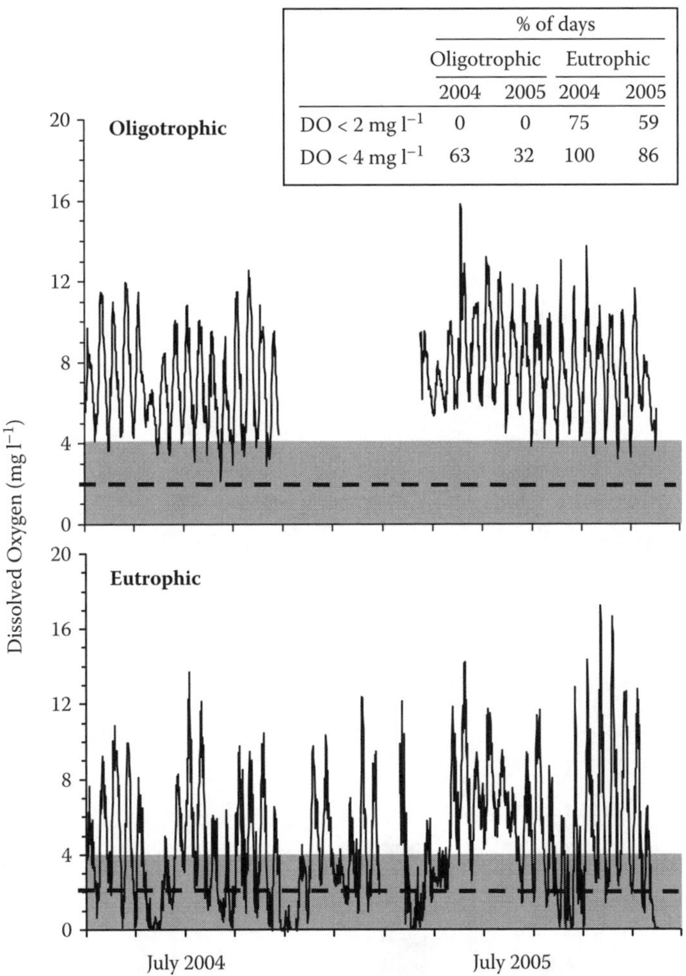

FIGURE 9.4 Continuous dissolved oxygen concentrations in an oligotrophic (Sage Lot Pond) and a eutrophic (Childs River) subestuary of Waquoit Bay during the month of July, 2004 and 2005 (data courtesy of Waquoit Bay National Estuarine Research Reserve). The data were collected with YSI data loggers deployed continuously at a representative location in each of the two estuaries at a depth of 1 m (in the estuarine bottom waters). Dashed line indicates dissolved oxygen (DO) concentration defined as hypoxia (<2 mg L^{-1}), and shaded area indicates DO concentrations below which there are negative effects on benthic fauna (<4 mg L^{-1}). Inset: table of percentage of days in July 2004 and 2005 DO concentrations below the dashed line and within the shaded area in each estuary. (Adapted from Fox, S. E. et al. 2009. *Mar. Ecol. Progr. Ser.* 380: 43–57.)

To understand the relative influence of nutrients and herbivory controls in structuring macroalgal communities in the estuaries of Waquoit Bay, we carried out two experiments to examine the growth response of macroalgae to nutrients and grazing. First, to identify the nutrients that effectively control growth in the different macroalgal taxa, we assessed responses of the common green and red bloom-forming species *Ulva lactuca* and *Gracilaria tikvahiae* to short-term enrichment with nitrate (NO_3^-), ammonium (NH_4^+), and phosphate (PO_4^{3-}) in a high and a low nitrogen loaded estuary (Teichberg et al. 2008). The macroalgal growth response to nutrient enrichment varied by species, estuary, and nutrient treatment (Figure 9.7). Growth rates of *U. lactuca* tended to be higher overall than those of *G. tikvahiae*. Mean *U. lactuca* growth was higher in controls in the high than in the low nitrogen loaded estuary, and growth was significantly higher with the addition of either NO_3^- or NH_4^+ in the low nitrogen loaded estuary (Figure 9.7). In the high nitrogen loaded estuary,

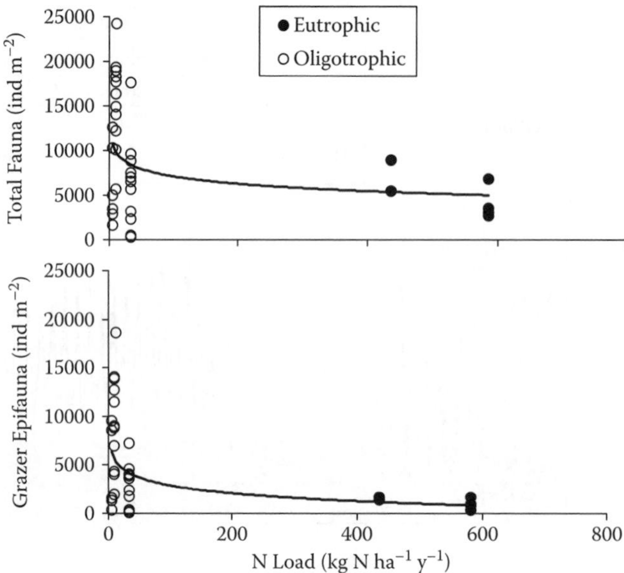

FIGURE 9.5 Abundance of animals and grazers in eutrophic and oligotrophic estuaries of Waquoit Bay with different land-derived nitrogen loads. Lines represent significant regressions, for total fauna $F = 5.1$, $p = 0.029$, and for epifaunal grazers $F = 5.0$, $p = 0.031$.

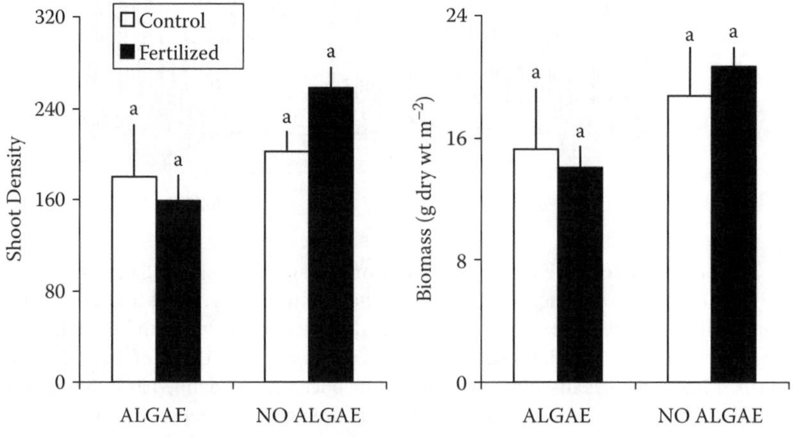

FIGURE 9.6 Seagrass biomass and shoot density in response to nutrient enrichment and macroalgae removal in a low nitrogen load estuary in Waquoit Bay. Letters correspond to statistically homogeneous groups determined by post hoc Tukey's test for each response variable.

U. lactuca growth was not significantly different in nitrogen-enriched than in the control treatments. Growth of *G. tikvahiae* tended to be higher overall in the low nitrogen loaded estuary, and did not respond significantly to nitrogen enrichment in either estuary, although growth tended to be higher in the NH_4^+ enrichments (Figure 9.7). Neither species grew in response to PO_4^{3-} enrichment (relative to the control) or to N + P enrichment (relative to nitrogen enrichment).

These experimental results show that the macroalgae were primarily nitrogen limited, with no phosphorus limitation during the experiment, as suggested by the results of other studies in temperate estuaries (Howarth 1988; Pedersen and Borum 1996; Teichberg et al. 2007). There were species-specific differences in growth rate responses to NO_3^- or NH_4^+, and *U. lactuca* may be more

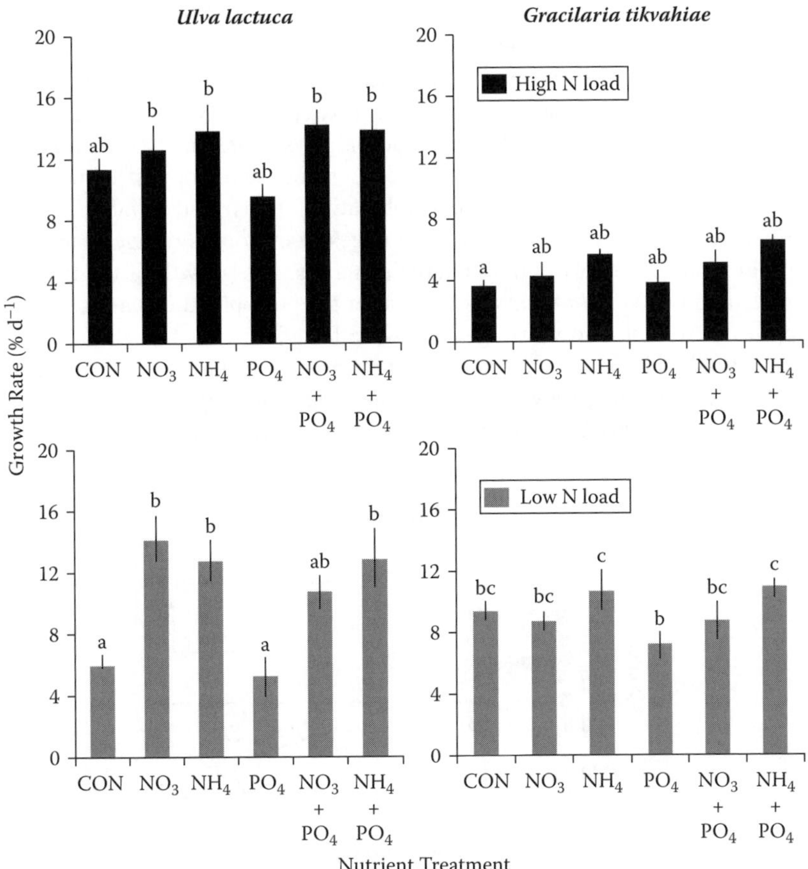

FIGURE 9.7 *Ulva lactuca* and *Gracilaria tikvahiae* growth rates in response to nutrient enrichment treatments in a high (top) and low (bottom) nitrogen loaded estuary of Waquoit Bay. Letters correspond to statistically homogeneous groups determined by post hoc Tukey test between nutrient treatments within a species. (From Teichberg, M. et al. 2008. *Mar. Ecol. Prog. Ser.* 368: 117–126. With permission.)

nitrogen limited under low nitrogen conditions, where it responds to increased supply of both NO_3^- and NH_4^+. *Gracilaria tikvahiae* exhibited lower growth rates and, thus, may be less nitrogen limited than *U. lactuca*. Studies of *Ulva* and *Gracilaria* spp. growth support these findings (DeBoer et al. 1978; Lapointe and Tenore 1981). Species-specific differences in growth responses may also be controlled by the prior nutrient history of the fronds, the long-term ambient nutrient supply, and species-specific differences in nutrient storage capacity (DeBoer et al. 1978; Peckol et al. 1994; Aguiar et al. 2003; Fong et al. 2003; Teichberg et al. 2007, 2008).

To examine the effect of long-term nutrient forcing on the relative influence of nutrients and herbivory as controls on macroalgal growth, we ran an *in situ* cage experiment in subestuaries of Waquoit Bay, where nitrogen loads, eutrophication status, and grazer abundance differed (Fox 2008; Fox et al. submitted). To evaluate the importance of bottom-up control of growth of macroalgae, we ran the cage experiments in three subestuaries of Waquoit Bay (i.e., Childs River, Quashnet River, and Sage Lot Pond) (Table 9.1, Figure 9.1), because we wanted to capture the effects of longer-term, regional-scale forcings, like nutrient loading rates and annual mean nitrogen concentrations, rather than the short-term responses to experimental nutrient enrichment. Within each of the three subestuaries, we employed cages with four different sized mesh openings to create four grazer treatments, where different numbers of consumers were present inside the cages. Four replicate cages of each of the mesh treatments were placed in each subestuary for a total of 48 cages. The use of these three subestuaries with different long-term nutrient regimes was particularly important because we knew that the macrophyte taxa and biomass and grazer communities present in the Waquoit Bay estuaries were the product of chronic nutrient loading (Fox 2008; Fox et al. 2008, 2009) (Figures 9.2 and 9.5). We measured the growth response of three macroalgal species (i.e., *U. lactuca, Cladophora vagabunda,* and *G. tikvahiae*) and growth differed in response to both different long-term nutrient regimes (nitrogen loading rates and eutrophication status) and number of grazers inside cages (Figure 9.8). There were fewer grazers in cages in eutrophic (range, 0 to 27 individuals) than oligotrophic (range, 29 to 169 individuals) estuaries. Net growth of the two green macroalgae, *U. lactuca* and *C. vagabunda*, was highest in the eutrophied estuaries, where long-term nutrient supplies were high and where grazers were fewer ($F = 29.2$, $p < 0.001$, for *U. lactuca* and $F = 25.6$, $p < 0.001$, for *C. vagabunda*). Only in the oligotrophic estuary did grazers control the growth of the green algae, shown by numerous cages with negative growth (Figure 9.8). In contrast, the red alga *G. tikvahiae* had high growth rates, and there was no significant effect of grazing ($F = 1.4$, $p > 0.05$).

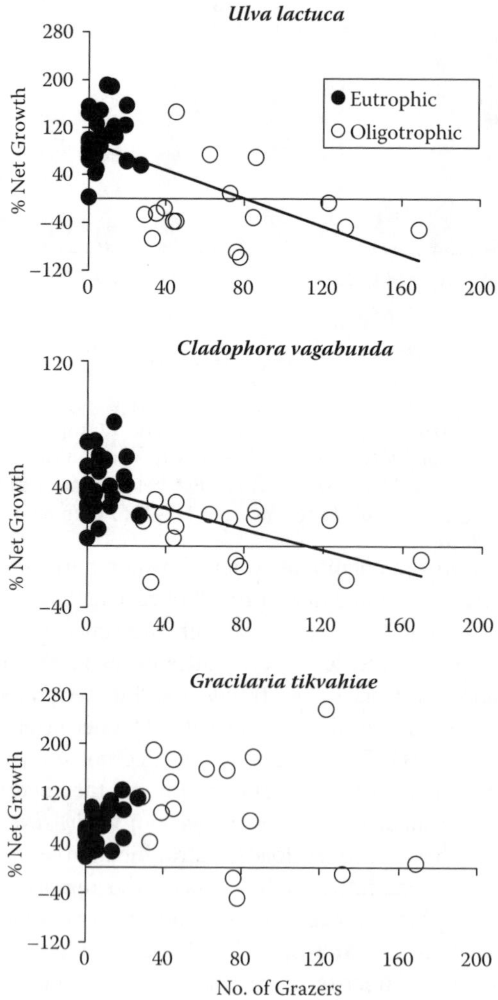

FIGURE 9.8 Number of grazers per experimental cage vs. percent net growth of macroalgae after a 14-day incubation in eutrophic and oligotrophic subestuaries of Waquoit Bay. Lines represent significant regressions. (Data from Fox 2008; Fox et al. submitted.)

In the subestuaries of Waquoit Bay, both nutrients and grazers can control growth of benthic macrophytes. For both seagrasses and macroalgae, the long-term nutrient regime in surrounding waters most likely determined their growth responses. For seagrasses, growth was not stimulated by short-term sediment fertilization in the presence of free-floating macroalgae, while growth tended to increase where macroalgae were removed. These results suggest that macroalgal biomass accumulation limits growth of the eelgrass, *Z. marina*, in these estuaries. Macroalgae were predominantly controlled from the bottom up by short-term and long-term nutrient supplies, since macroalgal growth rates increased in response to nitrogen enrichment and were highest in eutrophied estuaries with higher nitrogen loads (Figures 9.7 and 9.8). When growth rates of macroalgae were high, as they were for *G. tikvahiae* in the oligotrophic estuary, grazers, even in very high numbers, were not able to compensate for the rapid growth, and herbivory control was overwhelmed by the bottom-up control by nutrients. Accumulations of macroalgal mats in the oligotrophic estuary contributed to reduced *Z. marina* growth in experimental plots (Figure 9.6), and are likely to reduce seagrass biomass throughout the estuary since herbivory is not controlling macroalgal growth.

In the eutrophied estuaries, nutrients not only increased macroalgal biomass, but also increased frequency and duration of hypoxia (Figure 9.4) that may control the grazer abundance and community composition (Figure 9.5) (Fox et al. 2009). The abundance of grazers (predominantly small crustacean amphipods and isopods, and gastropods) was significantly lower in the eutrophic estuaries relative to oligotrophic estuaries, with almost a 25-fold decrease from low to high nitrogen loads (Figure 9.5). Eutrophication and the associated hypoxia were most likely responsible for the differences in the grazer communities between estuaries (Fox et al. 2009). These studies support the notion that nutrients predominate as controls on benthic macrophyte growth in these temperate estuaries, by controlling both producer growth and herbivore abundance.

9.4.2 Jobos Bay, Puerto Rico: Bottom-Up and Top-Down Controls

Jobos Bay is a tropical estuary located on the south coastal plain of Puerto Rico (Table 9.1, Figure 9.9). The land cover on the Jobos Bay watershed is a mix of land uses, including extensive agriculture and urban development (Bowen and Valiela 2008). The bay is fringed by mangroves, and the benthic macrophyte community is dominated by the seagrass *Thalassia testudinum*, with smaller stands

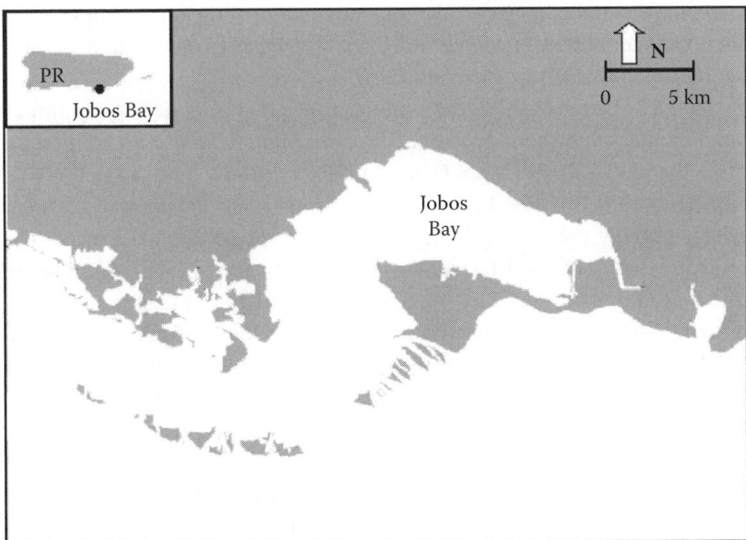

FIGURE 9.9 Map of Jobos Bay, Puerto Rico. Inset shows location of Jobos Bay on the south coast of Puerto Rico.

of the seagrasses *Syringodium filiforme* and *Halodule wrightii*. Macroalgae are rare and patchily distributed. In contrast to the Waquoit Bay system, the land uses on the Jobos Bay watershed appear to be decoupled from receiving estuarine waters so that, although nitrogen loads from the watershed may be high, nutrient concentrations in the bay are low (Bowen and Valiela 2008). In these relatively pristine waters, seagrass cover and biomass have remained largely unchanged over the past 35 years (Kolehmainen 1973; Vicente 1975; Y. Olsen unpublished).

Jobos Bay is a good site to study bottom-up and top-down controls on benthic macrophytes in the tropics. It is representative of many tropical estuaries because of its low nutrient concentrations, oligotrophic status, and the presence of many potential grazers of both seagrasses and macroalgae. Grazers include manatees, sea turtles, herbivorous fish, and small crustaceans. Sea urchins, a common grazer of seagrass and macroalgae, do inhabit the benthos, but were rarely observed during our study period (Y. Olsen unpublished). In the Jobos Bay estuary, we carried out a set of *in situ* experiments to examine the relative influence of nutrient supply and herbivory as controls on seagrass and macroalgal biomass.

To test the combined effects of nutrients and grazers on seagrass growth, we simultaneously enriched the sediments (with N + P) and excluded mega-grazers with fences in a *T. testudinum* meadow (Olsen and Valiela 2010). Since grazing mega-herbivores exert a significant grazing pressure on seagrasses in tropical estuaries (Kenworthy et al. 2007), we constructed mesh fences around plots to exclude manatees and sea turtles. Herbivorous fish are common in the seagrass beds of Jobos Bay, and were able to move freely in and out of the fenced plots through large holes in the mesh. The experimental design consisted of control (C) and fertilized (F) plots, in which we excluded (–G) or allowed (+G) mega-grazers (for a total of 12 plots with three replicates of each of the four treatments, C – G, C + G, F – G, F + G). Although the experimental plots were in an area where manatees often fed, there was little evidence of manatee grazing during the study period (from April 2005 to May 2006). Fish, however, were attracted to the fences and were abundant in the fenced plots, so that the fenced plots fortuitously created an experimental treatment that markedly elevated abundance of fish of all trophic levels (Olsen and Valiela 2010). Bite marks on leaves provided evidence of fish grazing on the seagrasses. In contrast, open plots (no fences) had fewer fish and showed few signs of grazing. The cages, thus, created significant differences in grazing pressure, primarily by herbivorous fish, between treatments, with higher grazing pressure in caged plots relative to open plots. Further, there were significantly higher abundances of fish in fertilized caged plots compared to unfertilized caged plots and open plots. We took advantage of these conditions to assess the influence of different porewater nutrient regimes and grazing pressures on the biomass and shoot density of *T. testudinum* (Olsen 2008; Olsen and Valiela 2010).

The seagrass biomass response showed an interactive effect of nutrient enrichment and grazing. Both aboveground biomass and shoot density of *T. testudinum* decreased with increasing number of herbivorous fish in the plots ($F = 16.2$, $p = 0.002$, for aboveground biomass and $F = 51.3$, $p < 0.001$, for shoot density) (Figure 9.10). The lowest biomass and shoot densities were found in fertilized fenced plots, where grazing pressure was highest. There were clear differences in seagrass cover between the control caged plots and the nutrient-enriched caged plots, with a significantly lower percentage of the fenced plots covered by seagrass with fertilization (Figure 9.11).

Increased grazing pressure with fertilization has previously been recorded for seagrasses (McGlathery 1995, Heck et al. 2000, Goecker et al. 2005). Previous studies have attributed higher fish grazing rates to increased palatability of the seagrasses due to lower C:N; however, this is not a likely explanation in our study since the percent nitrogen in the seagrass tissues from fertilized caged plots was similar to control caged plots, where substantially less grazing occurred (Olsen and Valiela 2010). The increase in porewater nutrient supply may have also changed chemical grazing deterrent concentrations in the *T. testudinum* tissues and, thus, altered palatability of the seagrasses to grazers. Several studies have shown reduced concentrations of grazing deterrents in response to higher nitrogen availability (Buchsbaum et al. 1990; Yates and Peckol 1993), and *T. testudinum* leaves that were high in nitrogen had significantly lower levels of phenolics (Goecker et al. 2005).

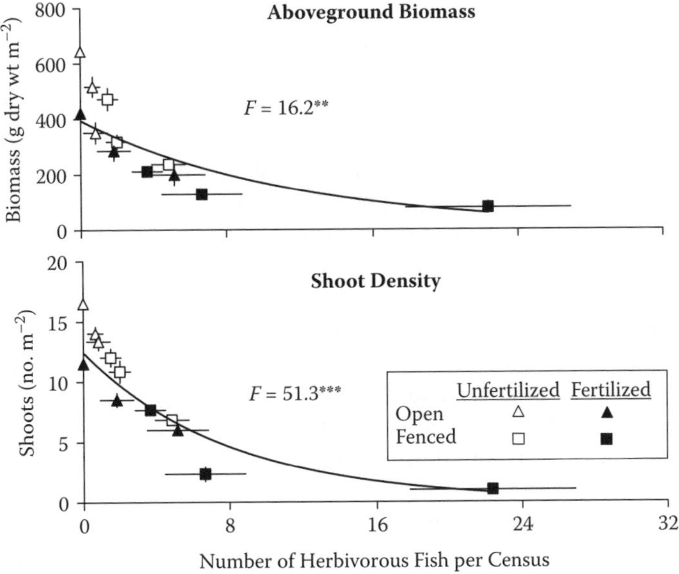

FIGURE 9.10 Aboveground biomass and shoot density of *Thalassia testudinum* at the end of the manipulative field experiment as a function of the mean number of grazing fish in Jobos Bay, Puerto Rico. Lines represent significant regressions. (From Olsen, Y.S. and Valiela, I., *Estuaries Coasts*, DOI 10.1007/s12237-009-9256-7, 2010. With permission from Springer Science & Business Media.)

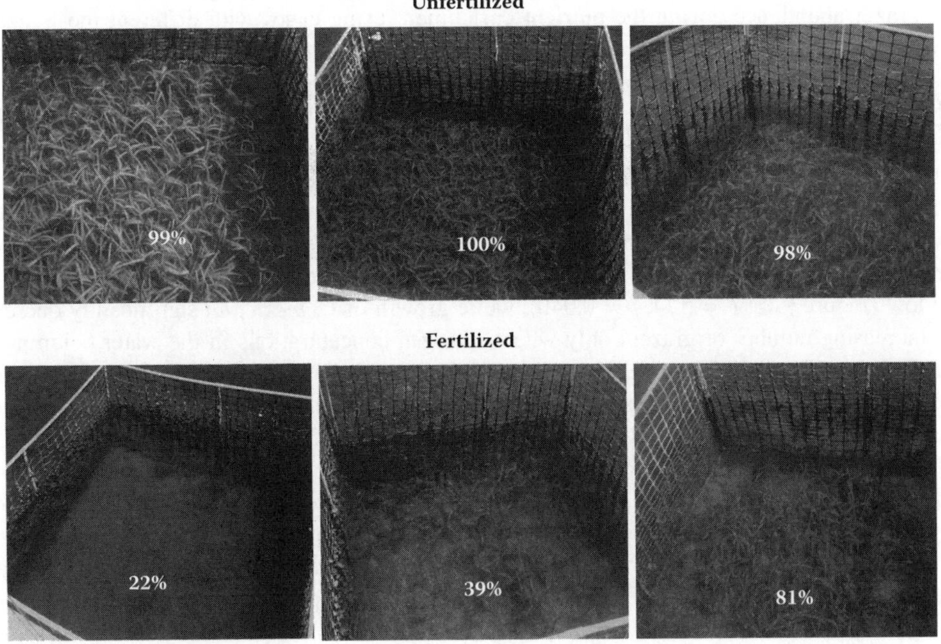

FIGURE 9.11 Seagrass percent cover in unfertilized (top) and fertilized (bottom) fenced plots after 12 months of experimental treatment in Jobos Bay, Puerto Rico. (From Olsen, Y.S. and Valiela, I., *Estuaries Coasts*, DOI 10.1007/s12237-009-9256-7, 2010. With permission from Springer Science & Business Media.)

With increased supply of nutrients, carbon could also have been allocated to growth rather than increased storage of defensive compounds (M. Teichberg, unpublished). It therefore seems likely that the chemical deterrents of the seagrasses were in some way altered by the sediment fertilizer application, making the seagrass more palatable to the numerous grazers that were attracted to the fences.

In Jobos Bay, herbivorous fish were able to control seagrass biomass, and grazing pressure was mediated by nutrient enrichment. This experiment provides evidence for significant interactions between nutrients and grazing, such that bottom-up mechanisms control the degree of control by herbivores.

To understand the relative influence of nutrient supply and herbivore controls in structuring macroalgal communities in tropical estuaries, we carried out a field experiment to assess the growth response of macroalgae to nutrients and grazing in Jobos Bay. To identify the nutrient that effectively controls macroalgal growth, we examined the responses of the bloom-forming species *U. lactuca, G. tikvahiae,* and *Caulerpa mexicana,* to short-term enrichment with N or P. The experimental design consisted of three nutrient treatments, Control (C), NO_3^- (N) , and PO_4^{3-} (P), and two herbivory treatments, excluding (–G) or allowing (+G) crustacean grazers for a total of 24 cages with four replicates of each of six treatments (C + G, C – G, N + G, N – G, P + G, and P – G). All three macroalgal species tended to increase growth in response to enrichment with NO_3^- (Figures 9.12a and 9.12b), although only significantly so in *C. mexicana* ($F = 10.7$, $p = 0.004$). Growth rates of *U. lactuca* were significantly higher overall than those of *G. tikvahiae* and *C. mexicana* for all treatments. None of the macroalgal species grew in response to enrichment with PO_4^{3-}, with all species tending to grow less when PO_4^{3-} was added than in control and NO_3^- treatments (Figure 9.12c). These data suggest that Jobos Bay macroalgae were nitrogen limited, with no phosphorus limitation during the experiment, contrary to the results of many studies examining nutrient limitation of macroalgae in tropical waters (Lapointe et al. 1992; McGlathery et al. 1994). There were also species-specific differences in growth rates with NO_3^- fertilization, with significantly higher growth of *U. lactuca* than the other species.

To test the interactive effects of nutrients and grazers on macroalgal growth, we also manipulated grazer abundance during the nutrient enrichment using cages with different mesh sizes to allow (+G) or restrict (–G) grazer access. The macroalgal growth response to nitrogen enrichment and grazing pressure differed among the different macroalgal species (Figure 9.13). Grazers in this study were small crustaceans and gastropods, and herbivorous fish were excluded from all of the cages. *U. lactuca* tended to have higher growth with higher nitrogen concentrations in the water column, and grazers were unable to control growth at low or high nitrogen concentrations. In contrast, *G. tikvahiae* and *C. mexicana* showed interactive effects of nutrients and grazing (Figure 9.13). However, the treatments to which each species responded differed. The growth of *G. tikvahiae* decreased in response to increased grazing pressure only where water nitrogen concentrations were low (Figure 9.13; $F = 5.14$, $p = 0.040$), while growth of *C. mexicana* significantly decreased with increasing number of grazers only where nitrogen concentrations in the water column were high (Figure 9.13; $F = 7.53$, $p = 0.034$). The *G. tikvahiae* growth response to grazing followed a pattern that has been shown for macroalgae under low nutrient conditions in temperate estuaries (Geertz-Hansen et al. 1993; Hauxwell et al. 1998; Lotze et al. 2001; Worm and Lotze 2006; Fox 2008; Fox et al. submitted). The mechanisms controlling growth of macroalgae in Jobos Bay change markedly depending on the species involved. Increased nitrogen supply led to growth in all species, and there were clear interactions of bottom-up and herbivore controls. The degree of response to controlling mechanisms would likely have differed if herbivorous fish were allowed in the experimental cages. Given the highly species-specific nature of the responses, however, we cannot speculate as to the results if herbivorous fish had been included in our study.

In Jobos Bay, nutrients and grazers were both important in controlling benthic macrophytes, and there were complex interactions between nutrients and grazers. The relative influence of herbivory and nutrients and the response to changes in these controls differed among producers. For some

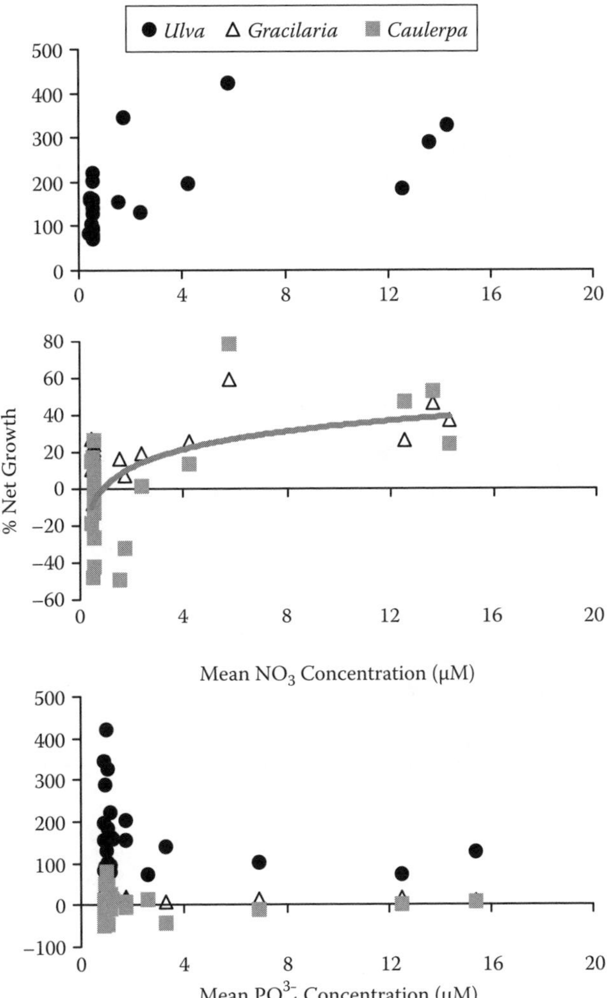

FIGURE 9.12 Mean nitrate and phosphate concentration vs. percent net growth of macroalgae after a 14-day incubation in Jobos Bay, Puerto Rico. Lines represent significant regressions.

species of macroalgae, grazer control of biomass was absent or only present at low nutrient concentrations, when growth was limited by nutrient supply. For other producers, such as the seagrass *T. testudinum* and the macroalga *C. mexicana*, grazing overwhelmed growth only when nutrient concentrations were high. The experimental nutrient additions may have increased the nutritional value of the macrophytes (McGlathery 1995; Boyer et al. 2004), and may have overridden the effect of chemical deterrents (Valiela and Rietsma 1984) leading to increased grazing pressure (Bjorndal 1980; McGlathery 1995; Preen 1995; Heck et al. 2000).

9.5 CONCLUSIONS

In Waquoit Bay, seagrass biomass was controlled by macroalgal shading, where increased light availability by removal of the overlying macroalgal canopy tended to lead to higher *Z. marina* biomass and shoot density. The macroalgal biomass in these estuaries was nitrogen limited, and *U. lactuca* had the largest growth response to fertilization. Bottom-up mechanisms dominated control of macroalgal biomass, so that where nutrients stimulated algal growth, grazers even in very

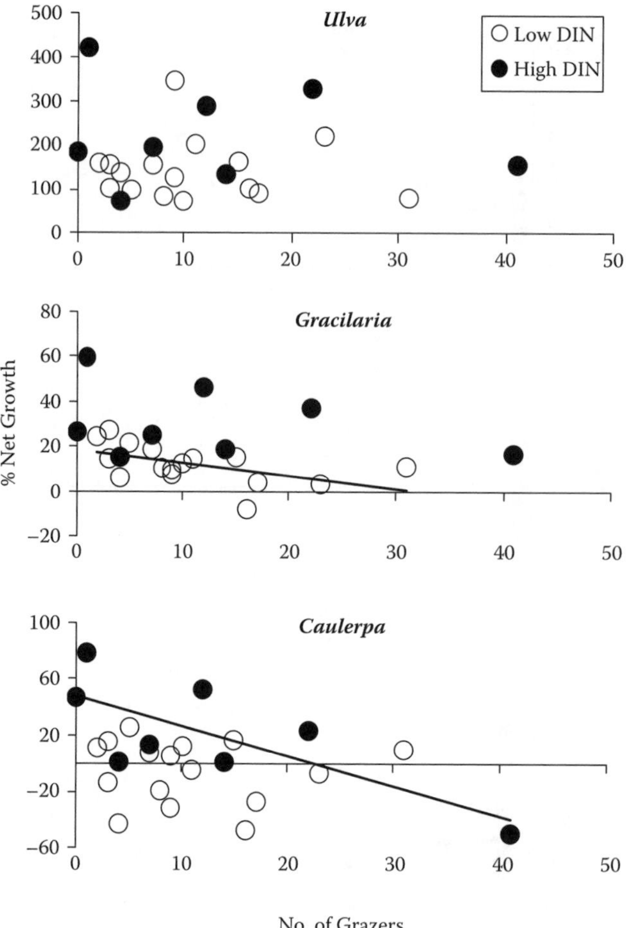

FIGURE 9.13 Number of grazers per experimental cage vs. percent net growth of macroalgae after a 14-day incubation under different dissolved inorganic nitrogen (DIN) concentrations in Jobos Bay, Puerto Rico.

high numbers could not control biomass. Bottom-up control mechanisms structured the benthic macrophyte communities in the Waquoit Bay estuaries.

In Jobos Bay, macroalgal growth was nitrogen limited, so that nitrogen supply stimulated an increase in macroalgal biomass. This study contributes new data to the unresolved arguments regarding the primacy of nitrogen or phosphorus limitation of macroalgae in tropical waters (Larned 1998; Fong et al. 2001). The macroalgal responses to grazing pressure depended on nutrient supply and macroalgal species. Seagrasses responded to a different combination of bottom-up and top-down controlling mechanisms. Seagrass biomass and shoot density were controlled by an interaction between porewater nutrients and herbivory, where increased nutrients led to higher grazing pressure on seagrasses. Despite previous experimental assessments of interactive effects of bottom-up and top-down controls in seagrass meadows, we do not have a clear, predictable understanding of how seagrasses respond to simultaneous nutrient enrichment and grazing (McGlathery 1995; Heck et al. 2000; Hillebrand and Kahlert 2002; Posey et al. 2002; Armitage et al. 2005; Heck and Valentine 2007), probably owing to the high natural variability in seagrass meadows.

The experimental results highlight some of the many complex interactions taking place in seagrass meadows. In Waquoit Bay, nutrients led to lower grazer abundance through hypoxia, and decreased herbivore pressure, while in Jobos Bay, nutrients stimulated herbivory, increasing grazing pressure. Nutrients, light, producers, and consumers interact continuously and simultaneously, and these connections result in myriad combinations, many of which can control growth and biomass on different spatial and temporal scales.

In these studies, we experimentally examined the effects of fertilization on seagrass standing stocks, and our results did not show significant direct effects of porewater nutrient enrichment on seagrass biomass or shoot density. A dynamic measure (productivity) rather than a static measure (standing stock) of seagrass response might have yielded an effect of fertilization (Peterson et al. 2007), since biomass turnover rates of seagrasses are high, particularly in the tropics (Martinez-Daranas et al. 2005).

The results of these experimental manipulations examining the mechanisms of control for seagrasses and macroalgae overwhelmingly emphasize that the relative primacy of nutrients vs. herbivory effects is more nuanced than has previously been thought. There are various mechanisms involved (quality of food, cover, predator and prey sizes), and in any one setting, the food web may be governed by combinations of factors, and different compartments (plants, grazers, predators) could respond differently to the control mechanisms. Although the specific mechanisms and outcomes may differ from one environment to another, we should expect a complex interplay of controlling forces in structuring benthic producer communities.

ACKNOWLEDGMENTS

We would like to thank the staff at the Waquoit Bay and the Jobos Bay National Estuarine Research Reserves for assistance and logistical support. We would also like to thank Leanna Heffner, Carolina Aguila, and Laurie Hofmann for assistance in the lab and the field. This research was supported by the following funding sources: NOAA National Estuarine Research Reserve Systems Graduate Research Fellowships to S.E.F and Y.S.O., EPA STAR Graduate Research Fellowship to S.E.F., Palmer McCleod Fellowship awarded to M.T., Lerner-Gray Award from the American Museum of Natural History and a Sounds Conservancy Grant from the Quebec Labrador Foundation awarded to Y.S.O., Woods Hole Marine Science Consortium, and NOS/ECOHAB Grant #NA16OP2728.

REFERENCES

Aguiar, A. B., J. A. Morgan, M. Teichberg, S. Fox, and I. Valiela. 2003. Transplantation and isotopic evidence of the relative effects of ambient and internal nutrient supply on the growth of *Ulva lactuca*. *Biol. Bull.* 205: 250–251.

Alcoverro, T. and S. Mariani. 2002. Effects of sea urchin grazing on seagrass (*Thalassodendron ciliatum*) beds of a Kenyan lagoon. *Mar. Ecol. Prog. Ser.* 226: 255–263.

Armitage, A. R., T. A. Frankovich, K. L. Heck, and J. W. Fourqurean. 2005. Experimental nutrient enrichment causes complex changes in seagrass, microalgae, and macroalgae community structure in Florida Bay. *Estuaries* 28: 422–434.

Baden, S. P., L. O. Loo, L. Pihl, and R. Rosenberg. 1990. Effects of eutrophication on benthic communities including fish: Swedish west coast. *Ambio* 19: 113–122.

Bjorndal, K. A. 1980. Nutrition and grazing behavior of the green turtle *Chelonia mydas*. *Mar. Biol.* 56: 147–154.

Blaabjerg, V. and K. Finster. 1998. Sulfate reduction associated with roots and rhizomes of the marine macrophyte *Zostera marina*. *Aquat. Microb. Ecol.* 15: 311–314.

Bologna, P. A. X. and K. L. Heck. 1999. Differential predation and growth rates of bay scallops within a seagrass habitat. *J. Exp. Mar. Biol. Ecol.* 239: 299–314.

Bowen, J. L. and I. Valiela. 2001. The ecological effects of urbanization of coastal watersheds: Historical increases in nitrogen loads and eutrophication of Waquoit Bay estuaries. *Can. J. Fish. Aquat. Sci.* 58: 1489–1500.

Bowen, J. L. and I. Valiela. 2008. Using $\delta^{15}N$ to assess coupling between watersheds and estuaries in temperate and tropical regions. *J. Coastal Res.* 24: 804–813.

Boyer, K. E., P. Fong, A. R. Armitage, and R. A. Cohen. 2004. Elevated nutrient content of tropical macroalgae increases rates of herbivory in coral, seagrass, and mangrove habitats. *Coral Reefs* 23: 530–538.

Bricker, S. B., B. Longstaff, W. Dennison, A. Jones, K. Boicourt, C. Wicks, and J. Woerner. 2007. *Effects of nutrient enrichment in the nation's estuaries*. NOAA, Coastal Ocean Program Decision Analysis Series No. 26, National Centers for Coastal Ocean Science, Silver Spring, Maryland.

Buchsbaum, R. N., F. T. Short, and D. P. Cheney. 1990. Phenolic-nitrogen interactions in eelgrass, *Zostera marina* L.: Possible implications for disease resistance. *Aquat. Bot.* 37: 291–297.

Burkepile, D. E. and M. E. Hay. 2006. Herbivore vs. nutrient control of marine primary producers: Context-dependent effects. *Ecology* 87: 3128–3139.

Carmichael, R. H. and I. Valiela. 2005. Coupling of near-bottom seston and surface sediment composition: Changes with nutrient enrichment and implications for estuarine food supply and biogeochemical processing. *Limnol. Oceanogr.* 50: 97–105.

Cebrián, J. 1999. Patterns in the fate of production in plant communities. *Amer. Nat.* 154: 449–468.

Cebrián, J. and C. M. Duarte. 1998. Patterns in leaf herbivory on seagrasses. *Aquat. Bot.* 60: 67–82.

Cebrián, J. and C. M. Duarte. 2001. Detrital stocks and dynamics of the seagrass *Posidonia oceanica* (L.) Delile in the Spanish Mediterranean. *Aquat. Bot.* 70: 295–309.

D'Avanzo, C. and J. N. Kremer. 1994. Diel oxygen dynamics and anoxic events in a eutrophic estuary of Waquoit Bay, Massachusetts. *Estuaries* 17: 131–139.

DeBoer, J. A., H. J. Guigli, T. L. Israel, and C. F. D'Elia. 1978. Nutritional studies of two red algae. I. Growth rate as a function of nitrogen source and concentration. *J. Phycol.* 14: 261–266.

den Hartog, C. 1994. Suffocation of a littoral *Zostera* bed by *Enteromorpha radiata*. *Aquat. Bot.* 47: 21–28.

Dennison, W. C., R. C. Aller, and R. S. Alberte 1987. Sediment ammonium availability and eelgrass (*Zostera marina*) growth. *Mar. Biol.* 94: 469–477.

Domning, D. P. 2001. Sirenians, seagrasses, and Cenozoic ecological change in the Caribbean. *Palaeogeogr. Palaeoclimatol. Palaeoecol.* 166: 27–50.

Dorenbosch, M., M. C. van Riel, I. Nagelkerken, and G. van der Velde. 2004. The relationship of reef fish densities to the proximity of mangrove and seagrass nurseries. *Estuar. Coastal Shelf Sci.* 60: 37–48.

Duarte, C. M. and C. L. Chiscano. 1999. Seagrass biomass and production: A reassessment. *Aquat. Bot.* 65: 159–174.

Duffy, J. E. and A. M. Harvilicz. 2001. Species-specific impacts of grazing amphipods in an eelgrass-bed community. *Mar. Ecol. Progr. Ser.* 223: 201–211.

Duffy, J. E. and M. E. Hay. 1994. Herbivore resistance to seaweed chemical defense: The roles of mobility and predation risk. *Ecology* 75: 1304–1319.

Erftemeijer, P. L. A., J. Stapel, M. J. E. Smekens, and W. M. E. Drossaert. 1994. The limited effect of in-situ phosphorus and nitrogen additions to seagrass beds on carbonate and terrigenous sediments in South Sulawesi, Indonesia. *J. Exp. Mar. Biol. Ecol.* 182: 123–140.

Ferdie, M. and J. W. Fourqurean. 2004. Responses of seagrass communities to fertilization along a gradient of relative availability of nitrogen and phosphorus in a carbonate environment. *Limnol. Oceanogr.* 49: 2082–2094.

Fong, P., K. E. Boyer, K. Kamer, and K. A. Boyle. 2003. Influence of initial tissue nutrient status of tropical marine algae on response to nitrogen and phosphorus additions. *Mar. Ecol. Progr. Ser.* 262: 111–123.

Fong, P., K. Kamer, K. E. Boyer, and K. A. Boyle. 2001. Nutrient content of macroalgae with differing morphologies may indicate sources of nutrients to tropical marine systems. *Mar. Ecol. Progr. Ser.* 220: 137–152.

Fox, S. E. 2008. Ecological effects of nitrogen loading to temperate estuaries: Macrophyte and consumer community structure and food web relationships. Ph.D. Thesis, Boston University, Boston, Massachusetts.

Fox, S. E., E. Stieve, I. Valiela, J. Hauxwell, and J. McClelland. 2008. Macrophyte abundance in Waquoit Bay: Effects of land-derived nitrogen loads on seasonal and multi-year biomass patterns. *Estuaries Coasts* 31: 532–541.

Fox, S.E., M. Teichberg, L. Heffner, and I. Valiela. Submitted. The relative role of nutrients, grazing, and predation as controls on macroalgal growth in a temperate estuary. *Estuaries Coasts*.

Fox, S. E., M. Teichberg, Y. S. Olsen, L. Heffner, and I. Valiela. 2009. Restructuring of benthic communities in eutrophic estuaries: Lower abundance of prey leads to trophic shifts from omnivory to grazing. *Mar. Ecol. Progr. Ser.* 380: 43–57.

Frankovich, T. A. and J. C. Zieman. 2005. A temporal investigation of grazer dynamics, nutrients, seagrass leaf productivity, and epiphyte standing stock. *Estuaries* 28: 41–52.

Geertz-Hansen, O., K. Sand-Jensen, D. F. Hansen, and A. Christiansen. 1993. Growth and grazing control of abundance of the marine macroalga, *Ulva lactuca* L. in a eutrophic Danish estuary. *Aquat. Bot.* 46: 101–109.

Goecker, M. E., K. L. Heck, and J. F. Valentine. 2005. Effects of nitrogen concentrations in turtlegrass *Thalassia testudinum* on consumption by the bucktooth parrotfish *Sparisoma radians*. *Mar. Ecol. Prog. Ser.* 286: 239–248.

Graham, S., J. Davis, L. A. Deegan, J. Cebrian, J. Hughes, and J. Hauxwell. 1998. Effect of eelgrass (*Zostera marina*) density on the feeding efficiency of mummichog (*Fundulus heteroclitus*). *Biol. Bull.* 195: 241–243.

Green, E. P. and F. Short. 2003. *World atlas of seagrasses*. University of California Press, Berkeley, California.

Gruner, D. S., J. E. Smith, E. W. Seabloom, S. A. Sandin, J. T. Ngai, H. Hillebrand, W. S. Harpole, J. J. Elser, E. E. Cleland, M. E. S. Bracken, E. T. Borer, and B. M. Bolker. 2008. A cross-system synthesis of consumer and nutrient resource control on producer biomass. *Ecol. Lett.* 11: 740–755.

Harlin, M., M. B. Thorne-Miller, and J. Boothroyd. 1982. Seagrass-sediment dynamics of a flood-tidal delta in Rhode Island (USA). *Aquat. Bot.* 14: 127–138.

Hauxwell, J., J. Cebrián, C. Furlong, and I. Valiela. 2001. Macroalgal canopies contribute to eelgrass (*Zostera marina*) decline in temperate estuarine ecosystems. *Ecology* 82: 1007–1022.

Hauxwell, J., J. Cebrián, and I. Valiela. 2003. Eelgrass *Zostera marina* loss in temperate estuaries: Relationship to land-derived nitrogen loads and effect of light limitation imposed by algae. *Mar. Ecol. Prog. Ser.* 247: 59–73.

Hauxwell, J., J. McClelland, P. J. Behr, and I. Valiela. 1998. Relative importance of grazing and nutrient controls of macroalgal biomass in three temperate shallow estuaries. *Estuaries* 21: 347–360.

Heck, K. L., G. Hays, and R. J. Orth. 2003. Critical evaluation of the nursery role hypothesis for seagrass meadows. *Mar. Ecol. Prog. Ser.* 253: 123–136.

Heck, K. L., J. R. Pennock, J. F. Valentine, L. D. Coen, and S. A. Sklenar. 2000. Effects of nutrient enrichment and small predator density on seagrass ecosystems: An experimental assessment. *Limnol. Oceanogr.* 45: 1041–1057.

Heck, K. L. and J. F. Valentine. 1995. Sea-urchin herbivory: Evidence for long-lasting effects in subtropical seagrass meadows. *J. Exp. Mar. Biol. Ecol.* 189: 205–217.

Heck, K. L. and J. F. Valentine. 2007. The primacy of top-down effects in shallow benthic ecosystems. *Estuaries Coasts* 30: 371–381.

Heck, K. L., J. F. Valentine, J. R. Pennock, G. Chaplin, and P. M. Spitzer. 2006. Effects of nutrient enrichment and grazing on shoalgrass *Halodule wrightii* and its epiphytes: Results of a field experiment. *Mar. Ecol. Prog. Ser.* 326: 145–156.

Hemmi, A. and V. Jormalainen. 2002. Nutrient enhancement increases performance of a marine herbivore via quality of its food alga. *Ecology* 83: 1052–1064.

Hillebrand, H. and M. Kahlert. 2002. Effect of grazing and water column nutrient supply on biomass and nutrient content of sediment microalgae. *Aquat. Bot.* 72: 143–159.

Holmer, M. and R. M. Nielsen. 2007. Effects of filamentous algal mats on sulfide invasion in eelgrass (*Zostera marina*). *J. Exp. Mar. Biol. Ecol.* 353: 245–252.

Howarth, R.W. 1988. Nutrient limitation of net primary production in marine ecosystems. *Annu. Rev. Ecol. Syst.* 19: 89–110.

Hughes, A. R., K. J. Bando, L. F. Rodriguez, and S. L.Williams. 2004. Relative effects of grazers and nutrients on seagrasses: A meta-analysis approach. *Mar. Ecol. Prog. Ser.* 282: 87–99.

Hughes, T., A. M. Szmant, R. Steneck, R. Carpenter, and S. Miller. 1999. Algal blooms on coral reefs: What are the causes? *Limnol. Oceanogr.* 44: 1583–1586.

Ibarra-Obando, S. E., K. L. Heck, and P. M. Spitzer. 2004. Effects of simultaneous changes in light, nutrients, and herbivory levels, on the structure and function of a subtropical turtlegrass meadow. *J. Exp. Mar. Biol. Ecol.* 301: 193–224.

IUCN (International Union for Conservation of Nature). 2008. IUCN World Conservation Congress. Barcelona. October 5–14.

Jackson, J. B. C., M. X. Kirby, W. H. Berger, K. A. Bjorndal, L. W. Botsford, B. J. Bourque, R. H. Bradbury, R. Cooke, J. Erlandson, J. A. Estes, T. P. Hughes, S. Kidwell, C. B. Lange, H. S. Lenihan, J. M. Pandolfi, C. H. Peterson, R. S. Steneck, M. J. Tegner, and R. R. Warner. 2001. Historical overfishing and the recent collapse of coastal ecosystems. *Science* 293: 629–638.

Jephson, T., P. Nyström, P. O. Moksnes, and S. P. Baden. 2008. Trophic interactions in *Zostera marina* beds along the Swedish coast. *Mar. Ecol. Progr. Ser.* 369: 63–76.

Jernakoff, P., A. Brearley, and J. Nielsen. 1996. Factors affecting grazer-epiphyte interactions in temperate seagrass meadows. *Oceanogr. Mar. Biol. Annu. Rev.* 34: 109–162.

Kenworthy, W. J., G. DiCarlo, J. P. Reid, and M. Merello. 2007. Can grazing manatees alter species composition and abundance in tropical seagrass communities in Puerto Rico? Abstract, Estuarine Research Federation Meeting, Providence, Rhode Island, November 4–8.

Kirsch, K. D., J. F. Valentine, and K .L. Heck. 2002. Parrotfish grazing on turtlegrass *Thalassia testudinum*: Evidence for the importance of seagrass consumption in food web dynamics of the Florida Keys National Marine Sanctuary. *Mar. Ecol. Prog. Ser.* 227: 71–85.

Kolehmainen, S. 1973. *Ecology of turtle grass (*Thalassia testudinum*) beds in Jobos Bay*. Aguirre Power Project, Environmental Studies 1972. Annual Report. Puerto Rico Nuclear Center.

Lapointe, B. E., M. M. Littler, and D. S. Littler. 1992. Nutrient availability to marine macroalgae in siliciclastic versus carbonate rich coastal waters. *Estuaries* 15: 75–82.

Lapointe, B. E. and J. D. O'Connell. 1989. Nutrient-enhanced growth *of Cladophora prolifera* in Harrington Sound, Bermuda: Eutrophication of a confined, phosphorus limited marine ecosystem. *Estuar. Coast. Shelf Sci.* 28: 347–360.

Lapointe, B. E. and K. R. Tenore. 1981. Experimental outdoor studies with *Ulva fasciata* Delile. I. Interaction of light and nitrogen on nutrient uptake, growth, and biochemical composition. *J. Exp. Mar. Biol. Ecol.* 53: 135–152.

Larned, S. T. 1998. Nitrogen-versus phosphorus-limited growth and sources for coral reef macroalgae. *Mar. Biol.* 132:409–421.

Lavery, P. S., R. J. Lukatelich, and A. J. McComb. 1991. Changes in the biomass and species composition of macroalgae in a eutrophic estuary. *Estuar. Coastal. Shelf Sci.* 33: 1–22.

Lee, K. S. and K. H. Dunton. 2000. Effects of nitrogen enrichment on biomass allocation, growth, and leaf morphology of the seagrass *Thalassia testudinum. Mar. Ecol. Progr. Ser.* 196: 39–48.

Lewis, S. M. 1986. The role of herbivorous fishes in the organization of a Caribbean reef community. *Ecol. Monogr.* 56: 183–200.

Lotze, H. K., H. S. Lenihan, B. J. Bourque, R. H. Bradbury, R. G. Cooke, M. C. Kay, S. M. Kidwell, M. X. Kirby, C. H. Peterson, and J. B. C. Jackson. 2006. Depletion, degradation, and recovery potential of estuaries and coastal seas. *Science* 312: 1806–1809.

Lotze, H. K. and B. Worm. 2000. Variable and complementary effects of herbivores on different life stages of bloom forming macroalgae. *Mar. Ecol. Prog. Ser.* 200: 167–175.

Lotze, H., K. B. Worm, and U. Sommer. 2001. Strong bottom-up and top-down control of early life stages of macroalgae. *Limnol. Oceanogr.* 46: 749–757.

Lubbers, L., W. R. Boynton, and W. M. Kemp. 1990. Variations in structure of estuarine fish communities in relation to abundance of submersed vascular plants. *Mar. Ecol. Prog. Ser.* 65: 1–14.

Martinez-Daranas, B., P. M. Alcolado, and C. M. Duarte. 2005. Leaf production and shoot dynamics of *Thalassia testudinum* by a direct census method. *Aquat. Bot.* 81: 213–224.

Mateo, M. A., J. Cebrián, K. Dunton, and T. Mutchler. 2006. Carbon flux in seagrass ecosystems. Pp. 159–192, In: A. W. D. Larkum, R. J. Orth, and C. M. Duarte (eds.), *Seagrasses: Biology, ecology, and conservation.* Dordrecht, the Netherlands: Springer.

McGlathery, K. J. 1995. Nutrient and grazing influences on a subtropical seagrass community. *Mar. Ecol. Prog. Ser.* 122: 239–252.

McGlathery, K. J. 2001. Macroalgal blooms contribute to the decline of seagrass in nutrient-enriched coastal waters. *J. Phycol.* 37: 453–456.

McGlathery, K. J., R. Marino, and R. W. Howarth. 1994. Variable rates of phosphate uptake by shallow marine carbonate sediments: Mechanisms and ecological significance. *Biogeochemistry* 25: 127–146.

Menge, B. A., B. A. Daley, P. A. Wheeler, E. Dahlhoff, E. Sanford, and P. T. Strub. 1997. Benthic-pelagic links and rocky intertidal communities: Bottom-up effects on top-down control? *Proc. Natl. Acad. Sci. USA* 94: 14530–14535.

Moksnes, P. O., M. Gullstrom, K. Tryman, and S. Baden. 2008. Trophic cascades in a temperate seagrass community. *Oikos* 117: 763–777.

Mumby, P. J., C. P. Dahlgren, A. R. Harborne, C. V. Kappel, F. Micheli, D. R. Brumbaugh, K. E. Holmes, J. M. Mendes, K. Broad, J. N. Sanchirico, K. Buch, S. Box, R. W. Stoffle, and A. B. Gill. 2006. Fishing, trophic cascades, and the process of grazing on coral reefs. *Science* 311: 98–101.

Murray, L., W. C. Dennison, and W. M. Kemp. 1992. Nitrogen versus phosphorus limitation for growth of an estuarine population of eelgrass (*Zostera marina* L). *Aquat. Bot.* 44: 83–100.

Nagelkerken, I., C. M. Roberts, G. van der Velde, M. Dorenbosch, M. C. van Riel, E. C. de la Moriniere, and P. H. Nienhuis. 2002. How important are mangroves and seagrass beds for coral-reef fish? The nursery hypothesis tested on an island scale. *Mar. Ecol. Prog. Ser.* 244: 299–305.

Neckles, H. A., R. L. Wetzel, and R. J. Orth. 1993. Relative effects of nutrient enrichment and grazing on epiphyte-macrophyte (*Zostera marina* L.) dynamics. *Oecologia* 93: 285–295.

Nixon, S. W., C. A. Oviatt, J. Frithsen, and B. Sullivan. 1986. Nutrients and the productivity of estuarine and coastal marine ecosystems. *J. Limnol. Soc. S. Africa* 12: 43–71.

Oesterling, M. and L. Pihl. 2001. Effects of filamentous green macroalgal mats on benthic macrofaunal functional feeding groups. *J. Exp. Mar. Biol. Ecol.* 263: 159–183.

Olsen, Y.S. and I. Valiela. 2010. Effect of sediment nutrient enrichment and grazing on turtlegrass *Thalassia testudinum* in Jobos Bay Puerto Rico. *Estuaries Coasts* DOI 10.1007/s1223-7-009-9256-7.

Orth, R. J. 1977. Effect of nutrient enrichment on growth of the eelgrass *Zostera marina* in the Chesapeake Bay, Virginia, USA. *Mar. Biol.* 44:187–194.

Orth, R. J., T .J. B. Carruthers, W. C. Dennison, C. M. Duarte, J. W. Fourqurean, K. L. Heck, W. J. Kenworthy, S. Olyarnik, F. T. Short, M. Waycott, and S. L. Williams. 2006. A global crisis for seagrass ecosystems. *Bioscience* 56: 987–996.

Orth, R. J. and J. V. van Montfrans. 1984. Epiphyte–seagrass relationships with an emphasis on the role of micrograzing: A review. *Aquat. Bot.* 18:43–69.

Peckol, P., B. DeMeo-Anderson, and J. Rivers. 1994. Growth, nutrient uptake capacities, and tissue constituents of the macroalgae *Gracilaria tikvahiae* and *Cladophora vagabunda*, related to site-specific nitrogen loading rates. *Mar. Biol.* 121: 175–185.

Pedersen, M. F. and J. Borum. 1996. Nutrient control of algal growth in estuarine waters: Nutrient limitation and the importance of nitrogen requirements and nitrogen storage among phytoplankton and species of macroalgae. *Mar. Ecol. Progr. Ser.* 142: 261–272.

Peterson, B. J., T. A. Frankovich, and J. C. Zieman. 2007. Response of seagrass epiphyte loading to field manipulations of fertilization, gastropod grazing and leaf turnover rates. *J. Exp. Mar. Biol. Ecol.* 349: 61–72.

Posey, M. H., T. D. Alphin, L. B. Cahoon, D. G. Lindquist, M. A. Mallin, and M. B. Nevers. 2002. Top-down versus bottom-up limitation in benthic infaunal communities: Direct and indirect effects. *Estuaries* 25: 999–1014.

Powell, G. V. N., W. J. Kenworthy, and J. W. Fourqurean. 1989. Experimental evidence for nutrient limitation of seagrass growth in a tropical estuary with restricted circulation. *Bull. Mar. Sci.* 44: 324–340.

Preen, A. 1995. Impacts of dugong foraging on seagrass habitats: Observational and experimental evidence for cultivation grazing. *Mar. Ecol. Prog. Ser.* 124: 201–213.

Ralph, P. J., M. J. Durako, S. Enriquez, C. J. Collier, and M. A. Doblin. 2007. Impact of light limitation on seagrasses. *J. Exp. Mar. Biol. Ecol.* 350: 176–193.

Risgaard-Petersen, N., T. Dalsgaard, S. Rysgaard, P. Christensen, J. Borum, K. McGlathery, and L. Nielsen. 1998. Nitrogen balance of a temperate eelgrass *Zostera marina* bed. *Mar. Ecol. Prog. Ser.* 174: 281–291.

Ruiz, J. M., M. Perez, and J. Romero. 2001. Effects of fish farm loadings on seagrass (*Posidonia oceanica*) distribution, growth and photosynthesis. *Mar. Poll. Bull.* 42: 749–760.

Sfriso, A. and A. Marcomini. 1997. Macrophyte production in a shallow coastal lagoon. I: Coupling with the chemico-physical parameters and nutrient concentrations in waters. *Mar. Environ. Res.* 44: 351–375.

Sfriso, A., B. Pavoni, and A. Marcomini. 1989. Macroalgae and phytoplankton standing crops in the central Venice lagoon: Primary production and nutrient balance. *Sci. Tot. Environ.* 80: 139–159.

Short, F. T., D. M. Burdick, and J. E. Kaldy. 1995. Mesocosm experiments quantify the effects of eutrophication on eelgrass, *Zostera marina*. *Limnol. Oceanogr.* 40: 740–749.

Short, F. T., W. C. Dennison, and D. G. Capone. 1990. Phosphorus-limited growth of the tropical seagrass *Syringodium filiforme* in carbonate sediments. *Mar. Ecol. Prog. Ser.* 62: 169–174.

Teichberg, M., S. E. Fox, C. Aguila, Y. S. Olsen, and I. Valiela. 2008. Macroalgal responses to experimental nutrient enrichment in shallow coastal waters: Growth, internal nutrient pools, and isotopic signatures. *Mar. Ecol. Prog. Ser.* 368: 117–126.

Teichberg, M., L. Heffner, S. Fox, and I. Valiela. 2007. Nitrate reductase and glutamine synthetase activity, internal N pools, and growth of *Ulva lactuca*: Responses to long- and short-term N supply. *Mar. Biol.* 151: 1249–1259.

Tewfik, A., J. B. Rasmussen, and K. S. McCann. 2005. Anthropogenic enrichment alters a marine benthic food web. *Ecology* 86: 2726–2736.

Thayer, G. W., K. A. Bjorndal, J. C. Ogden, S. L. Williams, and J. C. Zieman. 1984. Role of larger herbivores in seagrass communities. *Estuaries* 7: 351–376.

Thorne-Miller, B., M. M. Harlin, G. B. Thursby, M. M. Brady-Campbell, and B. A. Dworetsky. 1983. Variations in the distribution and biomass of submerged macrophytes in five coastal lagoons in Rhode Island, U.S.A. *Bot. Mar.* 26: 231–242.

Thybo-Christensen, M., M. B. Rasmussen, and T. H. Blackburn. 1993. Nutrient fluxes and growth of *Cladophora sericea* in a shallow Danish Bay. *Mar. Ecol. Prog. Ser.* 100: 273–281.

Udy, J. W., W. C. Dennison, W. J. L. Long, and L. J. McKenzie. 1999. Responses of seagrass to nutrients in the Great Barrier Reef, Australia. *Mar. Ecol. Prog. Ser.* 185: 257–271.

Valentine, J. and J. E. Duffy. 2006. The central role of grazing in seagrass ecology. Pp. 463–501, In: W. W. D. Larkum, R. J. Orth, and C. M. Duarte (eds.), *Seagrasses: Biology, ecology and conservation.* Dordrecht, the Netherlands: Springer.

Valentine, J. F. and K. L. Heck. 1999. Seagrass herbivory: Evidence for the continued grazing of marine grasses. *Mar. Ecol. Prog. Ser.* 176: 291–302.

Valentine, J. F., K. L. Heck, D. Blackmon, M. E. Goecker, J. Christian, R. M. Kroutil, K. D. Kirsch, B. J. Peterson, M. Beck, and M. A. Vanderklift. 2007. Food web interactions along seagrass-coral reef boundaries: Effects of piscivore reductions on cross-habitat energy exchange. *Mar. Ecol. Prog. Ser.* 333: 37–50.

Valiela, I. 2006. *Global coastal change*. Oxford, U.K.: Blackwell Publishing.

Valiela, I., J. L. Bowen, and J. K. York. 2001. Mangrove forests: One of the world's threatened major tropical environments. *Bioscience* 51: 807–815.

Valiela, I. and M. L. Cole. 2002. Comparative evidence that salt marshes and mangroves may protect seagrass meadows from land-derived nitrogen loads. *Ecosystems* 5: 92–102.

Valiela, I., K. Foreman, M. LaMontagne, D. Hersh, J. Costa, P. Peckol, B. DeMeo-Anderson, C. D'Avanzo, M. Babione, C. Sham, J. Brawley, and K. Lajtha. 1992. Couplings of watersheds and coastal waters: Sources and consequences of nutrient enrichment in Waquoit Bay, Massachusetts. *Estuaries* 15: 443–457.

Valiela, I., J. McClelland, J. Hauxwell, P. J. Behr, D. Hersh, and K. Foreman. 1997. Macroalgal blooms in shallow estuaries: Controls and ecophysiological and ecosystem consequences. *Limnol. Oceanogr.* 42: 1105–1118.

Valiela, I. and C. S. Rietsma. 1984. Nitrogen, phenolic acids, and other feeding cues for salt marsh detritivores. *Oecologia* 63: 350–356.

van Lent, F., J. M. Verschuure, and M. L. J. Vanveghel. 1995. Comparative study on populations of *Zostera marina* L (eelgrass): In-situ nitrogen enrichment and light manipulation. *J. Exp. Mar. Biol. Ecol.* 185: 55–76.

Vicente, V. P. 1975. *Seagrass bed communities of Jobos Bay*. Aguirre Environmental Studies, Jobos Bay, Puerto Rico. Final Report. Puerto Rico Nuclear Center.

Worm, B. and H. K. Lotze. 2006. Effects of eutrophication, grazing, and algal blooms on rocky shores. *Limnol. Oceanogr.* 51: 569–579.

Worm, B., H. K. Lotze, and U. Sommer. 2000. Coastal food web structure, carbon storage, and nitrogen retention regulated by consumer pressure and nutrient loading. *Limnol. Oceanogr.* 45: 339–349.

Yates, J. L. and P. Peckol. 1993. Effects of nutrient availability and herbivory on polyphenolics in the seaweed *Fucus vesiculosus*. *Ecology* 74: 1757–1766.

Zieman, J. C., R. L. Iverson, and J. C. Ogden. 1984. Herbivory effects on *Thalassia testudinum* leaf growth and nitrogen content. *Mar. Ecol. Prog. Ser.* 15: 151–158.

10 Phase Shifts, Alternative Stable States, and the Status of Southern California Lagoons

Peggy Fong and Rachel L. Kennison*

CONTENTS

Abstract 227
10.1 Introduction 228
10.2 Geology and Climate 229
10.3 Hydrology and Nutrient Supply: What Is Natural? 230
10.3.1 Open Estuarine/Lagoon Systems 232
10.3.2 Frequently/Permanently Closed Lagoon Systems 233
10.3.3 Lagoon Systems with Limited Riverine and Oceanic Influence 233
10.4 Nature of Change in Southern California Lagoon Communities 234
10.4.1 Phase Shifts vs. Alternative Stable States: Some Theory 234
10.4.2 How Primary Producer Communities in Southern California Lagoons Change: Bottom-Up Forcing by Nutrients 236
10.4.3 Positive Feedbacks Stabilize the Macroalgal-Dominated State: Support for Lagoons as Alternative Stable States 238
10.5 Have Community Shifts Resulted in Loss of Ecosystem Function? 240
10.5.1 Upward Cascading Negative Effects on Trophic Support 240
10.5.2 Case Study of Nutrient Retention and Recycling in Five Southern California Lagoons 241
10.6 Future Challenges: Climate Change—How Will it Affect Southern California Lagoons? 245
10.7 Conclusions 247
Acknowledgments 247
References 248

ABSTRACT

Our overall objective was to summarize current knowledge of the ecological status or present state of primary producer communities in southern California lagoons and the physical and ecological processes producing changes in these states. To accomplish this, we address a number of difficult questions such as what is the natural state, what target state is "desirable" or "healthy," what is the nature and trajectory of historical change, and can we use this history to predict future changes. The answers to these questions were based on a combination of scientific evidence from field surveys, experiments, and modeling, as well as a good deal of speculation based on this evidence. As scientists, we were uncomfortable with speculation; therefore, in each case, we tried to clearly state the source of our conclusions and our evaluation of their strength.

* Corresponding author. Email: pfong@biology.ucla.edu

We first discussed the prevalence and importance of coastal lagoons in southern California, and their unique character that is derived from a combination of physical setting (geology and oceanography) and Mediterranean climate. We provided strong evidence that bottom-up forcing (nutrients) in these systems has shifted the state of primary producer communities to dominance by opportunistic green macroalgae. Further, several feedback loops were identified that may stabilize the shifted state, suggesting that southern California communities may exist as alternative stable states, where two or more communities may exist under the same environmental conditions (nitrogen supply rates) and restoration of the "desirable" state may be difficult. We explored how community shifts may result in loss of vital ecosystem functions such as nutrient retention, nutrient recycling and trophic support. Finally, we speculated on how climate change, especially sea level rise and increased sedimentation, may impact southern California lagoons, and we identified knowledge gaps and future research needs.

Key Words: southern California, coastal lagoons, phase shifts, alternative stable states, eutrophication, harmful macroalgal blooms, climate change

10.1 INTRODUCTION

Coastal estuaries, the transition zones where marine and fresh water mix, are some of the most diverse and productive ecosystems in the world, as they intercept land-derived nutrients and organic matter that support both autotrophic primary producers and heterotrophic organisms (Day et al. 1989). Due to a unique combination of physical setting and climate, most southern California estuaries can be considered lagoonal systems, defined as shallow brackish or seawater bodies separated from the ocean by a barrier island, spit, or sand bank, and connected at least intermittently to the open ocean by restricted tidal inlets (Kennish and Paerl, Chapter 1, this volume). Although small in area and set in isolated river valleys, lagoons in southern California are numerous, with about 30 of them occurring between Point Conception and the Mexican border (Figure 10.1). A diverse set of migratory, resident, and endemic organisms have adapted to the wide ranges of salinity, temperature, water flow, nutrient levels, and productivity typical of southern California lagoons (Zedler et al. 2001), while potentially an even more diverse suite of human impacts threatens their sustainability. Southern California lagoons are subject to multiple stresses, including hydrological modifications, watershed development, contaminants, invasive species, and declining diversity (Sutula et al. 2007; Stein et al. 2007; Zedler and West 2008). In this chapter, we focus on one of these important stresses, increased nutrient supplies that have resulted in eutrophication.

Coastal lagoons worldwide support key ecosystem functions that include biogeochemical cycling and retention of nutrients, and they sustain a diverse array of primary producers and consumers (Ferguson et al. 2004). When healthy and functioning as productive ecosystems, coastal lagoons provide vital goods and services to human society, including flood abatement, nurseries for fish, control of coastal erosion, protection of near-shore water quality, and recreation and tourism (Costanza et al. 1997; Kennish 2002; Paerl 2006). Although far less studied, the little evidence that exists suggests southern California lagoons provide these same local goods and services (e.g., Zedler et al. 2001; Kennison 2008). In addition, southern California lagoons are located along the Pacific Flyway, a migratory bird route, providing critical stopover habitat (Lafferty 2001). Our lagoons, therefore, function as a critical local and crucial global resource.

Southern California lagoons have a long history of physical and hydrological modification that continues to threaten their vital ecosystem functions (Zedler 1982). To date, approximately 90% of southern California wetland area has been lost to development (Zedler et al. 2001). In much of the remainder, water quality is severely degraded by eutrophication (Boyle et al. 2004; Kennison 2008), which is the buildup of excessive organic matter usually as a result of nutrient enrichment (Nixon 1995). An exponentially increasing coastal human population will cause further development of watersheds and accelerate nutrient loading, placing the remaining ecosystems and the services they

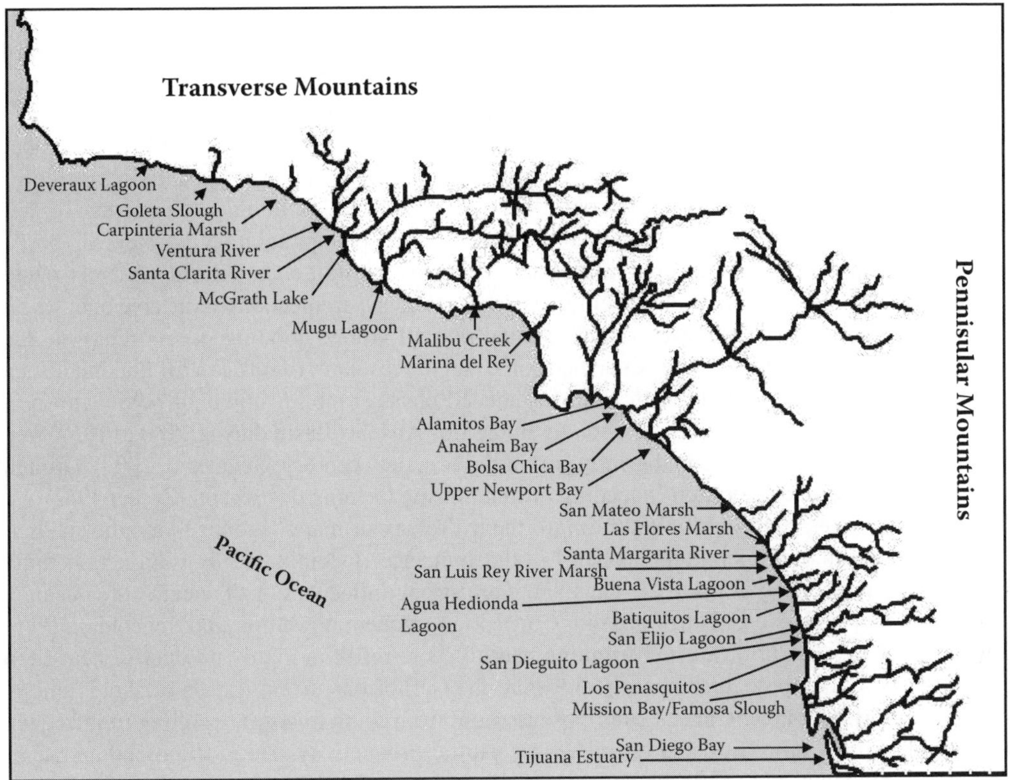

FIGURE 10.1 Map of southern California estuaries and lagoons, with inflowing rivers and the proximity of the coastal mountain ranges. Point Conception at the upper left delimits the northern boundary while the Mexican border forms the southern limit. (Figure drawn by Kendal Fong.)

provide at risk of future degradation and losses. If so, the key functions of biogeochemical cycling and retention of nutrients that support vegetation and sustain macrofauna in lagoons may be overloaded by severe modification to and degradation of their linked terrestrial systems (Zedler et al. 2001). Thus, it is important to further our understanding of the current state of lagoon functions, as well as the effects of both natural and anthropogenic changes on these functions over time, in order to ensure the systems continue to supply essential local and global services in a world being transformed by the effects of increasing population growth and global climate change.

10.2 GEOLOGY AND CLIMATE

California is divided from north to south by the San Andreas Fault, with the largest eastern portion on the leading edge of the North American Plate and the most western portion on the Pacific Plate (Press and Siever 1997). These two plates form a transform tectonic zone, where the plates move past each rather than converging or diverging. As they move, they rub and grind along their edges, producing very wide zones of deformation that are extremely tectonically active. The result is that the entire coast of California has many mountainous welts, down-dropped valleys, and limited area of flat topography to support bays, estuaries, and coastal lagoons.

Southern California, delineated to the north by Point Conception and to the south by a political boundary with Mexico, is also characterized by coastal mountains and limited area of coastal plain. Point Conception is where the north–south coastal mountain range of central California turns east to form the Transverse Range; the coastline also turns east and runs close and parallel to the mountains (Figure 10.1). The meeting of the Transverse Range and the Peninsular Range, a group of mountains that run 1500 km from southern California to the tip of Baja, Mexico, forms the Los Angeles Basin, the only extent of coastline with a significant area of relatively flat topography. As a result, in contrast to other U.S. lagoons on the East Coast and Gulf of Mexico with their broad coastal plains, watersheds in California typically are small and isolated.

California's climate is Mediterranean, characterized by warm dry summers (May to October) and cooler wet winters (November to April). In southern California, average daily temperatures vary from about 15°C in the north to about 21°C in the south. Rainfall, although universally low, also varies along the coast from a low of 20 cm (mean annual precipitation) at the Mexican border to almost 40 cm near Point Conception (Zedler 1982). Most rainfall occurs during discreet winter storm events, and the frequency and magnitude of these events is tremendously variable both within and between years. Ironically, even though annual rainfall is low, flooding occurs relatively frequently. Sequential storms saturate watershed soils, and the resultant runoff can fill floodplains and inundate lagoons; in fact, many of the human-induced physical modifications of both watersheds and lagoons of southern California, especially in the Los Angeles Basin during 1930 to 1950, were in response to extreme flooding in the early part of the twentieth century (Stein et al. 2007). Drought is also common, with evaporation exceeding rainfall during the long dry season (Zedler 1982)

With small watersheds and limited rainfall, the hydrology of many lagoons along the southern California coast that are perennially open to tidal influence is dominated by tidal action rather than river input for much of the year. Thus, there are strong influences of the near-shore ocean on southern California lagoons. North of Point Conception frequent upwelling and the cold, nutrient-rich waters of the California Current running near to shore result in highly productive near-shore environments. In contrast, in southern California, the California current travels offshore, and the warmer California Counter Current runs close to shore from south to north, resulting in infrequent and episodic spring upwelling and overall lower natural productivity. Tides are mixed and semi-diurnal, meaning that the two high and two low daily tides are of different height. Mean tidal amplitude is ~1.1 m, while spring tides usually exceed 1.6 m (Zedler 1982). Although these systems are generally strongly driven by tides, longshore movement of sand coupled with wave action result in a seasonal cycle of beach building and erosion that historically caused migration of the lagoon mouth (connection to the ocean) and may have caused seasonal or even permanent lagoon closure in some systems (Ambrose and Orme 2000). The duration and frequency of closure under natural conditions is unknown, but most likely was extremely variable, as it would have been linked to a combination of physical drivers such as tides, waves, rainfall, riverflow, and groundwater influence.

10.3 HYDROLOGY AND NUTRIENT SUPPLY: WHAT IS NATURAL?

Because lagoons in southern California have a long history of major modification, it is difficult to state with any certainty the natural hydrological regime in these ecosystems. With that caveat, given the unique physical setting and climate, it is likely that even the natural hydrology of this region was highly variable over space and time, with conditions ranging from permanently closed to continuously open (Figure 10.2). Humans have had myriad effects on watersheds over time that directly affected river and lagoon hydrology (Zedler 1996). For example, inflowing streams were dammed to create a year-round water supply, thereby reducing seasonality of inflow while channelization and diversion of rivers directly to the ocean for flood control reduced the duration of peak flows. Within lagoons, large areas have been filled for airports and urban development or dredged to create deep water for marinas and other recreational uses. Hydrological connections to the ocean were severely restricted by roads, railroad tracks, and finally freeways. All of these modifications most likely enhanced the naturally high variability in hydrology, and this increased variability between systems and years resulted in the extreme environmental conditions that characterize southern California lagoons today.

The current hydrology of each southern California lagoon depends on its unique history of human modifications to watershed and lagoon. Therefore, these lagoons should be considered as

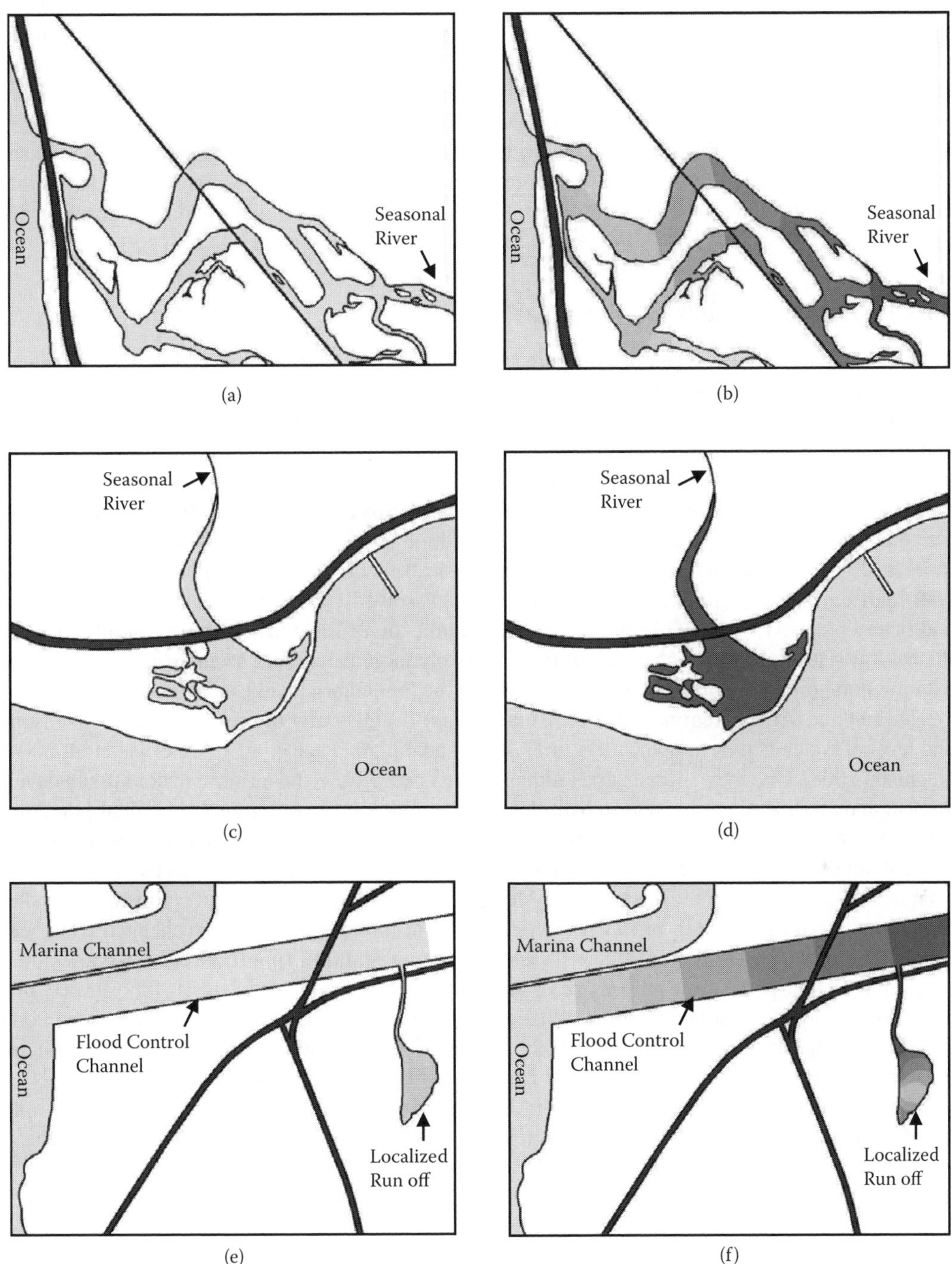

FIGURE 10.2 Examples of the diversity of lagoons of southern California: open estuarine/lagoon systems with a largely continuous connection to both a seasonal river and the ocean in (a) dry and (b) wet seasons; frequently/permanently closed lagoon systems with seasonal connection to a river but usually closed to tidal influence in (c) dry and (d) wet seasons; lagoon systems with limited river and oceanic influence due to the diversion of the inflowing river in (e) dry and (f) wet seasons. Shading indicates nutrient supply, ranging from low (lightest) to high (darkest). Dark thick lines show roads that cross these lagoons. (Figure drawn by Kendal Fong.)

a set of systems along a very wide spectrum of conditions. Significant progress has been made toward developing a classification scheme for California lagoons with respect to their eutrophication (Sutula et al. 2007). We have not attempted to re-create those efforts here, but rather we will discuss three types of lagoons commonly found in southern California that serve to demonstrate their extreme diversity. We discuss systems with largely seasonal river inflow that are usually open to the ocean, those with seasonal river inflow that are generally closed off from direct ocean flushing, and those that have had the main inflowing river diverted. However, it is important to stress that lagoons may shift along the condition spectrum, over time scales that may vary from occasionally to seasonally.

10.3.1 Open Estuarine/Lagoon Systems

Many of the relatively large lagoons in southern California are tidally influenced year round via an open connection to the ocean. Highly seasonal river flow results in estuarine conditions during the short, wet season and systems that function as marine embayments in the dry season (Figures 10.2a and 10.2b). While some maintain an ocean connection naturally due to episodic erosion by river water during the wet season and tidal scouring during the dry season, many are maintained open by dredging. Management to maintain an open condition is based both on the belief that this may be the "natural" state and the practicality that eutrophication may severely impact ecological functioning if flushing ceases even for a short time. For example, brief closure of San Elijo Lagoon combined with high dry season temperatures resulted in low dissolved oxygen and mass mortality of fish and invertebrates (Douglas Gibson personal communication). Lagoons that were mainly open in the natural state were likely characterized by relatively large catchment basins with capacities for extreme flood events, such as those that occurred in the San Gabriel and Los Angeles rivers.

Nutrient and salinity regimes in open estuaries vary dramatically between wet and dry seasons and level of watershed development (Figures 10.2a and 10.2b) (Page et al. 1995; Boyle et al. 2004; Kennison 2008). Given the climate and unique geology, we believe that prior to major human development and modification, dry season hydrology in open systems was dominated by tidal pumping, salinities were oceanic, and the primary source of nutrients was the ocean. There is evidence that nutrient import from the ocean drove productivity of Tijuana Estuary as recently as the 1970s (Winfield 1980). At present, even in open systems salinity can range from hypersaline in areas that have muted tidal influence to brackish in systems where inflowing rivers have changed from seasonal to perennial by addition of treated wastewater or agricultural runoff (Greer and Stow 2003; Kennison 2008). For the same reasons, dry season nutrients can also range widely; for example, in a study comparing five southern California lagoons, NO_3^- was extremely high in the dry season in one system with adjacent strawberry fields (maximum mean 1700 μM NO_3^-), over an order of magnitude greater than the four other estuaries (range 2 to 100 μM NO_3) (Kennison 2008).

Wet season hydrology in open systems is a result of both river inflow and tidal flushing (Figure 10.2b). Thus, salinity ranges from freshwater at the river entrance to oceanic at the ocean mouth, as in a typical estuarine system (Day et al. 1989). However, upstream dams and urban development in most of these watersheds reduce peak flows and extend the duration of freshwater influence beyond individual storm events. Developed watersheds destabilize terrestrial soils that then enter lagoons, while reduced peak floods increase sediment deposition within the lagoons. As a result, many lagoons have little shallow subtidal habitat (albeit many have been dredged to create "deep" water habitat as mitigation for dredging in ports and marinas, a habitat that may have no natural analog in southern California) and large areas of intertidal mudflat and saltmarsh vegetation that may not have been as extensive in the natural state. In both natural and modified lagoons, the river supplies nutrients from the watershed, with the magnitude of loading dependent on the nature and degree of watershed nutrient sources. One study found water column nutrients to be highest during the wet season in four of five open estuaries, with NO_3 ranging from 50 to 1100 μM (Kennison 2008). In addition, spatial patterns of higher concentrations at the upstream river end demonstrated a watershed source and dilution down estuary by tidal oceanic water (Page et al. 1995; Boyle et al. 2004; Kennison 2008). Overall, water column nutrients during the wet season in southern California lagoons have an extremely wide range compared to other open estuaries (Valiela et al. 1992; Nixon 1995; Taylor et al. 1995; Svensson et al. 2000).

10.3.2 Frequently/Permanently Closed Lagoon Systems

Many lagoons in southern California have seasonal river flow, yet they are frequently or even permanently cut off from direct oceanic influence by sand bars (Figures 10.2b and 10.2c). Few studies have been conducted in these types of systems (but see Ambrose and Orme 2000; Sutula et al. 2004), so our understanding of present and past conditions is limited. Most local geologists agree that at least seasonal closure, especially of smaller lagoons, is one of several natural conditions that occurred historically (Zedler 1982; Ambrose and Orme 2000). There is little doubt, however, that human modifications on watersheds and within lagoons have increased the frequency and duration of closure. Management of these systems is mixed, with some lagoons left closed, some opened to tidal flushing by dredging during winter to prevent flooding and/or excessive eutrophication, and others with one-way gates that maintain a largely freshwater system. At present, most lagoons that remain closed for extensive periods are characterized by small watersheds or watersheds with major upstream diversion of freshwater.

Nutrient and salinity regimes in closed lagoons of southern California are poorly understood due to a relative paucity of research. The few studies that exist suggest that these systems vary even more dramatically than open systems between wet and dry seasons and among levels of watershed development and within-lagoon modification. Their current biogeochemical state, however, depends heavily on the recent history of their connection to the ocean (Ambrose and Orme 2000; Sutula et al. 2004).

Lagoons that close naturally (i.e., no weirs, tide gates, etc.) in the dry season can vary in salinity from freshwater to hypersaline. This depends, in part, on time of closure, the elevation of each lagoon with respect to sea level, the duration and magnitude of anthropogenic freshwater flows in the dry season, and the importance of groundwater inputs. Once closed, high rates of evaporation during the hot dry season may result in higher salinities (Zedler 1982); however, any salinity regime along that gradient is also possible. Based on this same reasoning, dry season nutrient regimes can also be extremely variable and depend on the level of watershed development. In lagoons with little watershed development and low river inflow in the dry season, we would expect low nutrient supplies and oligotrophic conditions. In lagoons with watersheds subject to continuous river flow due to agriculture or where rivers are used for disposal of treated wastewater, we would expect extremely high supplies and severe eutrophication. In addition, sediments transported from the watershed and deposited in the lagoon during the wet season can regenerate considerable nutrients, supporting blooms of primary producers in the dry season (Sutula et al. 2004)

In systems that remain closed even during the wet season, variability in nutrients and salinities will be amplified. Where watersheds are small and undeveloped, wet season river inflow supplies the only external nutrients to these systems and must result in extreme seasonality in productivity. In contrast, where watersheds are developed, high nutrient supplies support eutrophic conditions that can only worsen in the wet season as there is little or no capacity for dilution or washout of nutrients from the lagoon. More research on these systems is essential to understand present conditions. Even more key is to understand what may be the historical or "natural" hydrological state of these systems in order to make decisions for future management.

10.3.3 Lagoon Systems with Limited Riverine and Oceanic Influence

Some lagoons in southern California have restricted, episodic, or even complete and permanent diversion of riverine influence, essentially uncoupling these systems from their watersheds

(Figures 10.2e and 10.2f). In many cases, these systems are characterized by frequent and even permanent reduction of direct oceanic influence due to the lack of significant river flows scouring a lagoon mouth. This state is purely a result of human modification of the course of rivers, usually due to channelization for flood control. However, historical records of the Los Angeles and San Gabriel rivers, which drain large portions of the mountain ranges surrounding the Los Angeles Basin and the San Fernando Valley, show they dramatically switched course repeatedly during extremely wet years and thus created, abandoned, and reconnected with several lagoons. This suggests there may have been a natural counterpart to these types of modifications (Stein et al. 2007). Most likely these natural events were limited to lagoons with very large watersheds with significant rainfall and strong flooding potential. At present, systems with diverted rivers are remnants of much larger lagoons that have been transformed into deep-water bays for recreational purposes. These small remnant systems are often connected to a diverted and channelized main river, and therefore are only indirectly connected to the ocean. Connections are often by a small creek with circulation restricted by culverts that are well above mean sea level. Thus, freshwater and nutrient supplies are either from localized runoff or only during extreme flood events when the river level exceeds the culverts (Fong and Zedler 2000). Similarly, oceanic exchange is very muted, limited to tides that exceed culvert elevations.

The degree to which the river and the ocean remain connected to these isolated lagoons determines the nature of the salinity and nutrient regimes. These systems may function as muted open systems, with estuarine conditions in the wet season and tidal lagoon conditions in the dry season (Figures 10.2e and 10.2f). In the wet season, there may be very local inflowing nutrients from culverts and storm drains as well as river-derived sources (Fong and Zedler 2000). Depending on local land use as well as extent of watershed development, wet season nutrient supplies can be very high. As these systems are essentially "dead ends" hydrologically, nutrients supplied during the wet season accumulate and eutrophication can be extreme. In the dry season, freshwater flow is low yet nutrient rich; however, low volume flows may never make it into these systems, and eutrophication in the dry season may be less extreme. During the dry season, recycled sources may be very important; studies in Famosa Slough suggested that the flux of nutrients from enriched sediments fueled massive macroalgal and cyanobacterial blooms in summer (Fong and Zedler 2000; Fong unpublished data).

10.4 NATURE OF CHANGE IN SOUTHERN CALIFORNIA LAGOON COMMUNITIES

10.4.1 Phase Shifts vs. Alternative Stable States: Some Theory

Dramatic shifts in populations, communities, and ecosystems worldwide in response to natural and anthropogenic alterations in environmental conditions have focused much attention on the nature of change and the processes that drive them (Scheffer et al. 2001; Beisner et al. 2003; Didham and Watts 2005). Ecological theory predicts many possible patterns of change for populations and communities. Some studies found ecological communities shifted in a relatively simple linear manner, where a given change in a controlling environmental variable produced a predictable community or population-level response (Figure 10.3a). Much recent attention, however, has focused on nonlinear shifts, with extreme nonlinearities or phase shifts occurring in response to both natural and anthropogenic changes in environmental forcing functions (Figure 10.3b). Examples span a diversity of ecosystems, including the collapse of most major marine fisheries (Roughgarden and Smith 1996), desertification of the sub-Sahara (Xue and Shukla 1993), and shifts from coral- to algal-dominated tropical reefs (Hughes et al. 2007).

The rising frequency of rapid collapses from one community state to another, often a less desirable state (e.g., Jackson et al. 2001), has renewed interest in whether this process is a simple and reversible phase shift, or if these shifts represent alternative stable states (ASS) (Scheffer et al. 2001;

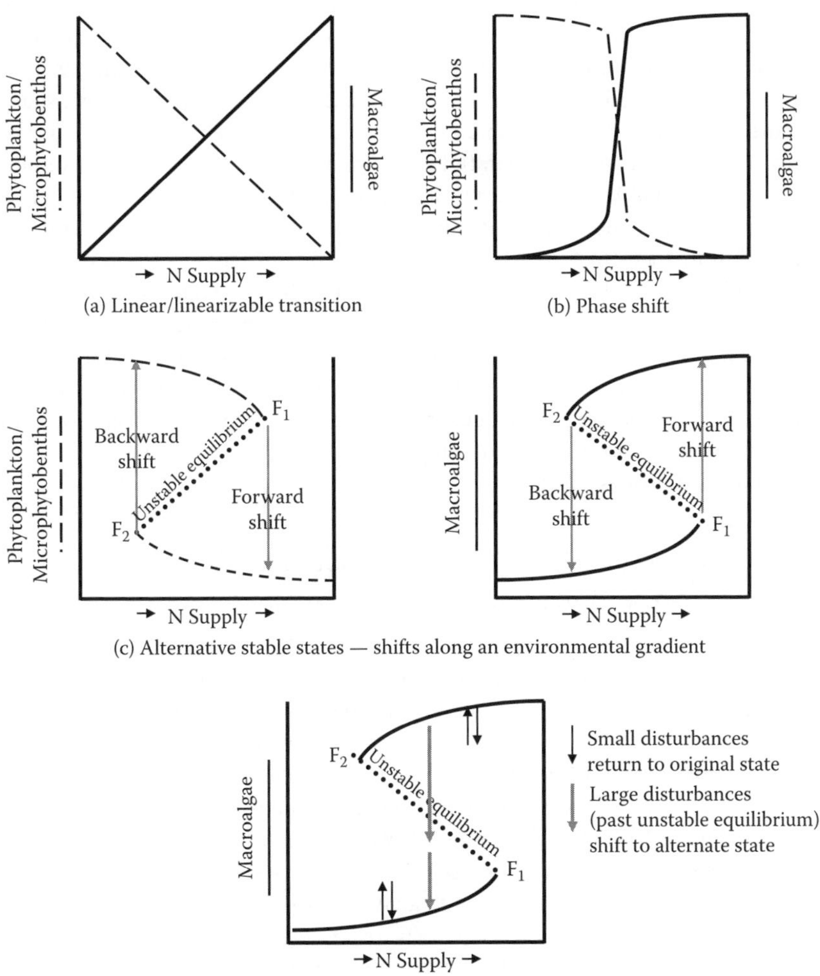

FIGURE 10.3 Conceptual diagram showing three stages along a continuum of possible patterns of change that lagoon primary producer communities may undergo over time in response to nitrogen supply. Nitrogen supply is the amount of nitrogen available to primary producers and is a function of both nitrogen loading and removal via tidal flushing. Dashed lines represent the microphytobenthic community, solid lines the macroalgal community. (a) In the simplest case, a community shift can occur in a linear manner, with an incremental change in an environmental condition (nitrogen supply) resulting in an incremental, predictable change in the relative abundance of the two dominant communities. This relationship allows prediction of future change based on past history of the condition. (b) Communities can also undergo phase shifts, or rapid, catastrophic shifts from one dominant community to another over a very short range of environmental change. In this case, incremental changes in nitrogen supply result in little change until a critical threshold is reached; thus, the past history of environmental change will not aid prediction of future change. (c) Communities can also exist as alternative stable states, where shifts between states are also rapid and catastrophic along a nutrient-loading gradient. However, forward transitions occur at different points along the gradient than backward transitions or reversals, resulting in a range of conditions (between F_1 and F_2) where two different states can occur. In this case, there is limited ability to predict the community state based on the condition of the environment. (d) Shifts between stable states can also occur within the range F_1 and F_2 if disturbances are large enough to push community abundance beyond the unstable equilibrium. (Figure drawn by Kendal Fong.)

Beisner et al. 2003; Didham and Watts 2005). It is important to distinguish between these two methods of community change because management strategies must be very different for each, if the goal is to promote and stabilize the initial, often "desirable," state. Two key aspects of ASS theory with important management implications involve the processes that cause transitions between states and those mechanisms that stabilize them. Theory predicts forward shifts from one state to another (e.g., F_1 in Figure 10.3c) occur at very different points or conditions of the environmental driver than backward shifts (e.g., F_2) that restore the initial state. Thus, there is a range of environmental conditions, between points F_1 and F_2, where either of two stable states can occur. At any environmental condition between F_1 and F_2, theory predicts that small disturbances will result in a return to a stable state; only large disturbances that push the state beyond the unstable equilibrium (represented by the dotted lines in Figures 10.3c and 10.3d) will result in shifts from one state to another. The only other mechanism to restore the initial state is to reverse the environmental conditions far beyond the forward shift, to F_2, often a very difficult and expensive management option. Another key prediction from theory is that, in order for ASS to be maintained, positive feedback mechanisms must exist to stabilize each state. These can include abiotic or biotic processes, but must be strong enough to buffer these states across a range of environmental conditions (between F_1 and F_2) and provide resilience to small disturbances.

10.4.2 How Primary Producer Communities in Southern California Lagoons Change: Bottom-Up Forcing by Nutrients

There is empirical, experimental, and theoretical evidence that bottom-up forcing has shifted the composition of primary producer communities through at least two states along a nutrient-loading gradient in shallow southern California lagoons (Fong et al. 1993a, 1993b; Fong et al. 1994). Results of large-scale experiments have demonstrated that as nitrogen loading from highly developed watersheds increased to southern California's lagoons, producer communities dominated by low abundances of phytoplankton and microphytobenthos (MPB) were replaced by blooms of opportunistic green macroalgae (Figure 10.4). These community replacements were not smooth, linear transitions along the nutrient gradient; rather, changes were abrupt, with communities dominated by either one state or another, with few transient transitional states. This pattern of change could be a result of either phase shifts (Figure 10.3b) or shifts between stable states (Figures 10.3c and 10.3d), but with this type of evidence we are unable to distinguish between these two processes. However, transitions to macroalgal-dominated states have been documented worldwide, with blooms of green or red algae dominating many shallow estuaries and lagoons subject to nitrogen enrichment (e.g., Sfriso

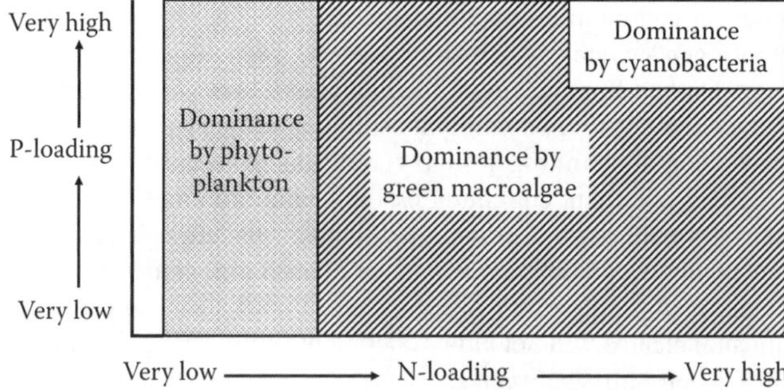

FIGURE 10.4 Results of a field mesocosm experiment modeling lagoons in southern California subject to a wide range of nutrient loading. There were 20 treatments varying both nitrogen and phosphorus with five-fold replication. Communities demonstrated rapid, catastrophic shifts along the nitrogen loading axis, transitioning from domination by phytoplankton to opportunistic green macroalgae. When both nutrients were extremely high, cyanobacteria dominated. (Data from Fong et al. 1993b.)

et al. 1987, 1992; Raffaelli et al. 1989; Valiela et al. 1992, 1997; Geertz-Hansen et al. 1993; Peckol et al. 1994; Marcomini et al. 1995; Page et al. 1995; Hernández et al. 1997; Hauxwell et al. 1998; Kamer et al. 2001). Thus, rapid and catastrophic community change driven by bottom-up forces may be characteristic of estuarine systems worldwide.

There is mounting evidence that an additional phase shift may be in the process of occurring in southern California lagoons presently dominated by macroalgal blooms as nitrogen loading continues to rise (Figure 10.4). In systems outside of California, the driving mechanism that limited macroalgal blooms was self-shading of macroalgal mats once they reach a critical density (e.g., Peckol et al. 1994; Krause-Jensen et al. 1999; Fox et al. 2008); decomposition of these mats leads to release of nitrogen to the water for use by other producers. In estuaries with low to moderate tidal amplitude, macroalgae are replaced by blooms of phytoplankon (Duarte 1995; Valiela et al. 1997), and this transition is thought to be dependent on water residence time (Valiela et al. 2000). In contrast, results of the only *in situ* enrichment experiment on the West Coast of the United States suggest that this final ecosystem transition along a eutrophication gradient will be to benthic cyanobacterial mats (Armitage and Fong 2004) rather than to phytoplankton. We hypothesize that dominance by benthic cyanobacteria may be attributed to the higher tidal amplitudes regularly flushing out phytoplankton and the existence of broad areas of intertidal mudflats providing suitable habitat for benthic algae in southern California lagoonal systems. These mats of cyanobacteria are unpalatable or even toxic to the dominant herbivores (Armitage and Fong 2004).

A competition model has been developed and tested empirically where the outcome of nutrient competition among phytoplankton, macroalgae, and cyanobacterial mats varies along a resource (nitrogen) gradient (Figure 10.5a) (Fong et al. 1994). Phytoplankton, with efficient nutrient uptake and growth at low nutrient loading rates dominates oligotrophic lagoons subject to low levels of

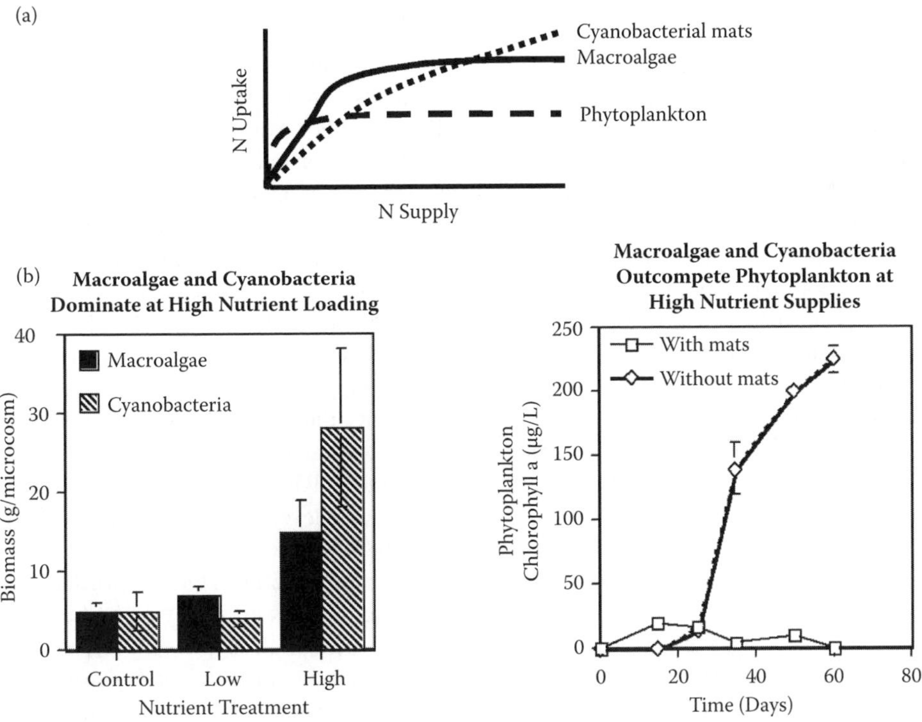

FIGURE 10.5 A competition model predicts that the outcome of competition will vary based on nitrogen supply, with phytoplankton dominating at low, macroalgae at medium and high, and cyanobacteria at very high supplies of nitrogen. (Modified from Fong, P. et al. 1994. *Ecol. Monogr.* 64: 225–247.)

nitrogen loading. As nitrogen loading increases, phytoplankton uptake rates saturate and macroalgae and cyanobacteria, with higher maximum uptake rates, take up a larger percentage of the nitrogen and out-compete phytoplankton. Competition experiments along a resource gradient conducted in experimental mesocosms modeling southern California lagoons supported this competition model (Figure 10.5b; Fong et al. 1993b). At low loading rates relative to natural systems, phytoplankton dominated producer communities. However, opportunistic green macroalgae and cyanobacteria with rapid growth capacities dominated experimental units at high nitrogen loading rates. Macroalgae and cyanobacteria suppressed phytoplankton via indirect, exploitative competition when grown together at high nitrogen loading rates.

We speculate that the transition from MPB and phytoplankton to macroalgae may not have been the first community shift in open lagoons of southern California. We believe that seagrasses may have dominated at least the larger systems prior to scientific documentation in this region. Our reasoning is that remnant seagrass populations occur in a few of the more open, well-flushed systems of southern California, such as Lower Newport Bay (personal observations) and San Diego Bay (Stewart 1991), where a direct connection with the ocean limits nutrients and turbidity and enhances water clarity. In addition, seagrasses dominate large lagoons and estuaries in northern California (e.g., Huntington and Boyer 2008), as well as many systems around the world (e.g., Hauxwell et al. 2003). Ecological processes that are driving current shifts from seagrass to other primary producers usually involve nutrient-mediated competition for light (e.g., Short and Wyllies-Echeverria 1996; Hauxwell et al. 2003); we hypothesize that this same mechanism is responsible for the loss of seagrass in southern California lagoons.

10.4.3 Positive Feedbacks Stabilize the Macroalgal-Dominated State: Support for Lagoons as Alternative Stable States

Field surveys and experiments that demonstrate rapid transitions among community states are often unable to distinguish between phase shifts and ASS because limited temporal and/or spatial scales usually restrict observations of transitions to single events. For example, in southern California, we infer from present conditions in other regions that a transition occurred from seagrass to phytoplankton and MPB. However, this most likely happened long enough ago that there is little scientific documentation of this event, and we have yet to observe reversals to seagrass domination. The types of evidence that can distinguish between these patterns of change and support ASS include: (1) persistent dominance by macroalgae across a wide nutrient supply gradient and (2) the existence of strong and consistent positive feedbacks that stabilize the macroalgal state. Below we summarize the evidence supporting the concept that southern California lagoons exist as alternative stable states driven by the bottom-up force of changing nutrient supply.

The present state of every primary producer community that has been quantitatively surveyed in southern California lagoons can be characterized, at least seasonally, as dominated by opportunistic green macroalgae (e.g., Rudnicki 1986; Page et al. 1995; Kamer et al. 2001; Sutula et al. 2004; Kennison 2008). Although always associated with some level of enhanced supply of nitrogen from the watershed and enriched sediments (e.g., Fong 1986; Ambrose and Orme 2000; Boyle et al. 2004; Sutula et al. 2004; Kennison 2008), temporal and spatial patterns in these characteristics vary widely within and between lagoons. In some lagoons, water NO_3^- is extremely high year-round due to closely associated agricultural fields (Figure 10.6a). In others, high supplies of NH_4^+ occur in the wet season (Figure 10.6b). Algal blooms, however, may occur in either wet or dry seasons and do not directly correlate to either water column (NO_3^- or NH_4^+) or sediment (organic content) sources of nitrogen (Figure 10.6). The clear disconnect between water nutrient concentration and magnitude and timing of blooms (Kennison 2008) suggests that complex direct and indirect feedback mechanisms may support macroalgal blooms, even during times of relatively low external supplies.

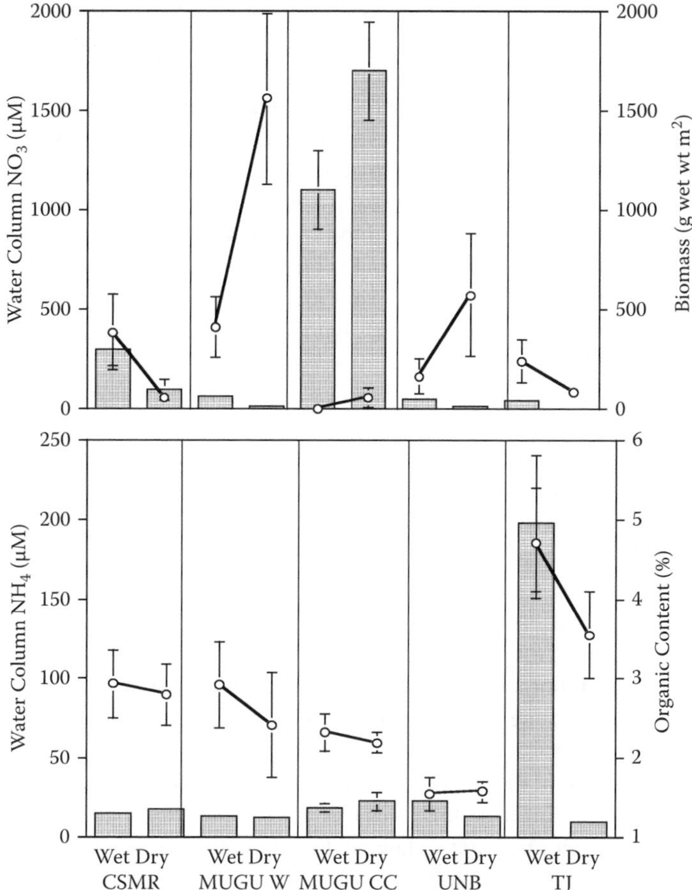

FIGURE 10.6 Seasonal patterns in five southern California lagoons of water column nutrients, biomass, and sediment organic content. Wet season averaged values from December 2001, February and December 2002, and February 2003. Dry season averaged values from June and September 2002. Bars represent standard error. Lagoons are Carpinteria Marsh (CSMR), Mugu West (Mugu W), Mugu Calleguas Creek (Mugu CC), Upper Newport Bay (UNB), and Tijuana Estuary (TJ).

One of the positive feedbacks that stabilizes the present macroalgal-dominated state is the presence of enriched sediments that sustain supplies of nitrogen to macroalgae beyond direct supply of nutrients from the watershed during episodic rainfall and river flow events. Lagoon sediments accumulate higher concentrations of organic nutrients during the short rainy season via deposition of soils and organic matter (Figure 10.6b) eroded from the watershed and by adsorbing high concentrations of inorganic nutrients carried by freshwater flows (Sutula et al. 2004). During the dry season, mineralization of organic matter and flux of inorganic nutrients from the sediment result in seasonal decreases in sediment nutrient content, providing a source of nitrogen during times when macroalgal biomass is highest and water column sources often lowest (Fong and Zedler 2000; Kamer et al. 2001; Boyle et al. 2004; Sutula et al. 2004, 2006). This shifts nitrogen cycling from the processes of nitrification and denitrification, pathways of permanent loss, to one of recycling of nitrogen into algal biomass (Tyler et al. 2003; Sutula et al. 2006).

In one manipulative experiment using dry season water nutrient concentrations, macroalgae grew four times as rapidly with estuarine sediments than with inert sands (Kamer et al. 2004b). In addition, sediment sources of nutrients greatly reduced and even eliminated strong bottom-up nitrogen limitation of macroalgae during the dry season (Kamer et al. 2004a). Thus, the sediments act as a slow release fertilizer to support macroalgal blooms when water column supplies are low, providing a strong positive feedback stabilizing the macroalgal community state.

Another positive feedback stabilizing macroalgal dominance is that higher nitrogen supplies enhance tolerance of bloom-forming algae to a variety of environmental extremes. First, high nitrogen supply ameliorates the negative effects of lowered salinity on growth. For example, in a laboratory microcosm experiment, growth of algae subjected to lowered salinity and ambient nutrients were suppressed by 50%, while algae subjected to lowered salinity and higher nutrients were suppressed by only 15% (Kamer and Fong 2000). This trait is especially adaptive, as pulses of high concentrations of nitrogen from the watershed are always associated with lower salinity because of the freshwater transport mechanism. Second, ontogenetic shifts from benthic juvenile to adult free-floating stages expand suitable habitat for macroalgal proliferation (Kennison 2008). Benthic stages are limited by light to intertidal or very shallow subtidal zones, while floating rafts avoid light limitation by occupying the upper layers of water throughout lagoons. Field experiments demonstrated that this ontogenetic shift is controlled, in part, by thallus size, and the rate at which this size is achieved is directly dependent on nutrient supply. Both higher salinity tolerance and ontogenetic shifts in habitat usage stabilize the macroalgal-dominated community state in the extreme environmental conditions typical of southern California lagoons.

A major herbivore in southern California lagoons, the mud snail *Cerithidia californica*, stabilizes the dominant macroalgal state through its trophic interactions. Snails exhibited top-down control of the microphytobenthos, thereby removing the competitor that may have replaced macroalgae when nutrient supply was reduced (Armitage and Fong 2004). In addition, snails facilitate dominance by macroalgae by releasing nitrogen sequestered in sediments (Fong et al. 1997). This may be through consumption and subsequent release of nutrients contained in the microphytobenthos or through physical disturbance of the sediment enhancing flux of sediment nutrients.

The rapid and highly nonlinear nature of transitions to macroalgal dominance, the prevalence and persistence of the present macroalgal state, and the multitude of positive feedbacks stabilizing their dominance all support the hypothesis that southern California lagoons exist as alternative stable states. Ultimately, however, we will need to quantify the nature of the community response to reduction in nutrient supplies to levels below a threshold that caused the forward transition to confirm this hypothesis. Will the phytoplankton/MPB state return at the same point it disappeared (F_1 in Figure 10.3)? Or will nutrient supply need to be reduced even further (to F_2) to reestablish the original state? The answers to these questions are essential if we want to manage these systems to restore or even conserve the present status of the valuable ecosystem functions they provide.

10.5 HAVE COMMUNITY SHIFTS RESULTED IN LOSS OF ECOSYSTEM FUNCTION?

10.5.1 Upward Cascading Negative Effects on Trophic Support

One key ecosystem function of coastal lagoons is to support the abundant and diverse community of consumers that rely on the high level of lagoon productivity. Macroalgal blooms reduce diversity of primary producers via competition for nutrients and light, thus changing the base of this important food web. A shift from a microphytobenthic community to macroalgae or cyanobacteria may have repercussions that cascade up the food web. In addition, severe blooms may deplete the water column and sediments of oxygen (Sfriso et al. 1987; Valiela et al. 1992; Young et al. 1998), resulting in fish and invertebrate mortality, producing subsequent changes in both trophic and community structure (Raffaelli et al. 1989; Bolam et al. 2000). Several experiments have demonstrated that macroalgal mats may have significant negative impacts on the benthic infaunal community (Everett 1994; Norkko and Bonsdorff 1996; Osterling and Pihl 2001; Paalme et al. 2002; Fox et al. 2009), an important food source for secondary consumers including crabs, fish, and resident and migratory shorebirds. A few studies have directly connected macroalgal blooms to negative impacts on higher trophic functioning such as shorebird foraging (Stenzel et al. 1976; Lopes et al. 2006).

There is strong evidence that nutrients have upward cascading negative effects in southern California lagoons by altering primary producer community structure. *In situ* experiments manipulating top-down (consumers) and bottom-up (nutrients) processes resulted in shifts in benthic producer communities to dominance by cyanobacteria and purple sulfur bacteria with nutrient addition (Armitage and Fong 2004). This shift toward cyanobacteria in field communities supported earlier predictions from theoretical and experimental work (Fong et al. 1993a, 1993b, 1994.). Nonpredation mortality of the major herbivore, the California mud snail, was up to three times higher in nutrient-enriched plots, suggesting toxicity by bacteria (Armitage and Fong 2004). Another experiment manipulating the third trophic level demonstrated that crabs ameliorated this bottom-up toxicity effect, most likely due to the physical disturbance produced by crabs moving across the sediments rather than direct consumption (Armitage and Fong 2006). Snails are an important link in the trophic structure of California lagoons, supporting both crab and bird populations, making these upward cascades of great management concern.

Nutrient-driven macroalgal blooms also have negative upward cascading effects directly on consumers, which may be attributed to changing sediment biogeochemistry. Field surveys showed that densities of benthic infauna are one to two orders of magnitude lower under mats of macroalgae than they were where mats were absent (Green and Fong, in preparation). Field experiments manipulating macroalgal mat thickness suggested that low and moderate abundances of macroalgae supported a larger and more diverse infaunal assemblage, but that abundances were severely reduced under thick mats (Green and Fong in preparation). Most estuaries surveyed in southern California, however, have thick mats at least seasonally (Kamer et al. 2001; Kennison 2008). Recent studies of bird behavior link these changes in infauna to negative cascades further up the trophic structure as many resident and migratory shorebirds change foraging behavior with macroalgal abundance. For example, sandpipers spend more time probing in their search for food when they are on top of macroalgal mats, but more time repeatedly pecking when mats are absent (Green and Fong, in preparation). As infauna are a major food source for birds and other secondary consumers, macroalgal impacts that alter this link of the food chain may signify a significant impairment of this vital ecosystem function.

10.5.2 Case Study of Nutrient Retention and Recycling in Five Southern California Lagoons

The biogeochemical cycling of nutrients is one of the primary ecological functions of estuaries and lagoons and is essential on a global scale to maintain the health of coastal waters (Conley et al. 2000; Cloern 2001; Ferguson et al. 2004). Two aspects of biogeochemical cycling, nutrient retention (uptake of "new" nitrogen entering the system) and nutrient cycling (transfer of nitrogen among biotic and abiotic pools), protect other near-shore coastal ecosystems from excessive nutrient supplies (Schaefer and Alber 2007). Lagoon biota and sediments retain nitrogen and phosphorus from nutrient-rich terrestrial sources while microbial communities cycle nitrogen via denitrification and remineralization of organic compounds (Thybo-Christesen et al. 1993; Eyre and Ferguson 2002; Sundbäck et al. 2003). However, if nutrient loads exceed critical thresholds, reduction and even loss of these ecosystem functions may result (Cloern 2001; Kemp et al. 2005). The extent to which systems stave off loss of function depends on attributes such as hydrology, sediment characteristics, oxygen depletion, and biotic filtration (Schramm, 1999; Cloern 2001).

The natural or historical capacity of southern California lagoons to retain and recycle nutrients is largely unknown as they have been subjected to such a long history of human modification. That they were, in general, linked to small watersheds with highly seasonal rainfall and were semi-enclosed by sand bars with muted connectivity to the ocean suggests that residence times were long and thus nutrient processing capacities relatively high. The exception may have been during times of peak rainfall and flooding, when export to near-shore waters likely dominated. Long-term studies in one southern California estuary suggest that cycles of catastrophic storms, flooding, and sedimentation may be linked to the Pacific Decadal Oscillation, producing benign and stormy decades (Zedler and West 2008) that could greatly alter nutrient retention and recycling rates in these systems. In the following case study, we summarize results of a field survey quantifying present ecological functions of nutrient retention and recycling (Kennison 2008), and speculate on reasons why we think significant loss of these functions has occurred.

Nutrient retention and cycling capacities varied from highly efficient to nonmeasurable among five southern California lagoon systems (Kennison 2008), and appeared to be largely independent of nutrient loading rate. Mixing diagrams relate salinity, a nonbiologically active substance, to inorganic nitrogen concentration in the water column (e.g., Figure 10.7). A straight line with a negative slope indicates that nitrogen is diluted by tidal water but not biologically processed; a line that lies below this straight line indicates net uptake (retention) and a line above the straight line indicates net generation (recycling) within the lagoon. Two of the five lagoons studied, Carpinteria Marsh and Upper Newport Bay, showed considerable retention of nitrate (the dominant nitrogen species in all but Tijuana Estuary) despite high to very high inflowing concentrations spanning an order of magnitude (Figures 10.7a and 10.7g). At least some southern California lagoons have the capacity to absorb anthropogenic nutrient sources, and serve the ecological function of protecting ocean water quality. Both of these systems are open lagoons with direct connections to their watersheds, have a large proportion of area comprised of intertidal mudflat and salt marsh vegetation, and are ecological reserves (Figures 10.8a and 10.8b). Previous studies in Newport Bay showed that large amounts of nutrients are taken up and retained by macroalgal blooms and the sediments (Kamer et al. 2001; Boyle et al. 2004). The dilemma we face is that, although these processes may support nutrient retention functions, excessive macroalgal blooms have strong negative trophic effects and enriched sediments may fuel blooms during any season (Figure 10.6). Thus, one key to managing our lagoons is balancing within-system health with whole system ecological functions.

Of the five lagoons, only Carpinteria Marsh demonstrated significant nitrogen cycling by generating ammonium within the lagoon, although the relationship is weak (Figure 10.7b); this function was also identified in an earlier study and attributed to sediment nutrient flux to the water (Page et al. 1995). Uncovering the processes and mechanisms supporting a recycling function in this system may be key to maintaining and ultimately restoring this function to other southern California lagoons.

In two of five lagoons (Mugu Lagoon West and Tijuana Estuary), nitrate was diluted and exported directly to the ocean with no measurable retention within the system (Figures 10.7c and 10.7i). In Tijuana Estuary, ammonium, the dominant nitrogen species in inflowing water in this lagoon, was also exported with no measurable uptake (Figure 10.7j). We believe that these two lagoons previously supported the functions of retention and recycling based on the current functioning of other southern California systems, and had a naturally high retention capacity based on climate and geomorphology. That these functions have been reduced below measurable amounts or even completely lost over time is most likely a result of each lagoon's unique history of human modifications, and therefore the mechanisms of loss are likely to be different for each system.

Investigating the different mechanisms that have acted to limit and even eliminate nutrient retention in Mugu West and Tijuana Estuary is key to understanding how we may restore this ecological function. Mugu West is an example of a lagoon where the river has been diverted from flowing directly through the system, although the disconnection is not as severe as in other systems (Figure 10.8c). Calleguas Creek, the main river, was dredged into a channel 10 m deep that carries agricultural runoff directly from strawberry fields along its banks. River water only enters Mugu West when flood tides back up the nutrient-rich water, displacing it into the western arm. Within Mugu West, roads, buildings and airstrips create drastically muted tidal influence, increasing water

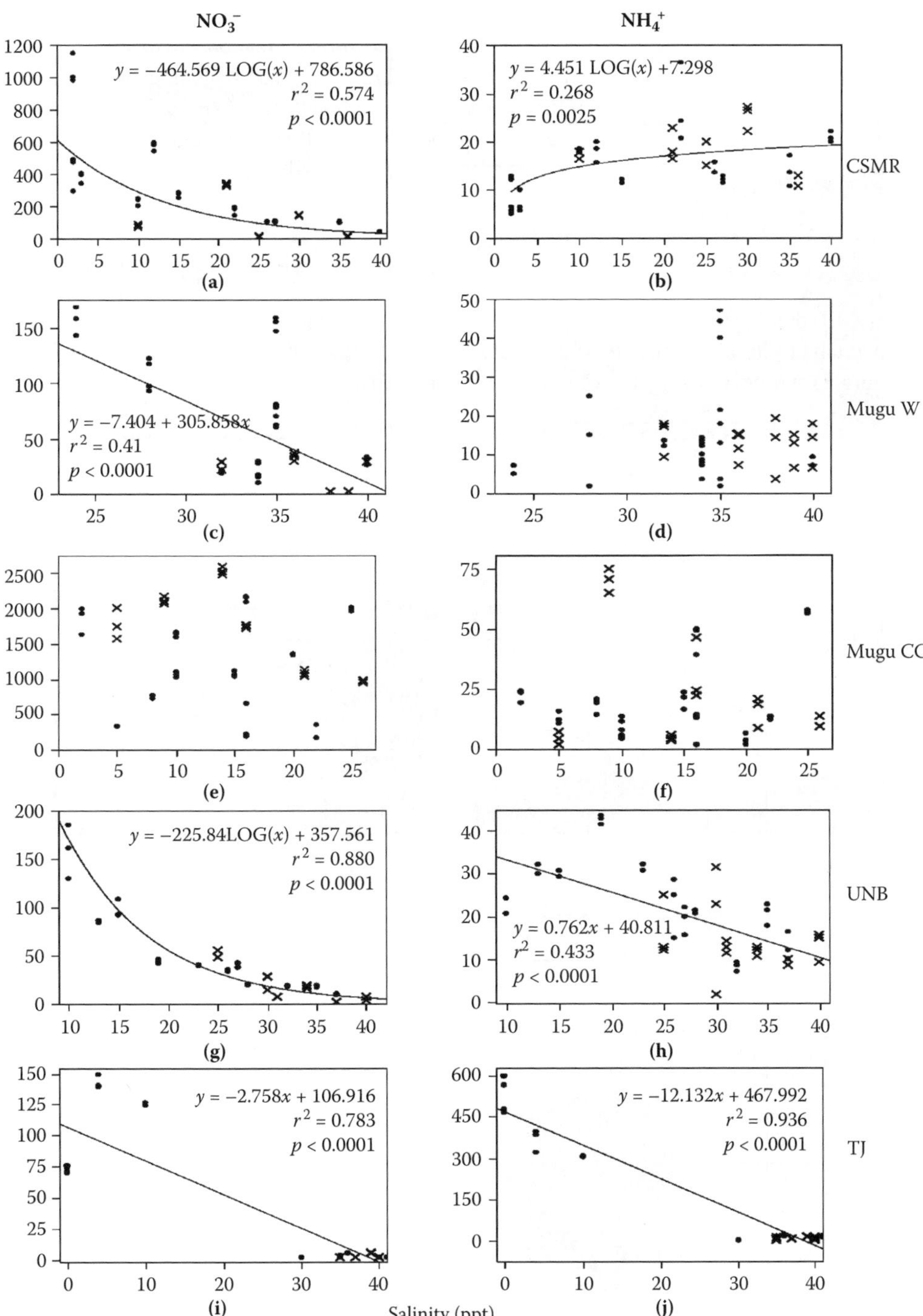

FIGURE 10.7 Mixing diagrams compare salinity, a nonbiologically active substance, with inorganic nitrogen. Symbols: • = wet season, x = dry season. A straight line with a negative slope indicates that nitrogen is diluted by tidal water but not biologically processed. A line that lies below this straight line indicates uptake and a line above the straight line indicates generation within the lagoon. Lagoons are Carpinteria Marsh, Mugu West, Mugu Calleguas Creek, Upper Newport Bay, and Tijuana Estuary.

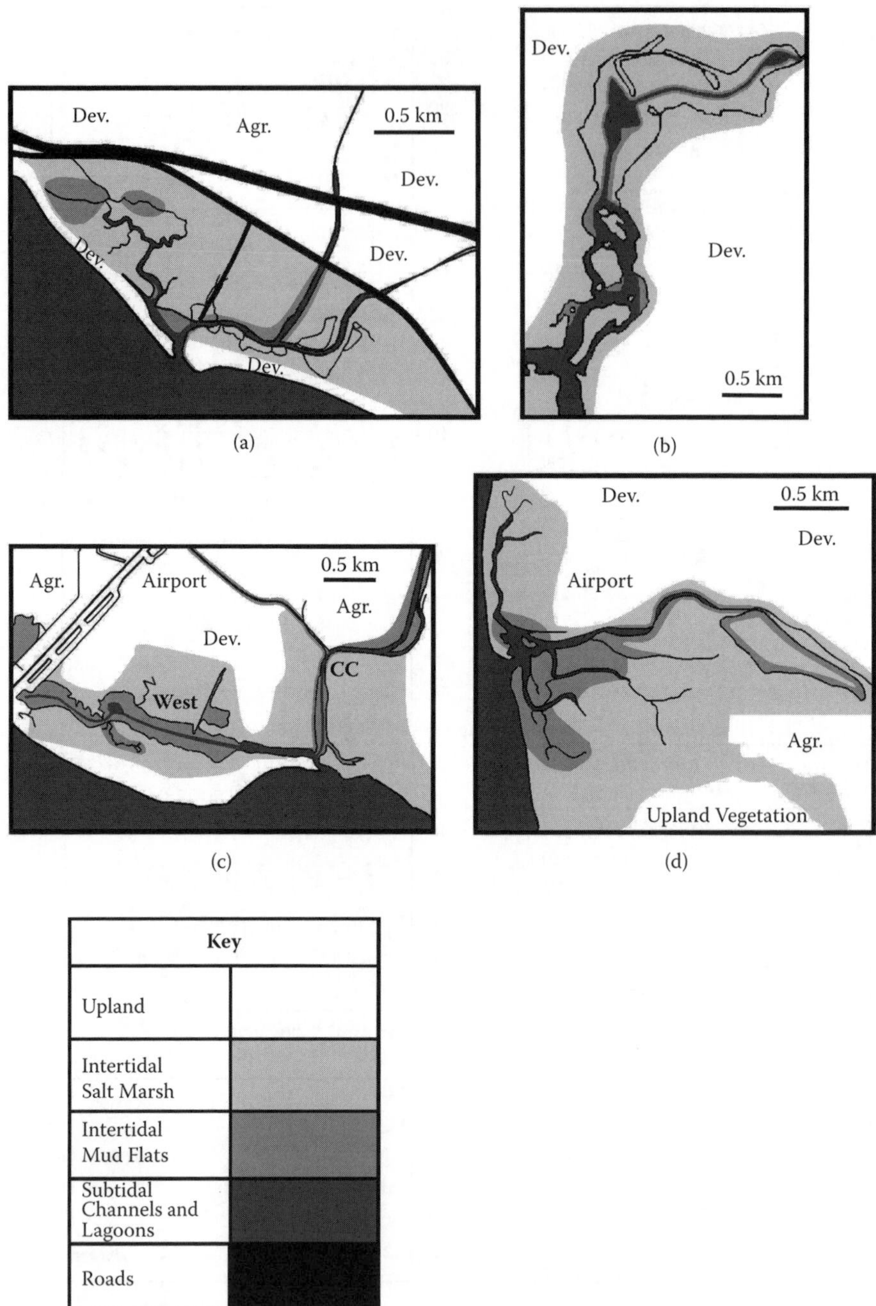

FIGURE 10.8 Maps of (a) Carpinteria Marsh; (b) Upper Newport Bay; (c) Mugu Lagoon West and Calleguas Creek; and (d) Tijuana Estuary with different habitat types delineated as subtidal, intertidal mudflat, and intertidal salt marsh. In the upland community, usage has been indicated as development (Dev.), agriculture (Agr.), airports, and upland vegetation. (Figure drawn by Kendal Fong.)

residence time. Due to these hydrological changes, we hypothesize that both macroalgal bloom potential and sediments became nutrient saturated from internal cycling sources, reducing and ultimately eliminating the ability to process chronic "new" anthropogenic inputs, and facilitating dilution and export of nitrogen to the ocean.

Tijuana Estuary is a large and open system designated as a reserve, with broad intertidal mudflats and saltmarsh vegetation (Figure 10.8d). Thus, it is hydrologically more similar to Newport and Carpinteria than Mugu West. Yet, like Mugu West, Tijuana Estuary functions primarily as a conduit of nitrogen from the watershed to the sea. We hypothesize high concentrations of NH_4^+ (Figure 10.7j) from pulsed release of treated wastewater directly upstream result in ammonium toxicity rather than uptake and cycling. Previous studies suggest this change in ecosystem function may be relatively new. During the 1970s, Winfield (1980) determined that Tijuana Estuary was a net sink for oceanic nutrients, not a source of terrestrial nutrients to coastal waters. In the 1980s, when tidal flushing in this estuary was reduced due to sedimentation (Zedler and West 2008), prolific macroalgal mats were found to be taking up large amounts of nutrients from the watershed (Fong 1986; Rudnicki 1986), implying that the estuary at that time was retaining nutrients. After tidal flushing was restored, recent increases in storms and flooding (Zedler and West 2008) most likely resulted in the net export of nitrogen from this system.

Mixing diagrams showed no clear patterns of retention or uptake in Mugu Calleguas Creek, a system reduced to a deep channel draining agricultural fields with little connectivity to intertidal mudflat or vegetation. We hypothesize that the dredging of this channel resulted in complex tidal and freshwater mixing patterns, high turbidity, and high flushing rates. Additional complexity is added by direct nutrient-rich freshwater inflows from the adjacent strawberry fields timed to individual water cycles for each field rather than larger-scale patterns generated by rainfall and subsequent river flow. While systems with short water residence times tend to export nutrients and have lower nutrient cycling capacity (Nixon et al. 1996; Conley et al. 2000), this is not the only process that is controlling nutrient retention in the highly modified Mugu Callguas Creek.

It is critical that future work assesses the degree of natural or historic functionality, focusing on the history of water flow and land use for each system, in order to adequately hindcast loss of ecological functions. Macroalgae proliferate, take up, retain, and recycle large amounts of nutrients in response to enrichment and hydrodynamic forces. However, in some systems their capacity to "keep up" with supply has been exceeded, and these estuaries are not functioning to retain a significant portion of the nutrient load. In order to retain invaluable ecosystem services provided by estuaries in southern California, managers cannot only look to nutrient reduction but also must consider secondary feedback mechanisms, habitat availability for nutrient processing, and both the historic context and current condition of hydrodynamic processes within each estuary.

10.6 FUTURE CHALLENGES: CLIMATE CHANGE—HOW WILL IT AFFECT SOUTHERN CALIFORNIA LAGOONS?

Our changing climate is producing a suite of changes in global environmental drivers, including sea-level rise, increased CO_2 in the air and water, ocean acidification, rising temperatures, and changes in weather patterns (Parmesan and Yohe 2003). These global changes will affect overall productivity of the ocean as well as species distributions, abundances, and diversity. Predictions that have been made for California are tenuous, but include wetter, warmer winters with more rain than snow, and therefore more winter and less summer riverflow, resulting in more extreme climatic events such as floods, landslides, droughts, and fires (Field et al. 1999). The ocean is expected to be warmer and therefore less productive due to nutrient limitation resulting from a more stable thermocline, with subsequent cascades up the food web (Doney 2006). Declines in fish abundances in the southern California Bight over the last two decades suggest these changes are well under way (Brooks et al. 2002). However, more rainfall combined with even greater development of coastal watersheds may bring more nutrients and sediments into coastal lagoons, and productivity may increase even further in these systems (Field et al. 1999).

One of the most serious climate-related concerns for southern California lagoons is sea-level rise, which is expected to continue even if we drastically reduce emissions today (IPCC 2007). The ocean absorbs more than 80% of the heat added to the global climate system, causing thermal expansion of seawater and melting of ice sheets and snow from Greenland and Antarctica. Sea-level rise averaged 1.8 mm yr^{-1} from 1961 to 2003, with more recent rates (1993 to 2003) accelerated to 3.1 mm yr^{-1}. The Intergovernmental Panel on Climate Change (IPCC) projected a range of sea-level rise from 0.18 to 0.59 m by the end of the twenty-first century. Uncertainties were generated by the lack of studies of ice flow from melting ice sheets, and the high variability of these systems. However, with continued warming, snow cover is projected to decrease and ice is projected to shrink, and even with a stabilization of these factors, by 2300, sea level is expected to rise by 0.3 to 0.8 m by thermal expansion alone. As the IPCC reports, if the Greenland ice sheet is eliminated in the coming millennia, sea level would rise 7 m, which is comparable to the last interglacial period (125,000 years ago) when sea level rose 4 to 6 m.

Mean sea level is predicted to rise between 1.0 and 1.4 m along the California coast by 2100 (Heberger et. al. 2009). However, it is important to note that this prediction does not include Greenland or Antarctic ice melt, so it is likely an underestimate. Key risks include flooding and coastal erosion, in some cases of barrier dunes (Heberger et. al. 2009). Sea-level rise will flood areas of low land adjacent to lagoons which, if undeveloped, could add to their area. However, in southern California, most lagoons are encircled by high-density development. In addition, these lagoons are tightly constrained between the ocean and the mountains, limiting the possibility of these systems naturally "migrating" upstream. Erosion of barrier dunes may fundamentally change lagoon hydrology; one result may be continual tidal flushing, transitioning these lagoons into open embayments.

A second important threat to southern California lagoons that may be related to climate change is increased supply of sediments in response to increased storminess. The fourth assessment report of the IPCC (2007) predicted an increase in cyclone intensity, with a pole-ward shift of extratropical storm tracks. Thus, these more intense storms may reach southern California more frequently. This trend was confirmed by a long-term study (1858 to 2000) of tidal gauge data in San Francisco Bay that detected decadal scale "pulses" in storminess, with heightened storminess in the last two decades that peaked during the 1997 to 1998 El Niño Southern Oscillation (Bromirski et al. 2003). However, the authors did not relate these patterns to climate change, as values for the current storminess were not unprecedented in the data record. Whether related to climate change or not, storms interact with watershed development to determine rates of sedimentation in lagoons. Thus, these more intense, and possibly more frequent, storms may increase sedimentation beyond the already high rates currently measured in southern California lagoons. Two studies have demonstrated that present sedimentation rates are higher than sea-level rise in some southern California lagoons (Greer and Stowe 2003; Zedler and West 2008). Sediments from the dunes and rivers filled lagoons and channels and elevated marsh plains, with major effects on producer and consumer communities including lowered diversity and increased brackish marsh habitat.

More research into the effects of climate change on southern California lagoons is essential for their long-term conservation. Key will be determining the balance in each system between sea-level rise and sedimentation. Methods that have proven effective include coring of salt marshes to quantify accretion rates, analysis of historic data on elevation and tide gauges, application of shorebird surveys and zonation studies, and experimental studies on hydrological and physical processes (Moorhead and Brinson 1995; Day et al. 1999; Najjar et al. 2000; Wijnen and Bakker 2001; Galbraith et al. 2002; Erwin et al. 2006). In addition, a more comprehensive understanding of factors affecting ecosystem structure and function is needed, including historical and current hydrological regimes, a historical context of the geomorphology of the system, and a degree of current degradation. Research on impacts of sea-level rise should be conducted in this ecological context in order to strengthen future predictions.

10.7 CONCLUSIONS

Southern California lagoons are set in tectonically active valleys of the coastal mountain ranges, resulting, in general, in small and isolated ecosystems. Each has a unique history of human modifications both to the watershed and the lagoon, making it difficult to advance generalizations about their functioning. One key to predicting the future may be to understand their historical condition. Many believe that most of these lagoons were closed off from the ocean seasonally or episodically. A Mediterranean climate with winter rains produced river flows that, depending on their magnitude and timing, could scour open the ocean connection. During the long, dry season longshore drift of sand resulted in building of the barrier sand dune, migration of this connection southward, and ultimately could cause lagoon closure. It is likely the historical condition of these systems contained more shallow water and less intertidal mudflat, and that deeper water (>l m depth) was rare. Seagrasses may have dominated primary producer communities.

With human modification, ocean connections became restricted to one possible opening, sedimentation increased infilling of subtidal areas and elevating marsh plains, and nutrient supplies increased dramatically. Although timing and trajectories most likely varied widely, seagrass communities were replaced first by microphytobenthos and then by blooms of macroalgae and cyanobacteria, with concurrent changes in consumer communities. At present, southern Caifornia lagoons are experiencing upward cascading negative effects of macroalgal blooms. Macroalgae draw down sediment oxygen content, killing benthic infauna. High nutrient supplies cause blooms of toxic cyanobacteria, which increases mortality of important epibenthic invertebrate grazers. Both of these change food supplies to resident and migratory birds. At present, some lagoons may no longer provide the important ecosystem function of nutrient retention, thus no longer serving to protect water quality of other near-shore marine habitats. Climate change will result in sea-level rise flooding lagoons and adjacent areas with oceanic water. Climate change may also increase storminess, which will further accelerate sedimentation from the watershed as well as cause erosion of barrier dunes.

The ultimate fate of southern California lagoons is highly uncertain. However, they are clearly at risk. Southern California lagoons are currently subjected to multiple stresses that continue to cause major changes in populations, communities, and trophic structure, and threaten ecosystem functioning. These stresses and resultant changes can only accelerate in the future with continued coastal and watershed development combined with climate change. If there is any hope of conserving southern California lagoons, it is essential that we continue efforts to understand their historical or "natural state," study current ecological processes controlling their present status in order to predict future responses, and formulate active, adaptive management plans to maintain and restore their functioning in the future.

ACKNOWLEDGMENTS

We thank California Sea Grant, the U.S. Environmental Protection Agency, the University of California's Center for Water Resources, the University of California's Academic Senate, the Southern California Coastal Water Research Project, the Mildred E. Mathias Graduate Student Research Grant, UCLA Department of Ecology and Evolutionary Biology and UCLA Graduate Division for the funding provided over the last 20 years to support this research. We also thank all of the graduate students whose dissertations were invaluable to advancing our knowledge of California coastal lagoons: Karleen Boyle, Krista Kamer, Anna Armitage, Kathy Boyer, Lauri Green, Tonya Kane, and Risa Cohen. Thanks to others who contributed their time selflessly to both our field and lab efforts: Jay Smith, Alina Corcoran, Matt Lurie, Brittany Huntington, Dan Reineman, Vanessa Gonzalez, and countless undergraduate students. We thank Kendal Fong, who drew the maps and conceptual models. Finally, many thanks to Rich Ambrose, who was a constant source of information and support.

REFERENCES

Ambrose, R. F. and A.R. Orme. 2000. *Lower Malibu Creek and lagoon resource enhancement and management.* Final Report to the California State Coastal Conservancy.

Armitage, A. R. and P. Fong. 2004. Upward cascading effects of nutrients: Shifts in a benthic microalgal community and a negative herbivore response. *Oecologia* 139: 560–567.

Armitage, A. R. and P. Fong. 2006. Predation and physical disturbance by crabs reduce the relative impacts of nutrients in a tidal mudflat. *Mar. Ecol. Prog. Ser.* 313: 205–213.

Beisner, B. E., D. T. Haydon, and K. Cuddington. 2003. Alternative stable states in ecology. *Front. Ecol. Environ.* 1: 376–382.

Bolam, S. G., T. F. Fernandes, P. Read, and D. Raffaelli. 2000. Effects of macroalgal mats on intertidal sandflats: An experimental study. *J. Exp. Mar. Biol. Ecol.* 249: 123–137.

Boyle, K. A., P. Fong, and K. Kamer. 2004. Spatial and temporal patterns in sediment and water column nutrients in an eutrophic southern California estuary. *Estuaries* 27: 254–267.

Bromirski, P. D., R. E. Flick, and D. R. Cayan. 2003. Storminess variability along the California coast: 1858–2000. *J. Climate* 16: 982–993.

Brooks A. J., R. J. Schmitt, and S. J. Holbrook. 2002. Declines in regional fish populations: Have species responded similarly to environmental change? *Mar. Freshw. Res.* 53: 189–198.

Cloern, J. E. 2001. Our evolving conceptual model of the coastal eutrophication problem. *Mar. Ecol. Prog. Ser.* 210: 223–253.

Conley, D. J., H. Kaas, F. Mohlenberg, B. Rasmussen and J. Windolf. 2000. Characteristics of Danish estuaries. *Estuaries* 23: 820–837.

Costanza, R., R. d'Arge, R. de Groot, S. Farber, M. Grasso, B. Hannon, K. Limburg, S. Naeem, R. V. O'Neill, J. Paruelo, R. G. Raskin, P. Sutton, and M. van den Belt. 1997. The value of the world's ecosystem services and natural capital. *Nature* 387: 253–260.

Day, J. W., C. A. S. Hall, W. M. Kemp, and A. Yáñez-Arancibia. 1989. *Estuarine ecology*. New York: John Wiley & Sons.

Day, J. W., J. Rybczyk, F. Scaron, A. Rismondo, D. Are, and G. Cecconi. 1999. Soil accretionary dynamics, sea-level rise and the survival of wetlands in Venice Lagoon: A field and modeling approach. *Estuar. Coastal Shelf Sci.* 49: 607–628.

Didham, R. K. and C. H. Watts. 2005. Are systems with strong underlying abiotic regimes more likely to exhibit alternative stable states? *Oikos* 110: 409–416.

Doney, S. C. 2006. Oceanography-plankton in a warmer world. *Nature* 444: 695–696.

Duarte, C. M. 1995. Submerged aquatic vegetation in relation to different nutrient regimes. *Ophelia* 41: 87–112.

Erwin, R. M., D. R. Cahoon, D. J. Prosser, G. M. Sanders, and P. Hensel. 2006. Surface elevation dynamics in vegetated *Spartina* marshes versus unvegetated tidal ponds along the Mid-Atlantic coast, USA, with implications to waterbirds. *Estuaries Coasts* 29: 96–106.

Everett, R. A. 1994. Macroalgae in marine soft-sediment communities: Effects on benthic faunal assemblages. *J. Exp. Mar. Biol. Ecol.* 175: 253–274.

Eyre, B. D. and A. J. P. Ferguson. 2002. Comparison of carbon production and decomposition, benthic nutrient fluxes and denitrification in seagrass, phytoplankton, benthic microalgae and macroalgae-dominated warm-temperate Australian lagoons. *Mar. Ecol. Prog. Ser.* 229: 43–59.

Ferguson, A. J. P., B. D. Eyre, and J. M. Gay. 2004. Benthic nutrient fluxes in euphotic sediments along shallow sub-tropical estuaries, northern New South Wales, Australia. *Aquat. Microb. Ecol.* 37: 219–235.

Field, C. B., G. C. Daily, F. W. Davis, S. Gaines, P. A. Matson, J. Melack, and M. L. Miller. 1999. *Confronting climate change in California: Ecological impacts on the golden state.* Union of Concerned Scientists, Cambridge, Massachusetts and Ecological Society of America, Washington, D.C.

Fong, P. 1986. Monitoring and manipulation of phytoplankton dynamics in a southern California estuary. M.S. Thesis, San Diego State University, San Diego, California.

Fong, P., J. S. Desmond, and J. B. Zedler. 1997. The effect of a horn snail on *Ulva expansa* (Chlorophyta): consumer or facilitator of growth? *J. Phycol.* 33: 353–359.

Fong, P., R. M. Donohoe, and J. B. Zedler. 1993a. Competition with macroalgae and benthic cyanobacterial mats limits phytoplankton abundance in experimental microcosms. *Mar. Ecol. Prog. Ser.* 100: 97–102.

Fong, P., T. C. Foin, and J. B. Zedler. 1994. A simulation model of lagoon algal communities based on competition for nitrogen and internal storage. *Ecol. Monogr.* 64: 225–247.

Fong, P. and J. B. Zedler. 2000. Sources, sinks, and fluxes of nutrients (N + P) in a small highly-modified estuary in southern California. *Urban Ecosys.* 4: 125–144.

Fong, P., J. B. Zedler, and R. M. Donohoe. 1993b. Nitrogen versus phosphorus limitation of algal biomass in shallow coastal lagoons. *Limnol. Oceanogr.* 38: 906–923.

Fox, S. E., E. Stieve, I. Valiela, J. Hauxwell, and J. McClelland. 2008. Macrophyte abundance in Waquoit Bay: Effects of land-derived nitrogen loads on seasonal and multi-year patterns. *Estuaries Coasts* 31: 532–541.

Fox, S. E., M. Tichberg, Y. S. Olsen, L. Heffner, and I. Valiela. 2009. Restructuring of benthic communities in eutrophic estuaries: Lower abundance of prey leads to trophic shifts from omnivory to grazing. *Mar. Ecol. Prog. Ser.* 380: 43–57.

Galbraith, H., R. Jones, R. Park, J. Clough, S. Herrod-Julius, B. Harrington, and G. Page. 2002. Global climate change and sea level rise: Potential losses of intertidal habitat for shorebirds. *Waterbirds* 25: 173–183.

Geertz-Hansen, O., K. Sand-Jensen, D. F. Hansen, and A. Christiansen. 1993. Growth and grazing control of abundance of the marine macroalga, *Ulva lactuca* L. in a eutrophic Danish estuary. *Aquat. Bot.* 46: 101–109.

Greer, K. and D. Stow. 2003. Vegetation type conversion in Los Peñasquitos Lagoon, California: An examination of the role of watershed urbanization. *Environ. Manage.* 31: 489–503.

Hauxwell, J., J. Cebrián, C. Furlong, and I. Valiela. 1998. Relative importance of grazing and nutrient controls of macroalgal biomass in three temperate shallow estuaries. *Estuaries* 21: 347–360.

Hauxwell, J., J. Cebrián, and I. Valiela. 2003. Eelgrass *Zoster marina* loss in temperate estuaries: Relationship to land-derived nitrogen loads and effect of light limitation imposed by algae. *Mar. Ecol. Prog. Ser.* 247: 59–73.

Heberger, M., H. Cooley, P. Herrera, P. H. Gleick, and E. Moore. 2009. *The impacts of sea-level rise on the California coast.* California Climate Change Center. http://www.pacinst.org/reports/sea_level_rise/.

Hernández, I., G. Peralta, J. L. Pérez-Lloréns, J. J. Vergara, and F. X. Niell. 1997. Biomass and dynamics of growth of *Ulva* species in Palmones River estuary. *J. Phycol.* 33: 764–772.

Hughes, T. P., D. R. Bellwood, C. S. Folke, L. J. McCook, and J. M. Pandolfi. 2007. No-take areas, herbivory, and coral reef resilience. *Trends Ecol. Evol.* 22: 1–3.

Huntington, B. E. and K. E. Boyer. 2008. Effects of red macroalgal (*Gracilariopsis* sp.) abundance on eelgrass *Zostera marina* in Tomales Bay, California, USA. *Mar. Ecol. Prog. Ser.* 367: 133–142.

IPCC (Intergovernmental Panel on Climate Change). 2007. Climate change 2007: The physical science basis. In: S. Solomon, D. Qin, and M. Manning (eds.), Contribution of Working Group I to the Fourth Assessment Report of the Intergovernmental Panel on Climate Change. Geneva, Switzerland.

Jackson, J. B. C., M. X. Kirby, W. H. Berger, K. A. Bjorndal, L. W. Botsford, B. J. Bourque, R. H. Bradbury, R. Cooke, J. Erlandson, J. A. Estes, T. P. Hughes, S. Kidwell, C. B. Lange, H. S. Lenihan, J. M. Pandolfi, C. H. Peterson, R. S. Steneck, M. J. Tegner, and R.R. Warner. 2001. Historical overfishing and the recent collapse of coastal ecosystems. *Science* 293: 629–638.

Kamer, K., K. A. Boyle, and P. Fong. 2001. Macroalgal bloom dynamics in a highly eutrophic southern California estuary. *Estuaries* 24: 623–635.

Kamer, K. and P. Fong. 2000. A fluctuating salinity regime mitigates the negative effect of reduced salinity on the estuarine macroalga, *Enteromorpha intestinalis* (L.) link. *J. Exp. Mar. Biol. Ecol.* 254: 53–69.

Kamer, K., P. Fong, R. L. Kennison, and K. Schiff. 2004a. Nutrient limitation of the macroalga, *Enteromorpha intestinalis* collected along a resource gradient in a highly eutrophic estuary. *Estuaries* 27: 201–208.

Kamer, K., P. Fong, R. L. Kennison, and K. Schiff. 2004b. The relative importance of sediment and water column supplies of nutrients to the growth and tissue nutrient content of the green macroalga *Entermorpha intestinalis* along an estuarine resource gradient. *Aquat. Ecol.* 38: 45–56.

Kemp, W. M., W. R. Boynton, J. E. Adolf, D. F. Boesch, W. C. Boicourt, G. Brush, J. C. Cornwell, T. R. Fisher, P. M. Gilbert, J. D. Hagy, L. W. Harding, E. D. Houde, D. G. Kimmel, W. D. Miller, R. I. E. Newell, M. R. Roman, E. M. Smith, and J. C. Stevenson. 2005. Eutrophication of Chesapeake Bay: Historical trends and ecological interactions. *Mar. Ecol. Prog. Ser.* 303: 1–29.

Kennish, M. J. 2002. Environmental threats and environmental future of estuaries. *Environ. Conserv.* 29: 78–107.

Kennison, R. L. 2008. Evaluating ecosystem function of nutrient retention and recycling in excessively eutrophic estuaries. Ph.D. Thesis, University of California, Los Angeles.

Krause-Jensen, D., P. G. Christensen, and S. Rysgaard. 1999. Oxygen and nutrient dynamics within mats of the filamentous macroalga *Chaetomorpha Linum*. *Estuaries* 22: 31–38.

Lafferty, K. D. 2001. Birds at a southern California beach: Seasonality, habitat use and disturbance by human activity. *Biol. Conserv.* 10: 1949–1962.

Lopes, R. J., M. A. Pardal, T. Mu´ Rias, J. A. Cabral, and J. C. Marques. 2006. Influence of macroalgal mats on abundance and distribution of dunlin *Calidris alpina* in estuaries: A long-term approach. *Mar. Ecol. Prog. Ser.* 323: 11–20.

Marcomini, A., A. Sfriso, B. Pavoni, and A. A. Orio. 1995. Eutrophication of the Lagoon of Venice: Nutrient loads and exchanges. Pp. 59–80, In: A. J. McComb (ed.), *Eutrophic shallow estuaries and lagoons*. Boca Raton, Florida: CRC Press.

Moorhead, K. K. and M. M. Brinson. 1995. Response of wetlands to rising sea level in the lower coastal plain of North Carolina. *Ecolog. Appl.* 5: 261–271.

Najjar, R. G., H. A. Walker, P. J. Anderson, E. J. Barron, R. J. Bord, J. R. Gibson, V. S. Kennedy, C. G. Knight, J. P. Megonigal, R. E. O'Conner, C. D. Polsky, N. P. Psuty, B. A. Richards, L. G. Sorenson, E. M. Steele, and R. S. Swanson. 2000. The potential impacts of climate change on the mid-Atlantic coastal region. *Climate Res.* 14: 219–233.

Nixon, S. W. 1995. Coastal marine eutrophication: A definition, social causes, and future concerns. *Ophelia* 41:199–219.

Nixon, S. W., J. W. Ammerma, L. P. Atkinson, V. M. Berounsky, G. Billen, W. C. Biocourt, W. R. Boynton, T. M. Church, D. M. DiToro, R. Elmgren, J. H. Garber, A. E. Giblin, R. A. Jahnke, N. J. P. Owens, M. E. Q. Pilson, and S. P. Seitzinger. 1996. The fate of nitrogen and phosphorus at the land-sea margin of the North Atlantic Ocean. *Biogeochemistry* 35: 141–180.

Norkko, A., and E. Bonsdorff. 1996. Altered benthic prey-availability due to episodic oxygen deficiency caused by drifting algal mats. *Mar. Ecol.* 17: 355–372.

Osterling, M., and L. Pihl. 2001. Effects of filamentous green algal mats on benthic macrofaunal functional feeding groups. *J. Exp. Mar. Biol. Ecol.* 263: 159–183.

Paalme, T., H. Kukk, and H. Orav. 2002. "In vitro" and "in situ" decomposition of nuisance macroalgae *Cladophora glomerata* and *Pilayella littoralis*. *Hydrobiologia* 475/476: 469–476.

Paerl, H. W. 2006. Assessing and managing nutrient-enhanced eutrophication in estuarine and coastal waters: Interactive effects of human and climatic perturbations. *Ecol. Eng.* 26: 40–54.

Page, H. M. 1995. Variation in the natural abundance of ^{15}N in the halophyte, *Salicornia virginica*, associated with groundwater subsidies of nitrogen in a southern California salt marsh. *Oecologia* 104: 181–188.

Page, H. M., R. L. Petty, and D. E. Meade. 1995. Influences of watershed runoff on nutrient dynamics in a southern California salt marsh. *Estuar. Coastal Shelf Sci.* 41: 163–180.

Parmesan, C. and G. Yohe. 2003. A globally coherent fingerprint of climate change impacts across natural systems. *Nature* 421: 37–42.

Peckol, P., B. DeMeo-Anderson, J. Rivers, I. Valiela, M. Maldonado, and J. Yates. 1994. Growth, nutrient uptake capacities and tissue constituents of the macroalgae *Cladophora vagabunda* and *Gracilaria tikvahiae* related to site-specific nitrogen loading rates. *Mar. Biol.* 121: 175–185.

Press, F. and R. Siever. 1997. *Understanding earth,* 2nd ed. New York: W.H. Freeman and Company.

Raffaelli, D., S. Hull, and H. Milne. 1989. Long-term changes in nutrients, weed mats and shorebirds in an estuarine system. *Cahiers Biol. Mar.* 30: 259–270.

Roughgarden, J. and F. Smith. 1996. Why fisheries collapse and what to do about it. *Proc. Natl. Acad. Sci. USA* 93: 5078–5083.

Rudnicki, R. M. 1986. Dynamics of macroalgae in Tijuana Estuary: Response to simulated wastewater addition. M.S. Thesis, San Diego State University, San Diego, California.

Schaefer, S. C. and M. Alber. 2007. Temperature controls a latitudinal gradient in the proportion of watershed nitrogen exported to coastal ecosystems. *Biogeochemistry* 85: 333–346.

Scheffer, M., S. Carpenter, J. A. Foley, C. Folkes, and B. Walker. 2001. Catastrophic shifts in ecosystems. *Nature* 413: 591–596.

Schramm, W. 1999. Factors influencing seaweed responses to eutrophication: Some results from EU-project EUMAC. *J. Appl. Phycol.* 11: 69–78.

Sfriso, A., A. Marcomini, and B. Pavoni. 1987. Relationships between macroalgal biomass and nutrient concentrations in a hypertrophic area of the Venice Lagoon Italy. *Mar. Environ. Res.* 22: 297–312.

Sfriso, A., B. Pavoni, A. Marcomini, and A. A. Orio. 1992. Macroalgae, nutrient cycles, and pollutants in the Lagoon of Venice. *Estuaries* 15: 507–528.

Short, F. T. and S. Wyllies-Echeverria. 1996. Natural and human-induced disturbance of seagrasses. *Environ. Conserv.* 23: 17–27.

Stein, E. D., S. Dark, T. Longcore, N. Hall, M. Beland, R. Grossinger, J. Casanova, and M. Sutula. 2007. Historical ecology and landscape change of the San Gabriel River and floodplain. Southern California Coastal Water Research Project Technical Report #499, Westminster, California. www.sccwrp.org.

Stenzel, L. E., H. R. Huber, and G. W. Page. 1976. Feeding behavior and diet of the long-billed curlew and willet. *Wilson Bull.* 88: 324–332.

Stewart, J. G. 1991. *Marine algae and seagrasses of San Diego County*. California Sea Grant College, University of California, La Jolla, California.

Sundbäck, K., A. Miles, S. Hulth, L. Pihl, P. Engström, E. Selander, and A. Svenson. 2003. Importance of benthic nutrient regeneration during initiation of macroalgal blooms in shallow bays. *Mar. Ecol. Prog. Ser.* 246: 115–126.

Sutula, M., C. Creager, and G. Wortham. 2007. Technical Report 516. Technical approach to develop nutrient numeric endpoints for California estuaries. Southern California Coastal Research Project. Westminster, California. www.sccwrp.org.

Sutula, M., K. Kamer, J. Cable. 2004. Technical Report 441. Sediments as non-point source of nutrients to Malibu Lagoon, California (USA). Southern California Coastal Research Project, Westminster, California. www.sccwrp.org.

Sutula, M., K. Kamer, J. Cable, H. Berelson, and J. Mendez. 2006. Technical Report 482. Sediments as an internal source of nutrients to Upper Newport Bay, California. Southern California Coastal Research Project, Westminster, California. www.sccwrp.org.

Svensson, J. M., G. M. Carrer, and M. Bocci. 2000. Nitrogen cycling in sediments of the Lagoon of Venice, Italy. *Mar. Ecol. Prog. Ser.* 199: 1–11.

Taylor, D. I., S. W. Nixon, S. L. Granger, and B. A. Buckley. 1995. Nutrient limitation and the eutrophication of coastal lagoons. *Mar. Ecol. Prog. Ser.* 127: 235–244.

Thybo-Christensen, M., M. B. Rasmussen, and T. H. Blackburn. 1993. Nutrient fluxes and growth of *Cladophora sericea* in a shallow Danish bay. *Mar. Ecol. Prog. Ser.* 100: 273–281.

Tyler, A. C., K. J. McGlathery, and I. C. Anderson. 2003. Benthic algae control sediment-water column fluxes of organic and inorganic nitrogen compounds in a temperate lagoon. *Limnol. Oceanogr*. 48: 2125–2137.

Valiela, I., K. Foreman, M. LaMontagne, D. Hersh, J. Costa, P. Peckol, B. DeMeo-Andreson, C. D'Avanzo, M. Babione, C. H. Sham, J. Brawley, and K. Lajtha. 1992. Couplings of watersheds and coastal waters sources and consequences of nutrient enrichment in Waquoit Bay Massachusetts. *Estuaries* 15: 443–457.

Valiela, I., M. Geist, J. McClelland, and G. Tomasky. 2000. Nitrogen loading from watersheds to estuaries: Verification of the Waquoit Bay nitrogen loading model. *Biogeochemistry* 49: 277–293.

Valiela, I., J. McClelland, J. Hauxwell, P. J. Behr, D. Hersh, and K. Foreman. 1997. Macroalgal blooms in shallow estuaries: Controls and ecophysiological and ecosystem consequences. *Limnol. Oceanogr.* 42: 1105–1118.

Wijnen, H. J. and J. P. Bakker. 2001. Long-term surface elevation change in salt marshes: A prediction of marsh response to future sea-level rise. *Estuar. Coastal Shelf Sci.* 52: 381–390.

Winfield, P. 1980. Dynamics of carbon and nitrogen in a southern California salt marsh. Ph.D. Thesis, University of California Riverside and San Diego State University, San Diego, California.

Young, D. R., D. T. Specht, P. J. Clinton, and H. I. Lee. 1998. Use of color infrared aerial photography to map distributions of eelgrass and green macroalgae in a non-urbanized estuary of the Pacific Northwest, USA. *Proceedings of the Fifth International Conference on Remote Sensing of Marine and Coastal Environments* 2: 37–45.

Xue, Y. and J. Shukla. 1993. The influence of land surface properties on Sahel climate. I. Desertification. *J. Climate* 6: 2232–2245.

Zedler, J. B. 1982. The ecology of Southern California coastal salt marshes: A community profile. U.S. Fish and Wildlife Service, Biological Services Program, Washington, D.C., FWS/085-81/54.

Zedler, J. B. 1996. Tidal wetland restoration: A scientific perspective and Southern California focus. California Sea Grant College System, University of California, La Jolla, California.

Zedler, J. B., J. C. Callaway, and G. Sullivan. 2001. Declining biodiversity: Why species matter and how their functions might be restored in Californian tidal marshes. *BioScience*. 51: 1005–1017

Zedler, J. B. and J. M. West. 2008. Declining diversity in natural and restored salt marshes: A 30-year study of Tijuana Estuary. *Restoration Ecol.* 16: 249–262.

11 Lagoons of the Nile Delta

Autumn J. Oczkowski and Scott W. Nixon*

CONTENTS

Abstract 253
11.1 Introduction 254
11.2 The Nile Delta 255
11.3 Physical, Chemical, and Biological Characteristics of the Delta Lagoons 257
11.3.1 Burullus 258
11.3.2 Edku 260
11.3.3 Manzalah 262
11.3.4 Maryut 266
11.4 Lagoon Substrates 268
11.5 Pollutants 268
11.6 Nutrient Dynamics 270
11.7 Primary Producers 272
11.8 Zooplankton and Benthic Invertebrates 274
11.9 Lagoon Fisheries 275
11.10 Linking Nutrients and Fisheries 277
11.11 Future Directions 278
Acknowledgments 278
References 279

ABSTRACT

Burullus, Edku, Manzalah, and Maryut are four large (50 to 500 km^2), shallow (~1 m deep) lagoons on Egypt's Nile Delta that intercept great amounts of agricultural drainage, sewage, and industrial effluents before they are discharged to the oligotrophic eastern Mediterranean Sea. The lagoons have been rapidly decreasing in size as fish pond aquaculture, agriculture, and urbanization encroach on them. Following the completion of the Aswan High Dam on the Nile River in 1965, Egypt's freshwater resources became fully regulated and the intensity of agricultural irrigation and synthetic fertilizer consumption increased, leading to a rise in the amount of agricultural drainage received by the lagoons. These factors, combined with increases in upstream population and improvements in sewerage infrastructure, have had significant and often dramatic ecological effects. Salinity, as well as ecological indicators like Secchi depth (representing turbidity levels) and dissolved oxygen, have decreased over time while nutrient concentrations, particularly inorganic nitrogen, have been increasing. Commercial fish landings in Burullus, Edku, and Manzalah lagoons have been rising since the 1960s and today constitute almost half of Egypt's national fish catch. In the most altered and industrialized lagoon, Maryut, the fishery collapsed by around 1980, when sewage from the city of Alexandria was first collected and discharged to the lagoon's main basin. Nutrient concentrations

* Corresponding author. Email: ajo@gso.uri.edu

in this lagoon are also exceedingly high, with recent mean annual dissolved inorganic nitrogen (DIN) concentrations far greater than 100 μM. In fact, the Maryut fishery collapsed when DIN concentrations reached 100 μM, suggesting that rising inorganic nitrogen inputs may have contributed to the system's breakdown and that even the most productive systems characterized by resistant fish species have a limited capacity to assimilate and process nutrients.

Key Words: lagoons, Egypt, Nile Delta, fisheries, nutrients, water quality

11.1 INTRODUCTION

Of all coastal water bodies, lagoons are thought to be among the most sensitive to environmental changes. So-called "choked" lagoons (Kjerfve 1986) are particularly sensitive to human-induced alterations such as eutrophication and climate change. With most of the world's population living along the coast, and often adjacent to these systems, most lagoons are already greatly altered (Nixon 1982). This was particularly true in the United States and western Europe, where the advent of the industrial revolution in the late eighteenth and early nineteenth centuries brought about rapid enrichment of coastal systems with sewage and industrial effluents prior to widespread reliable and thorough ecological studies which might have documented such changes. In contrast, the coastlines of developing countries have been receiving the effects of rapid development and urbanization much more recently. Unfortunately, few of these systems are actively being monitored and the effects of any changes are again going undocumented. Egypt's coastal lagoons are unique as the anthropogenic impacts on these systems have been accelerating since the mid-1960s and they have been the subject of extensive research since the 1950s.

The famous Greek historian Herodotus once wrote that Egypt was the "Gift of the Nile." He was not exaggerating, for without the world's longest river flowing through it, all of Egypt would be a desert. Egypt is an environment of extremes, and the Nile Delta lagoons are no exception. Egyptians have understood their dependence on the seasonal variability of the Nile for feast or famine since ancient times and have been modifying and extending the river's floodplains to take better advantage of this necessary resource for more than 5000 years. Up until the mid-1900s, heavy summer rains in the Ethiopian headwaters caused an autumn flood in Egypt, irrigating and fertilizing the floodplain with nutrient-rich silt. These floodwaters would eventually overflow a large delta wetland and flow out into the Mediterranean Sea, supporting both fertile lands for crops and a productive fishery along the northern edge of the delta.

Starting in the nineteenth century, small dams built on the delta were able to retain Nile water at levels above summer low flows and, along with other dams built further upstream, they were able to partially control the hydrology (Dumont and El-Shabrawy 2007). However, it was not until 1965 that the Aswan High Dam was finally completed, reducing the fall flood by about 90% and giving Egyptians full control of their water resources for the first time. By 1967, virtually all of the water reaching the delta was used for irrigation (Dumont and El-Shabrawy 2007). The new dam allowed farmers to grow three crops a year instead of one, but with the nutrient-rich silt now trapped behind the dam, farmers came to rely on synthetic fertilizers to support increased food production and an ever-growing national population. From 1965 to 2002, Egypt's national fertilizer consumption increased almost fourfold, from 3.4×10^5 t yr^{-1} to 13×10^5 t yr^{-1} (FAO 2008). During this time, population more than doubled (1965 to 2000) and per capita calorie and total protein consumption also increased markedly (FAO 2008). The nutrient-rich runoff from agricultural fields and much of the human sewage from Cairo, Alexandria, and towns on the delta are released into large drains which eventually discharge directly into the Mediterranean or, more commonly, into the four lagoons on Egypt's delta coast (Burullus, Edku, Manzalah, and Maryut). Because water is such a precious resource in this country, as much of the land drainage as possible has been diverted to the lagoons.

Burullus, Edku, Manzalah, and Maryut are large (50 to 500 km^2), shallow (~1 m deep) lagoons along Egypt's northern coast that lie at the intersection of the fertile Nile Delta and the extremely

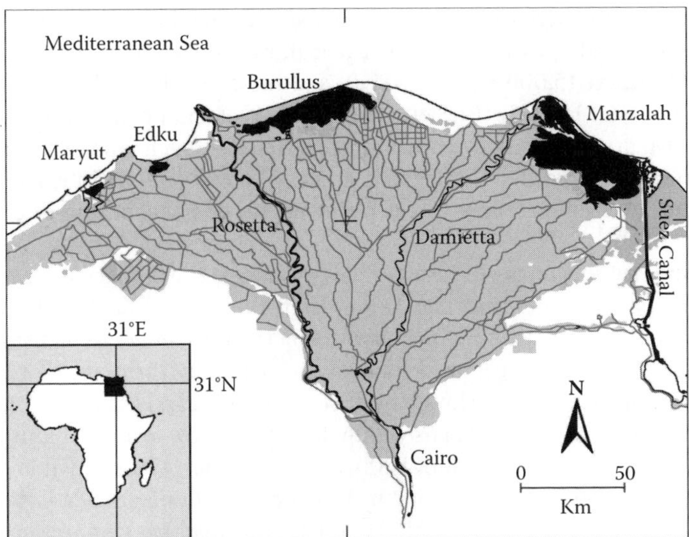

FIGURE 11.1 Egypt's Nile Delta with agricultural areas indicated in gray (source data, Digital Chart of the World [DCW], Environmental Systems Research Institute, Inc. [ESRI] 1:1,000,000 scale). The Nile flows north through Cairo and then splits into the Rosetta and Damietta branches (black lines). Dark gray lines represent some of the major drains and canals on the delta (data from DCW). The crosshairs in the center of the image correspond to 31°0′E, 31°0′N.

oligotrophic eastern Mediterranean Sea. They have been changing, both naturally and from human activities, since ancient times (Figure 11.1; Butzer 1976). However, it seems that the most rapid changes are or "as the result of" related to increasing inputs of nutrient-rich agricultural drainage and sewage, and area loss to urbanization and aquaculture have been occurring since the middle of the twentieth century, well within the timeframe in which reliable measurements of the chemistry and ecology of the systems have been made and allowing us to assess how some of these changes may be impacting the productivity of the lagoons (EA Engineering 1997; El-Rayis 2005; Shakweer 2005, 2006; Okbah and Hussein 2006). As the information presented in this chapter will demonstrate, the increased drainage, and the nutrients therein, as well as lagoon area loss to urbanization, aquaculture, and agriculture, have caused a series of dramatic changes, such as increased eutrophication, in the delta lagoons.

11.2 THE NILE DELTA

Egypt's fertile Nile Delta lies at the intersection of three great deserts: the eastern and western deserts on land and an offshore marine desert, the ultra-oligotrophic eastern Mediterranean Sea (Figure 11.1). In stark contrast, the delta itself is a 25,000 km^2 wedge of intense productivity, with densely packed villages and cities set among fertile agricultural fields, fish farms, and the four large brackish lagoons. The delta constitutes less than 3% of Egypt's total land area but provides for about 40% of the nation's agriculture, 50% of its industry, and 60% of its national fish catch (Mikhailova 2001). To accommodate the approximately 34 million inhabitants (half of Egypt's population), and their industry and food sources, the Nile Delta is highly engineered. Agricultural fields have tile drainage systems and are connected to a dense network of canals carrying clean Nile water to the fields and drains removing field runoff. The density and size of the drains increase downstream toward the Mediterranean Sea. Most of the drains are routed to the coastal lagoons, which serve as secondary treatment basins for the complex mixture of agricultural runoff, sewage (from the cities and villages), and industrial effluents (Figure 11.1).

Aquaculture ponds that encroach on the lagoons and intercept drainage waters have also become widespread in recent decades. Egypt's national aquaculture production increased more than 30-fold from 1984 to 2004 (from 15,000 t to 471,535 t), and most of this production occurs in earthen ponds around the coastal lagoons (El-Sayed 2007). Most of the ponds are semi-intensively cultured, where the fish are fed a mixture of commercial feed, bycatch, and agricultural byproducts. Aquaculture wastes discharge into the lagoons at an unknown rate, and their impact is not well known because the ponds are privately operated and records do not appear to be centrally collected (Ishak and Shafik 1982).

The sizes of the delta lagoons (Burullus, Edku, Manzalah, and Maryut) have been changing for hundreds of years, especially since the Aswan High Dam closed in 1965. Between land reclamation for agriculture and aquaculture, the encroachment of urban areas, and siltation associated with increased agricultural drainage, the delta lagoons are rapidly shrinking. For example, Kassim (2005) reported that Maryut Lagoon shrank by 50% in just 35 years. Manzalah Lagoon is decreasing at a rate of 5 km^2 yr^{-1}, while the islands within the lagoon grow at about 3 km^2 yr^{-1}, a total open water loss of just under 2% yr^{-1} (Mikhailova 2001). Using our best approximation of current lagoon areas, Burullus and Manzalah (500 and 450 km^2), are an order of magnitude larger than Edku and Maryut at 50 and 63 km^2, respectively (Table 11.1). The two larger lagoons are similar in size to the Lagoon of Venice (550 km^2) and about twice the size of Long Island's Great South Bay (245 km^2).

The lagoons are not of marine origin, but were formed by the pooling of Nile flood waters behind the coastal sand dunes in this low lying delta. As early as 6500 years ago, extensive sand ridges began to form along the outer margins of the delta creating a barrier behind which marshes and lagoons developed (Stanley and Warne 1993). Humans settled in the area at about 7000 B.P. and by 2000 B.P. were dramatically influencing the delta landscape through intensified irrigation, wetland drainage, and the artificial excavation of the Damietta and Rosetta branches of the Nile. Other naturally occurring distributary channels were rerouted for irrigation purposes and by the nineteenth century, the area of the lagoons began to decline (Stanley and Warne 1993). At some point, either naturally or through human manipulation, the lagoons were connected to the sea via breachways. The first record of a lagoon–sea connection for Edku Lagoon was drawn on a map in 1798 by French scientists (Samman 1974). After the last normal Nile flood in 1964, the lagoons lost their seasonally replenishing freshwater pulse. To compensate for this and maintain the already important fishery, drainage canals were designed to discharge into the lagoons rather than directly to the Mediterranean Sea (Al Sayes et al. 2007). Thus, it was intended that each lagoon receive very large amounts of agricultural drainage, although it is unclear whether the potential impact of increasing amounts of fertilizer runoff to the lagoon was considered. It is also impossible to assess the impact and importance of agricultural runoff relative to human sewage and industrial effluents as they appear not to have been described quantitatively or qualitatively.

TABLE 11.1
Some Basic Characteristics of the Nile Delta Lagoons

Lagoon	Size[a] (km^2)	Salinity[a]	Agricultural Drainage Inflow[b] (10^9 m^3 yr^{-1})	Direct Sewage Inflow[b] (10^9 m^3 yr^{-1})
Burullus	500	0.5–21	2.5	–
Edku	50	2–28	2	–
Manzalah	450	2–15	7	0.033
Maryut	63	2–5	0.3	0.22–0.35

[a] Data from Toews (1988); El-Shenawy (1994); EA Engineering (1997); Zaghloul and Hussein (2000); Dewidar and Khedr (2001); Okbah and El-Gohary (2002); Okbah and Hussein (2006).

[b] Data from EA Engineering (1997); El-Rayis (2005); Shakweer (2005); Okbah and Hussein (2006); Shakweer (2006); Al Sayes et al. (2007); Dumont and El-Shabrawy (2007); Mageed (2007).

The physical and chemical characteristics of some lagoons are far better documented than others. Of the four lagoons, Maryut Lagoon is the best understood, probably because it lies just inshore of Alexandria. Alexandria is Egypt's second largest city and home to a government research agency (National Institute of Oceanography and Fisheries) and a well-regarded oceanography program at the University of Alexandria. Many of the earliest studies of lagoon chemistry and biology were carried out in Maryut (Wahby 1961; Aleem and Samaan 1969; El-Wakeel and Wahby 1970; El-Zarka et al. 1970). In the mid-1990s, an environmental consulting firm, EA Engineering, Science and Technology, Inc., conducted an extensive review of both historical data and then current conditions of the lagoon. Over the period of a year, the consulting firm measured water and sediment composition and quality, including nutrients, heavy metals, and pollutants, as well as documented the flora and fauna in the lagoon (EA Engineering 1997). Manzalah was thoroughly studied in the 1960s and 1970s. While we have a fairly clear picture of lagoon conditions prior to 1980, more recent data are sparse, and we have only a poor sense of how different parameters have changed in the last quarter century. One exception is that the sedimentology of Manzalah Lagoon is exceptionally well documented in a series of papers (e.g., Stanley and Warne 1993; Benninger et al. 1998; Randazzo et al. 1998) which represent a large and important body of work concerning our understanding of the evolution, not only of Manzalah, but of the entire Nile Delta. There are a number of papers addressing different aspects of Burullus and Edku lagoons, but, with a few notable exceptions (e.g., Dumont and El-Shabrawy 2007), many of them are quite narrow in scope and published over a wide range of time, such that our understanding of the dynamics of these systems is incomplete.

11.3 PHYSICAL, CHEMICAL, AND BIOLOGICAL CHARACTERISTICS OF THE DELTA LAGOONS

The delta lagoons are close to one another and share a similar origin and certain physical and climactic characteristics. Despite their wide range in size (50 to 500 km^2), the mean depth of each of the lagoons is ~1 m (Table 11.1). Because they are so shallow, lagoon water temperatures closely track air temperatures and vary little spatially. Data from Maryut Lagoon show a typical seasonal profile, with air and water temperatures ranging from about 15°C in the winter to a high of 30°C in the summer (Figure 11.2). Annual rainfall is about 10 to 20 cm, with almost no rainfall from April to October (Figure 11.3). Warm temperatures in this dry climate contribute to high rates of evaporation on the order of 160 to 170 cm yr^{-1}, or about 8 to 17 times the amount of rainfall. Depending on

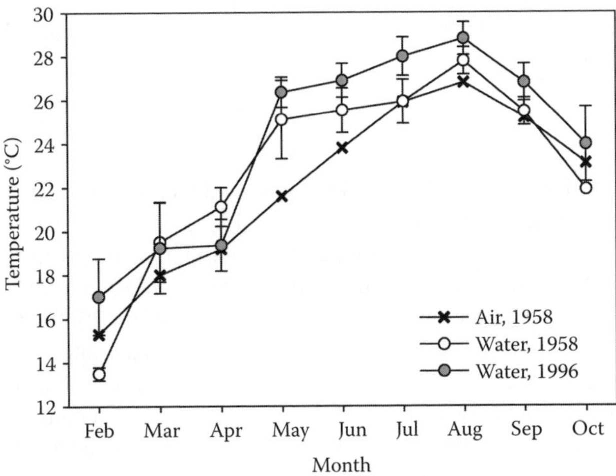

FIGURE 11.2 Maryut Lagoon mean monthly air and water temperatures (°C, ±SD) from 1958 (data from Wahby 1961) and mean water temperatures (±SD) in 1996. (Data from EA Engineering 1997.)

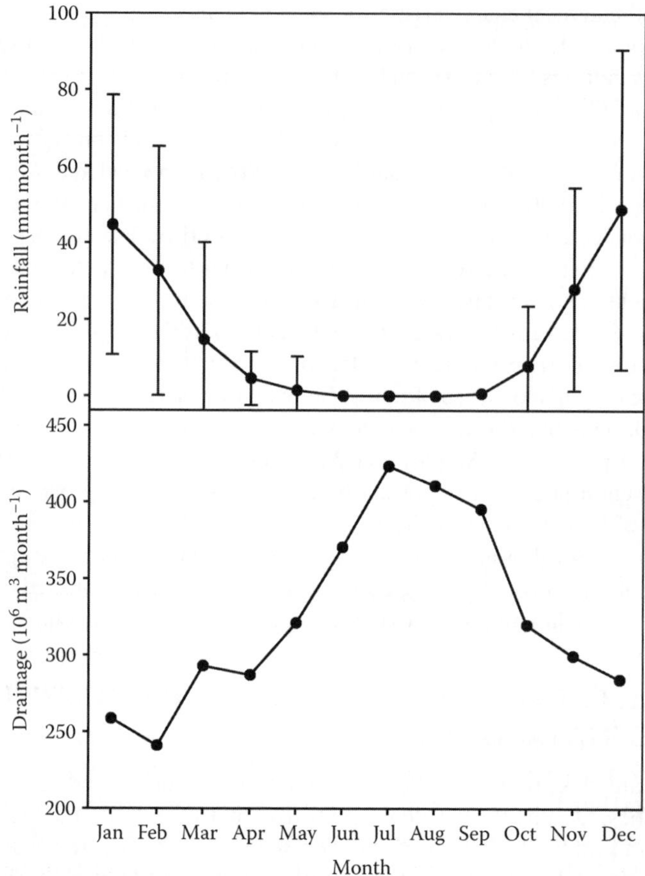

FIGURE 11.3 Top panel: Burullus Lagoon mean monthly rainfall (±SD). Bottom panel: Monthly agricultural drainage to Burullus Lagoon, averaged from 1997 to 2000. (Data from Al Sayes, A. et al. 2007. *Egyptian J. Aquat. Res.* 33: 312–351.)

the lagoon area and season, this translates to evaporation rates ranging from 10 m^3 s^{-1} to more than 50 m^3 s^{-1} (Aleem and Samaan 1969; El-Zarka et al. 1970; Toews and Ishak 1984; EA Engineering 1997; Mikhailova 2001).

11.3.1 Burullus

Burullus is currently the largest coastal lagoon (500 km^2) and is centrally located on the delta between the Rosetta and Damietta branches of the Nile (Figure 11.1). Separated from the Mediterranean Sea by sand dunes to the north, Burullus is surrounded by ever-encroaching fish farms along its southern margin (Radwan 2005). It is estimated that more than 35% of the lagoon area has been lost by conversion to fish farms and agricultural lands (Okbah and Hussein 2006). There are seven major drains discharging a mix of agricultural water, fish farm drainage, and mixed upstream sewage effluent into the lagoon (Figure 11.4). While these drains are distributed along the southern and eastern edges of the lagoon, there is also a large canal (Brimbal Canal) that connects the Rosetta branch of the Nile to the western edge of the lagoon. This canal used to provide about 50% of the water to the lagoon but, after the Aswan High Dam closed, this was reduced to 5% as flow in the Rosetta also declined rapidly (Dumont and El-Shabrawy 2007). Today, all of the drains combined provide about 97% of the incoming freshwater (the remaining 3% is precipitation and groundwater), discharging

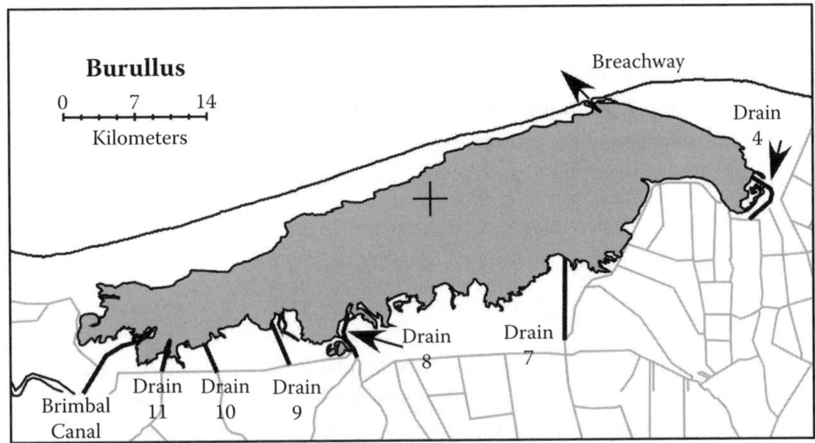

FIGURE 11.4 Burullus Lagoon with drain entrances shown in bold black lines (data from DCW and ESRI). Brimbal Canal connects Burullus with the Rosetta branch of the Nile. Crosshairs in the center of the image correspond to 30°50′E, 31°30′N.

an average of 2.5×10^9 m^3 yr^{-1}, with flows ranging more than threefold, from 80×10^6 m^3 mo^{-1} in the winter (January) to 272×10^6 m^3 mo^{-1} in the summer (July) (Table 11.1, Figure 11.3) (Okbah and Hussein 2006; Dumont and El-Shabrawy 2007). For some context, the Lagoon of Venice, which is of similar size (500 km^2) and depth (~1 m) to Burullus, is a well-known example of a heavily impacted lagoon. The freshwater inflows to this system are an order of magnitude less, ~0.5×10^9 m^3 yr^{-1} (Morin et al. 2000).

Although Burullus Lagoon is connected to the sea in its northeast corner through a breachway (locally called Boughaz El-Burullus), the lagoon surface is 25 to 60 cm above sea level (Dumont and El-Shabrawy 2007). A noticeable drop in lagoon level, down to 20 cm below sea level, is generally observed in both the summer and winter. The former drop is associated with high levels of evaporation and the latter with decreases in agricultural discharge (Al Sayes et al. 2007; Dumont and El-Shabrawy 2007)). In general, about 84% of the total lagoon freshwater input is discharged through the breachway, while the rest is lost to evaporation (Dumont and El-Shabrawy 2007).

The spatial distributions of water chemistry and water quality parameters are related to the seasons and the hydrologic gradients between drainage and sea water. In general, salinity in this lagoon ranges widely from 0.5 in the westernmost corner in the fall to 21 near the breachway in the winter (Table 11.1) (Okbah and Hussein 2006). In addition, the drainage water is not completely fresh, and salinity in the drains can range from 0 to 3 (Okbah and Hussein 2006). Data from 2003 show that both total suspended matter and transparency also varied by season, with total suspended matter an order of magnitude greater in the summer than in the winter (149 mg L^{-1} and 18 mg L^{-1}, respectively) (Radwan 2005; Okbah and Hussein 2006; Al Sayes et al. 2007). Transparency (Secchi depth) typically ranges from 20 to 40 cm, with greatest transparency in the fall, before winter wind-mixing and after the summer growing period (Radwan 2005). These convert roughly to vertical light attenuation coefficients of over 4 to almost 8 m^{-1}. Even in such a shallow system, very little light (<2%) reaches the bottom. While these seasonal variations in total suspended matter may be related to seasonal changes in productivity due to temperature variations, upstream growing seasons may also play an important role in these shifts. Not surprisingly, the total suspended matter in the drains is much greater than in the lagoon, and also shows seasonal variation from 225 mg L^{-1} during summer to 85 mg L^{-1} in the winter (Okbah and Hussein 2006). It seems likely that this allochthonous material contributes in some way to the suspended organic matter in the lagoon. In Egypt, as in most of the world, summer is the largest growing season when much greater amounts of water and fertilizers are used. Dissolved oxygen (DO) is higher in the winter than the summer (about 10 to 12 mg L^{-1}

vs. 5 to 6 mg L^{-1}). Presumably this increase is due to greater wind-mixing, which may also cause the observed decrease in winter water clarity, and an increase of oxygen solubility with decreasing water temperatures (Radwan 2005; Okbah and Hussein 2006; Al Sayes et al. 2007). While the suspended matter data do not distinguish between organic and inorganic fractions, the available data suggest that increased summertime runoff leads to high system productivity and phytoplankton growth, and winter wind-mixing may resuspend benthic organic matter and sediments.

11.3.2 Edku

Edku Lagoon lies just west of Rosetta and is currently about 50 km^2, less than half of its original size (120 to 130 km^2) (Table 11.1, Figures 11.1 and 11.5) (Banoub 1983; Zaghloul and Hussein 2000; Okbah and El-Gohary 2002). Unlike Burullus, where much of the lagoon has been lost to aquaculture, agricultural land reclamation is primarily responsible for the decline of Edku's surface area. Water movement is more dynamic along the western side of the lagoon, near the breachway (locally called Boughaz El-Maadiya). In contrast, the eastern half of the lagoon is heavily vegetated and much shallower (<1 m). It is generally believed that the flow through the breachway is stratified, with lagoon water flowing out along the surface, and marine water in along the bottom (Shakweer 2006). However, like Burullus, this lagoon is not in equilibrium with the Mediterranean as the water level of the lagoon is ~70 cm above sea level. This is reflected in a marked salinity gradient within the lagoon during all seasons except spring (Figure 11.6). Most of the seawater intrusion is thought to occur during the winter months, when wind-mixing is greatest (Shakweer 2006). There are two main drains that discharge into the eastern side of the lagoon, the Bersik and Kom Belag (Okbah and El-Gohary 2002; Shakweer 2006). The Kom Belag drain releases only about a sixth of the Bersik Drain's effluent. Together they discharge about 2×10^9 m^3 yr^{-1} and exhibit a similar seasonal pattern to the Burullus drains, although with a peak in discharge in the fall and not late summer as with Burullus (Table 11.1, Figure 11.7) (Shakweer 2006; Al Sayes et al. 2007).

Seasonal patterns in salinity are fairly consistent with our understanding of the hydrodynamics in Edku (Figure 11.6). El-Shenawy (1994) conducted seasonal measurements of salinity at five locations in the lagoon and found that salinity values near the breachway and in the easternmost portion of the lagoon (furthest from the breachway) are nearly identical in the spring, when agricultural

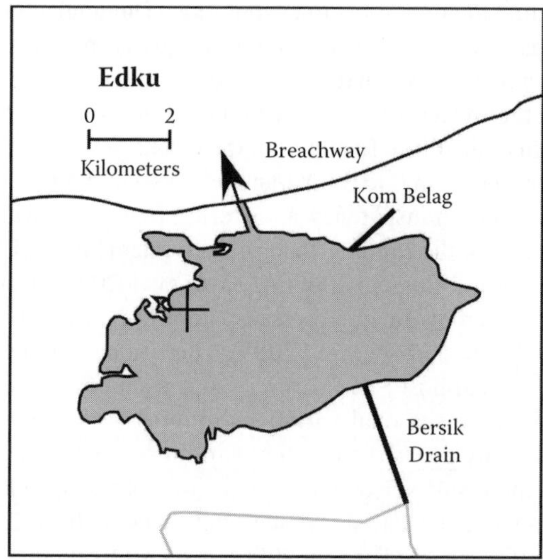

FIGURE 11.5 Edku Lagoon with the locations of the Kom Belag and Bersik drains shown in bold (data from DCW and ESRI). Crosshairs in the upper left corner of the lagoon correspond to 30°10′E, 31°15′N.

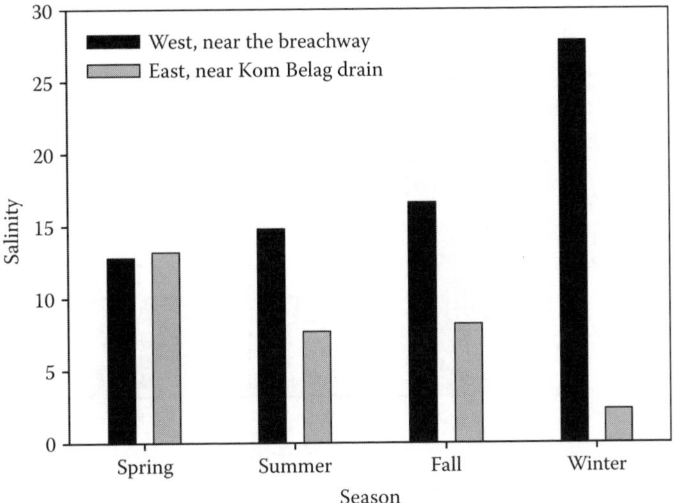

FIGURE 11.6 Edku Lagoon seasonal salinity from two stations (data from El-Shenawy 1994). Measurements were made twice per season in 1992.

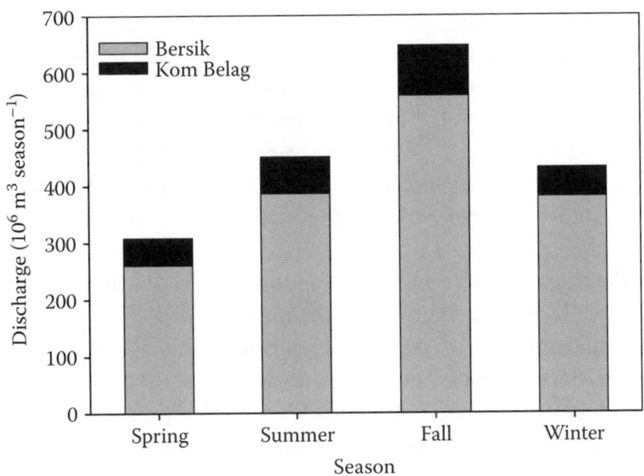

FIGURE 11.7 Edku Lagoon mean seasonal drainage in 2003 from the two main drains, Bersik and Kom Belag. (From Shakweer, L. M. 2006. *Egyptian J. Aquat. Res.* 32: 264–282. With permission.)

drainage is lowest. Throughout the rest of the year, salinity is greater near the lagoon–sea connection than in the east (Table 11.1; Figure 11.6). In the winter, when wind-mixing is presumably high, salinity near the breachway is an order of magnitude greater than the rest of the lagoon (27.8 vs. about 2.5 for the other four stations). While stormier weather may have driven more Mediterranean water into the lagoon, it does not explain why this high salinity water did not mix with the rest of the lagoon. Salinity measurements were made at least twice per season (exact frequency is not known), so a single storm event cannot explain this difference (El-Shanawy 1994). Further, one might expect large differences in water temperature to correspond to the winter differences in salinity as the Mediterranean seawater is likely cooler than the very shallow, productive lagoon waters. However, temperature variations are extremely small within Edku throughout the year (≤1°C), but especially

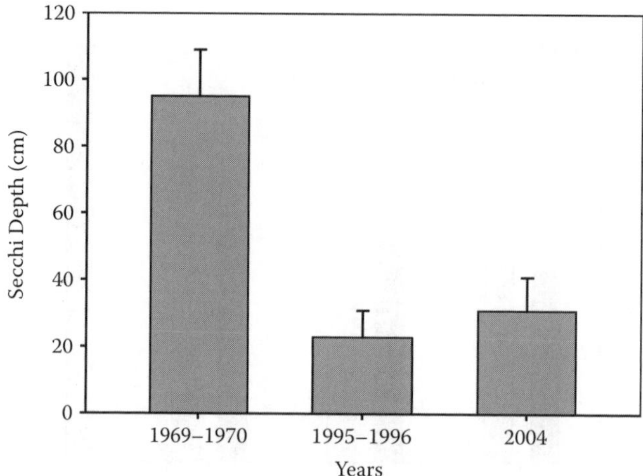

FIGURE 11.8 Mean Secchi depth (±SD) in Edku Lagoon between 1969 and 2004. Data were collected monthly from five stations between June 1969 and May 1970, monthly at three stations from June 1995 to April 1996, and from 20 stations in 2004 at an unknown frequency. (Data from Samman 1974; Soliman 2005; Shakweer 2006.)

in the winter when the station near the breachway was only 0.2°C cooler than the other stations (17.8°C vs. 18°C) (El-Shanawy 1994).

Based on five reports published from 1992 to 2004, variations in DO in this lagoon are great (up to fivefold within a given season), with no chronological patterns evident. Data from 1992 and 2000 ranged from 0.9 to 4.5 mg L^{-1} and were much lower than 1999 and 2004, which ranged from 6.7 to 9.9 mg L^{-1} (El-Shenawy 1994; Siam and Ghobrial 2000; Abbas et al. 2001; Okbah and El-Gohary 2002; Shakweer 2006). While some of these differences may arise from different sampling frequencies and/or number of sampling stations, they are not acknowledged in any of the publications.

While we cannot determine if DO concentrations have changed over time, there is some preliminary evidence that water clarity has declined significantly since 1970 (Figure 11.8). From June 1969 to May 1970, the bottom (at about 1 m) was visible throughout most of the year (Samman 1974). By the mid-1990s, however, Secchi depth ranged from an average of 30 cm near the breachway to 15 cm in the east near the agricultural drains (Soliman 2005). Similar values were observed in 2004, with slightly poorer water clarity in the east and poorest water clarity throughout the lagoon in the spring (~22 cm) (Shakweer 2006). Surprisingly, and unlike Burullus Lagoon, the greatest seasonal water clarity was in the summer (~36 cm), and nearly identical in the winter (~31 cm) (Shakweer 2006). The authors attribute low spring clarity to phytoplankton blooms and deeper summer and fall values to a lack of wind-mixing and general decrease in turbidity during these seasons.

11.3.3 Manzalah

Manzalah has the most complicated morphology of all the lagoons (Figure 11.9). It is also the subject of the widest body of literature, particularly pertaining to fisheries and sedimentology. Although most of this information is from before 1990, it provides an excellent picture of the physical and hydrodynamic characteristics of this ecosystem. Like all of the lagoons, Manzalah is shallow, but it is also mottled with over 1000 islands that have a total area of about 13 km^2 (Shakweer 2005). The islands fall into three basic morphological categories. The first group is in the north, parallel to the coast, and contains islands that are large, sandy, and flat (Randazzo et al. 1998). These are thought to represent ancient shorelines (El-Wakeel and Wahby 1970). The second category is in

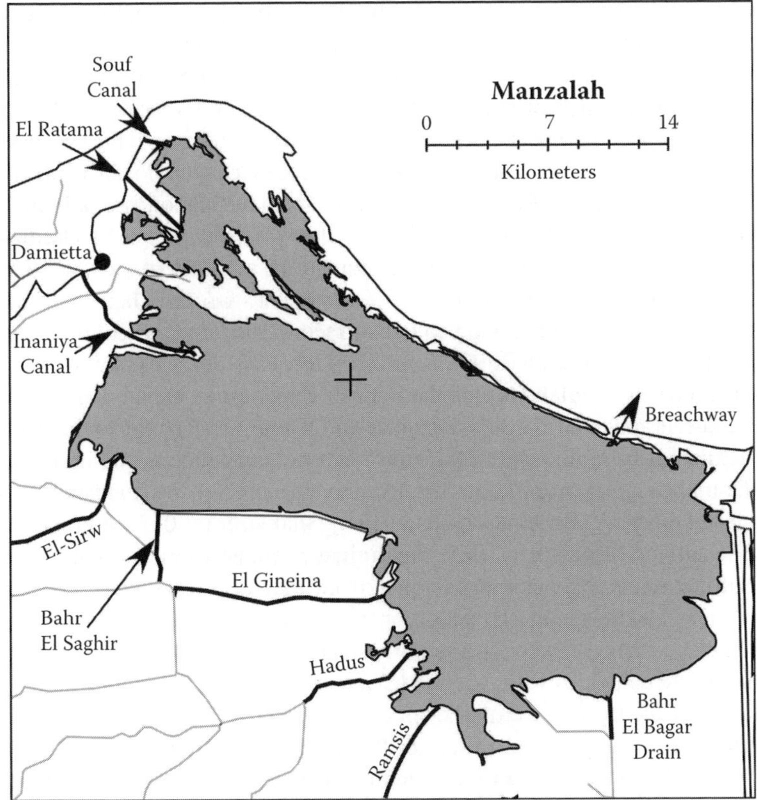

FIGURE 11.9 Manzalah Lagoon with drain entrances marked with bold black lines (data from DCW and ESRI). Note that spellings of drain names are approximate, as many different spellings are available in the literature. The Inaniya, El Ratama, and Souf canals connect the Damietta branch of the Nile with the lagoon. The location of the city of Damietta is also shown as a closed circle. The crosshairs at the center of the image correspond to 32°0′E, 31°20′N.

the central and eastern parts of the lagoon and oriented either north–south or northwest–southeast. The third type is in the south and smaller than the others. Islands in the latter two categories are mud rich and may be the depositional remnants of former branches of the Nile that used to flow through this lagoon (Randazzo et al. 1998). Shaheen and Yousef (1980) also report some small islands of mollusk shells in Manzalah comparable to Native American Shell Middens. There is archaeological evidence that these islands have been occupied for a long time, and most are still inhabited by fishermen and farmers (Randazzo et al. 1998; Shakweer 2005). Many of the islands appear to be continuously increasing in size as a result of increasing influx of agricultural drainage water and silt (Dewidar and Khedr 2001).

Manzalah Lagoon is bordered to the north by the Mediterranean Sea, on the west by the Damietta branch of the Nile River, and on the east by the Suez Canal (Figure 11.1). The connection to the sea, locally called the Boughaz El-Gameel, is in the northeast corner of the lagoon. There are three canals connecting the lagoon with the Damietta, all of which were originally constructed to decrease the salinity in the western part of the lagoon (Figure 11.9). The canals were opened prior to 1965 and used to supply freshwater only seasonally, during the Nile flood. Today, in the absence of the flood, they provide a permanent source of drainage water to the area (El-Wakeel and Wahby 1970). Manzalah is also connected with the Suez Canal a few kilometers south of the coastal city of Port Said, although it is not clear in which direction the water flows at this canal–lagoon interface.

Prior to the closure of the Aswan High Dam in 1965, groundwater and the Nile flood were the primary sources of freshwater to the lagoon. Since then, with the careful year-round regulation of Nile water and tile drained fields, virtually all of the freshwater inputs to the lagoon are from the Damietta canals and five main drains (Table 11.1, Figure 11.9) (Shakweer 2005). The Bahr El-Bagar and Hadus drains discharge 75% of the drainage water to the southeasternmost corner of the lagoon and provide about 3.3×10^9 m^3 yr^{-1} and 1.7×10^9 m^3 yr^{-1}, respectively. Bahr El-Bagar released 92% of the total nitrogen and 25% of the total phosphorus loads to the lagoon annually in 1979 and 1980 (6200 and 1700 t yr^{-1}, respectively; Toews and Ishak 1984). While the city of Damietta discharges about 90,000 m^3 d^{-1} of secondary treated sewage into the western side of the lagoon, it is the southeastern corner that is infamous for its pollution and noxious odors (Mageed 2007). Some have observed hydrogen sulfide (H_2S) bubbling to the surface in this area, where the average H_2S concentration is about 15 mL L^{-1} (Samir 2000). Approximately 65% of Cairo's municipal and industrial wastes are discharged to Manzalah through the Bahr El-Bagar drain, explaining the drain's nutrient-rich effluent (El-Sherif and Gharib 2001). Dewidar and Khedr (2001) refer to this southeast region, known as the Ghinka sub-basin, as a "black spot" rich in heavy metals and nutrients. Despite the apparent pollution, the lagoon overall does not seem to have problems with anoxia or hypoxia, with the lowest dissolved oxygen concentration observed by Shakweer (2005) reported at 6.3 mg L^{-1}.

There is some early evidence that, while the drainage inflows to the lagoon have been increasing over time, the amount of water evaporated from the lagoon's surface has been declining (Figure 11.10) (Toews and Ishak 1984). In addition, Cairo's sewage must have increased markedly in the 1980s with the expansion of water and sewer systems in the city (Nixon 2003). Not surprisingly, the increased drainage coincides with a general freshening of lagoon waters (Figure 11.11) (Toews 1988). The salinity of the lagoon declined from approximately 20 in the 1920s and 1930s to about 3 in 1980. For reference, the average salinity of the drains was approximately 1.3 during the later time (Toews 1988). Macrofossil, diatom, and sediment core data also suggest that barrages (small dams) on the Damietta branch of the Nile significantly increased freshwater inflows to this lagoon as early as 1912 (Appleby et al. 2001; Flower et al. 2001). However, what might not be immediately intuitive is the observed decline in the amount of water evaporated. As in the case of the other three lagoons, the surface area of Manzalah has been decreasing rapidly, resulting in less surface evaporation (Figures 11.10 and 11.12) (Toews 1988; Randazzo et al. 1998; El-Sherif and Gharib 2001).

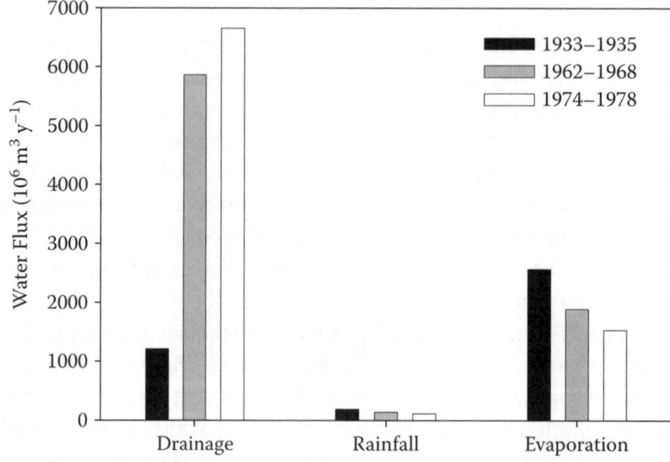

FIGURE 11.10 Manzalah Lagoon mean annual drainage, rainfall, and evaporation from the lagoon surface (data from Toews 1988). The earliest data from 1933 to 1935 are shown as black bars. Gray bars represent water fluxes from 1962 to 1968 and the open bars from 1974 to 1978.

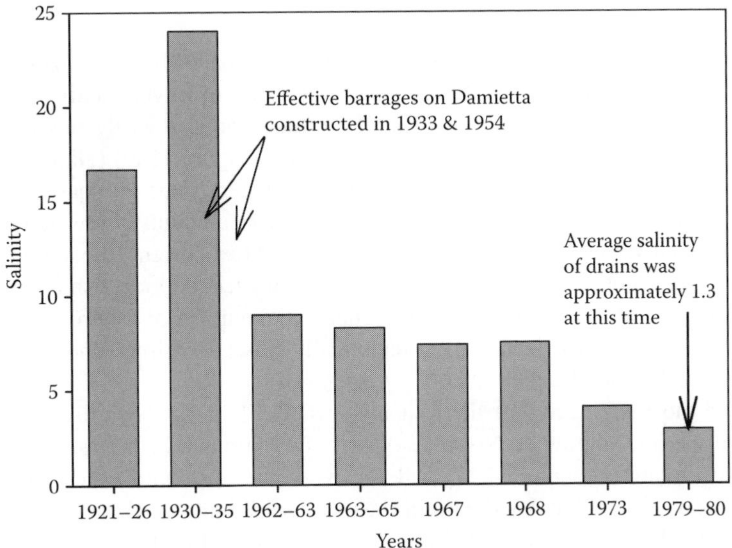

FIGURE 11.11 Decline in Manzalah Lagoon mean salinity between 1921 and 1980 (data from Toews 1988). Dams built in 1933 and 1951 diverted Nile water to agricultural fields, increasing the amount of agricultural drainage discharging into the delta lagoons (data from Appleby et al. 2001). The drops in salinity from 1933 to 1935 and in 1962 may reflect the increasing effectiveness of the dams. It is unlikely that the lagoon will ever become entirely fresh as the primary source of water to the lagoon (drainage) has a salinity >0.

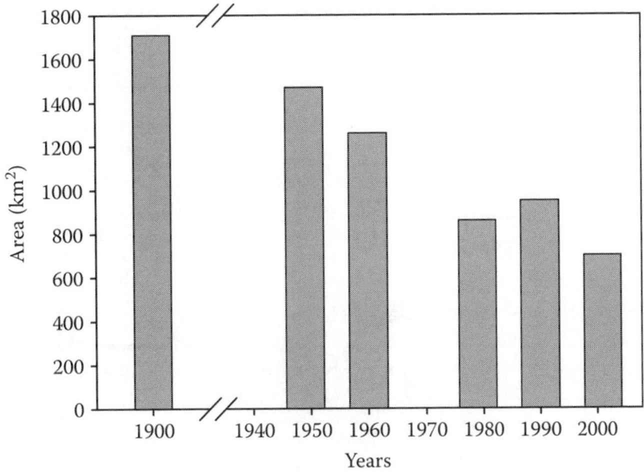

FIGURE 11.12 Manzalah Lagoon open water area decreased by about 60% from 1900 to 2000. (Data from Toews 1988; Randazzo et al. 1998; El Sherif and Gharib 2001.)

The loss of lagoon area was estimated at 5.2 $km^2\ yr^{-1}$ between 1922 and 1995, with the greatest loss along the western and southern borders (Dewidar and Khedr 2001). Others have estimated more recent lagoon area losses of 15 $km^2\ yr^{-1}$, suggesting that evaporation rates are even lower today (Reinhardt et al. 2001). As with the other lagoons, conversions of lagoon area to aquaculture ponds and agricultural lands, as well as the building of roads, are primarily responsible for the decline in area (Dewidar and Khedr 2001).

11.3.4 Maryut

Historically, Maryut is the most studied of all the delta lagoons. This is probably because it lies just inshore of the city of Alexandria, the second largest city in Egypt (behind Cairo), making it an important resource for more than 3.3 million Egyptians and also easily accessible to scientists at the local universities and government research agencies (Figure 11.13) (Nixon 2003). However, this lagoon has received much less scientific attention in recent years and has been dismissed by many as "entirely polluted" and a "dead lake" due to the large amounts of raw and primary treated sewage and industrial wastes it receives from Alexandria. Most current research has shifted away from ecology to studies of heavy metal and/or pesticide impacts on local fish populations. We are not so quick to dismiss Maryut as a lost cause. A substantial number of fishermen still fish from the lagoon's main basin, despite the fact that it is technically closed to fishing, and this region provides both a livelihood and habitat for a number of Egyptians.

Maryut is the most engineered of the four lagoons. While it has been divided into four main basins (Main, Fisheries, Southwest, Northwest), only the Main and Fisheries basins are addressed in the scientific literature (Figure 11.13). All receive some upstream agricultural drainage and untreated industrial effluent, but it is the Main Basin, which lies along Alexandria's southern shore, that is considered polluted and "dead." Prior to 1993, there were four large outfalls along this city–sea interface that discharged substantial amounts of raw sewage and untreated industrial wastes from more than 1200 manufacturing plants into the lagoon (Abdel-Shafy and Aly 2002; El-Rayis 2005). After 1993, almost all of the sewage was transferred to one of two treatment plants for primary treatment. The Western Treatment Plant discharges about 1.8×10^5 m^3 d^{-1} of primary effluent directly into the northeast corner of the Main Basin, while the Eastern Treatment Plant discharges 4.1×10^5 m^3 d^{-1} of primary effluent into an open drain (Kalaa) where it mixes with agricultural drainage, raw sewage, and industrial effluents. The mix then travels about 12 km before discharging into the southeastern corner of the Main Basin (as of 1997, discharge was 8.0×10^5 m^3 d^{-1}) (Table 11.1; EA Engineering 1997). Today, almost all of the sewage effluent is treated to some

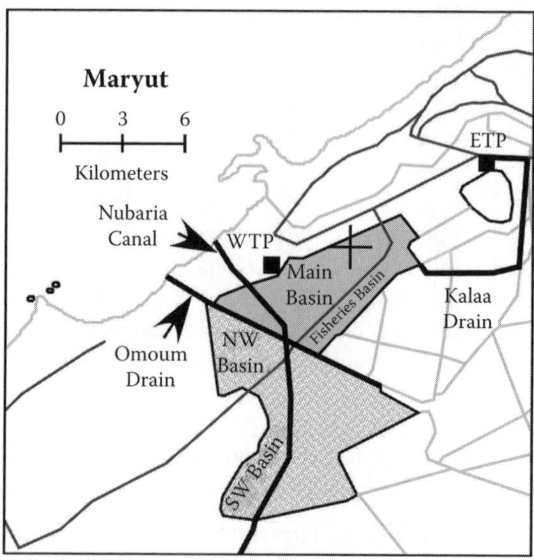

FIGURE 11.13 Maryut Lagoon with the Main and Fisheries basins indicated in gray and the northwest and southwest basins stippled (data from DCW and ESRI). The three main drains that release water into (Kalaa and Nubaria) or exchange water with (Omoum) Maryut are represented by bold lines. The West Treatment Plant (WTP) and East Treatment Plant (ETP) are also shown. The ETP discharges its effluent into Kalaa Drain before discharging into the southeastern corner of the Main Basin. The coastline and some of the smaller regional drains are shown as gray lines. The crosshairs correspond to 29°55′E, 31°10′N.

degree, although small local sources of untreated domestic and industrial wastewater continue to enter the lagoon directly at other locations (EA Engineering 1997).

None of Maryut's basins have a direct connection to the sea, and the water level in the Main and Fisheries basins is carefully maintained at 3.7 m below sea level to allow land drainage to flow into them. The very large Omoum drain divides the eastern two basins from the western two. This drain is both a source of nutrient-rich drainage water and a sink for Maryut Lagoon water. It is unclear exactly how much exchange of water occurs between Maryut and this drain, although numerous reports mention Omoum as a significant source of drainage water to the lagoon. A map in an early report shows underground pipes leading water from the drain to the lagoon, which the caption describes as "... underground pipe connecting the upper limit of the fresh water canal with Lake Mariut" (Figure 11.3 from Aleem and Samaan 1969). In addition, the sides of this open canal rise only about 0.5 m or less above the water surface, and local fishermen frequently create openings in these walls to allow the "fertilizing drain waters" to flow into the lagoon. At the northwestern-most corner of the Main Basin, surplus lagoon water flows back into the drain at a rate of approximately 7.2×10^5 m^3 d^{-1}. This mix of lagoon and drainage water is subsequently pumped out into the Mediterranean Sea at a rate of about 6×10^6 m^3 d^{-1} (El-Rayis and Abdallah 2005). The lagoon water going into Omoum drain is about 70% wastewater but, after mixing with the drainage, the sewage component dilutes to about 10% (El-Rayis and Abdallah 2005).

Because there is no direct sea connection, salinity is low throughout Maryut Lagoon, ranging from approximately 2 to 5 (Table 11.1) (EA Engineering 1997). This measurable salinity may be attributable to either marine groundwater intrusion or agricultural drainage and subsequent upstream desalinization of fields. Maryut was actually open to the sea at various points in its history, the most recent being in 1801, when the French, led by Napoleon, reopened a dyke to the sea (El-Wakeel and Wahby 1970; EA Engineering 1997). Despite dramatic changes in the spatial distribution and volume of drainage to the lagoon, the transparency of the lagoon has not changed much, with Secchi depths generally less than about 50 cm in both 1960 and in 1992, the most recent measurements reported (Aleem and Samaan 1969; EA Engineering 1997). In contrast, there was a drop in DO concentrations from the 1950s to the 1990s. We compiled all of the available mean annual DO measurements into decadal bins, and the results show a clear drop in concentration from 8 mg L^{-1} in the 1950s to a lagoon-wide average of less than 4 mg L^{-1} in the 1990s (Figure 11.14).

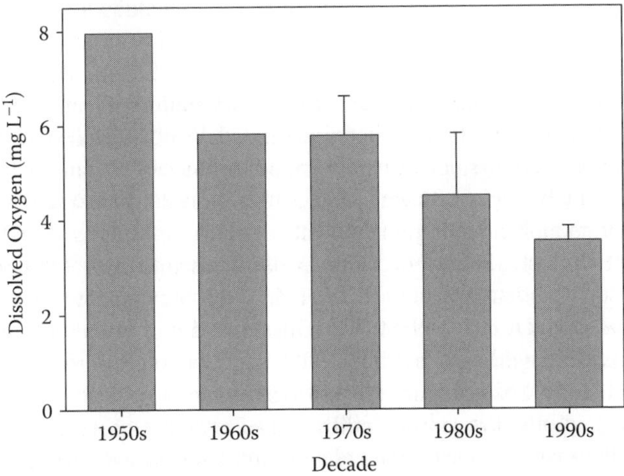

FIGURE 11.14 Maryut Lagoon mean decadal dissolved oxygen (DO) concentrations (±SD). The number of measurements varies among decades as do the number of stations and frequency of measurements. (Data from Wahby 1961; Aleem and Samman 1969; Saad 1983; Saad et al. 1984; EA Engineering 1997; Abdel Aziz and Aboul Ezz 2004; El-Rayis 2005.)

There is a clear east to west gradient in the Main Basin, with completely anoxic water near the sewage outfalls. Studies have reported corresponding H_2S odors along this eastern edge for decades (Saad et al. 1984; El-Rayis 2005), which we can also personally confirm.

11.4 LAGOON SUBSTRATES

Sediments in the three lagoons with an open exchange with the sea (Burullus, Edku, and Manzalah) range from coarse sands near the breachways to fine silty clays further inshore along the southern edges. The poorly sorted sands contain mollusk shells of marine origin, particularly *Cardium* (Randazzo et al. 1998; Al Sayes et al. 2007). Remains of barnacles and the tube worm *Mercierella* are also abundant (Samman 1974). Carbonate ($CaCO_3$) constitutes at least 50% of the total sediments, and organic matter is generally between 2% and 6% (Abdel-Moati and El-Sammak 1997). In Burullus Lagoon, $CaCO_3$ also comprises a substantial portion of the fine-grained sediments (12% to 34%), with the highest concentrations near the drain mouths. The silty clays in the drains themselves are 7% to 8% $CaCO_3$ (Khalil et al. 2007).

While Maryut Lagoon is now composed entirely of fine-grained sediments, the composition of its substrate has changed over time. In 1968, the lagoon bottom was covered in a poorly sorted sand–silt–clay mix, where the sand size fraction was made up entirely of shell fragments, mostly pieces of lamellibranches, gastropods, and tubeworms. About 40% of the sediment was composed of $CaCO_3$ and organic matter was 8% (El-Wakeel and Wahby 1970; El-Rayis 2005). By 1987, just two decades later, the organic matter content more than doubled to ~20%. With time, the suspended matter coming in from the former drains along the Main Basin's northern shore, and from Kalaa drain in the southeast, had formed a fine-grained, nutrient-rich sludge bed on the bottom of the Main Basin (El-Rayis 2005). These northern shore drains closed in 1993 and their effluent was routed to either the East or West Treatment Plants for primary treatment. By 1996, the organic matter content of the sediments adjacent to these former outfalls had decreased considerably to about 10%. Organic matter concentrations increased to >20% near the Western Treatment Plant, which discharges directly to the lagoon (El-Rayis 2005). Overall, the mean organic matter content for the whole lagoon was ~14% (El-Rayis 2005). Even with recent declines, overall organic matter content in Maryut is still at least twice that of the other lagoons, and Maryut is the only lagoon where the percent organic matter exceeds the percent carbonate (Abdel-Moati and El-Sammak 1997).

11.5 POLLUTANTS

With large amounts of agricultural drainage, sewage, and industrial waste entering the lagoons, associated pollutants such as pesticides, herbicides, and heavy metals could have a significant impact on the ecology of these systems. Despite the importance of agriculture in the region and the number of crops grown per year, pesticides are not heavily used. In the mid-1990s, Egypt applied pesticides at a rate of about 1.5 kg ha^{-1}, or half that applied in the United States at the same time (Pimentel et al. 1991; Dinham 1995; FAO 2008). Pesticide use is largely (65%) fungicides (Badawy 1998) and does not seem to be adversely affecting the lagoonal ecosystems (El-Hoshy and Mousa 1994; Abdel-Moati and El-Sammak 1997; EA Engineering 1997). A recent study of organochlorine insecticides and polychlorinated biphenyls (PCBs) in sediments from nine lagoons along the North African coast found low to undetectable levels of pesticides in surficial sediments from Burullus, Edku, and Manzalah lagoons (Peters et al. 2001). However, Peters et al. (2001) point out that their measured concentrations were at least three orders of magnitude lower than previously reported for other river and lagoon sediments. They attribute their very low levels to sample location, and lagoon heterogeneity. The sampling sites were not characterized by heavy human impact.

There have been many studies of various heavy metals in the sediments, water, and fish of the lagoons. Overall, Maryut and Manzalah clearly have the highest levels of heavy metals, while Edku and Burullus have low levels typically observed elsewhere in the region (Table 11.2) (Abdel-Moati

TABLE 11.2
Concentrations (μg g⁻¹, ± SD) of Various Metals in Sediments from the Delta Lagoons

Lagoon	Cd (μg g^{-1})	Cr (μg g^{-1})	Cu (μg g^{-1})	Fe (μg g^{-1})	Mn (μg g^{-1})	Pb (μg g^{-1})	Zn (μg g^{-1})
Burullus	5.2 ± 2.1		25 ± 17	17.9 ± 5.1	85 ± 22	14 ± 6	90 ± 79
Edku	7.3 ± 2.9		19 ± 9	23.6 ± 3.8	115 ± 80	20 ± 8	317 ± 179
Manzalah	11.8 ± 5.6		74 ± 89	35.9 ± 12.3	847 ± 644	79 ± 59	164 ± 238
Ghinka Sub-basin[a]	15 ± 4	221 ± 24	105 ± 11			99 ± 34	188 ± 73
Rest of Basin[a]	7 ± 2	148 ± 33	73 ± 25			65 ± 24	129 ± 55
Maryut	10.8 ± 3.4		574 ± 220	31.9 ± 9.5	598 ± 221	114 ± 55	229 ± 179
ERL[b]	1.2	81	34			46.7	150
ERM[b]	9.6	370	270			218	410

Note: Unless otherwise noted, data are from Abdel-Moati and El Sammak (1997). Note that metal concentrations in the Ghinka sub-basin of Manzalah lagoon and the rest of Manzalah lagoon are italicized. The Ghinka region receives effluent from two large drains, one of which discharges more than half of Cairo's sewage and industrial waste, and metal concentrations are much higher there than in the rest of the lagoon (El-Sherif and Gharib 2001; see text for further discussion). Effects Range-Low and Effects Range-Median concentrations published by the National Oceanic and Atmospheric Administration's (NOAA) National Status and Trends Program are included for reference.

[a] Data from Samir (2000).

[b] When concentrations are below the Effects Range-Low (ERL), adverse effects are not likely to occur. When concentrations exceed the Effects Range-Median (ERM), adverse effects frequently occur (Sediment Quality Guidelines developed for the National Status and Trends Program, 6/12/1999, http://response.restoration.noaa.gov/book_shelf/121_sedi_qual_guide.pdf).

and El-Sammak 1997; Samir 2000). Only copper (Cu) and cadmium (Cd) levels in Maryut sediments, and Cd in Manzalah sediments, exceeded the Effects Range-Median (ERM) concentration established for the United States by NOAA as a guideline beyond which adverse effects frequently occur (Long and MacDonald 1998). Some regions of the lagoons may show a slight enrichment in Cu associated with the use of $CuSO_4$ as an algicide to control massive blooms of floating plants and macroalgae in the Nile and large drains. However, while mean Cu concentrations in Maryut sediments were 574 ± 220 μg g^{-1}, values in fish were much lower, at about 12 ± 12 μg g^{-1} (Abdel-Moati and El-Sammak 1997; Adham et al. 1999). Although below ERM concentrations, zinc (Zn), lead (Pb), and chromium (Cr) levels in Manzalah and Maryut do exceed the Effects Range-Low (ERL), and adverse effects may occur (Long and MacDonald 1998).

Edku and Burullus receive only drainage water and have no large cities or industrial centers along their banks. With the exception of Zn in Edku, metal concentrations were below the ERL and adverse effects are not likely (Long and MacDonald 1998). Overall, concentrations of Pb, Cu, and Zn in sediments in all four lagoons were generally less than those observed in Boston Harbor between 1970 and 1995, when sludge and sewage effluent from the city of Boston were being discharged into the system (Zn 224 μg g^{-1}, Cu 111 μg g^{-1}, Pb 136 μg g^{-1}; Bothner et al. 1998). One exception to this was Maryut Lagoon, where Cu concentrations were more than five times those observed in Boston Harbor sediments (Abdel-Moati and El-Sammak 1997; Bothner et al. 1998).

In the two lagoons that have the highest metal concentrations, Manzalah and Maryut, the distribution of metals in the sediments is spotty and metal contamination does not appear to have clear adverse impacts on commercial fish species. Alexandria is Egypt's industrial center, and the highest concentrations of metals were observed in sediments in front of the old outfalls closed in 1993 and off the West Treatment Plant, which discharges its effluent directly into the northeastern corner of Maryut Lagoon (Abdel-Moati and El-Sammak 1997; El Rayis 2005). While some of these metals have been observed in detectable to low levels in fish near the sediment "hot spots," factors like low DO are probably more responsible for fish disease and death (Adham et al. 2001). Serum analyses suggest that, while heavy metal contamination may have some sublethal effects on fish, the lagoon fish were healthier and less stressed than those from nearby fish farms, which have to contend with competition for food and overcrowding (Adham et al. 2001). Catfish (*Clarias Lazera*) caught in the western portion of Maryut's Main Basin, away from these hotspots, often had lower metal concentrations than catfish caught in Edku Lagoon (Adham et al. 1999). Sediments from the northwestern corner of this basin were also low (El-Rayis and Abdallah 2005). As with Maryut, sediments near the Bahr El-Baquar drain (which discharges into Manzalah's Ghinka sub-basin) had much higher metal concentrations than the rest of Manzalah Lagoon (Table 11.2; Figure 11.9). As noted earlier, this area is anoxic and H_2S has been observed bubbling to the water surface.

Fine-grained sediments near current or recent industrial outfalls and drains that are known to discharge large amounts of industrial and urban waste are likely enriched in heavy metals. However, the regions near these outfalls are generally inhospitable to higher-trophic-level organisms (including commercially important fish), as they also tend to be anoxic with high levels of H_2S. While metal concentrations in sediments in these hot spots may exceed permissible levels, concentrations quickly decrease to acceptable levels with distance (Abdel-Moati and El-Sammak 1997; Samir 2000; El Rayis 2005).

11.6 NUTRIENT DYNAMICS

Dissolved inorganic nitrogen (ammonium, NH_4^+; nitrite, NO_2^-; and nitrate, NO_3^-; or DIN) and phosphorus (PO_4^-, DIP) concentrations in all four lagoons have been rising since the late 1950s, especially in Maryut Lagoon. Prior to about 1970, combined $NO_3^- + NO_2^-$ was generally less than 10 µM in the lagoons. However, the most recent data from the 1990s to 2003 reported DIN concentrations more than double earlier observations and greater than 10 µM (Figure 11.15). In Maryut Lagoon, concentrations increased more than 50 times to very high values (>500 µM DIN). Concentrations exceeding 100 µM are extremely rare for coastal systems, and we know of only one other lagoon that has had DIN concentrations this high (Nixon 1983; Nixon et al. 2007; Oczkowski and Nixon 2008). Korle Lagoon in Ghana drains the capital city of Acraa and has DIN concentrations on the order of 1000 µM (Nixon et al. 2007). Recent DIP concentrations range from <4 µM in Burullus Lagoon to greater than 50 µM in Maryut (EA Engineering 1997; El-Rayis 2005; Okbah 2005; Radwan 2005; Al Sayes et al. 2007). DIP concentrations are less than DIN, and a comparison of mean annual concentrations suggests that primary production in all four lagoons may be nitrogen limited (Oczkowski and Nixon 2008).

The high concentrations of DIN in Maryut are probably due to a number of factors. First, Maryut receives all of the city of Alexandria's sewage. While there is one report that sewage was first introduced into this lagoon in 1981, it seems possible that lesser amounts of sewage were entering the lagoon far earlier, as Alexandria lies on the narrow strip of land between the lagoon and the sea. Alexandria's population has grown exponentially over the past five decades, but wastewaters were not collected and treated widely until the early 1990s (Nixon 2003; EA Engineering 1997; El-Rayis 2005). Through USAID sponsored efforts, sewerage infrastructure, including two primary sewage treatment plants (East and West Treatments Plants), was established by 1993 and the effluent was temporarily discharged into Maryut Lagoon (EA Engineering 1997). Unfortunately, no agreement was reached as to where the sewage should be discharged before funding ran out, and Maryut has remained the recipient for over a decade. This has had a profound effect on the lagoon water quality. In regions close to the drains, dissolved oxygen is essentially zero, and hydrogen sulfide concentrations are high. The only inputs of water to Maryut Lagoon are from sewage and some agricultural drainage, both nutrient-rich sources, and outputs are denitrification, evaporation, and whatever water is pumped out to sea.

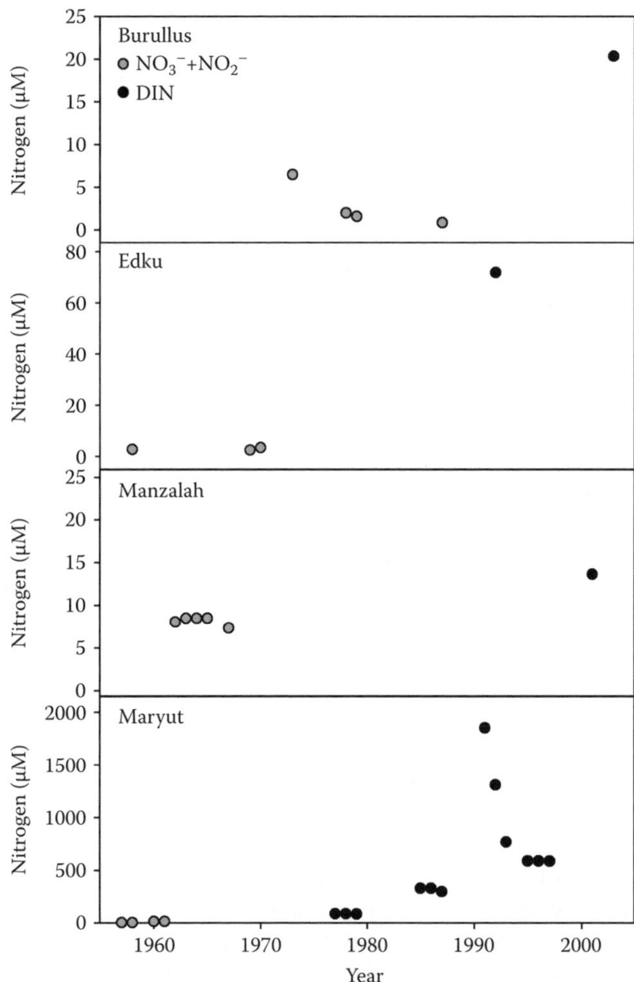

FIGURE 11.15 Mean annual inorganic nitrogen concentrations in the delta lagoons from 1955 to 2005. Unfortunately, ammonium (NH_4^+) was not widely measured prior to 1990. Combined nitrate and nitrite (NO_3^- + NO_2^-) concentrations are shown as gray circles and dissolved inorganic nitrogen (DIN, or NO_3^- + NO_2^- + NH_4^+) concentrations are black circles. Note the different scales on the Y-axes. (Data from Wahby 1961; El-Wakeel and Wahby 1970; Wahby et al. 1972; Samman 1974; Shaheen and Yousef 1980; Halim and Guerguess 1981; Banoub 1983; Saad et al. 1984; Toews and Ishak 1984; Saad 1987; El-Sherif 1993; El-Shenawy 1994; EA Engineering 1997; Abdallah 2003; El-Rayis 2005; Okbah 2005; Shakweer 2005.)

While Maryut certainly receives the most direct sewage effluent, all of the lagoons receive some waste, either directly or indirectly. Small villages and towns on the shores of the lagoons, such as Baltim on Burullus and El-Mataraya on Manzalah, discharge some untreated waste directly into the lagoons (El-Sherif and Gharib 2001; Okbah and Hussein 2006; Al Sayes et al. 2007). In addition, cities and towns upstream have little to no sewage treatment infrastructure, and virtually all of this sewage is released into septic tanks or drainage canals. Unfortunately, not much is known about the amount and nutrient concentrations of these discharges. The annual load of nitrogen and phosphorus (N and P, presumably total) from drains to Burullus lagoon are about 2300 and 560 t yr^{-1}, respectively, and these loads have quadrupled DIP concentrations between 1970 and 1987 (Okbah and Hussein 2006). Drains to Edku supply about 1100 t of DIN and 650 t of DIP annually (Shakweer 2006). To provide some context for these values, Italy's Venice Lagoon, a well-known eutrophied system, receives N and P loads of ~8000 t and 2500 t annually from rivers and urban sewage (Morin et al. 2000). This lagoon is also shallow (~1 m) and of similar size (550 km^2; Table 11.1) to Manzalah, where the Bahr El-Bagar drain alone released 6200 t of N and 1700 t of P to the lagoon annually in 1979 and 1980 (Toews and Ishak 1984).

The delta lagoons appear to act as tertiary treatment plants, removing some unknown amount of N and P introduced through drains and sewage before discharging lagoon water out to sea. A recent study using N and carbon (C) stable isotopes ($\delta^{15}N$, $\delta^{13}C$) to examine the potential influence of anthropogenic nutrients on higher trophic levels found a correlation between $\delta^{15}N$ values in fish tissues and estimated lagoon residence time, where lagoons with longer residence times had heavier $\delta^{15}N$ values (Oczkowski et al. 2008). This suggests that N removal processes, such as denitrification and uptake, may be important as these processes preferentially select for the lighter N isotope (^{14}N), leaving the remaining bioavailable N enriched in ^{15}N.

Further evidence was seen in Maryut, a lagoon with one of the longest estimated residence times (~45 days). Extraordinary nitrite (NO_2^-) concentrations (>40 μM), more than 10 times greater than ammonium (NH_4^+) concentrations, were measured in a few discrete water samples. Such high NO_2^- concentrations may occur when the reduction of nitrate (NO_3^-) to NO_2^- occurs more quickly than the conversion of NO_2^- to N_2 gas, or when the oxidation of NO_2^- to NO_3^- occurs more slowly than nitrification (the oxidation of NH_4^+ to NO_2^-). The potential importance of either process suggests that this system is probably a highly dynamic one for N (Oczkowski et al. 2008).

11.7 PRIMARY PRODUCERS

The delta lagoons appear to provide a hospitable environment for phytoplankton with optimal conditions for growth throughout the year. As described by Halim and Guerguess in 1981, p. 160, "Blooms occur at all times in the lake [lagoon], irrespective of variations in temperature or incident light. Neither the reduction in the rate of drainage inflow … nor its increase … appear to reduce or induce the blooms." Phytoplankton composition in the delta lagoons appears to be diverse, with at least three times the number of taxa present and higher cell densities than in other coastal lagoons on the North African coast in Morocco and Tunisia (Fathi et al. 2001; Flower 2001). Overall the phytoplankton composition of the four lagoons is fairly similar, where about a third to half of the phytoplankton are diatoms (Bacillariophyceae dominated by *Nitzschia* sp.) and a third to half are green algae (Chlorophyceae, mostly *Carteria* sp.) (El Sherif 1993; Zhagloul and Hussein 2000; El Sherif and Gharib 2001). The dominance of these various groups shifts both seasonally and spatially. For example, although the main phytoplankton species are ubiquitous throughout Manzalah Lagoon, there is an east to west variation in phytoplankton composition from diatoms in the most eutrophic region (east) to green and blue green algae characteristic of Nile phytoplankton in the west (Halim and Guerguess 1981). There are spatial gradients in both composition and cell counts in all four lagoons associated with proximity to cities and their waste, agricultural drains, and breachways. Regions with the highest cell counts are close to cities that discharge their waste directly into the lagoons (El-Sherif 1993; Okbah and Hussein 2006), while the lowest phytoplankton counts and chlorophyll concentrations are frequently found near the agricultural drains (Radwan 2005; Okbah and Hussein 2006).

Cell abundance data are sparse and variable, but counts from Edku and Manzalah appear to be increasing over time, while the other two lagoons appear to have remained the same or have declined (Table 11.3). The lagoon with the highest recorded phytoplankton counts (Maryut) also has nutrient concentrations two orders of magnitude greater than the others. Cell counts in the lagoons are comparable to more temperate systems, such as Narragansett Bay, Rhode Island (USA), where counts are generally less than 5×10^6 cells L^{-1} for most of the year, but can exceed 25×10^6 cells L^{-1} during the spring or fall blooms (Nixon et al. 2009). In at least two of the lagoons (Manzalah and Maryut), species diversity seems to have decreased (Abdalla et al. 1991; El-Sherif and Gharib 2001). The number

TABLE 11.3
Phytoplankton Cell Counts in the Nile Delta Lagoons over Time

Year	Burullus	Edku	Manzalah	Maryut
	(10^6 cells L^{-1})			
1976–1977		0.02		
1978	2.7			
1979	3.4			
1982				56
1986–1987			2.3	
1987–1988	1.04			
1990			3.9	
1992–1993			12.4	
1995–1996		7.5		>7.6
2000	0.7	4.5		
2003	1.9			

Note: Data from Ibrahim (1989); El-Sherif (1993); Gharib (1999); Zaghloul and Hussein (2000); Abdalla (1991); EA Engineering (1997); El-Sherif and Gharib (2001); Radwan (2005); Okbah and Hussein (2006).

of species in Manzalah Lagoon has declined from 170 in 1986–1987 to 140 in 1992–1993 (El-Sherif and Gharib 2001). Similarly, the number of taxa in Maryut Lagoon declined from 95 in 1960–1961 to 65 in 1982, and the plankton assemblage shifted from primarily diatoms to green algae (Abdalla et al. 1991). Many of the dominant species present year round in 1982 are characterized as pollution tolerant and are frequently used as indicators of organic pollution (e.g., *Merismopedia tenuissima*, *Planktosphaeria gelatinosa*, and *Crucigenia tetrapedia*) (Abdalla et al. 1991).

Unlike other lagoons described in this volume, these systems are not completely open water systems; rather, they are a complex continuum of shallow open water, floating and submerged aquatic vegetation, wetlands, and *Phragmites* (*australis* or *communis*) and *Typha australis* fringed islets (Samman 1974; Zagholoul and Hussein 2000; Dewidar and Khedr 2001). At the edges of the lagoons, channels are cut through the dense *Phragmites* for boats to pass through to more open waters. It has been our experience in Manzalah Lagoon to observe many of the small islands, often smaller than 1 km^2, populated with fishermen, their families, and a variety of livestock. There is almost no information available as to how many islands there are in each of the lagoons, but one source estimates over 1000 in Manzalah alone (Shakweer 2005).

Phragmites, *Typha*, and, to a lesser extent, *Arundo* and *Cyperus* (papyrus) cover areas shallower than about 50 cm (Samman 1974; Shaheen and Yousef 1980), beyond which submerged and floating plants dominate. In Burullus, emergent macrophytes form a band up to 200 to 330 m wide along the lagoon's northern shore (Ramdani et al. 2001). The more abundant submerged plants are *Potamogeton pectinatus* and *Ceratophyllum demersum* (Samman and Aleem 1972; Halim and Guerguess 1981; Aboul Ezz and Soliman 2000; Shakweer 2006). Perhaps the most widespread floating plant in the lagoons, and in all of Egypt, is *Eichornia crassipes*, the Nile Lily. This plant is notorious for clogging agricultural drains and canals and for making navigation very difficult in the lagoons. A more recent addition to the floating plant community of Edku Lagoon (and perhaps others) is the nitrogen-fixing *Azolla filiculoides*. It was observed for the first time in Edku in 1991 and in 1992, when it covered more than half of the lagoon area (El-Shenawy 1994).

While the lagoons are extremely shallow, albeit with Secchi depths frequently less than one meter and high suspended organic matter concentrations, the possibly important role of epibenthic algae appears not to have been studied.

11.8 ZOOPLANKTON AND BENTHIC INVERTEBRATES

Despite their physical separation, the lagoons appear to have similar zooplankton populations. Rotifers, especially species of *Brachionus* (*B. angularis, B. plicatilis, B. calyciflorus, B. urceolaris*), are most abundant (Aboul Ezz and Soliman 2000; Mageed 2007). These small, typically freshwater zooplankton comprise at least 70%, and often greater than 90%, of the total population in the lagoons (Aboul Ezz and Soliman 2000; Abdel Aziz and Aboul Ezz 2004; Mageed 2007). Other small animals that are present in smaller amounts include Cladocera (mainly *Moina micrura*), Copepoda, larvae of benthic invertebrates, and Ostracoda. In their study of zooplankton communities in Edku Lagoon, Aboul Ezz and Soliman (2000) found that, of the remaining zooplankton not identified as *Brachionus* (~25%), 16% were Copepoda, and most of these were nauplii of the genus *Acanthocyclops*.

General zooplankton abundance has increased in the lagoons. Gharib and Soliman (1998) report a 44-fold increase in zooplankton counts (from 6.1×10^4 to 2.7×10^6 individuals m^{-3}) in Edku Lagoon from the mid-1970s to the mid-1990s. In Manzalah, counts increased about 19-fold from 63×10^3 individuals m^{-3} in 1977 to 936×10^3 individuals m^{-3} in 2000–2001 to 1212×10^3 individuals m^{-3} in 2003 (Mageed 2007). Counts in Burullus have also increased about 10-fold, with seasonal counts from 2001 to 2004 rarely falling below 4×10^5 individuals m^{-3} and peaking at 2.5 to 3.0×10^6 individuals m^{-3} (Dumont and El Shabrawy 2008). In the mid-1990s, Maryut had a population of about 3.6×10^5 individuals m^{-3} despite tremendous amounts of sewage input and chronically low DO (annual average of 3.9 mL O_2 L^{-1}; Abdel Aziz and Aboul Ezz 2004). Zooplankton populations are quite high in all of these lagoons, particularly when compared to the heavily impacted Venice Lagoon where, despite similar physical characteristics, mean seasonal zooplankton counts peak at about 1.9×10^4 individuals m^{-3} in the summer and typically do not exceed 1×10^3 individuals m^{-3} throughout the rest of the year (Bianchi et al. 2000).

While zooplankton population data are inherently very dynamic, there have been some clear shifts in the community composition over time. Soliman (2005) has reported a rise in rotifers from 56% of the total counts in 1976–1977 to 92% in 1995–1996, as well as a decrease in copepods (from 32% to 7%) during this same time in Edku Lagoon. Dumont and El Shabrawy (2008) have thoroughly documented similar shifts in the lagoons, and in particular Burullus lagoons, from the 1930s to the present. These authors describe how, in the 1930s, the zooplankton were marine for most of the year and then shifted to freshwater assemblages during the four months during and following the fall flood. Since then, the marine species have disappeared from most of the lagoons, except from areas close to the breachways. Further, all large-bodied freshwater species have disappeared and small rotifers and cladocerans have increased by a factor of 10 or more (Dumont and El Shabrawy 2008). The authors attribute these changes to the shift from a marine fishery to a freshwater one dominated by omnivorous tilapia, and to increased agricultural drainage. With the closure of the dam, the increased nutrient-rich agricultural drainage and associated eutrophication favored smaller species. Increased turbidity was likely detrimental to the large filter feeders, including crustaceans, and favored microfiltrators like rotifers (see Dumont and El Shabrawy 2008 for a more thorough review).

Much less is known about the benthic invertebrate populations. One study conducted in Maryut in 1960 listed some of the most common species as *Nereis diversicolor* (a polychaete), *Melania tuberculata* (gastropod), the amphipods *Corophium orientalis, C. volutator, Gammis locusta, G. oceanicus,* and *Chironomid* larvae (Samaan and Aleem 1972). We do not know which, if any, of these species are present today. In the 1930s, typically marine species of benthic invertebrates, including mysids, two polychaetes (*Nereis* sp.), three amphipods (including *Corophium volutator*), and a serpulid (*Mercierella enigmatica*) were reported in Burullus Lagoon (Dumont and El-Shabrawy 2007). More recently, Reinhardt et al. (2001) provided a list of benthic infaunal taxa from surficial sediment samples collected from Manzalah in 1990, and it included only two of the species described above, notably *M. enigmata* and *Melania tuberculata* (as *Melanoides tuberculata*). Tables of molluscan species and their abundances in Edku, Burullus, and Manzalah lagoons are given in Bernasconi and Stanley (1994). Unfortunately, living mollusks were not described, although the distribution of shells indicated distinct marine, freshwater, and lagoonal influences (Bernasconi and Stanley 1994).

11.9 LAGOON FISHERIES

The Nile Delta lagoons produce between 30% and 60% of Egypt's national fish catch (El-Karachilly 1991; Radwan and Shakweer 2004; GAFRD 2007; Oczkowski and Nixon 2008), and the two largest lagoons (Burullus and Manzalah) provide about 90% of the total lagoonal landings (Figure 11.16).

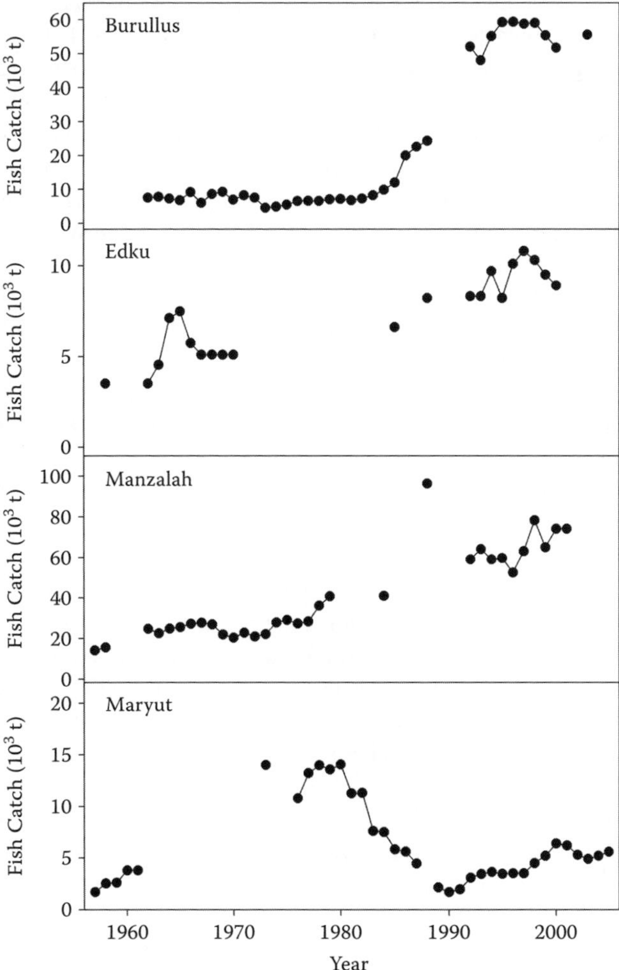

FIGURE 11.16 Mean annual commercial fish landings for each of the lagoons from 1957 to 2005. Note the different scales on the Y-axes. (Data from El-Zarka et al. 1970; Hashem et al. 1973; Toews and Ishak 1984; Fattouh 1990; El-Karachilly 1991; Thomas et al. 1993; EA Engineering 1997; Megapesca 2001; Abdallah 2003; Hoevenaars and Slootweg 2004; Alsayes 2005; WADI Project 2006.)

While Edku and Maryut are an order of magnitude smaller in size and today contribute little to the overall commercial fish catch, this has not always been the case. Maryut, in particular, contributed about 30% of the total lagoonal catch in 1980, but today provides just 5%. This change results from a decline of Maryut's fishery and the increase of the landings in the other lagoons (Figure 11.16). While total lagoonal landings have been increasing in the Burullus, Edku, and Manzalah lagoons, Maryut had a rise in its landings from 1960 to 1980 and then a dramatic decline from 1980 to the present. This decline does not appear to be related to fishing effort, as the number of fishermen in the lagoon increased only slightly during the period of major fishery decline—from 3200 in 1980 to 3900 in 1988 (EA Engineering 1997; Rashed 2001). Maryut is the most depressed fishery, with yields of ~0.4 t of fish per person, or an order of magnitude less than in the 1970s, before the fishery collapsed (EA Engineering 1997). Fishing in Manzalah appears to be the most profitable, with fishermen catching 4 t per person in the late 1980s, while Burullus and Edku yielded about half as much (1.8 t per person) during this same time (El-Karachilly, 1991).

Fishing in the lagoons remains largely artisanal. Few boats are mechanized and most fishermen still use oars, poles, or sails, and fish are primarily caught using barrier traps, seines, submerged traps, trammel nets, and longlines for catfish (Toews and Ishak 1984; EA Engineering, 1997). Tilapia dominate the lagoonal fisheries with four species present (*Oreochromis aureus, O. niloticus, Sarotherodon galilaeus*, and *Tilapia zillii*). Nile tilapia (*O. niloticus*) is the most abundant (Toews and Ishak 1984; Shakweer and Abbas 2005; Al Sayes et al. 2007) and comprises between 75% and 94% of the total catch in these lagoons (Toews and Ishak 1984; EA Engineering 1997; Al Sayes et al. 2007). Tilapia are herbivorous/omnivorous feeders with flexible feeding habits and wide environmental tolerances, which probably gives them an advantage in stressed systems and makes them the dominant aquaculture fish in Egypt (El-Sayed 2006).

Mullet and catfish also contribute in a small way to the lagoonal landings. In Burullus Lagoon, 2% and 4% of the catch are mullet and catfish, respectively (Al Sayes et al. 2007). Toews and Ishak (1984) observed similar contributions of 2% and 5% to the overall catch (by weight) of mullet and catfish in Manzalah Lagoon. Landings consist of four types of mullet (*Liza auratus, L. ramada, L. saliens,* and *Mugil cephalus*) and two main species of catfish (*Bagrus bayad* and *Clarias Lazera*). While mullet and tilapia are both omnivorous feeders, mullet are catadromous, often living in estuarine or near-shore environments for much of their life cycle, but spawning at sea (Blaber 1997; Froese and Pauly 2007). Catfish are generally omnivorous, but feed higher on the food chain than tilapia and mullet (Froese and Pauly 2007).

While three of the lagoons have open exchange with the sea, recent stable isotope data ($\delta^{13}C$) suggest that the catfish and tilapia are consuming carbon originating from freshwater phytoplankton, terrestrial C_3 plant material, and little to no marine phytoplankton (Oczkowski et al. 2008). Unfortunately, contributions from benthic algae are unknown. The $\delta^{13}C$ results are consistent with our conceptual understanding that the main source of carbon to these fisheries is from freshwater. The $\delta^{15}N$ values in the fish were heavy and indistinguishable among genus, suggesting that freshwater phytoplankton, rather than synthetically fertilized agricultural detritus, is probably an important food source. Tissue $\delta^{13}C$ values of mullet from the lagoons sometimes reflected freshwater and terrestrial sources, but on other occasions had values similar to offshore Mediterranean fish, which is consistent with their catadromous lifestyle. The mullet $\delta^{15}N$ values were also consistent with these observations (Oczkowski et al. 2008).

Other species (mainly marine or euryhaline) are occasionally recorded in the lagoon catches, but in extremely low quantities. These include carp, gilthead seabream (Denis, *Sparus aurata*), European seabass (Karous, *Dicentrarchus labrax*), sole (Samak Mousa, *Solea* spp.), meagure (Loot, *Argyrosomus regius*), grouper (Wakkar, *Epinephelus* spp.), and eel (Hensahn, *Anguilla anguilla*) (GAFRD 2007; Oczkowski et al. 2009; Abdel-Fattah El-Sayed personal communication).

11.10 LINKING NUTRIENTS AND FISHERIES

Despite, or perhaps because of, tremendous alterations to these lagoon ecosystems, they remain highly productive. Each is losing surface area rapidly, each receives large amounts of agricultural drainage water and, in some cases, direct sewage inputs. Probably in response to nutrient enrichment, phytoplankton cell counts, in at least some of the lagoons, are increasing (Table 11.3), and fish yields (per unit area) are much greater than capture fisheries in marine systems. While capture fisheries typically do not exceed yields of 40 t km^{-2} yr^{-1}, these lagoons produce between 90 and 160 t km^{-2} yr^{-1}, or levels comparable to intensive aquaculture (Nixon 1988; Megapesca 2001; Oczkowski and Nixon 2008). In fact, Maryut, the most intensively fertilized system, is the only lagoon that has shown a decline in fish landings, dropping precipitously from a high of 14,000 t in 1980 to about 4,000 t in recent decades (Figure 11.16). While we can only speculate about the reasons for this decline, we do know that the discharge of raw sewage into the lagoon began in 1981, just before the decline began (EA Engineering 1997). If we plot fish landings against available information on inorganic nitrogen (IN, which is $NO_3^- + NO_2^-$ or, when available, DIN) concentrations for all of the lagoons, there is an increase in fish yields with increasing IN up to 100 μM. Only Maryut was higher, and at that point the fishery declined (Figure 11.17) (Oczkowski and Nixon 2008). As increases in IN concentrations and landings in the lagoons are approximately chronological, N and associated P appear to stimulate primary and then secondary productivity in these systems, at least up until a point, as has been seen in other cross-system comparisons (Nixon and Buckley 2002).

The rise and fall of the Maryut fishery closely follows a theoretical curve described by Caddy (1993), who cautioned that adding nutrients to coastal systems will initially increase system productivity, but may eventually cause its decline (Figure 11.17). Alexandria's direct sewage inputs may have been the major contributing factor to the collapse of the fishery, but nutrient loads from agricultural drains and nearby cities and towns on the delta are increasing in the other three lagoons as well. While *in situ* nutrient concentrations appear to be rising at a much slower rate than Maryut experienced, the other lagoons are morphologically and biologically similar enough to suggest that, without management intervention, the trends observed in Maryut may be repeated in the future.

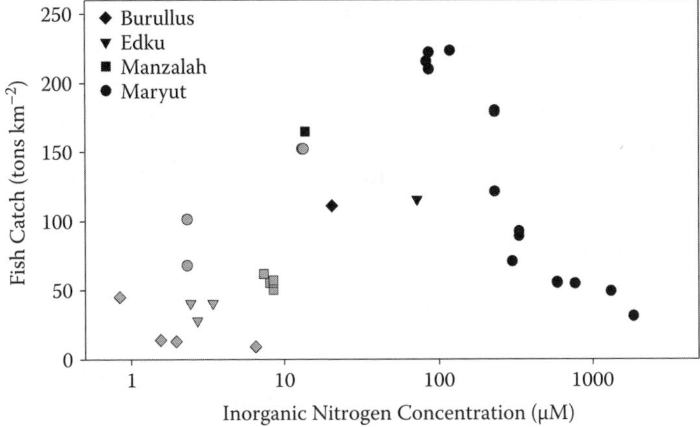

FIGURE 11.17 Mean annual fish catch plotted against available mean annual inorganic nitrogen concentrations for Burullus (♦), Edku (▾), Manzalah (▪), and Maryut (●) lagoons. Gray shapes represent NO_3^- + NO_2^- concentrations and DIN concentrations are shown as black symbols (references listed in Figures 11.15 and 11.16). Note that only Maryut, the lagoon with recent DIN concentrations >500 μM, shows the full progression of the curve. (From Oczkowski, A. and S. Nixon. 2008. *Estuar. Coastal Shelf Sci.* 77: 309–310. With permission.)

Dramatically increasing nutrient loads are not unique to these Egyptian lagoons, but are likely common in coastal systems of many developing nations. For example, DIN exports to the coast of Africa are expected to triple between 1990 and 2050 (Seitzinger et al. 2002), and lagoons in Ghana have shown a positive relationship between mean annual DIN concentrations and population density, where the most polluted lagoon has DIN concentrations of up to 1000 μM and a now nonexistent fishery (Nixon et al. 2007). We have been fortunate to observe ecological changes in the Egyptian coastal lagoons through the extensive documentation of these systems over the past 50 years. Their experience may provide an important cautionary tale of what the lagoons of other rapidly developing countries may expect if management actions are not implemented.

11.11 FUTURE DIRECTIONS

In compiling the data presented in this review, we noticed a few important gaps in our understanding of these systems that we would like to highlight and suggest as areas for future study. First and foremost, there are no recent water or nutrient budgets available for any of these lagoons. It would be helpful to have detailed quantitative information available on the recent monthly flows from individual drains into each lagoon, as well as some information on nitrogen and phosphorus loads, when trying to evaluate the impact of agricultural and aquaculture runoff on the productivity of the systems. Aquaculture is playing an increasingly important role in local food production and water dynamics in the regions surrounding the lagoons, yet details of pond water consumption and drainage, and the nutrient concentrations in these effluents, are unknown, as is how much drainage water is discharged into the lagoons and offshore. Again, this information is critical in understanding how nutrient loads from these different sources may be impacting (either positively or negatively) the lagoonal fisheries. Also, more quantitative information on the primary productivity (chlorophyll *a*) of these systems is needed. Finally, an assessment of the benthic infaunal community and benthic primary producers would be useful to compare to earlier studies conducted in the 1930s and 1960s and could also provide further insight into how eutrophication may be impacting these lagoons.

In reviewing hundreds of papers published over a span of some 50 years, we have observed that some recent work occasionally lacks context. We would encourage scientists to discuss their results in the framework of what has been done before in the same lagoon as well as to compare their data to conditions in the other lagoons. From an outside perspective, it was often difficult to understand why two or more studies conducted in the same system within a few years of one another and measuring some of the same parameters (for example, nutrients, Secchi depth, dissolved oxygen) yielded different results when there was no discussion of the other work. Occasionally we omitted studies from our review when they lacked information on sampling dates, frequency, or units.

ACKNOWLEDGMENTS

We thank Stephen Granger for his assistance with the lagoon maps and much of our work in Egypt, as well as Dr. Abdel-Fattah El-Sayed from the University of Alexandria for his helpful information on fish and fishing. Our understanding of these systems was greatly improved through conversations with Dr. El-Sayed, Dr. Youssef Halim, Dr. Wahid Moufaddal, and many of the scientists at the National Institute of Oceanography and Fisheries in Alexandria. We would also like to thank three anonymous reviewers for their helpful comments. This work was supported by NOAA's Dr. Nancy Foster Scholarship Program, the NSF Biological Oceanography Program (award no. OCE 0526332), and the U.S.-Egypt Joint Board through a Junior Scientist Development Visit Grant to A.J.O. The statements, findings, conclusions, and recommendations are those of the authors and do not necessarily reflect the views of NSF, NOAA, or the Department of Commerce.

REFERENCES

Abbas, M. M., L. M. Shakweer, and D. H. Youssef. 2001. Ecological and fisheries management of Edku Lake. I. Hydro-chemical characters of Edku Lake. *Bull. Natl. Inst. Oceanogr. Fish. A.R.E.* 27: 65–93.

Abdalla, R. R., A. A. Samaan, and M. G. Ghobrial. 1991. Eutrophication in Lake Mariut. *Bull. Natl. Inst. Oceanogr. Fish. A.R.E.* 17: 157–166.

Abdallah, M. A. M. 2003. The chemical changes in Lake Maryout after the construction of two wastewater treatment plants. Ph.D. Thesis, University of Alexandria, Alexandria, Egypt.

Abdel Aziz, N. E. and S. M. Aboul Ezz. 2004. The structure of zooplankton community in Lake Maryout, Alexandria, Egypt. *Egyptian J. Aquat. Res.* 30: 160–170.

Abdel-Moati, M. A. R and A. A. El-Sammak. 1997. Man-made impact on the geochemistry of the Nile Delta lakes. A study of metals concentrations in sediments. *Water Air Soil Pollut.* 97: 413–429.

Abdel-Shafy, H. I. and R. O. Aly. 2002. Water issue in Egypt: Resources, pollution, and protection endeavors. *Cent. Eur. J. Occupation Environ. Med.* 8: 1–21.

Aboul Ezz, S. M. and A. M. Soliman. 2000. Zooplankton community in Lake Edku. *Bull. Natl. Inst. Oceanogr. Fish., A.R.E.* 26: 71–99.

Adham, K. G., S. S. Hamed, H. M. Ibrahim, and R. A. Saleh. 2001. Impaired functions in Nile tilapia, *Oreochromis niloticus* (Linnaeus, 1757), from polluted waters. *Acta Hydrochim. Hydrobiol.* 29: 278–288.

Adham, K. G., I. F. Hassan, N. Taha, and T. H. Amin. 1999. Impact of hazardous exposure to metals in the Nile and delta lakes on the catfish, *Clarias Lazera. Environ. Monit. Assess.* 54: 107–124.

Aleem, A. A. and A. A. Samaan. 1969. Productivity of Lake Mariut, Egypt. I. Physical and chemical aspects. *Int. Rev. Ges. Hydrobiol.* 54: 313–355.

Alsayes, A. 2005. Environmental and fishery investigation on Lake Borollus. V. Fishing gears and methods used at Lake Borollus: Their effects on fish populations at the lake. *Egyptian J. Aquat. Res.* 31: 410–446.

Al Sayes, A., A. Radwan, and L. Shakweer. 2007. Impact of drainage water inflow on the environmental conditions and fishery resources of Lake Borollus. *Egyptian J. Aquat. Res.* 33: 312–351.

Appleby, P. J., H. H. Birks, R. J. Flower, N. Rose, S. M. Peglar, M. Ramadani, M. M. Kraïem, and A. A. Fathi. 2001. Radiometrically determined dates and sedimentation rates for recent sediments in nine North African wetland lakes (the CASSARINA Project). *Aquat. Ecol.* 35: 347–367.

Badawy, M. I. 1998. Use and impact of pesticides in Egypt. *Int. J. Environ. Pollut.* 8: 223–239.

Banoub, M. W. 1983. Comparative study of the nutrient-salts of Lake Edku (Egypt) before and after the construction of Asswan's high dam in 1958 and 1969. *Arch. Hydrobiol.* 99: 106–117.

Benninger, L. K., I. B. Suayah, and D. J. Stanley. 1998. Manzala lagoon, Nile Delta, Egypt: Modern sediment accumulation based on radioactive tracers. *Environ. Geol.* 34: 183–193.

Bernasconi, M. P. and D. J. Stanley. 1994. Molluscan biofacies and their environmental implications, Nile Delta lagoons, Egypt. *J. Coast. Res.* 10: 440–465.

Bianchi, F., F. Acri, L. Alberighi, M. Bastianini, A. Boldrin, B. Cavalloni, F. Cioce, A. Comaschi, S. Rabitti, G. Socal, and M. M. Turchetto. 2000. Biological variability in the Venice lagoon. Pp. 97–126, In: P. Lasserre and A. Marzollo (eds.), *The Venice Lagoon ecosystem inputs and interactions between land and sea.* Man and the Biosphere Series Vol. 25. Paris: UNESCO and the Parthenon Publishing Group.

Blaber, S. J .M. 1997. *Fish and fisheries of tropical estuaries.* London: Chapman & Hall.

Bothner, M. H., M. Buchholtz ten Brink, and F. T. Manheim. 1998. Metal concentrations in surface sediments of Boston Harbor: Changes with time. *Mar. Environ. Res.* 45: 127–155.

Butzer, K. W. 1976. *Early hydraulic civilization in Egypt: A study in cultural ecology.* Chicago: University of Chicago Press.

Caddy, J. F. 1993. Toward a comparative evaluation of human impacts on fishery ecosystems of enclosed and semi-enclosed seas. *Rev. Fish. Sci.* 1: 57–95.

Dewidar, K. and A. Khedr. 2001. Water quality assessment with simultaneous Landsat-5 TM at Manzala Lagoon, Egypt. *Hydrobiologia* 457: 49–58.

Dinham, B. 1995. Egyptian studies confirm pesticide related health effects in farm and factory. *Pesticide News,* no. 30. Online, http://www.pan uk.org/pestnews/Pn30/pn30p10.htm.

Dumont, H. J. and G. M. El Shabrawy. 2007. Lake Borullus of the Nile Delta: A short history and an uncertain future. *Ambio* 36: 677–682.

Dumont, H. J. and G. M. El Shabrawy. 2008. Seven decades of change in the zooplankton (s.l.) of the Nile Delta Lakes (Egypt), with particular reference to Lake Borullus. *Int. Rev. Hydrobiol.* 93:44–61.

EA Engineering, Science and Technology Inc. 1997. *Environmental Technical Report 8: Chemical and biological characterization of Lake Maryout.* Final Report, Alexandria Wastewater Project Phase II, USAID Project No. 263-0100, Prepared for Metcalf & Eddy International, Roushdie, Alexandria.

El-Hoshy, S. M. and M. Mousa. 1994. Association between water pollution with insecticides and their residues in fish. *Alexandria J. Veterinary Sci.* 10: 55–59.

El-Karachilly, A. F. 1991. Production functions analysis in Lake Burollos fisheries. *Bull. Natl. Inst. Oceanogr. Fish. A.R.E.* 17: 37–43.

El-Rayis, O. 2005. Impact of man's activities on a closed fishing-lake, Lake Maryout in Egypt, as a case study. *Mitigation Adapt. Strat. Global Change* 10: 145–157.

El-Rayis, O. A. and M. A. Abdallah. 2005. Contribution of some trace elements from an Egyptian huge drain to the Mediterranean Sea, west of Alexandria. *Egyptian J. Aquat. Res.* 31: 120–129.

El-Sayed, A. F. M. 2006. *Tilapia culture*. Wallingford, U.K.: CABI Publishing.

El-Sayed, A. F. M. 2007. Analysis of feeds and fertilizers for sustainable aquaculture development in Egypt. Pp. 401–422, In: M. R. Hasan, T. Hecht, S. S. De Silva and A.G. J. Tacon (eds.), *Study and analysis of feeds and fertilizers for sustainable aquaculture development*. FAO Fish. Tech. Paper. 497. Rome: Food and Agriculture Organization.

El-Shenawy, M. A. 1994. *Azollla filiculoides*, a new effective dinitrogen fixer in Lake Edku, Egypt. *Bull. Natl. Inst. Oceanogr. Fish. A.R.E.* 20: 83–97.

El-Sherif, Z. M. 1993. Phytoplankton standing crop, diversity and statistical multispecies analysis in Lake Burollus, Egypt. *Bull. Natl. Inst. Oceanogr. Fish., A.R.E.* 19: 213–233.

El-Sherif, Z. M. and S. M. Gharib. 2001. Spatial and temporal patterns of phytoplankton communities in Manzalah Lagoon. *Bull. Natl. Inst. Oceanogr. Fish.* 27: 217–239.

El-Wakeel, S. K. and S. D. Wahby. 1970. Hydrography and chemistry of Lake Manzalah, Egypt. *Archiv. Hydrobiol.* 67: 173–200.

El-Zarka, S., A. H. Shaheen, and A. A. El-Aleem. 1970. Tilapia fisherie[s] in Lake Mariut., age, and growth of *Tilapia rilotica* [*Tilapia nilotica*] L. in the alke [Lake]. *Bull. Natl. Inst. Oceanogr. Fish. A.R.E.* 1: 150–182.

Fathi, A. A., H. M. A. Abdelzaher, R. J. Flower, M. Ramdani, and M. M. Kraïem. 2001. Phytoplankton communities of North African wetland lakes: The CASSARINA Project. *Aquat. Ecol.* 35: 303–318.

Fattouh, S. A. L. 1990. Analysis of fishery exploitation in Lake Burollus. *Bull. Natl. Inst. Oceanogr. Fish., A.R.E.* 16: 109–118.

Flower, R. J. 2001. ChAnge, Stress, Sustainability, and Aquatic ecosystem Resilience in North African wetland lakes during the 20th century: An introduction to integrated biodiversity studies within the CASSARINA Project. *Aquat. Ecol.* 35: 261–280.

Flower, R. J., S. Dobinson, M. Ramdani, M. M. Kraïem, C. Ben Hamza, A. A. Fathi, H. M. A. Abdelzaher, H. H. Birks, P. G. Appleby, J. A. Lees, E. Shilland, and S. T. Patrick. 2001. Recent environmental change in North African wetland lakes: Diatom and other stratigraphic evidence from nine sites in the CASSARINA Project. *Aquat. Ecol.* 35: 369–388.

FAO (Food and Agriculture Organization of the United Nations). 2008. FAOSTAT online statistical service. Rome: FAO. http://faostat.fao.org/default.aspx. Accessed April 16, 2008.

Froese, R. and D. Pauly. Eds. 2007. FishBase. www.fishbase.org, version (10/2007). Accessed April 16, 2008.

GAFRD. 2007. *Fisheries statistics yearbook 2006*. General Authority for Fish Resources Development (GAFRD), Cairo, Egypt.

Gharib, S. M. 1999. Phytoplankton studies in Lake Edku and adjacent waters. *Egypt. J. Aquat. Biol. Fish* 3: 1–23.

Gharib, S. M. and A. M. Soliman. 1998. Some water characteristics and phyto-zooplankton relationship in Lake Edku (Egypt) and adjacent sea. *Bull. Fac. Sci. Alex. Univ.* 38: 25–44.

Hashem, M. T., A. M. El-Maghraby, and H. M El-Sedfy. 1973. The grey mullet fishery of Lake Borollus. *Bull. Natl. Inst. Oceanogr. Fish., A.R.E.* 3: 31–54.

Halim, Y. and S. K. Guerguess. 1981. *Coastal lakes of the Nile Delta. Lake Manzalah*. UNESCO Technical Papers in Marine Science 33: Coastal lagoon research present and future, pp. 135–172.

Hoevenaars, J. and R. Slootweg. 2004. Rapid assessment study towards integrating planning of irrigation and drainage in Egypt—natural resources perspective. http://www.iwmi.cgiar.org/Assessment/files/Synthesis/LowQWater/IIIMP-DrainframeAnalyses.pdf. Accessed November 13, 2008.

Ibrahim, E. A. 1989. Studies on phytoplankton in some polluted areas of Lake Manzalah. *Bull. Natl. Inst. Oceanogr. Fish. A.R.E.* 15: 1–19.

Ishak, M. M. and M. M. Shafik. 1982. The utilization of coastal areas for aquaculture in Eygpt. CIFA Technical Paper (FAO), no. 9, pp. 25–30.

Kassim, T. A. 2005. Forensic analysis and source partitioning of aliphatic hydrocarbon contamination in Lake Maruit aquatic sediments. *Egyptian J. Aquat. Res.* 31: 166–182.

Khalil, M. K., A. M. Radwan, and K. M. El-Moselhy. 2007. Distribution of phosphorus fractions and some of heavy metals in surface sediments of Burullus Lagoon and adjacent Mediterranean Sea. *Egyptian J. Aquat. Res.* 33: 277–289.

Kjerfve, B. 1986. Comparative oceanography of coastal lagoons. Pp. 63–81, In: D.A. Wolfe (ed.), *Estuarine variability*. New York: Academic Press.

Long, E. R. and D. D. MacDonald. 1998. Recommended uses of empirically derived, sediment quality guidelines for marine and estuarine ecosystems. *Human Ecol. Risk. Assess.* 4: 1019–1039.

Mageed, A. A. 2007. Distribution and long-term historical changes of zooplankton assemblages in Lake Manzala (South Mediterranean Sea, Egypt). *Egyptian J. Aquat. Res.* 33: 183–192.

Megapesca. 2001. Marine aquaculture in Egypt. Megapesca, Rua Gago Coutinho 11 Valado de Santa Quitéria 2460-207 Alfeizerão Portugal. http://www.megapesca.com.

Mikhailova, M. V. 2001. Hydrological regime of the Nile Delta and dynamics of its coastline. *Water Resour.* 28: 526–539.

Morin, P., P. Lasserre, C. Madec, P. Le Corre, E. Macé, and B. Cavalloni. 2000. Pelagic nitrogen fluxes in the Venice Lagoon. Pp. 143–188, In: P. Lasserre and A. Marzollo (eds.), *The Venice Lagoon ecosystem inputs and interactions between land and sea*. Man and the Biosphere Series Vol. 25. Paris: UNESCO and the Parthenon Publishing Group.

Nixon, S. W. 1982. Nutrient dynamics, primary production, and fisheries yields of lagoons. *Oceanol. Acta* Special Edition, 357–371.

Nixon, S. W. 1983. *Estuarine Ecology: A comparative and experimental analysis using 14 estuaries and the MERL microcosms*. Final report to the U.S. Environmental Protection Agency, Chesapeake Bay Program under Grant No. X-003259-01, Graduate School of Oceanography, University of Rhode Island, Kingston, Rhode Island.

Nixon, S. W. 1988. Physical energy inputs and the comparative ecology of lake and marine ecosystems. *Limnol. Oceanogr.* 33: 1005–1025.

Nixon, S. W. 2003. Replacing the Nile: Are anthropogenic nutrients providing the fertility once brought to the Mediterranean by a great river? *Ambio* 32: 30–39.

Nixon, S. W. and B. A. Buckley. 2002. "A strikingly rich zone": Nutrient enrichment and secondary production in coastal marine ecosystems. *Estuaries* 25: 782–796.

Nixon, S. W., B. A. Buckley, S. L. Granger, M. Entsua-Mensah, O. Ansa-Asare, M. J. White, and R. A. McKinney. 2007. Anthropogenic enrichment and nutrients in some tropical lagoons of Ghana, West Africa. *Ecol. Appl.* 17: S144–S164.

Nixon, S. W., R. W. Fulweiler, B. A. Buckley, S. L. Granger, B. L. Nowicki, and K. M. Henry. 2009. The impact of changing climate on phenology, productivity, and benthic-pelagic coupling in Narragansett Bay. *Estuar. Coastal Shelf Sci.* 82: 1–18.

Oczkowski, A. and S. Nixon. 2008. Increasing nutrient concentrations and the rise and fall of a coastal fishery: A review of data from the Nile Delta, Egypt. *Estuar. Coastal Shelf Sci.* 77: 309–310.

Oczkowski, A., S. Nixon, S. Granger, A. M. El-Sayed, M. Altabet, and R. McKinney. 2008. A preliminary survey of the nitrogen and carbon isotope characteristics of fish from the lagoons of Egypt's Nile Delta. *Estuaries Coasts* 31: 1130–1142.

Oczkowski, A. J., S. W. Nixon, S. L. Granger. A. M. El-Sayed, and R. A. McKinney. 2009. Anthropogenic enhancement of Egypt's Mediterranean fishery. *Proc. Natl. Acad. Sci. USA* 106: 1364–1367.

Okbah, M. A. 2005. Nitrogen and phosphorus species of Lake Burullus water (Egypt). *Egyptian J. Aquat. Res.* 31: 186–199.

Okbah, M. A. and S. El-Gohary. 2002. Physical and chemical characteristics of Lake Edku water, Egypt. *Mediterranean Mar. Sci.* 3/2: 27–39.

Okbah, M. A. and N. R. Hussein. 2006. Impact of environmental conditions on the phytoplankton structure in Mediterranean Sea Lagoon, Lake Burullus, Egypt. *Water Air Soil Pollut.* 172: 129–150.

Peters, A. J., K. C. Jones, R. J. Flower, P. G. Appleby, M. Ramdani, M. M. Kraïem, and A. A. Fathi. 2001. Recent environmental change in North African wetland lakes: A baseline study of organochlorine contaminant residues in sediments from nine sites in the CASSARINA Project. *Aquat. Ecol.* 35: 449–459.

Pimentel, D., L. McLaughlin, A. Zepp, B. Lakitan, T. Kraus, P. Kleinman, F. Vancini, W. J. Roach, E. Graap, W. S. Keeton, and G. Selig. 1991. Environmental and economic effects of reducing pesticide use. *BioScience* 41: 402–409.

Radwan, A. M. 2005. Some factors affecting the primary production of phytoplankton in Lake Burullus. *Egyptian J. Aquatic Res.* 31: 72–88.

Radwan, A. M. and L. M. Shakweer. 2004. On the present status of the environmental and fishery of Lake Borollus. 2A. Distribution and concentrations of trace elements in the water and bottom sediments of Lake Borollus. *Egyptian J. Aquatic Res.* 30: 43–98.

Ramdani, M., R. J. Flower, N. Elkhiati, M. M. Kraïem, A. A. Fathi, H. H. Birks, and S. T. Patrick. 2001. North African wetland lakes: Characterization of nine sites included in the CASSARINA Project. *Aquat. Ecol.* 35: 281–302.

Randazzo, G., D. J. Stanley, S. I. Di Geronimo, and C. Amore. 1998. Human-induced sedimentological changes in Manzala Lagoon, Nile Delta, Egypt. *Environ. Geol.* 36: 235–258.

Rashed, D. 2001. Killing waters. *Al-Ahram Weekly Online*, 6–12 September 2001. Issue No. 550. http://weekly.ahram.org.eg/2001/550/fe2.htm.

Reinhardt, E. G., D. J. Stanley, and H. P. Schwarcz. 2001. Human-induced desalinization of Manzala Lagoon, Nile Delta, Egypt: Evidence from isotopic analysis of benthic invertebrates. *J. Coastal Res.* 17: 431–442.

Saad, M. A. H. 1983. Influence of pollution on Lake Mariut, Egypt II. Nutrients. *Rapport Commission Internationale pour la Mer Méditeranée* 28: 209–210.

Saad, M. A. H. 1987. Limnological studies on the Nozha Hydrodome, Egypt, with special reference to the problems of pollution. *Sci. Total Environ.* 67: 195–214.

Saad, M. A. H., O. A. El-Rayis, and H. H. Ahdy. 1984. Status of nutrients in Lake Mariut, a delta lake in Egypt suffering from intensive pollution. *Mar. Pollut. Bull.* 15: 408–411.

Samir, A.M. 2000. The response of benthic foraminifera and ostracods to various pollution sources: A study from two lagoons in Egypt. *J. Foram. Res.* 30: 83–98.

Samman, A. A. 1974. Primary production of Lake Edku. *Bull. Natl. Inst. Oceanogr. Fish., A.R.E* 4: 260–317.

Samman, A. A. and A. A. Aleem. 1972. Quantitative estimation of bottom fauna in Lake Mariut. *Bull. Natl. Inst. Oceanogr. Fish. A.R.E* 2: 376–397.

Seitzinger, S. P., C. Kroeze, A. F. Bouwman, N. Caraco, F. Dentener, and R. V. Styles. 2002. Global patterns of dissolved inorganic and particulate nitrogen inputs to coastal systems: Recent conditions and future projections. *Estuaries* 25: 640–655.

Shaheen, A. H. and S. F. Yousef. 1980. Physico-chemical conditions, fauna and flora of Lake Manzala, Egypt. *Water Supply Manage.* 4: 103–113.

Shakweer, L. M. 2005. Ecological and fisheries development of Lake Manzalah (Egypt). I. Hydrography and chemistry of Lake Manzalah. *Egyptian J. Aquatic Res.* 31: 251–269.

Shakweer, L. M. 2006. Impacts of drainage water discharge on the water chemistry of Lake Edku. *Egyptian J. Aquat. Res.* 32: 264–282.

Shakweer, L. M. and M. M. Abbas. 2005. Effect of ecological and biological factors on the uptake and concentration of trace elements by aquatic organisms at Edku Lake. *Egyptian J. Aquat. Res.* 31: 271–287.

Siam, E. and M. Ghobrial. 2000. Pollution influence on bacterial abundance and chlorophyll-*a* concentraton: Case study at Idku Lagoon, Egypt. *Sci. Mar.* 64: 1–8.

Soliman, A. M. 2005. Zooplankton structure in Lake Edku and adjacent waters (Egypt). *Egyptian J. Aquat. Res.* 31: 239–252.

Stanley, D. J. and A. G. Warne. 1993. Nile Delta: Recent geological evolution and human impact. *Science* 260: 628–634.

Thomas, G. H., I. A. El-Karyony, and A. H. El-Din Hassan. 1993. Phosphorus-nitrogen loading and trend of fish catch as index of Lake Mariut eutrophication. *J. Egyptian Pub. Health Assoc.* 68: 593–615.

Toews, D. R. 1988. Fishery transformation on Lake Manzala, Egypt during the period 1920 to 1980. Pp. 25–50, In: N. W. Schmidtke (ed.), *Impact of toxic contaminants on fisheries management.* Boca Raton, Florida: CRC Press.

Toews, D. R. and M. M. Ishak. 1984. Fishery transformation on Lake Manzala, a brackish Egyptian delta lake in response to anthropological and environmental factors during the period 1920–1980. *Gen. Fish. Council Mediterranean Stud. Rev.* 64: 347–402.

WADI Project. 2006. Maryuit Lake: "a preliminary assessment" from WADI Project, Sustainable management of coastal fresh water bodies: a socio-economic and environmental analysis to enhance and sustain stakeholders' benefits. http://www.wadi.unifi.it/study_sites_egypt.htm.

Wahby, S. D. 1961. Chemistry of Lake Maryut. *Alexandria Inst. Hydrobiol. Notes Mem.* 65: 25.

Wahby, S. D., S. F. Youssef, and N. F. Bishara. 1972. Further studies on the hydrography and chemistry of Lake Manzalah. *Bull. Natl. Inst. Oceanogr. Fish. A.R.E.* 2: 400–422.

Zaghloul, F. A. and N. R. Hussein. 2000. Impact of pollution on phytoplankton community structure in Lake Edku, Egypt. *Bull. Natl. Inst. Oceanogr. Fish. A.R.E.* 26: 297–318.

12 Origins and Fate of Inorganic Nitrogen from Land to Coastal Ocean on the Yucatan Peninsula, Mexico

Troy Mutchler, Rae F. Mooney, Sarah Wallace, Larissa Podsim, Stein Fredriksen, and Kenneth H. Dunton*

CONTENTS

Abstract 283
12.1 Introduction 284
12.2 Methods 286
12.2.1 Study Area 286
12.2.2 Sample Collection and Analysis 286
12.2.3 Statistical Analysis 288
12.3 Results 288
12.3.1 Water Chemistry 288
12.3.2 Seagrass Biomass 290
12.3.3 Plant Isotope Values and Tissue Nutrient Content 291
12.4 Discussion 291
12.4.1 Water Column Processes in Cenotes and Bays 291
12.4.2 Nutrient Sources for Algae and Seagrass 295
12.4.3 Nutrient Impacts on Plants and Algae 298
Acknowledgments 302
References 302

ABSTRACT

Land use changes and poor wastewater management threaten valuable coastal resources by augmenting nutrient delivery to near-shore ecosystems. Oligotrophic, tropical systems are particularly vulnerable because relatively small changes in nutrient load can cause shifts in community composition and ecosystem function. In order to examine nutrient dynamics along a rapidly developing coastline, we characterized water chemistry in a series of cenotes and bays on the Caribbean coast of Mexico. We also measured the isotopic and tissue nutrient content of resident seagrasses and algae to examine possible nitrogen (N) sources and determine their influence on the nutrient status of primary producers in seagrass and coral reef ecosystems. Ranked water column NO_3^- concentrations were significantly different in cenotes (NO_3^-, 62.6 ± 27.7 μM) vs. bays (NO_3^-, 2.8 ± 2.6 μM); however, NH_4^+ concentrations were not significantly different (NH_4^+, 0.8 ± 1.2 μM and

* Corresponding author. Email: tmutchle@kennesaw.edu

NH_4^+, 1.7 ± 2 μM, respectively). NO_3^- behaved nonconservatively as it moved through the aquifer and discharged into coastal systems. Vertical profiles within the Sian Ka'an Cenote revealed increasing NH_4^+ concentrations with depth, while NO_3^- and dissolved oxygen declined. These patterns suggested that denitrification removes NO_3^- within the aquifer. Elevated $\delta^{15}N$ values for several algal species provided further evidence for nitrogen transformation in the aquifer. At Laguna Lagartos, $\delta^{15}N$ values of *Chara* sp. (9‰) and *Cladophora* sp. (12‰) suggested that intrusions of untreated wastewater from nearby residential and tourism developments may enter the aquifer. Seagrass tissue nutrient content was similar to other locations in the Caribbean characterized by high nutrient inputs, and indicated that nutrient loading to marine ecosystems in the area may be increasing along with tourism and development. Based on these observations, land use changes associated with the burgeoning tourism industry are altering the coastal nutrient dynamics and threatening the health of economically and ecologically valuable seagrass and coral reef ecosystems.

Key Words: nutrients, eutrophication, seagrass, Mexico, Yucatan, wastewater, $\delta^{15}N$, coral reefs, groundwater, cenote, watersheds

12.1 INTRODUCTION

Cultural eutrophication is widely recognized as a serious threat to diverse coastal ecosystems around the globe, including coral reefs, mangrove forests, seagrass meadows, and salt marshes (Hughes et al. 2003; Pandolfi et al. 2003; Duarte et al. 2008). Concerns regarding cultural eutrophication have inspired intensive and ongoing efforts both to quantify nutrient dynamics in coastal systems (e.g., Smith et al. 1978; Valiela et al. 1978; Capone and Bautista 1985), as well as to better understand the impacts of anthropogenic nutrient enrichment on system structure and function. Among the findings of these efforts was recognition that submarine groundwater discharge (SGD) (Johannes 1980) is a significant source for nutrients in coastal ecosystems, particularly in locations where limestone bedrock and karst topographies are common (Valiela et al. 1978; Johannes 1980; D'Elia et al. 1981; Lapointe et al. 1990).

Karst topography is found where rainwater infused with organic and carbonic acids from the soil dissolves limestone bedrock and creates porous, heterogeneous substrate that may contain large caves and conduits (Worthington et al. 2000). The porous nature of the bedrock facilitates extensive horizontal movement of groundwater within the aquifer and can alter the conventional density-stratified circulation that occurs in more homogeneous aquifers. For instance, Beddows et al. (2007) showed that in the Yucatan Peninsula (Mexico) large conduits allowed extensive landward flow of seawater up to 10 km inland from the coastline. Such landward flow of seawater contrasts with the parallel coastward flow of fresh and salt water observed in conventional density stratified circulation (Cooper 1959). The morphology of the conduits, particularly size and geometry, determines the location and degree of fresh and salt water mixing in the aquifer (Beddows et al. 2007). These conduits, therefore, greatly influence the amount and chemical composition of SGD delivered to coastal systems (Beddows et al. 2007).

In some locations, this SGD constitutes a substantial portion of freshwater and nutrient inputs to near-shore environments. Moore (1996) estimated that SGD into coastal waters of the South Atlantic Bight in South Carolina represented ~40% of the river water flux, whereas SGD accounted for 99% of freshwater flux to the Caribbean along the karstic coastline of the Yucatan Peninsula (Worthington et al. 2000). Such large fluxes of SGD have been shown to significantly affect coastal ecology by affecting the nutrient availability in these systems (e.g., Valiela et al. 1978; Johannes 1980; D'Elia et al. 1981). For instance, studies throughout the Caribbean have shown that SGD from systems with carbonate sediments typically have high levels of dissolved inorganic nitrogen (DIN), particularly in the form of NO_3^-, but little soluble reactive phosphorus (SRP; D'Elia et al. 1981; Lapointe et al. 1990; Lapointe and Clark 1992). In Discovery Bay, Jamaica, various processes, including terrestrial nitrogen fixation, precipitation, concentration via evapotranspiration, and decomposition in the vadose zone, generated levels of DIN in groundwater in excess of 80 μg L^{-1} (D'Elia et al. 1981). Elsewhere in the Caribbean region, changes in land use and wastewater inputs led to significant nutrient enrichment of groundwater (Simmons et al. 1985; Lapointe et al. 1990; Finkl and Charlier 2003). For example, groundwater adjacent to on-site sewage disposal systems in the Florida Keys experienced up to 5000-fold enrichment of DIN compared to unimpacted groundwater (Lapointe et al. 1990). In some cases reduced forms of nitrogen (N) contribute to the DIN pool (e.g., NH_4^+; Lapointe et al. 1990), but aerobic soil conditions frequently facilitate nitrification that results in high NO_3^- concentrations in the groundwater (Lapointe et al. 1990). NO_3^- is highly mobile within soil and groundwater and therefore can easily move through the aquifer and export in SGD.

Despite these high DIN concentrations, groundwater in systems with carbonate sediments typically has relatively low SRP content (D'Elia et al. 1981; Lapointe et al. 1990, 1992). Previous studies have demonstrated that SRP is removed from groundwater in carbonate sediments because it adsorbs to carbonate surfaces, reacts to form apatite, and adsorbs to iron and aluminum oxides (De Kanel and Morse 1978; Berner 1981). Experimental additions of phosphate to groundwater in the Florida Keys demonstrated that the removal of SRP can be quite rapid, with 95% removal of injected PO_4^{3-} occurring within 20 to 50 h (Corbett et al. 2000). Numerous studies have shown that, because of these SRP removal processes, primary production in coastal lagoons with carbonate sediment tends to be limited by P availability (Lapointe 1987, 1997; Littler et al. 1988; Lapointe and O'Connell 1989). This condition can be exacerbated in systems where SGD with high DIN concentrations removes any potential N limitation (Lapointe and Clark 1992).

Despite the prevalence of P limitation in Caribbean lagoons, several studies raised concerns that SRP enrichment is occurring in some locations. Lapointe (1997) reported apparent increases in DIN and SRP concentrations in Discovery Bay, Jamaica, throughout the 1980s. Although the origin of SRP was not identified, wastewater from nearby developments was recognized as a potential source (Lapointe 1997). Wastewater contamination of groundwater was more definitively established in the Florida Keys, where on-site sewage disposal systems contribute to both DIN and SRP enrichment (Lapointe et al. 1990). A similar process has been documented in coastal lagoons of the Yucatan Peninsula (Carruthers et al. 2005b). Human wastewater, however, is not the only potential source of SRP to coastal systems. For instance, agricultural runoff also has contributed to nutrient enrichment in the Florida Keys (Lapointe and Clark 1992).

Such increases in DIN and SRP loading to coastal systems are of concern because they can lead to dramatic shifts in the structure and function of coastal ecosystems. A common response to nutrient enrichment is the development of harmful algal blooms (HABs). In Puerto Rico, Diaz et al. (1990) and Corredor et al. (1992) documented changes in seagrass community structure as a result of extensive mat formation by the cyanophyte *Microcoleus lyngbyaceus*. Similar algal blooms have occurred in coral reefs throughout the Caribbean region, including Belize (Lapointe et al. 1992), Florida (Lapointe and Clark 1992; Lapointe 1997; Lapointe et al. 2005a, 2005b), Jamaica (Lapointe 1997), and Martinique (Littler et al. 1992). In some instances, the development of these algal blooms was directly linked to inputs of land-derived nutrients (Lapointe 1997; Lapointe et al. 2005a), although other factors, including reduced grazing intensity, likely contributed to the increase in algal biomass (Lessios 1988). Because of the link between nutrient enrichment and altered community composition, cultural eutrophication is a main cause of declines in seagrass meadows and coral reefs around the globe (Pandolfi et al. 2003; Duarte et al. 2008).

Based on this previous work demonstrating the significance of SGD and nutrient inputs to coastal lagoons, we examined groundwater coastal interactions in several bay and cenote systems on the Yucatan Peninsula, Mexico. The Caribbean coast of the Yucatan Peninsula is characterized by karst topography, and bays and lagoons receive significant inputs of SGD (Worthington et al. 2000). We hypothesized that groundwater in cenotes near developed areas would have high nutrient concentrations from inputs of anthropogenic N. In addition, we predicted that algal and seagrass tissue nutrient content would reflect these elevated nutrient concentrations and provide evidence that anthropogenic nutrient sources were impacting near-shore ecosystems. By better understanding the connections between development, freshwater inflow, and nutrient cycling within these coastal systems, we can better assess the risks to these valuable environments.

12.2 METHODS

12.2.1 Study Area

This study was conducted in May and June 2007 along the Caribbean Coast of the Yucatan Peninsula in Quintana Roo, Mexico (Figure 12.1). Samples for this study were taken from 13 sites: six bay sites and seven cenotes (Table 12.1). These sites allow for a comparison of nutrient levels across a broad range of salinities and land use. In the past 5 years, the areas around Xaak and Akumal Bay have experienced a dramatic increase in development. Prior to 2005, Xaak had no development and now is the location of a large resort. The Sian Ka'an bay sites are located in Ascensión Bay within the protected Sian Ka'an Biosphere Reserve. Sian Ka'an cenote is in Ascension Bay, where a large sinkhole has formed and freshwater upwells through rock into the bay. While development occurs around the perimeter, the interior of the reserve remains largely undeveloped. The bay sites are predominantly marine, whereas the cenote sites include both fresh and brackish waters, with salinities ranging between 2 and 19 psu (Table 12.1). Laguna Lagartos is an enclosed inland water body with a surface covered by *Cladophora* sp., a green alga that forms thick mats. Yal Kú Lagoon is connected to Laguna Lagartos through subsurface groundwater flow.

12.2.2 Sample Collection and Analysis

Data were collected during May and June 2007. Water column depth, temperature, salinity, dissolved oxygen, and pH were measured *in situ* with a YSI 600 XLM data sonde (YSI Inc., Yellow Springs, Ohio). One to six replicate water samples were collected per site for nutrient analysis (Table 12.2). Water samples were retrieved from the surface and at depth at Xaak, Akumal Bay, Sian Ka'an Bay mouth, Bat Cave Cenote, Pueblo Cenote, Beach Cenote, and Sian Ka'an Cenote. Because of a sharp halocline at Yal Kú Chico, a surface sample was taken where the groundwater emerges at the landward end of Yal Kú Chico, and an additional sample was collected at depth in the mouth of the embayment where salinity approached that of seawater. Only surface samples were taken at Casa Cenote, Laguna Lagartos, Yal Kú Lagoon mouth, and Sian Ka'an Bay middle due to shallow depths and/or well mixed water columns.

Water samples collected for nutrient analysis were placed immediately in the dark on ice and analyzed the day of collection or frozen at –20°C until analysis. Nitrate + nitrite (NO_3^-) concentrations were determined by standard colorimetric analyses after cadmium reduction to nitrite (NO_2^-; Strickland and Parsons 1972a). Ammonium (NH_4^+) concentrations were determined colorimetrically by the indophenol method (Strickland and Parsons 1972b).

In Akumal, Xaak, and Sian Ka'an bays, we collected between two and four replicate samples of three seagrass species (*Thalassia testudinum* Banks ex König, *Halodule wrightii* Ascherson, and *Syringodium filiforme* Kützing). At sites where low salinity excluded seagrasses, samples of representative algal species were sampled and identified. Seagrass and algal tissue samples were immediately placed on ice following collection. In the laboratory, samples were rinsed in deionized water and then scraped to remove epiphytes. Algal species were acidified in 10% HCl to remove carbonates prior to analysis. All tissues were dried at 60°C and pulverized by hand or using a Wig-L-Bug (Rinn Corp., Elgin, Illinois). $\delta^{15}N$, %C, and %N were measured using a Carlo Erba 2500 elemental analyzer coupled to a Finnigan MAT Delta+ isotope ratio mass spectrometer at the University of Texas Marine

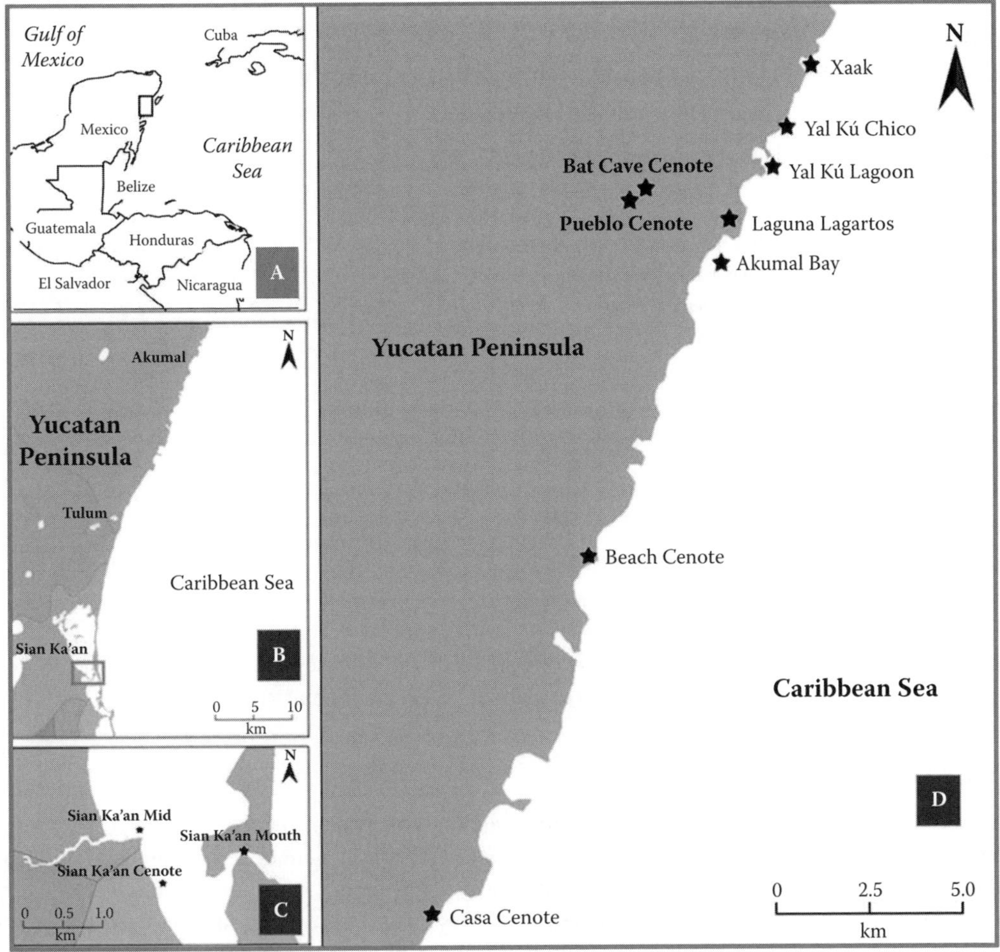

FIGURE 12.1 The study area and sampling sites on the Yucatan Peninsula in Quintana Roo, Mexico. Geographical information systems (GIS) data from Sistema Nacional de Información Estadística y Geográfica (http://www.inegi.org.mx/).

Science Institute. Stable isotope ratios are presented as $\delta^{15}N$ relative to atmospheric N_2. Seagrass sample dry weight was used to calculate the elemental content and to determine C:N ratios.

Tissue phosphorus (P) content was determined using a modified version of methods described in Fourqurean et al. (1992a) and Solorzano and Sharp (1980). Dried samples with known weights were combusted at 500°C, and P was extracted through acid hydrolysis. SRP concentrations were determined colorimetrically and converted to sample %P based on the dry weight, and used to determine C:N:P concentrations of sample tissue.

Three replicate biomass cores of seagrass species were collected at Xaak, Akumal Bay, and Sian Ka'an to estimate above- and belowground biomass and root:shoot ratio. A 15-cm-diameter corer was used to sample *T. testudinum,* and a 9 cm diameter corer was used to sample *H. wrightii* and *S. filiforme*. Samples of each species present were collected at each site. Cores were sieved in the field to remove sediment from the roots and rhizomes. After collection, seagrass samples were placed in plastic bags and refrigerated upon arrival to the laboratory. Processing of all biomass samples was completed within 3 days of their collection. In the laboratory, biomass cores were

TABLE 12.1
Sampling Sites and Habitat/Land Use Characteristics of the Study Sites

Study Site	Type of System/Habitat Description	Salinity	Land Use Characteristics
Xaak Bay	Bay, shallow reef and seagrass beds	37	Large resort
Yal Kú Chico	Bay, salt wedge estuary fed by cenote	Surface: 12 Depth: 38	Undeveloped forest
Yal Kú Lagoon	Bay, salt wedge estuary fed by cenote	37	S developed, N undeveloped
Akumal Bay	Bay, shallow reef and seagrass beds	37	Large resort
Sian Ka'an Bay mouth	Bay, inlet to Caribbean from bay	20	Nature reserve
Sian Ka'an Bay middle	Open bay between tidal channel and inlet	24	Nature reserve
Sian Ka'an Cenote	Cenote, submarine groundwater discharge	19	Nature reserve
Bat Cave Cenote	Cenote	3	Forested NW of development
Pueblo Cenote	Cenote	2	Forested NW of development
Laguna Lagartos sinkhole	Cenote	11	Forested NE of development
Laguna Lagartos	Cenote, open lagoon	11	Forested NE of development
Beach Cenote	Cenote	5	Undeveloped forest
Casa Cenote	Cenote, inlet fed by cenote	11	Forested inland of small resorts

sorted by species, and the aboveground portions of the shoots were separated from the belowground portions. Samples were then dried at 60°C to estimate seagrass biomass (g m^{-2}).

12.2.3 Statistical Analysis

Statistical analyses were performed using Statistical Analysis System version 9.1 (SAS, Cary, North Carolina). One-way and two-way analyses of variance (ANOVA) were done on elemental and isotopic composition of seagrass species with site and year as factors. Species and site were factors in ANOVAs performed on seagrass biomass data. Residuals of the fitted models were tested for homogeneity of variance and departures from normality. Kolmogorov–Smirnov tests indicated significant departures from normality for the variables above- and belowground biomass and %P of *Thalassia testudinum* tissue ($p < 0.05$). These variables were $\log_{10}$-transformed to satisfy the normality assumption. For all other seagrass variables, no transformations were necessary. Post hoc Tukey Honestly Significantly Different tests were performed to identify differences among treatment combinations. For all analyses, alpha was set at 0.05.

For water column NO_3^- and NH_4^+ concentrations, comparisons were made between bays and cenotes. Sites were assigned to bay or cenote classification as presented in Tables 12.1 and 12.2. Normality assumptions were not met for the nutrient concentrations; therefore, one-way ANOVA was performed on ranked nutrient data. Post hoc Tukey Honestly Significantly Different tests were performed on the ranks to identify differences among treatment combinations. For all analyses, alpha was set at 0.05.

12.3 RESULTS

12.3.1 Water Chemistry

Water column NO_3^- differed between cenote and bay sampling sites; however, NH_4^+ concentration did not (NO_3^-: $n = 27$, df = 1, $F = 48.76$, $p < 0.0001$; NH_4^+: $n = 23$, df = 1, $F = 2.85$, $p > 0.11$). An overall trend of decreasing NO_3^- concentrations with increasing salinity is apparent (Table 12.2). The cenote sites had an average salinity of 10 ± 8.6 psu ($\bar{x}$ ± SD), and the bay sites had salinity of 32.8 ± 7.2 psu, with average NO_3^- concentrations of 62.6 ± 27.7 μM and 2.8 ± 2.6 μM, respectively.

TABLE 12.2
Water Column Parameters at All Sites

Location	Temperature (°C)	Salinity (psu)	DO (mg L^{-1})	pH	NO_3^- (μM)	NH_4^+ (μM)	Chl (μg L^{-1})
Cenote Systems							
Yal Ku Chico	26.0 (1)	11.6 (1)	3.3 (1)	7.3 (1)	74.9 (1)	n.d.	0.2 (1)
Bat Cave	25.0 ± 0 (2)	2.7 ± 0 (2)	3.1 ± 1.3 (2)	7.6 ± 0.1 (2)	65.3 ± 44.3 (2)	0.1 ± 0.1 (2)	0.3 ± 0.1 (2)
Pueblo Cenote	25.9 ± 0.5 (4)	2.4 ± 0.1 (4)	3.9 ± 1.2 (4)	7.7 ± 0.2 (4)	65.0 ± 18.2 (5)	2.1 ± 1.1 (4)	2.9 ± 1.5 (4)
Laguna Lagartos	26.4 ± 0.7 (3)	10.5 ± 0.1 (3)	4.0 ± 1.5 (3)	7.5 ± 0.2 (3)	58.1 ± 26.6 (4)	3.7 ± 2.5 (3)	4.4 ± 6.9 (3)
Beach Cenote	25.3 ± 0 (2)	5.2 ± 0.1 (2)	2.6 ± 0.4 (2)	7.4 ± 0 (2)	62.8 ± 24.9 (2)	0.2 ± 0.1 (2)	0.3 ± 0.1 (2)
Casa Cenote	25.8 ± 0 (4)	11.1 ± 0.2 (4)	2.9 ± 1.7 (4)	7.3 ± 0.1 (4)	85.1 ± 10.1 (6)	0.0 ± 0 (2)	0.4 ± 0.4 (6)
Sian Kaan Cenote	25.8 ± 1.7 (2)	19.4 ± 0.1 (2)	2.4 ± 2.5 (2)	7.3 ± 0.1 (2)	10.9 ± 10.8 (6)	2.6 ± 3.0 (3)	0.4 ± 0.4 (3)
Bay Systems							
Xaak Bay	28.3 ± 0.5 (3)	37.0 ± 1.3 (3)	5.1 ± 0.7 (3)	8.1 ± 0.1 (3)	2.2 ± 0.9 (5)	0.0 ± 0 (4)	0.2 ± 0 (4)
Yal Ku Chico Mouth	28.8 (1)	37.6 (1)	6.6 (1)	8.3 (1)	1.0 (1)	0.0 (1)	0.2 (1)
Yal Ku Lagoon Mouth	29.1 (1)	37.0 (1)	6.6 (1)	8.3 (1)	0.6 (1)	0.0 (1)	0.3 (1)
Akumal Bay	28.8 ± 0.4 (2)	37.1 ± 0.9 (2)	5.5 ± 0.5 (2)	8.4 ± 0.1 (2)	1.7 ± 2.2 (4)	0.0 ± 0 (4)	0.3 ± 0.3 (4)
Sian Ka'an Mouth	28.3 ± 0.4 (2)	24.0 ± 1.6 (2)	6.5 ± 0.4 (2)	8.0 ± 0.1 (2)	6.8 ± 3.0 (4)	2.2 ± 0.4 (6)	0.6 ± 0.2 (2)
Sian Ka'an Mid	28.3 (1)	19.8 (1)	6.5 (1)	8.0 (1)	2.5 (1)	2.5 (1)	0.5 (1)
Average							
Cenotes	26.0 (±0.9)	10.0 (±8.6)	3.4 (±1.5)	7.5 (±0.3)	62.6 (±27.7)*	1.7 (±2.0)	1.5 (±2.9)
Bays	28.5 (±0.4)	32.8 (±7.2)	5.9 (±0.8)	8.2 (±0.2)	2.8 (±2.6)	0.8 (±1.2)	0.3 (±0.2)

Note: Values are $\overline{x} \pm SE$ (n). * indicates significant differences between bay and cenote sites. $\alpha = 0.05$. n.d. = not determined.

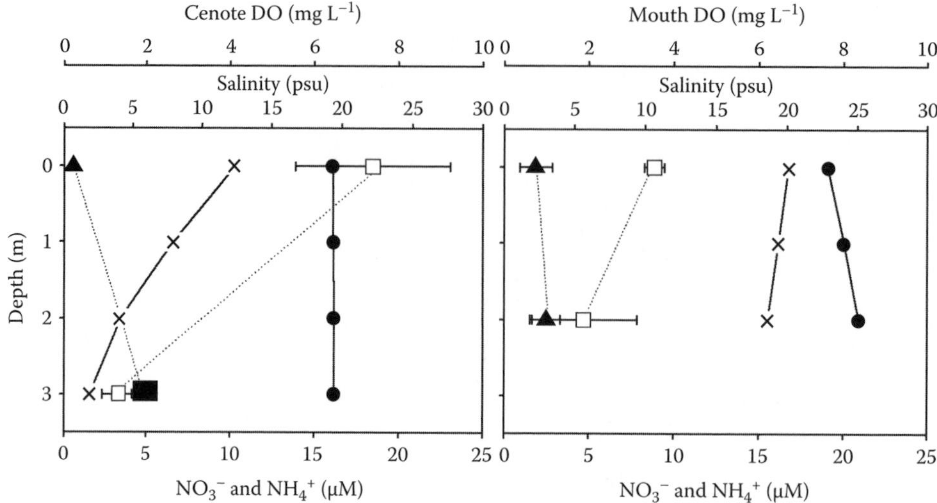

FIGURE 12.2 Vertical profiles of Sian Ka'an Cenote and Sian Ka'an mouth (opening from ocean to Ascension Bay). Maximum depth at mouth site is 2 meters. Temperature and chlorophyll *a* are constant at both sites throughout water column. Note the precipitous drop in NO_3^- that occurs in concert with a large decline in DO within the cenote. Symbols represent the following: salinity ●, nitrate (NO_3^-) □, ammonium (NH_4^+) ▲, and dissolved oxygen (DO) ×.

Within the cenote sites, NO_3^- concentrations were highest at Casa Cenote and lowest at Sian Ka'an cenote (Table 12.2). At the bay sites, the mouth of Sian Ka'an had the highest NO_3^- concentrations (Table 12.2). NH_4^+ concentrations in cenote and bay sites were 1.7 ± 2 μM and 0.8 ± 1.2 μM, respectively. NH_4^+ concentrations were higher at Laguna Lagartos than at all other sites (Table 12.2). To a lesser extent, the Pueblo Cenote and Sian Ka'an had relatively elevated NH_4^+ concentrations.

Vertical profiles from the Sian Ka'an cenote and mouth of the bay contrasted greatly (Figure 12.2). Cenote NO_3^- and DO concentrations decrease rapidly at depth along with increasing NH_4^+. At the mouth site, NO_3^- concentrations are higher at the surface with slightly higher salinity. NH_4^+ and DO concentrations are constant through the water column.

12.3.2 Seagrass Biomass

ANOVA indicated significant main effects of location and species for $\log_{10}$-transformed aboveground biomass; however, the location × species interaction term was not significant. (Table 12.3; Figure 12.3). Aboveground seagrass biomass differed significantly at all three sites ($p < 0.05$ in all comparisons), with the highest values in Sian Ka'an (410 ± 172 g m^{-2}), intermediate values in Akumal Bay (184 ± 177 g m^{-2}), and lowest values in Xaak. Also, *T. testudinum* (332 ± 228 g m^{-2}) had significantly greater aboveground biomass than *H. wrightii* (124 ± 146 g m^{-2}) and *S. filiforme* (101 ± 64 g m^{-2}). For $\log_{10}$-transformed belowground biomass, the location × species interaction term was significant (Table 12.3). Belowground biomass for *T. testudinum* tended to be high at all sites, while the belowground biomass of *H. wrightii* was highest at Sian Ka'an (Figure 12.3). Belowground biomass of *S. filiforme* was relatively low at all sites and was not collected at Sian Ka'an.

The root:shoot ratio also exhibited significant main effects of location and species but no significant interaction term (Table 12.3). The root:shoot ratios of seagrass were significantly lower ($p < 0.05$) in Akumal Bay (3.8 ± 1.7) and Sian Ka'an (3.1 ± 4.1) than Xaak (10.9 ± 6.4), reflecting the greater aboveground biomass at this location. Similarly, the large amounts of aboveground biomass

TABLE 12.3
Results of Two-Way ANOVA Examining Effects of Location (Akumal Bay, Sian Ka'an, and Xaak) and Species (*H. wrightii*, *S. filiforme*, and *T. testudinum*) on Seagrass Biomass Variables

Response Variable	Factor	df	MS	*F*	*p*-value
Log_{10} aboveground biomass	Location	2	3.1	28.6	<0.0001*
	Species	2	1.4	13.5	0.0001*
	Location × species	3	0.2	2.1	0.1
Log_{10} belowground biomass	Location	2	1.8	23.3	<0.0001*
	Species	2	3.4	44.6	<0.0001*
	Location × species	3	0.3	3.4	0.04*
Root:shoot	Location	2	60.4	6.7	0.01*
	Species	2	79.6	8.9	0.001*
	Location × species	3	26	2.9	0.06

Note: $n = 4$ for all location and species combinations except *T. testudinum* ($n = 3$) and *S. filiforme* ($n = 0$) from Sian Ka'an. * indicates significance at $\alpha = 0.05$.

for *T. testudinum* led to significantly greater root:shoot ratios (12.6 ± 5.3) when compared to those for *H. wrightii* (2.9 ± 2.4) and *S. filiforme* (4.6 ± 2.7).

12.3.3 Plant Isotope Values and Tissue Nutrient Content

$\delta^{15}N$ values for algae were elevated at several sites, especially those collected from Laguna Lagartos (9‰ and 12‰ for *Chara* sp. and *Cladophora* sp., respectively; Table 12.4) and the Sian Ka'an Mouth (13‰ for *Batophora* sp.; Table 12.3). $\delta^{13}C$ values, on the other hand, were exceptionally depleted, as exemplified by *Boodleopsis pusilla* (–40‰), an aggregate mixture of *Derbesia* sp. and *Polysiphonia* sp. (–41‰), and *Caulerpa verticillata* (–45‰).

Isotope values and tissue nutrient content for seagrass species showed few spatial trends (Table 12.4). $\delta^{13}C$ values for *T. testudinum* were significantly lower at Xaak compared to Akumal Bay, and *H. wrightii* had higher %P content in Akumal Bay compared to Xaak.

No site differences in $\delta^{15}N$ values, %C, or %N were detected for any of the three seagrass species.

12.4 DISCUSSION

Results of this study led to three conclusions relative to the nutrient dynamics in the coastal lagoons of the Caribbean coast of the Yucatan Peninsula: (1) groundwater was characterized by high nutrient concentrations; (2) isotopic signatures of algae suggested the presence of anthropogenic N sources to coastal systems; and (3) seagrass tissue nutrient contents indicated possible changes in the nutrient loads to bay systems.

12.4.1 Water Column Processes in Cenotes and Bays

The high concentration of NO_3^- in groundwater was consistent with reported values for SGD in the Caribbean region (Table 12.5) and other locations with carbonate substrata (Marsh 1977; Johannes 1980). At all cenote sites, NO_3^- was the dominant form of N in the DIN pool, and NH_4^+ was relatively low in most cases. However, somewhat elevated NH_4^+ concentrations were present in Laguna Lagartos (3.7 ± 2.5 μM) and, to a lesser extent, Pueblo Cenote (2.1 ± 1.1 μM) and Sian Ka'an Cenote

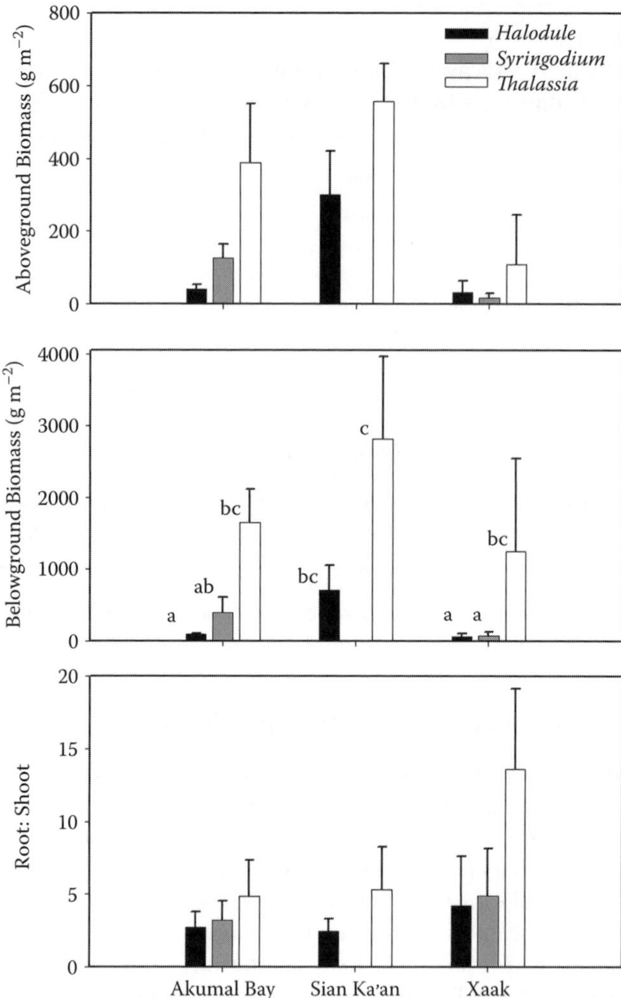

FIGURE 12.3 Aboveground biomass (top panel), belowground biomass (middle panel), and root:shoot ratios (bottom panel) for the three seagrass species collected at Akumal Bay, Sian Ka'an, and Xaak. Values are ± SD. In the middle panel, values with the same letter are not significantly different ($p > 0.05$). See text for description of significant differences for top and bottom panels.

(2.6 ± 3.0 μM). These values were slightly higher than NH_4^+ concentrations in groundwater in the unimpacted Dzilam Lagoon (1.69 ± 0.4 μM; median ± SE) on the northern coast of the Yucatan Peninsula, Mexico (Medina-Gómez and Herrera-Silveira 2003), but they were at the low end of the range of values for impacted groundwater in the Florida Keys (0.77 to 2579 μM; Lapointe et al. 1990). Although NH_4^+ was higher at the Sian Ka'an Bay site (Table 12.2), values were lower than seasonal (3.3 to 11.1 μM) and annual means (5.06 to 8.66 μM) for sites along the northern coast of the Yucatan Peninsula, where shrimp farming, urban wastewater, and SGD are believed to contribute to nutrient loading (Álvarez-Góngora and Herrera-Silveira 2006). Based on these patterns of DIN concentrations alone, there was no strong signal of anthropogenic nutrient inputs during the sampling period.

Despite the lack of measurements of SRP in this study, collections from 2005 at some of these same locations showed that concentrations were <0.5 μM (Table 12.5; Mutchler et al. 2007). Fresh- and brackish water systems that were characterized by high NO_3^- and NH_4^+, such as Casa Cenote and Laguna Lagartos, did not have significantly higher SRP concentrations than did bay

TABLE 12.4
Elemental and Isotopic Composition of Algae and Seagrasses Collected at Study Sites in 2007

Site	Species	*n*	$\delta^{15}N$ (‰)	$\delta^{13}C$ (‰)	%C	%N	%P	C:N	C:P	N:P	C:N:P
Algal Species											
Beach Cenote	*Cladophora* sp.	2	6.2 ± 0.9	–32.4 ± 2.4	26.6 ± 2.7	2.4 ± 1.1	0.2 ± 0.0	14 ± 5	420 ± 64	33 ± 17	420:33:01
Casa Cenote	*Derbesia* & *Polysiphonia* aggregate	1	5.5	–41.2	34.7	4.7		9			
Bat Cave	*Cladophora*	1	6.1	–21.6	18.7	1.3		17			
Pueblo Cenote	*Boodleopsis pusilla*	1	7.8	–40.3	29.1	2.9		12			
Laguna Lagartos	*Chara* sp.	1	9	–26.9	24.2	2.6	0.2	11	300	27	300:27:01
	Cladophora sp.	1	12	–20.4	24	2	0.1	14	500	35	500:35:01
Sian Ka'an mouth	*Batophora* sp.	1	12.5	–24.1	26.9	1.9	0	16	1929	117	1929:117:1
Yal Kú Lagoon	*Caulerpa verticillata*	2	3.9 ± 0.0	–45.1 ± 0.2	22.4 ± 26.1	3.3 ± 3.9		8 ± 1			
Seagrass Species											
Akumal Bay	*Halodule wrightii*	2	1.5 ± 2.5[a]	–10.0 ± 1.0[a]	29.2 ± 3.2[a]	1.6 ± 0.5[a]	0.2 ± 0.0[a]	19 ± 7[a]	419 ± 65[a]	24 ± 5[a]	419:24:01
	Syringodium filiforme	2	2.4 ± 0.5[a]	–7.1 ± 3.7[a]	24.1 ± 0.3[a]	1.6 ± 0.0[a]	0.2 ± 0.1[a]	17 ± 0.1[a]	434 ± 175[a]	25 ± 10[a]	437:25:01
	Thalassia testudinum	2	4.6 ± 1.0[a]	–7.9 ± 0.7[a]	30.9 ± 1.7[a]	2.2 ± 0.2[a]	0.2 ± 0.1[a]	17 ± 1[a]	425 ± 241[a]	26 ± 15[a]	425:26:01
S. Akumal Bay	*Thalassia testudinum*	2	6.1 ± 1.9[a]	–11.0 ± 0.5[ab]	30.9 ± 1.5[a]	2.2 ± 0.4[a]	0.12 ± 0.0[a]	16 ± 2[a]	462 ± 9[a]	28 ± 4[a]	462:28:01
Sian Ka'an mouth	*Halodule wrightii*	1	6.1	–18.2	35.4	2.4		17			
	Thalassia testudinum	2	5.8 ± 0.9[a]	–8.7 ± 0.1[ab]	32.34 ± 3.3[a]	1.8 ± 0.1[a]	0.1 ± 0.0[a]	21 ± 1[a]	625 ± 64[a]	30 ± 2[a]	625:30:1
Xaak	*Halodule wrightii*	2	2.5 ± 1.3[a]	–9.4 ± 0.8[a]	31.4 ± 0.7[a]	1.8 ± 0.1[a]	0.1 ± 0.0[b]	21 ± 1[a]	737 ± 25[b]	36 ± 2[a]	737:36:01
	Syringodium filiforme	2	–0.7 ± 1.4[a]	–6.5 ± 2.0[a]	22.2 ± 4.9[a]	1.7 ± 0.5[a]	0.1 ± 0.0[a]	15 ± 1[a]	478 ± 223[a]	32 ± 17[a]	478:32:01
	Thalassia testudinum	2	4.2 ± 0.8[a]	–11.4 ± 2.1[b]	32.1 ± 0.2[a]	2.3 ± 0.6[a]	0.1 ± 0.0[a]	17 ± 5[a]	630 ± 10[a]	38 ± 12[a]	630:38:01

Note: For seagrasses, means ($\bar{x} \pm SD$) with the same letter are not significantly different than means for the same species at a different site ($p > 0.05$).

TABLE 12.5
Representative Water Chemistry (μM) of SGD throughout the Caribbean Region ($\bar{x}$ ± SD Unless Otherwise Specified)

Location	Year (Season)	Water Type	Salinity (‰)	NH_4^+	$NO_3^- + NO_2^-$	SRP	Study
Discovery Bay, Jamaica	1987–1996	Springs	28	0.22 ± 0.07	27.86 ± 5.80	0.33 ± 0.06	Lapointe 1997
Florida Keys, USA	1987 (summer)	Groundwater control	0.5 ± 0.5	1.4 ± 0.48	0.2 ± 0.23	0.14 ± 0.11	Lapointe et al. 1990
		Groundwater impacted	13.4 ± 2.5	346 ± 222	125 ± 133	4.0 ± 3.35	
Celestún Lagoon, Mexico	1989–1990	Groundwater discharge	2.8	1.75	62.95	0.62 ± 0.3	Calculated from data in Herrera-Silveira 1996
Caribbean coast, Mexico	2007 (dry)	Cenotes	10 ± 8.6	1.7 ± 2	62.6 ± 27.7	n.r.	This study
Akumal vicinity, Mexico	2005 (dry)	Cenotes	32	2.0 ± 1.9	45.1 ± 29	0.4 ± 0.1[a]	Calculated from data presented in Mutchler et al. 2007
Northern coast of Yucatan Peninsula, Mexico	2000	Groundwater	n.r.	4.7[b]	120.52[b]	0.75[b]	Calculated from data in Aranda–Cirerol et al. (2006)
Dzilam Lagoon, Mexico	1998–1999	Groundwater	2.3[b]	1.69[b]	85.3[b]	0.6[b]	Calculated from data in Medina-Gómez and Herrera-Silveira (2003)

Note: n.r. = not reported.

[a] Indicates measurement of inorganic P as ortho-phosphate.

[b] Indicates median values.

sites with low DIN availability (Mutchler et al. 2007). The low SRP availability probably resulted from the adsorption of SRP to carbonate surfaces as well as Fe and Al oxides (De Kanel and Morse 1978; Berner 1981). As noted previously, experimental additions in the Florida Keys indicated that 95% of PO_4^{3-} could be removed within 20 to 50 h of addition, and that removal may occur within short distances of the SRP source (Corbett et al. 2000). Assuming that conditions during this study were similar to those encountered in 2005, we would expect that the groundwater DIN:SRP ratio was characteristically high for carbonate systems (Lapointe et al. 1992).

Comparing these nutrient concentrations to benchmarks for eutrophic conditions in the Caribbean indicated that the bay systems in this study may be slightly enriched. Lapointe et al. (1993) examined natural nutrient gradients in coral reefs of the Belize Barrier Reef (which is contiguous with the barrier reef along the eastern Yucatan Peninsula) and determined that concentrations of ~1 μM DIN and 0.1 μM SRP were sufficient for the development of macroalgal blooms. Further examination of studies conducted around the globe showed that these limits were generally applicable. Coral reefs where macroalgal overgrowth occurred were associated with nutrient concentrations above these thresholds (Lapointe 1997). If these thresholds were reliable predictors of eutrophic conditions in the Caribbean, average concentrations of DIN (3.5 ± 3.4 μM) and possibly SRP (0.3 ± 0.04 μM; Mutchler et al. 2007) in the bay systems have the potential to support extensive growth of macroalgae. Casual observations made during sampling in Akumal Bay appear to support this contention as the area of seagrass covered by patches of *Ceramium* sp. seemed to increase from 2005 to 2007.

12.4.2 Nutrient Sources for Algae and Seagrass

Patterns of nutrient distribution similar to those observed in this study have been attributed to both natural (D'Elia et al. 1981) and anthropogenic causes (Lapointe et al. 1990; Herrera-Silveira 1996; Lapointe 1997; McClelland and Valiela 1998; Lapointe et al. 2005a). In Discovery Bay, Jamaica, D'Elia et al. (1981) concluded that high NO_3^- and low PO_4^{3-} concentrations resulted from natural processes such as precipitation, release during decomposition, concentration via evapotranspiration, N fixation, and adsorption of P to carbonate surfaces. However, Lapointe et al. (1990) demonstrated N enrichment of groundwater by on-site sewage disposal systems in developed areas of the Florida Keys. Wastewater contamination of groundwater resulted in up to 5000-fold N enrichment of the groundwater. Because hypoxic conditions in the drainage fields inhibited nitrification, most of the enrichment was due to additions of NH_4^+ (Lapointe et al. 1990). Numerous additional studies over the past decade have used stable isotope analyses to highlight the importance of land-derived N sources to coastal ecosystems (McClelland and Valiela 1998; Umezawa et al. 2002; Lapointe et al. 2005a).

Stable isotope analyses of algae in this study provided evidence that anthropogenic N sources contributed to nutrient loads in several cenote systems. Wastewater-derived N sources tend to have high $\delta^{15}N$ values (5‰ to 9‰ for untreated wastewater; 10‰ to 22‰ for treated wastewater) as a result of fractionation during nitrogen transformations (Kreitler 1975; Aravena et al. 1993). For example, effluent from septic tanks in Florida had $\delta^{15}N$ values that ranged from 4.6‰ to 19.5‰ (Lapointe and Krupa 1995a, 1995b). These elevated isotope signatures of wastewater N have been used to examine whether anthropogenic N represented a significant source for groundwater and coastal systems (McClelland and Valiela 1998; Lapointe et al. 2005a; Cole et al. 2006). Cole et al. (2005) also demonstrated that $\delta^{15}N$ values of macrophytes were effective indicators of anthropogenic N in aquatic systems, observing a linear relationship between $\delta^{15}N$ values of macrophytes and wastewater contributions to nutrient loading in Cape Cod, Massachusetts. In this study, $\delta^{15}N$ for *Cladophora* sp. (12.0‰) and *Chara* sp. (9.0‰) from Laguna Lagartos as well as *Boodleopsis pusilla* (7.8‰) from Pueblo Cenote clearly suggested the presence of substantial wastewater inputs (Table 12.3). These measurements reinforced earlier values for *Cladophora* sp. (10‰) and $\delta^{15}N$ of NO_3^- (7‰ ± 0.4‰) obtained in 2005 at Laguna Lagartos (Mutchler et al. 2007). In fresh and estuarine water bodies of Cape Cod, $\delta^{15}N$ values of similar magnitude were obtained for macrophytes in Ashumet Pond (6.4‰ to 13.8‰) and Great Pond (7.6‰ to 9.9‰; Cole et al. 2005). Wastewater contributions to these ponds were estimated to be 80% and 60% of the total N load. Similarly, macroalgae in coral reefs near sewage outfalls in Boca Raton, Florida, had an average $\delta^{15}N$ value of 8.5‰ ± 1.3‰ (Lapointe et al. 2005a). Although we did not quantify the contribution of wastewater N to nutrient loads in Laguna Lagartos, comparisons with these other locations suggest that wastewater contamination is significant. Laguna Lagartos is located between Akumal Pueblo and a series of resorts that line the nearby beaches. Nutrients from injection wells or septic systems associated with these developments may easily move through the karst bedrock and into the groundwater that supplies Laguna Lagartos (Whitney et al. 2003).

Wastewater contamination has also been observed in water quality monitoring that is conducted by the Centro Ecologico Akumal, a local nonprofit organization in the area (CEA 2008). Counts of *Escherichia coli* in Laguna Lagartos, Yal Kú Lagoon, and several other freshwater springs near Akumal (including SGD inputs to Akumal Bay) previously exceeded thresholds established by the Mexican government (CEA 2008), and spikes in *E. coli* were correlated with the summer tourist season and rainfall events. This second line of evidence provides strong corroboration that widespread input of nutrients from wastewater infiltrates the groundwater around Akumal Pueblo.

Interestingly, exceptionally high $\delta^{15}N$ values were also observed in samples of *Batophora* sp. collected from the Sian Ka'an Biosphere Reserve. Such high values, suggestive of anthropogenic N sources, were surprising since the reserve is relatively isolated from any nearby development. The high algal $\delta^{15}N$ values suggested that despite the protected status and remoteness of this area, anthropogenic nutrient sources may be traveling longer distances than expected. If wastewaters were in fact entering this system, this would have serious implications for the management of the Biosphere Reserve and potential human impacts on the ecosystem; however, elevated $\delta^{15}N$ values could result from other processes.

Nitrogen transformations such as denitrification also lead to progressive ^{15}N enrichment of the DIN pool (Mariotti et al. 1988). It is possible that the spatial differences in $\delta^{15}N$ values observed in this study arose from variability in denitrification rates among our sites rather than inputs of anthropogenic N. For instance, the thickness of the vadose zone and length of residence times for groundwater in the biogeochemically active upper aquifer may be important determinants for the relative importance of denitrification and isotopic signatures of DIN (Cole et al. 2006). Sites with longer residence times and thick vadose zones would be expected to have higher $\delta^{15}N$ values. Without robust estimates of N transformations, it is difficult to assess the relative importance of denitrification and anthropogenic N sources in determining algal tissue $\delta^{15}N$ values; however, it is likely that both processes are involved.

To further examine this issue, we plotted salinity vs. NO_3^- to characterize the mixing and removal of NO_3^- within the aquifer along the Yucatan Peninsula (Figure 12.4). Although this approach has typically been applied to linearly connected systems (Fisher et al. 1988), we included all study sites in the diagram. The aquifer in our study area extends from Puerto Morelos in the north to the Sian Ka'an Biosphere Reserve in the south and consists of a highly porous cavernous zone with ~700 km of anastomosing caves (Beddows et al. 2007). These interconnected underground caves permit large-scale movement of groundwater over long distances and make identification of linear connections of study sites difficult. For example, 39% of the 30 cm amplitude semidiurnal tidal signal is transmitted to the free water surfaces of cenotes 6 km inland (Beddows 1999). These characteristics of the aquifer suggest a distinct possibility that distant study sites are, in fact, hydrologically linked, although perhaps not linearly. In addition, we reasoned that spatially heterogeneous N mixing should be apparent in the graph as departures from expected trends associated with conservative and nonconservative mixing. For example, lateral inputs of sewage from New York City generated

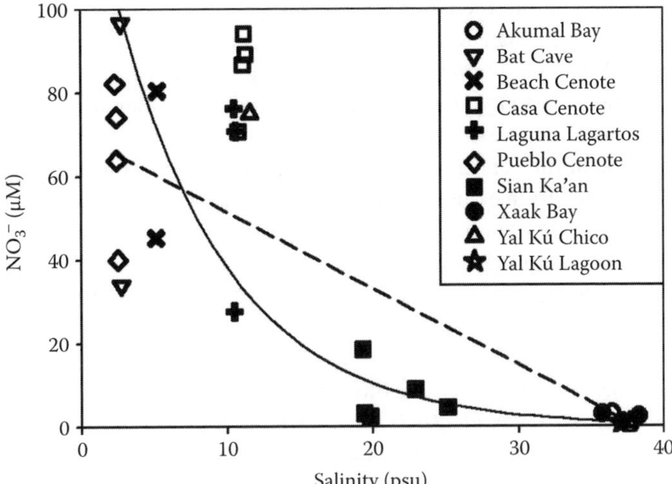

FIGURE 12.4 The relationship between NO_3^- concentration and salinity at all sampling sites. Dashed line is the conservative mixing line created by connecting the average NO_3^- concentration for freshwater (salinity < 3 psu; $\bar{x}$ = 65.1 μM) and marine (salinity > 35 psu; $\bar{x}$ = 1.7 μM) end members in the study area. The solid line is the exponential fit ($y = 139.92*e^{(-0.13x)}$) to all data ($R^2 = 0.80$) and shows nonconservative mixing with N removal.

a convex curve for NO_3^- and salinity mixing in the Hudson River (Fisher et al. 1988). Thus, we considered this approach to be valid for identifying N removal processes and inputs to help resolve ambiguities in algal tissue $\delta^{15}N$ values.

Examination of the salinity vs. NO_3^- relationship indicated that differences in N mixing may exist among our study sites (Figure 12.4). Assuming that the relatively fresh waters of Pueblo Cenote and Bat Cave were representative end-members for the study area as a whole (for comparison to groundwater in other areas of the Caribbean see Table 12.5), Casa Cenote, Yal Kú Chico, and possibly Laguna Lagartos had higher NO_3^- values than would be expected for either conservative mixing or nonconservative mixing with N removal. This pattern is very similar to values obtained previously from these study sites (Mutchler et al. 2007); however, in that instance, Laguna Lagartos served as the freshwater end-member, and interpretations of N mixing for the site were ambiguous.

Several potential explanations exist for the mixing patterns observed in Casa Cenote, Yal Kú Chico, and Laguna Lagartos. First, conservative mixing occurred but a temporal decrease in end-member NO_3^- concentrations resulted in a convex mixing diagram (Loder and Reichart 1981). Alternatively, lateral inputs of NO_3^- from other sources caused localized elevation in NO_3^- along the salinity gradient as was observed in mixing diagrams for the Hudson River mentioned above (Fisher et al. 1988). Without having a better understanding of temporal variability and flushing time of the aquifer near these sites, it was impossible to rule out temporal variation as a factor controlling the shape of the mixing diagram. However, the fact that $\delta^{15}N$ values of macroalgae and NO_3^- concentrations in Laguna Lagartos were both higher than the freshwater end-member suggested that anthropogenic N with elevated $\delta^{15}N$ signatures entered the system. If the elevated $\delta^{15}N$ values of macroalgae resulted from denitrification, we would have expected to see a decline in NO_3^- concentrations. Unfortunately, we did not have isotope data from Casa Cenote or Yal Kú Chico to evaluate the source of N in those systems.

The N mixing diagram indicated that a different scenario occurred at the Sian Ka'an study sites. In this location, a concave, nonconservative mixing line best describes the relationship between NO_3^- and salinity (Figure 12.4). Although this could have arisen from a temporal increase in end-member NO_3^- concentration, the pattern was consistent with N removal. In this case, the high $\delta^{15}N$ values of macroalgae in Sian Ka'an would have resulted from denitrification as opposed to anthropogenic inputs. A water column profile of the cenote in Sian Ka'an showed that NO_3^- and DO both declined with depth. These hypoxic conditions in the cenote may be conducive to denitrification (Figure 12.2). In addition, we noted distinct sulfurous odors and sulfur bacteria at depths >2 m in the cenote. Outside of the cenote, Ascension Bay is a shallow, well-mixed estuary with inputs from the tidal channels that drain extensive mangrove forests. Denitrification may occur in the sediments and contribute to further N removal as water mixes with marine inputs from the Caribbean. At the Sian Ka'an mouth sampling location where the marine influence was greatest, NO_3^- levels were nearly identical to those of Xaak, the marine location at the northern end of the study area (Table 12.2).

The $\delta^{15}N$ values of seagrasses provided no clear indication of extensive inputs of wastewater or terrestrial N to the marine locations. Values for all three species were within the ranges reported elsewhere for these species (Fry et al. 1987; Moncreiff and Sullivan 2001; Carruthers et al. 2005a, 2005b). It is worth noting, however, that values for *T. testudinum* (Table 12.4) were similar to values for the species collected in Nichupte Lagoon (5.49 ± 0.77 at Nichupte south; 9.06 ± 0.73 at Nichupte north) where evidence suggests inputs of wastewater may occur (Carruthers et al. 2005b). Although it is conceivable that the $\delta^{15}N$ values of *T. testudinum* reflected the influence of nitrogen sources with elevated $\delta^{15}N$, this seems unlikely given a lack of similar trends for *H. wrightii* and *S. filiforme*. Interestingly, $\delta^{15}N$ values of *H. wrightii* collected in 2007 actually showed a significant decline from 2005 (Table 12.5). This trend was apparent for the other two seagrass species but was not statistically significant. Such a reduction could be an indication that nitrogen sources or the nitrogen loading rates are temporally variable at these sites.

12.4.3 Nutrient Impacts on Plants and Algae

One of the objectives of this study was to determine if elevated nutrients in the cenote systems generated eutrophic conditions and altered seagrass growth dynamics. Despite the relatively high NO_3^- concentrations in many of these systems, the elevated nutrient levels were not accompanied by extensive phytoplankton or macroalgal biomass characteristic of advanced eutrophication. Overall, chlorophyll *a* (chl *a*) concentrations were generally low (Table 12.2). Chlorophyll *a* concentrations at Sian Ka'an (0.5 ± 0.2 μg L^{-1}), Xaak (0.2 ± 0.0 μg L^{-1}), and Akumal Bay (0.3 ± 0.3 μg L^{-1}) were much lower than average values (2.7 μg L^{-1}) reported for the dry season in lagoons that experience a range of nutrient inputs along the northern coast of the Yucatan (Álvarez-Góngora and Herrera-Silveira 2006). However, chlorophyll *a* for our sites were close to threshold values for eutrophication (0.3 to 0.5 μg L^{-1}) in the Great Barrier Reef Lagoon (Bell and Elmetri 1995) and were slightly lower than average concentrations (0.55 to 1.85 μg L^{-1}) at sites in the Florida Keys that receive land-based nutrient enrichment by seasonal rain events (Lapointe et al. 2004). These comparisons suggested that chlorophyll *a* concentrations were at or near levels associated with eutrophication. Laguna Lagartos and Pueblo Cenote, where NH_4^+ availability was greatest, had distinctly higher chlorophyll *a* concentrations (4.4 ± 6.9 μg L^{-1} and 2.9 ± 1.5 μg L^{-1}, respectively; Table 12.2). For many algae, NH_4^+ is the preferred form of nitrogen (McCarthy et al. 1977), and low NH_4^+ or SRP availability may have limited phytoplankton growth at most locations we studied.

In addition to having higher nutrient and chlorophyll *a* concentrations, Laguna Lagartos also differed from other locations in that *Cladophora* sp. grows extensively at the site. In fact, mats of the algae were removed from the surface in August 2006 (CEA 2008). Although we do not know how frequently such removal of algae occurs, it is apparent that the nutrient load delivered to the lagoon supported much greater biomass than was observed in our phytoplankton estimates. The prevalence of N in Laguna Lagartos was also apparent from the elevated %N content of macroalgal tissues at the site. N content of tissues for *Cladophora* sp. and *Chara* sp. were 2% and 2.6%, respectively. These values were approximately double those reported by Lapointe et al. (2005b) in their broad-scale study throughout the Caribbean region. Because of higher %C contents, however, the C:N ratios of algae from Laguna Lagartos (11 and 14 for *Chara* sp. and *Cladophora* sp., respectively) were similar to those reported for green algae in south Florida (12.4 ± 3.2). Chlorophyta from the Caribbean region had a mean C:N of 20.5 ± 4.9. Based on these comparisons (and those of other algae in Table 12.4), macroalgae from sites with relatively high NO_3^- (Casa Cenote, Pueblo Cenote, and Laguna Lagartos) were less N limited than algae from other locations in the Caribbean region and had tissue nutrient contents more like macroalgae from south Florida where nutrient enrichment and episodic upwelling reduce the likelihood of nutrient limitation (Lapointe et al. 2005b).

Although we had limited data on the P content of macroalgal tissues, a few trends were apparent. Tissue P content for algae from Beach Cenote and Laguna Lagartos (Table 12.4) was nearly double the values reported for green algae from south Florida (0.08 ± 0.03%) and *Codium isthmocladum* from the Caribbean region (0.04 ± 0.01%). However, N:P ratios (Table 12.4) were similar to those of green algae from south Florida (35.5 ± 8.41; Lapointe et al. 2005b). Typically, N:P ratios >30:1 are interpreted as P limited (Smith 1984; Lapointe et al. 1992). Using this benchmark, the data suggested that algal growth in Beach Cenote and Laguna Lagartos were limited slightly by P availability; whereas *Batophora* sp. at the mouth of Sian Ka'an showed signs of severe P limitation. These observations are consistent with the notion that groundwater near Akumal Pueblo experienced nutrient enrichment, but anthropogenic nutrient sources were absent from Sian Ka'an.

Seagrass tissue nutrient contents suggest that inputs of high nutrient SGD from the cenote systems may be impacting near-shore marine systems. For all three species, %C declined significantly from 2005 to 2007, dropping approximately 2% to 6% (Table 12.5). The %N of *T. testudinum* was significantly higher in 2007, and the %P increased significantly for *T. testudinum* in Akumal Bay. In addition, *H. wrightii* had significantly higher %P content in Akumal Bay compared to Xaak. Together, these trends implied that nutrient availability in coastal waters increased over time and was more pronounced in areas adjacent to human development.

Seagrasses are reported to have median C:N:P values of 474:24:1 (Duarte 1990), and macrophytes in general have a ratio of 550:30:1 (Atkinson and Smith 1983). Based on these benchmarks, it appears that all three seagrass species in Akumal Bay were at or near a balance in nutrient content and even may have been slightly elevated in N. Comparisons with other studies in the Caribbean region indicated that nutrient content of *T. testudinum* in Akumal Bay was exceeded only by plants from Bocas del Toro in Panama, where high freshwater inputs resulted from rainfall across mountainous terrain (Tables 12.6 and 12.7). *T. testudinum* in Akumal Bay has very similar C:N:P ratios to plants collected from the Bojórquez Lagoon sites (Table 12.6; sites BJ3 and BJ4) in Nichupte Lagoon on the northeast coast of the Yucatan Peninsula (van Tussenbroek et al. 1996). Bojórquez Lagoon is considered an impacted site that is surrounded by tourism development and has elevated P in the sediments (van Tussenbroek et al. 1996). This similarity coupled with the changes in nutrient content since 2005 suggested that Akumal Bay is undergoing nutrient enrichment. The average change in the C:P and N:P ratios for the three seagrass species in Akumal Bay was a substantial decline of 52% and 39%, respectively. The C:N:P ratios of seagrasses in Xaak also changed in the last 2 years; however, not nearly to the extent that was observed in Akumal Bay. The average C:P ratio for all three seagrasses in Xaak declined 22%, but there was a slight average increase in N:P of 2%. The difference in magnitude of change between Akumal Bay and Xaak demonstrated that the two embayments are on different trajectories regarding nutrient availability. It will be important to monitor this situation closely because Akumal Bay is one of the few places where green sea turtles regularly forage on seagrass, and a large beach front hotel was constructed at Xaak between 2005 and 2007. Therefore, changes in nutrient loading of these bays with continued coastal development may affect the seagrass populations and hence the supply of forage for sea turtles.

TABLE 12.6
Comparisons of Stable Isotope and Tissue Nutrient Content of Seagrasses Collected in 2005 and 2007

Species	Site	Year	*n*	$\delta^{15}N$ (‰)	%C	%N	%P	C:N	C:N:P
	Akumal Bay	2005	3	5 ± 1^{a}	36 ± 0.6^{a}	2 ± 0.2^{a}	0.16 ± 0.06^{a}	20 ± 2^{a}	626:32:1
H. wrightii		2007	2	1.53 ± 2.54^{b}	29.16 ± 3.16^{b}	1.56 ± 0.53^{a}	0.18 ± 0.01^{a}	19 ± 7^{a}	419:24:1
	Xaak	2005	3	4 ± 0.2^{a}	36 ± 2^{a}	2 ± 0.2^{a}	0.1 ± 0.01^{b}	25 ± 2^{a}	910:36:1
		2007	2	2.47 ± 1.30^{b}	31.43 ± 0.73^{b}	1.77 ± 0.09^{a}	0.11 ± 0.001^{b}	21 ± 1^{a}	737:36:1
	Akumal Bay	2005	3	3 ± 1^{a}	29 ± 0.6^{a}	2 ± 0.2^{a}	0.08 ± 0.02^{a}	21 ± 3^{a}	955:46:1
S. filiforme		2007	2	2.40 ± 0.47^{a}	24.06 ± 0.34^{b}	1.61 ± 0.01^{a}	0.15 ± 0.06^{a}	17 ± 0.1^{b}	437:25:1
	Xaak	2005	3	1 ± 3^{a}	28 ± 0.5^{a}	2 ± 0.2^{a}	0.13 ± 0.02^{a}	20 ± 3^{a}	582:29:1
		2007	2	-0.73 ± 1.43^{a}	22.18 ± 4.91^{b}	1.69 ± 0.52^{a}	0.13 ± 0.03^{a}	15 ± 1^{b}	478:32:1
	Akumal Bay	2005	3	7 ± 1^{a}	32 ± 2^{a}	1 ± 0.3^{a}	0.06 ± 0.01^{a}	29 ± 5^{a}	1336:47:1
		2007	2	4.64 ± 0.95^{a}	30.92 ± 1.65^{b}	2.16 ± 0.18^{b}	0.22 ± 0.11^{b}	17 ± 1^{b}	425:26:1
T. testudinum	Xaak	2005	3	6 ± 1^{a}	36 ± 0.2^{a}	2 ± 0.2^{a}	0.11 ± 0.03^{ab}	22 ± 2^{a}	867:40:1
		2007	2	4.24 ± 0.82^{a}	32.07 ± 0.23^{b}	2.27 ± 0.64^{b}	0.13 ± 0.003^{ab}	17 ± 5^{b}	630:38:1
	S. Akumal Bay	2007	2	6.10 ± 1.91^{a}	30.94 ± 1.51^{b}	2.22 ± 0.40^{b}	0.17 ± 0.01^{b}	16 ± 2^{b}	462:28:1
	Sian Ka'an mouth	2007	2	5.81 ± 0.88^{a}	32.37 ± 3.33^{b}	1.80 ± 0.10^{b}	0.13 ± 0.0001^{ab}	21 ± 1^{b}	625:30:1

Source: Data from Mutchler et al. 2007.

Note: Means ($\bar{x} \pm$ s.d.) with the same letter are not significantly different from means for the same species at a different site ($p > 0.05$).

TABLE 12.7
Comparison of Mean Tissue Nutrient Content for *T. testudinum* from Several Studies in the Caribbean Region

Location	Year (season)	Site	%C	%N	%P	C:N	C:P	N:P	C:N:P	Study
Florida Bay, USA	1987–1989	Porjoe Key	34.6	2.2	0.095	18.5	1,070	58.6		Fourqurean et al. 1992b
Nichupte Lagoon, Mexico	1991	CN1	38.53	1.871	0.11				851:41:1	van Tussenbroek et al. 1996
		CN2	37.5	2.31	0.15				594:36:1	
		BJ3	38.0	1.95	0.18				488:25:1	
		BJ4	38.2	2.14	0.2				476:26:1	
	2002 (dry)	North	33.49	2.93	0.17				541:42:1	Carruthers et al. 2005b
		South	34.35	2.5	0.13				794:50:1	
Puerto Morelos Lagoon, Mexico	2002 (dry)	Lagoon	35.25	1.80	0.13				740:32:1	
		Springs	35.60	2.11	0.18				528:26:1	
Bocas del Toro, Panama	2003	Bahía Almirante	31.5	2.46	0.26	15.1	313.9	20.9		Carruthers et al. 2005a
		Laguna de Chiriquí	32.6	2.38	0.21	16.4	419.1	25.7		
		Outer Lagoon	33.4	2.45	0.27	16.0	326.8	20.3		
Akumal vicinity, Mexico	2005 (dry)	Akumal Bay	32	1	.06	29	1336	47	1336:47:1	Mutchler et al. 2007
		Xaak	36	2	0.11	22	867	40	867:40:1	
	2007 (dry)	Akumal Bay	30.92	2.16	0.22	17	425	26	425:26:1	This study
		Xaak	32.07	2.27	0.13	17	630	38	630:38:1	
		Sian Ka'an mouth	32.37	1.80		21				

ACKNOWLEDGMENTS

Funding for this project was provided through the University of Texas at Austin Executive Vice President and Provost's 2007 Initiative for International Studies Maymester Abroad Program to Ken Dunton. P. Sánchez-Navarro, Director of Centro Ecológico Akumal (CEA), provided logistical support and laboratory facilities. We are also very grateful to B. van Tussenbroek at the Puerto Morelos Marine Laboratory for her long-term friendship and invaluable assistance in the procurement of chemicals needed for some of our on-site nutrient analyses. Karli Dunton and L. Bush provided meals and housing, respectively, during the study. We thank E. Sosa Bravo from CEA for sharing her local knowledge of sampling sites and land use patterns. K. Jackson provided assistance with field logistics, sample collection, and analysis. Our colleagues at Texas A&M University-Corpus Christi (led by W. Tunnell and G. Stunz) provided valuable travel information and vessel support for sampling in Sian Ka'an. Finally, we thank Maymester students D. Allen, M. Barajas, N. Butler, C. Cook, W. Dela Cruz, K. Garibay, M. Gil, A. Idlett, R. McBryde, N. Morgan, R. Murray, N. Raheja, K. Robertson, and H. Vigander, who assisted with sample collection, processing, and analysis. We are also grateful to S. Schonberg, who provided useful comments that improved the quality of the manuscript.

REFERENCES

Álvarez-Góngora, C. and J. A. Herrera-Silveira. 2006. Variations in community structure related to water quality trends in a tropical karstic coastal zone. *Mar. Pollut. Bull.* 52: 48–60.

Aranda-Cirerol, N., J. A. Herrera-Silveira, and F. A. Comín. 2006. Nutrient water quality in a tropical coastal zone with groundwater discharge, northwest Yucatán, Mexico. *Estuar. Coastal Shelf Sci.* 68: 445–454.

Aravena, R., M. L. Evans, and J. A. Cherry. 1993. Stable isotopes of oxygen and nitrogen in source identification of nitrate from septic systems. *Ground Water* 31: 180–186.

Atkinson, M. J. and S. V. Smith. 1983. C:N:P ratios of benthic marine plants. *Limnol. Oceanogr.* 28: 568–574.

Beddows, P. A. 1999. *Conduit hydrogeology of a tropical coastal carbonate aquifer: Caribbean coast of the Yucatan Peninsula*. M.S. Thesis, McMaster University, Hamilton, Ontario, Canada.

Beddows, P. A., P. L. Smart, F. F. Whitaker, and S. L. Smith. 2007. De-coupled fresh-saline groundwater circulation of a coastal carbonate aquifer: Spatial patterns of temperature and specific electrical conductivity. *J. Hydrol.* 346: 18–32.

Bell, P. R. F. and I. Elmetri. 1995. Ecological indicators of large-eutrophication in the Great Barrier Reef Lagoon. *Ambio* 24: 208–215.

Berner, R. A. 1981. Authigenic mineral formation resulting from organic matter decomposition in modern sediments. *Fortschr. Miner.* 59: 117–135.

Capone, D. G. and M. F. Bautista. 1985. A groundwater source of nitrate in nearshore marine sediments. *Nature* 313: 214–216.

Carruthers, T. J. B., P. A. G. Barnes, G. E. Jacome, and J. W. Fourqurean. 2005a. Lagoon-scale processes in a coastally influenced Caribbean system: Implications for the seagrass *Thalassia testudinum. Caribb. J. Sci.* 41: 441–455.

Carruthers, T. J. B., B. I. van Tussenbroek, and W. C. Dennison. 2005b. Influence of submarine springs and wastewater on nutrient dynamics of Caribbean seagrass meadows. *Estuar. Coastal Shelf Sci.* 64: 191–199.

CEA. 2008. Global Characterization of Akumal Coastal Aquifer. *Water quality monitoring 2006–2007*. Internal Institutional Report, Centro Ecologico Akumal, Centro de Investigación Científica de Yucatán, Unidad Quintana Roo.

Cole, M. L., K. D. Kroeger, J. W. McClelland, and I. Valiela. 2005. Macrophytes as indicators of land-derived wastewater: Application of a $\delta^{15}N$ method in aquatic systems. *Water Resour. Res.* 41: W01014.

Cole, M. L., K. D. Kroeger, J. W. McClelland, and I. Valiela. 2006. Effects of watershed land use on nitrogen concentrations and δ^{15} nitrogen in groundwater. *Biogeochemistry* 77: 199–215.

Cooper, H. H. 1959. A hypothesis concerning the dynamic balance of fresh water and salt water in a coastal aquifer. *J. Geophys. Res.* 64: 461–467.

Corbett, D. R., L. Kump, K. Dillon, W. Burnett, and J. Chanton. 2000. Fate of wastewater-borne nutrients under low discharge conditions in the subsurface of the Florida Keys, USA. *Mar. Chem.* 69: 99–115.

Corredor, J. E., J. M. Morell, and M. R. Díaz. 1992. Environmental degradation, nitrogen dynamics and proliferation of the filamentous cyanophyte *Micocoleus lyngbyaceus* in nearshore Caribbean waters. In: Amato, E. (ed.), *Meiterraneo e Caraibe due mari in pericolo? Sversamenti accidentali di idrocarburi ed emergenze causate dalle alghe*. ICRAM/IFREMER. Atti del Convegno Internazionale. Genova, Italy.

De Kanel, J. and J. W. Morse. 1978. The chemistry of orthophosphate uptake from seawater on to calcite and aragonite. *Geochim. Cosmochim. Acta* 42: 1335–1340.

D'Elia, C. F., K. L. Webb, and J. W. Porter. 1981. Nitrate-rich groundwater inputs to Discovery Bay, Jamaica: A significant source of N to local coral reefs? *Bull. Mar. Sci.* 31: 903–910.

Diaz, M. R., J. E. Corredor, and J. M. Morell. 1990. Inorganic nitrogen uptake by *Microcoleus lyngbyaceus* mat communities in a semi-eutrophic marine community. *Limnol. Oceanogr.* 35: 1788–1795.

Duarte, C. 1990. Seagrass nutrient content. *Mar. Ecol. Progr. Ser.* 67: 201–207.

Duarte, C. M., W. C. Dennison, R. J. W Orth, and T. J. B. Carruthers. 2008. The charisma of coastal ecosystems: Addressing the imbalance. *Estuaries Coasts* 31: 233–238.

Finkl, C. W. and R. H. Charlier. 2003. Sustainability of subtropical coastal zones in southeastern Florida: Challenges for urbanized coastal environments threatened by development, pollution, water supply, and storm hazards. *J. Coastal Res.* 19: 934–943.

Fisher, T. R., L. W. Harding, Jr., D. W. Stanley, and L. G. Ward. 1988. Phytoplankton, nutrients, and turbidity in the Chesapeake, Delaware, and Hudson estuaries. *Estuar. Coastal Shelf Sci.* 27: 61–93.

Fourqurean, J. W., J. C. Zieman, and G. V. N. Powell. 1992a. Relationships between porewater and seagrasses in a subtropical carbonate environment. *Mar. Biol.* 114: 57–65.

Fourqurean, J. W., J. C. Zieman, and G. V. N. Powell. 1992b. Phosphorus limitation of primary production in Florida Bay: Evidence from C:N:P ratios of the seagrass *Thalassia testudinum*. *Limnol. Oceanogr.* 37: 162–171.

Fry, B., S. A. Macko, and J. C. Zieman. 1987. Review of stable isotopic investigations of food webs in seagrass meadows. Pp. 117–138, In: M. J. Durako, R. C. Phillips and R. R. Lewis (eds.), *Proceedings of a symposium on subtropical-tropical seagrasses, southeast United States*. Florida Marine Resources Publication No. 42, Florida Department of Natural Resources, Bureau of Marine Research, St. Petersburg, Florida.

Herrera-Silveira, J. A. 1996. Salinity and nutrients in a tropical coastal lagoon with groundwater discharges into the Gulf of Mexico. *Hydrobiologia* 321: 165–176.

Hughes, T. P., A. H. Baird, D. R. Bellwood, M. Card, S. R. Connolly, C. Folke, R. Grosberg, O. Hoegh-Guldberg, J. B. C. Jackson, J. Kleypas, J. M. Lough, P. Marshall, M. Nyström, S. R. Palumbi, J. M. Pandolfi, B. Rosen, and J. Roughgardern. 2003. Climate change, human impacts, and the resilience of coral reefs. *Science* 301: 929–933.

Johannes, R. E. 1980. The ecological significance of the submarine discharge of groundwater. *Mar. Ecol. Prog. Ser.* 3: 365–373.

Kreitler, C. W. 1975. *Determining the source of nitrate in ground water by nitrogen isotope studies*. Report of Investigations 83. Bureau of Economic Geology, University of Texas, Austin, Texas.

Lapointe, B. E. 1987. Phosphorus and nitrogen-limited photosynthesis and growth of *Gracilaria tikvahiae* (Rhodophyceae) in the Florida Keys: An experimental field study. *Mar. Biol.* 93: 561–568.

Lapointe, B. E. 1997. Nutrient thresholds for bottom-up control of macroalgal blooms on coral reefs in Jamaica and southeast Florida. *Limnol. Oceanogr.* 42: 1119–1131.

Lapointe, B. E., P. J. Barile, M. M. Littler, and D. S. Littler. 2005a. Macroalgal blooms on southeast Florida coral reefs. II. Cross-shelf discrimination of nitrogen sources indicates widespread assimilation of sewage nitrogen. *Harmful Algae* 4: 1106–1122.

Lapointe, B. E., P. J. Barile, M. M. Littler, D. S. Littler, B. Bedford, and C. Gasque. 2005b. Macroalgal blooms on southeast Florida coral reefs. I. Nutrient stoichiometry of the invasive green alga *Codium isthmocladum* in the wider Caribbean indicates nutrient enrichment. *Harmful Algae* 4: 1092–1105.

Lapointe, B. E., P. J. Barile, and W. R. Matzie. 2004. Anthropogenic nutrient enrichment of seagrass and oral reef communities in the Lower Florida Keys: Discrimination of local versus regional nitrogen sources. *J. Exp. Mar. Biol. Ecol.* 308: 23–58.

Lapointe, B. E. and M. W. Clark. 1992. Nutrient inputs from the watershed and coastal eutrophication in the Florida Keys. *Estuaries* 15: 465–476.

Lapointe, B. E. and S. Krupa. 1995a. *Jupiter Creek septic tank water quality investigation*. Final Report to the Loxahatchee River Environmental Control District, Jupiter, Florida.

Lapointe, B. E. and S. Krupa. 1995b. *Tequesta Peninsula septic tank water quality investigation*. Final Report to the Loxahatchee River Environmental Control District, Jupiter, Florida.

Lapointe, B. E., M. M. Littler, and D. S. Littler. 1992. Nutrient availability to marine macroalgae in siliciclastic versus carbonate-rich coastal waters. *Estuaries* 15: 75–82.

Lapointe, B. E., M. M. Littler, and D. S. Littler. 1993. Modification of benthic community structure by natural eutrophication: The Belize Barrier Reef. *Proc. 7th Int. Coral Reef Symp*. University of Miami, Miami, Florida 1: 323–334.

Lapointe, B. E. and J. D. O'Connell. 1989. Nutrient-enhanced growth of *Cladophora prolifera* in Harrington Sound, Bermuda: Eutrophication of a confined, phosphorus-limited marine ecosystem. *Est. Coastal Shelf Sci.* 28: 347–360.

Lapointe, B. E., J. D. O'Connell, and G. S. Garrett. 1990. Nutrient couplings between on-site sewage disposal systems, groundwaters, and nearshore surface waters of the Florida Keys. *Biogeochemistry* 10: 289–307.

Lessios, H. A. 1988. Mass mortality of *Diadema antillarum* in the Caribbean: What have we learned? *Annu. Rev. Ecol. Syst.* 19: 371–393.

Littler, M. M., D. S. Littler, and B. E. Lapointe. 1988. A comparison of nutrient- and light-limited photosynthesis in psammophytic versus epilithic forms of *Halimeda* (Caulerpales, Halimedaceae) from the Bahamas. *Coral Reefs* 6: 219–225.

Littler, M. M., D. S. Littler, and B. E. Lapointe. 1992. Modification of tropical reef community structure due to cultural eutrophication: The southwest coast of Martinique. *Proc. 7th Coral Reef Symp.*, Guam 1: 335–343.

Loder, T. C. and R. P. Reichard. 1981. The dynamics of conservative mixing in estuaries. *Estuaries* 4: 64–69.

Mariotti, A., A. Landeau, and B. Simon. 1988. ^{15}N isotope biogeochemistry and natural denitrification process in groundwater: Application to the chalk aquifer of northern France. *Geochim. Cosmochim. Acta* 52: 1869–1878.

Marsh, J. A. 1977. Terrestrial inputs on nitrogen and phosphorus on fringing reefs of Guam. *Proc. 3rd Int. Coral Reef Symp.*, University of Miami, Miami, Florida 1: 331–336.

McCarthy, J. J., W. R. Taylor, and J. L. Taft. 1977. Nitrogenous nutrition of the plankton in the Chesapeake Bay. I. Nutrient availability and phytoplankton preferences. *Limnol. Oceanogr.* 22: 996–1011.

McClelland, J. W. and I. Valiela. 1998. Linking nitrogen in estuarine producers to land-derived sources. *Limnol. Oceanogr.* 43: 577–585.

Medina-Gómez, I. and J. A. Herrera-Silveira. 2003. Spatial characterization of water quality in a Kartic coastal lagoon without anthropogenic disturbance: A multivariate approach. *Estuar. Coastal Shelf Sci.* 58: 455–465.

Moncreiff, C. A. and M. J. Sullivan. 2001. Trophic importance of epiphytic algae in subtropical seagrass beds: Evidence from multiple stable isotope analyses. *Mar. Ecol. Prog. Ser.* 215: 93–106.

Moore, W. S. 1996. Large groundwater inputs to coastal waters revealed by ^{226}Ra enrichments. *Nature* 380: 612–614.

Mutchler, T., K. H. Dunton, A. Townsend-Small, S. Fredriksen, and M. K. Rasser. 2007. Isotopic and elemental indicators of nutrient sources and status of coastal habitats in the Caribbean Sea, Yucatan Peninsula, Mexico. *Estuar. Coastal Shelf Sci.* 74: 449–457.

Pandolfi, J. M., R. H. Bradbury, E. Sala, T. P. Hughes, K. A. Bjorndal, R. G. Cooke, D. McArdle, L. McClenachan, M. J. H. Newman, G. Paredes, R. R. Warner, and J. B. C. Jackson. 2003. Global trajectories of the long-term decline of coral reef ecosystems. *Science* 301: 955–957.

Simmons, J. A. K., T. Jickells, A. Knap, and W. B. Lyons. 1985. Nutrient concentrations in groundwaters from Bermuda: Anthropogenic effects. Pp. 383–398, In: D. E. Caldwell and J. A. Brierly (eds.), *Planetary ecology*. New York: Van Nostrand Reinhold Company.

Smith, S. V. 1984. Phosphorus versus nitrogen limitation in the marine environment. *Limnol. Oceanogr.* 29: 1149–1160.

Smith, S. V., R. E. Brock, J. T. Harrison, J. Hirota, P. L. Jokiel, W. Kimmerer, D. W. Kinsey, E. A. Laws, D. G. Redalje, S. Taguchi, and T. W. Walsh. 1978. *Kaneohe Bay sewage relaxation experiment: Pre-diversion report.* US EPA Report, Hawaii Institute of Marine Biology. Kaneohe, Hawaii.

Solorzano, L. and J. H. Sharp. 1980. Determination of total dissolved phosphorus and particulate phosphorus in natural waters. *Limnol. Oceanogr.* 25: 754–757.

Strickland, J. D. H. and T. R. Parsons. 1972a. Determination of reactive nitrate. Pp. 71–76, In: *A practical handbook of seawater analysis*. Bulletin 167 (2nd edition). Fisheries Research Board of Canada, Ottawa.

Strickland, J. D. H. and T. R. Parsons. 1972b. Determination of ammonia. Pp. 87–89, In: *A practical handbook of seawater analysis*. Bulletin 167 (2nd edition). Fisheries Research Board of Canada, Ottawa.

Umezawa, Y., T. Miyajima, M. Yamamuro, H. Kayanne, and I. Koike. 2002. Fine-scale mapping of land-derived nitrogen in coral reefs by $\delta^{15}N$ in macroalgae. *Limnol. Oceanogr.* 47: 1405–1416.

Valiela, I., J. M. Teal, S. Volkmann, D. Shafer, and E. J. Carpenter. 1978. Nutrient and particulate fluxes in a salt-marsh ecosystem: Tidal exchanges and inputs by precipitation and groundwater. *Limnol. Oceanogr.* 23: 798–812.

van Tussenbroek, B. I., K. Hermus, and T. Tahey. 1996. Biomass and growth of the turtle grass *Thalassia testudinum* (Banks ex König) in a shallow tropical lagoon system in relation to tourist development. *Caribb. J. Sci.* 32: 357–364.

Whitney, D., A. Rossman, and N. Hayden. 2003. Evaluating an existing subsurface flow constructed wetland in Akumal, Mexico. *Ecol. Eng.* 20: 105–111.

Worthington, S. R. H., D. C. Ford, and P. A. Beddows. 2000. Porosity and permeability enhancement in unconfined carbonate aquifers as a result of solution. Pp. 463–472, In: A. D. Klimchouk, D. C. Ford, A. M. Palmer, and W. Dreybrodt (eds.), *Speleogenesis: Evolution of karst aquifers*. National Speleological Society, Huntsville, Alabama.

13 Subtropical Karstic Coastal Lagoon Assessment, Southeast Mexico

The Yucatan Peninsula Case

Jorge A. Herrera-Silveira and Sara M. Morales-Ojeda*

CONTENTS

Abstract 307
13.1 Introduction 308
13.2 Putting Yucatan Coastal Lagoons in Context 309
13.3 Research and Monitoring Experience 311
13.4 Characterization and Status of Yucatan Coastal Lagoons 312
 13.4.1 Celestún 314
 13.4.2 Chelem 315
 13.4.3 Dzilam 317
 13.4.4 Rio Lagartos 318
 13.4.5 Holbox 320
 13.4.6 Chacmochuk 321
 13.4.7 Nichupte 324
 13.4.8 Bojorquez 325
 13.4.9 Ascención Bay 325
 13.4.10 Chetumal 326
13.5 What Have We Learned about This Karstic Ecosystem? 326
 13.5.1 Water Quality Characteristics 326
 13.5.2 Phytoplankton 328
 13.5.3 Submerged Aquatic Vegetation (SAV) 328
13.6 Conclusions 329
Acknowledgments 329
References 330

ABSTRACT

The Yucatan Peninsula of Mexico is characterized by numerous coastal lagoons, which are important both ecologically and economically. Wastewater discharge, groundwater pumping, and land use changes are modifying both the structure and function of these lagoons. Here, we assess the condition of 10 such lagoons, 5 on the Gulf of Mexico coast and 5 on the Caribbean coast of the peninsula, via multiple measures, including water quality variables, submerged aquatic vegetation, harmful phytoplankton species richness, and degree of eutrophication. Lagoons varied substantially

* Corresponding author. Email: jherrera@mda.cinvestav.mx

in terms of water quality, with much of this variation resulting from seasonal changes in lagoon salinity, which ranged from <10 to >40. Lagoon water quality generally suggested a low level of anthropogenic impact; however, two lagoons (Chelem on the Gulf side, and Bojorquez on the Caribbean side) showed signs of anthropogenic eutrophication, including high harmful phytoplankton species abundance and changes in seagrass cover. Additionally, lagoons on the Caribbean side of the peninsula differed considerably from those on the Gulf side. These results provide a baseline assessment of water quality in the Yucatan Peninsula's coastal lagoons and underscore the importance of accounting for natural geographic variation in water quality when comparing different lagoons and characterizing current and future human impacts on them.

Key Words: coastal lagoons, water quality, phytoplankton, seagrasses, management, ecosystem health, eutrophication

13.1 INTRODUCTION

The biological diversity, high productivity, and ecosystem services provided by tropical coastal systems are well recognized; indeed, these same characteristics may promote human settlement and urban development. Approximately 60% of the total human population lives in coastal areas (Costanza et al. 1997), and human impacts on coastal systems are increasing. Understanding how estuaries and coastal lagoons function ecologically is thus essential for effective management of these ecosystems.

Coastal lagoons are shallow, enclosed or semi-enclosed bodies of water that are typically oriented parallel to the coast, separated from the ocean by a sand barrier and connected to the ocean by one or more inlets, which remain open permanently or intermittently (Kjerfve 1994). Found on all continents except Antarctica, they are often highly productive systems that contain many types of habitat, supporting both high biodiversity and multiple human uses, including aquaculture, fisheries, tourism, and urban development. Coastal lagoons have innate differences in geomorphology and function, and some have also undergone significant changes as a result of human activity. Thus, it is necessary to establish specific research programs and management strategies to ensure continued ecosystem productivity and provision of valuable ecosystem services (Alongi 1998).

In order to assess how human activity has affected the health of a coastal lagoon, baseline information on ecosystem structure and function is required. Ecological indicators, such as water chemistry parameters or the abundance of certain species, can then be used to monitor the effects of management strategies, which may have myriad purposes: preserving lagoon function and services, maintaining water and habitat quality, and regulating specific ecosystem uses, such as fishing, aquaculture, or recreation. Assessing the condition of coastal lagoons is usually neither simple nor straightforward, and as a result numerous metrics, indices, frameworks and approaches for measuring coastal ecosystem health have been developed (Dennison et al. 1993; Borja et al. 2000, 2004; Amany and Dorgham 2003; Orfanidis et al. 2003; Borja 2005; Buchanan et al. 2005).

In the Yucatan case, a conceptual framework based on three core concepts was developed to guide ecosystem assessment and management. The first, connectivity, encompasses the biogeochemical, biological, and hydrological interactions between ecosystem components at different spatial and temporal scales (Seller and Causey 2005). The second concept is the dynamic interaction between land and sea (Twilley 1995), including freshwater inputs, tides, ocean currents, seasonal weather conditions, storms, and hurricanes, all of which can affect the system's productivity, biodiversity, responses to disturbance, and overall functioning. The third concept, ecological stability (Dayton et al. 1984), is the speed and process with which ecosystems return to their original structure and function after a disturbance, and how resistant they are to major disturbances such as exotic species or hurricanes.

The Yucatan Peninsula's human population has grown quickly over the past two decades, leading to intensified anthropogenic impacts on the coastal lagoons of the region. In 2005, 4.5 million people lived on the peninsula. Greater wastewater inputs and tourism infrastructure have led to increased inputs of nutrients and other pollutants, as well as the loss of mangrove forests and seagrass beds. These anthropogenic factors interact with site-specific geomorphologic conditions, submarine groundwater discharge, climate seasonality, and pulse events such as cold fronts and hurricanes to determine each lagoon's water quality.

In response to scientific and public interest in the ecological condition of the peninsula's lagoons, this chapter aims to characterize the status of 10 representative lagoons in the subtropical karstic ecosystems of the Yucatan Peninsula. Our characterization takes into account data that have already been collected on each lagoon, as well as new data on water quality, phytoplankton abundance and diversity, and seagrass community characteristics. The information presented here can be used to evaluate the present condition of these ecosystems and identify the processes, both natural and anthropogenic, that are most strongly linked to ecosystem health, thus contributing to the development of effective management strategies.

13.2 PUTTING YUCATAN COASTAL LAGOONS IN CONTEXT

This contribution included a total of 10 karst coastal lagoons located on the Yucatan Peninsula. Five (Celestún, Chelem, Dzilam, Rio Lagartos, and Holbox) are influenced by the Gulf of Mexico, and the other five (Chacmochuk, Nichupte, Bojorquez, Ascensión, and Chetumal) are influenced by the Caribbean Sea (Figure 13.1). The selected coastal lagoons differ in terms of structure, function, human use intensity, strength and type of impacts, and degree of vulnerability. Hence, this study constitutes an example of careful cross-ecosystem comparison and an overall assessment of the region's karst lagoon ecosystems.

Yucatan coastal lagoons vary in size, with surface areas ranging from 3 to 1500 km². They tend to be very shallow (depth range: 0.2 to 5 m) and have a small tidal range (0.06 to 1.5 m) and a high mean water temperature (>20°C). In addition, all receive freshwater inputs from springs or runoff from mangrove areas. Nutrient sources include wastewater discharges from urban sewage and

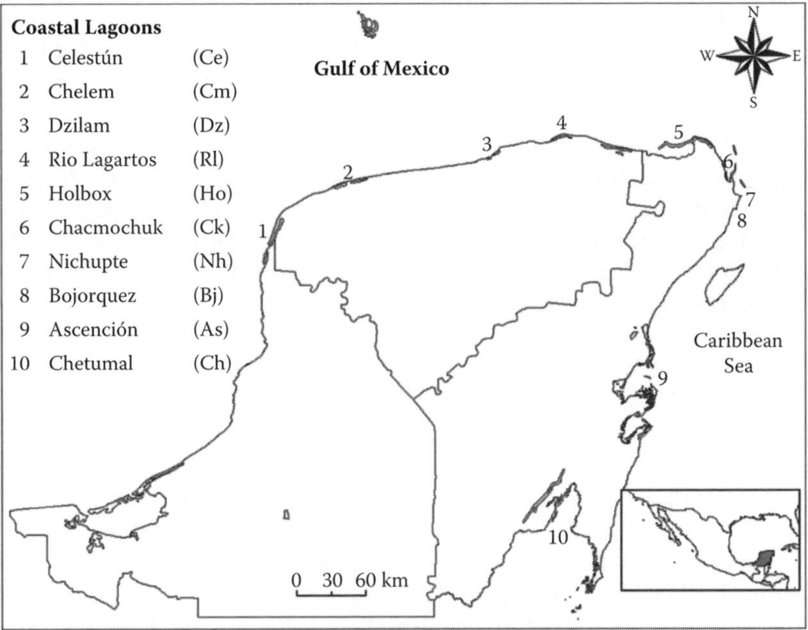

FIGURE 13.1 Map of the Yucatan Peninsula showing locations of the main coastal lagoons.

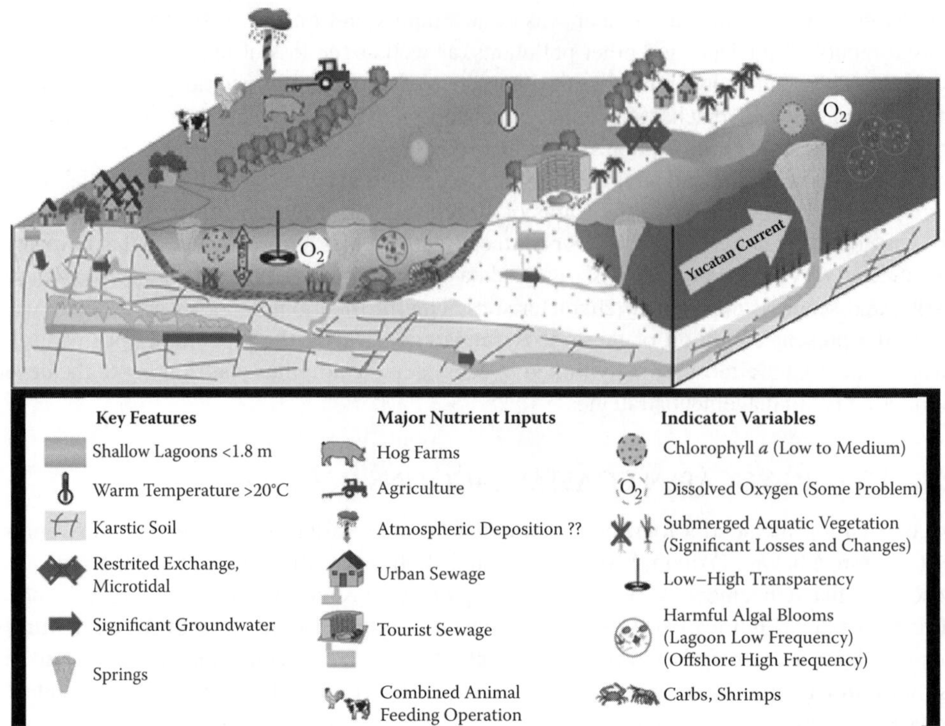

FIGURE 13.2 Conceptual model of the main characteristics of Yucatan Peninsula coastal lagoons.

agricultural activity (Figure 13.2). Although some of the lagoons are similar in terms of functional traits and nutrient inputs, they also exhibit important differences.

Water residence times of the lagoons studied vary from weeks to years depending on the number and size of inlets, wind- and tide-driven circulation patterns, and seasonal freshwater inflows. Some lagoons have exhibited changes in circulation patterns due to the modification of inlets connecting the lagoon to the sea (Chelem, Rio Lagartos, Nichupte, and Chetumal), or due to construction of bridges or embankments, which restrict water flow between portions of the lagoon (Celestún, Chelem, and Bojorquez).

Coastal ecosystems of the Yucatan Peninsula experience three well-defined seasons: dry (March to May), rainy (June to October), and "nortes" (November to February), which is dominated by cold fronts. Additionally, the hurricane season (from August to September) has a strong influence on coastal lagoon stability and disturbance regime (Herrera-Silveira 1994a; Herrera-Silveira et al. 1998).

The Yucatan Peninsula has unique hydrogeologic characteristics, including low relief, lack of rivers, highly permeable karst-derived soils, and substantial submarine groundwater discharge (SGD). In fact, SGD is the dominant connection between land and sea in Yucatan coastal lagoons, with the annual flux estimated at 8.6×10^6 m^3 km^{-1} (Hanshaw and Back 1980; Young et al. 2008). SGD is characterized by low salinity and high nitrate and silicate concentrations. However, its influence on coastal lagoon water chemistry depends on the magnitude of the discharge and how it occurs (i.e., discrete springs or diffuse seepage) (Herrera-Silveira 1994b; Young et al. 2005). Where submarine springs exist, they vary in number, size, chemistry, and flux rate. Rapid infiltration of rainwater results in very little surface runoff, making atmospheric precipitation a less important source of freshwater than in many other places. Only during the rainy season does surface runoff from mangrove fringe areas constitute an important source of freshwater to Yucatan coastal lagoons. Wind-generated and tidal currents also affect lagoon water chemistry. These currents are particularly intense during the "nortes" season and during hurricane events, often resulting in sediment resuspension, high turbidity, and more rapid movement of nutrients and organic matter from the lagoons to the open sea, resulting in fertilization of near-shore marine coastal areas (Herrera-Silveira 1994c; Herrera-Silveira et al. 2002a; Medina-Gomez and Herrera-Silveira 2003; Tapia et al. 2008).

A large number of human activities that take place on the coast of Yucatan—fisheries, salt extraction, port development, high- and low-impact tourism, and cattle ranching—have the potential to impact coastal lagoons by decreasing species richness and habitat diversity. In addition, some of the largest urban areas on the Yucatan coast, such as Cancun, are adjacent to lagoons, leading to high anthropogenic impacts in those water bodies. Progreso, Nichupte, Bojorquez, and Chetumal lagoons are all located near large urban centers. On the other hand, lagoons such as Chacmochuk and Ascensión are far from large cities and thus have experienced less human impact.

13.3 RESEARCH AND MONITORING EXPERIENCE

Beginning in 2000, 10 lagoons were sampled once in the middle of each season. Monitoring included the measurement of water quality variables, phytoplankton abundance and diversity, and seagrass community composition using standardized methods (Jeffrey and Humphrey 1975; Hasle 1978; Parson et al. 1984; Sournia 1987; Costanza et al. 1992; Hallegraeff et al. 1995; Durako and Duarte 1997; Bricker et al. 1999, 2003; Avery 2000; Boyer et al. 2000; Hemminga and Duarte 2000; U.S. Environmental Protection Agency 2001a, 2001b; Arhonditsis et al. 2003; Brander et al. 2003; Fourqurean et al. 2003; Jorgensen et al. 2005; Devlin et al. 2007). Here, we present hydrologic and phytoplankton data from 2001 to 2006 and seagrass data from 2001, 2006, and 2008. A variable number of stations were sampled at each lagoon, depending on the size and salinity gradient of each system (Herrera-Silveira 1994a, 2006; Herrera-Silveira et al. 1998; Medina-Gomez and Herrera-Silveira 2003; Arellano 2004; Sima-Morales 2004; Schumann et al. 2006; Tapia et al. 2008).

Water quality variables, including surface temperature (°C), salinity (expressed in adimensional units), dissolved oxygen (mg L^{-1}), and oxygen saturation (%) were measured *in situ* with a multiparameter probe. The concentrations of dissolved inorganic nutrients (ammonia (NH_4^+), nitrite (NO_2^-), nitrate (NO_3^-), soluble reactive phosphorus (SRP), and silica (SRSi), all expressed as μmol L^{-1}), were measured in surface water samples as described by Strickland and Parsons (1972). Chlorophyll *a* (Chl a) was used as an indirect measure of phytoplankton biomass. Because of the karstic geology of the study region, special emphasis was given to SRP and SRSi concentrations; the former is in some cases a limiting nutrient because it is lost in sediments due to precipitation in the presence of calcium carbonate (Coelho et al. 2004; Slomp and Van Cappellen 2004), and the latter has been used as SGD tracer (Smith et al. 1999; Young et al. 2008).

For phytoplankton analysis, the Utermöhl technique (Hasle 1978) was used and the taxonomic identification was done at genus and species level using the taxonomic methods of Chretiennot-Dinet et al. (1993) and Round et al. (1990). This methodology helps identify the number of harmful algal bloom (HAB) species, which was used as an indicator of the risk each system had of developing HAB events. The rationale behind this is that the more HAB species that are present, the more likely it is that a change in conditions will cause one of them to bloom (CEC 1991). This approach has been used previously in the Yucatan region (Morales-Ojeda 2007).

Submerged aquatic vegetation (SAV) cover was estimated *in situ* using a diver-assisted sampling design, which can obtain rapid and accurate estimates of SAV coverage. Surveys were accomplished using a modified version of the Braun-Blanquet scoring technique (Fourqurean et al. 2001) and 0.5-m^2 quadrats. Replicates were taken to determine species composition and estimate biomass (Zieman et al. 1999; Fourqurean et al. 2001). SAV cover is commonly used as an indicator of ecosystem health because decreases in SAV cover and shifts in species composition from seagrass to green macroalgae are associated with ecological stress (Orth and Moore 1983; Stevenson et al. 1993). In addition, if excessive growth of green macroalgae and epiphytes occurs, it can suffocate bivalves and cause SAV to die off (Dennison et al. 1993). There is currently no threshold above which green macroalgae or epiphytes are considered to be a threat to the overall community.

Based on water quality, phytoplankton, and SAV coverage, we classified each lagoon into one of four categories: "excellent," "good," "fair," or "poor." Classification criteria were based on published U.S. Environmental Protection Agency (U.S. EPA) and National Oceanographic and Atmospheric Administration (NOAA) recommendations (U.S. EPA 2001a, 2004), since Mexico lacks official criteria for assessing coastal ecosystem health status. We first established reference values for each of the three ecosystem health components (water quality, phytoplankton, and SAV) using the U.S. EPA (2005) guidelines. For parameters where higher values were associated with water quality impairment, values between the minimum and the first quartile were classified as "excellent." Values between the first quartile and the median were classified as "good," and values between the median and the third quartile were classified as "fair." Values higher than the third quartile were classified as "poor." For dissolved oxygen and phytoplankton diversity and richness, where higher values are associated with good water quality, the opposite classification scheme was used. For example, values between the minimum and the first quartile were classified as "poor."

13.4 CHARACTERIZATION AND STATUS OF YUCATAN COASTAL LAGOONS

The main features of the coastal lagoons studied are summarized in Table 13.1. All lagoons exhibited some spatial and temporal variation in water quality, which may be related to connectivity,

TABLE 13.1
Characteristics of Coastal Lagoons in the Yucatan Peninsula, Mexico

System	Coordinates	A (km²)	Z (m)	T (days)	Activities	Threat
Celestún (Ce)	20°46′–21°06′ 90°11′–89°25′	28	1.2	50–300 100	Fisheries Ecotourism Mineral extraction	Loss of SAV coverage Water level reduction by sediment refill Eutrophication
Chelem (Cm)	21°10′–21°19′ 88°34′–88°57′	14	0.8	300–750 400	Fisheries Ecotourism Urban development	Loss of SAV coverage Pollution Eutrophication
Dzilam (Dz)	21°19′–21°32′ 88°35′–88°58′	10	0.6	50–250 150	Fisheries Ecotourism	Pollution from its watershed
Rio Lagartos (Rl)	21°38′–21°20′ 88°15′–87°30′	91	0.5	450	Fisheries Ecotourism	Pollution from its watershed
Holbox (Ho)	21°26′–21°36′ 87°08′–87°29′	275	1.5	200–350 280	Fisheries Ecotourism	Pollution from its watershed
Chacmochuk (Ck)	21°10′–86°48′ 21°15′–86°51′	122	1	150–350 200	Fisheries Eutrophication	Pollution from its watershed
Nichupte (Nh)	21°08′–21°04′ 86°48′–86°46′	41	2.2	200–500 300	Massive tourism Loss of SAV coverage	Eutrophication
Bojorquez (Bj)	21°07′–21°08′ 86°46′–86°45′	3	1.7	360–800 400	Massive tourism Pollution Loss of SAV coverage	Eutrophication
Ascensión (As)	19°33′–19°56′ 87°28′–87°38′	740	2.5	200–400 150	Fisheries Ecotourism	Pollution from its watershed
Chetumal (Ch)	18°33′–18°51′ 88°02′–88°08′	1098	3	200–400 300	Fisheries Ecotourism Urban development	Pollution from its watershed

Note: A = surface area; Z = depth; T = residence time of water, upper-lower and average.

ocean-land/land-ocean controls, variable inputs of freshwater and seawater, springs, the intensity of marine currents, and/or the tidal range.

Lagoon salinities ranged from 10 to 40, reflecting differences in lagoon shape, area, water residence time, and seasonal variation in freshwater and saltwater inputs. Freshwater inputs are important in maintaining water chemistry gradients within a lagoon; thus, they promote biodiversity and ecological complexity (Newton and Mudge 2005). However, if freshwater sources are contaminated with nutrients or other pollutants, this can have a negative effect on ecosystem function.

Salinity and water residence time, which play key roles in determining ecosystem function, should be considered in characterizations of coastal lagoons. For instance, water residence time determines if a lagoon acts as a sink for pollutants or as a source of nutrients and organic matter to the open ocean (Medina-Gomez and Herrera-Silveira 2003). The water residence time of each lagoon was estimated following the LOICZ approach (Gordon et al. 1996). Large spatial and seasonal variations in residence time were observed; lagoons with longer residence times will probably be more vulnerable to pollution and eutrophication, while lagoons with shorter residence times will be less vulnerable. These rapidly exchanging lagoons would also act as a source of terrestrial organic matter to the ocean. In this study, the Chelem, Bojorquez, Holbox, and Chacmochuk lagoons have longer residence times, indicating that they are more vulnerable to pollution and eutrophication. In contrast, water in Celestún, Dzilam, Holbox, and Ascensión lagoons turned over more quickly, indicating lower vulnerability (Table 13.1).

The relations between salinity, DIN, and SRSi in the lagoons (Figure 13.3), suggest variability in freshwater input and circulation regime. Celestún, Nichupte, and Bojorquez lagoons exhibit large

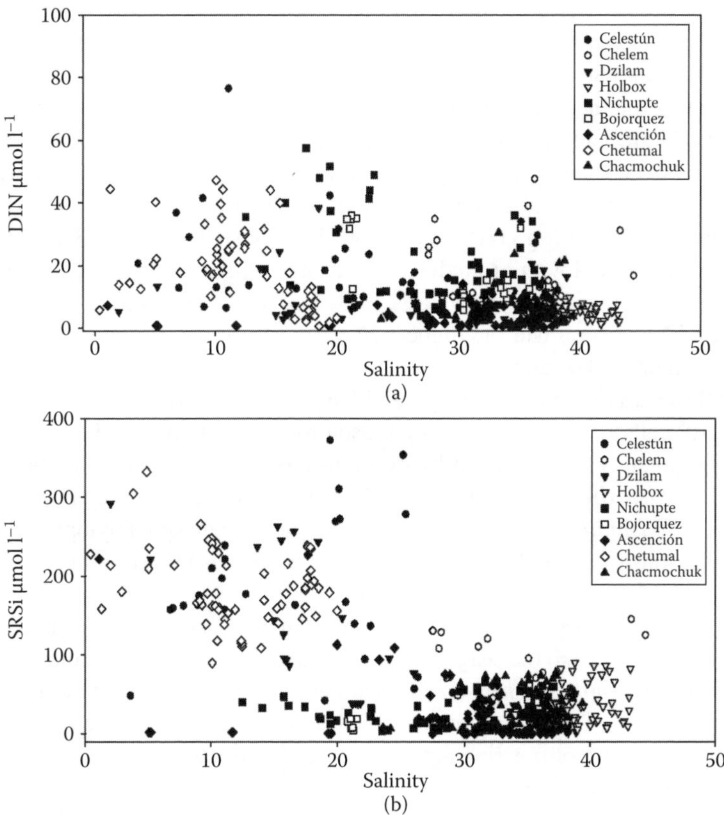

FIGURE 13.3 Relationship of (a) dissolved organic nitrogen and (b) soluble reactive silica concentrations to salinity, 2000 to 2008.

salinity gradients associated with SGD inputs. In other lagoons, such as Chetumal, Dzilam, and Ascención, less dilution of high-nutrient SGD by low-nutrient seawater was observed, and freshwater inputs, rather than mixing with seawater, determine nutrient concentrations. Other lagoons, such as Holbox, Chelem, and Chakmuchuc, are strongly influenced by ocean controls.

13.4.1 Celestún

The Celestún Lagoon (Table 13.1) is located on the western portion of the coast of Yucatan, on the edge of a coastal barrier island. It is within the Celestún Biosphere Reserve, and, since it is the most studied lagoon in Yucatan, much data about it are available. Celestún is the base site for the ECOPEY (acronym in Spanish of Coastal Ecosystems of Yucatan Peninsula) working group, which is part of the Red Mexicana de Estudios Ecologicos de Largo Plazo, or Mexican Long-Term Ecological Research (LTER) network. Salinity values in this system ranged from 5 in the inner zone to 37 in the outer region, which connects to the sea. The lagoon is surrounded by well-preserved mangrove forests; the dominant mangrove species are *Laguncularia racemosa* in the inner portions of the lagoon, *Rhizophora mangle* in the central region, and *Avicennia germinans* in the outside area that connects the inlet with the sea.

Based on water quality variables, the Celestún Lagoon can be divided into three zones: (1) the inner zone, which receives significant groundwater discharges and has low salinity, high NO_3^- and high SRSi concentrations; (2) the middle zone, which is characterized by high NH_4^+ and chlorophyll a concentrations, and (3) the euhaline outer zone, which is influenced by the sea and has low nutrient concentrations (Herrera-Silveira 1994a). Ammonium concentrations may be associated with natural processes of SAV decomposition and runoff from mangrove forests, as well as with wastewater discharge from tourism-related activities and infrastructure (Zaldivar-Jimenez et al. 2004; Tapia et al. 2008). The relatively high SRP concentrations in this system suggest an exogenous source and could be explained by the presence of birds inhabiting the lagoon, and/or by runoff from mangrove areas (Comin and Herrera-Silveira 2000). A comparison of the characteristics of this system in 2001 and 2006 showed no changes in salinity, NO_3^-, NO_2^-, or SRSi (Figures 13.4a, 13.5a, 13.6a, 13.6c, and 13.7a); however, in some cases differences were observed for variables related to anthropogenic eutrophication (e.g., NH_4^+, SRP, and chlorophyll a; Figures 13.6e, 13.8a, and 13.9a). Overall, variables associated with groundwater discharges (SRSi) were greater, while salinity and oxygen saturation values were lower (Figures 13.6a and 13.8a, respectively) during seasons with high SGD fluxes. This lagoon had the highest NO_3^- and SRSi concentrations of all the Gulf-side lagoons studied (Figures 13.6a and 13.7a).

A total of 150 species of phytoplankton were identified in Celestún Lagoon, with densities ranging from 2.8×10^5 cells L^{-1} during the rainy season to 6.3×10^7 during the "nortes" season. The phytoplankton community is dominated by cyanobacteria (*Chroomonas*) and diatoms (*Amphora, Navicula,* and *Thalassiosira*) (Figure 13.10). Only four HAB species were present (*Merismopedia glauca, Prorocentrum minimum, P. micans,* and *Protoperidinium* sp.). Phytoplankton blooms have been recorded previously during the late dry and early rainy seasons (Herrera-Silveira et al. 1999a) (Figure 13.11).

In terms of submerged vegetation, the inner zone of the lagoon was co-dominated by green algae such as *Chara fibrosa* and the seagrass *Ruppia maritima*. The mixing zone was dominated by *Halodule wrightii* and *Ruppia maritima*, and the outer zone was dominated by *H. wrightii* and several species of macroalgae. SAV cover has decreased significantly during recent years (Figure 13.12), mainly as a result of the use of trawling nets for fishing and boats for ecotourism (Herrera-Silveira et al. 2000c; Castrejon et al. 2005; Ascencio 2008).

Celestún Lagoon is considered to be in good condition according to most water quality variables; the only exception was turbidity, which indicated a fair condition (Table 13.3). The frequency and duration of phytoplankton blooms also suggest a fair condition. SAV cover in Celestún has

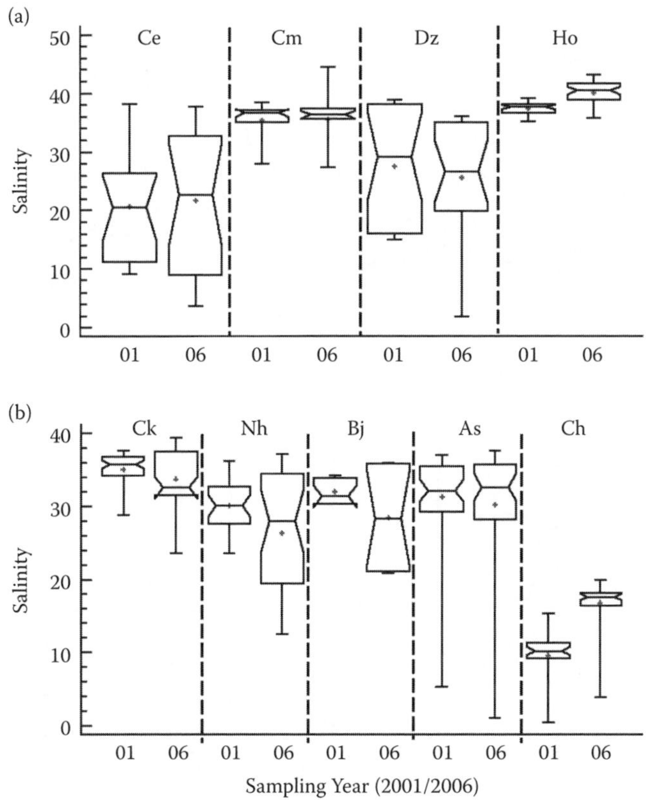

FIGURE 13.4 Salinity variation in the coastal lagoons over two sampling years. (a) Gulf of Mexico influence; (b) Caribbean influence.

decreased over the last 5 years, another indication of fair condition (Table 13.3). Thus, the overall health status of Celestún Lagoon is considered to be fair (Figure 13.13).

Given that the water residence time of this lagoon is relatively short (~100 days), the lagoon is likely a source of nutrients to the ocean, with a medium to low degree of vulnerability to human impacts. The main human activities conducted in the lagoon are fishing and ecotourism, and the main threats are habitat loss, the obstruction of water flux, eutrophication, pollution, and overfishing. We recommend continued monitoring of the impacts generated by ecotourism activities, restocking native species through aquaculture programs, and restoring natural water flow in the internal portions of the lagoon, which have been modified by a bridge and an embankment.

13.4.2 Chelem

Chelem Lagoon is located 35 km north of Mérida, the capital of the state of Yucatan, between the port of Progreso, the locality of Yucalpeten, and the town of Chelem (Figure 13.1). It is surrounded by a degraded mangrove forest dominated by *R. mangle* (Herrera-Silveira et al. 1998, 2000b). The lagoon is usually euhaline (Table 13.2, Figure 13.4a) and its hydrological characteristics have been modified due to the construction of roads and bridges, dredging, filling, and the creation of artificial connections with the sea, which increased water salinity in coastal wetlands, impacting mangrove vegetation and other ecological features of the lagoon (Herrera-Silveira et al. 2000a).

The highest nutrient and chlorophyll a concentrations in this lagoon occurred near the shipping channel of the canoeing trail, in the eastern portion of the lagoon. Such conditions, which appear to result from a combination of leaching from an old municipal dump, urban wastewater discharges,

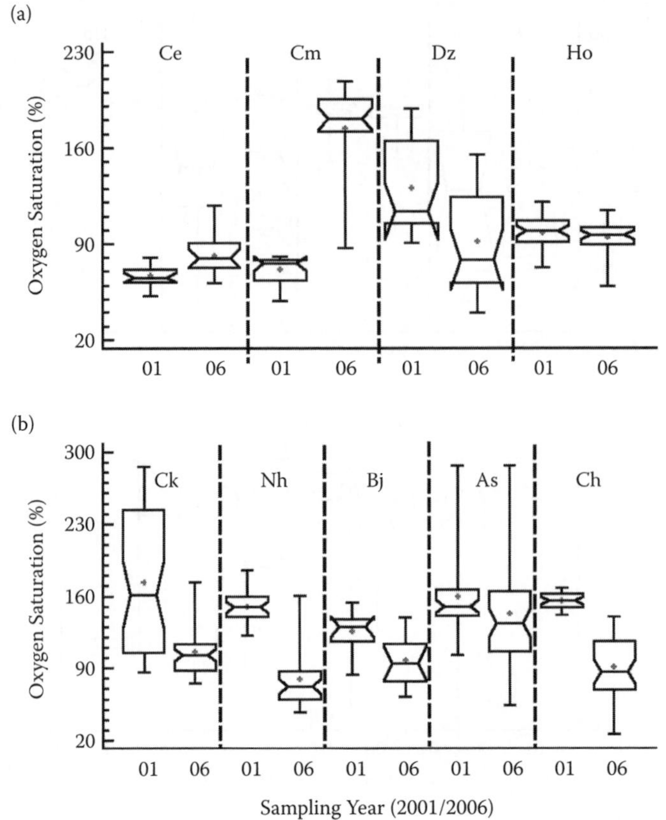

FIGURE 13.5 Oxygen saturation variation in the coastal lagoons over two sampling years. (a) Gulf of Meixco influence; (b) Caribbean influence.

and sluggish circulation in this part of the lagoon, are indicative of eutrophication (Tapia et al. 2008). Indicators associated with anthropogenic disturbance, such as NH_4^+, SRP, and chlorophyll a, increased between 2001 and 2006 (Figures 13.6a, 13.6c, 13.8a, and 13.9a), suggesting ongoing deterioration of this lagoon.

A total of 101 species of phytoplankton were identified, with densities ranging from 2.7×10^5 to 5.3×10^7 cell L^{-1}. Diatom taxa, especially *Cyclotella, Chaetoceros, Hannea,* and *Thalassiosira,* dominated (Figure 13.10). Eight HAB species, including *Cylindrotheca closterium*, *Nitzschia longissima,* and *Gambierdiscus toxicus*, were observed, which is high relative to other coastal lagoons in the Yucatan (Herrera-Silveira et al. 1999a) (Figure 13.11).

The dominant SAV species were *H. wrightii, Thalassia testudinum*, and green and red macroalgae. Changes in SAV community composition have been observed in this lagoon. It was originally dominated by freshwater algae and coraline macroalgae, but these have been largely replaced by seagrasses (Herrera-Silveira et al. 2000c). Jellyfish blooms have been observed in the shallow portions of the lagoon and seem to be increasing in frequency, suggesting a rapid deterioration of water quality. Additionally, there is public and private interest in infrastructure development related to port activities and tourism in the lagoon. Bridges, roads, and urban development, visible from the beach and in the adjacent coastal area, have contributed to SAV loss by increasing water turbidity and altering water circulation patterns in the lagoon (Ascencio 2008).

Overall, this lagoon's condition is classified as fair (Figure 13.13) according to water quality data (Table 13.3). Dissolved oxygen levels and water clarity were low, while nutrient and chlorophyll a

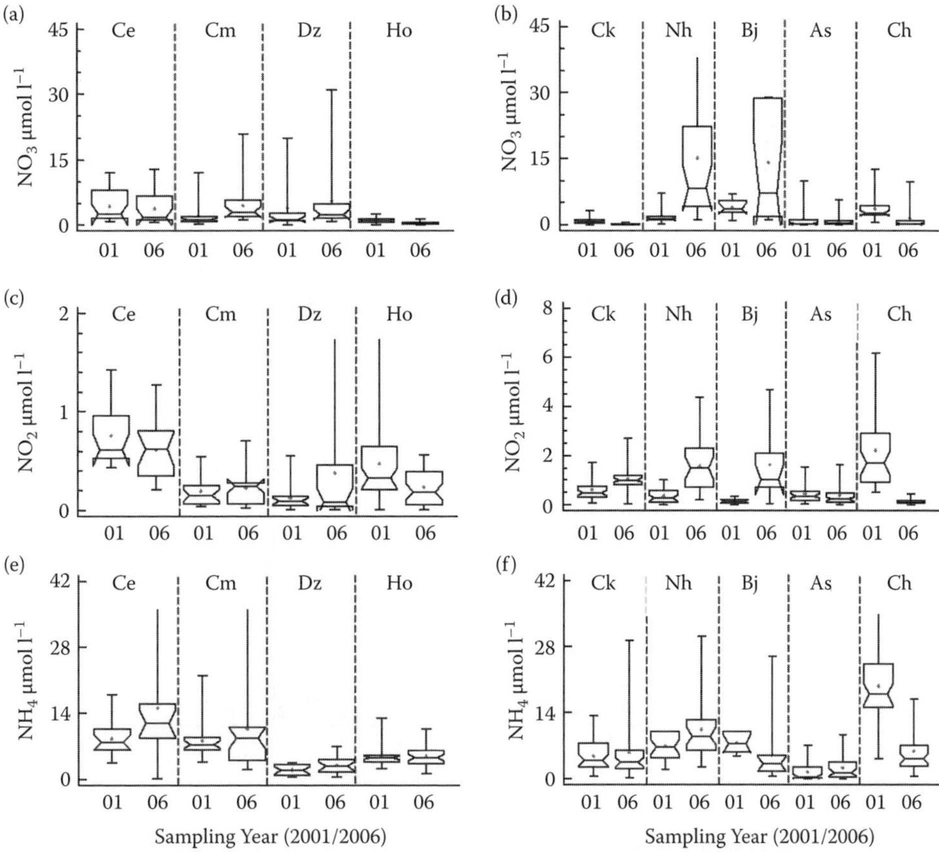

FIGURE 13.6 Variation in concentrations of dissolved inorganic nitrogen species in the coastal lagoons over two sampling years: (a) and (b) NO_3^-; (c) and (d) NO_2^-; (e) and (f) NH_4^+. (a), (c), (e) Gulf of Mexico influence; (b), (d) Caribbean influence.

concentrations were high (Table 13.2). The large number of HAB species, as well as the long and frequent phytoplankton blooms in the eastern area, indicated a fair condition in terms of phytoplankton. Both the cover and composition of SAV indicated a poor condition. Together, these results indicate that Chelem Lagoon has an overall poor condition. One of the most important variables contributing to this system's current condition is the high water residence time in the eastern portion of the lagoon (Table 13.1), making it highly vulnerable to both natural and anthropogenic impacts. Programs aimed at restoring water circulation and mangrove vegetation may help improve water quality.

13.4.3 Dzilam

This lagoon is located on the northern Yucatan coast, within a state-protected area, surrounded by mangrove forest (mainly *R. mangle* and *L. racemosa*). After it was hit by hurricane Isidore in 2002, this system has gradually shown signs of recovery (Teutli 2008). The lagoon is connected to the sea by an inlet located in its central region, and salinity levels are generally estuarine, although euhaline and hyperhaline conditions have been recorded occasionally (Table 13.2). Water residence time in this system is relatively long, possibly favoring the accumulation of SGD-derived pollutants, which may affect biota distribution and abundance patterns (Herrera-Silveira et al. 1999b). Nutrient

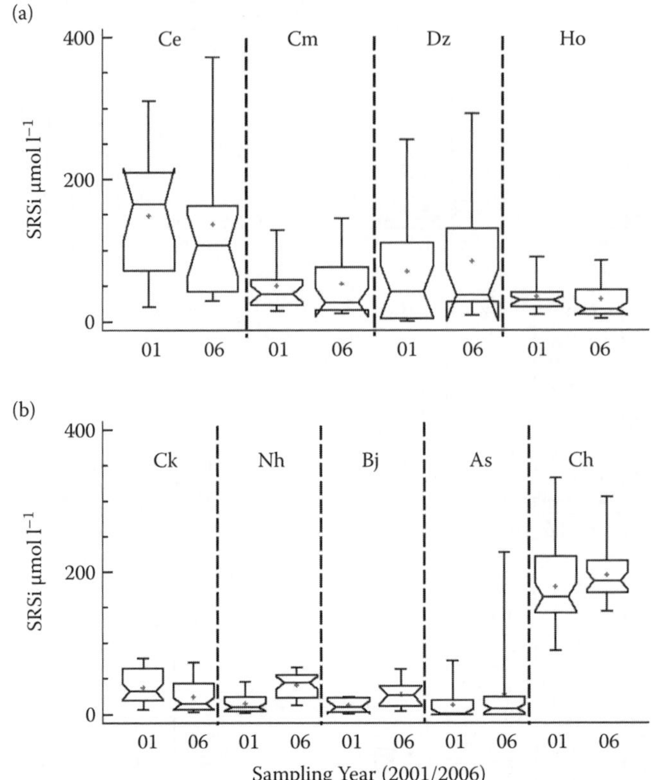

FIGURE 13.7 Soluble reactive silica variation in the coastal lagoons over two sampling years. (a) Gulf of Meixco influence; (b) Caribbean influence.

concentrations in the lagoon are high (Table 13.2) due to the presence of a large number of high-nitrate, high-silica springs in the inner zone of the lagoon, as well as runoff from the mangrove area. NO_3^- and NO_2^- increased between 2001 and 2006, whereas chlorophyll a and oxygen saturation decreased (Figures 13.5a, 13.6a, 13.6c, and 13.9a).

A total of 135 phytoplankton species were identified, with densities ranging from 3.1×10^5 to 4.8×10^7 cell L^{-1}. Few HAB species were observed (Figure 13.11). The most common HAB species was the diatom *Cylindrotheca closterium*, and members of the genera *Prorocentrum* and *Protoperidinium* were also observed (Figure 13.10) (Herrera-Silveira et al. 1999a).

Seagrasses (*H. wrightii*, *T. testudinum*, *Syringodium filiforme* and *R. maritima*) were the dominant type of SAV, although green and red macroalgae were also present. SAV cover did not change over the study period, probably because little urban development has occurred in the area (Figure 13.12). However, due to large inputs of groundwater, high nutrient concentrations and the presence of pollutants in sediments and SAV tissues have been reported in this system. The overall condition of the lagoon is good (Table 13.3). The water residence time is about 150 days, but it varies between different parts of the lagoon. The inner zones of both arms are more vulnerable to contaminants because they have residence times of up to 250 days, resulting in the storage of groundwater-borne pollutants.

13.4.4 Rio Lagartos

This lagoon, the largest in the state of Yucatan (Table 13.1), has three inlets where water is exchanged with the sea. It is long, shallow, and hyperhaline (Table 13.2), and its hydrochemistry is

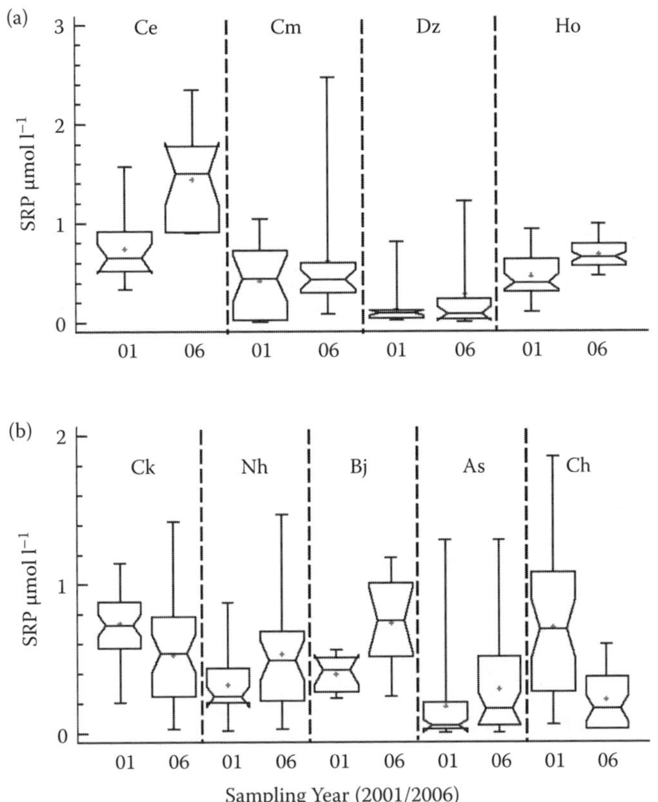

FIGURE 13.8 Soluble reactive phosphorus variation in the coastal lagoons over two sampling years. (a) Gulf of Meixco influence; (b) Caribbean influence.

quite different from the other lagoons. Since data for this lagoon are typically outside the normal range and may interfere with the interpretation of patterns observed for the other lagoons, we did not include hydrochemistry figures for this system. Rio Lagartos had high SRP and NO_2^- concentrations and low SRSi concentrations (Table 13.2). The hyperhaline region of the lagoon had high chlorophyll a concentrations (Herrera-Silveira et al. 1998, 2002a).

Because this system is located within a watershed influenced by agriculture and cattle farming, groundwater is contaminated, and early signs of eutrophication and pollution have been observed. The impact of SGD-derived pollutants may be magnified by the long water residence time (about 450 days). The lagoon is bordered by coastal dune vegetation and mangroves dominated by *R. mangle* and *A. germinans* which have been impacted by the local salt extraction industry and hurricanes.

A total of 80 species of phytoplankton were identified, varying in density from 3.5×10^5 to 4×10^7 cell L^{-1}. Cyanobacteria, such as *Gloeocapsa* sp. and *Gloeothece* sp., dominated (Figure 13.10). A small number of HAB species were observed (Figure 13.11), with *Scripsiella trochoidea* and *Nistzschia longissima* being the dominant species. There is no information on phytoplankton blooms for this system. Seagrasses (*H. wrightii, R. maritima,* and *T. testudinum*) and calcareous green algae were the dominant type of SAV in the western region, while the hyperhaline zone was dominated by microphytobenthos. Although this lagoon's overall condition is considered good, no information is available on temporal changes in ecological indicators, which limits the ability to make inferences based on the present data (Figure 13.12).

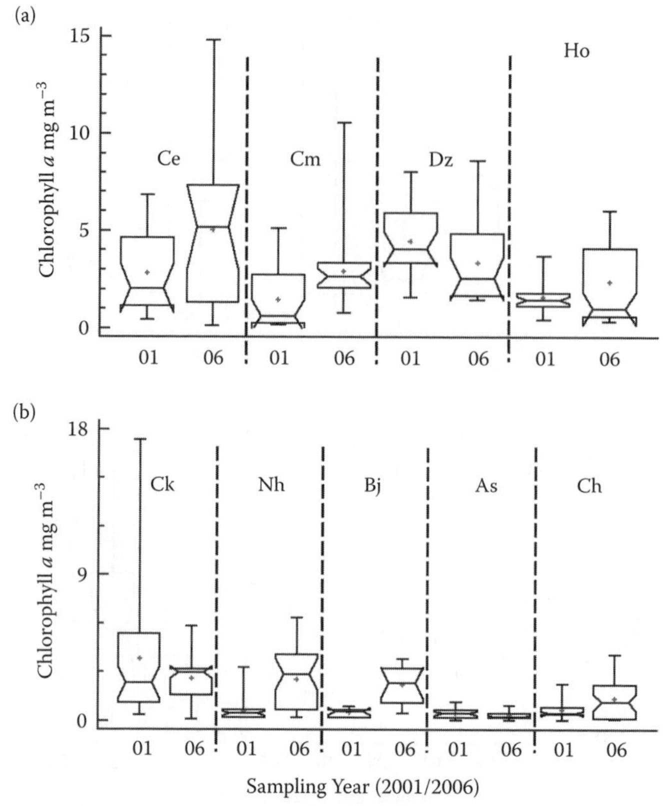

FIGURE 13.9 Chlorophyll *a* (Chl *a*) variation in the coastal lagoons over two sampling years. (a) Gulf of Meixco influence; (b) Caribbean influence.

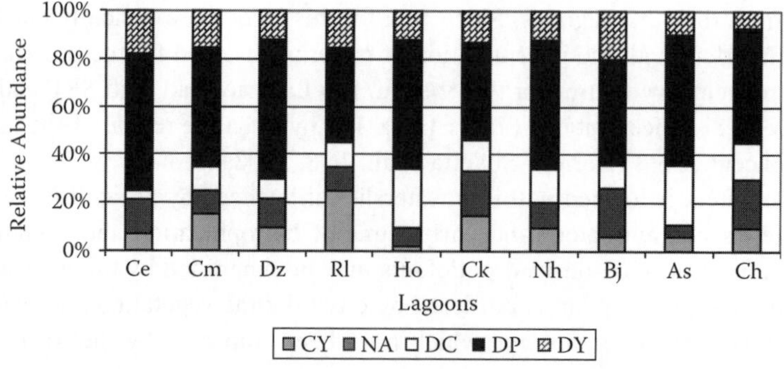

FIGURE 13.10 Phytoplankton community composition in the coastal lagoons. CY = Cyanophyta, NA = Nanoflagellates, DC = Centric Diatoms, DP = Pennate Diatoms, DY= Dinoflagellates.

13.4.5 Holbox

This lagoon is located on the northwestern Yucatan Peninsula coast, in the state of Quintana Roo, within the protected area of Yum Balam (Figure 13.1). Surrounded by mangrove forest, it is one of the largest coastal lagoons of northern Yucatan. Its overall condition cannot be assessed currently. The lagoon is typically euhaline, with lower salinities observed during the rainy season due to fresh groundwater inputs. Based on spatial differences in water chemistry, the lagoon can be divided into five zones. The inner zones are characterized by higher salinities, while those closer to urban areas

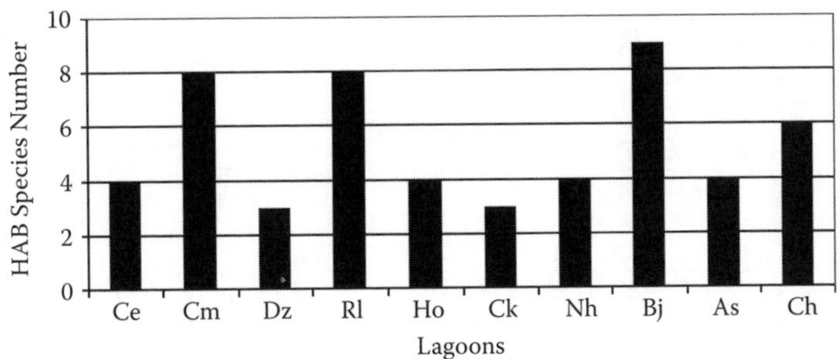

FIGURE 13.11 Number of harmful algae bloom (HAB) species in each coastal lagoon.

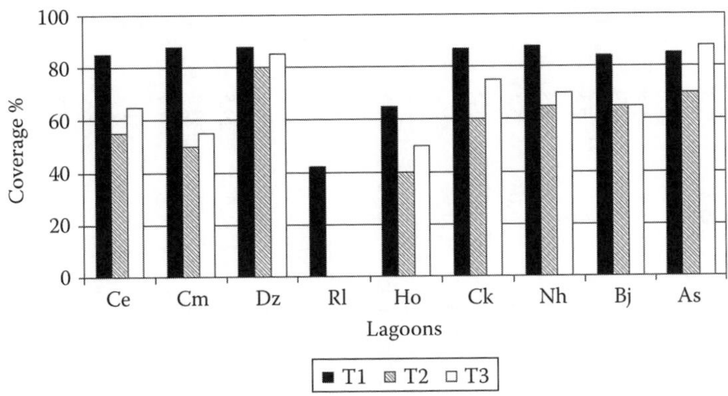

FIGURE 13.12 Submerged aquatic vegetation (SAV) coverage in the coastal lagoons at T1 (2001), T2 (2006), and T3 (2008).

have higher nutrient concentrations (Tran et al. 2002, 2005). Minor changes in NO_3^-, NO_2^-, NH_4^+, and SRSi occurred over the 5-year study period (Table 13.2, Figures 13.6a, 13.6c, 13.6e, and 13.7a); differences in salinity (Figure 13.3a), SRP, and chlorophyll a were somewhat larger (Figures 13.7a and 13.8a).

No information on phytoplankton blooms is available for this system. The limited data on phytoplankton community composition indicate that diatoms are the dominant group (Figure 13.10), and that the number of HAB species is low (Figure 13.11). Seagrasses (*H. wrightii, T. testudinum,* and *S. filiforme*) are the dominant type of SAV, and previous data indicate that there have been no changes in SAV cover (Figure 13.12). The presence of heavy metals in *T. testudinum* tissues (Solis et al. 2008) suggests pollution of this system, probably originating from the Cancun urban area.

Although SRP and chlorophyll a concentrations have increased during the last few years, water quality is considered good in this system. Lacking sufficient phytoplankton and SAV data, it is impossible to assess the lagoon's overall health, but its short water residence time suggests that its overall condition is probably good (Table 13.3).

13.4.6 Chacmochuk

This lagoon is located in the northeast corner of the Yucatan Peninsula, north of the city of Cancun. It is naturally connected to the Caribbean Sea by an inlet in its northern portion (Figure 13.2). Salinity data indicate that this system is euhaline, although salinity has increased over the last few years (Figure 13.4b). The lagoon is bordered by mangrove vegetation, which has been impacted by recent hurricanes and tourism development.

TABLE 13.2
Mean and Range of the Water Chemistry Parameters in Surface Waters of Yucatan Coastal Lagoons

	Celestún (Ce)			Chelem (Cm)			Dzilam (Dz)			Rio Lagartos (Rl)			Holbox (Ho)		
Variable	**Average**	**Min**	**Max**	**Average**	**Min**	**Max**	**Average**	**Min**	**Max**	**Average**	**Min**	**Max**	**Average**	**Min**	**Max**
T (°C)	27.8	25.5	33.3	26.7	23	31.6	31	28.1	35.5	27.3	20.7	35.8	28.1	25.2	29.8
Sal	21.1	3.6	38.2	35.4	27.5	44.5	26.5	2	38.9	57.6	23	186	38.8	35.2	43.2
O_2 (mg L^{-1})	3.6	2.4	5.4	5.6	2.4	9.2	5	1.9	8.3	5.8	2.8	9.9	4.2	2.6	5.4
SAT (%)	74	52	118	123	49	20.8	112	40	188	147	74	400	96	59	121
NO_3 (μmol L^{-1})	4.04	0.54	12.8	3.23	0.28	20.9	4.76	0.05	31	2.65	0.05	44	0.72	0.03	2.55
NO_2 (μmol L^{-1})	0.69	0.21	1.43	0.21	0.02	0.71	0.25	0.01	1.96	0.84	0.01	12	0.35	0.01	1.92
NH_4 (μmol L^{-1})	11.84	0.1	40.69	9.46	2.06	40.2	2.43	0.41	7.04	5.13	0.55	20	4.65	1.24	13
SRP (μmol L^{-1})	1.09	0.34	2.34	0.52	0.02	2.47	0.21	0.01	1.22	1.38	0.03	7.2	0.57	0.1	0.99
DIN:SRP	18	19	33	25	160	19	35	135	31	6	41	8	10	14	15
SiRS (μmol L^{-1})	143.1	22	372	51.8	11.5	144.9	77.4	1.6	292	33.3	0.1	90	33.2	5	90.1
Chlorophyll a (mg m^{-3})	3.93	0.13	14.81	2.17	0.15	10.5	3.91	1.45	8.62	4.7	0.01	15	1.99	0.36	6.05

	Chacmochuk (Ck)			Nichupte (Nh)			Bojorquez (Bj)			Ascención (As)			Chetumal (Ch)		
Variable	**Average**	**Min**	**Max**	**Average**	**Min**	**Max**	**Average**	**Min**	**Max**	**Average**	**Min**	**Max**	**Average**	**Min**	**Max**
T (°C)	31.5	29.6	33.9	28.6	23.2	32.3	28.9	26.2	31.6	29.5	23.5	34.9	29.2	27.3	32.1
Sal	34.4	23.6	39.4	28.2	12.5	37.1	30.2	20.8	36	30.7	1.1	37.7	13.2	0.4	20
O_2 (mg L^{-1})	5.1	3.2	8	5.8	2.8	7.9	5.1	2.7	7.2	7.1	2.8	12.8	6.2	1.3	8.6
SAT (%)	140	75	286	115	47	185	111	62	154	151	54	286	123	26	168
NO_3 (μmol L^{-1})	0.47	0.01	3.2	8.56	0.23	43.2	8.96	0.96	29	0.98	0.01	9.99	2.49	0.05	12.6
NO_2 (μmol L^{-1})	0.82	0.05	2.7	0.96	0.01	4.38	0.88	0.02	4.7	0.4	0.01	1.64	1.16	0.01	6.17
NH_4 (μmol L^{-1})	5.09	0.1	29	8.64	1.9	30.2	6.23	0.41	26	1.75	0.02	9.39	12.72	0.54	36.2
SRP (μmol L^{-1})	0.63	0.03	1.4	0.43	0.02	1.47	0.57	0.23	1.2	0.24	0.01	1.3	0.46	0.03	1.86
DIN:SRP	10	58	22	42	173	39	28	8	30	13	10	11	36	22	25
SiRS (μmol L^{-1})	31.2	3.6	78.3	38.1	2.8	66.3	20	2.1	63.3	20.8	0.1	227.1	187.3	89.8	332.4
Chlorophyll a (mg m^{-3})	3.22	0.1	17	1.57	0.2	6.34	1.36	0.22	3.8	0.41	0.02	1.16	1.04	0.03	4.1

TABLE 13.3
Assessment of Lagoon Condition Based on Individual Metrics and Overall (Using All Metrics Together)

	Water Variables					Water	Phytoplankton Variables		Phytoplankton	Habitat Variables Coverage Loss		Habitat	Overall
Lagoon	**DO**	**Tur**	**DIN**	**SRP**	**Chlorophyll a**	**Condition**	**HAB SP #**	**Blooms**	**Condition**	**SAV**	**Mangrove**	**Condition**	**Condition**
Celestún	●	◒	●	●	●	●	●	◒	◒	○	●	◒	◒
Chelem	◒	◒	○	◒	◒	◒	○	◒	◒	○	◒	○	○
Dzilam	●	●	●	●	●	●	●	●	●	●	◒	●	●
Rio Lagartos	●	□	●	◒	●	●	●	□	□	●	◒	●	●
Holbox	●	●	●	◒	●	●	●	□	□	●	□	□	●
Chacmochuk	●	●	●	◒	●	●	●	□	□	●	□	□	●
Nichupte	◒	◒	◒	◒	◒	◒	◒	◒	◒	◒	◒	◒	◒
Bojorquez	○	◒	○	○	◒	○	○	○	○	◒	○	○	○
Ascensión	●	●	●	●	●	●	●	●	●	●	●	●	●
Chetumal	●	◒	◒	◒	●	◒	●	□	□	□	□	□	◒

Note: ● Good; ◒ Fair; ○ Poor; □ No data available.

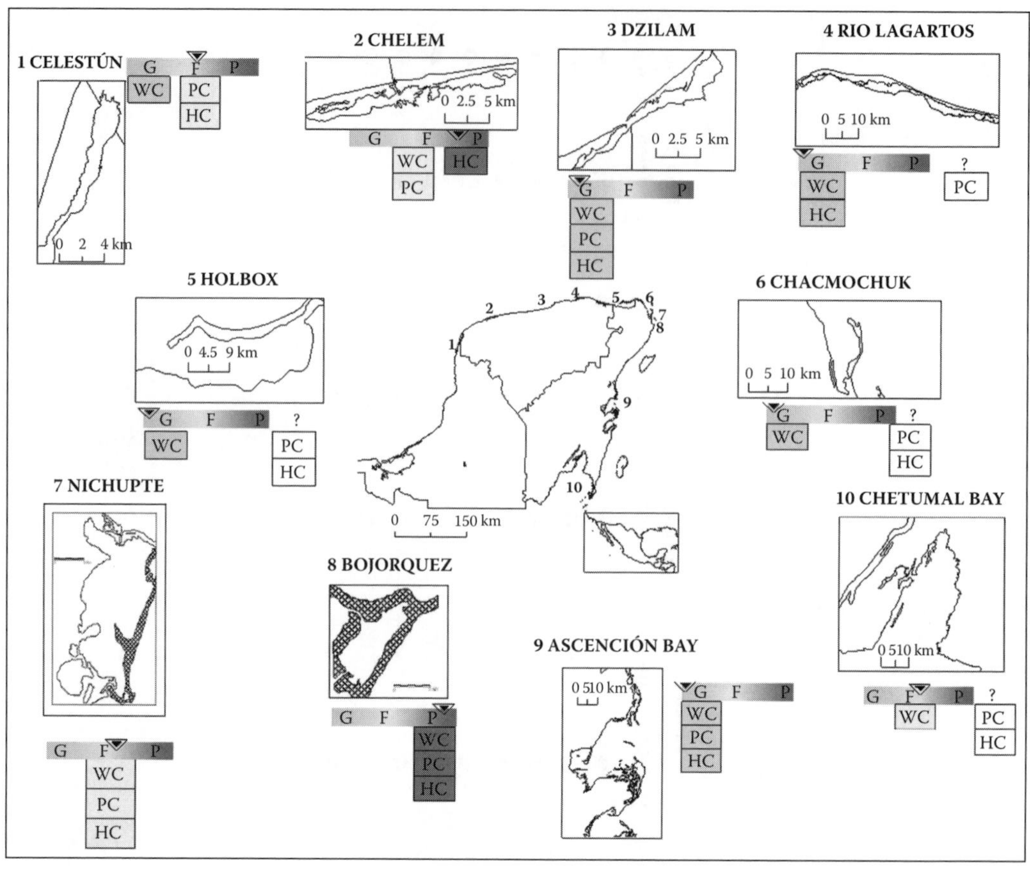

FIGURE 13.13 Overall condition of the Yucatan coastal lagoons.

Chlorophyll a concentrations in the lagoon are high, and, although it is influenced by the Caribbean Sea, it is quite similar to other lagoons on the northern Yucatan coast which are more influenced by the Gulf of Mexico (Figure 13.9b). The lagoon is located in the same watershed as the city of Cancun. Pollutants from the city's dumps may leach into groundwater, which then discharges into the lagoon.

Cyanobacteria and dinoflagellates dominate the phytoplankton community (Figure 13.10). There is no information available on algal blooms, although the number of HAB species reported is low (Figure 13.10). Seagrasses (*H. wrightii, T. testudinum, S. filiforme,* and *R. maritima*) are the dominant type of SAV, with green and red macroalgae also present. No temporal changes in SAV cover were observed (Figure 13.12).

The overall water quality in Chacmochuk was good (Table 13.3). Nutrient concentrations and turbidity were low. However, high SRP and chlorophyll a concentrations (Table 13.2) and proximity to the city of Cancun suggest that the lagoon may be vulnerable to eutrophication. Currently no data on phytoplankton or SAV are available. Human impacts on this system are expected to increase as Cancun expands to accommodate its thriving tourism industry and rapidly growing population.

13.4.7 Nichupte

Although the Nichupte and Bojorquez Lagoons are part of an interconnected lagoon system, their environmental conditions are different. Nichupte is connected to the Caribbean Sea by two small artificial inlets (Punta Cancun and Punta Nizuc), and it receives freshwater inputs from springs and surface runoff from the surrounding mangrove forest. Its salinity gradient resembles that of an estuary (Sima-Morales 2004).

Nitrate, nitrite, and ammonium concentrations were high, and changes in nearly all water quality variables over the past 5 years suggest degradation in water quality (Figures 13.6b, 13.6d, 13.6f, and 13.8b). Diatoms are the most abundant type of phytoplankton (Figure 13.10). Although the number of HAB species is low (Figure 13.11), annual algal blooms have been recorded in areas with lower water flow. Seagrasses (*H. wrightii* and *T. testudinum*) are the dominant type of SAV, with red and green macroalgae also present. Heavy boat traffic related to the tourism industry has damaged SAV (Figure 13.12). An increase in epiphyte abundance on *T. testudinum* leaves has also been observed over the last few years. Based on the available water quality, phytoplankton, and SAV data, this system's condition is classified as fair (Figure 13.13).

13.4.8 Bojorquez

Bojorquez, connected to Nichupte by two artificial channels, is the smallest lagoon included in this study (Table 13.1). Originally, it was completely surrounded by mangroves, but now most of that vegetation has been replaced by tourism infrastructure. Because of its limited connection with the Nichupte lagoon, changes in SGD and wastewater inputs have caused salinity changes over time (Figure 13.4b).

Bojorquez has the highest nitrate and SRP concentrations of all lagoons studied. However, its low silicate concentrations indicate low SGD fluxes into the lagoon, suggesting that wastewater is the main nutrient source. Concentrations of NO_3^-, NO_2^-, SRP, and chlorophyll a all increased between 2001 and 2006 (Figures 13.6b, 13.6d, 13.8b, and 13.9b).

Dinoflagellates comprise the dominant phytoplankton group in the lagoon. Many HAB species are present (Figure 13.11), and algal blooms have been recorded in areas with lower water flow. Seagrasses (*T. testudinum* and *H. wrightii*) were the dominant type of SAV, and small green macroalgae were also present. SAV cover has decreased over the last few years, perhaps due to high boat traffic and eutrophication (Figure 13.13). Disease symptoms and higher abundance of epiphytes have also been observed on leaves of *T. testudinum*. Total N and P content in *T. testudinum* leaves are higher here than at oligotrophic sites in the Caribbean (Carruthers et al. 2005). Based on these findings and the lagoon's 400-day residence time, it is considered to be in poor condition (Figure 13.13). Several studies conducted in Bojorquez and Nichupte lagoons have reported signs of eutrophication (Merino et al. 1992).

13.4.9 Ascención Bay

This lagoon, on the Caribbean coast, is part of the "Sian Ka'an" Biosphere Reserve in the state of Quintana Roo. It is one of the peninsula's largest lagoons, and it is separated from the sea by a barrier reef and mangrove islands. It receives seawater from the Caribbean Sea, as well as significant freshwater inputs from groundwater and mangrove runoff, resulting in a typical estuarine salinity gradient (Figure 13.4b) (Arellano 2004).

Water in this lagoon has high transparency, low nutrient concentrations, and the lowest chlorophyll a concentration of all lagoons included in this study (Figure 13.9b). Water quality varies from year to year, with generally high SRP and SRSi values (Figures 13.7b and 13.8b). Nitrogen compounds and salinity have not varied much between years (Figures 13.4b, 13.6b, 13.6d, and 13.6f).

The phytoplankton community is dominated by diatoms (Figure 13.10). Only a few HAB species have been reported for this system (Figure 13.11) and there have been no reports of phytoplankton blooms. Seagrasses (*H. wrightii, T. testudinum, S. filiforme,* and *R. maritima*) are the dominant type of SAV, with some species of green and red macroalgae also present. Probably due to the lagoon's environmental heterogeneity, high levels of morphological variation have been observed in *T. testudinum* leaves. No significant changes in SAV cover have been observed despite several recent hurricanes that have affected the area (Figure 13.12).

The lagoon's overall condition is considered good (Table 13.3, Figure 13.13). Ascensión Bay supports an important, environmentally responsible lobster fishery organized by local fishermen, which has probably contributed to the maintenance of a healthy ecosystem. However, despite the bay's relatively short (150 days) residence time, the potential impact of human activities, such as agriculture and urban development, should not be ignored.

13.4.10 Chetumal

Chetumal Bay, a large coastal lagoon (Table 13.1), is located in the southern portion of the state of Quintana Roo, close to the border with Belize. This system is different from the other lagoons studied because it receives significant freshwater input from a river (Rio Hondo), as well as groundwater discharges. As a result, it has the lowest mean salinity (<20) of all lagoons included in this study (Table 13.2, Figure 13.4b). It is connected to the sea by an inlet in its southern perimeter and an artificial channel, the Canal de Zaragoza, was recently opened, causing a salinity increase (Figure 13.4b).

Chetumal has the highest SRSi concentration of the lagoons in this study, suggesting that groundwater is its most important freshwater source. Previous studies indicated that, despite the oligotrophic lagoon's large size, it exhibits relatively low spatial variation. Nearby areas of the Rio Hondo and lagoon regions near the city of Chetumal have shown signs of eutrophication (Herrera-Silveira et al. 2002b). Salinity and chlorophyll a both increased over the study period (Figures 13.4b and 13.9b), while other parameters have remained relatively stable (Figures 13.6b, 13.6d, 13.6f, and 13.8b), suggesting that the increase in seawater inputs has had a positive impact on productivity. Overall, this lagoon's condition is classified as fair (Figure 13.13) because of human impacts that have altered this system's water quality.

Cyanobacteria and diatoms are the most abundant members of the phytoplankton community (Figure 13.10). No data are available on the duration and frequency of algae blooms, although the number of HAB species recorded is low (Figure 13.11). Future research should focus on filling this gap and assessing SAV coverage in the lagoon, in order to create a baseline characterization of the system. The overall condition of the lagoon is considered fair (Figure 13.13), although research and monitoring programs are urgently needed.

13.5 WHAT HAVE WE LEARNED ABOUT THIS KARSTIC ECOSYSTEM?

Ten coastal lagoons in the Yucatan Peninsula were characterized in terms of groundwater inputs, depth, SAV coverage, phytoplankton biomass and diversity, and the presence of harmful algae species. Dissolved inorganic phosphorus concentrations were generally low, while nitrate and silicate concentrations were generally high. This may be related to SGD-related inputs of nitrate and silicate. Water residence time was variable between and within lagoons, and some of the water bodies showed symptoms of eutrophication. In this highly permeable karst area, inland human activities have quickly contaminated groundwater with nutrients and, most likely, other pollutants. Groundwater affects coastal lagoon water quality via SGD, and the overall condition of Yucatan lagoons appears to be worsening due to land use changes and coastal human activities.

13.5.1 Water Quality Characteristics

Salinity gradients in the lagoons studied ranged from estuarine (Celestún, Dzilam, Nichupte, Bojorquez, Ascensión, and Chetumal) to marine (Chelem, Holbox, Chacmochuk) and hyperhaline

(Rio Lagartos) (Table 13.2). Moreover, all the systems studied were characterized by strong spatial and temporal variation in water chemistry, which would tend to favor habitat diversity and ecological complexity. Since freshwater inputs to these systems have an inland origin, human activities that impact the watershed represent a source of excess nutrients (NO_3^-, SRSi, SRP), which will modify nutrient dynamics and promote eutrophication, especially in lagoons with longer water residence times (Table 13.1). Specifically, the lagoons of Chelem, Nichupte, and Bojorquez are located in areas of high anthropogenic impact. The salinity gradients of these lagoons and the inverse relationship of salinity and nitrate concentration indicate freshwater is the main nitrate source (Herrera-Silveria 1994c; Tapia et al. 2008). In addition, high NO_3^- concentrations (Table 13.2) may reflect the influence of nearby tourism infrastructure and domestic wastewater inputs (Chumacero-Velazquez 2004; Sima-Morales 2004; Tapia et al. 2008).

The observed variability in concentrations of inorganic nitrogen species (NH_4^+, NO_3^-) suggests that the Yucatan groundwater system is contaminated with these constituents, and that the degree of contamination is variable. The main sources of nitrate to groundwater are excessive fertilizer use and inadequate manure disposal. On the other hand, the major source of NH_4^+ is organic matter decomposition within the aquifer. Due to the prevailing anoxic conditions, removal of NH_4^+ is inefficient (Spiteri et al. 2008). This is the case in Chelem, Rio Lagartos, Chacmochuk, Nichupte, and Bojorquez, where there are garbage dumps, urban areas, and housing developments within the lagoons' groundwater catchments. In constrast, nutrient inputs to Celestún, Dzilam, and Ascensión are dominated by natural processes, including the decomposition of SAV and mangrove vegetation (Herrera-Silveira et al. 2000c; Medina-Gomez and Herrera-Silveira 2003; Arellano 2004). However, it is possible that nutrient concentrations in the latter group may increase and eventually become harmful to aquatic life in the future if urban development continues.

Due to the karstic nature of the Yucatan Peninsula, coastal waters typically exhibit high alkalinity and low SRP concentrations (<0.1 μM). SRP precipitates and is stored in the sediment; thus, its concentration in the water column is low, limiting aquatic productivity. As a consequence, these lagoons are highly responsive to even small SRP additions (Phlips et al. 2002; Cox et al. 2005). In this study, SRP concentrations >1 μM were measured in Celestún and Rio Lagartos lagoons, suggesting exogenous phosphate sources. These sources may include the guano of local waterfowl and surface runoff from nearby mangrove forests (Comin and Herrera-Silveira 2000). Chelem, Holbox, Chacmochuk, Bojorquez, and Chetumal had moderately elevated SRP concentrations (0.5 to 0.7 μM), most likely due to an increase in wastewater discharges from urban development and tourism. Although SRP is an essential ingredient for aquatic productivity, large additions can lead to the negative outcome of eutrophication (Cloern 2001).

SRSi has generally received less attention, as a component of water quality, than N and P. However, Yucatan groundwater has high SRSi concentrations (>140 μM), making this species useful as a tracer of groundwater (Herrera-Silveira 1994b) and as an indicator of potential vulnerability to cultural eutrophication. Chetumal and Celestún lagoons had the highest average SRSi concentrations (143 to 187 μM), and in general, lagoons with low salinities had the highest concentrations of this nutrient (Table 13.2). The only exception was Bojorquez, which had low salinity and low SRSi concentrations (Table 13.2), suggesting that freshwater entering this lagoon originates from wastewater and stormwater rather than SGD.

Previous studies have linked residence time of water in coastal lagoons and vulnerability to human impacts. Residence time modulates the lagoon's behavior as a sink or source of pollutants and organic matter, and in this way determines the rate at which organic matter is exported to adjacent near-shore marine areas (Medina-Gomez and Herrera-Silveira 2003). Lagoons with longer residence times (e.g., Chelem, Bojorquez, Holbox, and Chacmochuk) are at greater risk of eutrophication and are generally less resistant to anthropogenic or natural impacts. Lagoons with shorter residence times (e.g., Celestún, Dzilam, Nichupte, and Ascensión Bay) may export organic matter and are less vulnerable.

Celestún, Rio Lagartos and Chacmochuk had the highest average chlorophyll a concentrations (Table 13.2), indicating potential eutrophication (Newton et al. 2003). In Celestún, nutrient concentrations and the water residence time suggest that the high chlorophyll a concentration may be natural, while Rio Lagartos and Chamuchuk appear to be affected by anthropogenic nutrient inputs. Based on nutrient levels in leaves and epiphyte abundance on seagrasses, human activity appeared to contribute nutrient subsidies to Chelem and Bojorquez, which had intermediate dissolved nutrient concentrations (Table 13.2). Both of these lagoons may be in the early stages of eutrophication.

13.5.2 Phytoplankton

Cyanobacteria, nanoflagellates, centric diatoms, pennate diatoms, and dinoflagellates comprise the bulk of phytoplankton at the study sites. Pennate diatoms were the dominant group, followed by dinoflagellates and centric diatoms. Because phytoplankton communities respond rapidly to changes in environmental conditions, they have been used extensively to characterize coastal ecosystem status (Phlips et al. 2002). The prevalence of HAB phytoplankton species and the frequency and intensity of algal blooms suggested that all the lagoons in this study are at risk of HAB events if nutrient concentrations continue to increase (Alvarez 2004). Shifts in phytoplankton community structure from communities dominated by diatoms to those dominated by harmful dinoflagellates have been reported in some of the lagoons, indicating environmental deterioration due to eutrophication (Bricker et al. 1999; Cloern 2001).

13.5.3 Submerged Aquatic Vegetation (SAV)

The lagoons studied are shallow, with good water clarity, promoting high SAV cover (>50%) (Figure 13.13). Seagrass morphology, species composition, and abundance are related to salinity, nutrient concentrations, turbidity, and physical changes such as the interruption of water flows; thus, these plants can serve as indicators of environmental change or degradation (Abal and Dennison 1996; Agostinia et al. 2003). SAV in the lagoons consists mainly of seagrasses (*T. testudinum, H. wrightii*, and *R. maritima*) and macroalgae (green and red algae). Temporal changes in SAV composition and abundance depend on specific hydrological and sediment conditions, as well as on anthropogenic activities.

The SAV communities of Dzilam, Rio Lagartos, Chacmochuk, and Ascensión Bay are dominated by *Halodule wrightii*, *Thalassia testudinum*, *Syringodium filiforme*, and *Ruppia maritima*, and to a lesser extent by green and red macroalgae. SAV in these lagoons did not change appreciably between 2001 and 2006, suggesting that these lagoons are in good condition, possibly because no major urban areas or tourism infrastructure is located near these lagoons.

However, all the lagoons receive contaminated groundwater inputs, resulting in some level of contamination. Evidence of this contamination comes from analyses of *T. testudinum* tissues from Bojorquez, Holbox, Ascensión, Chelem, and Dzilam (Carruthers et al. 2005; Solis et al. 2008) (Table 13.4). Monitoring of Celestún, Chelem, Dzilam, Chacmochuk, Nichupte, Bojorquez, and Ascensión indicate that only Dzilam, Chacmochuk and Ascensión have maintained or recovered their original SAV characteristics after hurricane events. The remaining lagoons have not exhibited such resilience; instead, SAV cover and biomass decreased after hurricanes and have not yet rebounded (Ascencio 2008). Trawling nets used for fishing and boat traffic in tourist areas can both damage SAV (Herrera-Silveira et al. 2000c; Castrejon et al. 2005), and, as a result, coastal ecosystems have lost some of their natural resilience to natural events such as hurricanes.

According to recent climate change projections (CICC 2007), the Yucatan Peninsula is likely to experience increased hurricane frequency and intensity, changes in climate seasonality, and sea

TABLE 13.4
Leaf Nutrient Content in *Thalassia testudinum* from Different Locations around Yucatan Peninsula

Location	Site	%N	%P	References
Caribbean	Akumal	1	0.06	Mutchler et al. (2007)
Caribbean	Xaak	2	0.11	Mutchler et al. (2007)
Caribbean	Puerto Morelos	1.8–2.11	0.13–0.38	Carruthers et al. 2005
Gulf of Mexico	Yucatan north coast	1.5–1.7	0.11–0.13	Herrera-Silveira (unpublished data)

level rise over the coming century. More research is necessary to identify vulnerable ecosystems and develop effective mitigation tactics, such as mangrove forest and seagrass restoration projects. In addition, watershed management policies and modern wastewater treatment systems are essential in order to avoid impacts on coastal ecosystems via contaminated groundwater.

13.6 CONCLUSIONS

The coastal lagoons studied are located along a 964-km long stretch of coast, and they cover a combined area of 19,000 km^2. Although they have many features in common, including the regional climate and the presence of groundwater inputs, each lagoon is affected by human activity in a different way.

The susceptibility of these coastal systems to eutrophication or other damage is directly related to the magnitude and quality of groundwater and wastewater inputs, water residence time and circulation, surface runoff from mangroves, and other exogenous nutrient sources such as bird feces. Based on their phytoplankton, SAV cover, and mangrove data, Chelem and Bojorquez lagoons seem to face the greatest threat. In contrast, Celestún, Dzilam, and Ascensión are currently in good condition because they are located relatively far from human development. Rio Lagartos, Holbox, Chacmochuk, and Chetumal, located in areas of more intense human activity, may also be at risk even though eutrophication is not obvious at present. Water circulation and residence time in the lagoon modulates the effects of pollution inputs, with long residence times associated with more severe effects.

We recommend management actions to restore natural water circulation and reduce pollution in wastewater and groundwater. New guidelines for zoning and infrastructure construction should be developed to account for the impact these activities have on coastal ecosystem productivity, diversity, nutrient dynamics, water quality, fishing, and human recreation. Long-term monitoring programs are necessary to assess ecological changes, differentiate impacts related to human activities from those related to natural events, and understand how natural and human disturbances interact.

ACKNOWLEDGMENTS

We thank all the students and laboratory staff of the Primary Production Laboratory of CINVESTAV-IPN, Unidad Mérida, for their assistance in the field and help with laboratory analyses. Special thanks to J. Ramirez, A. Zaldivar, C. Alvarez, L. Troccoli, F. Merino, I. Osorio, J. Trejo, F. Tapia, J. Sima, R. Colli, I. Medina, O. Cortes, and L. Arrellano. We especially thank Mike Kennish and Karen Knee for their advice and assistance in improving the manuscript, as well as the anonymous reviewers for their helpful comments. The financial support provided by CONABIO (B019; FQ004), CONANP, CONACYT (32356-T; D112-904672; 4147P-T; G34709; 2002-H602; YUC-2003-C02-046), PNUD (MEX/OP3/05/40), and CINVESTAV-IPN is gratefully acknowledged.

REFERENCES

Abal, E. G. and W. C. Dennison. 1996. Seagrass depth range and water quality in southern Moreton Bay, Queensland, Australia. *Mar. Freshwater Res.* 47: 763–771.

Agostinia, B. S., A. Capiomontc, B. Marchanda, and G. Pergentd. 2003. Distribution and estimation of basal area coverage of subtidal seagrass meadows in a Mediterranean coastal lagoon. *Estuar. Coastal Shelf Sci.* 56: 1021–1028.

Alongi, M. D. 1998. *Coastal ecosystem processes*. Boca Raton, Florida: CRC Press.

Alvarez, C. 2004. Cambios estructurales en el fitoplancton costero en un zona carstica de la peninsula de Yucatan con impacto natural y antrópico. M.S. Thesis, Marine Biology, CINVESTAV-IPN, Unidad Mérida, Mexico.

Amany, A. I. and M. M. Dorgham. 2003. Ecological indices as a tool for assessing pollution in El-Dekhaila Harbour (Alexandria, Egypt). *Oceanologia* 45: 121–131.

Arellano, L. U. 2004. Analisis espacio-temporal de las variables hidrologicas: Deteccion de la heterogeneidad a gran escala de los ecosistemas costeros. M.S. Thesis, Marine Biology, CINVESTAV-IPN, Unidad Mérida, Mexico.

Arhonditsis, G., M. Karydis, and G. Tsirtsis. 2003. Analysis of phytoplankton community structure using similarity indices: A new methodology for discriminating among eutrophication levels in coastal marine ecosystems. *Environ. Manage.* 31: 619–632.

Ascencio, J. M. 2008. Cambios en la Vegetacion Acuatica de las lagunas costeras de Celestún, Chelem y Dzilam 2000–2005. M.S. Thesis, Instituto Tecnologico de Conkal, Yucatan, Mexico.

Avery, W. 2000. Monitoring submerged aquatic vegetation in Hillsborough Bay, Florida. Pp. 137–145, In: S. A. Bortone (ed.), *Seagrasses: Monitoring, ecology, physiology, and management*. Boca Raton, Florida: CRC Press.

Borja, A., J. Franco, and V. Perez. 2000. A marine biotic index to establish the ecological quality of soft-bottom benthos within European estuarine and coastal environments. *Mar. Poll. Bull.* 40: 1100–1114.

Borja, A., V. Valencia, J. Franco, I. Muxika, J. Bald, M. J. Belzunce, and O. Solaun. 2004. The water framework directive: Water alone, or in association with sediment and biota, in determining quality standards? *Mar. Poll. Bull.* 49: 8–11.

Borja, A. 2005. The European water framework directive: A challenge for nearshore, coastal, and continental shelf research. *Cont. Shelf Res.* 25: 1768–1783.

Boyer, J. N., P. Sterling, and R. D. Jones. 2000. Maximizing information from a water quality monitoring network through visualization techniques. *Estuar. Coastal Shelf Sci.* 50: 3–48.

Brander, K. M., R. R. Dickson, and M. Edwards. 2003. Use of continuous plankton recorder information in support of marine management: Applications in fisheries, environmental protection, and in the study of ecosystem response to environmental change. *Prog. Oceanogr.* 58: 175–191.

Braun-Blanquet, J. 1932. *Plant sociology: The study of plant communities*. New York: McGraw-Hill.

Bricker, S. B., C. G. Clement, D. E. Pirhalla, S. P. Orlando, and D. R. G. Farrow. 1999. *National Estuarine Eutrophication Assessment: Effects of nutrient enrichment in the nation's estuaries*. NOAA, National Ocean Service, Special Projects Office and the National Centers for Coastal Ocean Science, Silver Spring, Maryland.

Bricker, S. B., J. G. Ferreira, and T. Simas. 2003. An integrated methodology for assessment of estuarine trophic status. *Ecol. Model.* 169: 39–69.

Buchanan, C., R. V. Lacouture, H. G. Marshall, M. Olson, and J. Johnson. 2005. Phytoplankton reference communities for Chesapeake Bay and its tidal tributaries. *Estuaries* 28: 138–159.

Carruthers, T. J. B., B. I. Van Tussenbroek, and W. C. Dennison. 2005. Influence of submarine springs and wastewater on nutrient dynamics of Caribbean seagrass meadows. *Estuar. Coastal Shelf Sci.* 64: 191–199.

Castrejon, H., R. Perez-Castañeda, and O. Defeo. 2005. Spatial structure and bathymetric patterns of penaeid shrimps in the southwestern Gulf of Mexico. *Fish. Res.* 72: 291–300.

CEC (Commission of the European Communities). 1991. Council Directive of 21 May 1991 concerning urban waste water treatment (91/271/EEC). *Official Journal of the European Communities*, L135 of 30.5.91: 40–52.

Chretiennot-Dinet, M. J., A. Sournia, M. Ricard, and C. Billard. 1993. A classification of the phytoplankton of the world from class to genus. *Phycologia* 32: 159–179.

Chumacero-Velazquez, M. 2004. Consumo de oxigeno del sedimento e intercambio con el mar del sistema lagunar Nichupte–Bojorquez. M.S. Thesis, Marine Biology, CINVESTAV-IPN, Unidad Mérida, Mexico.

CICC (Comision Intersecretarial del Cambio Climatico). 2007. *Estrategia nacional de cambio climatico*. Comision Intersecretarial de Cambio Climatico, Semarnat, Mexico.

Cloern, J. E. 2001. Our evolving conceptual model of the coastal eutrophication problem. *Mar. Ecol. Prog. Ser.* 210: 223–257.

Coelho, J. P., M. R. Flindt, H. S. Jensen, A. I. Lillebø, and M. A. Pardal. 2004. Phosphorus speciation and availability in intertidal sediments of a temperate estuary: Relation to eutrophication and annual P-fluxes. *Estuar. Coastal Shelf Sci.* 61: 583–590.

Comin F. A. and J. A. Herrera-Silveira. 2000. The role of birds on the trophic structure and nutrient cycles of aquatic ecosystems: A review. In: F. A. Comin, J. A. Herrera-Silveira, and J. Ramirez (eds.), *Proceedings of the limnology and waterfowl, monitoring, modeling, and management workshop*, Aquatic Birds Working Group. Societas Internationalis Limnologiae, Universidad Autonoma de Yucatan, Mexico.

Costanza, R., R. d'Arge, R. De Groot, S. Farber, M. Grasso, B. Hannon, K. Limburg, S. Naeem, R. O'Neill, J. Paruelo, R. G. Raskin, and P. Sutton. 1997. The value of the world's ecosystem services and natural capital. *Nature* 387: 253–260.

Costanza, R., B. G. Norton, and B. D. Haskell (eds.). 1992. *Ecosystem health: New goals for environmental management*. Washington, D.C.: Island Press.

Cox, M. E., A. Moss, and G. K. Smyth. 2005. Water quality condition and trend in North Queensland waterways. *Mar. Poll. Bull.* 51: 89–98.

Dayton, P. K., V. Currie, T. Gerrodette, B. D. Keller, R. Rosenthal, and D. V. Tresca. 1984. Patch dynamics and stability of some California kelp communities. *Ecol. Monogr.* 53: 253–289.

Dennison, W. C., R. J. Orth, K. A. Moore, J. C. Stevenson, V. Carter, S. Kollar, P. W. Bergstrom, and R. A. Batuik.1993. Assessing water quality with submerged aquatic vegetation. *Bioscience* 43: 86–94.

Devlin, M., M. Best, D. Coates, E. Bresnan, S. O'Boyle, R. Park, J. Silke, C. Cusack, and J. Skeats. 2007. Establishing boundary classes for the classification of U.K. marine waters using phytoplankton communities. *Mar. Poll. Bull.* 55: 91–103.

Durako, M. J. and C. M. Duarte. 1997. On the use of reconstructive aging techniques for assessing seagrass demography: A critique of the model test of Jensen et al. (1996). *Mar. Ecol. Prog. Ser.* 146: 297–303.

Fourqurean, J. W., J. N. Boyer, M. J. Durako, L. N. Hefty, and B. J. Peterson. 2003. Forecasting responses of seagrass distributions to changing water quality using monitoring data. *Ecol. Applic.* 13: 474–489.

Fourqurean, J. W., A. Willsie, C. D. Rose, and L. M. Rutten. 2001. Spatial and temporal pattern in seagrass community composition and productivity in south Florida. *Mar. Biol.* 138: 341–354.

Gordon, D. C., P. R. Boudreau, K. H. Mann, J. E. Ong, W. L. Silvert, S. V. Smith, G. Wattayakorn, F. Wulff, and T. Yanagi. 1996. *LOICZ biogeochemical modelling guidelines.* LOICZ Reports and Studies No. 5. Texel, the Netherlands.

Hallegraeff, G. M., D. M. Anderson, and A. D. Cembella (eds.). 1995. *Manual on harmful marine microalgae.* Paris: IOC-UNESCO.

Hanshaw, B. B. and W. Back. 1980. Chemical mass-wasting of the northern Yucatan Peninsula by groundwater dissolution. *Geology* 8: 222–224.

Hasle, S. 1978. The inverted microscope method. Pp. 88–96, In: A. Sournia (ed.), *Phytoplankton manual.* Paris: UNESCO.

Hemminga, M. A. and C. M. Duarte. 2000. *Seagrass ecology.* Cambridge, U.K.: University of Cambridge Press.

Herrera-Silveira, J. A. 1994a. Spatial and temporal patterns in a tropical coastal lagoon with groundwater discharges. *J. Coastal Res.* 10: 738–746.

Herrera-Silveira, J. A. 1994b. Nutrients from underground discharges in a coastal lagoon (Celestún, Yucatan, Mexico). *Verh. Internat. Verein. Limnol.* 25: 1398–1403.

Herrera-Silveira, J. A. 1994c. Phytoplankton productivity and submerged macrophytes biomass variation in a tropical coastal lagoon with groundwater discharge. *Vie Milieu* 44: 257–266.

Herrera-Silveira, J. A. 2006. Coastal lagoon of Yucatan (SE, Mexico): Research, diagnostic and management. *Ecotropicos* 19: 94–108.

Herrera-Silveira, J. A., M. B. Martin, and V. Diaz-Arce. 1999a. Variation of phytoplankton in four coastal lagoons of Yucatan Sate, Mexico. *Rev. Biol. Trop.* 47: 37–46.

Herrera-Silveira, J. A., I. Medina, and R. A. Colli. 2002a. Trophic status on nutrient concentration scales and primary producers community of tropical coastal lagoons influenced by groundwater discharges. *Hydrobiologia,* 475/476: 91–98.

Herrera-Silveira, J. A., J. Ramirez-Ramirez, N. Gomez, and A. Zaldivar. 2000a. Seagrass bed recovery after hydrological restoration in a coastal lagoon with groundwater discharges in the north of Yucatan. Pp. 123–135, In: S. Bortone (ed.), *Seagrass: Monitoring ecology, physiology and management.* Boca Raton, Florida: CRC Press.

Herrera-Silveira, J. A., J. Ramirez-Ramirez, and A. Zaldivar-Jimenez. 1998. Overview and characterization of the hydrology and primary producer community of selected coastal lagoons of Yucatan, Mexico. *Aquat. Ecosys. Health Manage.* 1: 353–372.

Herrera-Silveira, J. A., J. Ramirez, and A. Zaldivar. 2000b. Structure and soil salinity of two mangroves forests of Yucatan, SE Mexico. *Verh. Internat. Verein. Limnol.* 1707–1710.

Herrera-Silveira, J. A., L. Troccoli, J. Ramirez-Ramirez, and A. Zaldivar. 1999b. Yucatan groundwater-influenced systems: Dzilam Lagoon, Yucatan. Pp. 41–44, In: S. V. Smith, J. I. Crossland, and C. J. Crossland (eds.), *Mexican and Central American coastal lagoon systems: Carbon, nitrogen, and phosphorus fluxes.* Regional Workshop II, LOICZ Reports and Studies No. 13. Texel, the Netherlands.

Herrera-Silveira, J. A., A. Zaldivar, M. Aguayo, J. Trejo, I. Medina, F. Tapia, and O. Vazquez-Montiel. 2002b. Water quality of Chetumal Bay measured by indicators of trophic status. Pp. 185–196, In: F. J. Rosado-May, R. Romero Mayo, and A. D. Navarrete (eds.), *Science contribution to an integral coastal management of the Chetumal Bay and influence area.* Quintana Roo University, Chetumal, Q. Roo, Mexico.

Herrera-Silveira, J. A., A. Zaldivar, J. Ramirez, and D. Alonzo. 2000c. Habitat use of the American flamingo (*Phoenicopterus reber ruber*) in the Celestún Lagoon, Yucatan, Mexico. In: F. A. Comin, J. A. Herrera-Silveira, and J. Ramirez (eds.), *Proceedings of the limnology and waterfowl, monitoring, modeling, and management workshop.* Aquatic Birds Working Group, Societas Internationalis Limnology, November 24–27, 2000, Mérida, Yucatan Universidad Autonoma de Yucatán, Mexico.

Jeffrey, S. W. and G. F. Humphrey. 1975. New spectrophotometric equations for determining chlorophylls a, b in higher plants, algae, and natural phytoplankton. *Biochem. Physiol. Pflanzen* 167: 191–194.

Jorgensen, S., F. L. Xu, and R. Costanza. 2005. *Handbook of ecological indicators for assessment of ecosystem health.* Boca Raton, Florida: CRC Press.

Kjerfve, B. 1994. *Coastal lagoon processes.* New York: Elsevier.

Lewis, W. J., J. L. Farr, and S.D. Foster. 1980. The pollution hazard to village water supplies in eastern Bostwana. *Proc. Inst. Civil Eng.* 69 (Part 2): 281–293.

Medina-Gomez, I. and J. A. Herrera-Silveira. 2003. Spatial characterization of water quality in a karstic coastal lagoon without anthropogenic disturbance: A multivariate approach. *Estuar. Coastal Shelf Sci.* 58: 455–465.

Merino, M., A. Gonzalez, E. Reyes, M. E. Gallegos, and S. Czitrom. 1992. Eutrophication in the lagoons of Cancun, Mexico. *Sci. Total Environ. Suppl.* 861–870.

Morales-Ojeda, S. M. 2007. Diagnostico de las aguas costeras de Yucatan basado en las caraceristicas hidrologicas y del fitoplancton. M.S. Thesis, Marine Biology, CINVESTAV-IPN, Unidad Mérida, Mexico.

Mutchler, K., H. Dunton, A. Townsend-Small, S. Fredriksen, and M. K. Rasser. 2007. Isotopic and elemental indicators of nutrient sources and status of coastal habitats in the Caribbean Sea, Yucatan Peninsula, Mexico. Estuar., Coast. Shelf Sci. 74: 449-457.

Newton, A., J. D. Icely, M. Falcao, A. Nobre, J. P. Nunes, J. G. Ferreira, and C. Vale. 2003. Evaluation of eutrophication in the Ria Formosa coastal lagoon, Portugal. *Estuar. Coastal Shelf Sci.* 23: 1945–1961.

Newton, A. and S. M. Mudge. 2005. Lagoon-sea exchanges, nutrient dynamics and water quality management of the Ria Formosa (Portugal). *Estuar. Coastal Shelf Sci.* 62: 405–414.

Orfanidis, S., P. Panayotidis, and N. Stamatis. 2003. An insight to the Ecological Evaluation Index (EEI). *Ecol. Ind.* 3: 27–33.

Orth, R. J., and K. A. Moore. 1983. An unprecedented decline in submerged aquatic vegetation. *Science* 7: 51–53.

Parson, T. R., Y. Maita, and C. M. Lalli. 1984. *A manual of biological and chemical methods for seawater analysis*, 2nd Ed. Oxford, England: Pergamon Press.

Phlips, E. J., S. Badylak, and T. Grosskopf. 2002. Factors affecting the abundance of phytoplankton in a restricted subtropical lagoon, the Indian River Lagoon, Florida, USA. *Estuar. Coast. Shelf Sci.* 55: 385–402.

Round, F. E., R. M. Crawford, and D. G. Mann. 1990. The Diatoms. Biology and Morphology of the Genera. Cambridge University Press, USA, 747 pp.

Schumann, R., H. Baudler, A. Glass, K. Dumcke, and U. Karsten. 2006. Long-term observations on salinity dynamics in a tideless shallow coastal lagoon of the southern Baltic Sea coast and their biological relevance. *J. Mar. Syst.* 60: 330–334.

Seller, B. D. and B. D. Causey. 2005. Linkages between the Florida Keys National Marine Sanctuary and the South Florida Ecosystem Restoration Initiative. *Ocean Coastal Manage.* 48: 869–900.

Sima-Morales, A. L. 2004. *Hidrologia y estado trofico de las lagunas costeras del caribe Mexicano: Idrology and trophic status of the Caribbean coastal lagoon of Mexico—aporte de nutrientes de diferentes fuentes.* M.S. Thesis, Marine Biology, CINVESTAV-IPN, Unidad Mérida, Mexico.

Slomp, C. P. and P. Van Cappellen. 2004. Nutrient inputs to the coastal ocean through submarine groundwater discharge: Controls and potential impact. *J. Hydrol.* 295: 64–86.

Smith, S. V., V. Camacho-Ibar, J. A. Herrera-Silveira, D. Valdes, D. M. Merino, and R. W. Buddemeier.1999. Quantifying groundwater flow using water budgets and multiple conservative tracers. Pp. 96–105, In: S. V. Smith, J. I. Marshall Crossland, and C. J. Crossland (eds.), *Mexican and Central American coastal lagoon systems: Carbon, nitrogen, and phosphorus fluxes.* Regional Workshop II, LOICZ Reports and Studies No. 13. Texel, the Netherlands.

Solis, C., A. Martinez, E. Lavoisier, M. A. Martinez, and K. Isaac-Olive. 2008. Trace metal analysis in sea grasses from Mexican Caribbean Coast by particle induced X-ray emission (PIXE) *Rev. Mexicana Fisica* 54: 50–53.

Sournia, A. 1987. *Phytoplankton manual.* Paris: UNESCO.

Spiteri, C., C. P. Slomp, M. A. Charette, K. Tuncay, and Ch. Meile. 2008. Flow and nutrient dynamics in a subterranean estuary (Waquoit Bay, MA, USA): Field data and reactive transport modeling. *Geochim. Cosmochim. Acta* 72: 3398–3412.

Stevenson, J. C., L. W. Staver, and K. W. Staver. 1993. Water quality associated with survival of submersed aquatic vegetation along an estuarine gradient. *Estuaries* 16: 346–361.

Strickland, J. D. H. and T. R. Parsons. 1972. *Practical handbook of sea water analysis*, 2nd ed. Fisheries Research Board Canadian Bulletin, Ottawa, Canada.

Tapia, G. F., J. A. Herrera-Silveira, and L. Aguirre-Macedo. 2008. Water quality variability and eutrophic trends in karstic tropical coastal lagoons of the Yucatan Peninsula. *Estuar. Coastal Shelf Sci.* 76: 418–430.

Teutli, C. 2008. Regeneración de zonas de manglar bajo diferentes regimenes hidrológicos en sistemas cársticos-carbonatados. M.S. Thesis, Marine Biology, CINVESTAV-IPN, Unidad Mérida, Mexico.

Tran, Ch., D. Valdes, J. A. Herrera-Silveira, J. Euan, I. Medina-Gomez, and N. Aranda-Cirerol. 2002. Status of coastal water quality at Holbox Island, Quintana Roo State, Mexico. Pp. 331–340, In: C. A. Brebia (ed.), *Environment problems in coastal region IV.* Fourth International Conference on Environmental Problems in Coastal Regions. Boston: Wit-Press.

Tran, Ch., D. Valdes, J. A. Herrera-Silveira, and J. Euan. 2005. Coastal water quality monitoring: How can it contribute to sustainable development of Holbox Island, Mexico. Pp. 255–264, In: N. H. Afgan, Z. Bogdan, N. Duic, and Z. Guzovic (eds.), *Sustainable development of energy, water and environment systems, Volume II.* Conference on Sustainable Development of Energy, Water and Environment Systems, June 15–20, 2005, Faculty of Mechanical Engineering and Naval Architecture, Dubrovnik, Croatia.

Twilley, R. R. 1995. Properties of mangrove ecosystems and their relation to the energy signature of coastal environments. Pp. 43–61, In: C. A. S. Hall (ed.), *Maximum power: The ideas and applications of H. T. Odum.* Niwot, Colorado: University of Colorado Press.

U.S. Environmental Protection Agency. 2001a. *Nutrient criteria technical guidance manual. Estuarine and coastal marine waters.* EPA-822-B-01-003. Washington, D.C.: EPA, Office of Water.

U.S. Environmental Protection Agency. 2001b. *National coastal condition report.* EPA-620/R-01/005. Washington, D.C.: EPA, Office of Research and Development.

U.S. Environmental Protection Agency. 2004. *National coastal condition report.* EPA-620/R-03/002. Washington, D.C.: EPA, Office of Research and Development.

U.S. Environmental Protection Agency. 2005. *New indicators of coastal ecosystem condition.* EPA/600/S-05/004. Washington, D.C.: EPA, Office of Research and Development.

Young, M. B., M. E. Gonneea, D. A. Fong, W. S. Moore, J. A. Herrera-Silveira, and A. Payatan. 2008. Characterizing sources of groundwater to a tropical coastal lagoon in a karstic area using radium isotopes and water chemistry. *Mar. Chem.* 109: 377–394.

Young, M., M. E. Gonneea, J. A. Herrera-Silveira, and A. Paytan. 2005. Export of dissolved and particulate carbon and nitrogen from a mangrove-dominated lagoon, Yucatan Peninsula, Mexico. *Int. J. Ecol. Environ. Sci.* 31: 189–202.

Zaldivar-Jimenez, A., J. A. Herrera-Silveira., C. M. Coronado, and D. Alonso. 2004. Estructura y productividad de los manglares en la Reserva de la Biosfera Ria Celestún Yucatan (SE Mexico). *Maderas y Bosques* 2: 25–35.

Zieman, J. C., J. Fourqurean, and T. A. Frankovich. 1999. Seagrass dieoff in Florida Bay (USA): Long-term trends in abundance and growth of *Thalassia testudinum* and the role of hypersalinity and temperature. *Estuaries* 22: 460–470.

14 Seasonal and Interannual Variability of Planktonic Microbes in a Mesotidal Coastal Lagoon (Ria Formosa, SE Portugal)

Impact of Climatic Changes and Local Human Influences

*Ana B. Barbosa**

CONTENTS

Abstract ... 336
14.1 Introduction ... 336
14.2 The Ria Formosa Coastal Lagoon: An Overview ... 339
14.2.1 Climatic and Hydrologic Setting ... 339
14.2.2 Oceanographic Setting ... 339
14.2.3 Lagoon Morphology and Hydrodynamics ... 339
14.2.4 Biological Assemblages ... 340
14.3 Approach and Dataset ... 340
14.4 Recent Climate and Local Anthropogenic Changes ... 343
14.4.1 Climatic Changes ... 343
14.4.2 Local Anthropogenic Changes ... 344
14.4.3 Water Quality and Eutrophication ... 345
14.4.4 Changes in Hydrodynamics and Sedimentary Dynamics ... 346
14.4.5 Habitat Alterations ... 346
14.4.6 Exploitation of Biological Production ... 346
14.4.7 Introduced Species ... 347
14.5 Seasonal Dynamics of Planktonic Microbes ... 347
14.5.1 Environmental Setting ... 347
14.5.2 Phytoplankton ... 351
14.5.3 Heterotrophic Bacterioplankton ... 352
14.6 Interannual Dynamics of Planktonic Microbes ... 354
14.6.1 Environmental Setting ... 354
14.6.2 Phytoplankton ... 356

* Corresponding author. Email: abarbosa@ualg.pt

14.7 Conclusions ... 360
Acknowledgments ... 361
References ... 361

ABSTRACT

The Ria Formosa coastal lagoon is a multi-inlet mesotidal system, located along the south coast of Portugal. It is subjected to multiple anthropogenic influences and is situated in a region classified as very vulnerable to climate change. This study aimed to describe the seasonal and interannual variability of planktonic microbes in the lagoon, the driving forces underlying microbe variability, and the relative contribution of climate and local-human influences. Changes in anthropogenic activities or climatic processes affecting the Ria Formosa Lagoon were identified for the period 1967 to 2008 and linked to current knowledge on abiotic variables and planktonic microbes in the western subembayment of the lagoon. Phytoplankton and heterotrophic bacterioplankton at inner lagoon locations exhibited unimodal annual cycles with summer peaks, coupled to temperature and light availability, whereas phytoplankton at inlet areas displayed a bimodal annual cycle, probably more intensively controlled by nutrient availability. The effects of increased anthropogenic pressure, both on seasonal and interannual time scales, were evident only at sites close to major urban centers, probably due to the low water residence time inside the lagoon, and the relevance of physical variables (e.g., temperature and light) as drivers of the growth of planktonic microbes. Long-term climatic variability apparently reverberated into the lagoon's water column, leading to a generalized and significant winter warming trend, and strong interannual changes in the concentration of inorganic nutrients, particularly during the autumn–winter period. Phytoplankton biomass showed a generalized interannual declining trend during autumn and winter. This decline may be eventually related to the winter warming tendency, through its stimulatory effect upon phytoplankton grazers, or to decreases in nutrient concentrations associated with reductions in the autumn freshwater inflow. Long-term changes during spring and summer were less coherent across lagoon stations, indicating that more site-specific processes were driving phytoplankton variability. Despite overall reduced freshwater flows into the lagoon, interannual freshwater flow variability apparently exerted an important control on the biomass of phytoplankton, particularly during periods of relatively high flow rates and low light limitation (spring and autumn). Overall, synthesized data suggest that climate variability is an important driver of planktonic microbes in the lagoon, both on short-term, seasonal, and interannual time scales. Local anthropogenic impacts are apparently spatially restricted. More extensive time-series and dedicated experiments are needed to conclusively establish direct links between the dynamics of planktonic microbes and climatic and local anthropogenic alterations in the Ria Formosa Lagoon.

Key Words: coastal lagoon, Ria Formosa, phytoplankton, heterotrophic bacterioplankton, climate change, anthropogenic pressures, seasonal, interannual

14.1 INTRODUCTION

Coastal lagoons are shallow, semi-enclosed aquatic ecosystems partially separated from the adjacent coastal ocean by barrier islands or peninsular spits (see Kennish and Paerl, Chapter 1, this volume). Despite their low areal coverage (~13% of the world's coastline), coastal lagoons are among the most productive natural ecosystems on earth due to: (1) a high external supply of nutrients from land and, in the case of coastal upwelling areas, from the ocean; (2) the extension of the euphotic zone through most of the water column; and (3) rapid remineralization of nutrients (Kjerfve 1994). Coastal lagoons provide valuable goods and services, such as seafood, recreation opportunities, transformation and storage of organic matter and inorganic nutrients, and protection of the coastline (Nixon 1982; Levin et al. 2001).

Coastal lagoons comprise a large number of boundaries (e.g., water–sediment, land–lagoon, lagoon–ocean, and water–atmosphere) and are therefore susceptible to multiple natural and anthropogenic influences. In recent decades, coastal lagoons have been under increasing anthropogenic pressure (Levin et al. 2001; Viaroli et al. 2005; Lotze et al. 2006). Due to relatively restricted exchange with the ocean, coastal lagoons are considered particularly susceptible to eutrophication (Gamito et al. 2005), a serious and widespread problem often linked to human-induced nutrient overenrichment. Besides their susceptibility to multiple, local anthropogenic stressors, coastal lagoons are also considered very sensitive to climatic variations, including short-term perturbations (e.g., storms, floods, and heat waves) and long-term changes (Cloern et al. 2005; Paerl et al. 2006). The anticipated effects of global climate changes (e.g., increased air temperature, accelerated sea-level rise, and alterations in storm frequency, storm intensity, and precipitation) may reverberate through coastal lagoons, leading to increases in water temperature, dissolved carbon dioxide, and changes in salinity, turbidity, ultraviolet radiation, lagoon morphology, and hydrodynamics (IPCC 2007). These environmental alterations will have consequences for the structure and function of coastal lagoons (Lloret et al. 2008). There are growing concerns with coastal lagoons because of their exceptional ecologic and economic value and the serious threats posed by anthropogenic and global climate changes.

Ideally, evaluations of the impact of environmental alterations on coastal lagoons should be based on an integrated view of physical, chemical, geomorphological, and biological variables, associated with a wide range of communities, such as salt marshes, seagrass meadows, or plankton assemblages. Planktonic microbes, which include phytoplankton, heterotrophic bacteria, and phagotrophic protists, play vital roles in a wide range of ecosystem functions, such as primary production, nutrient cycling, organic matter decomposition, carbon sequestration, climate regulation, and water purification (Ducklow 2008). They exhibit high abundances and growth rates, and are therefore sensitive to environmental perturbations on short and long time scales. For these reasons, planktonic microbes are often used as gauges of ecological condition and change (Smetacek and Cloern 2008; Borkman et al. 2009).

The Ria Formosa coastal lagoon is a multi-inlet mesotidal system, located on the south coast of Portugal (Figure 14.1A), which has been classified as very vulnerable to climate change (IPCC 2007). The lagoon is a breeding and transit area for fish and birds, and supports a wide range of human uses, including tourism, fisheries, and shellfish farming. Despite the putative importance of planktonic microbes to the ecology of the lagoon (Ducklow 2008), published data are limited, particularly for heterotrophic microbes (Barbosa 1991; Morais et al. 2003). Published information on phytoplankton is more extensive and addresses spatial and seasonal variability of phytoplankton composition, chlorophyll *a* concentration (Chl *a*) and phytoplankton production (Assis et al. 1984; Falcão and Vale 2003; Newton et al. 2003; Loureiro et al. 2006; Pereira et al. 2007; Brito et al. 2009b), and the effects of nutrient enrichment (Loureiro et al. 2005; Edwards et al. 2005). However, most of the latter are confined to only 1- to 2-year study periods. No information on long-term changes in planktonic microbes is available for the lagoon.

The goals of this study were: (1) to describe seasonal and interannual variability of planktonic microbes in the Ria Formosa coastal lagoon; and (2) to understand the driving forces underlying microbial variability on these time scales, and the relative importance of climatic and local anthropogenic alterations. Seasonal and interannual changes in anthropogenic activities and climate in the Ria Formosa drainage basin were identified. Current knowledge on abiotic variables and planktonic microbes in the lagoon during the period 1967 to 2008 was synthesized, using published information, public datasets, and gray literature, in order to discern the relative contribution of climate and local human influences.

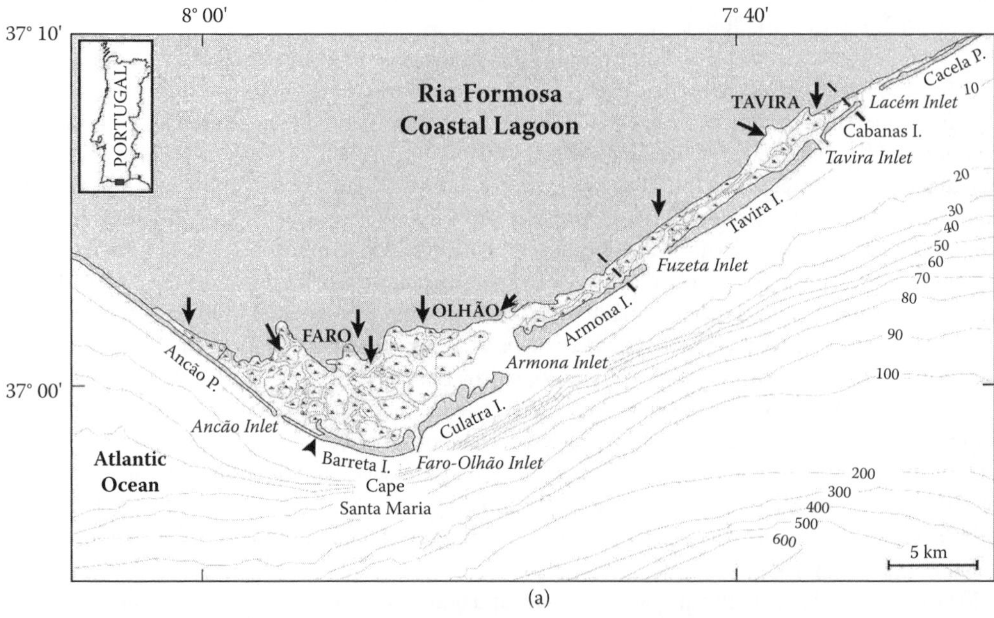

(a)

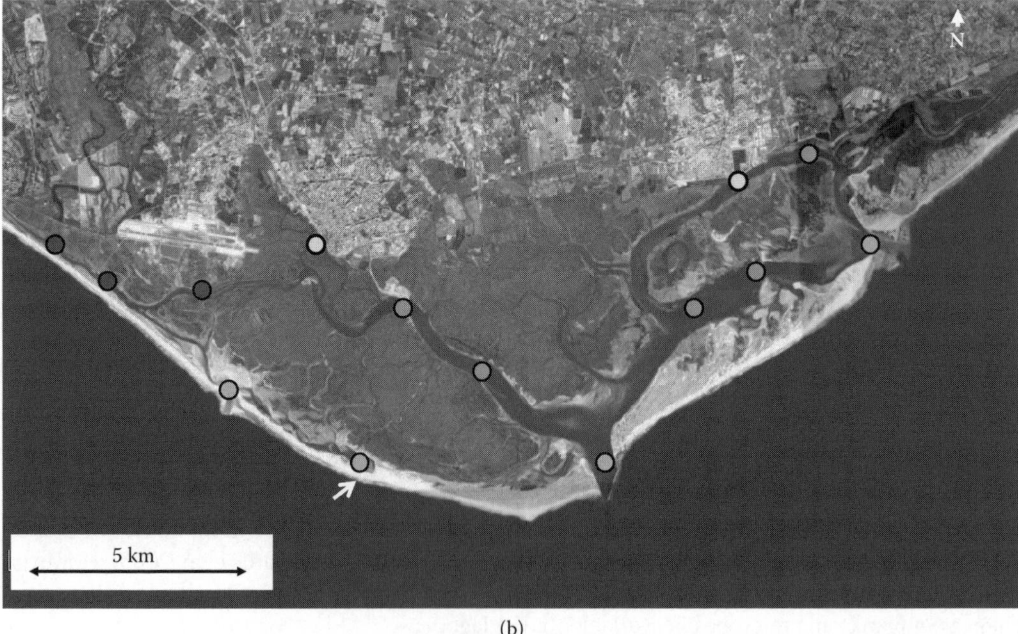

(b)

FIGURE 14.1 (a) Map of the Ria Formosa coastal lagoon and barrier island system (I., islands; P., peninsulas) showing divisions (dashed lines) between different lagoon subembayments (see Salles et al. 2005). Inlets, major urban centers, main freshwater courses (gray arrows), wastewater treatment plants (black arrows), and coastal bathymetry (in meters) are indicated. (b) Aerial view of the western subembayment of the Ria Formosa coastal lagoon with location of sampling stations. (1) Triangles represent stations located at lagoon inlets, and sea boundary conditions; squares represent stations located in the main navigational channels, and intermediate conditions; circles represent stations located at the inner lagoon areas, and landward boundary conditions; and diamonds represent stations located close to major urban centers, and urban boundary conditions (see text for details). The location of Ancão Inlet before 1997 is shown as an arrow.

14.2 THE RIA FORMOSA COASTAL LAGOON: AN OVERVIEW

The Ria Formosa is a highly dynamic, multi-inlet barrier island system located in southern Portugal, that protects a large mesotidal coastal lagoon. It extends ~55 km (E–W) and ~6 km (N–S) at its widest point (Figure 14.1A). The Ria Formosa coastal lagoon is a shallow, euryhaline ecosystem, with a wet surface area of 105 km^2 and an average depth, relative to mean sea level, of 2 m (Andrade et al. 2004; MCOA 2008). At present, the lagoon is separated from the Atlantic Ocean by a series of five barrier islands and two peninsular sand spits, and water exchanges with the adjacent coastal areas occur through six inlets: two artificially relocated inlets (Ancão and Fuseta inlets), two artificially opened and stabilized inlets (Faro-Olhão and Tavira inlets), and two natural inlets (Armona and Lacém inlets) (Matias et al. 2008).

14.2.1 Climatic and Hydrologic Setting

The Ria Formosa drainage basin has an area of 854 km^2. The climate is Mediterranean, with humid, moderate winters and hot, dry summers. Air temperature usually ranges from 8.0°C to 30.0°C, with annual mean values between 16.0°C and 20.0°C. Annual insolation and rainfall vary between 3000 and 3200 h and 400 and 600 mm, respectively. Rainfall is mostly concentrated between November and February (Serpa et al. 2005). There are 25 freshwater sources flowing into the lagoon, most of which flow during winter, but are dry during summer. The Gilão River (annual mean flow, 2.4 m^3 s^{-1}) is the only permanent watercourse, but it mainly affects the eastern regions of the lagoon. The most relevant freshwater inflows into the western regions of the lagoon (Rio Seco and Ribeira São Lourenço) have annual mean flow rates significantly smaller than the Gilão River (Serpa et al. 2005; MCOA 2008).

14.2.2 Oceanographic Setting

Coastal water features adjacent to the Ria Formosa Lagoon include the Gulf of Cadiz, a basin that connects the northeast Atlantic Ocean and the Mediterranean Sea. Tides in the area are semidiurnal, with tidal amplitudes averaging 2.1 m, and a range of 0.6 m and 3.6 m, for extreme neap and spring tides, respectively (Andrade 1990). Wave climate is usually moderate, with an offshore annual mean significant wave height of 1.0 m (Carrasco et al. 2008). This coastal region near the Ria Formosa is affected by regular upwelling events, related to local westerly winds, or advections from the southwestern coast. Upwelling events are most frequent from April to October, and relaxation of upwelling is usually associated with the development of a warm coastal countercurrent (Relvas et al. 2007).

14.2.3 Lagoon Morphology and Hydrodynamics

The Ria Formosa Lagoon is a strongly branched system of creeks and channels consisting of an extensive intertidal area, large salt marsh areas, and a complex network of natural and partially dredged subtidal channels, some of which are navigable to the ports of the main cities (Figures 14.1A and 14.1B). Dredged navigational channels can reach depths over 6 m, particularly in inlet channels. At mean sea level, the lagoon covers a flooded area of 49 km^2 (range: 16 to 84 km^2 for low and high tide of equinoctial spring tides), and its intertidal area represents 60% to 80% of the lagoon's flooded area during neap and spring tides, respectively (Andrade 1990).

The volume of water inside the lagoon at mean sea level is ~92 × 10^6 m^3 (range: 33 × 10^6 to 168 × 10^6 m^3), and at present, tidal exchanges occurs predominantly through the Faro-Olhão and Armona inlets (Figures 14.1A and 14.1B) (Salles et al. 2005). According to hydrodynamics, the lagoon may be divided into three subembayments: western, middle, and eastern embayments (see Figure 14.1A) (Salles et al. 2005). Tidal prism is clearly greater than the residual volume of the lagoon (Andrade

1990), suggesting high water renewal rates inside the lagoon and extremely low water residence times (0.5 to 2.0 d) (Tett et al. 2003). However, these high exchange rates should be applied only to the outer locations of the lagoon, since the inner regions, which are distant from inlets, may have significantly higher water residence times, up to 14 days (Duarte et al. 2008; Mudge et al. 2008). The lagoon is generally well mixed vertically, with no evidence of persistent or widespread haline or thermal stratification (Newton and Mudge 2003). Due to reduced freshwater inputs and elevated tidal exchanges, it is basically euryhaline with salinity values usually close to those observed at adjacent coastal ocean waters (Newton and Mudge 2003). As the lagoon is a fetch-limited system, the wind impact is relatively reduced, with wave heights usually on the order of a few centimeters (Carrasco et al. 2008). Tidal forcing is, therefore, the most relevant driver of water circulation inside the lagoon (Salles et al. 2005). The euryhaline nature of the lagoon and its low water residence time have been used to classify the Ria Formosa as a leakey lagoon (Sprung et al. 2001; Chubarenko et al. 2005), and as "coastal waters" (not "transitional waters") within the scope of the Water Framework Directive of the European Union (Bettencourt et al. 2003). However, the present absence of wide tidal passes, due to a significant decrease in total inlet width area since the 1940s (Vila-Concejo et al. 2002), the lack of sharp salinity and turbidity fronts, and the behavior of the lagoon during and after high rainfall periods (Duarte et al. 2008, see Section 5), are not in full accordance with these classifications. Indeed, these attributes are usually associated with restricted lagoons (Kjerfve and Magill 1989; Kjerfve 1994).

14.2.4 Biological Assemblages

The Ria Formosa coastal lagoon is a very productive ecosystem (Santos et al. 2004; Newton and Icely 2006; Cunha and Duarte 2007), with an average primary production of ~1400 g C m^{-2} yr^{-1} (Sprung et al. 2001). The subtidal area is colonized by the seagrasses *Cymodocea nodosa, Zostera marina,* and *Z. noltii* (2.4 km^2) (Cunha et al. 2009), and the higher and lower intertidal zones are occupied respectively by the salt marsh species *Spartina maritima* and extensive meadows of *Z. noltii* (~13 km^2 and. ~45% of the intertidal area, respectively) (Guimarães 2007). Salt marshes cover an area of 36 km^2 (MCOA 2008), and green macroalgae beds, mostly *Ulva* spp. and *Enteromorpha* spp., regularly form thick mats during winter (2.5 km^2) (Sprung et al. 2001). In addition, benthic microalgae attain concentrations up to 60 μg chlorophyll *a* cm^{-2} (Brito et al. 2009a). Data on the relative contribution of each primary producer inside the lagoon is somewhat conflicting. Some studies attribute a dominant role to salt marsh plants (Sprung et al. 2001), whereas others consider microphytobenthos (Santos et al. 2004) or phytoplankton (Duarte et al. 2008) as the dominant primary producers. The different habitats in the Ria Formosa Lagoon support diverse and abundant biological assemblages including holo- and merozooplankton (Sprung 1994; Chícharo and Chícharo 2001), benthic fauna (Gamito 2008), and fishes. The lagoon provides breeding, wintering, and staging sites for numerous species of water birds, including several threatened species, and serves as a breeding ground and nursery for economically important cephalopods, fishes, crustaceans, and bivalves (MCOA 2008; Ribeiro et al. 2008). The sediments of the lagoon are extensively used for shellfish farming, particularly the clam *Ruditapes decussatus.* Due to its environmental value, the Ria Formosa Lagoon and adjacent habitat, with a total area of 185 km^2, were designated a national Natural Park in 1987. It is also a wetland area of internationally recognized importance, designated as a Special Bird Protection Area under the Directive on the Conservation of Wild Birds, and a Site of Community Importance under the Habitat Directive, in the European Union (EU) Natura 2000 ecological network, and a Ramsar site.

14.3 APPROACH AND DATASET

Current data on abiotic variables and planktonic microbes (phytoplankton and heterotrophic bacterioplankton) in the Ria Formosa Lagoon were synthesized from published sources, public datasets, and gray literature (e.g., project and technical reports). Due to sparse sampling of the eastern and middle subembayments, only the western subembayment of the lagoon, controlled by tidal exchanges across the Ancão, Faro-Olhão, and Armona inlets (Salles et al. 2005), was considered (Figures 14.1A and 14.1B). Data derived from semi-artificial areas (e.g., reservoirs used for semi-intensive fish aquaculture, saline lakes), or sites in the immediate vicinity of wastewater treatment plants, were not included. Table 14.1 provides information on the dataset used in this study, with reference to variables, temporal coverage, and data sources. Most data were extracted directly from tables, but in some cases data digitized from figures was used (Table 14.1). Previously published data (Dionísio et al. 2000; Falcão and Vale 2003; Newton et al. 2003; Newton and Mudge 2003, 2005) were directly assessed from gray literature (Newton 1988, 1995; Falcão et al. 1991; Newton 1995; Dionísio 1996).

TABLE 14.1
Climatic, Hydrological, Physicochemical, and Biological Data Synthesized in This Study, with Information on Temporal Coverage, Sample Number (*n*), and Data Sources[a]

Variable	Period	*n*	Sources
NAO	1930–2008	78	http://www.cgd.ucar.edu/~jhurrell/indices.html2007-1
P	1930–2008	28,495	Water Institute (INAG, Portugal) public database, http://snirh.pt/
RF	1960–2008	17,248	INAG, http://snirh.pt/
T	1967–2004	1,539	1–7, 9–14, 16–21
S	1967–2004	1,144	1–7, 9–12, 14, 16–17, 19–21
SD/K_e	1976–2008	506	2–7, 11–13, 16, 19, 22
DO	1972–2004	460	3–7, 10–11, 14, 17, 20–21
I_m	1991–1993	395	19
NO_3^-	1980–2004	326	4–6, 11, 14, 17, 20, 21
NH_4^+	1980–2004	309	4, 6, 11, 14, 17, 20, 21
PO_4^{3-}	1976–2004	507	2, 4–7, 11, 12, 14, 16, 17, 20, 21
SiO_4^{4-}	1980–2004	517	4–7, 11, 12, 14, 16 17, 20, 21
BOD_5	1976–1988	143	2, 5, 7, 10, 12
Chlorophyll *a*	1967–2008	1,354	1, 3–6, 8–13, 16, 17–22
PP	1985–1993	86	8, 12, 19
BB	1988–1993	256	10, 15, 19
BP	1991–1993	7	19
CP	1988–2006	18	National Institute of Statistics (INE, Portugal), htpp:/www.ine.pt; Regional Directorate of Agriculture, Fisheries, and Aquaculture, Algarve (DRAPA, Portugal), for 2000–2006 data (M. Abreu, 2009, personal communication)

[a] See original sources for description of methods. Variables: BB, biomass of heterotrophic bacterioplankton; BOD_5, 5-day biochemical oxygen demand; BP, production of heterotrophic bacterioplankton; chlorophyll *a*, concentration of chlorophyll *a*; CP, annual clam farming production; DO, oxygen saturation; I_m, mean photosynthetic available radiation in the mixed layer; K_e, light extinction coefficient; NAO, North Atlantic Oscillation winter index; NH_4^+, ammonium; NO_3^-, nitrate; P, rainfall (measured at Loulé meteorological station, 37.146°N, 8.005°W); PO_4^{3-}, orthophosphate; PP, phytoplankton production; RF, river flow (measured at Coiro da Burra hydrometric station, 37.10°N, 7.903°W, or estimated using precipitation data); S, salinity; SD, Secchi depth; SiO_4^{4-}, silicate; T, water temperature.

Data sources: 1. Silva and Assis (1970); 2. Lima and Vale (1977); 3. Assis et al. (1984); 4. Benoliel (1984); 5. Cunha and Massapina (1984); 6. Benoliel (1985); 7. Falcão et al. (1985); 8. Marques (1988); 9. Newton (1988); 10. Barbosa (1989); 11. Benoliel (1989); 12. Falcão et al. (1991); 13. Thiele-Gliesche (1992); 14. Newton (1995); 15. Dionísio (1996); 16. Chícharo et al. (2000); 17. Loureiro et al. (2005); 18. Marques (2005);[†] 19. Barbosa (2006); 20. Loureiro et al. (2006);[†] 21. Water Institute database, Portugal (http://snirh.pt/); 22. Brito et al. (2009b).[†]

[†] Indicates data partially or totally extracted from digitized figures.

Light extinction coefficient (k_e) was estimated from Secchi disc data using an empirical model previously established for the Ria Formosa Lagoon ($R^2 = 0.67$, $n = 106$, $p < 0.001$; Barbosa 2006). Total daily radiation (W m^{-2}) was used to estimate photosynthetically active radiation (PAR) at the surface (I_0), considering that PAR constitutes 45% of the total radiation reaching the water surface and a 4% reflection at the surface (Baker and Froiun 1987). I_0 values were divided by the length of the light period (9.5 to 14.5 h) and subsequently converted using 4.587 μmol photons s^{-1} W^{-1} (Morel and Smith 1974). Mean light intensity in the mixed layer (I_m, μmol photons m^{-2} s^{-1}) was estimated according to Jumars (1993), and the water column was assumed to be homogeneously mixed to the bottom. A 5-day biochemical oxygen demand (BOD_5) was used as a proxy for the concentration of labile dissolved organic matter, and chlorophyll *a* was used as a proxy for phytoplankton biomass. It is worth mentioning, however, that alterations in carbon to chlorophyll *a* ratios, caused by changes in phytoplankton composition or its physiological state, may eventually lead to chlorophyll *a* changes without concurrent or parallel biomass alterations (Domingues et al. 2008). Potential nutrient limitation for phytoplankton growth was assessed using comparisons of *in situ* dissolved inorganic nutrient concentrations (dissolved inorganic nitrogen, DIN; silicate Si-SiO_4^{4-}; and orthophosphate P-PO_4^{3-}), with phytoplankton half-saturation constants, and nutrient molar ratios (DIN:P, Si:P, Si:DIN), according to criteria proposed by Justic et al. (1995). The concentration of DIN was calculated as the sum of nitrate (NO_3^-), nitrite (NO_2^-), and ammonium (NH_4^+).

The information used in this study consisted of discrete surface or subsurface (depth, 0.1 to 1 m) sampling of 14 stations situated in the western sector or subembayment of the lagoon (Figure 14.1B), from 1967 through 2008, on a discontinuous basis (as available). Since the lagoon is generally well mixed vertically (Newton and Mudge 2003), surface or subsurface samples were considered representative of the whole water column. The 14 sampling stations were subsequently grouped into four ecologically distinct water bodies: (1) stations situated at lagoon inlets, hereafter referred to as inlets, represented the sea boundary conditions; (2) stations located at main navigational channels, hereafter referred to as channels, represented an intermediate lagoon domain; (3) stations located close to the major cities (Faro and Olhão), hereafter referred to as urban-impacted stations, represented the urban boundary; and (4) stations located at inner, more restricted locations, hereafter referred to as inner stations, represented the landward boundary (Figure 14.1B). This classification complies with spatial differences in the water quality, human pressures (Newton 1995; Dionísio et al. 2000; Newton and Mudge 2003; J. G. Ferreira et al. 2006), and the composition of planktonic assemblages (Silva and Assis 1970; Assis et al. 1984; Cunha and Massapina 1984), and benthic assemblages (Gamito 2008) previously reported for the lagoon.

Seasonal and interannual variability in abiotic variables and planktonic microbes in the lagoon were analyzed for the four water bodies defined above. In order to facilitate the analyses of interannual variability and to reduce the impact of unevenly distributed sampling, sampling under different tidal and/or diel conditions, and extreme observations, for each year, all discrete data of each variable were grouped and averaged into monthly means. Subsequently, annual means were determined using all 12 monthly means of each year. Annual mean values were estimated only when data were available for 12 consecutive months (Cloern and Jassby 2008). Annual quarterly means were also calculated (first quarter, 1Q: January to March; second quarter, 2Q: April to June; third quarter, 3Q: July to September; fourth quarter, 4Q: October to December). Long-term linear trends for measured variables, based on annual mean values or annual quarterly mean values, were evaluated using linear regression analysis. Due to less frequent sampling of heterotrophic bacterioplankton, only phytoplankton was considered for the analysis of interannual variability. The strength of associations between variables, based on discrete or integrated data, was assessed using Spearman rank correlation coefficients (r_S). All statistical analyses were considered at a significance level of 0.05.

The dataset used for this study is based on the analysis of discrete surface or subsurface samples, using standard methods, and regular quality control with appropriate calibration standards. However, sampling under different tidal or diel conditions, and the use of different nonintercalibrated analytical methods (e.g., phytoplankton production, chlorophyll *a*), preparation techniques (e.g., filtration for nutrient analysis and chlorophyll *a*), or storage strategies prior to analysis, may have introduced a possible source of bias (Raabe and Wiltshire 2009). Furthermore, the analyses of mechanisms underlying plankton interannual variability is, of course, highly dependent on sampling frequency, series length (Jassby et al. 2002), and variables examined. The analysis of interannual trends in this study is clearly based on a limited and fragmented dataset; climate change effects, for instance, would be better evaluated by multi-decadal data.

14.4 RECENT CLIMATE AND LOCAL ANTHROPOGENIC CHANGES

The distribution patterns of planktonic microbes and their spatial and temporal variability reflect the interplay between *in situ* specific growth rates and death or loss rates. Growth rates are commonly controlled by resource availability (e.g., light, inorganic nutrients, dissolved organic matter), and other abiotic conditions (e.g., temperature, non-nutrient bioactive compounds), whereas loss rates are mostly regulated by pelagic and benthic predation, viral lyses, sedimentation, and horizontal advection (e.g., Cloern and Dufford 2005; Gasol et al. 2008). Consequently, both climate and local anthropogenic changes may control growth and loss rates of planktonic microbes, thereby influencing their seasonal and interannual variability in the Ria Formosa.

14.4.1 Climatic Changes

Long-term changes affecting the Ria Formosa drainage basin are apparently linked to the North Atlantic Oscillation (NAO) winter index, a global indicator of climate forcing over the area (Figure 14.2A). As previously reported for other southern Europe regions (Rodó et al. 1997), total annual rainfall in the Ria Formosa drainage basin is negatively correlated to the NAO winter index (Figure 14.2A). Annual freshwater flow into the western sector of the lagoon shows a marked interannual variability during the study period (0 to 0.8 $m^3 s^{-1}$), but without a discernable long-term trend (Figure 14.2B). Other climate change described for southern Portugal during the twentieth century include increases in the annual mean air temperature, particularly strong during the winter period and from the mid-1970s onward (~0.07°C yr^{-1}), the mean precipitation per rainy day, and the frequency of heat waves and extreme droughts (Klein Tank et al. 2002; Miranda et al. 2002).

Climate variability may also affect the Ria Formosa coastal lagoon indirectly, through its influence on oceanic processes. Indeed, during the twentieth century, a consistent increase in sea surface temperature (0.01 to 0.04°C yr^{-1}), together with a shallowing of the summer coastal thermocline depth, have been reported along the entire coast of Portugal, including the southern coast (Lemos and Sansó 2006; Relvas et al. 2009). Moreover, an intensification in upwelling during peak summer months (July to September) was recently reported for the southern Portuguese coast (Relvas et al. submitted). A major effect of global warming is an accelerated increase in sea level rise, which can have a profound impact on coastal lagoons. Although the Ria Formosa coastal lagoon is a very vulnerable system to accelerated sea level rise, the impacts of hard shore protection structures, sediment starvation due to river damming and sand exploitation, and episodic storms are currently considered an order of magnitude greater than those associated with the present rate of sea level rise (O. Ferreira et al. 2008).

Episodic high-energy events (e.g., a tsunami induced by the 1755 Lisbon earthquake; extreme storms in 1861, 1941, and 1961) are major drivers of change in the morphology of the lagoon, affecting the number of inlets, the width of the inlets, and their position (Salles 2001; Vila-Concejo et al. 2002, 2003). These extreme episodic events may also affect the water residence time inside the lagoon, light availability due to massive sediment resuspension, and nutrient concentrations, thereby affecting the dynamics of planktonic microbes (Guadayol et al. 2009). The current database

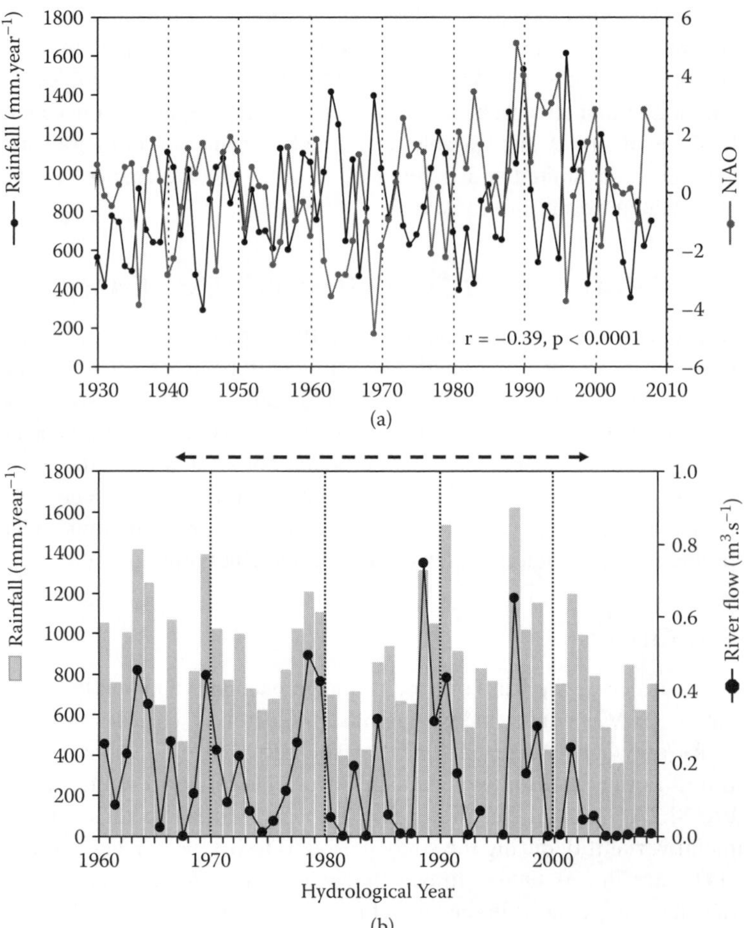

FIGURE 14.2 (a) Historical time series of the North Atlantic Oscillation (NAO) winter index and annual rainfall (mm yr^{-1}) in the Ria Formosa drainage basin, measured at Loulé, during the period 1930 to 2008. Spearman rank correlation and confidence level are indicated. (b) Time series of annual rainfall (mm yr^{-1}), and annual mean freshwater flow into the western subembayment of the Ria Formosa, measured at Coiro da Burra (m^3 s^{-1}), during 1960 to 2008. The horizontal line marks the study period, 1967 to 2004 (see Table 14.1 for data sources).

is insufficient for analysis of the evolution of the storminess in the southeastern Portuguese coast. However, overwash evolution, that integrates both the storminess regime and system susceptibility, indicates an overall decrease in washover occurrences in the Ria Formosa over the past 25 years (Matias et al. 2008).

14.4.2 Local Anthropogenic Changes

The number of residents inhabiting the Ria Formosa drainage basin has increased remarkably from ~99,950 to 159,530 individuals, from 1970 to 2001. The population is strongly concentrated in the littoral area, close to major urban centers. During the summer period, the population doubles due to a large influx of tourists (Serpa et al. 2005). This increasing population density leads to an intensification of anthropogenic pressures on the lagoon. The Ria Formosa drainage basin sustains different land uses including animal husbandry, industrial units mostly related to the food industry, and extensive agricultural areas intensively farmed with the use of fertilizers and pesticides (Serpa et al. 2005; Duarte et al. 2008). Some farmed areas overlie nitrate-contaminated aquifers, and are classified as Nitrate Vulnerable Zones under the EU Nitrate Directive (Stigter et al. 2006). The lagoon supports many human activities, namely finfish aquaculture, shellfish farming, salt production, fishing, recreation, and tourism (MCOA 2008). These different uses of the lagoon and its drainage basin are associated with multiple anthropogenic disturbances that can potentially affect planktonic microbes.

14.4.3 Water Quality and Eutrophication

The increased urban development, tourism, and agriculture in the Ria Formosa drainage basin are considered to be associated with a progressive deterioration in water quality. The discharge of untreated or partially treated domestic and industrial sewage, and agricultural runoff into the lagoon are commonly linked to decreased oxygen saturation (Newton et al. 2003; Mudge et al. 2007, 2008), and increased concentration of dissolved inorganic nutrients in the water column (Newton et al. 2003; Newton and Mudge 2005; Cabaço et al. 2008), fecal coliform and other indicators in water (Dionísio et al. 2000), and organic matter in water and sediments (Mudge et al. 1998; Santos et al. 2004). Due to human-induced nutrient over-enrichment, the eutrophication status of the Ria Formosa Lagoon, based on the European Environmental Agency criteria (nutrient concentrations), frequently scores as "poor" to "bad," especially in the eastern lagoon (Newton et al. 2003). Yet, the use of symptom-based evaluation approaches (e.g., the United States Estuarine Eutrophication Assessment) suggests the lagoon status is near pristine (Newton et al. 2003), or "good" (Nobre et al. 2005). High diversity of benthic macrofauna also points to an elevated ecological status (Gamito 2008). Relative to other coastal systems with restricted exchange, the robustness of the Ria Formosa Lagoon can be explained by its low water residence time. Nevertheless, the potential for episodic eutrophic conditions, particularly at inner lagoon locations with restricted tidal flushing, is clearly higher (Newton et al. 2003; Tett et al. 2003; Nobre et al. 2005; Newton and Icely 2006). In fact, eutrophication symptoms are expressed locally, in the immediate vicinity of discharges from wastewater treatments plants or industrial units, as strong depletion of dissolved oxygen, particularly during the night to early morning period (Mudge et al. 2008), increased phytoplankton biomass (A. Cravo, personal communication, 2009), presence of permanent mats of green macroalgae (Sprung et al. 2001), low physiological status of salt marsh plants (Padinha et al. 2000), alterations in population structure, morphology, and leaf nitrogen content in seagrasses (Cabaço et al. 2008), and decreases in the diversity and changes in the composition of macro- and meiofauna benthic communities (Chenery and Mudge 2005; Gamito 2008 and references therein). Toxic algal blooms, usually considered a symptom of eutrophication, are sometimes observed inside the lagoon (e.g., *Dynophysis* spp., *Gymnodinium catenatum*, *Pseudonitzschia* spp.), but they usually enter from adjacent coastal waters (Luis and Barbosa unpublished data).

Recently, a general improvement in the water quality of the western lagoon was attributed to the installation of urban wastewater treatment plants, in operation since the late 1990s, and to the enhancement in the water circulation caused by the artificial relocation of the Ancão Inlet, in 1997 (Figures 14.1A and 14.1B) (Vila-Concejo et al. 2003), and dredging of main channels (Newton and Icely 2002; Loureiro et al. 2006; Pereira et al. 2007; Brito et al. 2009b). Besides, the transposition of the EU Nitrate Directive has recently led to reduced nitrate concentrations in groundwater. Still, this effect may be countered, to some extent, by the recent sharp rise of the water table induced by the shift from locally extracted groundwater to regionally supplied surface water for urban water supply and irrigation (Stigter et al. 2006).

Anthropogenic activities have also introduced other inorganic and organic contaminants into the lagoon such as heavy metals, organochlorines, tributyltin (TBT), polychlorinated biphenyls, and polycyclic aromatic hydrocarbons. The Ria Formosa is considered poorly or slightly contaminated in respect to most contaminants (Barreira 2007), but the concentrations of TBT, copper, and cadmium are clearly higher (Bebianno 1995; Coelho et al. 2002). Sites close to major urban centers, metallic structures, and port facilities usually show higher contamination levels (Padinha et al. 2000; Coelho et al. 2002). Current information suggests a two- to fourfold increase in Cu and Pb contamination in the Ria Formosa between the 1980s and the 1990s (Padinha et al. 2000).

14.4.4 Changes in Hydrodynamics and Sedimentary Dynamics

Anthropogenic disturbances in the Ria Formosa have changed considerably the water fluxes and sedimentary dynamics inside the lagoon. The opening and stabilization of artificial inlets (Faro-Olhão and Tavira inlets) and the relocation of Ancão Inlet (June 1997; 3.5 km west of its previous position, Figures 14.1A and 14.1B) and Fuseta Inlet clearly affected the migration and width of natural inlets, and the hydrodynamic behavior of different lagoon basins (Salles 2001; Vila-Concejo et al. 2002, 2003). Furthermore, dredging of main channels, usually undertaken to ensure navigational safety, sediment extraction, and addition of coarser sediment to clam culture beds, may also affect water transparency and water residence time in the lagoon.

14.4.5 Habitat Alterations

Several human activities have contributed to the loss or degradation of natural habitats in the Ria Formosa Lagoon. These activities include

1. The artificial closure, in the early 1990s, of the Gondra and São Lourenço estuaries (western sector), as part of a reclamation campaign (Salles 2001).
2. Building of tourist and recreational resorts and golf courses.
3. Construction or amplification of infrastructure (e.g., airport, wastewater treatment plants, ports, and marinas).
4. Establishment of salt-extraction pans and aquaculture facilities in former salt marsh areas.
5. Establishment of shellfish culture beds in intertidal areas occupied by seagrass meadows.
6. Mechanical damage associated with dredging, use of inappropriate fishing gear, boating, anchoring, and shellfish farming and harvesting (Cabaço et al. 2005; MCOA 2008).

Despite these human disturbances, the evolution of the areal coverage of key lagoon habitats, such as salt marshes (1989: 32 km^2 [Fidalgo 1996]; 2001: 36 km^2 [MCOA 2008]) and intertidal *Zostera noltii* meadows (1989: 8 km^2 [Fidalgo 1996]; 2002: 13 km^2 [Guimarães 2007]), does not provide evidence of any declining trend.

14.4.6 Exploitation of Biological Production

Exploitation of marine living resources is an important human activity in the Ria Formosa Lagoon, and consists of fishing, finfish semi-intensive and intensive aquaculture, harvesting of benthic organisms for bait, recreational shellfish harvesting, harvesting of shellfish juveniles, and shellfish farming (Amaral 2008; MCOA 2008). Shellfish farming, especially of the clam *Ruditapes decussatus*, clearly is the most relevant type of exploitation. Clams are extensively cultured in sandy intertidal areas which are artificially transformed through the addition of coarser sediments, and seeded with clam juveniles, either transposed from natural banks of the lagoon or produced in hatcheries. Licensed areas used for shellfish farming inside the lagoon increased from 3 km^2 to 5 km^2 over the period from 1989 to 2001 (Fidalgo 1996; MCOA 2008).

The present official estimate of the annual production of cultivated clams is 2457 t during 2001 (DRAPA, 2009, personal communication) but unofficial estimates, considered more accurate, indicate values ranged from 10,000 to 15,000 t yr^{-1} (Amaral 2008). Despite this discrepancy, a major decline in clam production in the Ria Formosa was reported from 1983 onward. According to official estimates, the annual production of cultivated clams declined by ~72%, from ~7000 t to 2000 t, between the late 1980s and the 2000s, attaining a minimum value of 120 t during 1996 (INE, http://www.ine.pt). Decreasing bivalve harvests in the lagoon are related to a combination of several factors such as a general deterioration in the water quality, non-nutrient contaminants, and very high levels of infection with the protistan parasite *Perkinsus* sp. It is worth stating, however, that both natural climatic pressures and human pressures may have facilitated *Perkinsus* sp. infection and its progression (Leite et al. 2004).

14.4.7 Introduced Species

Human-mediated activities have contributed, whether intentionally or unintentionally (e.g., vessel ballast water, hull biofouling, or shellfish aquaculture import), to the introduction of several alien species in the Ria Formosa Lagoon (*Carpobrotus edulis*, *Spartina densiflora*, *Sargassum muticum*, *Ruditapes philippinarum*, and *Crassostrea gigas*) (MCOA 2008). Some of these alien species have caused significant economic and ecologic impacts in other estuaries and coastal lagoons, including the competitive displacement of native species and alterations in the water quality (e.g., *S. muticum* [Tweedley et al. 2008] and *R. philippinarum* [Pranovi et al. 2006]). However, no information is currently available on the impacts of alien species in the Ria Formosa.

14.5 SEASONAL DYNAMICS OF PLANKTONIC MICROBES

14.5.1 Environmental Setting

The western sector of the Ria Formosa Lagoon (see Figure 14.1) shows a pronounced seasonal variability, with subsurface water temperature (9.2°C to 30.0°C) and salinity (15.1 to 39.2) attaining minimum values during winter and peak values during summer (Figures 14.3a and 14.3b). Due

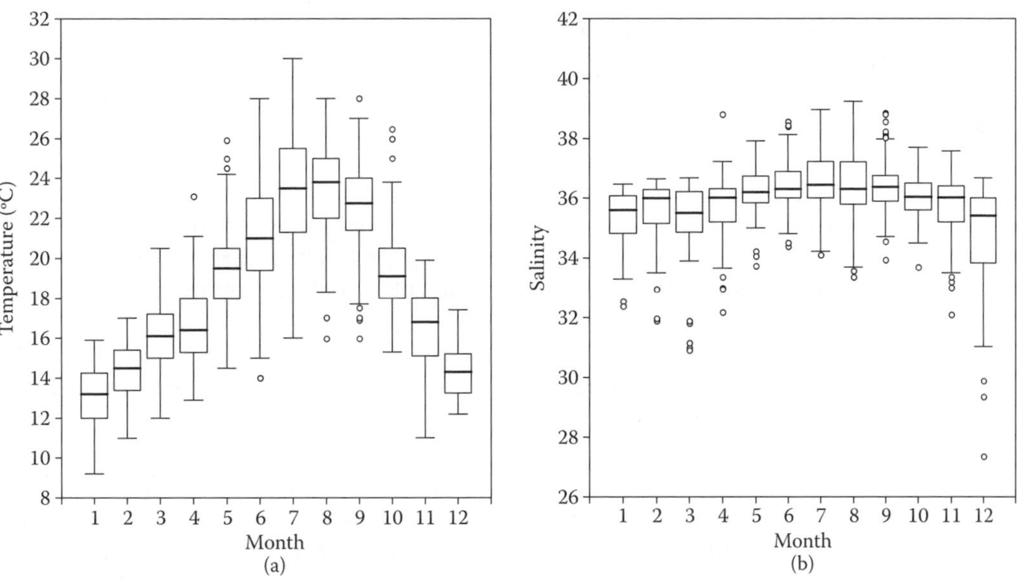

FIGURE 14.3 Box and whisker plots showing the distribution of: (a) subsurface water temperature (n = 1539); and (b) salinity (n = 1144) observations in the western subembayment of the Ria Formosa, grouped into months, for the period 1967 to 2004. Median value is represented by the line within the box: 25th to 75th percentiles are denoted by box edges; 5th to 90th percentiles are depicted by the error bars; and outliers are indicated by circles (see Table 14.1 for data sources).

to reduced freshwater inputs and elevated tidal exchanges, the lagoon is basically euryhaline. As previously reported (Newton and Mudge 2003; Mudge et al. 2008), heavy rainfall events during the autumn and winter are associated with major salinity decreases. Strong evaporation during the summer period leads to slightly hypersaline conditions, particularly at sites with restricted water exchange (average ± 1 SD: 36.96 ± 1.12; Figure 14.3b). Water temperature and salinity usually increase from inlets to inner stations during spring and summer, and the opposite occurs during autumn and winter (not shown).

Light extinction in the water column (0.4 to 5.3 m^{-1}), estimated using Secchi depth values (Barbosa 2006), is significantly lower at lagoon inlets (0.68 ± 0.18 m^{-1} vs. 0.82 ± 0.39 m^{-1}). However, as in other mixed transitional systems (Kocum et al. 2002), mean light intensity in the mixed layer (I_m) is higher in the more turbid, shallower inner lagoon than at the clearer, deeper outer locations. At non-inlet locations, the euphotic zone extends up to the bottom during most occasions (99%), and I_m values consistently surpass the critical value of Riley, ~45 µmol photons m^{-2} s^{-1}, below which net phytoplankton growth will not occur. During the autumn and winter, I_m values are usually below phytoplankton saturating light intensities (250 µmol photons m^{-2} s^{-1}) (Tang 1995), thus pointing to a potential light limitation for phytoplankton during that period (Barbosa 2006).

Seasonal changes in the concentration of inorganic nutrients are particularly marked, ranging from levels below detection to 163.1 µM and 97.1 µM for nitrate and silicate, respectively, and show clear spring–summer minima and winter maxima at all lagoon locations (Figure 14.4). This seasonal pattern mostly reflects nutrient enrichment associated with increased freshwater inflow during the rainy season, either via terrestrial runoff or torrential stream discharge (Newton et al. 2003; Newton and Mudge 2005), and increased nutrient consumption during the spring–summer period. At inlet stations, nitrate and silicate are further controlled by the seasonal stratification/mixing cycle and upwelling events (Loureiro et al. 2006). In contrast to silicate and nitrate, phosphate (0.01 to 9.8 µM) displays maximum values during summer (Figure 14.4b), possibly as a result of intensified decomposition and remineralization of organic matter in sediments, increased benthic bioturbation, increased desorption of phosphorus from sediment iron oxides (Falcão and Vale 1998), and greater phosphate loads associated with sewage discharges (Newton and Mudge 2005).

Seasonal distribution of ammonium at inlet stations shows a close relationship with phytoplankton, with maxima during late-spring and mid-autumn (Figure 14.4c). However, at other lagoon locations, ammonium peak concentrations are observed during the rainy season (Figure 14.4d), probably as a result of increased ammonium load associated with agricultural fertilizers or animal husbandry units. There appears to be little effect of increased ammonium release from the sediments (Falcão and Vale 1998; Newton et al. 2003), or increased population density observed in the lagoon during the summer period when increased uptake by planktonic and benthic primary producers and increased nitrification rates (Gurel et al. 2005) could, to some extent, explain the relatively low summer ammonium concentrations in the lagoon.

The major sources of inorganic macronutrients in the lagoon are considered to be urban effluents and agriculture-derived terrestrial runoff (Newton et al. 2003; Newton and Mudge 2005). Seasonal dynamics of inorganic nutrients in the western sector of the lagoon clearly suggest that the latter predominates over the former. Despite reduced freshwater flow into the lagoon, high nutrient concentrations in the freshwater courses apparently have significant effects on lagoon concentrations (Newton et al. 2003; Duarte et al. 2008). Indeed, notably higher nutrient concentrations, up to 326 µM nitrate, 32 µM ammonium, 4 µM phosphate, and 1201 µM silicate, are observed in the eastern lagoon, particularly close to the Gilão Estuary (Newton 1995; Newton et al. 2003). Sediments also represent an important nutrient source in the lagoon, potentially supplying most of the daily nitrogen and phosphorus requirements of phytoplankton (Falcão and Vale 1998; Serpa et al. 2007; Brito et al. 2009b). Other putative sources of nitrogen, phosphorus, and silicon for phytoplankton, not quantified so far, include atmospheric deposition, groundwater inflow, and import from the adjacent coastal ocean, particularly during upwelling events (Falcão

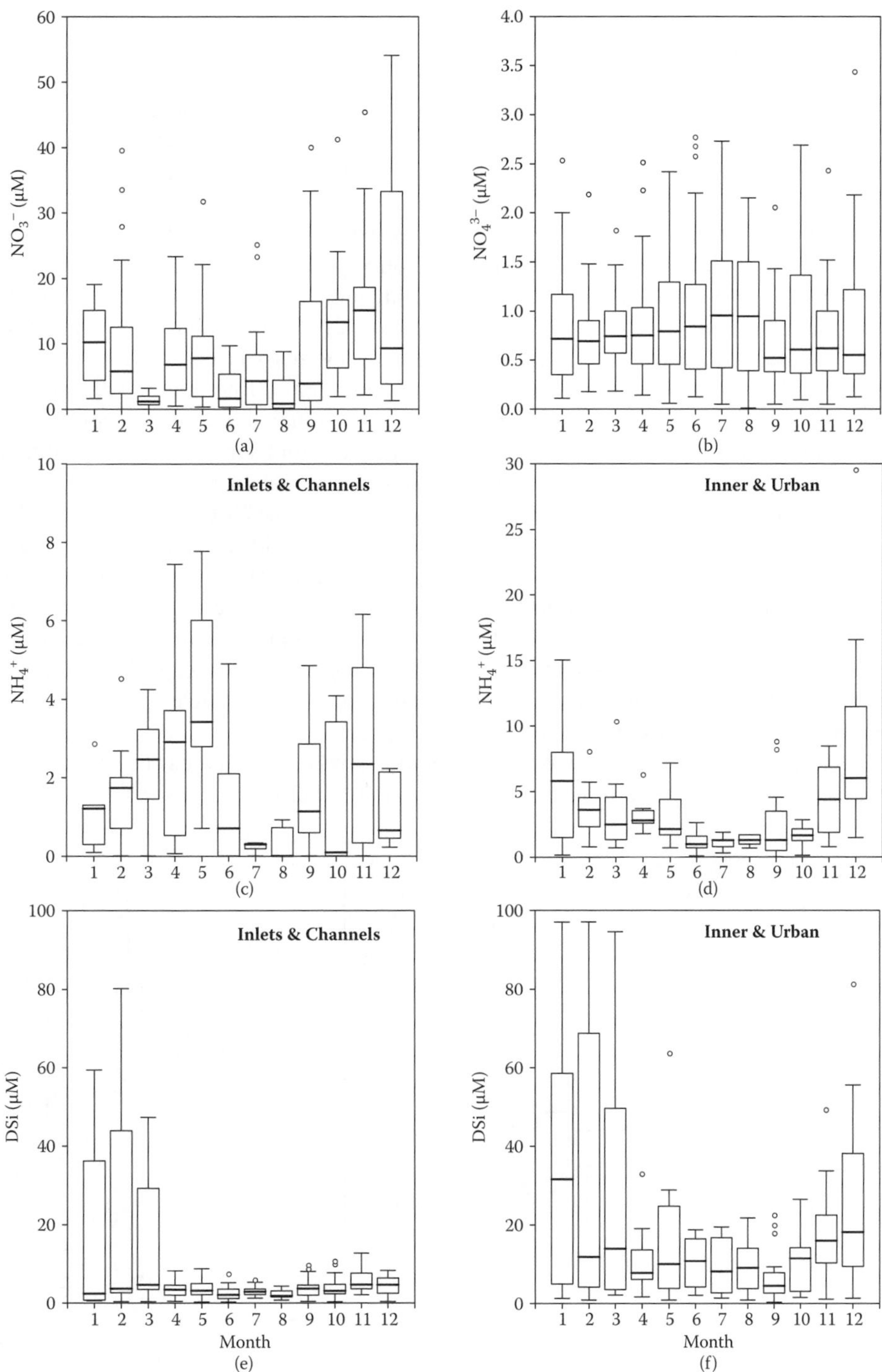

FIGURE 14.4 Box and whisker plots showing the distribution of dissolved inorganic nutrient concentrations in the western subembayment of the Ria Formosa, grouped into months, for the period 1967 to 2004. (a) nitrate (NO_3^-; n = 326); (b) phosphate (PO_4^{3-}; n = 507); (c) ammonium (NH_4^+) at inlets and channels (n = 160); (d) ammonium (NH_4^+) at inner and urban-impacted stations (n = 149); (e) silicate (DSi) at inlets and channels (n = 356); and (f) silicate (DSi) at inner and urban-impacted stations (n = 161). Median value is represented by the line within the box; 25th to 75th percentiles are denoted by box edges; 5th to 90th percentiles are depicted by the error bars; and outliers are indicated by circles (see Table 14.1 for data sources).

1996; Loureiro et al. 2006). In the case of mixotrophic taxa, dissolved organic matter or free living prey may additionally be used.

Apart from nitrate, which can be imported from coastal waters, dissolved inorganic nutrients usually display a strong spatial gradient inside the lagoon, with increased concentrations from inlets to inner areas (Falcão 1996; Newton and Mudge 2005). Nitrate is the dominant dissolved inorganic nitrogen (DIN) form at the western sector of the lagoon, but the proportion of ammonium increases from inlets (24% ± 29%) to urban locations (40% ± 29%). Potential nutrient limitation for phytoplankton growth, assessed according to criteria proposed by Justic et al. (1995), rarely occurs at inner lagoon sites (N limitation, 2% occasions), but appears 16% to 17% of the time in the main channels and inlets (8% to 9% Si limitation, 6% N limitation, and 1% to 3% P limitation). Urban-impacted stations mainly exhibit potential Si limitation (12% Si limitation, 1% N limitation, and 1% P limitation), as expected from increased anthropogenic nutrient enrichment (Cloern 2001). Recent data (2006 to 2008) derived from inner lagoon locations (Brito et al. 2009b) provide evidence of potential N limitation only. Although potential N limitation (n = 11) appears between late spring and mid-summer, P limitation (n = 3) emerges between mid-summer and early autumn, and potential Si limitation (n = 21) exhibits a more widespread seasonal distribution.

Seasonal changes in 5-d biochemical oxygen demand measured at 20°C (0.7 to 2.6 mg O_2 L^{-1}) and taken as a proxy for the concentration of labile organic matter provide evidence of an increase from late winter to late spring and a subsequent decline thereafter, possibly due to increased bacterial carbon demand (Figure 14.5a). Urban-impacted stations present significantly higher and sustained BOD_5 levels during summer (1.7 to 4.2 mg O_2 L^{-1}) due to increased inputs of allochthonous organic matter (Santos et al. 2004). The increased availability of organic matter, as well as increased concentration of inorganic nutrients at these locations, are probably responsible for the approximately fourfold higher microbial respiration rates (Barbosa unpublished data), and a local deterioration of the water quality, indicated by low oxygen saturation levels. Lagoon areas not directly affected by urbanization mainly exhibit oxygen saturations over 80% (92% of the cases; but see Brito et al.

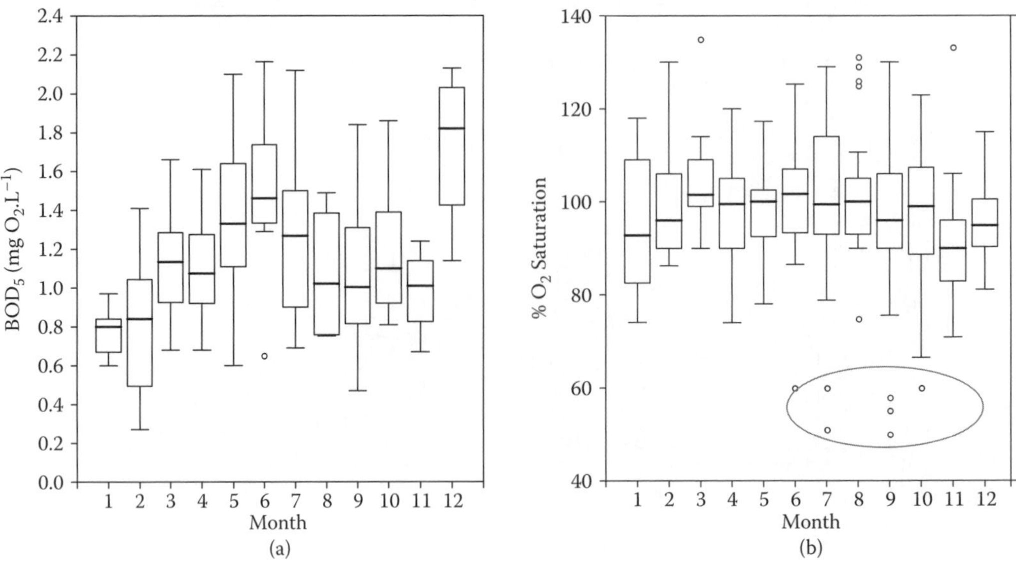

FIGURE 14.5 Box and whisker plot showing the distribution of (a) 5-d biochemical oxygen demand, BOD_5 (n = 143), and (b) oxygen saturation (n = 460) observations in the western subembayment of the Ria Formosa, binned into months, for the period 1967 to 2004. Median value is represented by the line within the box, 25th to 75th percentiles are denoted by box edges, 5th to 90th percentiles are depicted by the error bars, and outliers are indicated by circles (see Table 14.1 for data sources).

2009b), but at urban-impacted areas, this saturation level is attained in only 60% of the cases, with values below 40% occasionally observed mainly in summer (Figure 14.5b).

14.5.2 Phytoplankton

Chlorophyll *a* in the western sector of the Ria Formosa (0.1 to 23.5 μg L^{-1}) usually increases from the inlets to inner locations, probably due to higher nutrient concentrations, higher mean light intensity in the mixed layer (I_m), and lower tidal advection at the inner areas. Phytoplankton shows a clear seasonality, but seasonal patterns vary across lagoon locations. Lagoon inlets, directly impacted by adjacent coastal waters, exhibit a clear bimodal annual cycle, with peaks during early spring and mid-autumn (Figure 14.6a). Bimodal cycles, typically observed in temperate coastal areas, are driven

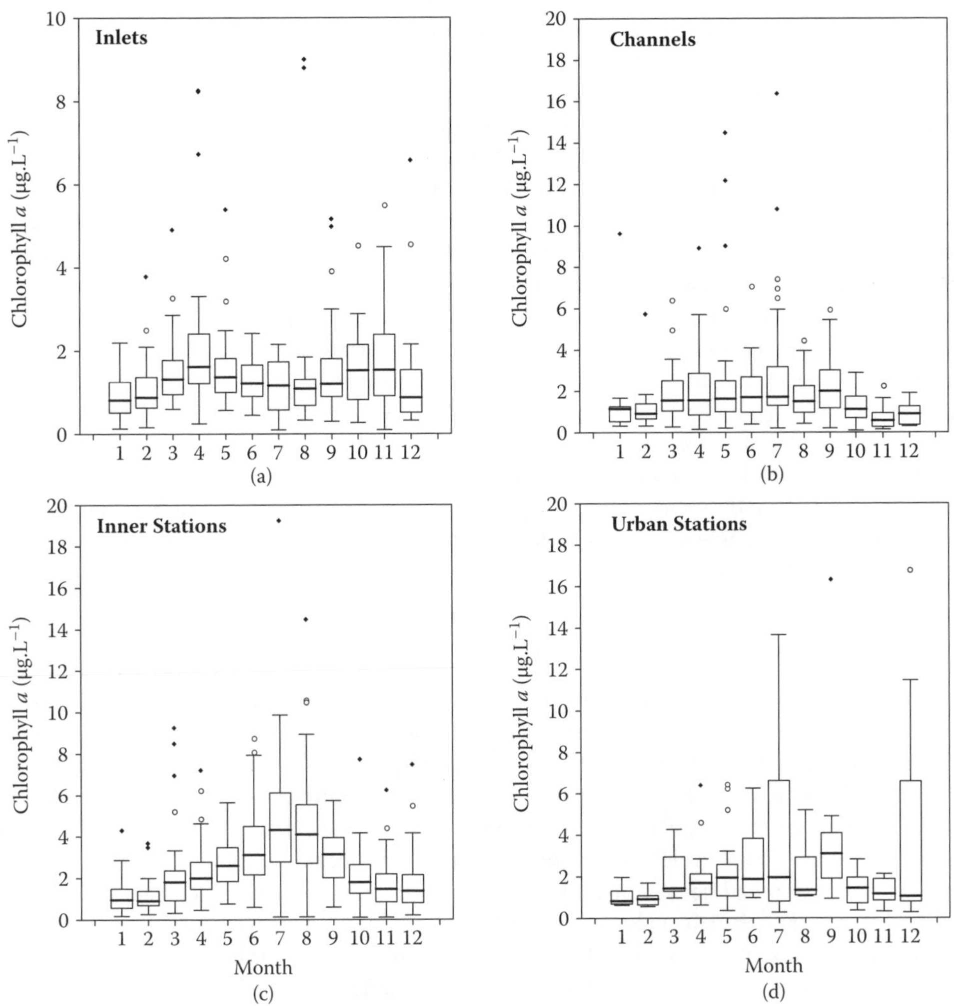

FIGURE 14.6 Box and whisker plots showing the distribution of chlorophyll *a* concentrations in different water bodies of the western subembayment of the Ria Formosa, grouped into months, for the period 1967 to 2008. (a) Inlets (n = 272); (b) channels (n = 322); (c) inner stations (n = 593); (d) urban-impacted stations (n = 167). Median value is represented by the line within the box, 25th to 75th percentiles are denoted by box edges; 5th to 90th percentiles are depicted by the error bars; outliers are indicated by circles, and extreme values by black diamonds (see Table 14.1 for data sources). Two extreme values (21.0 and 23.5 μg L^{-1}) were omitted for clarity.

by water column stratification-destratification cycles, which regulate light and nutrient availability (Cébrian and Valiela 1999). At more restricted lagoon locations, phytoplankton biomass shows a high-amplitude unimodal annual cycle, with clear summer peaks (Figures 14.6b through 14.6d). However, recent data (2006 to 2007) reveal a departure from this pattern (Brito et al. 2009b).

Unimodal annual cycles are frequently observed in shallow mixed estuaries and coastal lagoons, usually linked to growth limitation by light and/or temperature (Cébrian and Valiela 1999). In the Ria Formosa Lagoon, chlorophyll *a* is positively correlated with both water temperature ($r_S = 0.47$, $n = 950$, $p < 0.00001$) and I_m ($r_S = 0.50$, $n = 157$, $p < 0.00001$), apparently supporting this hypothesis. Indeed, light enrichment experiments induce a significant stimulation of phytoplankton net growth rates during autumn and winter (Barbosa unpublished data). Furthermore, instantaneous growth rates of phytoplankton at inner stations, measured using diffusion chambers incubated *in situ*, reveal a seasonal pattern apparently controlled by light during periods of I_m below 250 μmol photons m^{-2} s^{-1}, and by temperature during periods of higher I_m (Barbosa 2006). During late spring and summer, phytoplankton growth is probably limited by nutrient variability. Clearly, most of the potential nutrient limitation events (~72%), assessed according to criteria proposed by Justic et al. (1995), are concentrated between late spring and summer.

Nutrient enrichment experiments indicate that nitrogen is the most likely limiting nutrient for phytoplankton growth during summer in this system (Loureiro et al. 2005). Contrary to other taxa, *in situ* growth rates of diatoms in the Ria Formosa during the early 1990s do not provide evidence of nutrient limitation during summer (Barbosa 2006). Yet, nutrient addition bioassays undertaken in 2001 and 2002 show opposing results (Loureiro et al. 2005). Although these differences may indeed represent interannual variability, it is worth mentioning that bottle enclosures used in bioassays eliminate the most important sources of nutrients during summer (sediment fluxes) eventually generating an "apparent" nutrient-limiting state (Gallegos and Jordan 1997; Barbosa 2006). Processes reducing phytoplankton numbers inside the lagoon include pelagic and benthic grazing and tidal flushing. Microzooplankton grazing represents a major source of phytoplankton mortality. Microzooplankton remove, on a mean annual basis, 47% of daily phytoplankton production, and the impact is significantly greater for eukaryotic picophytoplankton and plastidic nanoflagellates. Diatoms are apparently under strong top-down control, probably exerted by benthic filter-feeders (Barbosa 2006).

14.5.3 Heterotrophic Bacterioplankton

Biomass of heterotrophic bacterioplankton in the western sector of the Ria Formosa shows a significant increase from inlets (7 to 79 μg C L^{-1}) to inner locations, attaining maximum values at urban-impacted stations (58 to 576 μg C L^{-1}). Spatial variability is probably regulated by the availability of labile organic matter, as suggested by positive and significant relationships between bacterial biomass and BOD_5 ($r_S = 0.87$, $n = 30$, $p < 0.00001$), and flushing rates. Bacterial biomass presents unimodal annual cycles, with summer peaks, at all locations (Figure 14.7). At the lagoon inlets, phytoplankton and heterotrophic bacterioplankton are apparently uncoupled (Figure 14.8), indicating that other pelagic sources of dissolved organic matter (e.g., grazing losses, excretion by phagotrophic protists, and viral lysis; Barbosa et al. 2001) are relevant for bacteria. In contrast, at inner stations, heterotrophic bacterioplankton is significantly correlated with phytoplankton ($r_S = 0.62$, $n = 115$, $p < 0.00001$; Figure 14.8), and also with water temperature ($r_S = 0.58$, $n = 127$, $p < 0.00001$). These relationships could be used as evidence for a bottom-up control of bacterial growth exerted by temperature and dissolved organic matter released by phytoplankton. However, these relationships should be interpreted with caution.

Extremely high ratios of bacterial carbon demand (88 to 506 μg C $L^{-1}d^{-1}$; Barbosa 2006) to phytoplankton production (8 to 2001 μg C L^{-1} d^{-1}) clearly indicate that heterotrophic bacterioplankton inside the lagoon is strongly supported by nonpelagic sources of organic carbon, particularly during winter (Figure 14.9). These may include allochthonous sources, such as upland plants and urban

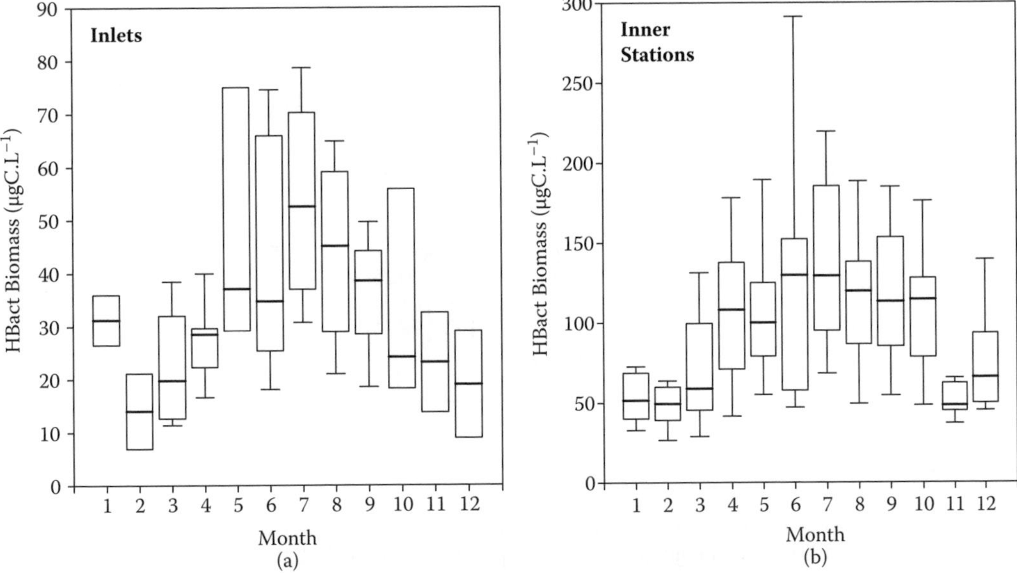

FIGURE 14.7 Box and whisker plot showing the distribution of the biomass of heterotrophic bacterioplankton in different water bodies of the western subembayment of the Ria Formosa, binned into months, for the period 1988 to 1993. (a) Inlets ($n = 42$); and (b) inner stations ($n = 152$). Median value is represented by the line within the box, 25th to 75th percentiles are denoted by box edges, 5th to 90th percentiles are depicted by the error bars, outliers are indicated by circles, and extreme values by black diamonds (see Table 14.1 for data sources).

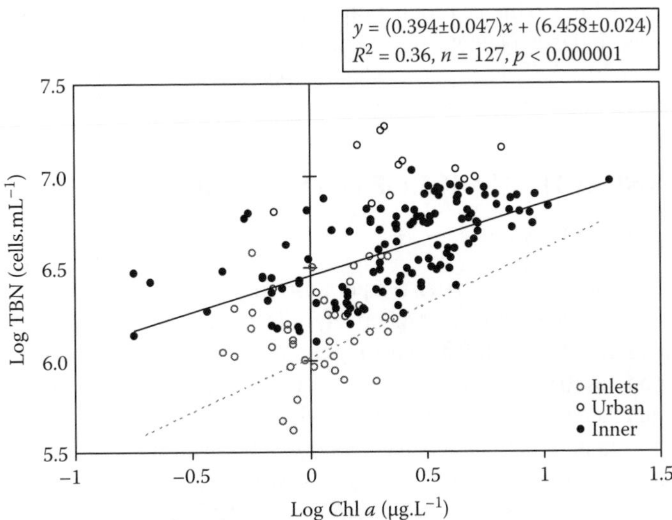

FIGURE 14.8 Relationship between chlorophyll *a* concentration (Chl *a*) and the abundance of heterotrophic bacterioplankton (TBN) in different water bodies of the western subembayment of the Ria Formosa Lagoon, during 1988 to 1993 (see Table 14.1 for data sources). △: inlets; ○: urban-impacted stations; ●: inner stations. Adjusted regression line and regression equation for the inner stations are shown (dashed line represents the empirical model of Cole et al. [1988] for comparison).

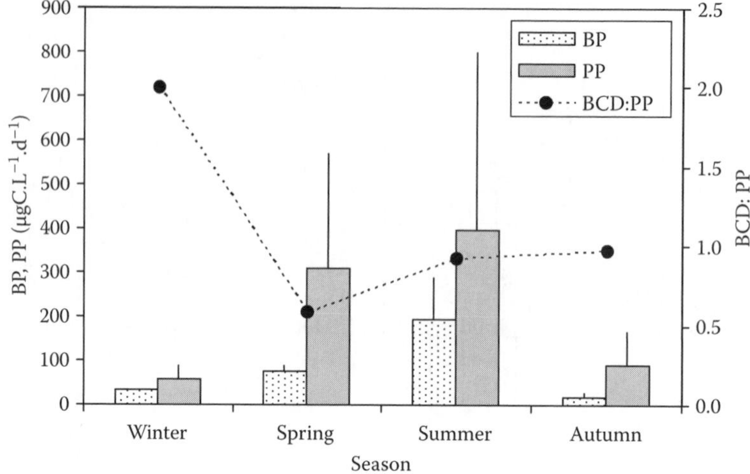

FIGURE 14.9 Seasonal variation of bacterial production (BP), phytoplankton production (PP), and the ratio between bacterial carbon demand and phytoplankton production (BCD:PP) at inner stations of the western subembayment of the Ria Formosa coastal lagoon (average ± 1 SD) during 1985 to 1993 (see Table 14.1 for data sources).

sewage discharges, as well as autochthonous sources such as microphytobenthos, benthic macroalgae beds, submerged seagrass meadows, and salt marshes (Machás and Santos 1999; Mudge et al. 1999; Machás et al. 2003; Santos et al. 2004; Mudge and Duce 2005). Very high and sustained production and leaf release rates of dominant seagrasses throughout the year (Machás et al. 2006; Cunha and Duarte 2007) support the potential role of macrophytes as sources of organic carbon. Similar to microbial processes in other coastal lagoons (e.g., Ziegler and Benner 1999; Grami et al. 2008), in the Ria Formosa Lagoon heterotrophic bacterioplankton and their predators possibly play a pivotal role in channelling primary production of macrophytes and allocthonous derived organic carbon to metazoans. Indeed, phagotrophic protists in the Ria Formosa Lagoon remove, on an annual average, 65% of the daily bacterial production per day, and constitute a potentially important food source for pelagic and benthic metazoans during most of the year (Barbosa 2006).

14.6 INTERANNUAL DYNAMICS OF PLANKTONIC MICROBES

14.6.1 Environmental Setting

During 1967 to 2002, annual mean water temperature in the lagoon (17.27°C ± 0.58°C to 18.63°C ± 0.95°C at inlets and inner locations, respectively) did not exhibit a generalized warming trend. However, a significant interannual linear warming trend during the winter period (1Q), was detected at all lagoon water bodies (0.07°C to 0.10°C yr^{-1}, $p < 0.01$; Figure 14.10). Since spatial differences in this winter warming tendency were not statistically significant, the importance of local processes as drivers of the winter temperature change (e.g., opening of Ancão Inlet during 1997, infilling of the lagoon, and changes in freshwater inflow) could be disregarded. Hence, the winter warming trend probably reflects the effect of global climate forcing on the Ria Formosa Lagoon. This winter warming rate is close to the increases in minimum air temperature reported for southern Portugal (about 0.07°C yr^{-1}; Miranda et al. 2002), and is higher than sea surface warming at near-shore locations off southern Portugal (~0.04°C yr^{-1}; Relvas et al. 2009). However, it is worth noting that the identification of trends and abrupt changes or shifts in time series is highly dependent on sampling frequency and length of the time series. In the case of Ria Formosa, a more continuous and extensive historic

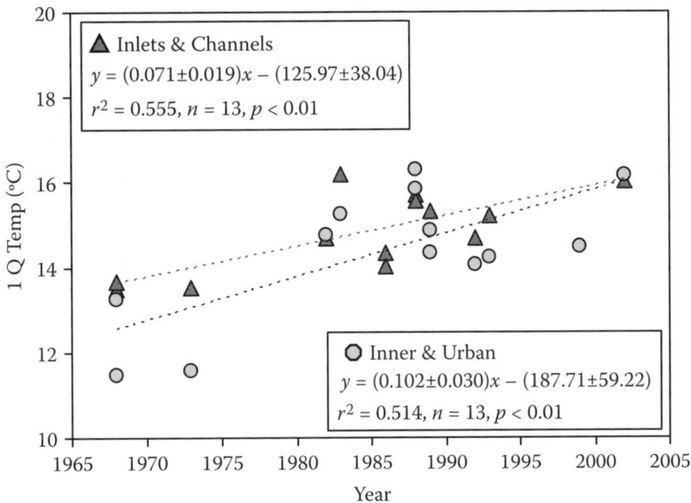

FIGURE 14.10 Time series of annual mean water temperature for the first quarter (January to March, 1Q) in different water bodies of the western subembayment of the Ria Formosa coastal lagoon (▲: inlets and channels; ●: inner and urban-impacted stations), from 1967 to 2002 (see Table 14.1 for data sources). Adjusted regression lines and regression equations are shown.

dataset would be needed to discriminate a linear winter warming trend, assumed in this study, from a step change in the winter mean temperature (Figure 14.10; Andersen et al. 2008).

The analysis of interannual trends in mean light intensity in the mixed layer, controlled by both incident irradiance and water turbidity, is hampered by limited data on water turbidity in the lagoon, namely at inlets and urban-impacted stations. However, existing data point to an overall reduction in turbidity from the 1980s to the late 1990s and 2000s, equivalent to an average reduction in the light extinction coefficient of 28% and 35% for channels and inner stations, respectively.

Annual mean concentrations of most inorganic nutrients in the lagoon did not exhibit any significant linear trends during 1967 to 2002. In the cases of nitrate (2.3 to 26.5 μM $N\text{-}NO_3^-$) and silicate (2.3 to 29.5 μM $Si\text{-}SiO_4^{4-}$), interannual variability was clearly driven by changes in rainfall and freshwater flow into the western sector of the lagoon. Periods with reduced annual precipitation and freshwater flow (1985–1986; 1998–1999) showed lower annual mean concentrations of both silicate and nitrate than periods of high rainfall levels (e.g., 1987–1988; Figure 14.11). Indeed, annual integrated freshwater flow was positive and significantly correlated with annual mean concentration of silicate ($r_s = 0.935$, $n = 12$, $p < 0.00001$) and marginally correlated with annual mean concentration of nitrate ($r_s = 0.691$, $n = 8$, $p = 0.058$). These relationships demonstrate that rainfall and freshwater flow should be considered in the analysis and interpretation of interannual trends of water quality in the Ria Formosa Lagoon. For instance, the general improvement in the water quality between 1987–1988 (272 $m^3\ s^{-1}$) and 2001–2002 (15 $m^3\ s^{-1}$) should not be exclusively attributed to the installation of urban wastewater treatments plants or increased lagoon hydrodynamics caused by the artificial opening of the New Ancão Inlet and extensive dredging of main channels, as hypothesized by others (Newton and Icely 2002; Loureiro et al. 2006; Pereira et al. 2007), but instead to changes in rainfall patterns (Figure 14.11). Data obtained at inner lagoon stations during 2006 to 2008, a period of low annual integrated rainfall and freshwater flow (0.7 to 1.9 $m^3\ s^{-1}$, INAG database), also reveal relatively low mean concentrations of silicate and dissolved inorganic nitrogen (2.2 μM N-DIN, and 8.5 μM $Si\text{-}SiO_4^{4-}$; Brito et al. 2009b). Overall, climatic changes apparently exert a strong influence on the interannual variability of bottom-up drivers of planktonic microbes inside the Ria Formosa, such as water temperature and concentrations of nitrate and silicate. Annual mean concentrations of phosphate (0.27 to 1.44 μM $P\text{-}PO_4^{3-}$) and ammonium (1.93 to 5.61 μM $N\text{-}NH_4^+$), usually classified as

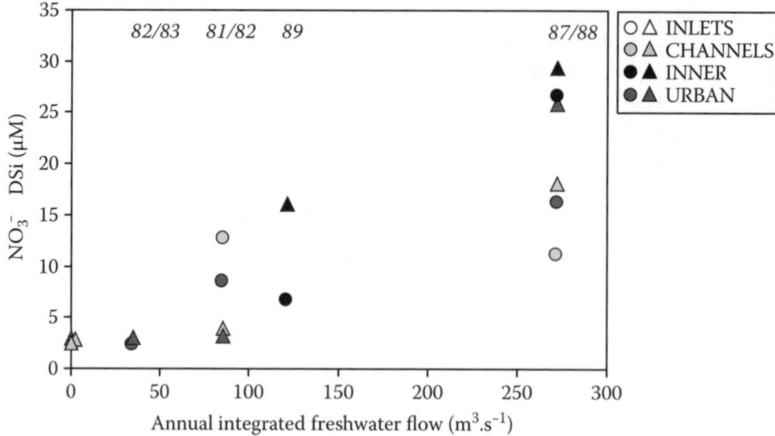

FIGURE 14.11 Relationship between annually integrated freshwater flow into the western subembayment of the Ria Formosa and mean annual concentrations of dissolved inorganic nitrate (NO_3^-), and silicate (DSi) for different lagoon water bodies (as available). Years are shown in italics, and data from 1985–1986 and 1998–1999 are compacted close to the origin (see Table 14.1 for data sources).

regenerated nutrients, are not related to rainfall or freshwater flow, thus suggesting the relevance of other factors, such as autochthonous and allochthonous biological sources. The effects of increased anthropogenic pressure in the Ria Formosa hinterland are evident only at urban-impacted stations, located close to major cities, where a sustained and elevated increase in the annual mean concentration of ammonium (0.446 ± 0.112 μM $N\text{-}NH_4^+$ yr^{-1}), and a significant linear increase in phosphate (0.126 ± 0.002 μM $P\text{-}PO_4^{3-}$ yr^{-1}, $p < 0.01$), were observed between 1981 and 1988.

14.6.2 Phytoplankton

During 1967 to 2008, chlorophyll *a* annual mean concentrations in the Ria Formosa Lagoon varied between 1.5 ± 0.3 μg L^{-1} and 2.5 ± 0.8 μg L^{-1} at inlets and inner stations, respectively. These chlorophyll *a* values are within the lower range reported for many other estuaries and lagoons (Cloern and Jassby 2008), possibly due to high phytoplankton losses associated with pelagic and benthic grazing, and tidal flushing (Barbosa 2006). Interannual trends observed over this period varied across lagoon sites. Chlorophyll *a* in urban-impacted stations exhibited an overall increase during 1969 to 1989, albeit not significant (Figure 14.12), whereas other lagoon locations exhibited an overall decrease throughout 1967 to 2008, albeit only significant at inner stations (–0.043 ± 0.013 (μg chlorophyll *a* L^{-1}).yr^{-1}, $R^2 = 0.62$, $n = 9$, $p < 0.01$; Figure 14.12).

In addition to interannual changes in biomass levels, phytoplankton seasonal patterns also varied among years, both in terms of amplitude and phasing. At inlets, phytoplankton showed a more variable phenology, with mean monthly chlorophyll *a* peaks (1.6 to 6.6 μg L^{-1}) distributed between April and December. The classic bimodal cycle, with spring and autumn blooms, is apparently "disturbed" by the enrichment effect of upwelling events occurring in the adjacent coastal area (Barbosa unpublished data). Inner stations and channels usually revealed unimodal annual cycles, with mean monthly chlorophyll *a* peaks (1.1 to 11.3 μg L^{-1}) consistently located between late spring and summer (not shown), a typical feature of shallow mixed ecosystems (Cébrian and Valiela 1999).

At each lagoon location, interannual chlorophyll *a* trends varied seasonally (Figures 14.13a and 14.13b). During autumn (1Q) and winter (4Q), significant linear declining trends, ranging from –0.025 ± 0.010 to –0.072 ± 0.015 (μg chlorophyll *a* L^{-1}) yr^{-1}, were detected at most lagoon locations (Figure 14.13a). During the autumn-winter period, neither nutrient concentrations and molar ratios (Justic et al. 1995) nor nutrient enrichment experiments (Loureiro et al. 2005) provide evidence of

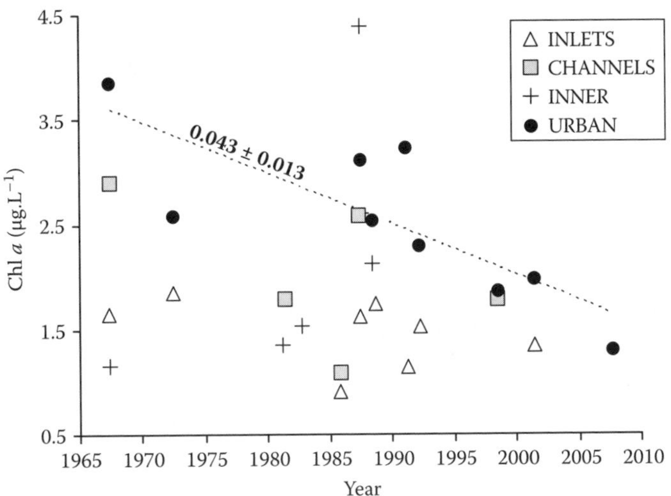

FIGURE 14.12 Time series of annual mean chlorophyll *a* concentration in different water bodies of the western subembayment of the Ria Formosa coastal lagoon, from 1967 to 2008 (see Table 14.1 for data sources). Adjusted regression line and slope (±1 SE) for inner lagoon locations are shown.

nutrient limitation for phytoplankton in the Ria Formosa Lagoon (see Section 14.5.2). Yet, I_m values lower than saturating light intensities typical for estuarine phytoplankton (Tang 1995; Kocum et al. 2002), and the stimulatory effect of light enrichment during this period of the year (Barbosa unpublished data), suggest that phytoplankton in the lagoon is probably light limited during autumn and winter. Since phytoplankton responses to temperature changes are usually not significant in light-limited assemblages (Underwood and Kromkamp 1999), the general decline of chlorophyll *a* observed in the lagoon during autumn and winter could be theoretically explained by two hypothesis, either alone or in combination: (1) a reduction in phytoplankton growth rate due to an overall reduction in light availability, (2) and an increase in the rate of phytoplankton losses. Because water transparency in the Ria Formosa, scarcely evaluated, do not provide evidence of a diminishing trend for this period (see Section 14.6.1), the increase in phytoplankton loss rates during autumn–winter is the more likely hypothesis.

The observed warming of the lagoon during winter (Figure 14.10) could have enhanced the activity of benthic and pelagic grazers, thus contributing to an overall decline in phytoplankton biomass during this period. Interannual declines of phytoplankton biomass, concomitant with increases in water temperature, were recently reported for estuaries and coastal lagoons (Borkman and Smayda 2009; van Beusekom et al. 2009). However, the decline of chlorophyll *a* during autumn was evidently driven by other mechanisms, unrelated to temperature changes. During autumn, positive correlations between mean chlorophyll *a* and freshwater inflow observed at inner stations and channels (Figure 14.14) may indicate that freshwater flow is an important source of inorganic nutrients (Figure 14.11) for phytoplankton during the autumn period, in those locations. These observations on temperature and freshwater inflow generally reflect the effects of long-term climatic variability upon planktonic communities. In the case of inner stations, the artificial opening and relocation of the Ancão Inlet 3.5 km west of its previous position (Figure 14.1b) could further explain chlorophyll *a* reductions after 1997 (Figures 14.13a and 14.13b). This human-induced structural change decreased the water residence time in the inner, western lagoon areas, probably leading to higher flushing rates and increased exportation of phytoplankton to coastal waters. For any fixed level of tidal flushing, the potential impact of tidal advection upon phytoplankton should be higher during periods of reduced phytoplankton growth, such as autumn and winter (Barbosa 2006).

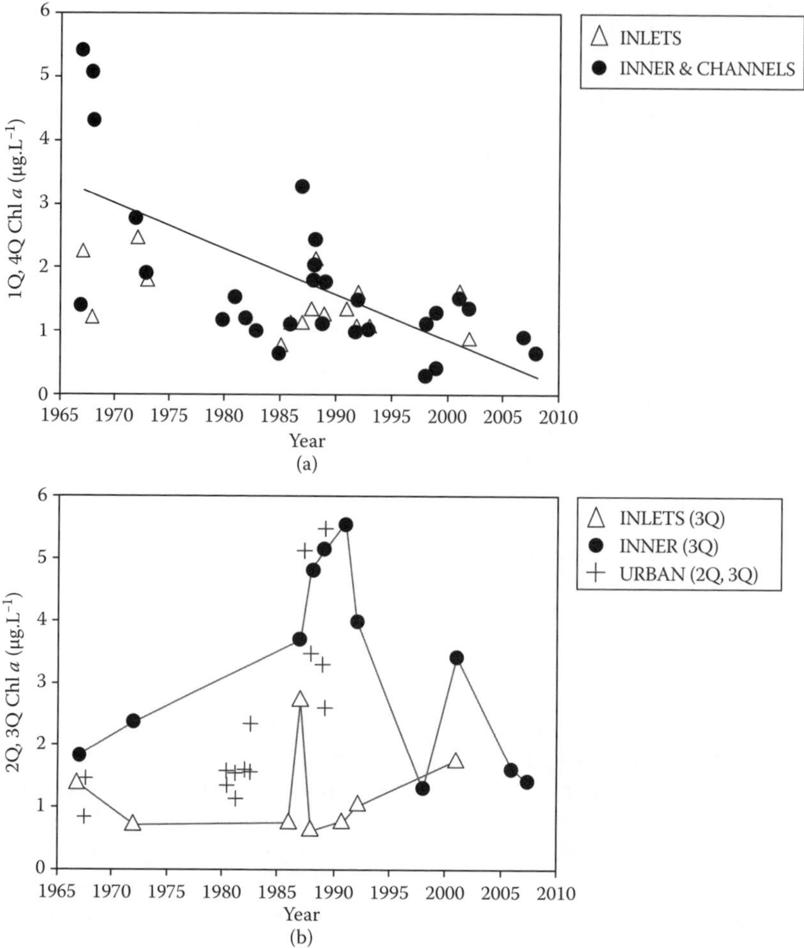

FIGURE 14.13 Time series of quarterly mean chlorophyll *a* concentration in different water bodies of the western subembayment of the Ria Formosa coastal lagoon, from 1967 to 2008 (see Table 14.1 for data sources). (a) Chlorophyll *a* mean concentrations during first (January to March, 1Q) and fourth (October to December, 4Q) quarters, and adjusted regression lines. △ – Inlets: $y = (-0.025 \pm 0.010)x + (51.210 \pm 20.847)$, $R^2 = 0.29$, $n = 16$, $p < 0.05$; ● – Channels and inner stations: $y = (-0.072 \pm 0.015)x + (144.822 \pm 30.284)$, $R^2 = 0.44$, $n = 31$, $p < 0.0001$). (b) Chlorophyll *a* mean concentrations during second (April to June, 2Q) and third quarters (July to September, 3Q), and adjusted regression line for urban-impacted stations [$y = (+0.144 \pm 0.044)x - (282.12 \pm 86.50)$, $R^2 = 0.47$, $n = 14$, $p < 0.01$].

Throughout spring and summer, chlorophyll *a* interannual trends in the lagoon were less coherent across stations, indicating that different processes, probably more site specific, were driving phytoplankton variability during the 1967 to 2008 period. During spring and summer, chlorophyll *a* at urban impacted stations exhibited a significant increase during 1969 to 1989 (+0.144 ± 0.044 [µg chlorophyll *a* L^{-1}].year^{-1}, $n = 14$, $p < 0.01$; Figure 14.13b). This tendency could be interpreted as a result of increased availability of inorganic nutrients, either derived from anthropogenic sources as in the case of phosphate and ammonium, or from increased riverine inputs, namely during 1987 to 1988, in case of silicate and nitrate (Figure 14.11). During spring, chlorophyll *a* showed significant declines at lagoon channels (−0.088 ± 0.031 [µg chlorophyll *a* L^{-1}] yr^{-1}, $R^2 = 0.61$, $n = 7$, $p < 0.05$), and no significant trends at inlets and inner stations. Summer mean chlorophyll *a* at lagoon inlets showed an increasing trend since the late 1980s (Figure 14.13b) that could be related to increased

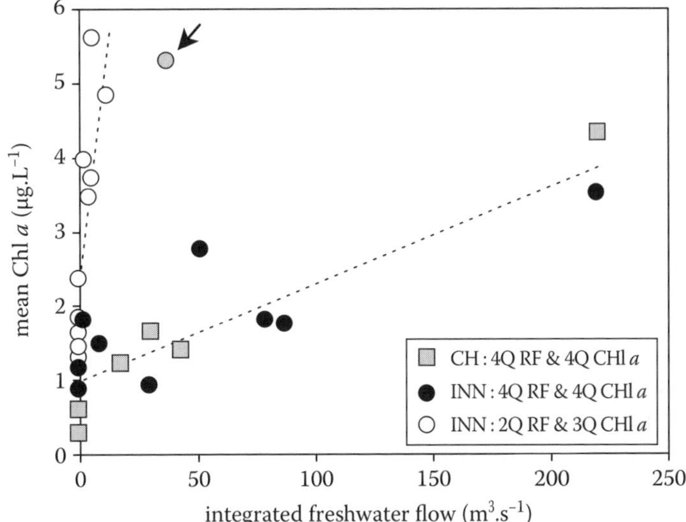

FIGURE 14.14 Relationship between quarterly integrated freshwater flow (RF) into the western sector of the Ria Formosa and quarterly mean chlorophyll *a* concentration (Chl *a*) for inner stations (INN), and channels (CH) during 1967 to 2007, as available (see Table 14.1 for data sources). Adjusted regression lines are shown. ○ – Inner stations, 2Q RF and 3Q Chl *a*: $y = (-0.249 \pm 0.070)x + (2.190 \pm 0.390)$, $R^2 = 0.612$, $n = 10$, $p < 0.01$; □ ● – Channels and inner stations, 4Q RF and 4Q Chl*a*: $y = (+0.013 \pm 0.002)x + (0.968 \pm 0.135)$, $R^2 = 0.850$, $n = 14$, $p < 0.00001$. The arrow marks a data point not included in the regression analysis.

nutrient availability due to the intensification of summer upwelling since 1985, along the southern Portuguese coast (Relvas et al. 2009). Summer mean chlorophyll *a* at inner stations exhibited an increase between 1967 and 1991, and a decrease thereafter (Figure 14.13B), and was positively correlated with integrated freshwater inflow during spring, the previous quarter ($r_S = 0.851$, $n = 10$, $p < 0.01$; Figure 14.14). This relationship indicates that freshwater inflow is probably also an important source of inorganic nutrients (Figure 14.11) for phytoplankton at inner lagoon stations during summer. It also demonstrates the role of climatic changes as drivers of phytoplankton interannual variability during the season of highest potential growth limitation in the lagoon (see Section 14.5.1). Overall, it is interesting to note that, despite low freshwater inflows into the lagoon, interannual variability of chlorophyll *a* during summer and autumn was correlated with, and possibly controlled by, freshwater flow occurring in previous periods (Figure 14.14). Interannual changes in the winter freshwater flow apparently had no effect on phytoplankton biomass, probably due to greater light limitation during this period.

Benthic filter feeders may exert a strong grazing pressure upon phytoplankton, and their role as key drivers of long-term variability of phytoplankton has been identified for other estuaries and coastal lagoons (Cadeé and Hegeman 1974; Cloern 1996; Mohlenberg et al. 2007; Strayer et al. 2008). No information on long-term changes in uncultured benthic communities is currently available for the Ria Formosa. However, the marked decrease in the production of cultivated clams since the early 1980s was not associated with an apparent increase in phytoplankton biomass, as might be expected. Still, bivalves may also exert a positive feedback on phytoplankton growth rates (Prins et al. 1998), and a strong top-down control on phagotrophic protists (Dupuy et al. 2000). Indeed, long-term changes in the abundance of loricate ciliates (tintinnids) were recently linked to changes in grazing exerted by benthic filter feeders (Strayer et al. 2008). Phagotrophic protists in the Ria Formosa, particularly ciliates, constitute a potentially important food source for benthic metazoans during most of the year (Barbosa 2006). Moreover, phytoplankton biomass in the lagoon is dominated by nanoplanktonic flagellates, namely cryptophytes (Barbosa 2006; Pereira et al. 2007), that are predominantly grazed by phagotrophic protists. Indeed, these grazers consume, on an annual average, 73% of the daily production of plastidic nanoflagellates (Barbosa 2006). Hence, decreases in benthic grazing pressure on phagotrophic protists may cascade down the food web, and eventually lead to a reduction in phytoplankton biomass.

Short-term episodic climate perturbations can also affect phytoplankton distribution inside the Ria Formosa. It is worth noting that, excluding urban-impacted areas, chlorophyll *a* values higher than 15.0 µg L^{-1} ($n = 2$) were associated with extreme perturbations, a heat wave (July 1991) and a flood (December 1987). The heat wave event (19.1 µg chlorophyll *a* L^{-1}) occurring during neap tides, was preceded by a 5-d period of extremely low wind velocities (<2 m s^{-1}), leading to exceptional air (42°C) and water temperatures (30°C) (Barbosa 2006). This scenario probably promoted a transient water column stratification that enhanced phytoplankton growth or decreased its consumption by benthic filter feeders (Cloern et al. 2005). The flood event (daily precipitation and freshwater flow of 118 mm d^{-1}, and 71.4 m^3 s^{-1}) led to significant increases in both nutrient concentrations and chlorophyll *a* (23.5 µg chlorophyll *a* L^{-1} (Newton 1988). In this case, improved light conditions for phytoplankton growth due to transient haline stratification of the water column, and not increased nutrient availability, was probably the underlying explanation.

14.7 CONCLUSIONS

During the period 1967 to 2008, the Ria Formosa Lagoon was subjected to seasonal and interannual environmental changes due to both climatic variability and local human activities. The effects of increased anthropogenic pressure, both on seasonal and interannual time scales, were evident only at sites close to major urban centers. These effects included decreased oxygen saturation, higher and sustained availability of labile organic substrates during summer, and long-term increases in the concentrations of dissolved inorganic nutrients, particularly phosphate and ammonium, and phytoplankton biomass during the spring–summer period. Anthropogenic effects apparently are not relevant drivers of planktonic microbes at other lagoon locations, probably due to low water residence times inside the lagoon, and the importance of physical variables (e.g., light and temperature) as growth-limiting factors for planktonic microbes.

At inlets, main navigational channels, and inner lagoon locations, seasonality of both heterotrophic bacterioplankton and phytoplankton was directly or indirectly coupled to changes in water temperature and light availability, both under climatic control. Long-term climatic variability (e.g., temperature and precipitation) apparently reverberated into the water column, leading to a generalized and significant winter warming trend, and strong interannual changes in the concentrations of inorganic nutrients, particularly during the autumn-winter period. Although the ultimate cause of changes in the nutrient loads associated with freshwater inflow is climate variability (rainfall), it is important to note that climate effects are superimposed on the effects of anthropogenic activities in the lagoon's drainage basin.

Despite overall reduced freshwater flows into the Ria Formosa, interannual variability of freshwater inflow apparently exerted a significant control on the biomass of phytoplankton in the lagoon, particularly during periods of relatively high flow rates and low light limitation (spring and autumn). Phytoplankton biomass showed a generalized interannual declining trend during autumn and winter. This decline may be eventually related to the winter warming tendency, through its stimulatory effect upon phytoplankton grazers, or to decreases in nutrient concentrations associated with reductions in the autumn freshwater inflow. Long-term changes in phytoplankton biomass during spring and summer were less coherent across stations, indicating that more site-specific processes were driving phytoplankton variability. After 1997, increased tidal exchanges associated with the Ancão Inlet relocation may further explain the decline of phytoplankton abundance at inner lagoon locations only.

Overall, synthesized data for the western subembayment of the Ria Formosa Lagoon (1967 to 2008) suggest that climate variability is an important driver of planktonic microbes in the lagoon, both on short-term, seasonal, and interannual time scales. Local anthropogenic impacts are apparently spatially restricted. More integrated monitoring and research programs, involving analysis of top-down controls, process-oriented studies, and dedicated experimental approaches, are clearly needed to test and conclusively establish direct links between the variability of planktonic microbes and climatic and local anthropogenic alterations in the Ria Formosa coastal lagoon.

ACKNOWLEDGMENTS

I thank Dr. A. Matias and Dr. T. Modesto for the figures of the Ria Formosa, Dr. A. Chícharo and Dr. H. Marques for permission to use unpublished data, and M. Abreu (DRAPA, Portugal) for providing access to recent data on production of cultivated clams. I also thank two anonymous reviewers for their valuable comments and suggestions.

REFERENCES

Amaral, A. 2008. Management of aquaculture of the clam *R. decussatus* (Linnaeus, 1758) in the Ria Formosa lagoon (South of Portugal): Effects on the ecosystem and species physiology. Ph.D. Thesis, Universidad de Santiago, Santiago de Compostela, Spain.

Andersen, T., J. Carstensen, E. Hernández-Garcia, and C. M. Duarte. 2008. Ecological thresholds and regime shifts: Approaches to identification. *Trends Ecol. Evol.* 24: 49–57.

Andrade, C. 1990. O ambiente de barreira da Ria Formosa (Algarve, Portugal). Ph.D. Thesis, Universidade Lisboa, Lisboa, Portugal.

Andrade, C., M. C. Freitas, J. Moreno, and S. C. Craveiro. 2004. Stratigraphical evidence of Late Holocene barrier breaching and extreme storms in lagoonal sediments of Ria Formosa, Algarve, Portugal. *Mar. Geol.* 210: 339–362.

Assis, M. E., M. A. Sampayo, and M. H. Vilela. 1984. A Ria Faro-Olhão. I. Pigmentos e formas planctónicas predominantes (Maio 1972–Maio 1973). *Cuaderno Area Ciencias Marinas, Seminario de Estudios Gallegos* 1: 217–236.

Baker, K. S. and R. Frouin. 1987. Relation between photosynthetically available radiation and total insolation at the ocean surface under clear skies. *Limnol. Oceanogr.* 32: 1370–1377.

Barbosa, A. M. B. 1989. Variação espaço-temporal da abundancia e biomassa bacterianas no sistema lagunar Ria Formosa. Graduation Thesis, Universidade do Algarve, Faro, Portugal.

Barbosa, A. M. B. 1991. Spatial and temporal variation of bacterioplankton abundance and biomass in a coastal lagoon (Ria Formosa, Southeastern Portugal). *Kieler Meeresforsch.* 8: 66–73.

Barbosa, A. M. B. 2006. Estrutura e dinâmica da teia alimentar microbiana na Ria Formosa. Ph.D. Thesis, Universidade do Algarve, Faro, Portugal.

Barbosa, A. B., H. Galvão, P. Mendes, X. Álvarez-Salgado, F. Figueiras, and I. Joint. 2001. Heterotrophic bacterioplankton dynamics off the NW Iberian margin. *Prog. Oceanogr.* 51: 339–359.

Barreira, L. P. V. A. 2007. Polycyclic aromatic hydrocarbons and oxidative stress markers in the clam Ruditapes decussatus from the Ria Formosa (Portugal). Ph.D. Thesis, Universidade do Algarve, Faro, Portugal.

Bebianno, M. J. 1995. Effects of pollutants in the Ria Formosa lagoon, Portugal. *Sci Total Environ.* 171: 107–115.

Benoliel, M. J. 1984. *Vigilância em contínuo da qualidade da água da Ria Formosa (Ria de Faro-Olhão) 1980*. Relatório Técnico Final REL-TF-QP-3/84, Divisão de Química e Poluição, Instituto Hidrográfico, Lisboa, Portugal.

Benoliel, M. J. 1985. *Vigilância da qualidade da água da Ria Formosa (Ria de Faro-Olhão) 1984*. Relatório Técnico Final REL-TF-QP-1/85, Divisão de Química e Poluição, Instituto Hidrográfico, Lisboa, Portugal.

Benoliel, M. J. 1989. *Vigilância da qualidade da água da Ria Formosa, 1981/1983*. Relatório Técnico Final REL-TF-QP-04/89, Divisão de Química e Poluição, Instituto Hidrográfico, Lisboa, Portugal.

Bettencourt, A. M., S. B. Bricker, J. G. Ferreira, A. Franco, J. C. Marques, J. J. Melo, A. Nobre, L. Ramos, C. S. Reis, F. Salas, M. C. Silva, T. Simas, and W. J. Wolff. 2003. Typology and reference conditions for Portuguese transitional and coastal waters. Development of Guidelines for the Application of the European Union Water Framework Directive, Instituto da Água, Lisboa, Portugal.

Borkman, D., H. Baretta-Bekker, and P. Henriksen. 2009. Introduction. *J. Sea Res.* 61: 1–2.

Borkman, D. and T. Smayda. 2009. Multidecadal (1959–1997) changes in *Skeletonema* abundance and seasonal bloom patterns in Narragansett Bay, Rhode Island. *J. Sea Res.* 61: 84–94.

Brito, A., A. Newton, P. Tett, and T. F. Fernandes. 2009a. Temporal and spatial variability of microphytobenthos in a shallow lagoon: Ria Formosa (Portugal). *Estuar. Coastal Shelf Sci.* 83: 67–76.

Brito, A., A. Newton, P. Tett, and T. F. Fernandes. 2009b. Understanding the importance of sediments to water quality in coastal shallow lagoons. *J. Coastal Res.* SI 56: 381–384.

Cabaço, S., A. Alexandre, and R. Santos. 2005. Population-level effects of clam harvesting on the seagrass *Zostera noltii. Mar. Ecol. Prog. Ser.* 298: 123–129.

Cabaço, S., R. Machás, V. Vieira, and R. Santos. 2008. Impacts of urban wastewater discharge on seagrass meadows (*Zostera noltii*). *Estuar. Coastal Shelf Sci.* 78: 1–13.

Cadée, G. C. and J. Hegeman. 1974. Primary production of phytoplankton in the Dutch Wadden Sea. *Neth. J. Sea Res.* 8: 240–259.

Carrasco, A. R., O. Ferreira, M. Davidson, and J. A. Dias. 2008. An evolutionary categorization model for backbarrier environments. *Mar. Geol.* 251: 156–166.

Cébrian, J. and I. Valiela. 1999. Seasonal patterns in phytoplankton biomass in coastal ecosystems. *J. Plankton Res.* 21: 429–444.

Chenery, A. M. and S. M. Mudge. 2005. Detecting anthropogenic stress in an ecosystem. III. Mesoscale variability and biotic indices. *Environ. Forensics* 6: 371–384.

Chícharo, L. M. Z. and M. A. Chícharo. 2001. Effects of environmental conditions on planktonic abundances, benthic recruitment, and growth rates of the bivalve mollusc *Ruditapes decussatus* in a Portuguese coastal lagoon. *Fish. Res.* 53: 235–250.

Chícharo, L. M. Z., M. A. Chícharo, I. Gouveia, S. Condinho, A. Amaral, C. Miguel, C. Graça, F. Alves, and F. Regala. 2000. *Biological assessment of bivalve stocks of* Ruditapes decussatus *and* Cardium edule *in the Ria Formosa*. Project DG XIV-97/106 Final Report, Faro, Portugal.

Chubarenko, B., V. G. Koutitonsky, R. Neves, and G. Umgiesser. 2005. Modelling concepts. Pp. 231–306, In: I. E. Gonenç and J. P. Wolfin (eds.), *Coastal lagoons: Ecosystem processes and modeling for sustainable use and development*. Boca Raton, Florida: CRC Press.

Cloern, J. E. 1996. Phytoplankton bloom dynamics in coastal ecosystems: A review with some general lessons from sustained investigation of San Francisco Bay, California. *Rev. Geophys.* 34: 127–168.

Cloern, J. E. 2001. Our evolving conceptual model of the coastal eutrophication problem. *Mar. Ecol. Prog. Ser.* 210: 223–253.

Cloern, J. E. and R. Dufford. 2005. Phytoplankton community ecology: Principles applied in San Francisco. *Mar. Ecol. Prog. Ser.* 285: 11–28.

Cloern, J. E. and A. D. Jassby. 2008. Complex seasonal patterns of primary producers at the land-sea interface. *Ecol. Lett.* DOI:10.1111/j.1461-0248.2008.01244.x

Cloern, J. E., T. S. Schraga, C. B. Lopez, and N. Knowles. 2005. Climate anomalies generate an exceptional dinoflagellate bloom in San Francisco Bay. *Geophys. Res. Lett.* 32: L14608.

Coelho, M. R., M. J. Bebianno, and W. J. Langston. 2002. Organotin levels in the Ria Formosa lagoon, Portugal. *Appl. Organomet. Chem.* 16: 384–390.

Cole, J. J., S. Findlay, and M. L. Pace. 1988. Bacterial production in fresh and saltwater ecosystems: A cross-system overview. *Mar. Ecol. Progr. Ser.* 43: 1–10

Cunha, A. H., J. Assis, and E. A. Serrão. 2009. Estimation of available seagrass meadow area in Portugal for transplanting purposes. *J. Coastal Res.* SI 56: 1100–1104.

Cunha, A. H. and C. M. Duarte. 2007. Biomass and leaf production of *Cymodocea nodosa* in the Ria Formosa lagoon, south Portugal. *Bot. Mar.* 50: 1–7.

Cunha, M. E. and M. C. Massapina. 1984. Contribution to the zooplankton community analysis of the Ria de Faro-Olhão. *Cuaderno Area Ciencias Marinas, Seminario de Estudios Gallegos* 1: 237–250.

Dionísio, L. A. P. C. 1996. Estudos microbiológicos da Ria Formosa. Qualidade sanitária de águas e moluscos bivalves. Ph.D. Thesis, Universidade do Algarve, Faro, Portugal.

Dionísio, L. A. P. C., G. Rheinheimer, and J. J. Borrego. 2000. Microbiological pollution of Ria Formosa (south of Portugal). *Mar. Poll. Bull.* 40: 186–193.

Domingues, R. B., A. Barbosa, and H. Galvão. 2008. Constraints on the use of phytoplankton as a biological quality element within the Water Framework Directive in Portuguese waters. *Mar. Pollut. Bull.* 56: 1389–1395.

Duarte, P., B. Azevedo, M. Guerreiro, C. Ribeiro, R. Bandeira, A. Pereira, M. Falcão, D. Serpa, and J. Reia. 2008. Biogeochemical modelling of Ria Formosa (south Portugal). *Hydrobiologia* 611: 115–132.

Ducklow, H. 2008. Microbial services: Challenges for microbial ecologists in a changing world. *Aquat. Microb. Ecol.* 53: 13–19.

Dupuy, C., A. Vaquer, T. Lam-Hoai, C. Rougier, N. Mazouni, J. Lautier, Y. Collos, and S. Le Gall. 2000. Feeding rate of the oyster *Crassostrea gigas* in a natural planktonic community of the Mediterranean Thau Lagoon. *Mar. Ecol. Prog. Ser.* 205: 171–184.

Edwards, V., J. Icely, A. Newton, and R. Webster. 2005. The yield of chlorophyll from nitrogen: A comparison between the shallow Ria Formosa lagoon and the deep oceanic conditions at Sagres along the southern coast of Portugal. *Estuar. Coastal Shelf Sci.* 62: 391–403.

Falcão, M. M. 1996. Dinâmica dos nutrientes na Ria Formosa: efeitos da interacção da laguna com as suas interfaces na reciclagem do azoto, fósforo e sílica. Ph.D. Thesis, Universidade do Algarve, Faro, Portugal.

Falcão, M. M., J. I. Pisarra, and M. H. Cavaco. 1985. Características físicas e químicas da Ria de Faro-Olhão: 1984. Relat. INIP, Instituto Nacional de Investigação das Pescas (INIP), Lisboa, Portugal. 61: 4–12.

Falcão, M. M., J. I. Pisarra, and M. H. Cavaco. 1991. Características químico-biológicas da Ria Formosa: análise de um ciclo anual (1985–1986). *Bol. Inst. Nac. Invest. Pescas*16: 5–21.

Falcão, M. and C. Vale. 1998. Sediment-water exchanges of ammonium and phosphate in intertidal and subtidal areas of a mesotidal coastal lagoon (Ria Formosa). *Hydrobiologia* 373/374: 193–201.

Falcão, M. and C. Vale. 2003. Nutrient dynamics in a coastal lagoon (Ria Formosa, Portugal): The importance of lagoon-sea water exchanges on the biological productivity. *Cienc. Mar.* 29: 425–433.

Ferreira, J. G., A. M. Nobre, T. C. Simas, M. C. Silva, A. Newton, S. B. Bricker, W. J. Wolff, P. E. Stacey, and A. Sequeira. 2006. A methodology for defining homogeneous water bodies in estuaries: Application to the transitional systems of the EU Water Framework Directive. *Estuar. Coastal Shelf Sci.* 66: 468–482.

Ferreira, O., J. A. Dias, and R. Taborda. 2008. Implications of sea-level rise for continental Portugal. *J. Coastal Res.* 24: 317–324.

Fidalgo, L. G. Z. 1996. Aplicação de um sistema de informação geográfica na conservação da biodiversidade, M.S. Thesis, Universidade do Algarve, Faro, Portugal.

Gallegos, C. L. and T. E. Jordan. 1997. Seasonal progression of factors limiting phytoplankton pigment biomass in the Rhode River estuary, Maryland (USA). I. Controls on phytoplankton growth. *Mar. Ecol. Prog. Ser.* 161: 185–198.

Gamito, S. 2008. Three stressors acting on the Ria Formosa lagoonal system (Southern Portugal): Physical stress, organic matter pollution and the land-ocean gradient. *Estuar. Coastal Shelf Sci.* 77: 710–720.

Gamito, S., J. Gilabert, C. M. Diego, and A. Pérez-Ruzafa. 2005. Effects of changing environmental conditions on lagoon ecology. Pp. 193–230, In: I. E. Gonenç and J. P. Wolfin (eds.), *Coastal lagoons: Ecosystem processes and modeling for sustainable use and development*. Boca Raton, Florida: CRC Press.

Gasol, J. M., J. Pinhassi, L. Alonso-Sáez, H. Ducklow, G. J. Herndl, M. Koblizek, M. Labrenz, Y. Luo, X. A. G. Morán, T. Reinthaler, and M. Simon. 2008. Towards a better understanding of microbial carbon flux in the sea. *Aquat. Microb. Ecol.* 53: 21–38.

Grami, B., N. Niquil, A. S. Hlaili, M. Gosselin, D. Hamel, and H. H. Mabrouk. 2008. The plankton food web of the Bizerte lagoon (South-western Mediterranean). II. Carbon steady-state modelling using inverse analysis. *Estuar. Coastal Shelf Sci.* 79: 101–113.

Guadayol, Ó., F. Peters, C. Marrasé, J. M. Gasol, C. Roldán, E. Berdalet, R. Massana, and A. Sabata. 2009. Episodic meteorological and nutrient-load events as drivers of coastal planktonic ecosystem dynamics: A time-series analysis. *Mar. Ecol. Prog. Ser.* 381: 139–155.

Gurel, M., A. Tanik, R. C. Russo, and I. E. Gonenç. 2005. Biogeochemical cycles. Pp. 79–192, In: I. E. Gonenç and J. P. Wolfin (eds.), *Coastal lagoons: Ecosystem processes and modeling for sustainable use and development*. Boca Raton, Florida: CRC Press.

Guimarães, M. 2007. Contribuição para a elaboração de um plano de gestão integrada dos campos de Z. noltii na Ria Formosa. M.S. Thesis, Universidade do Algarve, Faro, Portugal.

IPCC. 2007. *Climate change 2007: Synthesis report*. Contribution of Working Groups I, II and III to the Fourth Assessment Report of the Intergovernmental Panel on Climate Change, Geneva, Switzerland.

Jassby, A. D., J. E. Cloern, and B. E. Cole. 2002. Annual primary production: Patterns and mechanisms of change in a nutrient-rich tidal ecosystem. *Limnol. Oceanogr.* 47: 698–712.

Jumars, P. A. 1993. *Concepts in biological oceanography*. Oxford: Oxford University Press.

Justic, D., N. N. Rabalais, R. E. Turner, and Q. Dortch. 1995. Changes in nutrient structure of river-dominated coastal waters: Stoichiometric nutrient balance and its consequences. *Estuar. Coastal Shelf Sci.* 40: 339–356.

Kjerfve, B. 1994. Coastal lagoons. Pp. 1–8, In: B. Kjerfve (Ed.), *Coastal Lagoon Processes*, Amsterdam, Elsevier.

Kjerfve, B. and K. E. Magill. 1989. Geographic and hydrodynamic characteristics of shallow coastal lagoons. *Mar. Geol.* 88: 187–199.

Klein Tank, A. M. G., J. B.Wijngaard, G. P. Konnen, R. Bohm, and G. Demarée. 2002. Daily dataset of 20th-century surface air temperature and precipitation series for the European climate assessment. *Int. J. Climatol.* 22: 1441–1453.

Kocum, E., G. J. C. Underwood, and D. B. Nedwell. 2002. Simultaneous measurement of phytoplanktonic primary production, nutrient and light availability along a turbid, eutrophic UK east coast estuary (the Colne Estuary). *Mar. Ecol. Prog. Ser.* 231: 1–12.

Leite, R. B., R. Afonso, and M. L. Cancela. 2004. *Perkinsus* sp. infestation in carpet-shell clams, *Ruditapes decussatus*, along the Portuguese coast: Results from a 2-year survey. *Aquaculture* 240: 39–53.

Lemos, R. T. and B. Sansó. 2006. Spatio-temporal variability of ocean temperature in the Portugal current system. *J. Geophys. Res.* 111: C04010. DOI:10.1029/2005JC003051.

Levin, L. A., D. F. Boesch, A. Covich, C. Dahm, C. Erséus, K. C. Ewel, R. T. Kneib, A. Mildenke, M. A. Palmer, P. Snelgrove, D. Strayer, and J. M. Weslawski. 2001. The function of marine critical transition zones and the importance of sediment biodiversity. *Ecosystems* 4: 430–451.

Lima, C. and C. Vale. 1977. Ria de Faro-Olhão: aspectos físicos, químicos e bacteriológicos. *Bol. Inst. Nac. Invest. Pescas* 3: 5–25.

Lloret, J., A. Marín, and L. Marín-Guirao. 2008. Is coastal lagoon eutrophication likely to be aggravated by global climate change? *Estuar. Coastal Shelf Sci.* 78: 403–412.

Lotze, H. K., H. S. Lenihan, B. J. Bourque, R. H. Bradbury, R. G. Cooke, M. C. Kay, S. M. Kidwell, M. X. Kirby, C. H. Peterson, and J. B. C. Jackson. 2006. Depletion, degradation, and recovery potential of estuaries and coastal seas. *Science* 312: 1806–1809.

Loureiro, S., A. Newton, and J. D. Icely. 2005. Effects of nutrient enrichment on primary production in the Ria Formosa coastal lagoon (southern Portugal). *Hydrobiologia* 550: 29–45.

Loureiro, S., A. Newton, and J. D. Icely. 2006. Boundary conditions for the European water framework directive in the Ria Formosa lagoon, Portugal (physico-chemical and phytoplankton quality elements). *Estuar. Coastal Shelf Sci.* 67: 382–398.

Machás, R. and R. Santos. 1999. Sources of organic matter in Ria Formosa revealed by stable isotope analysis. *Acta Oecol.* 20: 463–469.

Machás, R., R. Santos, and B. Peterson. 2003. Tracing the flow of organic matter from primary producers to filter feeders in Ria Formosa lagoon, Southern Portugal. *Estuaries* 26: 846–856.

Machás, R., R. Santos, and B. Peterson. 2006. Elemental and stable isotope composition of *Zostera noltii* (Horneman) leaves during the early phases of decay in a temperate mesotidal lagoon. *Estuar. Coastal Shelf Sci.* 66: 21–29.

Marques, H. 1988. *Produtividade e poluição da Ria Formosa: fitoplâncton (análise qualitativa e quantitativa) e produção primária.* Relatório Final do Projecto "Produtividade e Poluição da Ria Formosa," Comissão de Coordenação da Região do Algarve/Universidade do Algarve, Faro, Portugal.

Marques, H. 2005. Distribution of the plankton community in the Ria Formosa, a coastal lagoon in South Eastern Portugal. Ph.D. Thesis, University of Wales, Menai Bridge, U.K.

Matias, A., O. Ferreira, A. Vila-Concejo, T. Garcia, and J. A. Dias. 2008. Classification of washover dynamics in barrier islands. *Geomorphology* 97: 655–674.

MCOA (Ministério das Cidades, Ordenamento do Território e Ambiente). 2008. *Revisão do plano de ordenamento do Parque Natural da Ria Formosa, estudo de caracterização.* Instituto de Conservação da Natureza, Parque Natural da Ria Formosa, Olhão, Portugal.

Miranda, P., F. E. S. Coelho, A. R. Tomé, M. A. Valente, A. Carvalho, C. Pires, H. O. Pires, V. C. Pires, and C. Ramalho. 2002. 20th century Portuguese climate and climate scenarios. Pp. 23–83, In: F. D. Santos, K. Forbes, and R. Moita (eds.), *Climate change in Portugal. Scenarios, impacts and adaptation measures.* SIAM project, Lisbon, Gradiva, Portugal.

Mohlenberg, F., S. Petersen, A. H. Petersen, and C. Gameiro. 2007. Long-term trends and short-term variability of water quality in Skive Fjord, Denmark: Nutrient load and mussels are the primary pressures and drivers that influence water quality. *Environ. Monitor. Assess.* 127: 503–521.

Morais, P., M. A. Chicharo, and A. B. Barbosa. 2003. Phytoplankton dynamics in a coastal saline lake (SE-Portugal). *Acta Oecol.* 24: S87-S96.

Morel, A. and R. C. Smith. 1974. Relation between total quanta and total energy for aquatic photosynthesis. *Limnol. Oceanogr.* 19: 591–600.

Mudge, S. M., M. J. Bebianno, J. A. East, and L. A. Barreira. 1999. Sterols in the Ria Formosa lagoon, Portugal. *Water Resour.* 33: 1038–1048.

Mudge, S. M. and C. E. Duce. 2005. Identifying the source, transport path and sinks of sewage derived organic matter. *Environ. Pollut.* 136: 209–220.

Mudge, S. M., J. E. East, M. J. Bebianno, and L. A. Barreira. 1998. Fatty acids in the Ria Formosa Lagoon, Portugal, *Org. Geochem.* 29: 963–977.

Mudge, S. M., J. D. Icely, and A. Newton. 2007. Oxygen depletion in relation to water residence times. *J. Environ. Monitor.* 9: 1194–1198.

Mudge, S. M., J. D. Icely, and A. Newton. 2008. Residence times in a hypersaline lagoon: Using salinity as a tracer. *Estuar. Coastal Shelf Sci.* 77: 278–284.

Newton, A. 1988. *Produtividade e poluição da Ria Formosa: nutrientes e clorofila a.* Relatório Final do Projecto "Produtividade e Poluição da Ria Formosa," Comissão de Coordenação da Região do Algarve/ Universidade do Algarve, Faro, Portugal.

Newton, A. 1995. The water quality of the Ria Formosa lagoon, Portugal. Ph.D. Thesis, University of Wales, Menai Bridge, U.K.

Newton, A. and J. D. Icely. 2002. Impact of coastal engineering on the water quality of the Ria Formosa lagoon, Portugal. Pp. 417–421, In: *Littoral 2002, The Changing Coast*, EUROCOAST/EUCC, Porto-Portugal.

Newton, A. and J. D. Icely. 2006. Oceanographic applications to eutrophication in a tidal, coastal lagoon, the Ria Formosa, Portugal. *J. Coastal Res.* 3 (SI 39): 1346–1350.

Newton, A., J. D. Icely, M. Falcão, A. Nobre, J. P. Nunes, J. G. Fereira, and C. Vale. 2003. Evaluation of eutrophication in the Ria Formosa coastal lagoon, Portugal. *Continent. Shelf Res.* 23: 1945–1961.

Newton, A. and S. M. Mudge. 2003. Temperature and salinity regimes in a shallow, mesotidal lagoon, the Ria Formosa, Portugal. *Estuar. Coastal Shelf Sci.* 57: 73–85.

Newton, A. and S. M. Mudge. 2005. Lagoon-sea exchanges, nutrient dynamics and water quality management of the Ria Formosa (Portugal). *Estuar. Coastal Shelf Sci.* 62: 405–414.

Nixon, S. W. 1982. Nutrient dynamics, primary production and fisheries yields in lagoons, *Oceanol. Acta* SP: 357–371.

Nobre, A. M., J. G. Ferreira, A. Newton, T. Simas, J. D. Icely, and R. Neves. 2005. Management of coastal eutrophication: Integration of field data, ecosystem-scale simulations and screening models. *J. Mar. Syst.* 56: 375–390.

Padinha, C., R. Santos, and M. T. Brown. 2000. Evaluating environmental contamination in Ria Formosa (Portugal) using stress indexes of *Spartina maritima*. *Mar. Environ. Res.* 49: 67–78.

Paerl, H. W., L. M. Valdes, J. E. Adolf, and L. W. Harding, Jr. 2006. Anthropogenic and climatic influences on the eutrophication of large estuarine ecosystems. *Limnol. Oceanogr.* 51: 448–462.

Pereira, M. G., J. D. Icely, S. Mudge, A. Newton, and R. Rodrigues. 2007. Temporal and spatial variation of phytoplankton pigments in the western part of Ria Formosa lagoon, Southern Portugal. *Environ. Forensics* 8: 205–220.

Pranovi, F., G. Franceschini, M. Casale, M. Zuchetta, P. Torricelli, and O. Giovanardi. 2006. An ecological imbalance induced by a non-native species: The Manila clam in the Venice lagoon. *Biol. Invasions* 8: 595–609.

Prins, T. C., A. C. Smaal, and R. F. Dame. 1998. A review of the feedbacks between bivalve grazing and ecosystem processes. *Aquat. Ecol.* 31: 349–359.

Raabe, T. and K. H. Wiltshire. 2009. Quality control and analyses of the long-term nutrient data from Helgoland Roads, North Sea. *J. Sea Res.* 61: 3–16.

Relvas, P., E. C. Barton, J. Dubert, P. B. Oliveira, Á. Peliz, J. C. B. da Silva, and A. M. P. Santos. 2007. Physical oceanography of the western Iberia ecosystem: Latest views and challenges. *Prog. Oceanogr.* 74: 149–173.

Relvas, P. and J. Luis. Importance of the mesoscale in the decadal changes observed in the northern Canary upwelling system. *Geophys. Res. Lett.* 36, L22601. doi: 10.1029/2009-1040504.

Ribeiro, J., C. C. Monteiro, P. Monteiro, L. Bentes, R. Coelho, J. M. S. Gonçalves, P. G. Lino, and K. Erzini. 2008. Long-term changes in fish communities of the Ria Formosa coastal lagoon (southern Portugal) based on two studies made 20 years apart. *Estuar. Coastal Shelf Sci.* 76: 57–68.

Rodó, X., E. Baert and F. A. Comin. 1997. Variations in seasonal rainfall in southern Europe during the present century: Relationships with the North Atlantic Oscillation and the El Niño-Southern Oscillation. *Climate Dynamics* 13: 275–284.

Salles, P., 2001. Hydrodynamic controls on multiple tidal inlet persistence. Ph.D. Thesis, Massachusetts Institute of Technology/Woods Hole Oceanographic Institution, Cambridge, Massachusetts.

Salles, P., G. Voulgaris, and D. G. Aubrey. 2005. Contribution of nonlinear mechanisms in the persistence of multiple tidal inlet systems. *Estuar. Coastal Shelf Sci.* 65: 475–491.

Santos, R., J. Silva, A. Alexandre, N. Navarro, C. Barrón, and C. Duarte. 2004. Ecosystem metabolism and carbon fluxes of a tidally-dominated coastal lagoon. *Estuaries* 27: 977–985.

Serpa, D., M. Falcão, P. Duarte, L. Cancela da Fonseca, and C. Vale. 2007. Evaluation of ammonium and phosphate release from intertidal and subtidal sediments of a shallow coastal lagoon (Ria Formosa, Portugal): A modelling approach. *Biogeochemistry* 82: 291–304.

Serpa, D., D. Jesus, M. Falcão, and L. Cancela da Fonseca. 2005. *Ria Formosa ecosystem: Socio-economic approach*. Relat. Cient. Téc. IPIMAR, Série digital, no. 28.

Silva, E.de S. and M. E. Assis. 1970. *Pigments and predominant plankton in sea water samples of "Ria de Faro-Olhão" from May 1967 to May 1968*. Notas e Estudos Inst. Biol. Marit. no. 38, Instituto de Biologia Maritima, Lisboa, Protugal.

Smetacek, V. and J. E. Cloern. 2008. On phytoplankton trends. *Science* 319:1346–1348.

Sprung, M. 1994. High larval abundances in the Ria Formosa (Southern Portugal): Methodological or local effect? *J. Plankton Res.* 16: 151–160.

Sprung, M., H. Asmus, and R. Asmus. 2001. Energy flow in benthic assemblages of tidal basins: Ria Formosa (Portugal) and Sylt-Romo Bay (North Sea) compared. Pp. 237–254, In: K. Reise (ed.), *Ecological comparisons of sedimentary shores*. Ecological Studies, Volume 151. Berlin: Springer-Verlag.

Stigter, T. Y., A. M. M. C. Dill, L. Ribeiro, and E. Reis. 2006. Impact of the shift from groundwater to surface water irrigation on aquifer dynamics and hydrochemistry in a semi-arid region in the south of Portugal. *Agricult. Water Manage.* 85: 121–132.

Strayer, D. L., M. Pace, N. F. Caraco, J. J. Cole, and S. E. G. Findlay. 2008. Hydrology and grazing jointly control a large-river food web. *Ecology* 89: 12–18.

Tang, E. P. Y. 1995. The allometry of algal growth rates. *J. Plankton Res.* 17: 1325–1335.

Tett, P., L. Gilpin, H. Svendsen, C. P. Erlandsson, U. Larsson, S. Kratzer, E. Fouilland, C. Janzen, J.-Y. Lee, C. Grenz, A. Newton, J. G. Ferreira, T. Fernandes, and S. Scory. 2003. Eutrophication and some European waters of restricted exchange. *Continent. Shelf Res.* 23: 1635–1671.

Thiele-Gliesche, D. 1992. Zur Ökologie pelagischer Ciliaten in der subtropischen Lagune "Ria Formosa" (Südkuste Portugal). Ph.D. Thesis, Christian-Albrechts Universität, Kiel, Germany.

Tweedley, J. R., E. L. Jackson, and M. J. Attrill. 2008. *Zostera marina* seagrass beds enhance the attachment of the invasive alga *Sargassum muticum* in soft sediments. *Mar. Ecol. Prog. Ser.* 354: 305–309.

Underwood, G. C. and J. Kromkamp. 1999. Primary production by phytoplankton and microphytobenthos in estuaries. *Adv. Ecol. Res.* 29: 93–153.

van Beusekom, J. E. E., M. Loebl, and P. Martens. 2009. Distant riverine nutrient supply and local temperature drive the long-term phytoplankton development in a temperate coastal basin. *J. Sea Res.* 61: 26–33.

Viaroli, P., G. Giordani, J. Martinez, Y. Collos, and J. M. Zaldivar. 2005. Ecosystem alteration and pollution in southern European coastal lagoon. *Chem. Ecol.* 21: 413–414.

Vila-Concejo, A., Ó. Ferreira, A. Matias, and J. A. Dias. 2003. The first two years of an inlet: Sedimentary dynamics. *Continent. Shelf Res.* 23: 1425–1445.

Vila-Concejo, A., A. Matias, Ó. Ferreira, C. Duarte, and J. A. Dias. 2002. Recent evolution of the natural inlets of a barrier island system in Southern Portugal. Proceedings of International Coastal Symposium, Templepatrick, Northern Ireland. *J. Coastal Res.* SI (36): 741–752.

Ziegler, S. and R. Benner. 1999. Dissolved organic carbon cycling in a subtropical seagrass-dominated lagoon. *Mar. Ecol. Prog. Ser.* 180: 149–160.

15 A Comparison of Eutrophication Processes in Three Chinese Subtropical Semi-Enclosed Embayments with Different Buffering Capacities

Kedong Yin, Jie Xu, and Paul J. Harrison*

CONTENTS

Abstract 368
15.1 Introduction 368
15.2 Geographic Locations and Geomorphologic Setting 369
15.3 Oceanographic Setting and Drivers of Change 369
15.4 Bays: Physical Processes and Anthropogenic Influences 371
  15.4.1 Port Shelter 371
  15.4.2 Tolo Harbour 372
  15.4.3 Deep Bay 377
15.5 Eutrophication Processes 377
  15.5.1 Chlorophyll *a* 380
  15.5.2 Dissolved Oxygen 383
  15.5.3 Nutrient Stoichiometry and Nutrient Limitation 383
15.6 Management Strategy and Perspectives 386
  15.6.1 Tolo Harbour 386
  15.6.2 Port Shelter 387
  15.6.3 Deep Bay 387
15.7 Climate Change Effects 388
15.8 Summary and Conclusions 389
Acknowledgments 392
References 394

* Corresponding author. Email: k.yin@griffith.edu.au

ABSTRACT

Phytoplankton biomass and bottom dissolved oxygen are controlled by processes such as residence time/exchange between a bay and the open ocean, vertical mixing/stratification, the most limiting nutrient/nutrient ratios, light penetration and grazing as well as climate variability. We examined the eutrophication processes and the drivers that regulate these processes by comparing three semi-enclosed bays (Port Shelter, Tolo Harbour, and Deep Bay) with different buffering capacities using long-term water quality data, including chlorophyll and dissolved oxygen along with salinity, temperature, and nutrients. The outer part of Port Shelter is wide open to southern coastal waters and has a residence time of about 20 days, but it is relatively pristine, with little input of sewage effluent. Nutrient concentrations are as low as chlorophyll *a* values. Tolo Harbour in northeast waters is a land-locked bay/inlet with high nutrients, a long 28-day residence time, and year-round stratification; hence, it is very vulnerable to eutrophication. Phytoplankton biomass has not reached the potential maximum that could be produced from the high ambient nutrient concentrations, suggesting that top-down control by grazing as well as turbidity is affecting the system. Deep Bay, in western waters between mainland China and Hong Kong, receives sewage input from the city of Shenzhen with up to 500 μM DIN at the head of the bay. However, there is no hypoxia and phytoplankton biomass is not as high as one would expect from such high nutrient levels. Turbidity, light limitation, and top-down control by benthic grazing likely play an important role in reducing the magnitude and frequency of phytoplankton blooms. Deep Bay receives the highest nutrient loads of the three bays, but eutrophication impacts are greatly reduced due to the aforementioned controls.

15.1 INTRODUCTION

Coastal eutrophication is a major environmental concern as the human population continues to grow rapidly in coastal areas (Ryther and Dunstan 1971; Rosenberg 1985; Nixon, 1995; Smith et al. 2006). One of the main symptoms of eutrophication is a long-term increase in chlorophyll *a* (chl *a*) in coastal regions due to accelerated phytoplankton growth in response to elevated concentrations of nutrients (Cloern 2001). Hypoxia (dissolved oxygen <2 mg L^{-1}) is another major symptom of eutrophication due to the increased production or input of organic matter (i.e., phytoplankton, benthic microalgae and macroalgae, and seagrass biomass) in coastal waters. Hypoxia and anoxia (dissolved oxygen $= 0$ mg L^{-1}) also cause the formation of "dead zones" in the sea.

The frequency and extent of dead zones are on the rise in coastal waters worldwide (Diaz and Rosenberg 2008), and governments have increased their efforts in managing this problem. However, there is extensive variability in the responses of ecosystems to the input of nutrients. Some ecosystems are more vulnerable than others to nutrient enrichment because their ecosystem buffering capacities do not cushion them against larger nutrient inputs (Le Pape et al. 1996; Cloern 1999, 2001). San Francisco Bay is an example, where the ecosystem appears to be well buffered against changes in nutrients due in part to top-down controls (Cloern 2001). Climate change may exacerbate the effects of nutrient enrichment (Justić et al. 1997). Recently, climate variability was reported to change the ecosystem community regardless of the efforts to reduce nutrients (Cloern et al. 2007). In order to manage the sustainability of coastal waters, therefore, an understanding of ecosystem functioning is important in controlling nutrient pollution and in predicting how coastal ecosystems may respond to the combined effects of eutrophication and climate variability.

Most studies on eutrophication processes have been conducted in temperate waters (Lillobø et al. 2005; Fisher et al. 2006; Howarth and Marino 2006), with only a few studies in subtropical and tropical waters (NRC 2000). Subtropical and tropical waters in southeast Asia are quite different ecosystems in terms of the high N:P ratios in riverine inputs into coastal waters and hence, phosphorus appears to be the most limiting nutrient in contrast to nitrogen in temperate waters. In addition, the Southeast Asian monsoon season is a pronounced feature that affects physical processes in

Hong Kong coastal waters and the South China Sea, and it produces a seasonal shift in upwelling and downwelling processes, which plays an important additional role in regulating eutrophication processes (Yin 2002).

In this study, we use an excellent 20-year dataset of water quality parameters to provide a comparative analysis of eutrophication processes in three contrasting environments in Hong Kong waters. We demonstrate how eutrophication impacts, such as the magnitude of phytoplankton blooms and the formation of hypoxia, differ in these three subtropical semi-enclosed bays, with different geomorphological settings and different physical processes, nutrient inputs, and nutrient ratios.

15.2 GEOGRAPHIC LOCATIONS AND GEOMORPHOLOGIC SETTING

Hong Kong waters are located on the south coast of China and connect to the northern part of the South China Sea (Figure 15.1). Situated in the subtropical region south of 23°N, Hong Kong waters cover an area of 1800 km^2 and harbor 230 islands. The west side of Hong Kong waters forms part of the Pearl River Estuary and receives freshwater outflow from the second largest river in China, especially during the summer wet season. The eastern waters mainly consist of Mirs Bay, which is semi-enclosed. The southern coastal waters of Hong Kong are open to the northern South China Sea. Most of Hong Kong waters are shallow with depths <20 m. Deep Bay, Port Shelter, and Tolo Harbour, are three relatively large semi-enclosed bays that are the focus of our eutrophication study (Figure 15.1).

Port Shelter is a semi-enclosed bay in the eastern part of Hong Kong (Figure 15.1) and is protected from the worst effects of adverse weather, except when the wind blows from the southeast. The whole area has relatively good water quality, and it has become one of Hong Kong's most important locations for water sports and recreational activities. Tolo Harbour and Tolo Channel (referred to as Tolo in this study) form a land-locked, semi-enclosed inlet in the northeastern part of Hong Kong with only a narrow eastern exit into semi-enclosed Mirs Bay. Tolo is about 16 km long, trending in an east-west direction with a total surface area of ~50 km^2. The mean water depth varies from about 21 m at the harbor entrance to 3 m inshore. Tolo is infamous for its history of extensive eutrophication which caused red tides and hypoxia before 1998. In the past decade, a dramatic improvement in water quality occurred in this system after the implementation of the Tolo Harbour Action Plan in 1998 (Broom et al. 2003). Finally, Deep Bay is a shallow (<5 m) semi-enclosed bay with a large area of intertidal mudflats, surrounded by a large mega-city, Shenzhen, to the north and the New Territories of Hong Kong to the south. The small Shenzhen River flows into the head of the bay and the mouth of the bay exits into the Pearl River Estuary. Deep Bay suffers from extensive anthropogenic inputs due to sewage discharge from Shenzhen, unsewered villages, and livestock farms (EPD 2004).

The area occupied by the coastal waters of Hong Kong is small, but there are large oceanographic gradients from the eutrophic freshwater input from the large Pearl River to the intrusion of oligotrophic oceanic waters from the South China Sea. Oceanographic processes of coastal Hong Kong waters are dynamic and exert different forcing on the three bays due to their different geographical locations and geomorphology.

15.3 OCEANOGRAPHIC SETTING AND DRIVERS OF CHANGE

The coastal waters of Hong Kong are profoundly influenced by three water regimes: the Pearl River discharge in the west, the oceanic waters from the South China Sea in the south, and coastal waters from the South China Coastal Current in the east (Chau and Abesser 1958; Chau and Wong 1960; Williamson 1970; Watts 1971a, 1971b, 1973). The oceanographic processes have strong seasonality with a dry season (October to March) and a wet season (April to September). In the winter dry season, Hong Kong waters are dominated by oceanic waters, while in the summer wet season, they are

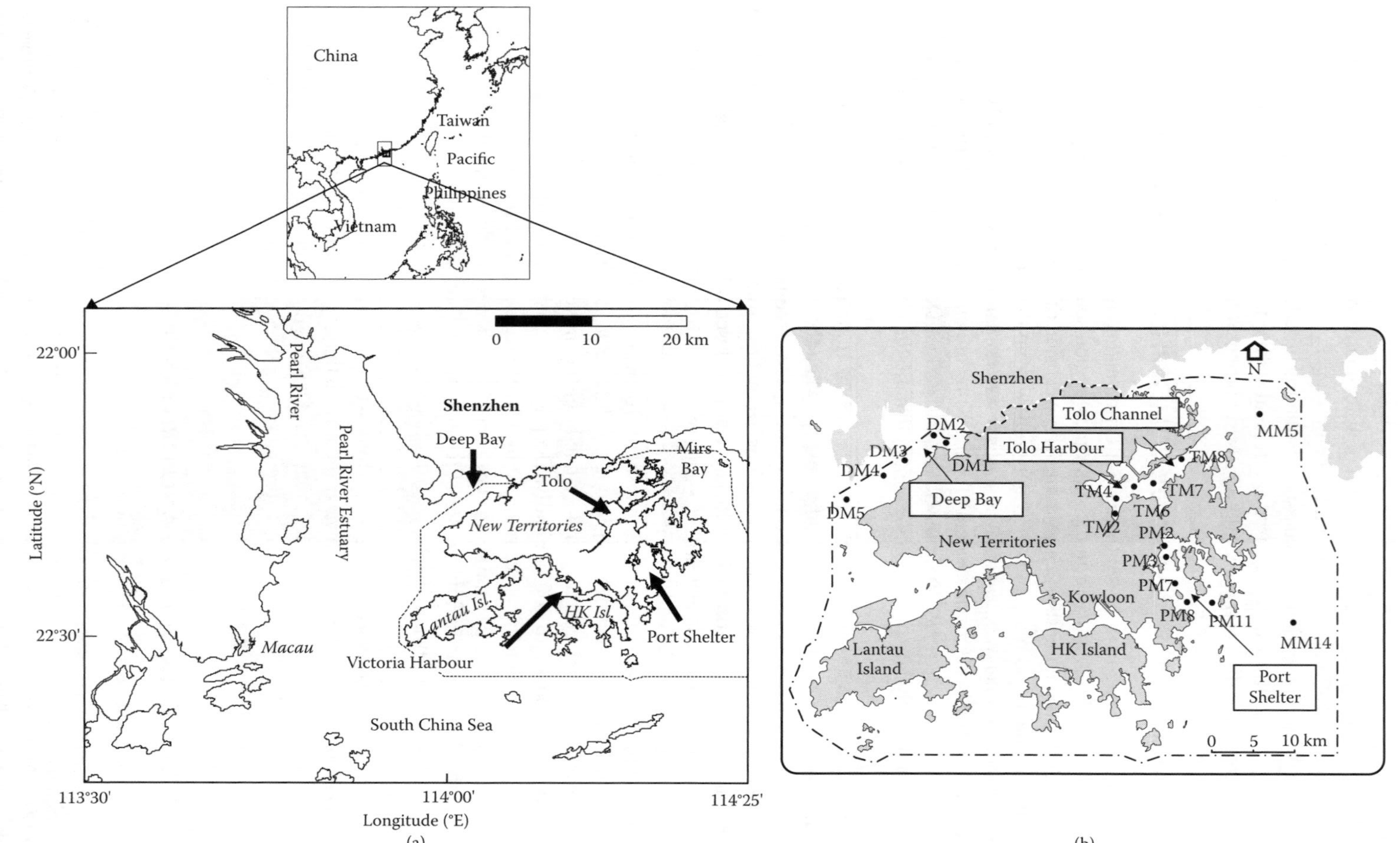

FIGURE 15.1 (a) Map of Hong Kong waters, showing the territorial water boundary (the dashed line) in relation to China, Port Shelter (PS), Tolo Harbour, and Tolo Channel (Tolo), and Deep Bay (DB). (b) Location of sampling stations monitored by EPD. Names of stations on the map are the same as those used by EPD: TM = Tolo, PM = Port Shelter, and DM = DB monitoring stations.

heavily influenced by the Pearl River discharge. Tidal cycles are mixed semidiurnal with an average tidal range of about 1 to 2 m.

The Pearl River Estuary has an area of over 2000 km^2. The Pearl River discharges into the South China Sea through eight distributaries. The four eastern-most distributaries, namely the Humen, Jiaomen, Hongqili, and Hengmen, discharge into the Pearl River Estuary. The average annual discharge through the estuary is about 5300 $m^3 s^{-1}$, or 53% of the total discharge of the Pearl River (Yin et al. 2000, calculated from Zhao 1990). Close to 80% of the total flow occurs between April and September, and the ratio of annual maximum to minimum discharge varies between 3 and 6 times interannually.

These oceanographic drivers (winds, river discharge, and tidal cycles) produce the dynamics of the physical oceanographic processes in Hong Kong waters. In the dry season, the Southeast Asian monsoon season winds prevail from the northeast, and the coastal circulation is dominated by downwelling due to the Coriolis effect. As a result, Hong Kong waters are mostly from offshore and residence times are long in the semi-enclosed bays (Yin 2002). During winter, the Pearl River discharge is minimal, and the river estuarine plume is small, largely confined to the river mouth and moving along the west shore away from Hong Kong. As a result, there is little input from the Pearl River Estuary into Hong Kong waters. In summer, Hong Kong experiences a south-southwest monsoon, and the Pearl River discharge reaches a maximum (22,190 $m^3 s^{-1}$) in July (Yin et al. 2000).

The Pearl River Estuary is dominated by typical two-layer estuarine circulation, with the seaward outflow of the estuarine plume at the surface and a landward inflow of the salt wedge in the bottom layer (Yin et al. 2001; Harrison et al. 2008). The estuarine plume flows out of the estuary, forming an estuarine coastal plume which spreads to the southern waters of Hong Kong and extends to the outer part of Mirs Bay on the eastern side (Yin 2002). Due to the southwest monsoonal wind, the coastal waters of Hong Kong are subjected to upwelling in the summer. The entrainment produced by the seaward flow of the surface estuarine coastal plume is superimposed on the upwelling process. This produces a two-layer surface outflow and bottom inflow circulation in the coastal shelf of Hong Kong where the bottom inflowing waters have oceanic properties (i.e., high salinity, cold temperatures, and relatively lower nutrients and oxygen than the surface estuarine plume). Tidal cycles produce relatively strong advection and frequent mixing due to tidal shoaling over shallow areas of Hong Kong waters (<15 m), which are extensive.

There are three nutrient sources for Hong Kong waters: (1) relatively oligotrophic oceanic waters from the South China Sea (SCS) and Southern China Coastal Current; (2) the eutrophic Pearl River discharge; and (3) the Hong Kong domestic sewage effluent discharge in the Victoria Harbour area. Nutrients in the Pearl River are up to two orders of magnitude higher than in the oligotrophic SCS (Yin 2002; Ning et al. 2004; He et al. 2009). In particular, dissolved inorganic nitrogen (DIN = $NO_3^- + NO_2^- + NH_4^+$) in the Pearl River is very high at about 100 μM, whereas PO_4^{3-} remains relatively low at about 1 μM (Yin 2002; Cai et al. 2004). Consequently, in the estuarine-influenced water, the DIN:P ratio is unusually high, resulting in potential P limitation in contrast to N limitation in oceanic waters (Yin et al. 2000, 2001; Xu et al. 2009). A population of about 7 million in Hong Kong produces 2.1 million tons of sewage effluent daily, of which about 70% is discharged into Victoria Harbour and its adjacent waters. The distribution of nutrients is closely coupled with the physical processes, and hence, they have strong seasonal and spatial gradients. Port Shelter, Tolo, and Deep Bay not only have different geomorphology and nutrient inputs, but also are affected differentially by oceanographic forcing and processes as discussed below.

15.4 BAYS: PHYSICAL PROCESSES AND ANTHROPOGENIC INFLUENCES

15.4.1 Port Shelter

Monitoring stations PM2, PM3, PM7, PM8, PM11, and MM14 along the transect from the inside bay to the open waters were used to illustrate spatial and temporal variability in Port Shelter. During most months of the year, the average surface salinity is >30 and salinity drops below 29.5 only during summer (Figure 15.2). The horizontal gradient of salinity at the surface is usually small except for summer when salinity decreases at stations PM2 and PM3, reflecting the influence of rainfall. This indicates that Port Shelter is not usually invaded by the Pearl River estuarine coastal plume at the surface from the south, although there appears to be some presence of the estuarine coastal plume at MM14 (outer Mirs Bay) as indicated by the decreased salinity (Figure 15.2). The average bottom salinity is between 32 and 33 in the dry season, but increases to 34 in the wet summer months. The average temperature in surface and bottom waters is 17°C in winter, but increases to 28°C–29°C in summer at the surface and only to 23°C–24°C at the bottom.

The Southwest Asian monsoon season in the summer causes upwelling with the surface outflow and high salinity, cold deep water inflow. The coupling of higher salinity and colder water at the bottom with the decrease in surface salinity induced by rainfall and land runoff indicates that Port Shelter is dominated by the typical estuarine-like, two-layer circulation with the surface water flowing out of the bay and bottom oceanic water flowing into the bay, especially in the inner bay. Outside the bay, the Pearl River outflow and monsoon-induced upwelling move waters seaward and, therefore, strengthens the two-layer circulation inside the bay. In winter, the entire water column is well mixed as shown by the small difference in salinity and temperature between the surface and bottom waters (Figure 15.2).

In Port Shelter, nutrient concentrations are relatively low due to the minor influence of anthropogenic nutrient inputs. Nitrate concentrations in the surface layer are <2 μM most of the year, even during the wet season. However, in July there is a patch of higher NO_3^- (up to 5 μM) at MM14 that is likely from the Pearl River estuarine coastal plume. Nitrate concentrations in summer are higher than in other months. However, these summer contour lines of higher NO_3^- form closed loops, suggesting that higher NO_3^- measurements have resulted from the regeneration of nutrients in the bottom waters or sediments (Figure 15.2) due to sinking algae from a summer bloom after a rainfall event, as observed in coastal waters outside of Port Shelter (Yin et al. 2000). The spatial and temporal distribution of SiO_4 is similar to that of NO_3^-, but the SiO_4 concentration is much higher than NO_3^- (Figure 15.2). Phosphate is <0.6 μM, with only small differences between seasons and between the surface and bottom waters (Figure 15.2).

15.4.2 Tolo Harbour

There are clear spatial and temporal variations in salinity in Tolo (Figure 15.3). Average surface salinity is low at station TM2 and increases all year toward station MM5 of Mirs Bay. There appears to be a lower salinity tongue at the surface extending from station TM2 all the way out to station MM5, with the lowest average surface salinity being about 27.5 between stations TM2 and TM4 in August. Thus, the largest salinity gradient occurs in August, demonstrating that Tolo Harbour is more greatly affected by freshwater in summer when rainfall is maximal. The average bottom salinity remains above 30 and increases from the inner to the outer harbor during the dry season. However, during the summer, there is a tongue of high salinity water >33 at the bottom between stations TM6 and MM5. Although stratification in Tolo appears to exist during all months, it is weaker in the winter because the topography shelters the bay from winds, leading to slow flushing of freshwater inputs out of the bay. The apparent vertical stratification starts in April when surface salinity decreases to 29.5 in the inner harbor (Figure 15.3), and the stratification becomes stronger in the summer when bottom salinity increases in the entire Tolo. Average temperature is uniformly distributed along Tolo at both the surface and bottom in all months, except in the summer when a cold water tongue occurs along the station TM6–MM5 transect (Figure 15.3). Coupled with the colder water tongue at the bottom is the higher salinity tongue, which occurs concurrently with the surface lower salinity tongue. This indicates that Tolo

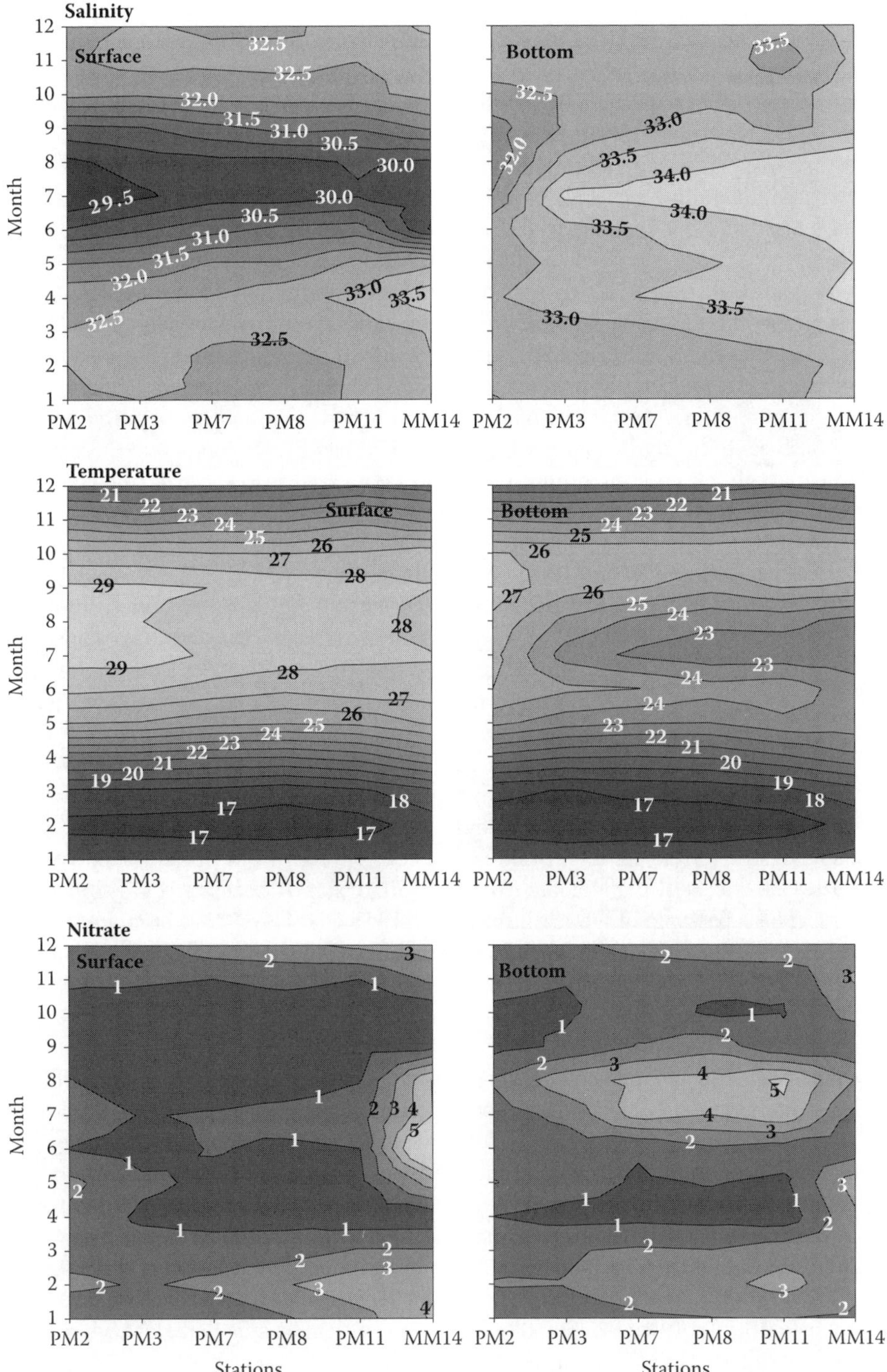

FIGURE 15.2 Port Shelter: Contour plots of average monthly (1986 to 2006) values for salinity, temperature, NO_3^-, NH_4^+, SiO_4^{2-}, and PO_4^{3-} along the transect from the inner bay at station PM11 to the open waters at station MM14. The methods of the measurement of these parameters are described in Yin (2002). Dissolved inorganic nitrogen = $NO_3^- + NO_2^- + NH_4^+$.

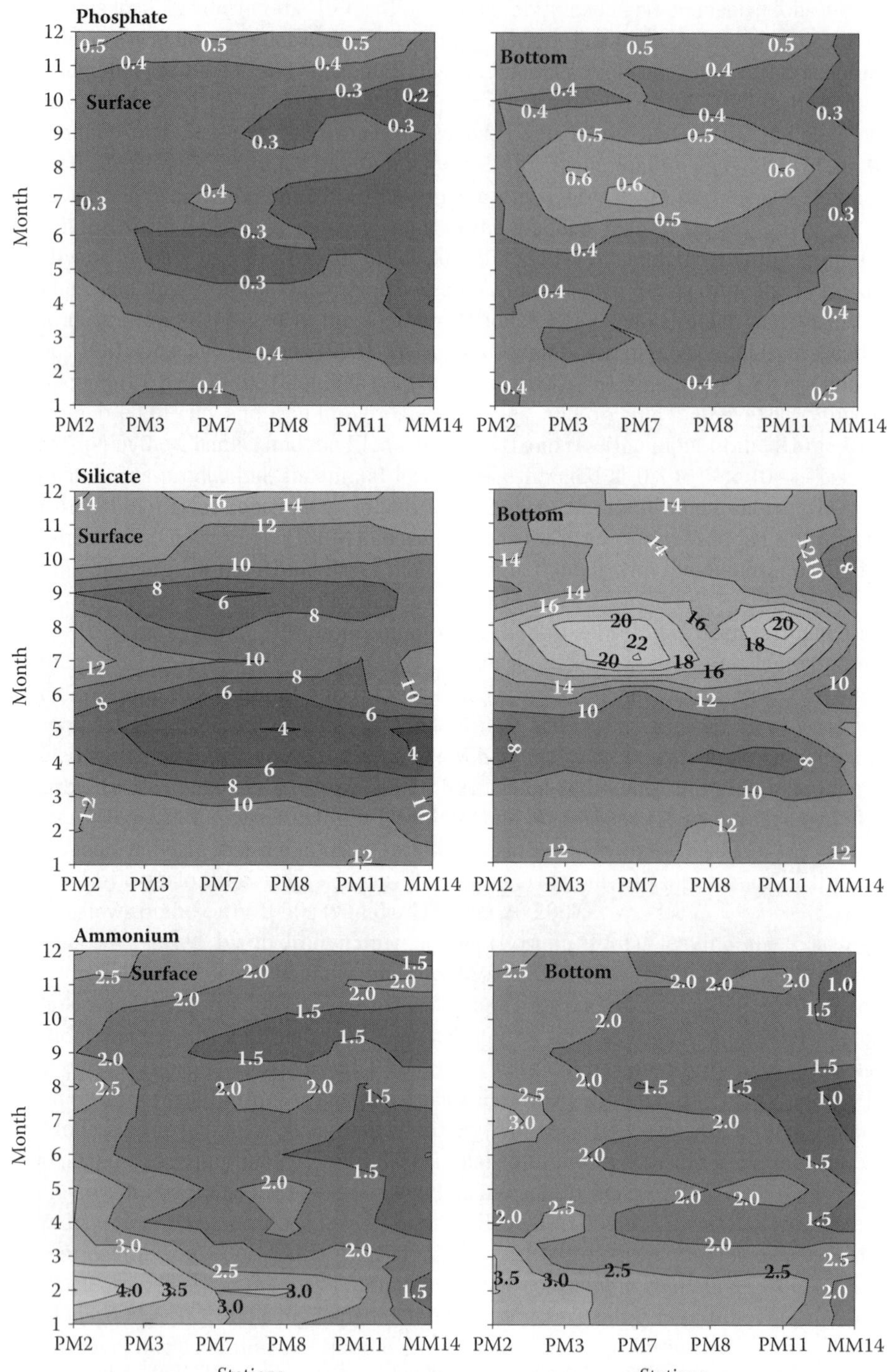

FIGURE 15.2 (continued).

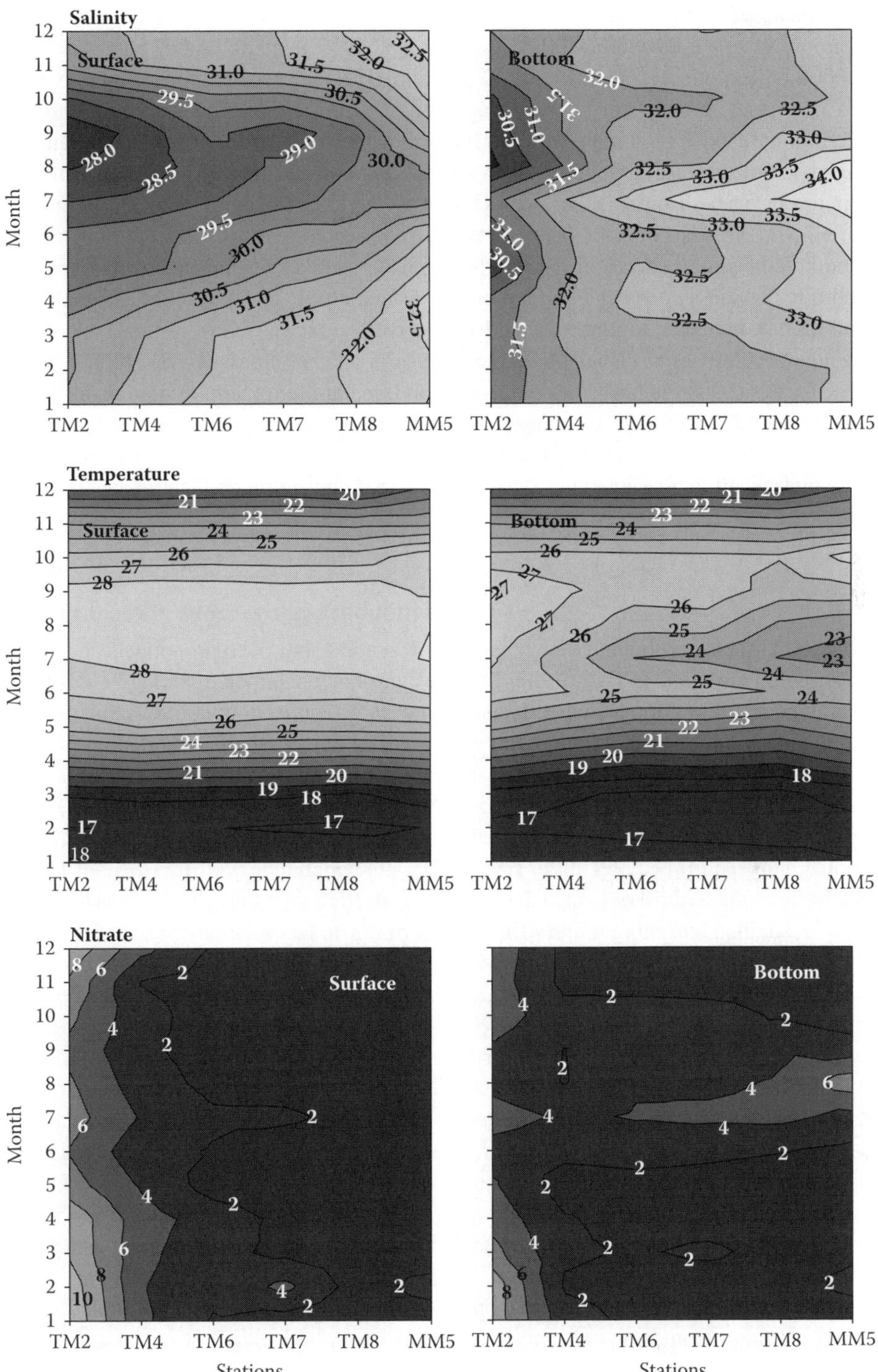

FIGURE 15.3 Tolo Harbour and Channel: Contour plots of average monthly (1986 to 2006) values for salinity, temperature, NO_3^-, NH_4^+, SiO_4^{2-}, and PO_4^{3-} along the transect from the inner bay at station TM1 to the open waters at MM5.

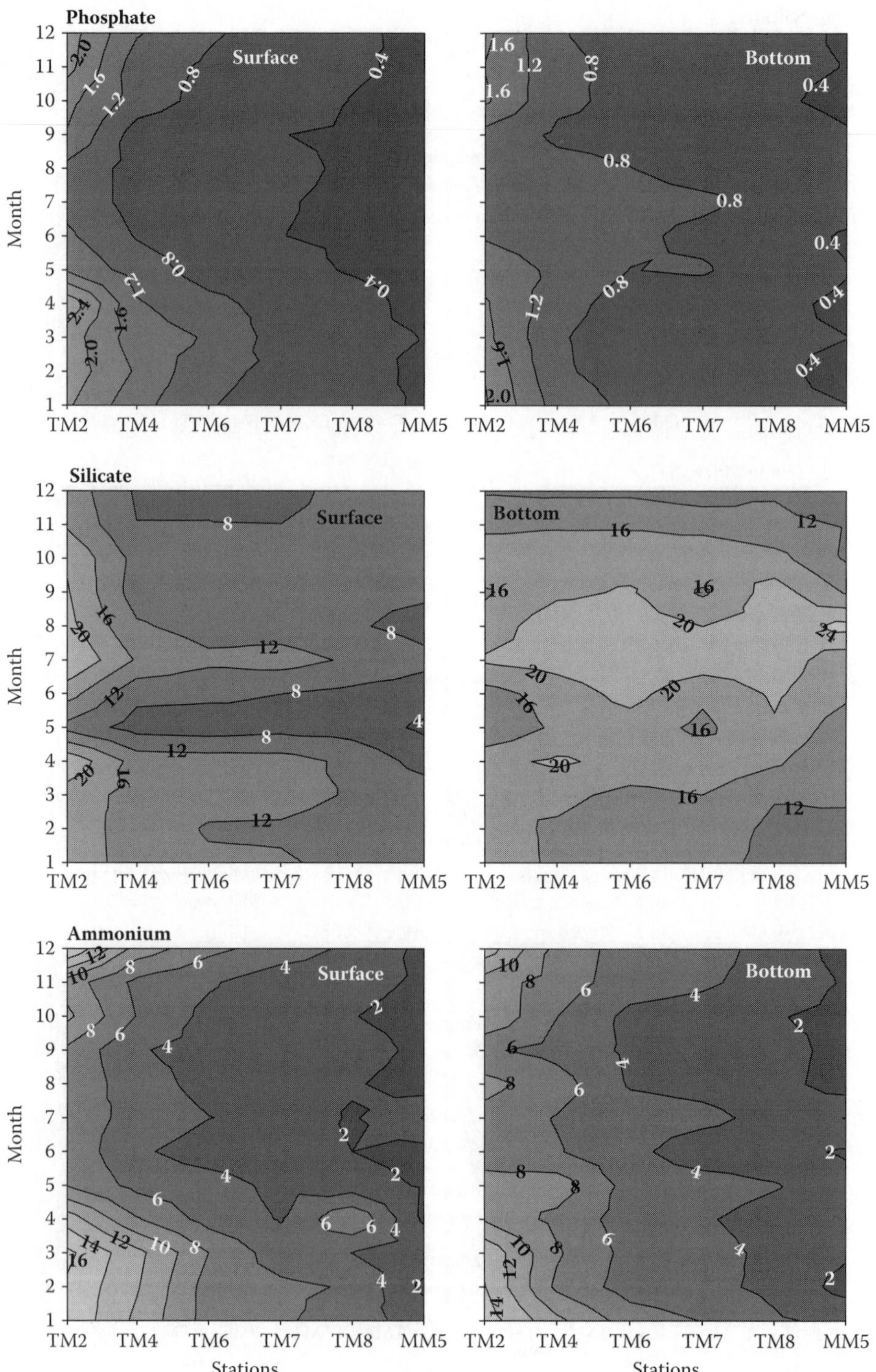

FIGURE 15.3 (continued).

is dominated by a two-layer opposite flow (estuarine-like circulation), which is strengthened during the rainy season.

Surface NO_3^- is >5 μM at station TM2 in the inner bay and decreases toward the outer bay, down to <2 μM at station TM8. Bottom NO_3^- is lower than surface NO_3^-. PO_4^{3-} at both the surface and bottom is >1 μM at station TM2 in the inner bay and decreases toward the outer bay; however, it remains above 0.4 μM. The spatial and temporal distribution of PO_4^{3-} appears to be opposite to salinity; the lower salinity tongue projects toward the outer bay, while the lower PO_4^{3-} tongue projects toward the inner bay at the surface. The higher salinity tongue projects from the outer bay toward the inner bay, and the 0.8 μM PO_4^{3-} tongue projects toward the outer bay. The opposite pattern at the surface suggests that freshwater is lower in PO_4^{3-}, whereas the pattern at the bottom suggests regeneration of organic matter in the bottom water or the sediment. NH_4^+ is >2 μM in the entire Tolo in all months. SiO_4^{2-} is >8 μM at the surface, but lower than that at the bottom (Figure 15.3), suggesting the regeneration of SiO_4^{2-} from the surface diatoms sinking into oceanic waters.

15.4.3 Deep Bay

Deep Bay receives freshwater from the Shenzhen River and opens into the Pearl River Estuary. The average salinity is rarely >30, being between 25 and 30 during the dry season and <20 during the wet season (Figure 15.4). Contour lines of salinity are nearly horizontal most of the year except for July when salinity drops to <10, indicating a small horizontal salinity gradient. There are only small differences in the average temperature between the surface and bottom, indicating that the bay water column is almost vertically mixed, which is typical for shallow waters. As a result, bottom contours of salinity are parallel to the surface contours; the tongue-shaped contour line of 25 intrudes from station DM5 toward station DM1 during the dry season, while during the wet season this contour line of <25 extrudes from station DM1 toward station DM5. This indicates that Deep Bay is dominated by freshwater flow from the Pearl River Estuary during the dry season. However, during the wet season, it is dominated by the Shenzhen River. These hydrodynamics determine the temporal and spatial distribution of nutrients.

Average NO_3^- at the surface and bottom is relatively low at 20 μM in the winter, but reaches 60 μM in the summer. However, NH_4^+ is very high, being extremely high near station DM1, reaching about 300 μM in winter. Ammonium decreases during the wet season, probably due to the dilution of more freshwater outflow from the river, and it quickly drops to 30 μM downstream between stations DM3 and DM4. Phosphate is also very high, with a maximum of 18 μM in winter at station DM1, declining to 2 μM at station DM4. Silicate is low during the winter and high during the summer; it reaches a maximum of 120 μM at station DM1 in June. The silicate concentrations are similar to those of the Pearl River (Cai et al. 2004).

15.5 EUTROPHICATION PROCESSES

Spatial and temporal variability of salinity, temperature, and nutrients in Port Shelter, Tolo Harbour, and Deep Bay are coupled to different oceanographic drivers. An array of environmental impacts may develop in response to increasing nutrient levels such as the magnitude of algal blooms, frequency of red tides, and the development of low dissolved oxygen concentrations.

High phytoplankton chlorophyll *a* concentrations are a good indicator of eutrophication (Pinckney et al. 1999, Paerl et al. 2006; Ho 2007; Wong et al. 2009). Phytoplankton biomass represents an accumulation of phytoplankton growth over time, and therefore, it is controlled by phytoplankton growth rate and grazing, and physical processes of vertical mixing and horizontal dilution or exchange between the bays and coastal ocean (Figure 15.5). The maximum biomass is determined by the concentration of the most limiting nutrient, typically either nitrogen or phosphorus. Light

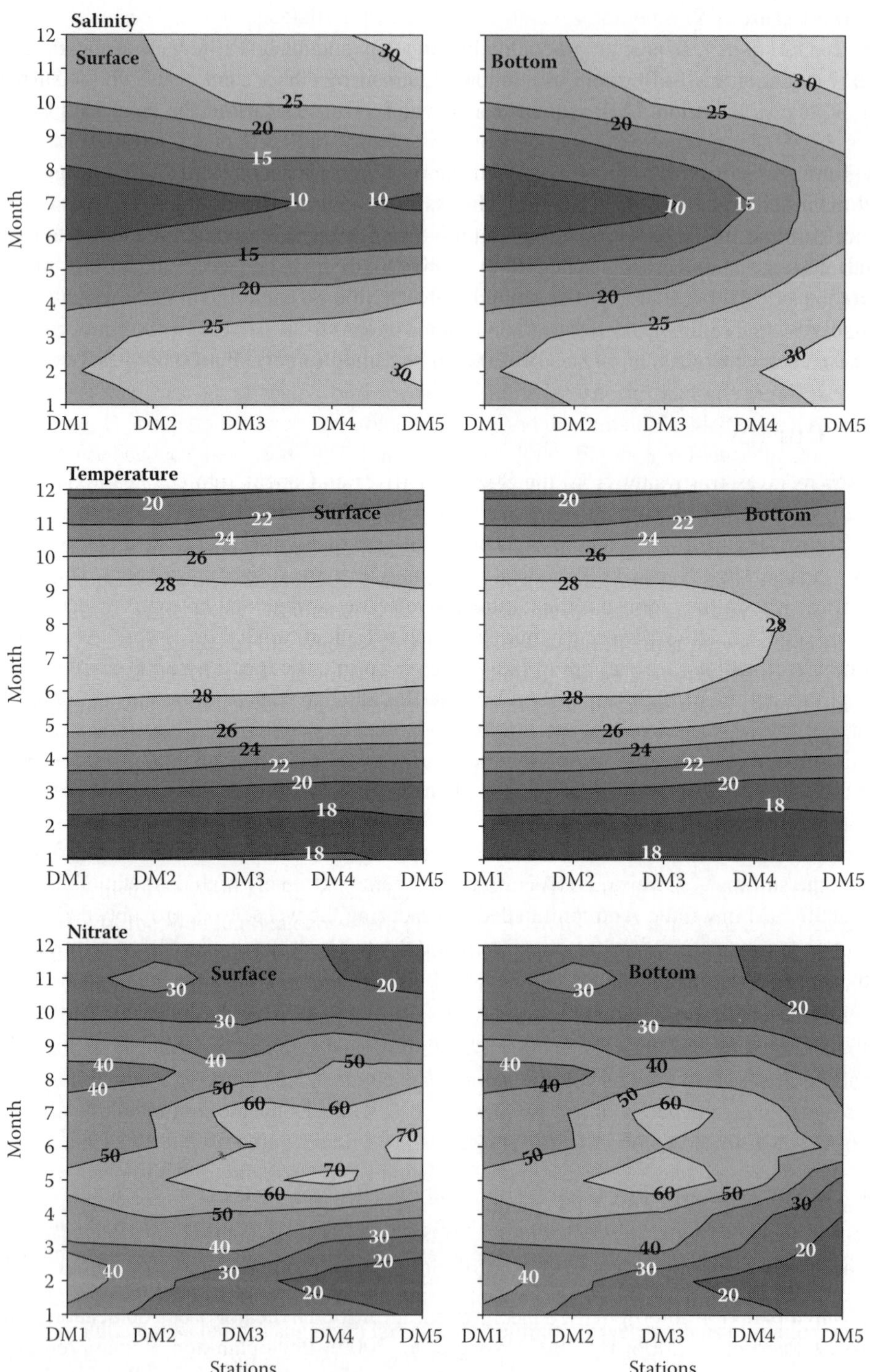

FIGURE 15.4 Deep Bay: Contour plots of average monthly (1986 to 2006) values for salinity, temperature, NO_3^-, NH_4^+, SiO_4^{2-}, and PO_4^{3-} along the transect from the inner bay at station DM1 to the open waters of Pearl River Estuary at station DM5.

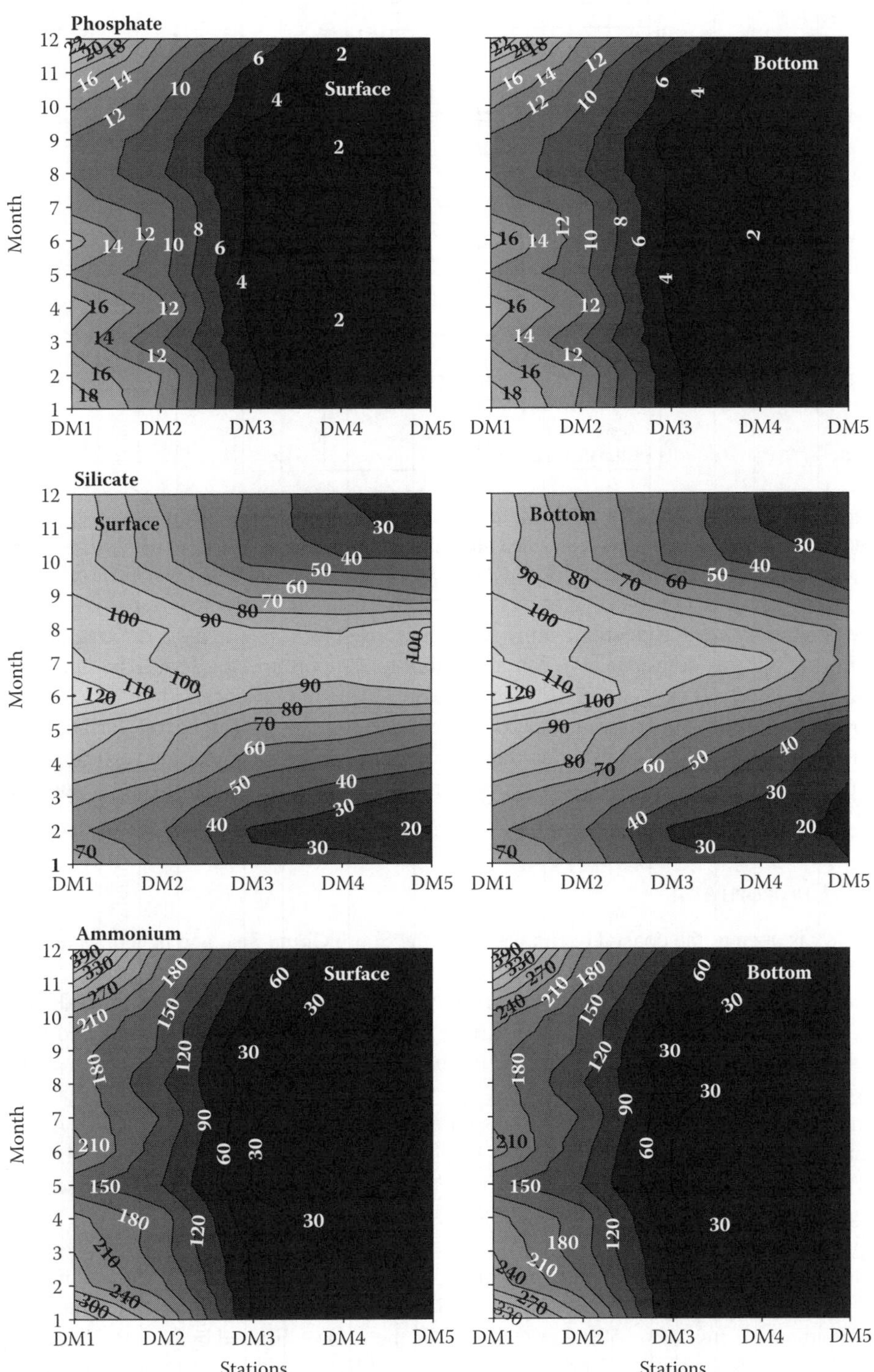

FIGURE 15.4 (continued).

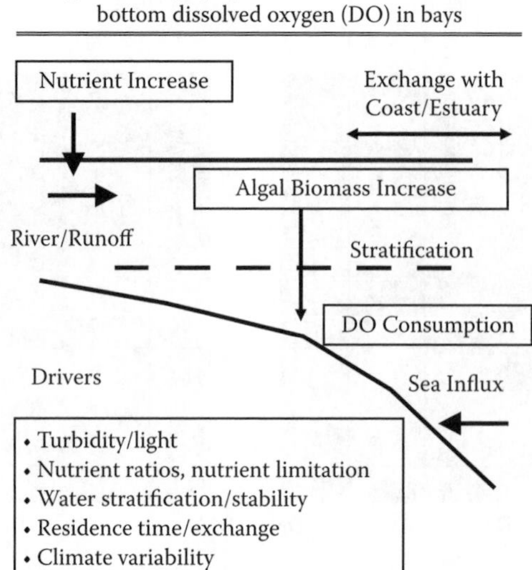

FIGURE 15.5 A conceptual model used to illustrate the regulation of phytoplankton biomass/blooms and dissolved oxygen (DO) in bays. Drivers include turbidity/light penetration, nutrient ratios/nutrient limitation, stratification/vertical mixing, residence time/exchange of a bay with open waters and climate effects.

regulates phytoplankton production, while nutrient ratios determine the most limiting nutrient and may influence species dominance. When algae sink to the bottom layer, bacterial decomposition will consume dissolved oxygen (DO). When excessive algae are produced in the surface layer due to an increase in the input of nutrients, DO may decrease to <2 mg/L (hypoxia). Variation in DO in the bottom water of a bay is subjected to the same driving processes, including vertical mixing and horizontal exchange with open ocean, light penetration, as well as organic matter production in the surface layer (Figure 15.5).

15.5.1 Chlorophyll *a*

Port Shelter is open to the coastal ocean waters of the South China Sea, and hence, water exchange is not restricted. However, Port Shelter is deeper than the other two bays, and water circulation is weak in the inner area, with a maximal current velocity of 0.15 m^{-1} (Yung et al. 2001). Vertical mixing and downwelling is strong in the winter due to northeast monsoon winds and the bay is dominated by oceanic waters from outside the bay. In the summer, the bay is dominated by the two-layer opposing flow circulation, and stronger stratification due to rainfall and surface heating. Modeling results show that the flushing times for eliminating 50% of a tracer are 9.1 days (>20 days for eliminating 90% of the tracer), with longer flushing times in the dry season (9.4 days) than in the wet season (8.9 days). The difference is greater between tidal phases than between seasons. The flushing time is 7.6 days during spring tides, and 11.5 days during neap tides in the wet season (Yin, unpublished results).

Phytoplankton biomass is low (<2 μg chlorophyll *a* L^{-1}) during the dry season and probably limited by vertical mixing, which causes light limitation. However, during the summer, phytoplankton biomass increases by 50% in the inner bay and is often limited by N or P (Xu et al. 2009). Nutrients are relatively high at depth (Figure 15.2), and light transmission reaches down to 20 m (Yin unpublished data). This results in higher biomass accumulation in the bottom layer than at the surface (Yin 2003). Port Shelter receives the least amount of nutrients among the three bays, and therefore, chlorophyll *a* levels are lowest. Even when there are blooms, chlorophyll *a* usually does not exceed 10 μg L^{-1} (EPD, unpublished data). There appears to be phytoplankton blooms outside of the bay at station MM4, probably due to the influence of the Pearl River estuarine coastal plume in the summer. It is unclear why Port Shelter is a hot spot for red tides (Yin 2003). Yin et al. (2008) hypothesized that physical aggregation by wind, plus vertical migration of phytoplankton, result in the accumulation of biomass leading to red tides.

Tolo is a land-locked bay with a relatively slow flushing rate that relies largely on tidal cycles. The average water velocity in the inner harbor is only 0.04 m s^{-1} and increases progressively to 0.08 m s^{-1} in the channel (EPD 1987). The residence time of water in Tolo Harbour is on average 28 days, being 38 and 14.4 days in the dry and the wet seasons, respectively (Lee et al. 2006). Therefore, Tolo Harbour appears to be most vulnerable to eutrophication impacts since there is little flushing to dilute the high anthropogenic nutrient inputs.

Compared to Port Shelter, Tolo receives much higher inputs of nutrients. The daily input of total nitrogen into Tolo Harbour, for example, was estimated to be over 6000 kg in 1986 and ~4000 kg in 1994 (Yung et al. 1997). Since 1998, sewage has been fully diverted to Victoria Harbour (Broom et al. 2003). The high input of nutrients resulted in higher chlorophyll *a*, especially in the inner harbor where chlorophyll *a* values were >10 μg L^{-1} all year, and algal blooms occurred every month. Even the average chlorophyll *a* at the bottom is very high (10 μg L^{-1}) (Yin et al. 2008). However, unlike other open waters of Hong Kong (e.g., southern waters and Victoria Harbour), maximal chlorophyll *a* concentrations in Tolo Harbour never occur in summer during the period of maximal rainfall, but instead the maximal chlorophyll *a* occurs over the winter–spring period. This may be explained by two distinct physical processes caused by the northeast and southwest monsoon winds. In winter–spring, the prevailing northeast monsoon winds produce downwelling of the coastal water, resulting in a longer residence time of surface water and a longer time for phytoplankton to respond to the local input of nutrients (Yin 2003). In contrast, in the summer, the more rapid outflow of surface water due to high rainfall and runoff reduces the residence time (Yin 2003). In addition, summer winds are from the southwest, and this helps force surface waters out of the harbor into Mirs Bay. As nutrients are generally abundant, phytoplankton may not become nutrient limited, although nutrients periodically drop to the limiting concentrations (Yin et al. 2008). In general, light does not appear to be limiting since water depths are shallow. This suggests that some additional factors other than bottom-up controls, such as physical dilution, nutrients, and light, may limit phytoplankton from achieving its potential maximal biomass. Few studies have been conducted on top-down effects (grazing) in Tolo; more work must be conducted on this subject in future research.

Tolo has the highest number of red tide events in all waters of Hong Kong (Yin 2003; Wong et al. 2009). Red tides in Tolo are dominated by dinoflagellates, particularly *Noctiluca scintillans*, which occurs from February to May (Yin 2003; AFCD 2005). This heterotrophic dinoflagellate grazes on other phytoplankton, notably diatoms (Fonda Umani et al. 2004) and may play a role in controlling chlorophyll *a*. *Noctiluca scintillans* disappears in May after the temperature increases to >22°C (Liu and Wong 2006). Tolo Harbour contains more red tide species than other Hong Kong waters, a total of 55 species, as compared with just 6 species in Deep Bay (EPD 2005). Many of these red tide species are heterotrophic and, therefore, may graze on diatoms. This aspect of certain red tide organisms acting as grazers on other phytoplankton has been overlooked.

Deep Bay receives freshwater inflow from the Shenzhen River, a small river that receives sewage from Shenzhen, one of the most rapidly growing cities in China with a population of 14 million residents (Chen 2008). Inner Deep Bay is heavily polluted with nutrients, particularly NH_4^+ (up to 200 μM), which is much higher than in Tolo. Chlorophyll *a* is high at stations DM1 and DM2 in January, reaching 25 μg L^{-1}, indicating large winter phytoplankton blooms. Chlorophyll *a* remains relatively low (<5 μg L^{-1}) from February to May in the entire bay and then increases to 20 μg L^{-1} in July. Phytoplankton blooms with chlorophyll *a* >10 μg L^{-1} occur in June–July between stations DM1 and DM3, but chlorophyll *a* is lower in the outer bay (Figure 15.6).

A modeling study has shown that there is spatial variability in hydrodynamic conditions throughout Deep Bay, with a high flushing rate (0.2 d^{-1} or a residence time of ~5 days) in the outer bay, but a low flushing rate (0.04 d^{-1} or a residence time of ~25 days) in the inner bay (Lee and Qian 2003).

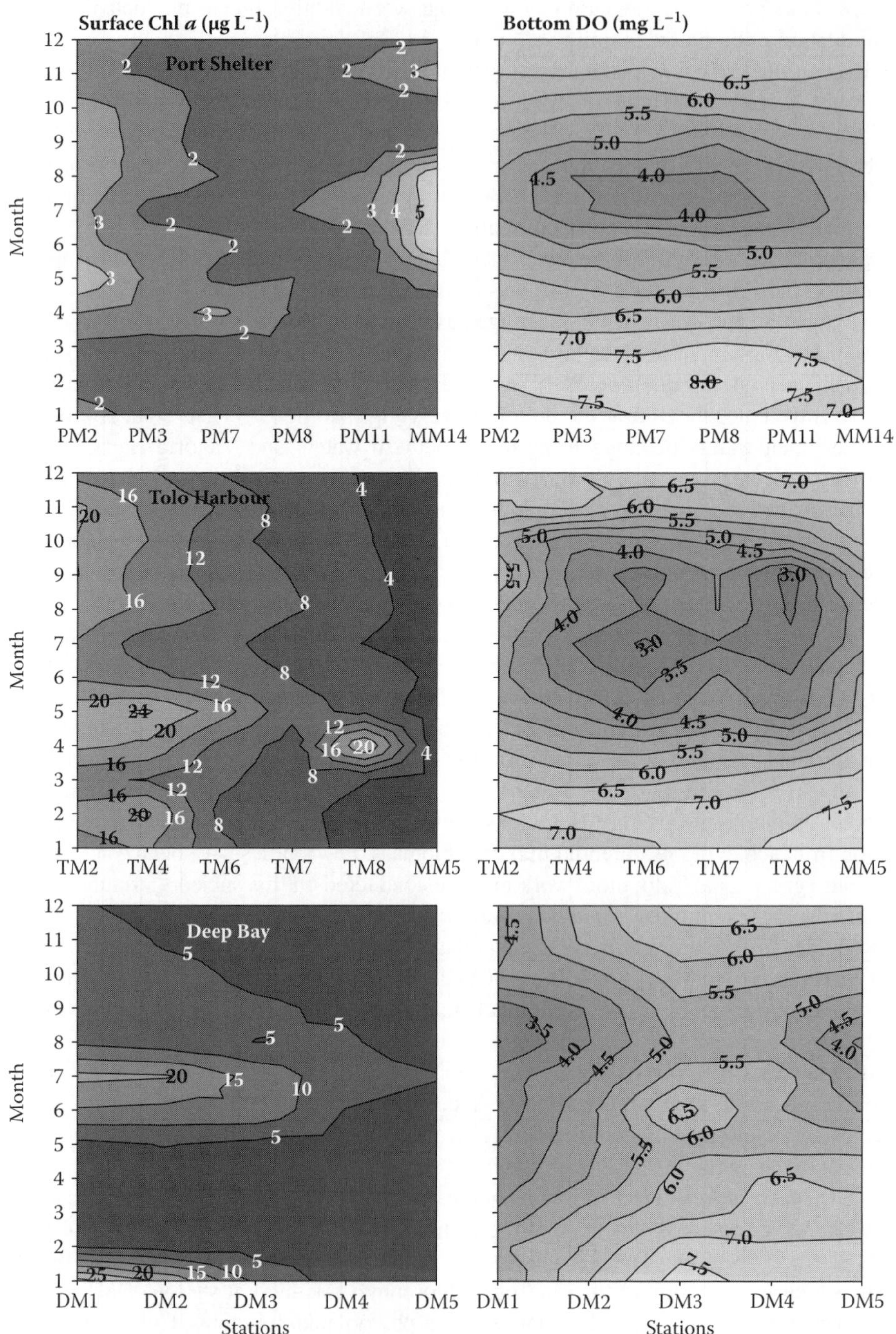

FIGURE 15.6 Contour plots of average monthly (1986 to 2006) values for surface chlorophyll *a* and bottom dissolved oxygen in Port Shelter, Tolo Harbour, and Deep Bay.

In spite of the low flushing rate and high concentrations of nutrients, phytoplankton biomass is low, except for high peaks of chlorophyll *a* in January and July, but the biomass peaks (20 to 25 μg L^{-1}) are comparable to Tolo. The phytoplankton biomass is much less than the potential biomass that could be produced from the high ambient nutrients; therefore, light is believed to be the limiting factor. There are large mud flats in Deep Bay, and water depths are shallower than in Tolo, which could increase sediment resuspension and light limitation, as the overall average (all data at five stations) Secchi disk depth is only 0.71 m (Yin et al. unpublished data). In addition, benthic grazing may be active and much of the water column volume may be filtered each day, as occurs in San Francisco Bay, where benthic feeding plays a key role in top-down control (Cloern and Dufford 2005). The weak stratification in the inner bay and frequent vertical mixing would help to make top-down control feasible, whereas in Tolo Harbour, the water column is stratified, which reduces the degree of the top-down control by benthic feeding. Furthermore, in Deep Bay there are oyster farms that filter algae and decrease chlorophyll *a* levels. Further studies are needed to investigate the hypothesis that a combination of light limitation and benthic feeding keeps phytoplankton biomass at relatively low levels. Surprisingly, few red tides occur in Deep Bay (Yin 2003; AFCD 2005), possibly due to high suspended loads, which tend to inhibit dinoflagellates (Anderson 1997).

15.5.2 Dissolved Oxygen

Dissolved oxygen in a water body is determined by physical processes of the air–sea exchange, vertical mixing in the water column, biological processes of community respiration, and sediment oxygen consumption. In general, longer residence times of a water body and stable stratification of the water column allow the formation of low oxygen waters in the bottom layer, but the level of low DO is determined by the supply of organic matter to the bottom layer. Higher phytoplankton productivity can result in lower DO if phytoplankton sink to the bottom.

Port Shelter experiences low DO in the bottom waters during the summer, averaging 4 mg L^{-1}. Before 2000, DO levels often dropped to hypoxic levels (2 mg L^{-1}) in August or September (EPD unpublished data). Chlorophyll *a* is low throughout the summer. The two-layer opposing flow may be partially responsible for the presence of low DO in the deep part of Port Shelter. The oceanic water invading Port Shelter comes from the continental shelf due to summer upwelling and DO in this bottom water at MM14 is already at 4.5 mg L^{-1} (Figure 15.6). During the movement of bottom water into the bay, phytoplankton at the surface sinks to the bottom layer and is decomposed by bacteria that further reduce DO levels. After rainfall events, near hypoxic waters can occur due to increased stratification and additional terrestrial organic inputs. In contrast, Tolo suffers from low oxygen in the deep bottom water between stations TM4 and TM7 in the wet season. The zone of low oxygen appears to occur downstream of the maximum chlorophyll *a* zone near station TM2. Station TM2 is shallow and light penetrates most of the water column, and therefore, DO consumption may be offset by phytoplankton photosynthesis in the water column.

In contrast, the average bottom DO in Deep Bay is low at station DM1 most of the year, and it increases downstream from stations DM2 to DM5. Hypoxia often occurs in Deep Bay, with only an occasional occurrence of anoxia, which would be expected from high nutrient concentrations. In this case, vertical mixing, as shown by similar salinities between the surface and bottom (Figure 15.4), would be responsible for the lack of anoxic waters in this shallow bay.

15.5.3 Nutrient Stoichiometry and Nutrient Limitation

Ambient ratios of nutrients have been used to indicate the most limiting nutrient based on the Redfield ratio of 16N:16Si:1P. Previous studies based on nutrient ratios have always agreed with the results from nutrient addition bioassays in Hong Kong waters (Table 15.1) (Xu et al. 2009). Ambient

TABLE 15.1
Comparison of the Potential Limiting Nutrient (N, P, Si or Co-Limitation) Derived from Two Methods (Ambient Inorganic Nutrient Ratios vs. Nutrient Enrichment Bioassay) at Four Stations in Hong Kong Waters during All Seasons of the Year

Stations	Seasons	Nutrient Ratios	Bioassays
NM2	Spring	P	P
	Summer	P	P
	Fall	P	P
	Winter	Si	N + Si
SM6	Spring	P	N + P
	Summer	P	P
	Fall	N	N
	Winter	N	N
VM5	Spring	Si	N
	Summer	P	P
	Fall	Si	Si
	Winter	N	N
PM7	Spring	N	N
	Summer	N	N + P, P
	Fall	N	N
	Winter	N	N

Source: Xu, J. et al. 2009. *Mar. Ecol. Prog. Ser.* With permission.

nutrient ratios have been robust enough to predict the most potentially limiting nutrient in Hong Kong waters, regardless of whether there is actual *in situ* nutrient limitation.

In general, nutrient stoichiometry in Hong Kong waters indicates low DIN:P ratios (~10:1) in marine dominated waters, and high DIN:P ratios (100:1) in Pearl River estuarine waters. As a result, the stoichiometry of nutrients and the most limiting nutrient vary temporally and spatially in waters at the outer edge of the three bays (Xu et al. 2009); consequently, this variation in nutrient stoichiometry can influence the bays during periods of nutrient exchange with outside waters.

In Port Shelter, the DIN:P ratio is below the Redfield ratio (= 16N:1P) most of the year (Figure 15.7). However, higher DIN:P ratios (>16:1) occur in summer during the rainy season as rain contains high DIN:P ratios (~50:1). Nutrient enrichment bioassays have shown that phytoplankton growth is potentially co-limited by N and P in Port Shelter (Xu et al. 2009). In the summer, the Pearl River discharge flows eastward due to the influence of the southwest monsoon wind and reaches the outer part of Port Shelter (Yin et al. 2001). As a result, P was the primary limiting nutrient (51% of all incidents) instead of N, in the freshwater-influenced outer part of the Port Shelter during summer (Figure 15.7). In contrast, N was most likely limiting phytoplankton growth in the inner bay due to the dominance of the coastal/oceanic water. P limitation occurred episodically in the inner bay during periods of high rainfall with a high N:P ratio (~50:1) in summer, as shown by nutrient enrichment bioassays (Table 15.1) (Xu et al. 2009).

Deep Bay is influenced by sewage from the Shenzhen River in the inner bay and by the Pearl River discharge with high NO_3^- and SiO_4^{2-} concentrations, respectively. Consequently, there is a shift in the most limiting nutrient from the inner bay to the outer bay in response to the dominance of different water masses, although no actual nutrient limitation occurs (Xu et al. 2010). In the inner bay, the nutrient ratios indicate that Si is potentially the most limiting nutrient (89% of all incidents)

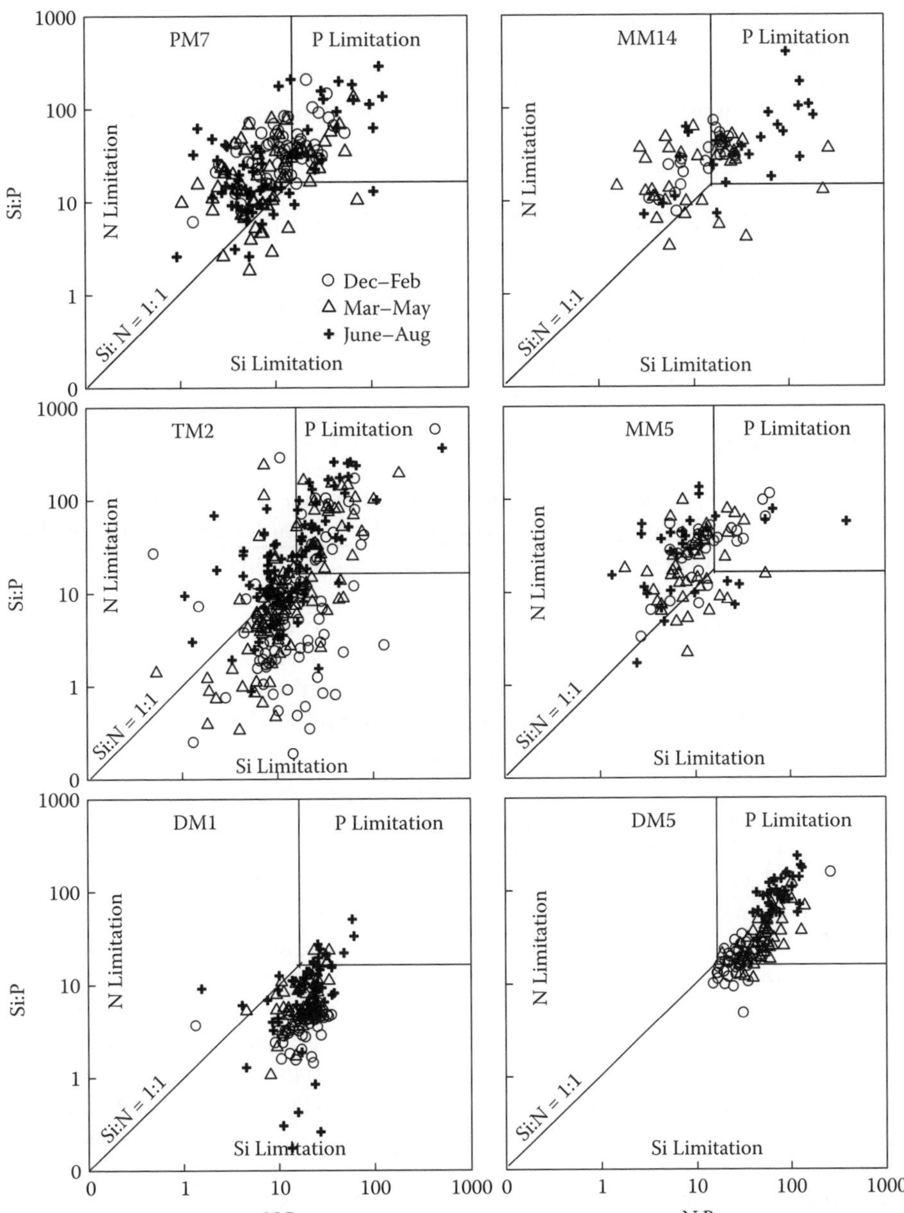

FIGURE 15.7 Scatter diagram of atomic nutrient ratios for the surface water during winter (Dec–Feb), spring (Mar–May), and summer (June–Aug) in three bays (Port Shelter, Tolo Harbour, and Deep Bay). N, P, and Si represent dissolved inorganic N, P, and Si. Stoichiometric (= potential) limitation for N, P, and Si is indicated by the number of data points in the various quadrants. (Data from EPD, Hong Kong.)

as a result of the input of the high NH_4^+ and PO_4^{3-} concentrations from sewage (Figure 15.7). In contrast, the input of the Pearl River discharge resulted in potential P limitation (85%) in the outer bay near the Pearl River Estuary (Figure 15.7). The degree of potential P limitation increases from winter to summer, as indicated by an increase in the DIN:PO_4^{3-} and SiO_4^{2-}:PO_4^{3-} ratios during this period in response to an increase in the influence of the Pearl River discharge.

Tolo Harbour is not influenced by the Pearl River discharge. In the outer part of Tolo Harbour, N was the most limiting nutrient (70%; Figure 15.7) because of the dominance of the coastal/oceanic water (Figure 15.3). In the inner Tolo Harbour, the reduction in nutrients had an important effect on the potential limiting nutrient after the diversion of the sewage. In the inner bay, before the full diversion sewage waste, Si was the potential limiting nutrient (63%; Figure 15.7). However, after the full diversion of the sewage, NH_4^+ and PO_4^{3-} concentrations dramatically decreased and P became the primary limiting nutrient (68%; Figure 15.7).

15.6 MANAGEMENT STRATEGY AND PERSPECTIVES

Hong Kong developed legislation on water quality criteria by dividing Hong Kong waters into 10 water quality control zones (WCZ) in the 1980s (EPD 2004). Port Shelter, Tolo, and Deep Bay are three different WCZs, and each zone usually has its own water quality standards that determine the pollution control and water quality management priorities.

15.6.1 Tolo Harbour

The most serious case of eutrophic impacts occurred in Tolo Harbour with the establishment of a relationship between the frequency of occurrence of red tides and the increase in population and nutrients (Holmes 1988; Lam and Ho 1989). Hypoxia also occurred frequently before 1998 (Broom et al. 2003).

Between 1986 and 2001, the population in the Tolo Harbour catchment area nearly doubled, from 0.5 million to 0.9 million, as the new towns of Sha Tin and Tai Po developed. The sudden rise in population was matched by an increase in wastewater generated in the area which was discharged into the harbor from the Sha Tin and Tai Po Sewage Treatment Works. Although much of the increased discharge into Tolo Harbour received secondary treatment at the sewage works, the increased levels of nutrients proved too much for the harbor's assimilation capacity. Red tides (or algal blooms) began to occur with increasing frequency in the 1980s and into the 1990s, with a record of 40 outbreaks in 1988 (EPD 2005). Frequent and serious hypoxic events (DO < 2 mg L^{-1}) occurred in the inner (stations TM2 and TM4) and middle (station TM6) harbor before 1990 (Chau 2007).

To halt the deterioration of the water quality in Tolo Harbour, the Hong Kong government drew up a Tolo Harbour Action Plan in 1986 consisting of initiatives to control and reduce pollution in the harbor. The plan was implemented in 1987. Livestock farming was prohibited or restricted in much of the harbor catchment area, and strict livestock waste controls were initiated. Meanwhile, sewage treatment works were improved, and pollution from an old landfill site was reduced. The plans also extended the public sewer network in rural areas. The plan to export and discharge the treated effluent from the Sha Tin and Tai Po Sewage Treatment Works into the better flushed Victoria Harbour was implemented in phases between 1995 and 1998. Thus, this plan reduced the nutrient loading in Tolo Harbour.

The Tolo Harbour Action Plan has achieved very positive results over the last two decades. Water quality in the harbor has improved substantially subsequent to sewage diversion. For example, 5-day biochemical oxygen demand (BOD_5), organic pollutants, and *Escherichia coli* have decreased significantly (EPD 2005). Total inorganic nitrogen and total phosphate between 1986–1997 and 1998–2006 have been reduced year-round throughout Tolo (Figure 15.8). Overall, sewage inflow into Tolo Harbour has decreased by 90%. As a result, total nitrogen (TN) and total phosphorus (TP) concentrations have decreased by 36% to 40% and 62% to 70%, respectively, in response to the reduction of sewage input. Chllorophyll *a* levels in Tolo have also decreased, and bottom DO has increased substantially (Figure 15.8). A 21% to 36% reduction in algal biomass and an 11% to 23% increase in bottom DO were observed in response to the nutrient reduction in Tolo Harbour, except in the narrow channel where eutrophication impacts were likely regulated by physical processes. Few hypoxic events (DO < 2 mg L^{-1}) occurred after sewage treatment, and red tide incidents have been reduced markedly from over 40 in 1988 to around 10 each year in the last decade. In addition, fish kills that were frequently seen in the 1980s and 1990s are currently rare (EPD 2005).

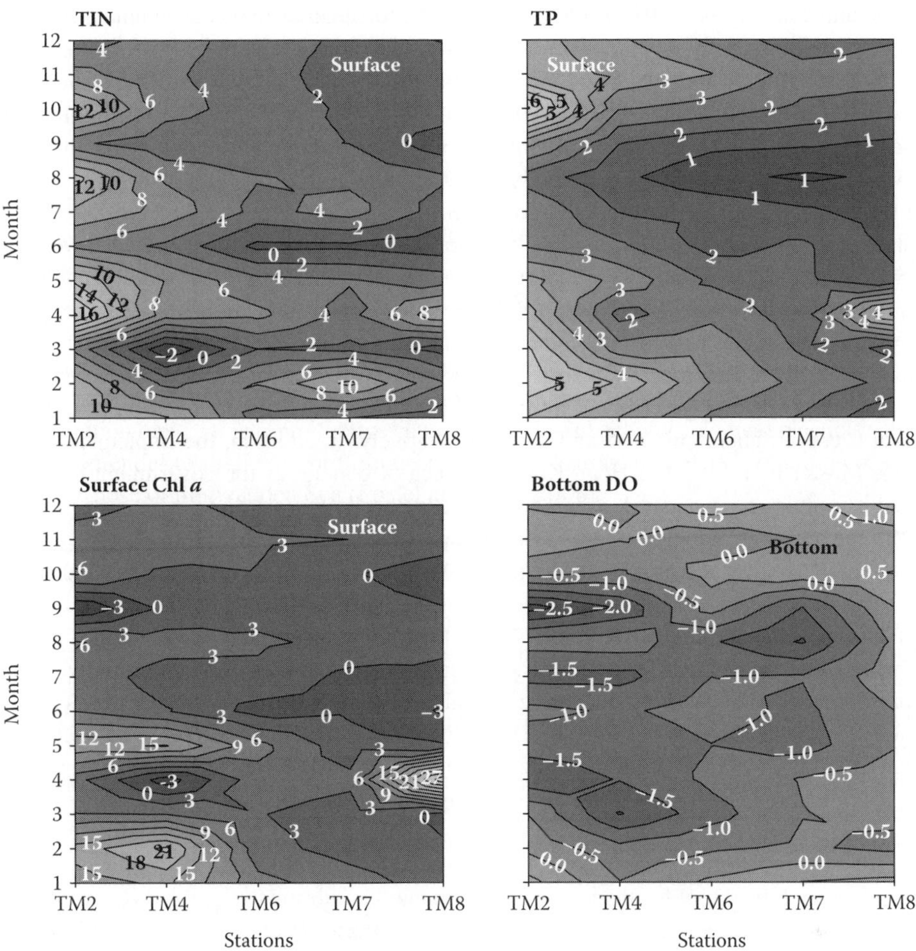

FIGURE 15.8 Differences in average monthly TIN, TP, chlorophyll *a*, and DO between 1986–1997 and 1998–2006. Note a reduction in TIN, TP, and chlorophyll *a*, but an elevation in DO during 1998–2006 after implementation of the Tolo Harbour Action Plan.

15.6.2 Port Shelter

The catchment around Port Shelter has no large urban developments within it, but there is a sizeable population scattered in small towns and villages, many of which are on the coast. Since 1990, local industrial and human wastes have largely been eliminated due to the implementation of sewage treatment and control of livestock wastes (Yung et al. 2001). An advanced secondary sewage treatment plant located at Sai Kung provides a high degree of sewage treatment, including UV disinfection and nutrient removal before discharge. This facility has improved water quality within Port Shelter, even as the population has increased during recent years. Most monitoring stations in Port Shelter have recorded decreasing trends in Kjeldahl nitrogen, total nitrogen, and total phosphorus in recent years (EPD 2005).

15.6.3 Deep Bay

Shenzhen is one of the most rapidly growing economies in China. The population in Shenzhen has grown from a fishing village of 30,000 people to a large complex with 10 million individuals (http://

english.people.com.cn/90002/95607/6532479.html). Unfortunately, rapid developments in Shenzhen and on the Hong Kong side of the northwestern New Territories have seriously affected the water quality in Deep Bay. In the 1980s and early 1990s, the levels of *E. coli* and BOD_5 increased due to sewage and livestock wastes. The bay began facing serious problems of increasing nutrient and organic enrichment, hypoxia, ammonium toxicity, and bacterial contamination which threatened its sensitive ecology and oyster culture industry (EPD 2005).

On the Hong Kong side, most of the pollution is the result of discharge from livestock farms and unsewered villages. Pollution control legislation was implemented for the northwestern New Territories, and the Livestock Waste Control Scheme, introduced in 1987 and amended in 1994, was particularly important. This control scheme bans livestock farming in areas designated as Prohibition Areas. It has significantly reduced the number of livestock farms in the New Territories and imposed effluent treatment standards on those remaining farms. To further reduce pollution from livestock farms, the government introduced the Voluntary Surrender of Poultry Scheme and Pig Farm License Scheme in 2005 and 2006, respectively. In addition, the implementation of the Sewerage Master Plan for all of Hong Kong has been of benefit, as the government has gradually extended its public sewer network to hundreds of previously unsewered villages, an operation that will continue to reduce the nutrient input over the next decade.

Due to increasing concerns of deteriorating water quality in 1992, a Deep Bay (Shenzhen Bay) Action Plan was formulated by the former Hong Kong–Guangdong Environmental Protection Liaison Group (EPLG) (renamed in 2000 as the Hong Kong–Guangdong Joint Working Group on Sustainable Development and Environmental Protection). The aim of this joint initiative from both sides of the border was to address the range of pollution problems threatening the bay's ecosystem. In 1998, EPLG declared a long-term goal to reduce pollution entering Deep Bay, and in 1999 a Deep Bay Water Pollution Control Joint Implementation Program was formulated. In 2000, both parties to the program agreed on a 15-year plan to clean up Deep Bay, which would reduce pollution loads from existing sources and control future pollution so that the water in the bay could regain its natural assimilative capacity by 2015. Both sides are now jointly developing a mathematical model that will act as an analytical tool for managing environmental condition in the Pearl River Estuary.

Environmental conditions remain problematic despite these efforts. For example, the water quality of Deep Bay remained generally poor based on a 2006 assessment, particularly in the inner bay area. The levels of dissolved inorganic nitrogen in Deep Bay are the highest of any recorded area in Hong Kong waters (Figure 15.4).

15.7 CLIMATE CHANGE EFFECTS

Large-scale climatic forcing can alter the physical processes and water quality that affect phytoplankton growth dynamics in oceanic ecosystems (Miller et al. 2004). A recent global-scale study of net oceanic primary productivity has shown that phytoplankton growth dynamics are tightly coupled with climatic variability (Behrenfeld et al. 2006). A number of mechanisms including changes in solar flux, temperature, horizontal advection, vertical mixing and upwelling, and freshwater outflow have been identified as the coupling processes between climatic variability and changes in phytoplankton biomass (Miller et al. 2004).

Hong Kong is linked to the South China Sea which, in turn, connects to the western Pacific Ocean through Luzon Strait between Taiwan and the Philippines. The dominant climatic forcing in the Pacific over interannual time scales is the El Niño Southern Oscillation (ENSO). ENSO can cause changes in the annual pattern in rainfall, winds, thermocline depth, and oceanic circulation (Fiedler 2002), and consequently, it may result in ecosystem variation from a change in primary productivity to changes in the population dynamics of fish, seabirds, and marine mammals (Fiedler 2002). In Hong Kong, ENSO has been found to lead to changes in temperature and rainfall patterns and also to influence the Southeast Asian monsoon season (Leung and Wu 2004). The air temperature in Hong Kong during the El Niño years is generally warmer than average, and there is higher rainfall, while the effects of La Niña are less pronounced (Chang 1999).

The intensity of El Niño in 1997 to 1998 was one of the strongest of the century. A series of red tides occurred in Hong Kong territorial waters between mid-March and mid-April 1998, resulting in a loss of HK $250 (US $32) million due to fish kills. The events consisted of the spatial and temporal progression of red tide outbreaks along the coast of southern China and Hong Kong and coincided with the dramatic change in the oceanographic conditions in the northern portion of the South China Sea between 1997 and 1998 from March to mid-April, as revealed in the weekly composite sea surface temperatures (SST) (Yin et al. 1999). The SST images showed a warm water tongue pointing north to the southern China coast in 1998 vs. a cold water tongue pointing south in 1997 in the middle of the South China Sea. The differences are believed to be due to El Niño and thought to be responsible for setting up the physical oceanographic conditions that were favorable for the formation of harmful algal blooms (HABs) along the south coast of China (Yin et al. 1999). The warm tongue in the SST images suggested that the warm water from the South China Sea might have been piling up on the south coast of China. Alternatively, downwelling of the south China Coastal Current along the coast due to the northeast monsoon during March might have been moving against the South China Sea warm water at the bottom. As a result, the coastal waters of the south China coast, including Hong Kong, became trapped along the coast. Given local eutrophic conditions of the China coast, the outbreak of HABs occurred over a coast-wide scale (~400 km) in winter 1997 and spring 1998 (Yin et al. 1999). The decreasing El Niño intensity in the South China Sea in 1998 lagged behind that of the Pacific Ocean (Wang et al. 2002). As a result, the SST observed from the satellite was higher than in 1997 on the continental shelf off the Hong Kong coast (Yin unpublished results). In response, the number of red tides in 1998 was somewhat higher than average (AFCD 2005).

The highest number of red tide events (88 incidents) among all years from 1983 to 2008 occurred in 1988. The record number of red tides during the 1988 El Niño year may have been due to the longer than usual residence time and the high input of nutrients from livestock and untreated domestic sewage into the Tolo. Tolo Harbour has the highest frequency of red tides, accounting for about 40% of all red tide occurrences among the water quality control zones in Hong Kong (EPD 2005; Wong et al. 2009).

Atmospheric temperature appears to be increasing in Hong Kong (Leung et al. 2004), and sea level is also rising (Yeung 2006). Water temperature has increased in Deep Bay (Xu et al. 2010) and other zones (Ho 2007) during the last several decades. Chlorophyll *a* in Port Shelter did not exhibit any significant long-term trend in temperature, but chlorophyll *a* significantly increased during 1986 to 2006 in Deep Bay at both stations DM1 and DM5 (Figure 15.9). However, bottom DO showed a small significant increase during 1986 to 2006 at station PM7 in Port Shelter, and a significantly larger increase at station TM2 in Tolo. A significant decrease in DO was recorded at station DM1 in Deep Bay (Figure 15.10). Outside of these bays, only DO at station DM5 decreased significantly. DO at station MM14 outside of Port Shelter and station MM5 outside of Tolo exhibited no significant trends (Figure 15.10). The correlation analysis did not reveal a significant relationship between chlorophyll *a* and temperature (data not shown), but the correlation between bottom DO and temperature is significant at all five stations (PM7-MM14, TM2-MM5, DM5) except for station DM1 in Deep Bay (Figure 15.11). This significant correlation can be an important indication of the potential future impacts that global warming may have on Hong Kong waters. Because it is a negative relationship, therefore, global warming would likely reduce the ecosystem buffering capacity against the formation of low oxygen waters due to a potential increase in stratification.

15.8 SUMMARY AND CONCLUSIONS

The management of coastal waters has become more urgent with the rapid population growth in the coastal zone. The input of nutrients and the accompanying eutrophication impacts remain a

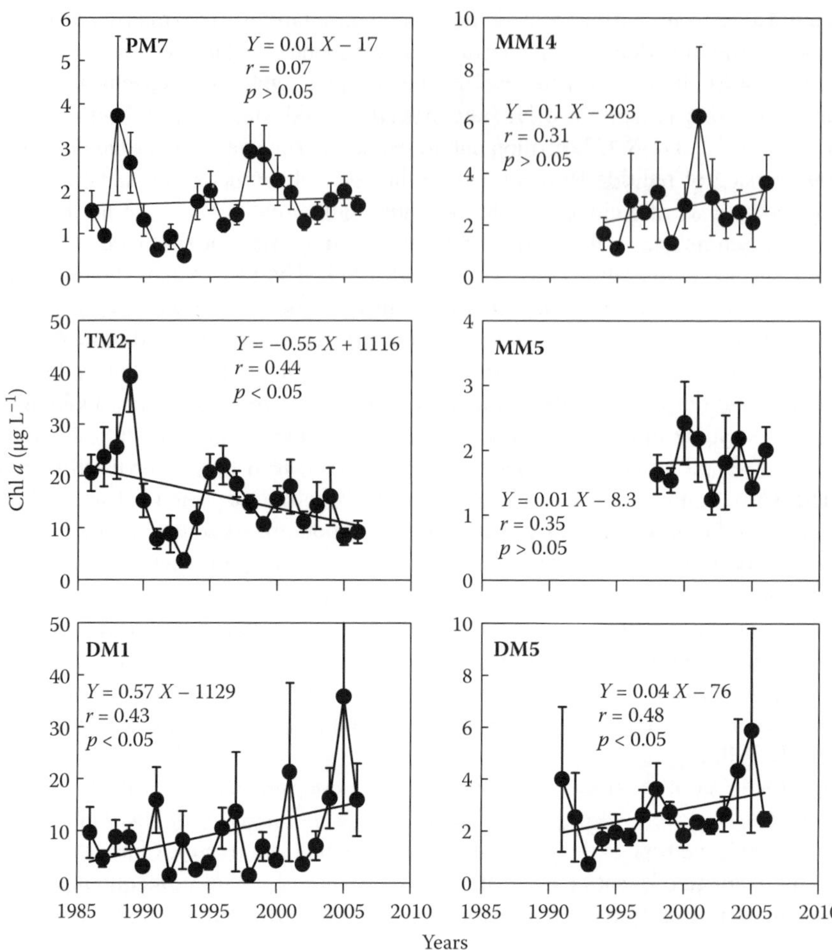

FIGURE 15.9 Chlorophyll *a* time series from 1986 to 2006 at stations PM7, MM5, DM1 and from 1994 to 2006 for MM14.

major concern. As the buffering capacity for eutrophication varies among coastal water bodies, it is necessary to understand the biotic and physicochemical processes that operate in these complex ecosystems in order to achieve sustainable development of the coastal zone.

The comparison of Port Shelter, Tolo, and Deep Bay demonstrates that each system has its own unique environmental characteristics (Table 15.2) and buffering capacity to accommodate the input of nutrients. Port Shelter receives freshwater input at the head of the bay, and it has a wide opening to coastal waters at the mouth of the bay. In the winter dry season, surface water outside of the bay is driven into the bay by northeast monsoon winds due to a downwelling mechanism, with vertical mixing dominating over stratification. In the summer rainy season, Port Shelter has estuarine-like circulation and strong stratification. Port Shelter is the deepest of the three bays. Nutrient concentrations limit the eutrophication impacts (Table 15.3). Low chlorophyll *a* concentrations in winter and low bottom oxygen levels in summer indicate that the buffering capacity against an increase in the input of nutrients may be high in winter but low in summer.

Tolo is shallow, with the longest residence times and year-round stratification of the three bays. Therefore, Tolo is the most vulnerable system to an increase in the input of nutrients and

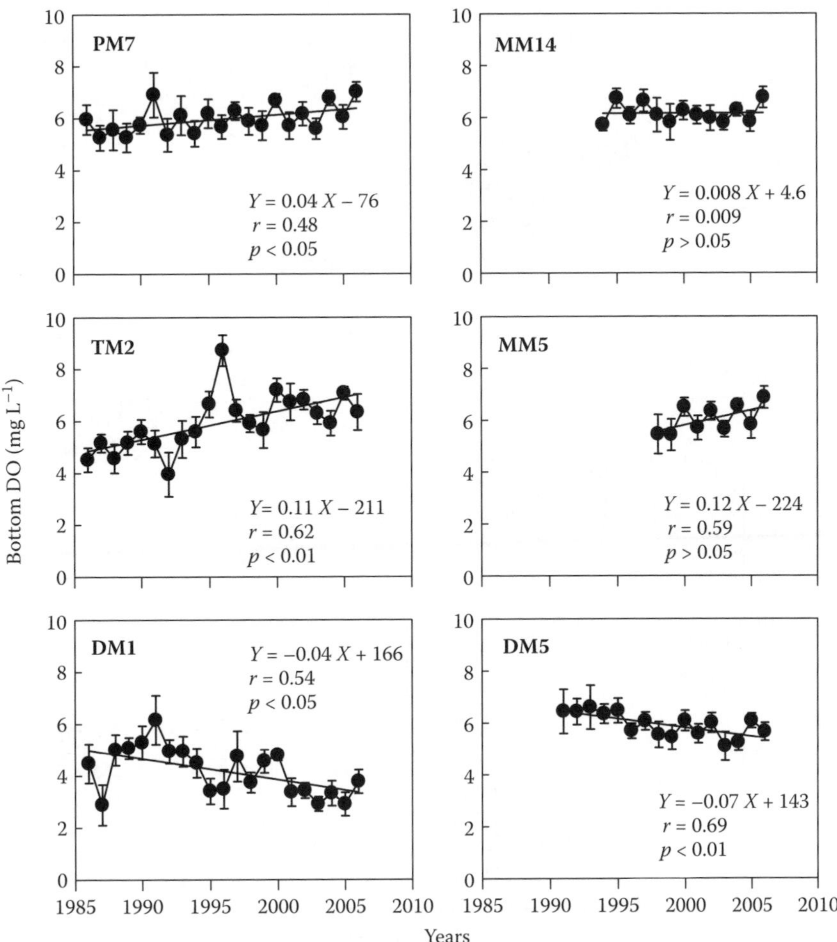

FIGURE 15.10 DO time series from 1986 to 2006 at stations PM7, MM5, DM1 and from 1994 to 2006 for MM14.

accompanying eutrophication impacts. The management actions implemented in 1996 were certainly appropriate, and produced a significant improvement in the water quality of Tolo.

Deep Bay is similar to Tolo with its shallow water column and long residence times. It has the highest nutrient concentrations among the three bays. However, there are large areas of shallow mudflats in the bay where sediment resuspension by wind and tidal mixing results in high turbidity. Therefore, light limits phytoplankton production, and benthic filter feeding may exert top-down controls of phytoplankton biomass. Its shallowness and possible top-down controls produce strong ecosystem buffering capacity because symptoms of eutrophication (i.e., algal blooms and low DO) are not as severe as expected from the high ambient nutrient concentrations.

The comparison of the three bays demonstrates the importance of physical and oceanographic driving factors and how these factors can lead to different eutrophication impacts. The ranking of the three bays in terms of eutrophication impacts (chlorophyll *a* concentrations, number of red tides, and hypoxia events), is Tolo > Deep Bay > Port Shelter. Therefore, high nutrient loads do not always produce the largest eutrophication impacts as observed for Deep Bay, where the impacts were reduced by turbidity and light limitation (bottom-up controls) and grazing (top-down controls) (Tables 15.2 and 15.3).

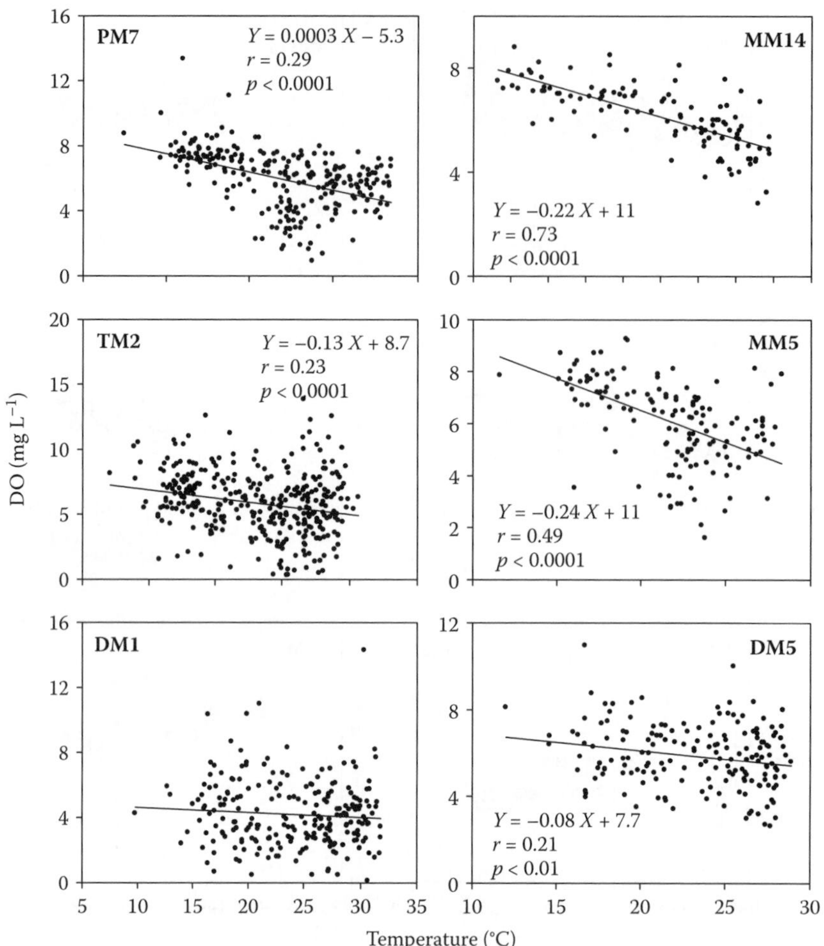

FIGURE 15.11 Correlation between bottom DO and temperature for the time series in Figure 15.10.

ACKNOWLEDGMENTS

This study was part of the AoE project UGC AoE/P-04/04-1, and the ARC linkage project LP088363. K. Yin acknowledges the support from NSFC Projects 40676074 and 40490264 and the CAS/SAFEA International Partnership Program for Creative Research Teams. We thank the Environmental Protection Department, Hong Kong Government, for providing data on Hong Kong marine waters and for permission to use and publish these data.

TABLE 15.2
Comparison of General Environmental Conditions in Port Shelter, Tolo Harbour, and Deep Bay

Condition	Season	Port Shelter	Tolo Harbour	Deep Bay
Depth (m)	Most Areas[a]	10–20	5–15	<5
Circulation	Dry	Downwelling/oceanic Intrusion	Downwelling/oceanic Intrusion	Oceanic Intrusion
	Wet	Estuarine-like two-layer	Estuarine-like two-layer	Freshwater ouflow
Residence times (days)	Avg	9.1[b]	28.5[c]	25[d]
	Dry	8.9	38	
	Wet	9.4	14.4	
Eutrophic conditions	Dry	Oceanic	Eutrophic	Eutrophic
	Wet	Rainfall and runoff	Eutrophic	Eutrophic
Chlorophyll *a* trend		No trend	Decreasing	Increasing
DO trend		Increasing	Increasing	Decreasing
Limiting factor (chlorophyll *a*)		Mixing/nutrients	Nutrients/grazing (?)	Light/benthic grazing (?)

[a] Most areas are estimated using the map with color-contoured depths.

[b] These values were the flushing time for eliminating 50% of the tracer obtained from the modeling results of the Delft 3D modeling run by CDM requested by the author. For eliminating 90% of the tracer, the flushing times were >500 h, the maximum the model output could produce.

[c] Data from Lee et al. (2006).

[d] Data from Arega and Lee (2000).

TABLE 15.3
Comparisons of Monthly Average Water Quality Parameters for Surface and Bottom Waters in Port Shelter, Tolo Harbour, and Deep Bay over a 21-Year Period (1986–2006)

		Port Shelter		Tolo Harbour		Deep Bay	
Parameters		**Surface**	**Bottom**	**Surface**	**Bottom**	**Surface**	**Bottom**
Salinity	Avg	31.8	33.1	30.5	32.3	20.6	22.3
	min	29.0	31.6	27.6	29.9	7.61	7.61
	max	33.8	34.5	32.7	34.3	30.9	32.1
Temperature	Avg	23.8	22.1	23.7	22.2	23.9	23.7
	min	16.5	16.4	16.8	16.6	17.0	17.0
	max	29.5	27.6	29.1	28.1	29.4	29.4
NO_3^- (μM)	Avg	1.59	1.99	2.80	2.64	39.4	35.5
	min	0.29	0.36	0.28	0.71	12.5	10.6
	max	7.76	5.17	11.8	9.27	75.6	67.4
PO_4^{3-} (μM)	Avg	0.36	0.42	0.81	0.79	7.01	6.96
	min	0.15	0.21	0.17	0.28	1.18	0.84
	max	0.59	0.62	2.69	2.05	25.3	25.3
SiO_4^{2-} (μM)	Avg	9.67	12.5	11.4	16.4	67.5	63.0
	min	3.43	6.81	3.26	8.19	18.9	17.5
	max	17.0	22.2	24.0	25.3	131	131
NH_4^+ (μM)	Avg	2.09	2.09	5.51	5.32	99.3	98.5
	min	0.92	0.78	0.91	1.04	4.84	5.24
	max	4.53	3.55	16.5	15.2	438	438
DO (mg L^{-1})	Avg	6.88	6.08	7.49	5.43	5.67	5.51
	min	5.93	3.61	5.95	2.78	2.97	2.97
	max	7.89	8.02	8.82	7.98	7.77	7.76
Chlorophyll *a* (μg L^{-1})	Avg	2.19	1.74	9.71	4.05	5.97	5.61
	min	1.09	0.91	1.00	0.96	1.41	1.09
	max	6.56	4.31	24.5	12.9	31.6	31.6

REFERENCES

AFCD (Agriculture, Fisheries and Conservation Department). 2005. *Red tides/harmful algal blooms in Hong Kong 2002 and 2003*. AFCD Report, Hong Kong SAR Print, Hong Kong.

Anderson, D. M. 1997. Turning back the harmful red tides. *Nature* 388: 513–514.

Arega, F. and J. H. W. Lee. 2000. Long-term circulation and eutrophication model for Tolo Harbor, Hong Kong. *Water Qual. Ecosys. Modeling* 1: 169–192.

Behrenfeld, M. J., R. T. O'Malley, D. A. Siegel, C. McClain, J. L. Samiento, G. C. Feldman, A. J. Milligan, P. G. Falkowski, R. M. Letellier, and E. S. Boss. 2006. Climate-driven trends in contemporary ocean productivity. *Nature* 444: 752–755.

Broom, M., G. Chiu, and A. Lee. 2003. Long term water quality trends in Hong Kong. Pp. 519–536, In: B. Morton (ed.), *Perspectives on marine environment change in Hong Kong and the South China Sea, 1977–2001*. Proceedings of an International Workshop Reunion Conference, October 21–26, 2001. Hong Kong: Hong Kong University Press.

Cai, W. J., M. Dai, Y. Wang, W. Zhai, T. Huang, S. Chen, F. Zhang, Z. Chen, and Z. Wang. 2004. The biogeochemistry of inorganic carbon and nutrients in the Pearl River estuary and the adjacent Northern South China Sea. *Continent. Shelf Res.* 24: 1301–1319.

Chang, W. L. 1999. *Some impacts of El Niño and La Niña events on the weather of Hong Kong*. Workshop on the Impact of the El Niño Southern Oscillation (ENSO)/La Niña on Meteorology and Hydrology in the Typhoon Committee Area, Macau. June 29 to July 1, 1999, Hong Kong Observatory, Hong Kong.

Chau, K. W. 2007. Integrated water quality management in Tolo Harbor, Hong Kong: A case study. *J. Clean. Prod.* 15: 1568–1572.

Chau, Y. K. and R. Abesser. 1958. A preliminary study of the hydrology of Hong Kong territorial waters. *Hong Kong Univ. Fish J.* 2: 37–42.

Chau, Y. K. and C. S. Wong. 1960. Oceanographic investigations in the northern region of the South China Sea off Hong Kong. *Hong Kong Univ. Fish J.* 3: 1–25.

Chen, H. 2008. Shenzhen to relax rules on residency. *China Daily*, 2008-08-06 06. http://www.chinadaily.com.cn/china/2008-08/06/content_6906707.htm.

Cloern, J. E. 1999. The relative importance of light and nutrient limitation of phytoplankton growth: A simple index of coastal ecosystem sensitivity to nutrient enrichment. *Aquat. Ecol.* 33: 3–16.

Cloern, J. E. 2001. Our evolving conceptual model of the coastal eutrophication problem. *Mar. Ecol. Prog. Ser.* 210: 223–253.

Cloern, J. E. and R. Dufford. 2005. Phytoplankton community ecology: Principles applied in San Francisco Bay. *Mar. Ecol. Prog. Ser.* 285: 11–28.

Cloern, J. E., A. D. Jassby, J. K. Thompson, and K. A. Hieb. 2007. A cold phase of the East Pacific triggers new phytoplankton blooms in San Francisco Bay. *Proc. Natl. Acad. Sci. USA* 104: 18,561–18,565.

Diaz, R. J. and R. Rosenberg. 2008. Spreading dead zones and consequences for marine ecosystems. *Science* 321: 926–929.

EPD (Environmental Protection Department). 1987. *Marine water quality in Hong Kong in 1985*. Hong Kong Government Printer, Hong Kong.

EPD (Environmental Protection Department). 2004. *Marine water quality in Hong Kong in 2003*. Hong Kong Government Printer, Hong Kong. http://www.epd.gov.hk/epd/english/environmentinhk/water.

EPD (Environmental Protection Department). 2005. *Marine water quality in Hong Kong in 2004. Hong Kong Government Printer, Hong Kong*. http://www.epd.gov.hk/epd/english/environmentinhk/water.

Fiedler, P. C. 2002. Environmental change in the eastern tropical Pacific Ocean: Review of ENSO and decadal variability. *Mar. Ecol. Prog. Ser.* 244: 265–283.

Fisher, T. R., J. D. Hagy III, W. R. Boynton, and M. R. Williams. 2006. Cultural eutrophication in the Choptank and Patuxent estuaries of Chesapeake Bay. *Limnol. Oceanogr.* 51: 435–447.

Fonda Umani, S., A. Beran, S. Parlato, D. Virgilio, T. Zollet, A. De Olazzarini, B. Lazzarini, and M. Cabrini. 2004. *Noctiluca scintillans* Macartney in the Northern Adriatic Sea: Long-term dynamics, relationships with temperature and eutrophication and role in the food web. *J. Plankton Res. 26*: 545–561.

Harrison, P. J., K. Yin, J. H. W. Lee, J. P. Gan, and H. B. Liu. 2008. Physical-biological coupling in the Pearl River Estuary. *Continent. Shelf Res.* 28: 1405–1415.

He, L., K. Yin, X. Yuan, D. Li, D. Zhang, and P. J. Harrison. 2009. Spatial distribution of viruses, bacteria, and chlorophyll in the northern South China Sea. *Aquat. Microb. Ecol.* 54: 153–162.

Ho, A. Y. T. 2007. Dynamics of nutrients and phytoplankton biomass and production in Hong Kong waters. Ph.D. Thesis, The Hong Kong University of Science and Technology, Hong Kong.

Holmes, P. R. 1988. Tolo Harbor: The case for integrated water quality management in a coastal environment. *Water Environ. J. (J. CIWEM)* 2: 171–179.

Howarth, R. W. and R. Marino. 2006. Nitrogen as the limiting nutrient for eutrophication in coastal marine ecosystems: Evolving views over three decades. *Limnol. Oceanogr.* 51: 364–376.

Justić, D., N. N. Rabalais, and R. E. Turner. 1997. Impacts of climate change on net productivity of coastal waters: Implications for carbon budgets and hypoxia. *Climate Res.* 8: 225–237.

Lam, C. W. Y. and K. C. Ho. 1989. Phytoplankton characteristics of Tolo Harbor. *Asian Mar. Biol.* 6: 5–18.

Le Pape, O., Y. Del Amo, A. Ménesguen, B. Quequiner, and P. Treguer. 1996. Resistance of a coastal ecosystem to increasing eutrophic conditions: The Bay of Brest (France), a semi-enclosed zone of Western Europe. *Continent. Shelf Res.* 16: 1885–1907.

Lee, J. H. W. and A. G. Qian. 2003. Three-dimensional modeling of hydrodynamics and flushing in Deep Bay. Pp. 814–821, In: *Proceedings of International Conference on Estuaries and Coasts*. Hangzhou: Zhejiang University Press.

Lee, J. H. W., P. J. Harrison, C. Kuang, and K. D. Yin. 2006. Eutrophication dynamics in Hong Kong coastal waters: Physical and biological interactions. Pp. 187–206: In: E. Wolanski (ed.), *The environment in Asia Pacific harbors*. Dordrecht: Springer.

Leung,Y. K. and M. C. Wu. 2004. The effect of ENSO and East Asian Monsoon on the annual rainfall in Hong Kong, China. Paper presented at Sixth Joint Meeting of Seasonal Prediction on East Asian Summer Monsoon. Guilin, China.

Leung, Y. K., K. H. Yeung, E. W. L. Ginn, and W. M. Leung. 2004. *Climate change in Hong Kong*. Hong Kong Observatory Technical Note No. 107, Hong Kong SAR.

Lillobø, A. I., J. M. Neto, I. Martins, T. Verdelhos, S. Leston, P. G. Cardoso, S. M. Ferreira, J. C. Marques, and M. A. Pardal. 2005. Management of a shallow temperate estuary to control eutrophication: The effect of hydrodynamics on the system's nutrient loading. *Estuar. Coastal Shelf Sci.* 65: 697–707.

Liu, X. J. and C. K.Wong. 2006. Seasonal and spatial dynamics of *Noctiluca scintillans* in a semi-enclosed bay in the northeastern part of Hong Kong. *Bot. Mar.* 49: 145–150.

Miller, A. J., F. Chai, S. Chiba, J. R. Moisan, and D. J. Neilsen. 2004. Decadal-scale climate and ecosystem interactions in the North Pacific Ocean. *J. Oceanogr*. 60: 163–188.

Ning, X., F. Chai, H. Xue, Y. Cai, C. Liu, and J. Shi. 2004. Physical–biological oceanographic coupling influencing phytoplankton and primary production in the South China Sea. *J. Geophys. Res.* 109: C1005.

Nixon, S. W. 1995. Coastal marine eutrophication: A definition, social cause, and future concerns. *Ophelia* 41:199–219.

NRC (National Research Council). 2000. *Clean coastal waters: Understanding and reducing the effects of nutrient pollution*. Washington, D.C.: National Academy Press.

Paerl, H. W., L. M. Valdes, M. F. Piehler, and C. A. Stow. 2006. Assessing the effects of nutrient management in an estuary experiencing climatic change: The Neuse River estuary, North Carolina. *Environ. Management* 37: 422–436.

Pinckney, J. L., H. W. Pearl, and M. B. Harrington. 1999. Responses of the phytoplankton community growth rate of nutrient pulses in variable estuarine environments. *J. Phycol.* 35: 1455–1463.

Rosenberg, R. 1985. Eutrophication: The future marine coastal nuisance? *Mar. Pollut. Bull.* 16: 227–231.

Ryther, J. H. and W. M. Dunstan. 1971. Nitrogen, phosphorus, and eutrophication in the coastal marine environment. *Science* 171: 1008–1013.

Smith, V. H., S. B. Joye, and R. W. Howarth. 2006. Eutrophication of freshwater and marine ecosystems. *Limnol. Oceanogr.* 51: 351–355.

Watts, J. C. D. 1971a. A general review of the oceanography of the northern sector of the South China Sea. *Hong Kong Fish. Bull.* 2: 41–50.

Watts, J. C. D. 1971b. The hydrology of the continental shelf area south of Hong Kong. I. Oceanographic observation for the year 1969. *Hong Kong Fish. Bull.* 2: 51–57.

Watts, J. C. D. 1973. Further observations on the hydrology of the Hong Kong territorial waters. *Hong Kong Fish. Bull.* 3: 9–35.

Wang, D. X, Q. Xie, Y. Du, W. Wang, and J. Chen. 2002. The 1997–1998 warm event in the South China Sea. *Chinese Sci. Bull.* 47: 1221–1227.

Williamson, G. R. 1970. The hydrography and weather of the Hong Kong fishing grounds. *Hong Kong Fish. Bull.* 1: 43–49.

Wong, K. T. M., J. H. W. Lee, and P. J. Harrison. 2009. Forecasting of environmental risk maps of coastal algal blooms. *Harmful Algae* 8: 407–420.

Xu, J., K. Yin, A. Y. T. Ho, J. H. W. Lee, D. M. Anderson, and P. J. Harrison. 2009. Nutrient limitation in Hong Kong waters inferred from comparison of nutrient ratios, bioassays and P turnover times. *Mar. Ecol. Prog. Ser.* 388: 81–97.

Xu, J., K. Yin, J. H. W. Lee, H. Liu, A. Y. T. Ho, X. Yuan, and P. J. Harrison. 2010. Long-term and seasonal changes in nutrients, phytoplankton biomass, and dissolved oxygen in Deep Bay, Hong Kong. *Estuar. Coasts.* 33: 399–416.

Yeung, K. H. 2006. Issues related to global warming: Myths, realities and warnings. Paper presented at 5th Conference on Catastrophe Insurance in Asia, June 20–21, 2006, Keynote Address, Hong Kong.

Yin, K. 2002. Monsoonal influence on seasonal variations in nutrients and phytoplankton biomass in coastal waters of Hong Kong in the vicinity of the Pearl River estuary. *Mar. Ecol. Prog. Ser.* 245: 111–122.

Yin, K. 2003. Influence of monsoons and oceanographic processes on red tides in Hong Kong waters. *Mar. Ecol. Prog. Ser.* 262: 27–41.

Yin, K., P. J. Harrison, J. C. Chen, W. Huang, and P. Y. Qian. 1999. Red tides during spring 1998 in Hong Kong waters: Is El Ninõ responsible? *Mar. Ecol. Prog. Ser.* 187: 289–294.

Yin, K., P. Y. Qian, J. C. Chen, D. P. H. Hsieh, and P. J. Harrison. 2000. Dynamics of nutrients and phytoplankton biomass in the Pearl River estuary and adjacent waters of Hong Kong during summer: Preliminary evidence for phosphorus and silicon limitation. *Mar. Ecol. Prog. Ser.* 194: 295–305.

Yin, K., P. Qian, M. C. S. Wu, J. C. Chen, L. Huang, X. Song, and W. Jian. 2001. Shift from P to N limitation of phytoplankton growth across the Pearl River estuarine plume during summer. *Mar. Ecol. Prog. Ser.* 221: 17–28.

Yin, K., X. X. Song, S. Liu, J. J. Kan, and P. Y. Qian. 2008. Is inorganic nutrient enrichment a driving force for the formation of red tides? A case study of the dinoflagellate *Scrippsiella trochoidea* in an embayment. *Harmful Algae* 8: 54–59.

Yung, Y. K., C. K. Wong, M. J. Broom, J. A. Ogden, S. C. M. Chan, and Y. Leung. 1997. Long-term changes in hydrography, nutrients and phytoplankton in Tolo Harbor, Hong Kong. *Hydrobiologia* 352: 107–115.

Yung, Y. K., C. K.Wong, K.Yau, and P. Y. Qian. 2001. Long-term changes in water quality and phytoplankton characteristics in Port Shelter, Hong Kong, from 1988–1998. *Mar. Pollut. Bull.* 42: 981–992.

Zhao, H. 1990. *Evolution of the Pearl River Estuary*. Beijing: Ocean Press (in Chinese).

16 The Wadden Sea
A Coastal Ecosystem under Continuous Change

Katja J. M. Philippart and Eric G. Epping*

CONTENTS

Abstract ... 399
16.1 Introduction ... 400
16.2 Physiography ... 400
16.3 Major Changes ... 403
16.3.1 Abiotic Conditions ... 404
16.3.2 Biotic Factors ... 407
16.3.2.1 Primary Producers and Productivity ... 407
16.3.2.2 Secondary Producers and Productivity ... 409
16.3.2.3 Higher Trophic Levels ... 409
16.4 Major Historical Drivers ... 410
16.4.1 Overexploitation ... 416
16.4.2 Habitat Change and Destruction ... 416
16.4.3 Pollution ... 417
16.4.4 Nutrients ... 417
16.4.5 Invasive Species ... 418
16.5 Climate Change Effects on Ecosystem Function ... 419
16.6 Future Research ... 422
Acknowledgments ... 424
References ... 424

ABSTRACT

The Wadden Sea is one of the largest coastal lagoons in the world, characterized by extensive wetlands, high biotic production, and valuable recreational and commercial fisheries. Based on physicochemical and biotic conditions, the Wadden Sea can be divided into three main regions: (1) the southern Wadden Sea (tidal range = 1.5 to 3 m) bounded by 12 elongated islands forming a sandy barrier 5 to 15 km parallel to the mainland; (2) the central Wadden Sea (tidal range >3 m) lacking barrier islands and ebb-deltas due to macrotidal conditions and the large volume of tidal water exchange; and (3) the northern Wadden Sea (with a tidal range decreasing from 3 to 1.2 m in a northward direction) consisting of 8 islands and high sand bars that form a seaward barrier some 5 to 25 km off the mainland coastline. Each region can be further divided into subareas, some of which are strongly influenced by freshwater inputs from large rivers such as the Rhine, IJssel, Ems, Weser, and Elbe, while others are considered to be dominated by coastal North Sea water. The Wadden Sea has a long history of human impacts. During the twentieth century, the exploitation

* Corresponding author. Email: katja.philippart@nioz.nl

of resources and coastal landscape transformation intensified and culminated in large-scale habitat destruction and overexploitation. After a large part of the southwestern Wadden Sea had been closed off in the early 1930s, the remaining area was heavily polluted by heavy metals, PCBs, and pesticides in the 1950s, and fertilizer use (nitrogen and phosphorus inputs) in the 1960s to 1970s. Although nutrient loads declined as the result of effective measures and policies implemented in the 1980s to 1990s, this reduction was not followed by a decrease in the biomass of phytoplankton. Warming climatic conditions have affected the system since the late 1980s and have the capacity to alter the present structure and function of the food web and other critical components of the system.

Key Words: Wadden Sea, primary production, secondary production, infrastructures, pollutants, eutrophication, fisheries, ecosystem functioning, climate change, scenarios

16.1 INTRODUCTION

The Wadden Sea, the largest nontropical barrier-island system with nonvegetated tidal flats in the world, is characterized by rich and diverse habitats (Wolff 1983). It is highly productive and provides an important nursery area for fish, foraging and resting habitat for seals, and foraging habitat for waders and other waterfowl (Reise 1995; CWSS 2008). Its rich resources attracted early settlers, who first started exploiting the Wadden Sea ~5000 years BP (Wolff 2000). With increasing coastal activities from peat excavation, salt mining, and tourism, the exploitation and transformation pressures increased, causing changes in the structure and function of the ecosystem (Lotze et al. 2005; Wolff 2000, 2005a). In order to protect and preserve some of the remaining unique natural services and functions of this system, the UNESCO World Heritage Committee inscribed the Dutch-German Wadden Sea on the List on June 26, 2009.

This chapter provides an overview of the abiotic and biotic characteristics of the Wadden Sea. Although this overview is not exhaustive, it demonstrates that the Wadden Sea is a system under long-term and continuous change due to a complex interplay of natural and anthropogenic drivers. We will discuss some of the important drivers of change, their impacts on the functioning of the Wadden Sea ecosystem, and the needs in terms of long-term field observations and research to assess these impacts. Finally, some potential future scenarios of ecosystem functioning are deduced. We will focus on the westernmost part of the Wadden Sea because this is historically one of the most intensively studied subareas of the system (Table 16.1).

16.2 PHYSIOGRAPHY

Bound by the North Sea and the Dutch, German, and Danish coastline, the Wadden Sea stretches for 450 km along the coast from Den Helder (the Netherlands) at 53°N to the peninsula of Skallingen (Denmark) at 55°30′ N. Using the 15 m isobath as the seaward boundary, the Wadden Sea presently covers a total area of ~15,000 km^2, 4700 km^2 of which covers intertidal flats and 3700 km^2 of which consists of subtidal areas and gullies within the chain of barrier islands (CWSS 2008) (Figure 16.1). The Elbe, Weser, Ems, Eider, and Varde Å rivers are the largest tributaries, discharging directly into the Wadden Sea (Harten and Vollmers 1978), whereas the Rhine indirectly drains via Lake IJssel and via the coastal North Sea along the Dutch coast (Zimmerman and Rommets 1974). Coastal water is transported from the English Channel to Denmark in a period of 1 year and is advected with Atlantic water from the British coast and river water. The residual current is strong and parallel to the Dutch coast, but diverts more offshore east of the barrier island of Ameland. As a consequence, the current field in the German Bight is rather weak, more subject to wind stress and therefore more variable, resulting in a rather slow renewal of water (Goedecke 1968). The tidal exchange of this coastal water leaves a clear imprint on the physics, chemistry, and biology of the Wadden Sea.

Tides are semidiurnal with a neap to spring tidal cycle having a range of 1.2 to >3 m. The effect of full and new moons on the tidal range is only 20%, whereas strong onshore winds may cause

TABLE 16.1
Wadden Sea Main Regions, Subareas, and Their Main Characteristics[a]

Region	Subarea Code	Name	Characteristics
Southwestern Wadden Sea			
	NL1	Western Dutch Wadden Sea	The area receives fresh water directly from Lake IJssel by the sluices of Den Oever and Kornwerderzand at an average rate of 16.3×10^9 m^3 yr^{-1}. The water mass that originates from the Rhine River passes Lake IJssel in about 50 days. The coastal North Sea water entering the Wadden Sea at the Marsdiep constitutes about 15% Rhine water.
	NL2	Eastern Dutch Wadden Sea	The area receives a minor freshwater source from Lake Lauwers and an industrial waste line. The area is considered to be dominated by coastal North Sea water.
	NL3	Ems-Dollard Estuary	The freshwater sources to the area are the Ems River (90%) and Westerwoldsche Aa (10%) at a total average rate of about 3.4×10^9 m^3 yr^{-1}. Industrial and harbor activities border the estuary at Emden, Delfzijl, and Eemshaven.
	NDS1	Ostfriesland Wadden Sea	The area is slightly influenced by local freshwater sources. Only small harbors are present. The area is considered to be dominated by coastal North Sea water.
Central Wadden Sea			
	NDS2	Jade Basin	The harbor of Wilhelmshaven is the main activity in the area. Virtually no fresh water enters the area. The area is considered to be dominated by coastal North Sea water.
	NDS3	Weser Estuary	The Weser River is the main freshwater source at an average rate of about 11.3×10^9 m^3 yr^{-1}. The river borders are densely populated. Harbor activities are present at the cities of Bremen and Bremerhaven.
	SH1	Elbe Estuary	The Elbe River is the main freshwater source into the area at an average rate of about 24.5×10^9 m^3 yr^{-1}, which is about 43% of the total freshwater input in the international Wadden Sea. The river is bordered by large cities (e.g., Hamburg), harbors, and industrial activities.
	SH2	Eider Estuary	The Eider River constitutes a relatively small freshwater source of about 0.9×10^9 m^3 yr^{-1} on average. The population density is moderate. Some small recreational and fisheries harbors are present.
Northeastern Wadden Sea			
	SH3	Halligen	Virtually no freshwater input and a low population density. The area is considered to be dominated by coastal North Sea water.
	DK1	List Tidal Basin	The area is physically bordered by the dams connecting the islands Sylt and Rømø to the mainland. The area is considered to be dominated by coastal North Sea water. The freshwater input from southern Jutland was about 0.8×10^9 m^3 yr^{-1} in 1990, which was about 1.3×10^9 m^3 yr^{-1} to DK1 and DK2 in the same year.
	DK2	Knudedyb and Juvredyb	The Knube, Konge, Rejsby, and Brøns Å are small rivers, thus constituting a small freshwater input. The input to DK1 and DK2 is about 1.2×10^9 m^3 yr^{-1} on average. The area is considered to be dominated by coastal North Sea water.
	DK3	Grådyb	The city of Esbjerg is the main center of population, and harbor and industrial activities. The area is considered to be dominated by coastal North Sea water.

[a] See Figure 16.2 for locations.
Source: Adapted from Essink et al. (2005).

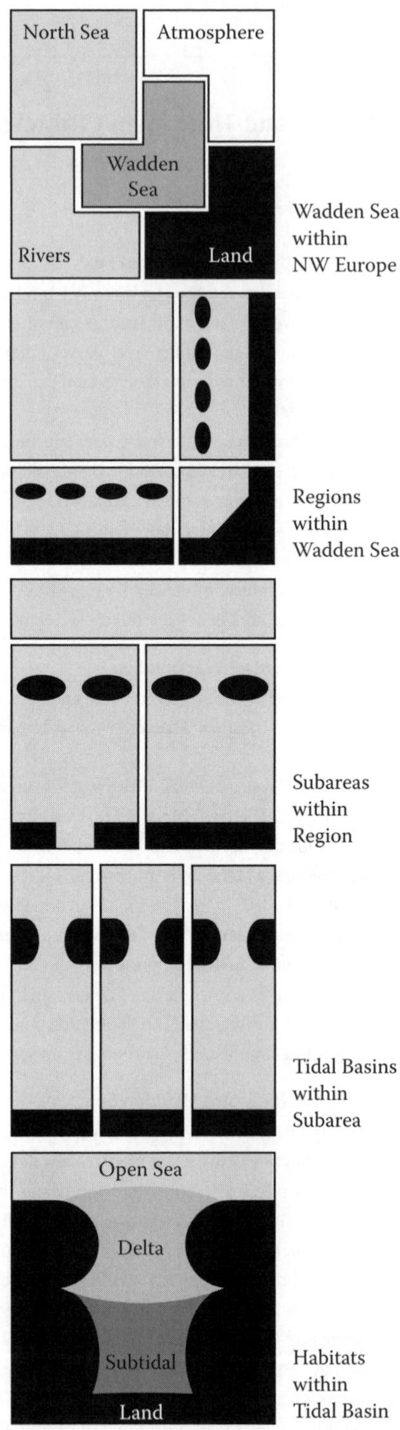

FIGURE 16.1 Basic features of the Waddden Sea at various scales. From top to bottom: the Wadden Sea within NW Europe, main regions within the Wadden Sea (i.e., southwestern, central and northeastern parts; Reise et al. 1995), main subareas within regions (adapted from Essink et al. 2005; see Table 16.2), tidal basins within subareas, and habitats within tidal basins, e.g., open sea, ebb- and flood delta, subtidal areas, intertidal areas, and islands and main land.

storm surges up to 4 m above normal high tide levels. Strong offshore winds on the other hand may dampen low tides to 1.5 m below normal low tide. This asymmetry and the predominance of western onshore winds keep the tidal flats submerged, and long-term emergence is rare. The flushing time of fresh water, as derived from freshwater contents within basins (Ridderinkhof et al. 1990), increases strongly with increasing distance between the freshwater source and the North Sea. Reported flushing times of fresh water amount to average values of 14 tides for the Marsdiep (Postma 1954; Zimmerman 1976) and Lister Tief basin (Hickel 1984), 31 tides for the Vlie basin (Zimmerman 1976), 80 tides for the Ems-Dollard Estuary (Dorrestein and Otto 1960), and up to 120 tides for the Jade Busen (Gillbricht 1956). The flushing time of sea water, as derived from one-dimensional tidally averaged box models, ranges within the Dutch Wadden Sea from 3 tides for the Eierlandse Gat to 25 tides for the Ems-Dollard Estuary (Ridderinkhof et al. 1990). Actual flushing times of fresh and sea water are considered to vary with wind, three-dimensional aspects of current fields, and freshwater discharge (Helder and Ruardij 1982; Ridderinkhof et al. 1990).

The climatic conditions prevailing in northwest Europe, including the Wadden Sea area, are strongly dominated by a permanent atmospheric low-pressure system near Iceland and a high-pressure system over the Azores. The relative strengths and positions of this pressure system may vary from year to year, and this so-called North Atlantic Oscillation correlates well with the variability in weather conditions in the North Atlantic region (Hurrell 1995; Hurrell and van Loon 1997). For the Wadden Sea region, the overall effect is a variable temperate climate, where westerly winds prevail; the mean annual air temperature is 8.5°C, and precipitation is moderate, between 700 and 800 mm yr^{-1}.

With regard to ecosystem processes, the Wadden Sea is not uniform, but can be divided into three main regions (Reise 1995; CWSS 2008) (Figures 16.1 and 16.2):

- Southwestern Wadden Sea—This region extends from the Marsdiep Inlet in the west to the Jade Inlet in the east. Twelve elongated islands, resulting from a tidal range of 1.5 to 3 m, form a sandy barrier 5 to 15 km parallel to the mainland. A large embayment, the former brackish Zuiderzee (3600 km^2), was separated by a causeway (Afsluitdijk) in 1932 and turned into a fresh water lake, Lake IJssel, and agricultural land. The southwest Wadden Sea receives fresh water from Lake IJssel via two sluices in the Afsluitdijk at an average rate of approximately 470 m^3 s^{-1} (van Raaphorst and de Jonge 2004) and from the Ems River at the Dutch-German border at a rate of about 125 m^3 s^{-1} (Helder and Ruardij 1982).
- Central Wadden Sea—This region extends from the Jade Inlet to the Eiderstedt Peninsula. Due to macrotidal conditions (>3 m mean tidal range) resulting in massive volumes of tidal water exchange, barrier islands and ebb-deltas are absent in this region (Ehlers 1988). Major estuaries developed in conjunction with the Weser, Elbe, and Eider rivers, which discharge fresh water at a rate of 310, 700, and 14 m^3 s^{-1}, respectively (Harten and Vollmers 1978). This freshwater input leads to highly variable salinities.
- Northeastern Wadden Sea—This region extends from the Eiderstedt Peninsula in the south to the Skallingen Peninsula in the north. Eight islands and major sand bars form a seaward barrier some 5 to 25 km off the mainland coastline. The tidal range for this region decreases from 3 to 1.2 m in a northward direction. Large estuaries are absent and the discharge of the Varde Å River (12.7 m^3 s^{-1}) only marginally affects the hydrography (Andersen and Pejrup 2001). Salinities range from 27 psu in late winter to 31 psu in summer (van Beusekom et al. 2008).

16.3 MAJOR CHANGES

Pristine conditions for the Wadden Sea are hard to define since human impact began ~5000 years BP (Wolff 2000). Until ~1000 BP, the Wadden Sea was bordered by very extensive salt, brackish,

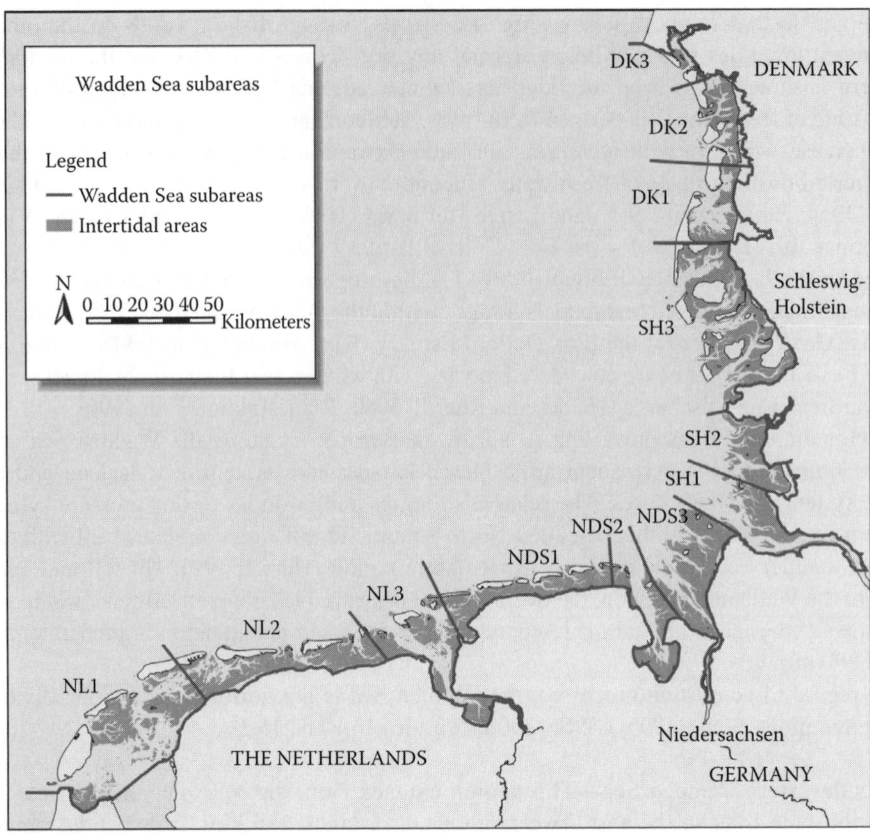

FIGURE 16.2 Map of the Wadden Sea, depicting subareas as described in Table 16.1. (From Essink, K. et al. (eds.). 2005. Wadden Sea Quality Status Report 2004. Wilhelmshaven, Germany: Common Wadden Sea Secretariat. With permission.)

and freshwater marshes and extensive peat bogs, altogether covering an area of the same order of magnitude as the present western Wadden Sea. These marshes were largely reclaimed during the period 1000 to 600 BP, which may have constituted the largest change in the history of the Wadden Sea (Wolff 1992). The exploitation of resources (e.g., hunting birds, mammals, shellfish, and fish) and coastal landscape transformation (e.g., embankments and reclamations of salt marshes) intensified as the coastal population increased, and culminated in large-scale habitat destruction and overexploitation during the twentieth century (Wolff 2000; Lotze et al. 2005). This history of increasing anthropogenic pressure is paralleled by abiotic, biotic, and functional changes. The major changes reported here are based on long-term observations on the system hydrography since the 1880s and on biotic components since the 1930s. Time series data enable us to assess human impacts (e.g., diking and river management) and recent climate change effects on the hydrography and functioning of the Wadden Sea.

16.3.1 Abiotic Conditions

Water temperatures in the Wadden Sea vary on tidal to centennial time scales (Van Aken 2008a). Time series measurements on sea surface temperatures (SST) in the westernmost inlet (Marsdiep) exhibit a clear annual trend in the daily mean SST following the coastal surface air temperature with a lag of a few days only. For the period 2002 to 2006, the daily mean sea surface temperature ranged from 0°C to 23°C, with typical mean amplitude of 7.9°C. The averaged monthly SST data in the Marsdiep over the period 1861 to 2006 indicate that July to September are the warmest months, whereas January to March are the coldest months, ranging from 3°C in February to 18°C in September. Most cold winters with SST <1.0°C occurred during the mid-1900s and warmest winters with SST >6.5°C occurred in 1989 and 1990. Analysis of the 10-year running mean shows a consistent increase in SST for all seasons from 1983 onward (van Aken 2008a).

A reconstruction of storminess over northwestern European starts with rather high levels in the 1880s, a decrease below average around 1930 till the 1960s, and then a pronounced rise to initial values until the mid-1990s. Since the mid-1990s, storminess has been average or below (Matulla et al. 2008). The results for moderate wind events (occurring an average 10 times per year) and strong wind events (that occur on average twice a year) indicate a decrease in storminess over the Netherlands between 5% and 10% per decade between 1962 and 2002 (Smits et al. 2005).

Salinity provides an excellent tracer for river discharge to the various basins. Long-term monthly mean values for the Marsdiep Inlet show a regular annual cycle, with a minimum of 28.85 psu in February and a maximum of 30.72 psu in September. Extremely high salinities can be observed during extended dry periods over large areas of western Europe, when river discharge is low, and the salinity of North Sea water increases as well. The interannual variability reflects the variation of the freshwater input caused by variations in precipitation and river management (e.g., building of dikes and sluices). For the western Wadden Sea, the past 150 years can be divided into three sub-periods: (1) the period 1861 to 1931 (decrease in salinity from 31.5 to 29.5 psu) when the Wadden Sea was the northern part of the Zuiderzee, an estuary with the IJssel River as its major freshwater source; (2) the period 1932 to 1970 (29.6 psu) after the closure of the Afsluitdijk with a natural discharge rate; and (3) the period 1972 to 2003 (28.7 psu) with an increased discharge rate for the IJssel River after the lock weirs in the Nederrijn River became operational (van Raaphorst and de Jonge 2004; van Aken 2008b).

A reconstruction of the total nitrogen (TN) discharge from Lake IJssel into the western Wadden Sea showed a gradual 12-fold increase (20 to 240 mol N s^{-1}) for the period 1935 to 1988, followed by a decrease to levels still five times the 1935 levels. Starting in the early 1960s, the discharge of total phosphorus (TP) increased by a factor of 10 (0.5 to 5 mol P s^{-1}) in 1983, and decreased strongly to values 2.5 times the 1965 levels. The TN:TP atomic ratio of discharge decreased from ~50 in the 1940s to values around 25 in the 1980s and increased steeply to values of 100 around 1995 due to more effective P reduction measures (van Raaphorst and de Jonge 2004). Since the Wadden Sea is a net heterotrophic system (de Jonge and Postma 1974; van Es 1977; Hoppema 1990), its nutrient status is not merely determined by direct nutrient discharge from freshwater sources, but also from the mineralization of autochthonous organic matter and of allochthonous organic matter imported from the coastal North Sea. The atmospheric deposition of nutrients is less important, especially for regions receiving considerable riverine nutrient input.

The import of allochthonous organic matter is considered to be proportional to the stock of organic matter in the coastal zone as modulated by riverine nutrient discharge and primary production in the North Sea. A reconstruction of coastal North Sea productivity suggests that the supply of organic matter and subsequent remineralization would have increased fivefold from 22 g C m^{-2} yr^{-1} in preindustrial times to current values of 110 g C m^{-2} yr^{-1} (van Beusekom et al. 1999, 2005). Earlier estimates for the Marsdiep area suggest an import of ~80 g C m^{-2} yr^{-1} for the 1950s increasing to 155 g C m^{-2} yr^{-1} in the 1970s (correction on de Jonge and Postma 1974 in Postma 1982). For the Ems Estuary, the total annual supply of particulate organic carbon from the coastal area to the estuary was calculated at 148 g C m^{-2} yr^{-1} in the late 1970s (de Jonge 1992).

The concentrations of major nutrients cannot be easily predicted from riverine loads and organic matter import due to the differences in geochemical cycling between N, P, and Si. An analysis of a 20-year (1974 to 1994) dataset on nutrient concentrations revealed that the annual loads of TP from Lake IJssel and concentrations of inorganic phosphate in the western Wadden Sea were significantly correlated in early spring. Phosphate concentrations were relatively high during the period of increased phosphorus loads until the mid-1980s, and declined in parallel with the decreasing phosphorus loadings during the 1990s. For silicon, both loads and concentrations significantly increased during (the last two decades of) the study period. For nitrogen, no significant relationship was found between loads and concentrations, and neither loads nor concentrations of nitrogen significantly differed for the various discharge periods (Philippart et al. 2007b). In contrast, the TN input to the North Sea from the Rhine and Meuse rivers before and during the phytoplankton growth season explained 50% of the interannual variability in ammonium and nitrite in the Wadden Sea during autumn (van Beusekom 2005). This may again signify the importance of organic matter import from the North Sea for the nutrient status of the Wadden Sea.

Annually, an amount of 8 to 21 × 10^6 m^3 or 20 to 55 × 10^6 t of sand, originating from the shore face, beaches, and dunes of the Dutch North Sea coast and North Sea sediments, is imported by bed-load transport and deposited in the southern Wadden Sea (Winkelmolen and Veenstra 1974; Anonymous 1981; Oost and de Haas 1993). These sands, composed of quartz, feldspars, carbonates, and micas, comprise 87% of the tidal flats (van Straaten 1964). The remaining 13% is a fine-grained fraction, composed of clay and iron minerals, carbonates, and organic matter and originates from seafloor erosion in the Channel and the Flemish Banks, from riverine input of the Rhine River, and from autochthonous production. Its annual loading is estimated at 0.8 to 3 × 10^6 t (Eisma 1981). Although the Rhine River contributes only 10% to 20% of the total input, short-term variations in annual mean suspended matter (SPM) concentrations for the western Wadden Sea appear to be related to the discharge of the Rhine River more strongly though with the disposal of dredge spoil in the outlet area of the Rhine in the North Sea (de Jonge and de Jong 2002).

The pre-1950s background SPM concentration, without disposal, has been estimated at <15 g m^{-3}. For the period 1970 to 2000, the overall annual mean SPM concentrations increased by a factor of 2 to 3, peaking to a factor of 5 to 6 for individual annual mean values. Based on unchanged dredging policies and expected changes in wind climate and river discharge, SPM concentrations are predicted to increase 20% above the present levels in the next 50 years and may increase water column turbidity (de Jonge and de Jong 2002). Suspended sediments and phytoplankton are the primary factors controlling water column turbidity (e.g., Postma 1961; van der Heide et al. 2007). Recent estimates indicate that the average turbidity of the westernmost basins of the Wadden Sea has increased by a factor of 5 from ~1.2 m^{-1} to 6.8 m^{-1} since the disappearance of the extensive seagrass meadows in the early 1930s (van der Heide et al. 2007).

Sediments are sorted by hydrodynamic action (Postma 1961), leaving the coarse-grained sands near the inlet and major channels, and progressively smaller fractions toward the mainland (Winkelmolen and Veenstra 1974). Fine materials are transported to shallow areas of the inner margins, tidal watersheds, and mainland coast chiefly by tidal asymmetry between the ebb and flood tidal current (Postma 1961; Groen 1967). For each tidal basin with its inlet, tidal flats, and channels, a dynamic equilibrium is sustained between the hydrodynamic energy and basin morphology. Moderate and temporal changes in hydrology can be compensated for by sediment redistribution and adaptation in morphology until the original equilibrium is restored. Permanent or severe changes may result in a transition toward a new dynamic equilibrium (CPSL 2005).

The Wadden Sea, however, does have a high resilience to counter local and moderate perturbations from equilibriums due to the high mobility potential of the sediment. Sea level rise and enhanced meteorological forcing may continuously and increasingly challenge this resilience. Due to morphological diversity and a high spatial and temporal variability in sediment transport potential and hydrodynamic conditions, deviations from the dynamic equilibrium are hard to detect in an early stage and various and even opposing trends may be detected within close vicinity.

Eight years of monitoring in the northern Wadden Sea showed that the short-term changes in bed level due to temporal deposition and erosion exceeded annual net deposition by two orders of magnitude. Mudflat deposition dominated in spring, summer, and early autumn, whereas erosion dominated during the remaining period of the year. For the northern Wadden Sea, long-term sedimentation has kept pace with sea level rise since the Holocene, and it is argued that the fine-grained tidal flats in this region are not seriously threatened by the expected sea level rise for the twenty-first century (Andersen et al. 2006).

For a nearby somewhat larger area, Dolch and Hass (2008) reported a steady increase in the proportion of sandy flats between the 1930s and the 2000s, whereas for the period 2003 to 2006 increased mud deposition in the innermost part and mud removal from the outer parts were observed. These changes were attributed to an increase in sea level, hindering mud deposition in the exposed areas, and to a decrease in storm events since the mid-1990s reducing resuspension of accumulated mud in the more sheltered inner areas. In the western Dutch Wadden Sea, almost every intertidal flat (>2 to 4 km from the coast) turned more sandy between the 1950s and the 1990s, while tidal flats to the east remained unchanged or became more silty when positioned along mainland borders (Zwarts 2004). Overall, the basins in the western Wadden Sea have shown net accretion since the closure of Lake IJssel in 1932 (Dronkers 2005), probably as a result of a decreased tidal flow and a change in the tidal wave asymmetry. The erosion anomaly during 1978 to 1982 correlates with an interannual fluctuation in tidal amplitude and tidal asymmetry (Dronkers 2005).

These results serve to illustrate that local trends in response to hydrological changes cannot be generalized for a single basin let alone for the entire Wadden Sea. They also demonstrate the highly variable physicochemical conditions and biotic components characteristic of the system.

16.3.2 Biotic Factors

16.3.2.1 Primary Producers and Productivity

The primary production in the Wadden Sea is dominated by phytoplankton and microphytobenthos and, to a lesser extent, by macroalgae and seagrasses (Table 16.2). The literature values for benthic and pelagic primary production suggest that, on average, the contribution of microphytobenthos to the primary production is about as high as that of the phytoplankton. With increasing depth, however, the productivity by microphytobenthos, decreases most probably due to more limiting light conditions (Table 16.2). Locally, the production by seagrasses (*Zostera noltii* and *Z. marina*) can still be high, but the contribution to the overall productivity of the Wadden Sea is presently low as the result of their overall low abundance (e.g., Philippart et al. 1992; Philippart and Cadée 2000; van Beusekom et al. 2005). Unfortunately, the production data are too limited, too scattered, and too different in regard to methodological origin to identify different regions and habitats.

Over the last decades, a two- to threefold increase in phytoplankton and microphytobenthos primary production has been reported throughout the entire Wadden Sea (Cadée 1984; Cadée and Hegeman 1993; de Jonge et al. 1996; Asmus et al. 1998). Subsequent nutrient reduction appears to have been followed by a more gradual decrease in primary production of phytoplankton while its biomass remained more or less constant, which implies that the P:B ratio of the phytoplankton was not constant over time (Philippart et al. 2007b). At present, the annual summed pelagic and benthic production is estimated at ~300 g C m^{-2} for the western Dutch Wadden Sea. By applying present-day relationships between winter nutrient concentration, annual primary production, and annual organic matter turnover rates, carbon budgets have been extrapolated to preindustrial times. For preindustrial conditions, the annual primary production for the western Wadden Sea has been estimated at 40 to 55 g C m^{-2} (Postma 1954; de Jonge 1990; van Beusekom 2005).

The phytoplankton community composition for the western Wadden Sea shifted between 1975 and 1978, and between 1987 and 1988, presumably in response to nutrient enrichment and nutrient reduction, respectively (Philippart et al. 2000, 2007b). During the first shift, phytoplankton biomass and overall cell abundance increased for most groups, except for large flagellates. During the second shift, however, overall phytoplankton biomass did not change; (smaller) flagellates further increased whereas diatom concentrations declined (Philippart et al. 2007b).

Analysis of long-term field observations (1974 to 2007) on total phytoplankton biomass in the western Wadden Sea showed no long-term trends in the timing of the wax and wane of phytoplankton

TABLE 16.2
Overview of Incidental Primary Production (g C m^{-2} yr^{-1}) Data in Various Regions of the Wadden Sea

Region	Period	Zone	Production	Additional Information	Source
			Phytoplankton		
SW	1960s	Subtidal	152	Average depth = 12 m	Postma and Rommets (1970)[a]
	1960s	Subtidal	93	Average depth = 4 m	Postma and Rommets (1970)[a]
	1970s	Intertidal	20		Cadée and Hegeman (1974b)
	1970s	Subtidal	70	Average depth = 3 m	Cadée and Hegeman (1974b)
	1970s	Subtidal	100		Cadée and Hegeman (1974a)
	1970s	Subtidal	120		Cadée and Hegeman (1974a)
	1970s	Subtidal	55		Cadée and Hegeman (1974a)
	1970s	Subtidal	13		Cadée and Hegeman (1974a)
	1970s	Subtidal	150		Cadée and Hegeman (2002)
	1970s	Subtidal	70	Inner parts	Colijn (1983)
	1970s	Subtidal	91	Middle parts	Colijn (1983)
	1970s	Subtidal	283	Outer parts	Colijn (1983)
	1980s	Subtidal	350		Cadée and Hegeman (2002)
	1980s	Subtidal	303		Veldhuis et al. (1988)
	1990s	Subtidal	300		Cadée and Hegeman (2002)
C	1990s	Subtidal	152		Tillman et al. (2000)
NW	1980s	Intertidal	27	*Arenicola*-flat	Asmus and Asmus (1985)
	1980s	Intertidal	90	*Zostera noltii* bed	Asmus and Asmus (1985)
	1990s	Subtidal	160		Asmus et al. (1998)
	2000s	Subtidal	146	Basin-wide	Loebl et al. (2007)
			Microphytobenthos		
SW	1970s	Intertidal	100	Balgzand and Vlieland	Cadée and Hegeman (1974b)
	1970s	Intertidal	130	Texel	Cadée (1980)
	1970s	Intertidal	50	Low (ca. 35% emersion)	Colijn and de Jonge (1984)
	1970s	Intertidal	250	High (ca. 80% emersion)	Colijn and de Jonge (1984)
	1970s	Intertidal	25	*Amphipleura*	van den Hoek et al. (1979)
	1970s	Intertidal	80		van den Hoek et al. (1979)
	1970s	Subtidal	10	Average depth = 3m	Cadée and Hegeman (1974b)
	1970s	Subtidal	0	Average depth = 20m	Cadée and Hegeman (1974b)
NE	1960s	Intertidal	430		van den Hoek et al. (1979)
	1980s	Intertidal	121	*Arenicola*-flat	Asmus and Asmus (1985)
	1980s	Intertidal	152	*Nereis-Corophium* belt	Asmus and Asmus (1985)
			Macroalgae		
SW	1970s	Intertidal	100	*Enteromorpha, Ulva*	van den Hoek et al. (1979)
			Seagrasses		
SW	1930s	Subtidal	300	*Zostera marina*	van den Hoek et al. (1979)
	1950s	Intertidal	14	*Zostera marina/noltii*	van den Hoek et al. (1979)
	1970s	Intertidal	6	*Zostera marina/noltii*	van den Hoek et al. (1979)
C	1930s	Intertidal	175	*Zostera noltii*	van den Hoek et al. (1979)
NE	1980s	Intertidal	459	*Zostera noltii*	Asmus and Asmus (1985)

Note: Abbreviations: SW = southwestern, C = central, NE = northeastern.

[a] Corrected data in Cadée and Hegeman (1974b).

spring blooms. There is weak evidence, however, that the autumn bloom has decreased from the 1970s to the 2000s. This fading of the autumn bloom may have had consequences for the carbon transfer to higher trophic levels, currently hampering primary consumer species that mostly rely on food supply during late summer (Philippart et al. 2010).

16.3.2.2 Secondary Producers and Productivity

For the Wadden Sea, microzooplankton grazing pressure is lower than the average 40% of primary production as reported by Duarte and Cebrian (1996) for coastal environments. Microzooplankton grazing does not impact the diatom fraction during phytoplankton blooms in spring, but may have a large impact on blooms of *Phaeocystis* and during summer blooms, implying efficient mass and energy conservation for the pelagic food web (Riegman et al. 1993; Brussaard et al. 1995; Loebl and van Beusekom 2008). For the western Wadden Sea, the mean abundance of the herbivorous pelagic copepod *Temora longicornis* increased from ~0.5 individual dm^{-3} in 1973 to 1976 to ~1 individual dm^{-3} in 1977 to 1983 to ~2.5 individuals dm^{-3} in 1990 to 1991 (Fransz et al. 1992). If not out-competing other pelagic grazers, this may have increased the flux of phytocarbon to primary consumers between the 1970s and the early 1990s. For the northeastern Wadden Sea, the herbivorous *Acartia* sp. occurred earlier in the season particularly after 1987, and the share of carnivorous zooplankton (mainly due to a sudden increase in *Obelia geniculata* after 1997) to total zooplankton biomass increased between 1984 and 2005 (Martens and van Beusekom 2008).

On the scale of estuaries, Herman et al. (1999) showed that between 5% and 25% of the annual primary production is consumed by macrozoobenthos, mostly deposit-feeding bivalves, gastropods, and polychaetes (Beukema et al. 2002). Pathways of food depend, however, on the trophic state of the estuary; high organic enrichment appears to favor filter feeders, while deposit-feeding organisms predominate in areas with a low supply of organic matter (Pearson and Rosenberg 1987). The macrozoobenthos community in the southwestern part of the Wadden Sea changed at the end of the 1970s by an increase in biomass of deposit- and mixed-feeding benthos and by a change in species composition, and at the end of the 1980s, when only a shift in species composition was observed, while biomass remained more or less at the same level (Philippart et al. 2007b). With regard to overall filtering capacity, the increase in relatively slow filtering sandgapers (*Mya arenaria*) during the last decades could not compensate for the concurrent decrease in the filtering capacity of cockles and mussels. This decrease in bivalve filtering capacity on the tidal flats since the late 1980s suggests that the top-down control of the filter-feeding bivalves on the phytoplankton was lower during the 1990s than during the 1980s (Philippart et al. 2007b).

Data on secondary production are scarce, and often restricted to a single species at a particular location. For cockles (*Cerastoderma edule*) living on a tidal flat near Spiekeroog Island, the secondary production amounted to 20 and 30 g C m^{-2} yr^{-1} at different stations along a tidal gradient in 1994 to 1995 (Ramón 2003, using conversion factors in Salonen et al. 1976). At the Balgzand tidal flats, the 31-year mean (1973 to 2003) of annual net secondary production (spring/summer minus subsequent autumn/winter) of this species varied from 0 to 15 g C m^{-2} yr^{-1}, with a 31-year average of 3.4 ± 0.6 g C m^{-2} yr^{-1} (Beukema and Dekker 2006). At the latter location and for the same study period, the among-cohort values of secondary production of *Macoma balthica* varied between less than 0.5 and more than 2.5 g C m^{-2} yr^{-1}, and the observed variation could be attributed to year-to-year variation in recruitment (van der Meer et al. 2001). Network analyses of the ecosystem of the Sylt-Rømø Bight, a semienclosed basin situated between the islands of Sylt (Germany) and Rømø (Denmark), resulted in an estimate for secondary production by zooplankton of 0.54 g C m^{-2} yr^{-1}, by benthic grazing macrozoobenthos (snails) of 2.74 g C m^{-2} yr^{-1}, and by suspension-feeding macrozoobenthos (bivalves and polychaetes) of 24.26 g C m^{-2} yr^{-1} (Baird et al. 2008).

16.3.2.3 Higher Trophic Levels

The Wadden Sea accommodates pelagic and benthic fish species that are (near) resident, or visit the area during a particular season, a particular phase of their life cycle, or during their migration between the sea and the rivers (Zijlstra 1978). It is an important nursery area for sole (*Solea solea*), dab (*Limanda limanda*), and especially for plaice (*Pleuronectes platessa*) (Zijlstra 1972; van Beek et al. 1989; Bolle et al. 1994). The abundance of juvenile flatfish was relatively high in the 1970s but has declined to low levels since 1980 (Vorberg et al. 2005). For plaice, the decline in local densities is in line with a large-scale offshore shift of juveniles to deeper waters in the adjacent North Sea (Pastoors et al. 2000; Grift et al. 2004).

The estuarine birds of the Wadden Sea may be grouped into species consuming filter-feeding bivalves, deposit- or mixed-feeding macrozoobenthos, or other prey (Leopold et al. 2004). For the Dutch Wadden Sea, the numbers of birds (oystercatchers, herring gulls, and eiders) feeding on shellfish increased between the mid-1970s and the end of the 1980s to early 1990s and decreased thereafter, while the numbers of bird species that mainly feed on polychaetes exhibited an opposite trend (Leopold et al. 2004; van Roomen et al. 2005). These shifts in species composition of estuarine avifauna appear to have been limited to the Dutch Wadden Sea (van Roomen et al. 2005). Recent trend analysis for the international Wadden Sea showed that out of the 33 bird species examined, 8 species had an overall moderate to strong increase, 11 species a moderate to strong decrease, and the remaining 14 species a more or less stable abundance between the period 1987–1988 and 2003–2004 (Figure 6 in Blew et al. 2007).

At present, the Wadden Sea is inhabited by harbor seals (*Phoca vitulina*), gray seals (*Halichoerus grypus*), and harbor porpoises (*Phocoena phocoena*). On average, the number of both species of seals has gradually increased during the last decades. Harbor seals, however, were struck by the phocine distemper virus in 1988 and 2002 which resulted in an instant reduction of 60% and 47% of the population, respectively (Reijnders et al. 1997; Essink et al. 2005). After four centuries of absence, gray seals returned to the Wadden Sea in the 1980s, and the first pups were born in 1985 (Reijnders et al. 1995). After two decades of absence, harbor porpoises returned to Dutch coastal waters, including the Wadden Sea, in the mid-1980s and their numbers have strongly increased (41% per year) since the mid-1990s, most probably the result of redistribution of these mammals in the North Sea (Camphuysen 2004).

16.4 MAJOR HISTORICAL DRIVERS

Throughout the history of Wadden Sea observations, several drivers have been identified that changed the functioning of the Wadden Sea ecosystem (Table 16.3). Some have acted for a restricted period of time and have been countered by effective measures and policies, whereas others continue to affect the functioning of the system. Most of the identified drivers, presented here in chronological order (cf. Wolff 2000), have been treated as isolated structuring factors. More likely, several drivers have been acting in concert which may have amplified or suppressed responses to individual drivers. The reduction in nutrient supply, for example, has not been followed by a decrease in productivity and biomass of phytoplankton. This lack of response may be due to concurrent ecosystem changes (e.g., lower stocks of filter-feeding bivalves) that prevent a restoration according to the original nutrient-primary production relationships (Figure 16.3).

Some of the observed changes for the Wadden Sea coincided with large-scale shifts in the North Sea (Weijerman et al. 2005) and the English Channel (Southward et al. 2005). Changes in weather conditions, temperature, and salinity preceded the shifts in 1979 and 1988. Since the hydrography of the Wadden Sea is intimately linked to the North Sea and the English Channel by coastal transport, these simultaneous changes suggest a large-scale common driver, such as climatic conditions. However, at present we may only speculate on the causality of these simultaneous shifts.

In future years, warming, sea level rise, invasions of introduced species, and variations in nutrient supplies will be the most pressing factors affecting the changes in the Wadden Sea (Reise and van Beusekom 2008). Because climate change and weather variability are intertwined with all other major drivers (e.g., fluxes in nutrient loads and proliferation of invasive species) and will be the major drivers of the future, we will discuss climate change and its associated drivers below (see Section 16.5).

TABLE 16.3
Overview of Indicators of Major Changes in Ecosystem Function in the Western Wadden Sea and Their Possible Causes from the 1930s to the 2000s

Year(s)	Observed Indicator	Possible Structuring Factors								References
		Infrastructure	Pollutants	Nutrients	Fisheries	Diseases	Invasions	Climate	Other	
1930s										
	Continuation of decrease in salinity	√								van Aken (2008a)
1932	Increase in tidal amplitudes	√								de Jonge et al. (1993)
1932	Increase (10%–25%) in current velocities	√								Thijsse (1972); de Jonge and de Jong (2002)
	Disappearance of subtidal seagrass beds	√				√				den Hartog (1987); Van der Heide et al. (2007)
	Decrease in transparency								√	van der Heide et al. (2007)
	Coarsening of sediment of central parts	√								Zwarts (2004)
	Sedimentation of silt near sheltered borders	√								Zwarts (2004)
1940s										
	Rapid accretion in parts of the western Wadden	√								Glim et al. (1987)
1960s										
	Decline of intertidal seagrass beds			√						den Hartog (1987); Philippart et al. (1992)
	Disappearance of beds of European flat oysters				√					Postuma and Rauck (1978)

(continued on next page)

TABLE 16.3 (continued)
Overview of Indicators of Major Changes in Ecosystem Function in the Western Wadden Sea and Their Possible Causes from the 1930s to the 2000s

Year(s)	Observed Indicator	Possible Structuring Factors								References
		Infrastructure	Pollutants	Nutrients	Fisheries	Diseases	Invasions	Climate	Other	
	Disappearance of bottle-nosed dolphins	√								Verweij (1975)
	Disappearance of harbor porpoises		√		√					Verweij (1975); Reijnders (1992); Camphuysen (2004)
	Reduced pup production of harbor seals		√							Reijnders (1976, 1978, 1980)
1970s										
Mid-1970s	Increase of herbivorous copepods			√						Fransz et al. (1992)
1977–78	Increase in phytoplankton biomass and production			√						Cadée and Hegeman (2002)
1977–78	Shift in phytoplankton species composition			√						Philippart et al. (2000, 2007b)
1979	Maximum tidal exchange with adjacent North Sea								√	Dronkers (2005)
Late 1970s	Increased biomass and production of microphytobenthos			√						Cadée (1984)
Late 1970s	Introduction of American razor clams									Beukema and Dekker (1995)
Late 1970s	Increase in biomass macrozoobenthos			√						Beukema et al. (2002)

1980s										
	Further increase in herbivorous copepods			√						Fransz et al. (1992)
	Habitat shift of cockles to higher tidal flats				√				√	Zwarts (2004)
Early 1980s	Further decrease in transparency	√								de Jonge et al. (1993)
Early 1980s	Decrease in juvenile flatfish numbers			√	√			√		Vorberg et al. (2005)
Early 1980s	Return of gray seals								√	Reijnders et al. (1995)
1983	Introduction and rapid increase of Pacific oysters				√		√	√		Dankers et al. (2004)
1985	Start of increase in water temperatures							√		van Aken (2008b)
1986	Return of harbor porpoises				√					Camphuysen (2004)
1987–88	Minimum tidal exchange with adjacent North Sea								√	Dronkers (2005)
1987–88	Shift in phytoplankton species composition			√				√		Philippart et al. (2000, 2007b)
1988	Mass mortality (60%) of harbor seals					√				Osterhaus and Vedder (1988)
Late 1980s	Shift in species composition of macrozoobenthos			√				√		Philippart et al. (2007b)
Late 1980s	Decrease in overall filtering capacity of macrozoobenthos			√				√		Philippart et al. (2007b)

(continued on next page)

TABLE 16.3 (continued)
Overview of Indicators of Major Changes in Ecosystem Function in the Western Wadden Sea and Their Possible Causes from the 1930s to the 2000s

		Possible Structuring Factors								
Year(s)	Observed Indicator	Infrastructure	Pollutants	Nutrients	Fisheries	Diseases	Invasions	Climate	Other	References
	Continuation of habitat shift of cockles to higher tidal flats				√				√	Zwarts (2004)
Early 1990s	Decrease in shellfish-eating birds			√	√			√	√	Leopold et al. (2004)
Early 1990s	Increase in worm-eating birds			√	√			√	√	Leopold et al. (2004)
1991–92	Decrease of intertidal beds of blue mussels				√			√		Beukema and Cadée (1996)
1991–92	Mortality shellfish-eating birds				√			√		Camphuysen et al. (1996, 2002); Ens et al. (2004)
1994	Start of strong increase of harbor porpoises				√			√		Camphuysen (2004)
1998	Maximum tidal exchange with adjacent North Sea								√	Dronkers (2005)
2000s										
	Disappearance of phytoplankton autumn blooms			√				√	√	Philippart et al. submitted
2002	Mass mortality (47%) of harbor seals					√				Härkönen et al. (2006)

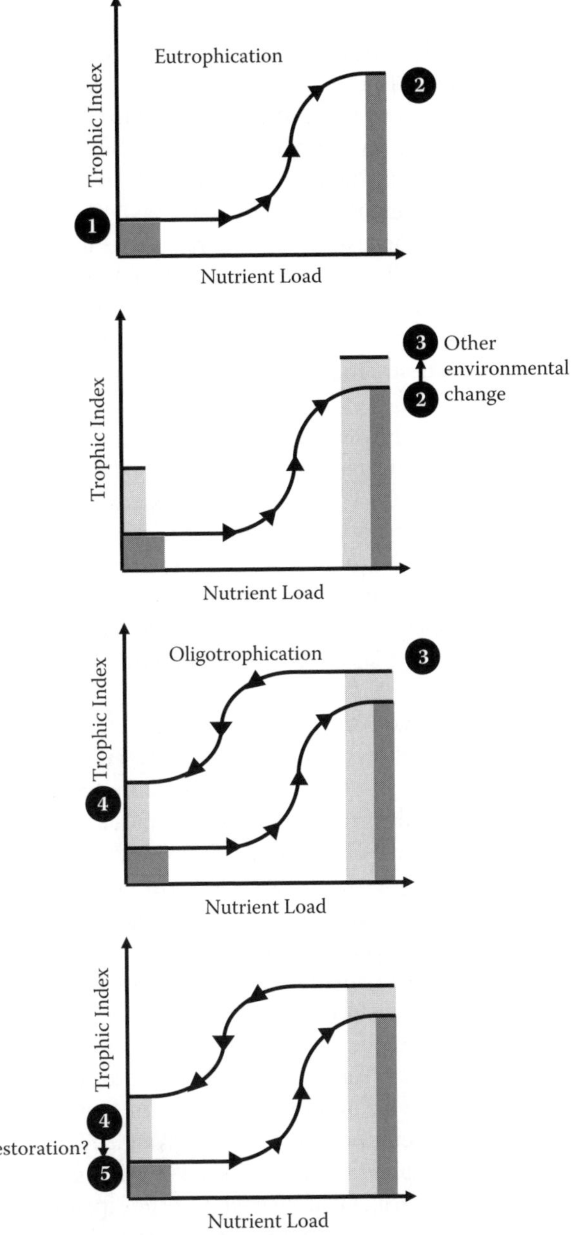

FIGURE 16.3 Conceptual model of environmental changes in the Wadden Sea following nutrient enrichment (eutrophication) and reduction (oligotrophication). From top to bottom: an increase in nutrient loads ("eutrophication") causes a change in a trophic index (e.g., primary productivity) from a low to a higher value, i.e., moving from state 1 (low loads, low trophic index) to state 2 (high loads, high trophic index). During this period of eutrophication, however, other environmental conditions may alter as well as illustrated by the shift from dark gray bars to light gray bars. This may result in a shift from state 2 (high loads, high trophic index) to state 3 (similarly high loads, higher trophic index). Such a state change may be induced, for example by a decline in stocks of filter-feeding bivalves as the result of overexploitation or a decrease in turbidity due to lowered wind stress. Reduction of nutrient loads to historical levels is followed by a reduction in trophic index, i.e., from state 3 to state 4. Because the Wadden Sea environment now differs from the historical one, the relationship between loads and trophic index might also have been changed, preventing a relaxation to initial conditions (i.e., from state 4 to state 5).

16.4.1 Overexploitation

A historical inventory of extirpations of marine and estuarine species in the Dutch Wadden Sea revealed that the decline in at least 17 of the 42 species could be attributed to hunting and fishing (Wolff 2000). Exploitation was too strong to be sustained for the European flat oyster, several large species of fish (e.g., thornback ray, sting ray, and sturgeon), a few species of coastal birds (e.g., Eider), and the gray seal (Wolff 2005a). In the northeastern Wadden Sea, comparisons of present species inventories with historical surveys in tidal channels suggest that a decline of nearly 50% of all epifaunal species within the last hundred years may be attributed to fishery disturbances (Buhs and Reise 1997). Here, the sessile and slow-moving animals have declined, while fast-moving decapod crustaceans have persisted (Buhs and Reise 1997). Apart from the direct effects on stock and recruitment of biota, the frequent reworking of surficial sediments may affect the sediment granulometry, resulting in habitat destruction and a further decline in recruitment success (Piersma et al. 2001).

At present, the main fisheries in the Wadden Sea are those for brown shrimp (*Crangon crangon*), cockles (*Cerastoderma edule*), and blue mussels (*Mytilus edulis*). Shrimp fishing takes place mainly in the subtidal areas of all regions of the Wadden Sea (Essink et al. 2005). Cockle fishing is neither allowed in the German part nor in most of the Danish part of the Wadden Sea (de Vlas et al. 2005). In the Dutch part, mechanical cockle fishing was closed in 2006. Hand-raking of cockles is still allowed and presently intensifying. While commercial-sized mussels are fished in Danish waters, mussel fisheries are mostly restricted to mussel spat from wild natural beds in Dutch and German parts of the Wadden Sea. These juveniles are then dispersed on culture plots to grow to a marketable size.

Shellfish fishing in Danish and Dutch waters was restricted after strong indications of overexploitation. In both areas, a combination of overfishing and extreme winter conditions (a series of severe winters in Denmark and mild winters in the Netherlands) caused a strong reduction in intertidal mussel beds, unprecedented low bivalve stocks, and the emigration and mass mortality of shellfish-eating birds in the 1980s in Denmark (Essink et al. 2005) and in the 1990s in the Netherlands (Beukema and Cadée 1996; Camphuysen et al. 1996, 2002; Ens et al. 2004).

The introduction of mussel culture started in the eastern Dutch Wadden Sea, following a decimation of the cultured mussels in the estuarine southwestern part of the Netherlands in 1949 after a parasitic infection by the copepod *Mytilicola intestinalis* (Korringa 1968; Wolff 2005a). Some years later, these mussel plots were transferred to the western part of the system, when the parasite invaded the German and eastern part of the Dutch Wadden Sea as well. Concurrently, new beds were introduced in the western Wadden Sea (Havinga 1960) in order to compensate for the loss of mussel cultivation plots due to the foreseen damming of all tidal inlets in the south of the Netherlands. Due to this concentration of mussel culture in the western Wadden Sea, *Mytilus edulis* is now one of the most abundant species in the subtidal area (Dekker 1987) and contributes substantially to the bivalve filtering potential of the western Dutch Wadden Sea (Philippart et al. 2007b). In addition, it may provide up to 20% of the food presently required by Eider ducks (Swennen 1991a).

16.4.2 Habitat Change and Destruction

The landscape of the Wadden Sea has been strongly modified through a long history of building dikes and polders, which has intensified over time (Wolff 1992, 2000; Lotze et al. 2005). In the Netherlands, two major dikes were built in 1919 and 1932, closing a large part of the southwestern Wadden Sea in order to prevent flooding of coastal regions surrounding the Zuiderzee and to reclaim land for agriculture and living. After the enclosure of the "Zuiderzee," this brackish inland sea turned into a freshwater lake, "IJsselmeer," due to the freshwater discharge of the IJssel River (the northern branch of the Rhine River), and it was partly impoldered in later years. The rate of flow of Rhine water via the IJssel River to Lake IJssel was 200 to 350 $m^3 s^{-1}$ between 1933 and

1970. Since 1971, the minimum rate of flow has been 400 $m^3 s^{-1}$ (van Raaphorst and de Jonge 2004). In 1971, a series of three lock weirs in the Nederrijn River (a southern branch of the Rhine River) became operational, which resulted in an increase of the IJssel discharge to 17% of the Rhine rate of flow, and a subsequent reduction of the transport of Rhine water toward the Wadden Sea along the Dutch coast (van Aken 2008b). The closure of the Lauwerszee, a bay in the eastern part of the Dutch Wadden Sea, was completed in 1969.

Through this diking and draining, the risk of floods was reduced, but the large-scale changes in coastal configuration such as embankments, elimination of flood plains, exploitation of peat lands, a straight dike line, and the transformation of estuaries into shipping canals also enhanced tidal height and the risk of floods (Reise 2005). The effects of large infrastructural works on the Wadden Sea are highly diverse and range from a change in timing of freshwater and nutrient supply to limiting the dynamic equilibration of sediments between the interior and exterior domains of a barrier basin. For example, the closure of the Zuiderzee was followed by a decline in subtidal seagrass beds and its associated flora and fauna, most probably due to wasting disease (Wolff 2000). Its failure to reestablish, however, was most likely the result of changed hydrographic conditions in the southwestern Wadden Sea, most notably an increase in tidal amplitude (de Jonge et al. 1993). In the southeastern and the northeastern Wadden Sea, the disappearance of typical habitats, such as beds of native oysters (*Ostrea edulis*) and reef-building polychaetes (*Sabellaria spinulosa*), was probably caused by bottom-trawling activities (Reise 1982; Wolff 2000).

16.4.3 Pollution

The rivers in the region are important sources of pollutants to the Wadden Sea, such as heavy metals, PCBs, pesticides, and xenobiotics. During the 1950s and 1960s, the concentrations of pollutants in marine organisms reached very high and even lethal levels in some cases (Koeman et al. 1969). In the Dutch Wadden Sea, the large decrease in the population size of harbor seals (*Phoca vitulina*) between the 1950s and the mid-1970s was attributed to elevated levels of contaminants (e.g., PCBs) that reduced reproduction rates (Reijnders 1980) and lowered immune responses (Brouwer et al. 1989). In the early 1960s, the number of eiders in the largest colony in the Dutch Wadden Sea was reduced by 75% (Swennen 1991b). Sandwich terns were falling dead from the sky in some areas of the western Wadden Sea, and numerous other species all died (Koeman 1971). This mass mortality was the result of the discharge of pollutants (e.g., aldrin, endrin, and telodrin) by a pesticide factory near Rotterdam (Koeman 1971).

Pollutant-induced reduction of the population size of (top) predators may have had cascading effects on lower trophic levels, but indications for such impacts are not recorded. In general, the concentration of pollutants has been significantly reduced during the past decades with major reductions in the input and concentration of metals from the late 1980s until the early 1990s, but targets of background concentrations in sediment and organisms have not been reached for all pollutants and for all regions of the Wadden Sea. However, many newly developed xenobiotics, such as hormone disruptors, have a wide occurrence in the Wadden Sea and may affect its ecosystem (Bakker et al. 2005).

16.4.4 Nutrients

Long-term field observations on phytoplankton dynamics in the western Wadden Sea revealed a steep increase in biomass and production for both phytoplankton (Philippart et al. 2000; Cadée and Hegeman 2002) and microphytobenthos (Cadée 1984) in response to nutrient enrichment during the 1970s. A subsequent reduction in nutrient discharge did not result in a proportional decrease in phytoplankton primary production, although its biomass remained more or less constant (Philippart et al. 2007b). Regrettably, no recent information is available on microphytobenthos due to termination of this time series of observations in the early 1980s (Cadée 1984).

Various indices for nutrient limitation suggest that phosphorus was the main regulator of primary production in the western part of the southern Wadden Sea for phytoplankton in general, while silicon limited diatom growth during the 1970s and early 1980s (Philippart et al. 2000, 2007b). In the 1990s and early 2000s, limitation patterns were dominated by potential PO_4^{-3} and Si limitation in the southern Wadden Sea, whereas potential DIN limitation was observed in the northern Wadden Sea near the island of Sylt (Loebl et al. 2009). The latter area was characterized by optimum light conditions, due to a combination of shallow depth and low turbidity, allowing excessive phytoplankton growth and a seasonal development.

When using nutrient concentrations as indices for their availability during the bloom, however, the relatively rapid and complete recycling of phosphorus compared to that of nitrogen is not taken into account. Although the results on relative nutrient concentrations indicated that phosphorus and silicon were most probably limiting phytoplankton growth in the Wadden Sea, it cannot be excluded that nitrogen inputs may still have driven primary productivity and algal species succession (Philippart et al. 2000, 2007b; Loebl et al. 2009).

In addition to riverine nutrient loads, primary production within the Wadden Sea is supported by nutrient supply and nutrient release from remineralized organic matter that originates from the North Sea (Postma 1961; Philippart et al. 2000; van Beusekom 2005). Elevated nutrient concentrations have been observed over the years in the English Channel and the Straits of Dover between the 1950s and the late 1980s (Laane et al. 1993) and in coastal zones of the Southern Bight of the North Sea between 1930 and 1970 (De Galan et al. 2004). Concentrations appear to be decreasing for some nutrients as a result of the stringent measures taken by different North Sea states to reduce river loads to the North Sea (De Galan et al. 2004).

Among other factors, the magnitude of the exchange of matter between the North Sea and the Wadden Sea is determined by the tidal amplitude, which follows an 18.6-year cycle. This cyclicality, however, may be obscured by the dominant wind field and storminess (Matulla et al. 2008). During the study period, maximum values occurred in 1979 and 1998, and minimum values around 1987 to 1988 (Dronkers 2005). Nutrient enrichment originating from the Rhine River, therefore, may have been strengthened by the relatively high supply of nutrient-rich organic matter from the North Sea in the mid-1970s, while nutrient reduction may resulted from the relatively low import of organic matter into the Wadden Sea at the end of the 1980s.

Since phosphorus and silicon were the least abundant (most limiting) nutrients at the start of the spring bloom, changes in their concentrations may have affected the species and size composition of the phytoplankton community and, not unlikely, also higher trophic levels. Due to the potential complexity of the relationships and a dearth of data on trophic fluxes, however, it cannot be determined conclusively if, and to what extent, nutrient enrichment and subsequent reduction contributed to some of the coinciding changes in microalgae, macrozoobenthos, and birds within the Wadden Sea ecosystem (Philippart et al. 2007b).

16.4.5 Invasive Species

In historical times, new species of plants and animals were introduced to the Wadden Sea via shipping and aquaculture. At least 52 nonindigenous species are now established in the Wadden Sea (Reise et al. 1999; Wolff 1999, 2005b). The soft sediment bivalve *Mya arenaria*, for example, was introduced 400 to 700 years ago (Reise et al. 1999; Strasser 1999) and is presently one of the most important producers of benthic biomass and calcimass in the Wadden Sea (Beukema 1982, 1992).

A few species appeared to strongly affect habitat properties and native biota (Wolff 2005b; van der Weijden et al. 2007). For example, the suspension-feeding potential in the coastal system has been strengthened by the invasion of the American razor clam (*Ensis americanus*), the American slipper limpet (*Crepidula fornicata*), and the Pacific oyster (*Crassostrea gigas*) (Armonies and Reise 1998; Reise 2008). As yet, there is no evidence that introduced species have caused the extinction of native species in the Wadden Sea (Wolff 2000, 2005b). A successful invader, such as the Pacific oyster, is a better competitor for food than are many of the native fauna. However, it is less edible for shellfish-eating crustaceans and birds than native filter-feeders. Moreover, it builds solid calcareous reefs which increase local sedimentation and biodiversity (Cadée 2007; Kochmann et al. 2008) and, given sufficient time, may induce ecosystem changes with an unpredictable future for native species.

A recent conspicuous invader is the gelatinous zooplankter *Mnemiopsis leidyi*, which has been expanding its geographic distribution to northern European waters since 2006, including the Wadden Sea (Faasse and Bayha 2006; Hansson 2006; Boersma et al. 2007). This invader has caused serious biotic problems in the Black and Caspian Seas, where its introduction was followed by a drastic reduction in the zooplankton mass and in a collapse of the fishing industries within 10 years (Shiganova et al. 2004; Daskalov and Mamedov 2007). Most probably, however, overfishing of the anchovy stock, together with climate-induced changes in the food web, enabled this invasive ctenophore to greatly increase in abundance (Bilio and Niermann 2004). As the result of its excessive food consumption rates, large reproductive capacity, and broad ecophysiological plasticity, this species has the potential to markedly proliferate in northern European waters, affecting coastal food webs such as those of the Wadden Sea (Postel and Kube 2008).

16.5 CLIMATE CHANGE EFFECTS ON ECOSYSTEM FUNCTION

The Northern Hemisphere has been warmer since 1980 than for any period during the last 2000 years. Climate models predict an increase for global mean air temperature of 1°C up to 6°C by the year 2100 relative to 1990 (IPCC 2007). In Europe, the average temperature increase will probably be slightly faster than the world average. Based on the most recent results from climate research, KNMI (Royal Netherlands Meteorological Institute) has developed four climate scenarios for the Netherlands (http://www.knmi.nl/climatescenarios/knmi06/) (see Table 16.4). Although the rate of change varies among these scenarios, a number of key characteristics of climate change in the Netherlands and bordering areas are envisioned: (1) increasing temperature levels, (2) increasing frequency of mild winters and hot summers, (3) increasing precipitation during winter, and (4) increasing sea level rise.

The following consequences of the aforementioned climate changes may occur in the Wadden Sea:

- With increasing temperature, the kinematic viscosity of water decreases. As a result, the mobility potential of noncohesive sediments decreases, whereas the particle settling velocities increase (Krögel and Flemming 1998). The overall effect would be a reduced resuspension of sediments and a faster clearance due to sediment settling, and thus a tendency toward lower water column turbidities. This may enhance pelagic and benthic primary production in systems that are light limited such as the southern Wadden Sea. Moreover, an increase in stability of the benthic habitat may result in a higher benthic biomass and production. The response of organisms and the ecosystem as a whole may not only depend on these absolute shifts in temperature, but also on the phasing of the new temperature regimen (tidally, daily, and seasonally) with other key variables as well.
- Enzymatic process rates, like respiration, on average increase by a factor of 2 to 3 with a rise in temperature of 10°C (Eppley 1972). Photosynthesis, however, is temperature sensitive only at light saturating conditions when the carboxylation rate is limited by RuBisCO activity (Crafts-Brandner and Salvucci 2000). At nonsaturating conditions, photosynthesis is far less sensitive to changes in temperature. Therefore, the predicted increase in temperature may reduce the overall growth yield and fitness of individual phototrophic species. As a consequence, the species composition of the phototrophic community may change.
- On the level of higher organisms, the demand for oxygen may exceed the capacity of oxygen supply to tissues at elevated temperatures, restricting whole-animal tolerance to thermal extremes (Portner and Knust 2007; Beukema et al. 2009). Mismatches in metabolic

TABLE 16.4
Climate Change in the Netherlands around 2100 Compared to the Baseline Year 1990, According to the Four KNMI'06 Climate Scenarios

1990–2100	Moderate	Moderate+	Warm	Warm+
Global temperature rise	+2°C	+2°C	+4°C	+4°C
Change in air circulation patterns	No	Yes	No	Yes
Winter				
Average temperature	+1.8°C	+2.3°C	+3.6°C	+4.6°C
Coldest winter day per year	+2.1°C	+2.9°C	+4.2°C	+5.8°C
Average precipitation amount	+7%	+14%	+14%	+28%
Number of wet days (≥0.1 mm)	0%	+2%	0%	+4%
10-day precipitation sum exceeded once in 10 years	+8%	+12%	+16%	+24%
Maximum average daily wind speed/year	–1%	+4%	–2%	+8%
Summer				
Average temperature	+1.7°C	+2.8°C	+3.4°C	+5.6°C
Warmest summer day per year	+2.1°C	+3.8°C	+4.2°C	+7.6°C
Average precipitation amount	+6%	–19%	+12%	–38%
Number of wet days (≥0.1 mm)	–3%	–19%	–6%	–38%
Daily precipitation sum exceeded once in 10 years	+27%	+10%	+54%	+20%
Potential evaporation	+7%	+15%	+14%	+30%
Sea Level				
Absolute increase	35–60 cm	35–60 cm	40–85 cm	40–85 cm

Note: The climate in the baseline year 1990 is described with data from the period 1976 to 2005. The seasons are defined as follows: *winter* stands for December, January, and February, and *summer* stands for June, July, and August. (http://www.knmi.nl/climatescenarios/knmi06/)

balances, food availability, and relationships between predators and prey affect mortality and reproduction rates, and have resulted in shifts from Arctic to Atlantic species in the more northern seas and from temperate to more subtropical species in southern waters (Philippart et al. 2007a). With increasing temperature, more shifts and the establishment of new trophic relationships are to be expected.

- On an ecosystem level, the differential effect of temperature increase on photosynthesis (i.e., very limited effects at nonsaturating light conditions; Crafts-Brandner and Salvucci 2000) and respiration (i.e., an average increase by a factor of 2 to 3 with a rise in temperature of 10°C; Eppley 1972) may also result in a more heterotrophic system with a concomitant reduction in net productivity and coinciding reductions in carrying capacity for higher trophic levels.
- Microphytobenthos can decrease the erodibility of sediment by producing extracellular polysaccharides that bind the sediment (Kromkamp et al. 2006). Warming is expected to result in a high peak diatom density because of the lower grazing by deposit feeders such as *Macoma balthica*, but low summer values of microphytobenthos biomass due to desiccation associated with high summer temperatures (Wood and Widdows 2003). Such circumstances may lead to reduced intertidal erosion and accelerated deposition of sediment advected from offshore (Wood and Widdows 2003).
- The combination of a decrease in oxygen solubility and an increase in sulfate reduction rate with increasing temperature may promote low redox conditions or even the formation

of sulfureta in sediments. This transition may induce major shifts in the geochemistry of sediments (e.g., the mobilization of metals and phosphate, formation of metal-sulfide precipitates, and a reduction in phosphorus adsorption capacity, suppression of nitrification and subsequent denitrification).

- A moderate downshift in redox conditions may force bivalves, such as *Macoma balthica*, to aggregate and reduce their burial depth, rendering them more vulnerable to resuspension and predation by crabs and birds (Epping et al. in preparation). Severe downshifts in redox conditions, as may occur in sheltered areas, may lead to complete surfacing of infauna and mass mortality of benthos. However, the high rates of water renewal for many basins reduce this risk and prevent the development of water column anoxia.
- For the Wadden Sea, an increased frequency of mild winters will favor macrobenthic species that are sensitive to cold winters, and result in greater weight loss in all bivalves during the winter and low reproductive success in the subsequent summer for various important (bivalve) species (e.g., *Mya arenaria, Mytilus edulis,* and *Cerastoderma edule*) (Beukema 2002; Beukema et al. 2009), possibly as a consequence of enhanced predation on bivalve spat by juvenile shrimps (Beukema 1992; Philippart et al. 2003). The reproductive success of warm-water species, such as *Crassostrea gigas*, may be favored by higher temperatures in (late) summer (Diederich et al. 2005).
- An increase in river runoff and subsequent lowering in salinity may lead to shifts from marine to more brackish species such as a shift within macrophytes from seagrasses (*Zostera noltii*) to widgeon grass (*Ruppia maritima*) (van Katwijk et al. 2005) and a shift within polychaetes from lugworms (*Arenicola marina*) to nereid polychaetes (*Nereis diversicolor* and *N. virens*) (Zipperle and Reise 2005).
- Rising sea level not only inundates low-lying coastal regions but also contributes to the redistribution of sediment along sandy coasts. Long-term beach erosion may increase due to accelerated sea level rise (SLR) and may eventually lead to the deterioration of barrier chains such as the Wadden Sea (Fitzgerald et al. 2008). Plans for the development of typical coastal features such as washovers to restore the historical transport of sand from the open sea to the lee sides of the islands are presently under discussion (Hillen and Roelse 1995). If SLR is not compensated by sedimentation, tidal exchange through inlets increases, which leads to sand sequestration in tidal deltas and (further) erosion of adjacent barrier shorelines.
- Higher annual river runoff will inevitably result in proportionally higher TN and TP outputs from Lake IJssel to the Wadden Sea, thereby counteracting ongoing de-eutrophication measures and stimulating coastal eutrophication (van Raaphorst and de Jonge 2004). Secondly, it may increase the input of suspended matter from the Rhine River, which comprised 10% to 20% of the total supply during the past decades. A reduction in the transparency of the Wadden Sea may affect the absolute and relative production by phytoplankton and microphytobenthos.
- A change from a rather constant to a strongly pulsed supply of riverine nutrients may drastically change the microalgal competition for nutrients and favor large phytoplankton species (Stolte and Riegman 1995; Roelke et al. 2003). Because of their larger storage capacity, these species are better competitors under high and fluctuating nutrient regimes than are smaller algae (Sommer 1984; Stolte et al. 1994; Grover 1997). Reduced grazing pressures and enhanced rates of sedimentation may result in enhanced carbon and energy flux to benthic communities (Thingstad and Sakshaug 1990; Riegman et al. 1993).
- Earlier seasonal onset of zooplankton grazing due to increased spring temperatures (Wiltshire and Manly 2004) may restrain the peak in microalgal biomass during the spring phytoplankton bloom and promote regenerative production by flagellates at the expense of diatoms.

- Changes in the balance between production by phytoplankton and microphytobenthos may affect the cycling of energy and matter (e.g., via pelagic zooplankton vs. filter-feeding macrozoobenthos), the food availability for macrozoobenthos (e.g., filter-feeding bivalves vs. deposit-feeding polychaetes), and the fate of energy and matter (e.g., export through pelagic communities vs. local retention in benthic communities).
- Changes in standing stocks of filter-feeding bivalves may affect algal stocks due to changes in grazing pressure, affect growth of other benthic species due to reduced or enhanced competition for food, and change the food availability for shell-fish eating fish and birds (Beukema 2002; Cadée 2008).
- Changes in reef-building bivalves (such as *Mytilus edulis* and *Crassostrea gigas*) will affect the balance between erosion and sedimentation and, as the result of supplying specific habitats for marine flora and fauna, the local biodiversity (Cadée 2007).
- Changes in macrozoobenthos communities will change the effects of this fauna on the sediment (e.g., by means of stabilization, destabilization, and bioturbation), subsequently affecting bio-irrigation of oxygen and nutrients, survival of resting stages of planktonic organisms such as copepods, diatoms, and dinoflagellates, and the burial rates of "solid particles" such as clay, organic matter, eggs, and shells (Reise 2002; Widdows and Brinsley 2002; Meysman et al. 2006).

In conclusion, warming may stress the present structure and function of the food web and may result in a cascade of yet unknown effects, in particular when interacting with other stressors such as invasions of introduced species, changes in nutrient supply, and sea level rise (Reise and van Beusekom 2008). Extreme scenarios involve major changes in the present balance between surface of tidal and subtidal areas, autrotrophy and heterotrophy, pelagic and benthic production, and import and export of energy and matter (Figure 16.4). These shifts may change present-day ecosystem functioning and have consequences for possibilities and limits of sustainable use and for the protection of natural ecosystems and their services. Resilience (i.e., the ability of ecosystems to absorb recurrent disturbances) may be achieved locally by limiting destructive human activities (e.g., overexploitation and pollution) and thereby reducing the likelihood of undesirable phase shifts. Such a resilience-based approach represents a fundamental change of focus, from reactive to proactive management, aimed at sustaining the socioeconomic and ecological values of marine systems in an increasingly uncertain world (Hughes et al. 2007).

16.6 FUTURE RESEARCH

Coastal waters are most vulnerable due to past and current activities and changes in climate settings. A Dutch, trilateral, and European-integrated coastal monitoring network is needed to keep close track of variations and trends in the coastal environment, as well as to examine the effects of human activities.

National and European policies tend to focus on state variables (such as nutrient concentrations and phytoplankton biomass), which may signal environmental changes, but are of limited use for identifying the causes of change (Philippart et al. 2007a). Thus, there is a need to extend conventional monitoring programs and assessment of key processes. Since ecosystem functioning often comes down to species and their interactions, we should encourage the detailed work of monitoring marine organisms to species level, including their larval and post-larval stages (early warning system approach). In order to project the consequences of environmental changes for biodiversity, we have to assess the plasticity of (key) species to environmental conditions.

This knowledge should be integrated in mechanistic ecosystem models as a tool for supporting the management of human activities in the marine environment (e.g., examining scenarios of nutrient reduction, protected areas, and exploitation of marine resources). Effort should focus on ensuring that the level of detail (complexity) of the models matches what can and will be measured,

FIGURE 16.4 Extreme scenarios of warming when interacting with other stressors such as invasions of introduced species, changes in nutrient supplies, and sea level rise. Top panel: Increase in grazing pressure on phytoplankton due to increase of herbivorous zooplankton and filter-feeding bivalves such as Pacific oysters suppressed coastal phytoplankton blooms, resulting in clear water and facilitating birds that prey on fish. Reduction of deposit-feeding bivalves such as the Baltic telling results in proliferation of microphytobenthos, reducing the resuspension of sediment and therefore enhancing the clarity of the water. Improved light conditions stimulate growth of seagrasses. Due to full compensation of sea level rise by sedimentation, the tidal flats and the associated flora and fauna (including waders such as Oystercatchers) remain in the area. Bottom panel: Because sedimentation cannot keep pace with sea level rise, the tidal flats drown and the associated flora and fauna is lost. Only flora and fauna that can live in permanently flooded areas (such as eiders) remain. Increased abundances of juvenile shrimp and crabs prey heavily upon young bivalves, reducing stocks of filter-feeding bivalves. Reduced grazing pressure promotes phytoplankton blooms, and subsequently reduces light and nutrient availability for microphytobenthos. Loss of stabilizing microphytobenthos and an increase in frequencies of storms promote resuspension of sediment, resulting in more turbid waters. Invasive jellyfish prey on fish larvae and subsequently reduce the food availability for fish-eating birds. (See text for references.)

including species-specific data. The latter will enable appropriate parameter values to be obtained and ensure that the models are realistic and not unnecessarily complicated.

As a summary of strategic research topics to adequately understand and project the consequences of climate change for biodiversity and ecosystem functioning of the sea, we need to (1) extend and further integrate our coastal long-term field observations, (2) extend our knowledge of sensitivities and adaptation capabilities of key species in the marine environment, and (3) optimize hydrodynamic-ecological models and end-to-end food web models to predict and evaluate responses of our coastal areas to managerial activities. More explicitly, we advocate including measurements of process rates to present monitoring efforts on state variables at several specific locations, reflecting the various habitats and large-scale variation throughout the Wadden Sea. These locations are to be designated as Long-Term Ecological Research sites, and studied coherently, not only for monitoring, but also as focus points for experiments and for teaching and learning about the ongoing changes in ecosystem function and the factors that are underlying these changes.

ACKNOWLEDGMENTS

This work was supported by the National Programme Sea and Coastal Research funded by The Netherlands Organization for Scientific Research (NWO). We greatly appreciated the valuable comments by Victor de Jonge, Michael Kennish, Karsten Reise, Wim Wolff, and one anonymous referee.

REFERENCES

Andersen, T. J. and M. Pejrup. 2001. Suspended sediment transport on a temperate microtidal mudflat, the Danish Wadden Sea. *Mar. Geol.* 173: 69–85.

Andersen, T. J., M. Pejrup, and A. A. Nielsen. 2006. Long-term and high-resolution measurements of bed level changes in a temperate, microtidal coastal lagoon. *Mar. Geol.* 226: 115–125.

Anonymous. 1981. *Zandwinning in de Waddenzee, resultaten van een hydrografisch-sedimentologisch onderzoek.* Stuurgroep, Rijkswaterstaat, Directie Friesland, Leeuwarden, the Netherlands.

Armonies, W. and K. Reise. 1998. On the population development of the introduced razor clam *Ensis americanus* near the island of Sylt (North Sea). *Helgoländer Mar. Res.* 52: 291–300.

Asmus, H. and R. Asmus. 1985. The importance of grazing food chain for energy flow and production in three intertidal sand bottom communities of the northern Wadden Sea. *Helgoländer wissenschaftlichen Meeresuntersuchungen* 39: 273–301.

Asmus, R., M. H. Jensen, D. Murphy, and R. Doerffer. 1998. Primary production of the microphytobenthos, phytoplankton and the annual yield of macrophytic biomass in the Sylt-Rømø Wadden Sea (In German, English abstract). Pp. 367–391, In: C. Gätje and K. Reise (eds.), *Ökosystem Wattenmeer—Austausch-, Transport- und Stoffumwandlungsprozesse*. Berlin: Springer-Verlag.

Baird, D., H. Asmus, and R. Asmus. 2008. Nutrient dynamics in the Sylt-Rømø Bight ecosystem, German Wadden Sea: An ecological network analysis approach. *Estuar. Coastal Shelf Sci.* 80: 339–356.

Bakker, J. F., M. van den Heuvel-Greve, D. Vethaak, K. Camphuysen, D. M. Fleet, B. Reineking, H. Skov, P. H. Becker, and J. Muñoz Cifuentes. 2005. Pollutants. Pp. 83–140, In: K. Essink, C. Dettmann, H. Farke, K. Laursen, G. Lüerßen, H. Marencic, and W. Wiersinga (eds.), *Wadden Sea Quality Status Report 2004*. Wadden Sea Ecosystem 19, Common Wadden Sea Secretariat, Wilhelmshaven, Germany.

Beukema, J. J. 1982. Annual variation in reproductive success and biomass of the major macrozoobenthic species living in a tidal flat area of the Dutch Wadden Sea. *Netherlands J. Sea Res.* 16: 37–45.

Beukema, J. J. 1992. Dynamics of juvenile shrimp *Crangon crangon* in a tidal-flat nursery of the Wadden Sea after mild and cold winters. *Mar. Ecol. Prog. Ser.* 83: 157–165.

Beukema, J. J. 2002. Expected changes in the Wadden Sea benthos in a warmer world: Lessons from periods with mild winters. *Netherlands J. Sea Res.* 30: 73–79.

Beukema, J. J. and G. C. Cadée. 1996. Consequences of the sudden removal of nearly all mussels and cockles from the Dutch Wadden Sea. *PSZNI Mar. Ecol.* 17: 279–289.

Beukema, J. J., G. C. Cadée, and R. Dekker. 2002. Zoobenthic biomass limited by phytoplankton abundance: Evidence from parallel changes in two long-term data series in the Wadden Sea. *J. Sea Res.* 48: 111–125.

Beukema, J. J. and R. Dekker. 1995. Dynamics and growth of a recent invader of European coastal waters: The American razor clam, *Ensis directus*. *J. Mar. Biol. Assoc. U.K.* 75: 351–362.

Beukema, J. J. and R. Dekker. 2005. Decline of recruitment success in cockles and other bivalves in the Wadden Sea: Possible role of climate change, predation on postlarvae and fisheries. *Mar. Ecol. Prog. Ser.* 287: 149–167.

Beukema, J. J. and R. Dekker. 2006. Annual cockle *Cerastoderma edule* production in the Wadden Sea usually fails to sustain both wintering birds and commercial fishery. *Mar. Ecol. Prog. Ser.* 309: 189–204.

Beukema, J. J., R. Dekker, and J. M. Janssen. 2009. Some like it cold: Populations of the tellinid bivalve *Macoma balthica* (L.) suffering in various ways from a warming climate. *Mar. Ecol. Prog. Ser.* 384: 135–145.

Bilio, M. and U. Niermann. 2004. Is the comb jelly really to blame for it all? *Mnemiopsis leidyi* and the ecological concerns about the Caspian Sea. *Mar. Ecol. Prog. Ser.* 269: 173–183.

Blew, J., K. Günther, K. Laursen, M. van Roomen, P. Südbeck, K. Eskildsen, and P. Potel. 2007. Trends of waterbird populations in the International Wadden Sea 1987–2004: An update. Pp. 9–31, In: B. Reineking and P. Südbeck (eds.), *Seriously declining trends in migratory waterbirds: Causes, concerns, consequences.* Proceedings of the International Workshop, Wadden Sea Ecosystem No. 23, August 31, 2006, Common Wadden Sea Secretariat, Wadden Sea National Park of Lower Saxony, Institute of Avian Research, Joint Monitoring Group of Breeding Birds in the Wadden Sea, Wilhelmshaven, Germany.

Boersma, M., A. M. Malzahn, W. Greve, and J. Javidpour. 2007. The first occurrence of the ctenophore *Mnemiopsis leidyi* in the North Sea. *Helgoländer Mar. Res.* 61: 153–155.

Bolle, L. J., R. Dapper, J. I. J. Witte, and H. W. van der Veer. 1994. Nursery grounds of dab (*Limanda limanda* L.) in the southern North Sea. *Netherlands J. Sea Res.* 32: 299–307.

Brouwer, A., P. J. H. Reijnders, and J. H. Koeman. 1989. Polychlorinated biphenyl (PCB)-contaminated fish induces vitamin A and thyroid hormone deficiency in the common seal (*Phoca vitulina*). *Aquat. Toxicol.* 15: 99–106.

Brussaard, C. P. D., R. Riegman, A. A. M. Noordeloos, G. C. Cadée, H. Witte, A. J. Kop, G. Nieuwland, F. C. Van Duyl, and R. P. M. Bak. 1995. Effects of grazing, sedimentation and phytoplankton cell lysis on the structure of a coastal pelagic food web. *Mar. Ecol. Prog. Ser.* 123: 259–271.

Buhs, F. and K. Reise. 1997. Epibenthic fauna dredged from tidal channels in the Wadden Sea of Schleswig-Holstein: Spatial patterns and a long-term decline. *Helgoländer wissenschaftlichen Meeresuntersuchungen* 51: 343–359.

Cadée, G. C. 1980. Reappraisal of the production and import of organic matter in the Wadden Sea. *Netherlands J. Sea Res.* 14: 305–322.

Cadée, G. C. 1984. Has input of organic matter increased during the last decades? In: *The role of organic matter in the Wadden Sea.* Netherlands Institute for Sea Research Publication Series, no. 10, pp. 71–82.

Cadée, G. C. 2007. Will the recent *Crassostrea gigas* reefs replace the former *Ostrea edulis* beds in the Wadden Sea? *De Levende Natuur* 108: 62–65.

Cadée, G. C. 2008. Scholeksters en Japanse oesters. *Natura* 105: 6–7.

Cadée, G. C. and J. Hegeman. 1974a. Primary production of phytoplankton in the Dutch Wadden Sea. *Netherlands J. Sea Res.* 8: 240–259.

Cadée, G. C. and J. Hegeman. 1974b. Primary production of the benthic microflora living on tidal flats in the Dutch Wadden Sea. *Netherlands J. Sea Res.* 8: 260–291.

Cadée, G. C. and J. Hegeman. 1993. Persisting high levels of primary production at declining phosphate concentrations in the Dutch coastal area (Marsdiep). *Netherlands J. Sea Res.* 31: 147–152.

Cadée, G. C. and J. Hegeman. 2002. Phytoplankton in the Marsdiep at the end of the 20th century: 30 years monitoring biomass, primary production, and *Phaeocystis* blooms. *J. Sea Res.* 48: 97–110.

Camphuysen, C. J. 2004. The return of the harbor porpoise (*Phocoena phocoena*) in Dutch coastal waters. *Lutra* 47: 113–122.

Camphuysen, C. J., B. J. Ens, D. Heg, J. B. Hulscher, J. van der Meer, and C. J. Smit. 1996. Oystercatcher *Haematopus ostralegus* winter mortality in the Netherlands: The effect of severe weather and food supply. *Ardea* 84A: 469–492.

Camphuysen, C. J., C. M. Berrevoets, H. J. W. M. Cremers, A. Dekinga, R. Dekker, B. J. Ens, T. M. van der Have, R. K. H. Kats, T. Kuiken, M. F. Leopold, J. van der Meer, and T. Piersma. 2002. Mass mortality of common eiders *Somateria mollissima* in the Dutch Wadden Sea, winter 1999/2000: Starvation in a commercially exploited wetland of international importance. *Biol. Conserv.* 106: 303–317.

Colijn, F. 1983. Primary production in the Ems-Dollard Estuary. Ph.D. Thesis, State University Groningen, Groningen, the Netherlands.

Colijn, F. and V. N. de Jonge. 1984. Primary production of microphytobenthos in the Ems-Dollard Estuary. *Mar. Ecol. Prog. Ser.* 14: 185–196.

CPSL. 2005. *Coastal protection and sea level rise: Solutions for sustainable coastal protection in the Wadden Sea region.* Wadden Sea Ecosystem 21, Common Wadden Sea Secretariat, Wilhelmshaven, Germany.

Crafts-Brandner, S. J. and M. E. Salvucci. 2000. Rubisco activase constrains the photosynthetic potential of leaves at high temperature and CO_2. *Proc. Natl. Acad. Sci. USA* 97: 13430–13435.

CWSS. 2008. *Nomination of the Dutch-German Wadden Sea as World Heritage Site.* Common Wadden Sea, Secretariat Common Wadden Sea Secretariat, Wilhelmshaven, Germany.

Dankers, N. M. J. A., E. M. Dijkman, M. L. de Jong, G. de Kort, and A. Meijboom. 2004. *De verspreiding en uitbreiding van de Japanse Oester in de Waddenzee.* Alterra-Rapport, 909, Alterra, Wageningen, the Netherlands.

Daskalov, G. M. and E. V. Mamedov. 2007. Integrated fisheries assessment and possible causes for the collapse of the anchovy kilka in the Caspian Sea. *ICES J. Mar. Sci.* 64: 503–511.

De Galan, S., M. Elskens, L. Goeyens, A. Pollentier, N. Brion, and W. Baeyens. 2004. Spatial and temporal trends in nutrient concentrations in the Belgian Continental area of the North Sea during the period 1993–2000. *Estuar. Coastal Shelf Sci.* 61: 517–528.

de Jonge, V. N. 1990. Response of the Dutch Wadden Sea ecosystem to phosphorus discharges from the River Rhine. In: D. S. McLusky, V. N. de Jonge, and J. Pomfret (eds.), North Sea–estuaries interactions. *Hydrobiologia* 195: 49–62.

de Jonge, V. N. 1992. *Physical processes and dynamics of microphytobenthos in the Ems Estuary (the Netherlands).* Ministry of Transport, Public Works and Water Management, The Hague, the Netherlands.

de Jonge, V. N., J. F. Bakker, and M. van Stralen. 1996. Recent changes in the contributions of River Rhine and North Sea to the eutrophication of the western Dutch Wadden Sea. *Netherlands J. Aquat. Ecol.* 30: 27–39.

de Jonge, V. N. and D. J. De Jong. 2002. "Global change" impact of inter-annual variation in water discharge as a driving factor to dredging and spoil disposal in the river Rhine system and of turbidity in the Wadden Sea. *Estuar. Coastal Shelf Sci.* 55: 969–991.

de Jonge, V. N., K. Essink, and R. Boddeke. 1993. The Dutch Wadden Sea: A changed ecosystem. *Hydrobiologia* 265: 45–71.

de Jonge, V. N. and H. Postma. 1974. Phosphorus compounds in the Dutch Wadden Sea. *Netherlands J. Sea Res.* 8: 139–153.

Dekker, R. 1987. The importance of the subtidal macrobenthos as a food source for the Wadden Sea ecosystem. Pp. 27–35, In: S. Tougaard and S. Asbirk (eds.), *Proceedings of the 5th International Wadden Sea Symposium.* National Forest and Nature Agency and Museum of Fishery and Shipping, Esbjerg, Denmark.

den Hartog, C. 1987. "Wasting disease" and other dynamic phenomena in *Zostera* beds. *Aquat. Bot.* 27: 3–14.

de Vlas, J., B. Brinkman, C. Buschbaum, N. Dankers, M. Herlyn, P. S. Kristensen, G. Millat, G. Nehls, M. Ruth, J. Steenbergen, and A. Wehrmann. 2005. Intertidal blue mussel beds. Pp. 190–200, In: K. Essink, C. Dettmann, H. Farke, K. Laursen, G. Lüerßen, H. Marencic, and W. Wiersinga (eds.), *Wadden Sea Quality Status Report 2004.* Wadden Sea Ecosystem 19, Common Wadden Sea Secretariat, Wilhelmshaven, Germany.

Diederich, S., G. Nehls, J. E. E. van Beusekom, and K. Reise. 2005. Introduced Pacific oysters (*Crassostrea gigas*) in the northern Wadden Sea: Invasion accelerated by warm summers? *Helgoländer Mar. Res.* 59: 97–106.

Dolch, T. and H. C. Hass. 2008. Long-term changes of intertidal and subtidal sediment compositions in a tidal basin in the northern Wadden Sea (SE North Sea). *Helgoländer Mar. Res.* 62: 3–11.

Dorrestein, R. and L. Otto. 1960. On the mixing and flushing of the water in the Ems Estuary. Verhandelingen van het Koninklijk Nederlands Mijnbouwkundig Genootschap. *Geologische Serie* 19: 83–102.

Dronkers, J. 2005. *Natural and human impacts on sedimentation in the Wadden Sea: An analysis of historical data.* RWS/RIKZ, The Haag, the Netherlands.

Duarte, C. M. and J. Cebrian. 1996. The fate of marine autotrophic production. *Limnol. Oceanogr.* 41: 1758–1766.

Ehlers, J. 1988. *Morphodynamics of the Wadden Sea.* Rotterdam: A. A. Balkema.

Eisma, D. 1981. Supply and deposition of suspended matter in the North Sea. *Special Publication of the International Association of Sedimentologists* 5: 415–428.

Ens, B. J., A. C. Smaal, and J. de Vlas. 2004. *The effects of shellfish fishery on the ecosystems of the Dutch Wadden Sea and Oosterschelde.* Wageningen: Alterra-rapport 1011, RIVO-rapport C056/04, RIKZ-rapport RKZ/2004.031.

Epping, H. G., E. Koning, R. Van Raaphorst, and C. J. M. Philippart. In prep. The effect of redox conditions on burrowing depth and behaviour of the Baltic Tellin Macoma balthica.

Eppley, R. W. 1972. Temperature and phytoplankton growth in the sea. *Fishery Bull.* 70: 1063–1085.

Essink, K., C. Dettmann, H. Farke, K. Laursen, G. Lüerßen, H. Marencic, and W. Wiersinga (eds.). 2005. *Wadden Sea Quality Status Report 2004.* Wadden Sea Ecosystem 19. Common Wadden Sea Secretariat, Wilhelmshaven, Germany.

Faasse, M. A. and K. M. Bayha. 2006. The ctenophore *Mnemiopsis leidyi* A. Agassiz 1865 in coastal waters of the Netherlands: An unrecognized invasion? *Aquat. Invasions* 1: 270–277.

FitzGerald, D. M., M. S. Fenster, B. A. Argow, and I. V. Buynevich. 2008. Coastal impacts due to sea-level rise. *Annu. Rev. Earth Planet. Sci.* 36: 601–647.

Fransz, H. G., S. R. Gonzales, G. C. Cadée, and F. C. Hansen. 1992. Long-term change of *Temora longicornis* (Copepoda, Calanoida) abundance in a Dutch tidal inlet (Marsdiep) in relation to eutrophication. *Netherlands J. Sea Res.* 30: 23–32.

Gillbricht, M. 1956. Die hydrography des jadebusens und der innenjade. *Veröff. Inst. Meeresf. Bremerhaven* 4: 153–170.

Glim, G. W., G. Kool, M. F. Lieshout, and M. de Boer. 1987. *Erosie en sedimentatie in de binnendelta van het zeegat van Texel 1932–1982*. Rijkswaterstaat, Directie Noord-Holland, Report ANWX 87.H201, Rijkswaterstaat, Hoorn, the Netherlands.

Goedecke, E. 1968. Über die hydrographische Struktur der deutschen Bucht im Hinblick auf die Verschmutzung in der Konvergenzzone. *Helgoländer wissenschaftlichen Meeresuntersuchungen* 17: 108–125.

Grift, R. E., I. Tulp, L. Clarke, U. Damm, A. McLay, S. Reeves, J. Vigneau, and W. Weber. 2004. *Assessment of the ecological effects of the Plaice Box*. Report of the European Commission Expert Working Group to evaluate the Shetland and Plaice boxes. Brussels, Belgium.

Groen, P. 1967. On the residual transport of suspended matter by an alternating tidal current. *Netherlands J. Sea Res.* 3–4: 564–574.

Grover, J. P. 1997. *Resource competition*. London: Chapman & Hall.

Hansson, H. G. 2006. Ctenophores of the Baltic and adjacent seas: The invader *Mnemiopsis* is here! *Aquat. Invasions* 1: 295–298.

Härkönen, T., R. Dietz, P. Reijnders, J. Teilmann, K. Harding, A. Hall, S. Brasseur, U. Siebert, S. J. Goodman, P. D. Jepson, T. Dau Rasmussen, and P. Thompson. 2006. The 1988 and 2002 phocine distemper virus epidemics in European harbor seals. *Diseases Aquat. Org.* 68: 115–130.

Harten, H. and H. Vollmers. 1978. The estuaries of the German North Sea coast. *Die Küste* 32: 50–65.

Havinga, B. 1960. *De mosselteelt in de Waddenzee*. Rapport aangeboden aan het College van Gedeputeerde Staten van de provincie Friesland, IJmuiden, the Netherlands.

Helder, W, and P. Ruardij. 1982. A one-dimensional mixing and flushing model of the EMS-Dollard estuary: Calculation of time scales at different river discharges. *Netherlands J. Sea Res.* 15: 293–312.

Herman, P. M. J., J. J. Middelburg, J. van de Koppel, and C. H. R. Heip. 1999. Ecology of estuarine macrobenthos. *Adv. Ecol. Res.* 29: 195–240.

Hickel, W. 1984. *Seston in the Wadden Sea of Sylt (German Bight, North Sea)*. Netherlands Institute of Sea Research Publication Series no. 10, pp. 113–131.

Hillen, R. and P. Roelse. 1995. Dynamic preservation of the coastline in the Netherlands. *J. Coastal Conserv.* 1: 17–28.

Hoppema, J. M. J. 1990. The distribution and seasonal variation of alkalinity in the Southern Bight of the North Sea and in the western Wadden Sea. *Netherlands J. Sea Res.* 26: 11–23.

Hughes, T. P., M. J. Rodrigues, D. R. Bellwood, D. Ceccarelli, O. Hoegh-Guldberg, L. McCook, N. Moltschaniwskyj, M. S. Pratchett, R. S. Steneck, and B. Willis. 2007. Phase shifts, herbivory, and the resilience of coral reefs to climate change. *Current Biol.* 17: 360–365.

Hurrell, J. W. 1995. Decadal trends in the North Atlantic Oscillation: Regional temperatures and precipitation. *Science* 269: 676–679.

Hurrell, J. W. and H. van Loon. 1997. Decadal variations in climate associated with the North Atlantic Oscillation. *Climate Change* 36: 301–326.

IPCC (Intergovernmental Panel on Climate Change). 2007. *Climate Change 2007: Synthesis Report*. Contribution of Working Groups I, II and III to the Fourth Assessment, Report of the Intergovernmental Panel on Climate Change [Core Writing Team, R. K. Pachauri, and A. Reisinger (Eds.)], IPCC, Geneva, Switzerland.

Kochmann, J., C. Buschbaum, N. Volkenborn, and K. Reise. 2008. Shift from native mussels to alien oysters: Differential effects of ecosystem engineers. *J. Exp. Mar. Biol. Ecol.* 364: 1–10.

Koeman, J. H. 1971. Het voorkomen en de toxicologische betekenis van enkele chloorkoolwaterstoffen aan de Nederlandse kust in de periode 1965 tot 1970. Ph.D. Thesis, University of Utrecht, the Netherlands.

Koeman, J. H., M. C. Ten Noever de Brauw, and R. H. de Vos. 1969. Chlorinated biphenyls in fish, mussels and birds from the River Rhine and the Netherlands coastal area. *Nature* 221: 1126–1128.

Korringa, P. 1968. On the ecology and distribution of the parasitic copepod *Mytilicola intestinalis*. *Bijdragen tot de Dierkunde* 38: 47–57.

Krögel, F. and B. W. Flemming. 1998. Evidence for temperature-adjusted sediment distributions in the back-barrier tidal flats of the East Frisian Wadden Sea (southern North Sea). In: C. R. Alexander, R. A. Davis, and V. J. Henry (eds.), *Tidalites: Processes and products*. *SEPM Spec. Publ.* 61: 31–41.

Kromkamp, J., J. F. C. de Brouwer, G. Blanchard, R. M. Forster, and V. Créach. 2006. *Functioning of microphytobenthos in estuaries: Proceedings of the Colloquium, Amsterdam, 21–23 August 2003*. Koninklijke Nederlandse Akademie van Wetenschappen Verhandelingen, Afdeling Natuurkunde (Tweede Reeks) 103, Amsterdam, the Netherlands.

Laane, R. W. P. M., G. Groeneveld, A. de Vries, A. J. van Bennekom, and J. S. Sydow. 1993. Nutrients (P, N, Si) in the Channel and the Dover Strait: Seasonal and year-to-year variation and fluxes to the North Sea. *Oceanol. Acta* 16: 607–616.

Leopold, M. F., C. J. Smit, P. W. Goedhart, M. van Roomen, E. van Winden, and C. van Turnhout. 2004. *EVA II Deelproject C2: Langjarige trends in aantallen wadvogels, in relatie tot de kokkelvisserij en het gevoerde beleid in deze*. Wageningen: Alterra Report 954, Alterra, the Netherlands.

Loebl, M., F. Colijn, J. E. E. van Beusekom, J. G. Baretta-Bekker, C. Lancelot, C. J. M. Philippart, V. Rousseau, and K. H. Wiltshire. 2009. Recent patterns in potential phytoplankton limitation along the Northwest European continental coast. *J. Sea Res.* 61: 34–43.

Loebl, M., T. Dolch, and J. E. E. van Beusekom. 2007. Annual dynamics of pelagic production and respiration in a shallow coastal basin. *J. Sea Res.* 58: 269–282.

Loebl, M. and J. E. E. van Beusekom. 2008. Seasonality of microzooplankton grazing in the northern Wadden Sea. *J. Sea Res.* 59: 203–216.

Lotze, H. K., K. Reise, B. Worm, J. van Beusekom, M. Busch, A. Ehlers, D. Heinrich, R. C. Hoffmann, P. Holm, C. Jensen, O. S. Knottnerus, N. Langhanki, W. Prummel, M. Vollmer, and W. J. Wolff. 2005. Human transformations of the Wadden Sea ecosystem through time: A synthesis. *Helgoländer wissenschaftlichen Meeresuntersuchungen* 59: 84–95.

Martens, P. and J. E. E. van Beusekom. 2008. Zooplankton response to a warmer northern Wadden Sea. *Helgoländer Mar. Res.* 62: 67–75.

Matulla, C., W. Schöner, H. Alexandersson, H. von Storch, and X.L. Wang. 2008. European storminess: Late nineteenth century to present. *Climate Dynamics* 31: 125–130.

Meltofte, H., J. Blew, J. Frikke, H.-U. Rösner, and C. J. Smit. 1994. *Numbers and distribution of waterbirds in the Wadden Sea: Results and evaluation of 36 simultaneous counts in the Dutch-German-Danish Wadden Sea 1980–1991*. IWRB Publication 34/Wader Study Group Bulletin 74, Special Issue, Wilhelmshaven, Germany.

Meysman, F. J. R., J. J. Middelburg, and C. H. R. Heip. 2006. Bioturbation: A fresh look at Darwin's last idea. *Trends Ecol. Evol.* 21: 688–695.

Oost, A. P. and H. de Haas. 1993. *Het Friese Zeegat, morfologisch-sedimentologische veranderingen in de periode 1927–1970, Deel 1&2*. Reports of the Faculty of Earth Sciences, University of Utrecht, the Netherlands.

Osterhaus, A. D. M. and E. J. Vedder. 1988. Identification of a virus causing recent seal death. *Nature* 335: 20.

Pastoors, M., L. Bolle, K. Groeneveld, P. Groot, P. van Leeuwen, G. Piet, A. Rijnsdorp, and G. Rink. 2000. *Evaluatie van de scholbox*. RIVO rapport C002/00. Rijksinstituut voor Visserijonderzoek, IJmuiden, the Netherlands.

Pearson, T. H. and R. Rosenberg. 1987. Feast and famine: Structuring factors in marine benthic communities. Pp. 373–395, In: J. H. R. Gee and P. S. Giller (eds.), *The 27th Symposium of the British Ecological Society*, Aberystwyth 1986. Oxford: Blackwell Scientific Publications.

Philippart, C. J. M., R. Anadón, R. Danovaro, J. W. Dippner, K. F. Drinkwater, S. J. Hawkins, T. Oguz, G. O'Sullivan, and P. Reid. 2007a. *Impacts of climate change on the European marine and coastal environment: Ecosystems approach*. ESF Marine Board Position Paper, 9. European Science Foundation, Marine Board, Strasbourg, France.

Philippart, C. J. M., J. J. Beukema, G. C. Cadée, R. Dekker, P. W. Goedhart, J. M. van Iperen, M. F. Leopold, and P. M. J. Herman. 2007b. Impact of nutrient reduction on coastal communities. *Ecosystems* 10: 96–119.

Philippart, C. J. M. and G. C. Cadée. 2000. Was total primary production in the western Wadden Sea stimulated by nitrogen loading? *Helgoländer Mar. Res.* 54: 55–62.

Philippart, C. J. M., G. C. Cadée, W. van Raaphorst, and R. Riegman. 2000. Long-term phytoplankton-nutrient interactions in a shallow coastal sea: Algal community structure, nutrient budgets, and denitrification potential. *Limnol. Oceanogr.* 45: 131–144.

Philippart, C. J. M., K. S. Dijkema, and J. van der Meer. 1992. *Wadden Sea seagrasses: Where and why?* Netherlands Institute for Sea Research Publication Series no. 20, pp. 177–191.

Philippart, C. J. M., H. M. van Aken, J. J. Beukema, O. G. Bos, G. C. Cadée, and R. Dekker. 2003. Climate-related changes in recruitment of the bivalve *Macoma balthica*. *Limnol. Oceanogr.* 48: 2171–2185.

Philippart, C. J. M., J. M. van Iperen, G. C. Cadée, and A. F. Zuur. Submitted. Long-term field observations on phytoplankton seasonality in a shallow coastal marine ecosystem, the Wadden Sea. *Estuaries Coasts*.

Piersma, T., A. Koolhaas, A. Dekinga, J. J. Beukema, R. Dekker, and K. Essink. 2001. Long-term indirect effects of mechanical cockle-dredging on intertidal bivalve stocks in the Wadden Sea. *J. Appl. Ecol.* 38: 976–990.

Portner, H. O. and R. Knust. 2007. Climate change affects marine fishes through the oxygen limitation of thermal tolerance. *Science* 315: 95–97.

Postel, L. and S. Kube. 2008. A matter of time and temperature: The spread of *Mnemiopsis leidyi*. *ICES Insight* 45: 16–19.

Postma, H. 1954. Hydrography of the Dutch Wadden Sea. *Arch. Néerlandaises Zool.* 10: 1–106.

Postma, H. 1961. Transport and accumulation of suspended matter in the Dutch Wadden Sea. *Netherlands J. Sea Res.* 1: 148–190.

Postma, H. 1982. *Hydrography of the Wadden Sea: Movement and properties of water and particulate matter.* Final report on "Hydrography" of the Wadden Sea Working Group, report 2. Rotterdam: A. A. Balkema.

Postma, H. and J. W. Rommets. 1970. Primary production in the Wadden Sea. *Netherlands J. Sea Res.* 4: 470–493.

Postuma, K. H. and G. Rauck, 1978. The fishery in the Wadden Sea. Pp. 139–157, In: N. Dankers, W. J. Wolff, and J. J. Zijlstra (eds.), *Fishes and fisheries of the Wadden Sea*. Rotterdam: A. A. Balkema.

Ramón, M. 2003. Population dynamics and secondary production of the cockle *Cerastoderma edule* (L.) in a backbarrier tidal flat in the Wadden Sea. *Scientia Mar.* 67: 429-443.

Reijnders, P. J. H. 1976. The harbor seal (*Phoca vitulina*) population in the Dutch Wadden Sea: Size and composition. *Netherlands J. Sea Res.* 10: 223–235.

Reijnders, P. J. H. 1978. Recruitment in the harbor seal (*Phoca vitulina*) population in the Dutch Wadden Sea. *Netherlands J. Sea Res.* 12: 164–179.

Reijnders, P. J. H. 1980. Organochlorine and heavy metal residues in harbor seals from Schleswig Holstein plus Denmark and the Netherlands: Their possible effects in relation to the reproduction in both populations. *Netherlands J. Sea Res.* 14: 30–65.

Reijnders, P. J. H. 1992. Harbor porpoises *Phocoena phocoena* in the North Sea: Numerical responses to changes in environmental conditions. *Netherlands J. Aquat. Ecol.* 26: 75–85.

Reijnders, P. J. H., E. H. Ries, S. Tougaard, N. Nørgaard, G. Heidemann, J. Schwarz, E. Vareschi, and I. M. Traut. 1997. Population development of harbor seals *Phoca vitulina* in the Wadden Sea after the 1988 virus epizootic. *J. Sea Res.* 38: 161–168.

Reijnders, P. J. H., J. van Dijk, and D. Kuiper. 1995. Recolonization of the Dutch Wadden Sea by the grey seal *Halichoerus grypus*. *Biol. Conserv.* 71: 231–235.

Reise, K. 1982. Long-term changes in the macrobenthic invertebrate fauna of the Wadden Sea: Are polychaetes about to take over? *Netherlands J. Sea Res.* 16: 29–36.

Reise, K. 1995. Predictive ecosystem research in the Wadden Sea. *Helgoländer wissenschaftlichen Meeresuntersuchungen* 49: 495–505.

Reise, K. 2002. Sediment mediated species interactions in coastal waters. *J. Sea Res.* 48: 127–141.

Reise, K. 2005. Coast of change: Habitat loss and transformations in the Wadden Sea. *Helgoländer Mar. Res.* 59: 9–21.

Reise, K., S. Gollasch, and W. J. Wolff. 1999. Introduced species in the Wadden Sea. *Helgoländer wissenschaftlichen Meeresuntersuchungen* 52: 219–234.

Reise, K., and J. E. E. van Beusekom. 2008. Interactive effects of global and regional change on a coastal ecosystem. *Helgoländer Mar. Res.* 62: 85–91.

Ridderinkhof, H., J. F. T. Zimmerman, and M. E. Philippart. 1990. Tidal exchange between the North Sea and Dutch Wadden Sea and mixing time scales of the tidal basins. *Netherlands J. Sea Res.* 25: 331–350.

Riegman, R., B. R. Kuipers, A. A. M. Noordeloos, and H. Witte. 1993. Size-differential control of phytoplankton and the structure of plankton communities. *Netherlands J. Sea Res.* 31: 255–265.

Roelke, D., S. Augustine, and Y. Buyukates. 2003. Fundamental predictability in multispecies competition: The influence of large disturbance. *Am. Nat.* 162: 615–623.

Rooth, J. and D. A. Jonkers. 1972. The status of some piscivorous birds in the Netherlands. *TNO-Nieuws* 27: 551–555.

Salonen, K., J. Sarval, I. Hakala, and M.-L. Viljanen (1976) The relation of energy and organic carbon in aquatic invertebrates. *Limnol. Oceanogr.* 21: 724–730.

Shiganova, T. A., H. J. Dumont, A. Mikaelyan, D. M. Glazov, Y. V. Bulgakova, E. I. Musaeva, P. Y. Sorokin, L. A. Pautova, Z. A. Mirzoyan, and E. I. Studenikina. 2004. Interactions between the invading ctenophores *Mnemiopsis leidyi* (A. Agassiz) and *Beroe ovata* Mayer 1912, and their influence on the pelagic ecosystem of the Northeastern Black Sea. *NATO Science Series IV: Earth and Environmental Sciences* 35: 33–70.

Smits, A., M. G. Klein Tank, and G. P. Können. 2005. Trends in storminess over the Netherlands, 1962–2002. *Int. J. Climatol.* 25: 1331–1344.

Sommer, U. 1984. The paradox of the plankton: Fluctuations of phosphorus availability maintain diversity in flow-through cultures. *Limnol. Oceanogr.* 29: 633–636.

Southward, A. J., O. Langmead, N. J. Hardman-Mountford, J. Aiken, G. T. Boalch, P. R. Dando, M. J. Genner, I. Joint, M. A. Kendall, N. C. Halliday, R. P. Harris, R. Leaper, N. Mieskowska, R. D. Pingree, A. J. Richardson, D. W. Sims, T. Smith, A. W. Walne, and S. J. Hawkins. 2005. Long-term oceanographic and ecological research in the western English Channel. *Adv. Mar. Biol.* 47: 1–105.

Stolte, W., T. McCollin, A. A. M. Noordeloos, and R. Riegman. 1994. Effect of nitrogen source on the size distribution within marine phytoplankton populations. *J. Exp. Mar. Biol. Ecol.* 184: 83–97.

Stolte, W. and R. Riegman. 1995. Effect of phytoplankton cell size on transient-state nitrate and ammonium uptake kinetics. *Microbiology* 141: 1221–1229.

Strasser, M. 1999. *Mya arenaria*: An ancient invader of the North Sea coast. *Helgoländer wissenschaftlichen Meeresuntersuchungen* 52: 309–324.

Swennen, C. 1991a. Ecology and population dynamics of the Common Eider in the Dutch Wadden Sea. Ph.D. Thesis, Rijksuniversiteit Groningen, Groningen, the Netherlands.

Swennen, C. 1991b. Fledgling production of eiders *Somateria mollissima* in the Netherlands. *J. Ornithol.* 132: 427–437.

Thijsse, J. Th. 1972. *Een halve eeuw Zuiderzeewerken 1920–1970*. Groningen, the Netherlands: Tjeenk Willink.

Thingstad, T. F. and E. Sakshaug. 1990. Control of phytoplankton growth in nutrient recycling ecosystems: Theory and terminology. *Mar. Ecol. Prog. Ser.* 63: 261–272.

Tillmann, U., K. J. Hesse, and F. Colijn. 2000. Planktonic primary production in the German Wadden Sea. *J. Plankton Res.* 7: 1253–1276.

Van Aken, H. M. 2008a. Variability of the water temperature in the western Wadden Sea on tidal to centennial time scales. *J. Sea Res.* 60: 227–234.

Van Aken, H. M. 2008b. Variability of the salinity in the western Wadden Sea on tidal to centennial time scales. *J. Sea Res.* 59: 121–132.

Van Beek, F. A., A. D. Rijnsdorp, and R. de Clerck. 1989. Monitoring juvenile stocks of flatfish in the Wadden Sea and coastal areas of the southeastern North Sea. *Helgoländer wissenschaftlichen Meeresuntersuchungen* 43: 461–477.

Van Beusekom, J. E. E. 2005. A historic perspective on Wadden Sea eutrophication. *Helgoländer Mar. Res.* 59: 45–54.

Van Beusekom, J., P. Bot, J. Göbel, M. Hanslik, H.-J. Lenhart, J. Pätsch, L. Peperzak, T. Petenati, and K. Reise. 2005. Eutrophication. Pp. 141–154, In: K. Essink, C. Dettmann, H. Farke, K. Laursen, G. Lüerßen, H. Marencic, and W. Wiersinga (eds.), *Wadden Sea Quality Status Report 2004. Wadden Sea Ecosystem* 19, Common Wadden Sea Secretariat, Wilhelmshaven, Germany.

Van Beusekom, J. E. E., U. H. Brockman, K.-J. Hesse, W. Hickel, K. Poremba, and U. Tillmann. 1999. The importance of sediments in the transformation and turnover of nutrients and organic matter in the Wadden Sea and German Bight. *Germ. J. Hydr.* 51: 245–266.

Van Beusekom, J. E. E., S. Weigelt-Krenz, and P. Martens. 2008. Long-term variability of winter nitrate concentrations in the Northern Wadden Sea driven by freshwater discharge, decreasing riverine loads and denitrification. *Helgoländer Mar. Res.* 62: 9–57.

Van den Hoek, C., W. Admiraal, F. Colijn, and V. N. de Jonge. 1979. The role of algae and seagrasses in the ecosystem of the Wadden Sea: A review. In: W. J. Wolff (ed.), *Flora and vegetation of the Wadden Sea*. Rotterdam: A. A. Balkema.

van der Heide, T., E. H. van Nes, G. W. Geerling, A. J. P. Smolders, T. J. Bouma, and M. M. van Katwijk. 2007. Positive feedbacks in seagrass ecosystems: Implications for success in conservation and restoration. *Ecosystems* 10: 1311–1322.

Van der Meer, J., J. J. Beukema, and R. Dekker. 2001. Long-term variability in secondary production of an intertidal bivalve population is primarily a matter of recruitment variability. *J. Animal Ecol.* 70: 159–169.

van der Weijden, W., R. Leewis, and P. Bol. 2007. *Biological globalisation: Bio-invasions and their impacts on nature, the economy, and public health*. Koninklijke Nederlandse Vereniging voor Veldbiologie, Utrecht, the Netherlands.

van Es, F. B. 1977. A preliminary carbon budget for a part of the Ems Estuary: The Dollard. *Helgoländer wissenschaftlichen Meeresuntersuchungen* 30: 283–294.

van Katwijk, M. M., A. R. Bos, and D. C. R. Hermus. 2005. *Klein zeegras en Snavelruppia op het Balgzand*. Report Afdeling milieukunde, Radboud Universiteit, Nijmegen, the Netherlands.

van Raaphorst, W. and V. N. de Jonge. 2004. Reconstruction of the total N and P inputs from the IJsselmeer into the western Wadden Sea between 1935–1998. *J. Sea Res.* 51: 109–131.

van Roomen, M., C. van Turnhout, E. van Winden, B. Koks, P. Goedhart, M. Leopold, and C. Smit. 2005. Trends in benthivorous waterbirds in the Dutch Wadden Sea 1975–2002: Large differences between shellfish-eaters and worm-eaters. *Limosa* 78: 21–38.

van Straaten, L. M. J. U. 1964. De bodem der Waddenzee. In: W. F. Anderson, J. Abrahamse, J. D. Buwalda, and L. M. J. U. van Straten (eds.), *Het Waddenboek*. Thieme Zutphen, the Netherlands.

Veldhuis, M. J. W., F. Colijn, L. A. H. Venekamp, and L. Villerius. 1988. Phytoplankton primary production and biomass in the western Wadden Sea (the Netherlands): A comparison with an ecosystem model. *Netherlands J. Sea Res.* 22: 37–49.

Verweij, J. 1975. The cetaceans *Phocoena phocoena* and *Tursiops truncatus* in the Marsdiep area (Dutch Wadden Sea) in the years 1931–1973. *Intern Verslag Nederlands Instituut voor Onderzoek der Zee, Texel* 1975-17a : 1–98; 1975-17b: 1–153.

Vorberg, R., L. Bolle, Z. Jager, and T. Neudecker. 2005. Fish. Pp. 219–236, In: K. Essink, C. Dettmann, H. Farke, K. Laursen, G. Lũerßen, H. Marencic, and W. Wiersinga (eds.), *QSR Wadden Sea*, Wadden Sea Ecosystem No. 19, Common Wadden Sea Secretariat, Wilhelmshaven, Germany.

Weijerman, M., H. Lindeboom, and A. F. Zuur. 2005. Regime shifts of the marine ecosystems of the North Sea and Wadden Sea. *Mar. Ecol. Prog. Ser.* 298: 21–39.

Widdows, J. and M. Brinsley. 2002. Impact of biotic and abiotic processes on sediment dynamics and the consequences to the structure and functioning of the intertidal zone. *J. Sea Res.* 48: 143–156.

Wiltshire, K. H. and B. F. J. Manly. 2004. The warming trend at Helgoland Roads, North Sea: Phytoplankton response. *Helgoländer Mar. Res.* 58: 269–273.

Winkelmolen, A. M. and H. J. Veenstra. 1974. Size and shape sorting in a Dutch tidal inlet. *Sedimentology* 21: 107–126.

Wolff, W. J. (ed.). 1983. *The ecology of the Wadden Sea: Final report of the Wadden Sea Working Group*. Rotterdam: A. A. Balkema.

Wolff, W. J. 1992. The end of a tradition. 1000-years of embankment and reclamation of wetlands in the Netherlands. *Ambio* 21: 287–291.

Wolff, W. J. 1999. Exotic invaders of the meso-oligohaline zone of estuaries in the Netherlands: Why are there so many? *Helgoländer Mar. Res.* 52: 393–400.

Wolff, W. J. 2000. Causes of extirpations in the Wadden Sea, an estuarine area in the Netherlands. *Conserv. Biol.* 14: 876–885.

Wolff, W. J. 2005a. The exploitation of living resources in the Dutch Wadden Sea: A historical overview. *Helgoländer Mar. Res.* 59: 31–38.

Wolff, W. J. 2005b. Non-indigenous marine and estuarine species in the Netherlands. *Zoologische Mededelingen Leiden* 79: 1–116.

Wood, R. and J. Widdows. 2003. Modelling intertidal sediment transport for nutrient change and climate change scenarios. *Sci. Total Environ.* 314–316: 637–649.

Zijlstra, J. J. 1978. The function of the Wadden Sea for the members of its fish-fauna. In: W. J. Wolff (ed.), *Fishes and fisheries of the Wadden Sea*. Rotterdam: A. A. Balkema.

Zijlstra, J. J. 1972. On the importance of the Wadden Sea as a nursery area in relation to the conservation of the southern North Sea fishery resources. *Symp. Zool. Soc. London* 29: 233–258.

Zimmerman, J. T. F. 1976. Mixing and flushing of tidal embayments in the western Dutch Wadden Sea. *Netherlands J. Sea Res.* 10: 149–191, 397–439.

Zimmerman, J. F. T. and J. W. Rommets. 1974. Natural fluorescence as a tracer in the Dutch Wadden Sea and the adjacent North Sea. *Netherlands J. Sea Res.* 8: 117–125.

Zipperle, A. and K. Reise. 2005. Freshwater springs on intertidal sand flats cause a switch in dominance among polychaete worms. *J. Sea Res.* 54: 143–150.

Zwarts, L. 2004. *Soil conditions and mechanical cockle fisheries in the Wadden Sea: report and data*. Lelystad, the Netherlands: RIZA.

17 The Patos Lagoon Estuary, Southern Brazil

Biotic Responses to Natural and Anthropogenic Impacts in the Last Decades (1979–2008)

Clarisse Odebrecht, Paulo C. Abreu, Carlos E. Bemvenuti, Margareth Copertino, José H. Muelbert, João P. Vieira, and Ulrich Seeliger*

CONTENTS

Abstract 434
17.1 History of the Patos–Mirim Lagoon System 434
  17.1.1 Geological Evolution 434
  17.1.2 Natural Protection Areas 435
17.2 The Microtidal Patos Lagoon Estuary 436
  17.2.1 Water Exchange, Salinity, and Temperature 436
  17.2.2 Dissolved Inorganic Nutrients 437
  17.2.3 Estuarine Habitats, Biological Components, and Processes 439
    17.2.3.1 Primary Producers 439
    17.2.3.2 Secondary Producers 440
17.3 Biotic Responses to Environmental Changes 442
  17.3.1 Impacts of El Niño Southern Oscillation (ENSO) 443
    17.3.1.1 Phytoplankton 443
    17.3.1.2 *Ruppia maritima* 444
    17.3.1.3 Macrozoobenthos 446
    17.3.1.4 Fish 446
    17.3.1.5 Fisheries 447
17.4 Human Influence 448
  17.4.1 Eutrophication 448
  17.4.2 Overfishing 449
  17.4.3 Harbor Dredging 449
  17.4.4 Bioinvasion 450
17.5 Concluding Remarks 451
References 451

* Corresponding author. Email: doclar@furg.br

ABSTRACT

The warm temperate Patos Lagoon Estuary receives waters from a 200,000 km^2 watershed shared by southern Brazil and northeastern Uruguay. The exchange of water with the Atlantic Ocean occurs through an 800-m-wide inlet, flanked by two 4-km-long jetties, at the southernmost part of the lagoon near the city of Rio Grande (32°S). The estuary is a river-dominated system. Peaks of river discharge are associated with El Niño episodes (negative El Niño Southern Oscillation; ENSO Index), when rainfall significantly increases in the region. Low discharge periods occur during La Niña when drought conditions are observed. These phenomena have direct influence on salinity variations in the estuary, with low values recorded during El Niño years (1994–1995, 1997–1998, and 2002–2003) and high saltwater intrusion during La Niña years (1988–1989 and 1999–2000). Changes in estuarine water and sediment dynamics as well as physicochemical water characteristics induce significant biological changes and ecological responses. For instance, phytoplankton growth and biomass accumulation in the estuary are closely related to rainfall and freshwater discharge. When rainfall exceeds 1500 mm year^{-1}, which is typical of El Niño years, phytoplankton biomass decreases mainly due to high freshwater runoff that flushes the phytoplankton biomass out of the estuary. However, low chlorophyll concentrations are also observed following La Niña periods when high ammonium concentration and long water residence occur. High rainfall drastically reduces the amount of *Ruppia maritima* on the lagoon floor due to increased erosion of margins and reduced bottom light caused by high water levels. Macrozoobenthic abundance in the estuary is also affected by El Niño, since recruitment gaps of species like the tanaidacean *Kalliapseudes schubartii* and the bivalve *Erodona mactroides* are observed. Similarly, high freshwater discharge during El Niño years decreases recruitment of resident and estuarine dependent fish, although species richness increases due to the greater number of freshwater or marine species during El Niño or La Niña conditions, respectively. ENSO effects on rainfall and on hydrodynamics in the Patos Lagoon Estuary cause significant variation in commercial fisheries. Changes in catch of the mullet *Mugil platanus* and the pink shrimp *Farfantepenaeus paulensis* have an important economic and social impact on the region. It is clear that species composition, abundance, and biomass of microalgae, fish, and macrobenthic flora and fauna strongly respond to ENSO events. Dredging activities and fisheries seem to interact synergistically with ENSO events, affecting the ecology of the highly dynamic Patos Lagoon Estuary. Moreover, freshwater discharge to the Atlantic Ocean has an overriding influence on ecological processes in the adjacent coastal ocean region.

Key Words: Southwest Atlantic Ocean, El Niño Southern Oscillation, human influence, phytoplankton, macrozoobenthos, macroalgae, seaweed, fish

17.1 HISTORY OF THE PATOS–MIRIM LAGOON SYSTEM

17.1.1 Geological Evolution

The warm temperate Patos Lagoon Estuary is part of the Patos–Mirim lagoon system, which receives waters from a 200,000 km^2 watershed (Seeliger 2001) shared between southern Brazil and northeastern Uruguay (Figure 17.1). The geological evolution of this system is marked by successive transgression-regression cycles (Calliari 1997; Seeliger et al. 2004). Early environmental, biological, and paleontological observations recognized the importance of sea level changes in shaping this region (von Ihering 1885). During the early Pleistocene (400,000 BP) when the Atlantic Ocean extended over the coastal plain, coastal waves and longshore currents progressively deposited sand bars that gradually formed the large water bodies of the Mirim and the Patos lagoons (120,000 BP). During the following glacial period (17,000 BP), the sea retreated to ~120 m below the present sea level. The two lagoons dried out and the shoreline extended about 70 km seaward. The Holocene marine transgression (5,500 BP) raised sea level 3 to 5 m above the

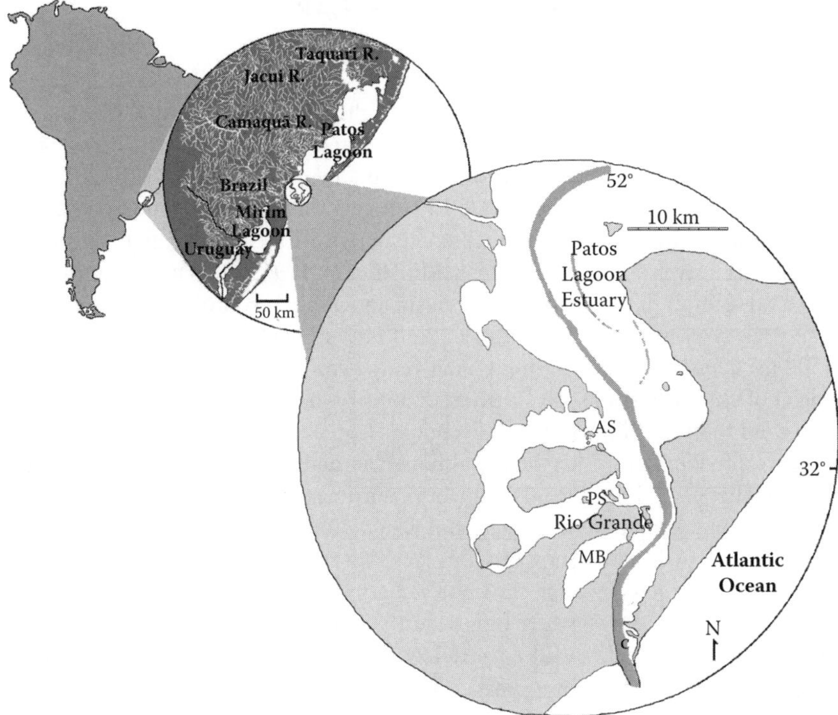

FIGURE 17.1 Geographical location showing the watershed and main tributaries of the Patos–Mirim lagoons in the Southwest Atlantic Ocean. Sampling areas in the zoom of the Patos Lagoon Estuary are indicated as the Principal Site (PS), Mangueira Bay (MB), the main channel (C), and additional site (AS).

present level, and subsequent regressions deposited external sand bars composed of beach ridges and dunes that shape the actual landscape. About 1000 years ago, the inlet connecting the Mirim Lagoon (3749 km^2) to the Atlantic Ocean closed, after which fresh water drained exclusively through a narrow channel (i.e., São Gonçalo; 75 km long, 250 m wide) into the Patos Lagoon (10,360 km^2).

A dam was constructed in 1977 to impede the penetration of seawater from the Patos Lagoon, thus rendering Mirim Lagoon waters available for agricultural irrigation (Seeliger and Costa 1997; Burns et al. 2006). The larger part of the Patos Lagoon is predominantly fresh to oligohaline (Möller et al. 1996; Odebrecht et al. 2005), and the exchange of water with the Atlantic Ocean is restricted to a 0.8 km-wide and 15 m-deep inlet, bounded by 4 km-long jetties that were constructed to stabilize the inlet at the beginning of the twentieth century near the city of Rio Grande (32°S).

17.1.2 Natural Protection Areas

To partially compensate for natural losses and to reduce environmental conflicts national protection areas were established in wetlands surrounding the Patos–Mirim lagoon system, namely the National Park Lagoa do Peixe (1980) and the Ecological Station Taim (1986) in Brazil and the World Biosphere Reserve Bañados del Este (1976) and RAMSAR area for migrating birds (1984) in Uruguay. Both, the Taim wetlands and the Patos Lagoon Estuary became sites (1999) of the Brazilian Long-Term Ecological Research Program (BRLTER) and the International Long-Term Ecological Research (ILTER), which aim to identify the main natural and anthropic impacts on each system (Seeliger et al. 2002). This chapter focuses on the temporal and spatial variations of biota in the Patos Lagoon Estuary (1979 to 2008) and evaluates biotic responses to environmental factors including natural phenomena and human influences.

17.2 THE MICROTIDAL PATOS LAGOON ESTUARY

17.2.1 Water Exchange, Salinity, and Temperature

The main freshwater sources of the Patos Lagoon are the Jacuí, Taquari, and Camaquã rivers and the São Gonçalo channel, which connects the Mirim and Patos lagoons (Figure 17.1). Rainfall in southern Brazil ranges from 3 to 500 mm per month (total annual mean 1200 to 1500 mm) (Figure 17.2). The highest values are related to El Niño periods of the El Niño Southern Oscillation (ENSO) phenomenon, while drought periods coincide with La Niña events (Grimm et al. 1998). The Patos Lagoon is a river-dominated system (Möller et al. 1996, 2001) with a significant relationship between total rainfall in the hydrographic basin and annual river discharge (mean 2400 $m^3 s^{-1}$) (Figure 17.2). Peak river discharge as high as 12,000 $m^3 s^{-1}$ occurs during El Niño, and low river discharge (500 $m^3 s^{-1}$) is associated with La Niña years (Vaz et al. 2006; Möller et al. 2009).

As is typical of choked lagoons (*sensu* Kjerfve 1986), water exchange with the coastal ocean is controlled by wind forcing and freshwater discharge (Abreu and Castello 1997; Möller et al. 1996, 2001). Winds are predominantly northeast, although in autumn and winter southwest winds are more frequent (Möller et al. 2001, 2009). In general, northeast and southwest winds generate a pressure gradient between lagoon and nearby coastal waters, forcing water outflow and inflow, respectively. During flood periods, the lagoon remains fresh for months since only very strong southwest winds tend to reverse outflow (Moller et al. 2001; Fernandes et al. 2002). The distance (250 km) between the rivers with large freshwater input and the mouth of the lagoon delays the influence of fresh water in the estuary by about 20 days (Niencheski and Windom 1994). In contrast, a water deficit in summer/autumn facilitates seawater penetration northward into the lagoon. When the southern reaches of the lagoon are marine and the large central lagoon body becomes oligohaline, salinity gradients may exceed 150 km (Möller et al. 1996; Odebrecht et al. 2005).

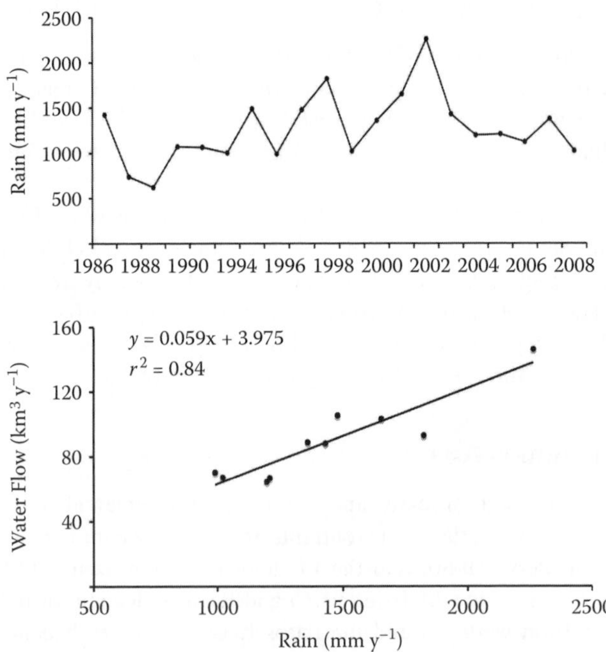

FIGURE 17.2 Total annual rainfall in Rio Grande (upper panel) and the relationship between the amount of annual rainfall in the hydrographic basin and water flow from river discharge (lower panel).

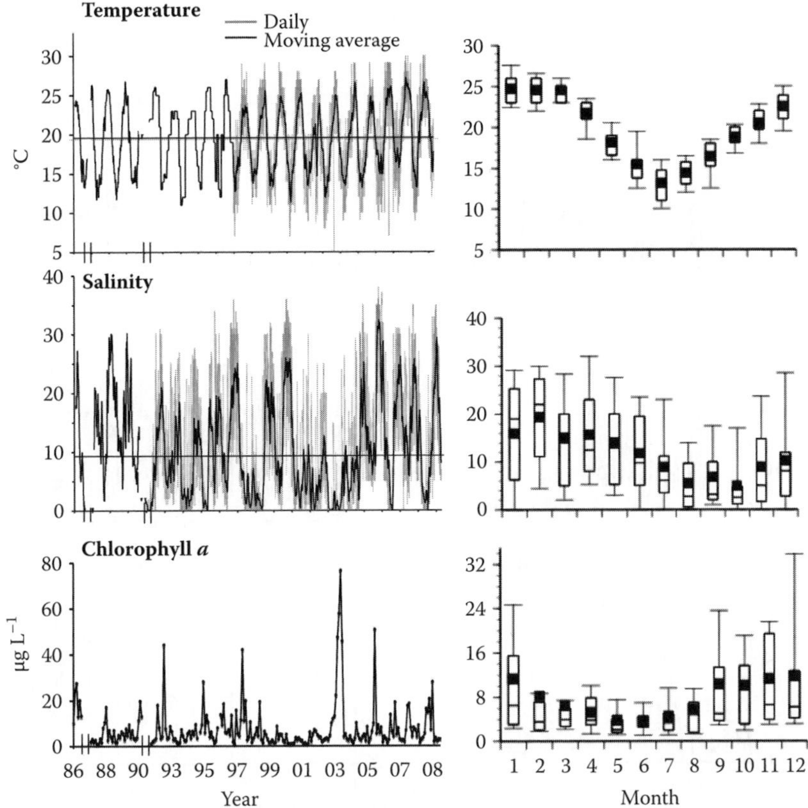

FIGURE 17.3 Left: Temperature, salinity and chlorophyll *a* variation at the principal sampling site in the Patos Lagoon Estuary since 1986 (except 1987 and 1991). Temperature and salinity are based on weekly (1986 to 1990) and daily sampling (mid-1992 to July 2008, gray bars), and the moving average (30 days) and mean value are presented. Chlorophyll (µg L^{-1}) is based on monthly sampling. Right: Box plot presenting the mean, median, and the 50% (vertical box) and the 75% (line end) percentiles per month of each parameter.

Water temperature variations are typical of warm-temperate systems, with lowest temperatures (10°C to 15°C) during the austral winter (July) and highest temperatures from January to March (22°C to 30°C) (Figure 17.3). Owing to winds, significant salinity oscillations occur on an hourly or daily scale (Fujita and Odebrecht 2007; Abreu et al. 2009). Despite the short-scale variability, seasonal salinity patterns occur in the shallow estuary (1986 to 2008), with brackish to euryhaline water tending to be more frequent in summer/autumn (January to June) and fresh to oligohaline conditions prevailing in winter/spring (July to October) (Figure 17.3). Seasonal patterns are related to the annual balance between rainfall and evaporation in the lagoon and to the horizontal advection of seawater inflow caused by winds. Interannual estuarine salinity variability, with high salinities in 1988–1989 and 1999–2000 and low values in 1997–1998 and 2001–2003, coincided with drought during La Niña periods and El Niño events following high discharge (~5000 m^{-2} s^{-1}; Möller et al. 2009) of tributaries in 2001.

17.2.2 Dissolved Inorganic Nutrients

Fluvial nutrients entering the upper lagoon decline significantly in concentration before reaching the estuary as a result of phytoplankton uptake (Niencheski and Windom 1994; Odebrecht et al.

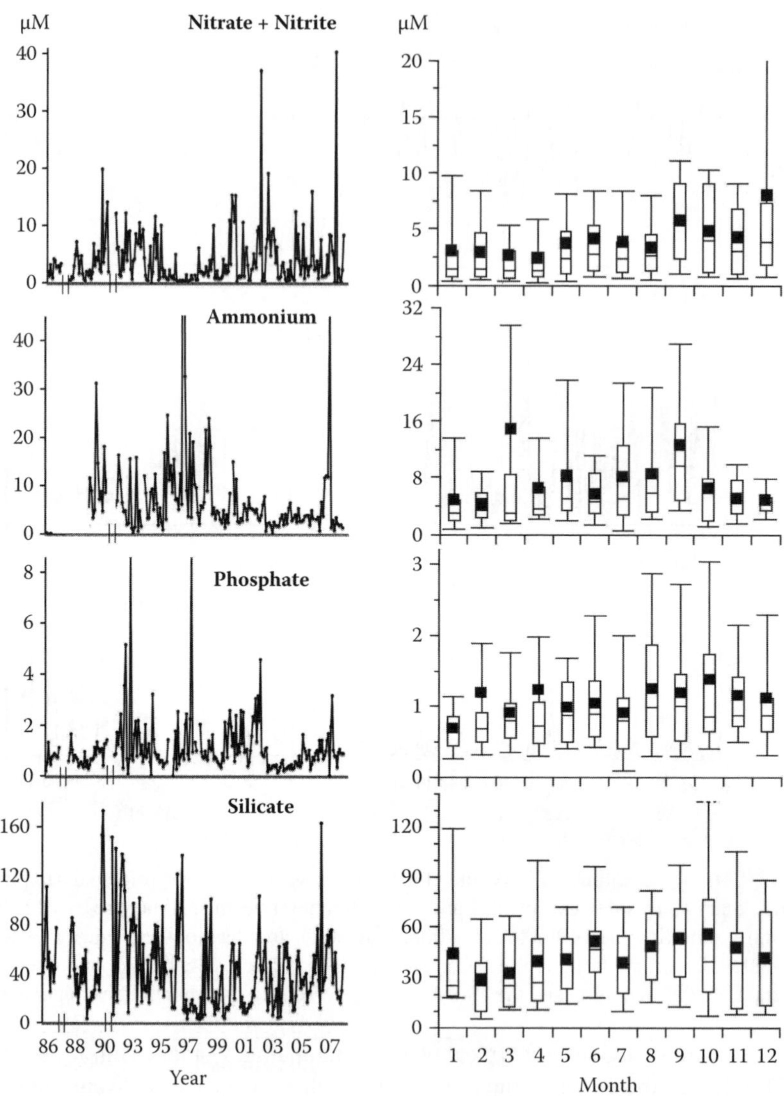

FIGURE 17.4 Left: Monthly variation between 1986 and 2007 (except 1987 and 1991) of the concentration (μM) of dissolved inorganic nutrients nitrate + nitrite, ammonium, phosphate, and silicate at the principal sampling site in the Patos Lagoon Estuary. Right: Box plot presenting the mean, median, and the 50% (vertical box) and the 75% (line end) percentiles per month of each parameter.

2005). The long-term (1986 to 2008) oligotrophication trend of dissolved inorganic silicate, ammonium, and phosphate in the water column (Figure 17.4) may result from eutrophication of the northern lagoon, where phytoplankton biomass increases (Odebrecht et al. 2005; Abreu et al. 2009), and phytoplankton uptake acts as a biofilter that removes excess nutrients from the water column. Freshwater discharge, local sewage input, and phytoplankton uptake also result in significant temporal variation of dissolved inorganic nutrients in the estuary, though seasonal patterns are less evident than for salinity and temperature. Nevertheless, concentrations of nitrate + nitrite (~0.1 to 40 μM; median 2.4 μM) and silicate (3.0 to 163.2 μM; median 35.0 μM) are slightly higher in spring, due to greater freshwater flow into the estuary. The random annual variation of dissolved inorganic phosphate (n.d. to 8.7 μM; median 0.8 μM) and ammonium (~0.1 to 40 μM; median 4.7 μM) is related to complex biogeochemical processes, diverse anthropogenic inputs, and sediment interactions (Niencheski and Baumgarten 1997).

17.2.3 Estuarine Habitats, Biological Components, and Processes

The funnel-shaped southern reaches of the Patos Lagoon represent ~10% (971 km^2) of the total lagoon area (Figure 17.1) and comprise the estuary proper (Asmus 1997). Both the geographic location near an amphidromic point and the long and narrow inlet attenuate the advance of tidal waves (mean tidal amplitude 0.47 m). The estuarine zone of Patos Lagoon is predominantly shallow (80% < 1.5 m) consisting of embayments (<1.5 m), open waters (1.5 to 5 m) and channels (>6 m). Unvegetated subtidal soft bottoms (300 km^2), seagrass beds (120 km^2), marginal marshes (40 km^2), and artificial hard substrates (0.18 km^2) are the main habitats for marine invertebrates and fish species (Seeliger 2001). Estuarine shoals are important nursery habitat for commercially and recreationally important fish and shrimp species (Chao et al. 1982, 1986; D'Incao 1991; Muelbert and Weiss 1991; Vieira and Castello 1997), and the development of these organisms in the estuary is vital for the maintenance of regional and local fisheries. In recent years, marine shrimp and fish aquaculture operations have been established along the estuarine margins (Cavalli et al. 2008).

17.2.3.1 Primary Producers

Microalgae are influenced by both horizontal water exchange and vertical interactions between sediments and the overlying water column. During periods of increased fluvial discharge and influx of seawater, respectively, freshwater and marine species of diatoms, cyanobacteria, chlorophytes, dinoflagellates, and other flagellates are observed in the water column. Periods of fresh or oligohaline water favor growth of planktonic coccoid and filamentous cyanobacteria, including toxic species of *Microcystis* (Yunes et al. 1998; Rosa et al. 2005). In estuarine shoals where wind action leads to the suspension of diatoms (*Surirella* spp., *Bacillaria paxillifer*, and *Cylindrotheca closterium*), the benthic microalgae biomass is three times higher than the phytoplankton biomass (Bergesch et al. 1995). In the channels, microalgae growth is generally light limited (Odebrecht et al. 2005) and has lower biomass and primary production rates than in the shoals. Nano- and picoplankton are the main contributors when chlorophyll *a* concentrations are ≤5 μg L^{-1}. Chlorophyll *a* concentrations increase in spring (10 to 70 μg L^{-1}), and high values occur in the estuary until summer (Figure 17.3), principally due to the growth of coastal euryhaline microplankton diatoms (Fujita and Odebrecht 2007). In summer, phytoplankton biomass appears to be controlled by nutrient concentrations and grazing pressure as indicated by the high abundance of secondary producers (Abreu et al. 1994a; Bemvenuti 1997a, 1997b; Montú et al. 1997).

In general, phytoplankton primary production (PPP) rates relate to incident irradiance and water temperature, while high seston loads and low nitrate concentrations appear to limit photosynthetic activity. Particulate PPP varies between 0.7 and 90.6 mg C m^{-3} h^{-1}, and dissolved organic carbon (DOC; 0.1 and 89.3 mg C m^{-3} h^{-1}) released by phytoplankton may reach 50% of the total PPP, though the annual mean DOC generated represents 22% of total (particulate + dissolved) carbon uptake. In general, high rates of absolute DOC excretion are positively related to particulate carbon production, as well as to high salinity and light in the water column, correlated with coastal phytoplankton species occurring in the estuary after the input of less turbid salt water in the system (Abreu et al. 1994b).

Shoals in the Patos Lagoon Estuary serve as suitable habitats for the settlement and development of seagrass meadows dominated by *Ruppia maritima* (Asmus 1984; Seeliger 1997a; Silva and Asmus 2001). These habitats provide shoreline stabilization; they also intercept nutrients and support ecologically and commercially important fisheries resources. The distribution and abundance of local *R. maritima* populations are related to natural cycles of temperature, salinity, and light attenuation (Costa and Seeliger 1998; Silva and Asmus 2001). Hydrodynamics and sediment conditions also influence the development of seagrass meadows during critical initial stages (Silva 1995), while light availability is important after plants become established, affecting both photosynthesis and intrinsic growth rates (Colares and Seeliger 2006). Net primary production estimates between spring and autumn are 50 to 300 g dry weight m^{-2} (Silva and Asmus 2001).

A high biomass of epiphytic microalgae occurs on the leaves and shoots of seagrasses, while benthic and unattached macroalgae grow entangled between the plants (Ferreira and Seeliger 1985; Silva 1995). Abundant macroalgae (*Ulva* spp., *Cladophora* spp., and *Rhizoclonium riparium*) grow on soft sediments in shallow areas, but are frequently suspended into the water column where they form dense masses of drift algae in the photic zone (0.5 to 1.2 kg C m^{-2} y^{-1}). While light, temperature, and nutrient availability control their biomass and community composition (Coutinho and Seeliger 1986), wind fetch and water currents are responsible for dispersing drift algae along shallow shoals or to open and deeper waters.

More than 20 macroalgal species (mostly Rhodophyta) that are absent in soft-bottom estuarine areas thrive on hard substrates provided by the rocky jetties along the access channel of the estuary (Seeliger 1997b), though they contribute little to primary production compared to drift macroalgae. The composition and diversity of macroalgae along the jetties of Patos Lagoon are indicative of warm-temperate environmental conditions (Coutinho and Seeliger 1986).

17.2.3.2 Secondary Producers

Zooplankton in the Patos Lagoon Estuary consists of marine and freshwater holoplanktonic and meroplanktonic species (Montú et al. 1997). Common holoplanktonic marine species are copepods (*Acartia tonsa*, *Euterpina acutrifrons*, *Oncea conifera*) and chaetognaths (*Sagitta tenuis*). Meroplanktonic marine organisms like cirripeds of *Balanus improvisus*, larvae of echinoderms, polychetes, mollusks, and crustaceans are also found. Freshwater zooplankton abundance is generally low, consisting of copepods (*Notodiaptomus carteri*, *N. incompositus*, *Mesocyclops annulatus*) and cladocera (*Diaphanosoma sarsi*, *Eubosmina tubicen*, *Simosa* sp., *Alona* sp.) (Montú and Gloeden 1986; Montú et al. 1997).

Zooplankton abundance and biomass are highest in summer (40,016 organisms m^{-3} and 0.2 to 11 mL m^{-3}). Abundance decreases in autumn (15,526 organisms m^{-3}), and lowest densities occur in winter (6251 organisms m^{-3}) and spring (14,180 organisms m^{-3}). Cladocera (*Pleopys polyphemoides*, *Podon intermedius*, *Penilia avirostris*), ctenophores, siphonophores, medusa, and chaetognaths (*Sagitta tenuis*) are abundant in the lower estuary during summer when subtropical species such as *Sagitta enflata* and *Centropages velificatus* may occur. The copepod *Acartia tonsa* and freshwater cladocera (*Moina micrura*, *Diaphanosoma sarsi*, *Ceriodaphnia cornuta*) are present the entire year (Montú et al. 1997).

Low diversity and abundance of macrozoobenthos occur on sandy and biodetritic bottoms of high current channels (Bemvenuti et al. 1992), but higher densities of the opportunistic gastropod *Heleobia australis* can be found in channels with fine sediments (Bemvenuti 1997a). During periods of greater marine influence, diversity increases in southern estuarine channels, particularly due to recruitment of marine polychetes which disappear during freshwater outflow (Bemvenuti et al. 2005). The channels are migration routes for species that use the estuary as feeding and nursery grounds. Furthermore, the deeper bottom layer (>6 m) of the channel between the lagoon mouth and Rio Grande Harbor may provide protection for some decapods (crabs, shrimp) during their austral winter migration (Bemvenuti 1987).

Abundant macrozoobenthic fauna in the extensive shoals are the main food for birds, juvenile decapod crustaceans, and fishes (Bemvenuti 1997b). The number of macrozoobenthic species that inhabit estuarine shoals is relatively low (15 to 20 species) due mainly to variable salinity. Some species occur at high density (>10,000 individuals m^{-2}) and frequency, including the gastropod *H. australis* that lives on sediments in unvegetated shoals and infaunal polychetes (*Laeonereis acuta*, *Nephtys fluviatilis*, *Heteromastus similis*), tanaidaceans (*Kalliapseudes schubartii*), and the bivalve *Erodona mactroides* that burrows in bottom sediments. Vegetated shoals that are dominated by macroalgae or submersed *Ruppia maritima* beds support high diversity and abundance of amphipods, isopods, and juvenile decapod crustaceans (crabs and shrimp) (Bemvenuti 1997a; Rosa and Bemvenuti 2006).

Macrozoobenthos in the shoals reproduce mainly during spring/summer following temperature increases and infaunal recruitment which lead to increased abundances (Bemvenuti 1987; Rosa and Bemvenuti 2006). Infaunal recruitment periods coincide with enhanced trophic activity of macropredators like crabs, shrimp, and macrobenthic feeding fishes that keep infaunal densities below the carrying capacity of the system. Predation pressure decreases in autumn when temperature decreases, and crabs and shrimp abandon the shoals. During this season, macrozoobenthos recruitment is important because it maintains the population density during winter (June to September) when reproduction rates of most infaunal species decline (Bemvenuti 1987; Rosa and Bemvenuti 2006).

Only a few fish species are abundant and frequent in the estuary, although eggs and larvae of 29 species are common (Sinque and Muelbert 1997), and about 150 species use the estuary for feeding and growth (Chao et al. 1982; 1985; Pereira 1994; Vieira and Castello 1995). The dominant species are the whitemouth croaker, mullets, and marine catfishes (Muelbert and Weiss 1991; Vieira and Castello 1997; Vieira et al. 2008). The shallow shoals favor the recruitment of a variety of species (Table 17.1). Ninety-five percent of juvenile and adult fish are small estuarine resident species of marine origin (Vieira and Castello 1997) represented by mullets (*Mugil platanus*, *M. curema*, and *M. gaimardianus*), silversides (*Atherinella brasiliensis* and *Odontesthes argentinensis*), clupeoids (*Brevoortia pectinata*, *Platanichthys platana,* and *Ramnogaster arcuata*), one-sided livebearer (*Jenynsia multidentata*), and the whitemouth croaker (*Micropogonias furnieri*). Freshwater visitors, like Characiforms and members of the families Poeciliidae and Cichlidae (Chao et al. 1985; A. M. Garcia et al. 2003a; Table 17.1), are also common. Fish eggs are almost nonexistent and larvae are not abundant in the shoal.

In contrast, deeper waters (Table 17.2) shelter larger individuals but also serve as migration routes for fish with greater mobility (Vieira and Castello 1997). Here juvenile and subadults of epibenthic or demersal species are important artisanal and industrial fishery resources, especially the abundant

TABLE 17.1
Temporal Variation of Dominant Fishes Sampled with Beach Seine at the Shallow Area of Patos Lagoon Estuary between 1979 and 2001

Species	1979	1980	1981	1982	1983	1984	1996	1997	1998	1999	2000	2001
Mugil platanus[a]	1	1	1	1	1	1	1	1	1	1	1	1
Atherinella brasillensis	3	1	1	1	1	1	1	1	1	1	1	1
Odontesthes argentinensis	1	3	1	1	1	1	1	1	1	1	3	3
Mugil curema[a]		3	1	1	1	1		1	1	1	1	1
Jenynsia multidentata	3	1	1	1	2	1	1	3	4	1	3	4
Brevoortia pectinata[a]	4	4	3	1	4	1	4	3	1	1	1	4
Micropogonias fumieri[a]	4	3	3	3	1	1	4	3	1	3	3	3
Mugil giamardianus		4	1	1	1	1		3	4	1	3	4
Lycengraulis grossidens	4	3	3	3	1	1	4	4	3	4	4	3
Platanichythys platana			4	4	1	1	3	3	3	4	4	3
Ramnogaster arcuata	4	4	4	4	4	1	4	2	4	4	4	4
Ctenogobius shufeldti		4		4	4		3	3	4	4	4	4
Parapimelodus nigribarbis				4	4	1			3	4	4	4

Note: (1) Abundant and frequent; (2) abundant but not frequent; (3) frequent but not abundant; and (4) present but not abundant or frequent. See Garcia and Vieira (2001) for category details.

[a] Commercially important species.

TABLE 17.2
Temporal Variation of Dominant Fishes Sampled with Bottom Trawl at the Channel in the Patos Lagoon Estuary between 1979 and 1998

Species	1979	1980	1981	1982	1983	1984	1995	1996	1998
Micropogonias fumieri[a]	1	1	1	1	1	1	1	1	1
Genidens barbus[a]	1	1	2	1	1	1	1	1	1
Genidens genidens	4	4	4	4	2	1	1	4	1
Lycengraulis grossidens	4	3	3	2	2		3	3	4
Paralonchurus brasiliensis	2	4	3	4	4	4		2	4
Menticirrhus americanus[a]	2	2	4	1	4		4	4	4
Achirus garmani	4	4	4	4	2		2	3	3
Cynoscion guatucupa[a]	3	3	3	4	4	3	4	4	
Parapimelodus nigribarbis	4			4	3	2	4		3
Genidens planifrons[a]	3	2	3	3	4	3	4	4	4
Porichthys poroissimus	4	4	3		4	4		3	
Umbrina canosai[a]	4		3		3			4	4
Anchoa marinii	4	4	3	3	4	4	4		4
Paralichthys orbignyana[a]	4	4	4	4	3		4	4	4
Prionotus punctatus[a]	4	4	4		4			2	4
Pimelodus maculatus					4	3	4	4	4
Ctenosciaena gracilicirrhus	3		4						
Symphurus jenynsi	4	4	4	3					
Citharichthys spilopterus								3	

Note: (1) Abundant and frequent; (2) abundant but not frequent; (3) frequent but not abundant; and (4) present but not abundant or frequent. See Garcia and Vieira (2001) for category details.

[a] Commercially important species.

whitemouth croaker, marine catfish, and flatfish (*Paralichthys orbignyanus*). Deep channel currents transport eggs and larvae in and out of the estuary (Sinque and Muelbert 1997). Saltwater intrusions in the channels are responsible for the recruitment of marine fish larvae into the estuary and for the abundance of eggs and larvae of the whitemouth croaker, which is controlled by variable water advection (Muelbert and Weiss 1991; Martins et al. 2007). Oceanic species (e.g., *Porichthys porosissimus, Prionotus punctatus, Parona signata,* and *Peprilus paru*) are restricted to the mouth of the estuary during intense saltwater intrusions (Sinque and Muelbert 1997; Vieira 2006).

Despite a pronounced decline in abundance of birds since the early nineteenth century, 13 species are still common in wetlands, open waters, and protected bays of the estuary (Vooren 1997). Feeding habits are diverse, including piscivorous birds, like the cormorant *Phalacrocorax olivaceus*, terns (*Sterna trudeaui, S. hirundo*), black skimmer (*Rynchops niger*), and the snowy egret (*Egretta thula*), while black-necked swan *Cygnus melancorhyphus* and the red-gartered coot *Fulica armillata* are herbivorous. The main scavenger in the estuary is the brown-hooded gull *Larus maculipennis.* Marine mammals are rare. Between 31 and 100 individuals of the bottlenose dolphin *Tursiops truncatus*, which tolerates low salinity water (2 to 3), enter the estuary throughout the year, while the sea lion *Otaria flavescens* is observed only occasionally (Pinedo 1997).

17.3 BIOTIC RESPONSES TO ENVIRONMENTAL CHANGES

There is a significant response of biota in the Patos Lagoon Estuary to eutrophication, predatory fishing, and dredging as well as to natural environmental effects caused by the El Niño Southern

Oscillation (ENSO) phenomenon. The retention of frontal systems due to the strength of atmospheric jetties during El Niño events typically increases rainfall in southern Brazil, while weaker atmospheric jetties accelerate the passage of frontal systems during La Niña events, generating reduced rainfall and drought conditions. Annual anomalies related to ENSO cycles affect freshwater discharge, and long-term salinity regimes seem to play a major role in the life cycle, growth, and biomass of most species, thus controlling the interannual variation in their composition and abundance.

17.3.1 Impacts of El Niño Southern Oscillation (ENSO)

17.3.1.1 Phytoplankton

Compared to coastal ecosystems elsewhere (Cloern and Jassby 2008), chlorophyll *a* values in the estuary are elevated (Figures 17.3 and 17.5). Significant short- and long-term temporal variability occurs largely due to horizontal water advection. Short-term wind-induced water exchange in the estuary is reflected by biotic community changes, alternating between freshwater and marine species at scales of hours to days (Fujita and Odebrecht 2007; Abreu et al. 2010). Changes in microalgae composition are closely related to the frequency of polar atmospheric fronts, which are important drivers of seawater flow into the estuary.

Phytoplankton biomass in estuarine shoals are strongly correlated with rainfall and freshwater discharge (1986 to 1990) (Figure 17.5). This relationship between mean annual chlorophyll *a* and the total amount of rainfall in the hydrographic basin, however, disappears when rainfall exceeds 1500 mm yr^{-1} because the large amount of fresh water discharged from the basin flushes the phytoplankton biomass out of the estuary (Abreu et al. 2010).

Phytoplankton biomass, mainly composed of freshwater cyanobacteria and oligohaline/euryhaline diatoms (*Skeletonema* spp., *Cylindrotheca closterium*, *Chaetoceros subtilis*), increases during prominent low salinity periods. The most conspicuous event was observed during the El Niño 2002–2003 (Figure 17.6), when freshwater discharge in spring (October to December 2002) resulted in high nitrogen inputs (10 to 40 μM) that were taken up by phytoplankton, leading to a biomass peak (chlorohyll *a* 40 to 80 μg L^{-1}; biovolume 30 to 80 mm^3 L^{-1}), low nitrogen levels, and N:P ratios (<5 μM TN; N:P <5). Nitrogen-fixing species, such as filamentous heterocytic cyanobacteria (*Anabaena oumiana*, *Anabaenopsis* sp., and *Aphanizomenon aphanizomenoides*) and the colony-forming *Microcystis* spp. were abundant during a 3-month period. In contrast, lowest chlorophyll concentrations, following the La Niña periods of 1988–1989 and 1999–2000, coincided with higher ammonium concentrations and longer water residence times (Abreu et al. 2010). During periods of

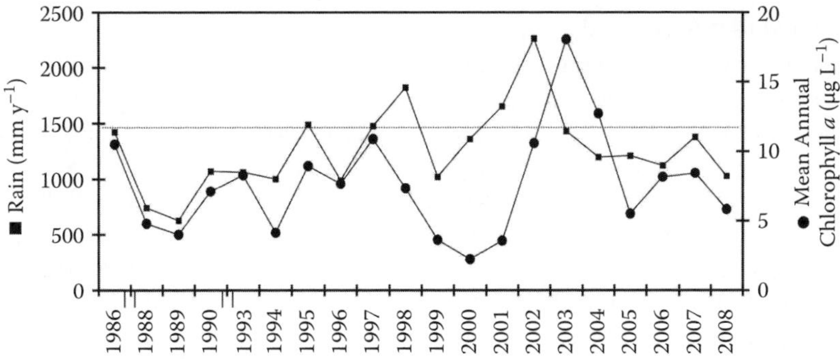

FIGURE 17.5 Relationship between the annual amount of rainfall and mean annual chlorophyll *a* based on a 20-year time series and indicating the 1500 mm y^{-1} threshold line. Chlorophyll *a* sampling was weekly until 1990 and monthly thereafter. (Adapted from Abreu, P. C. et al. 2010. *Estuaries Coasts* DOI 10.1007/s12237-009-9181-9.)

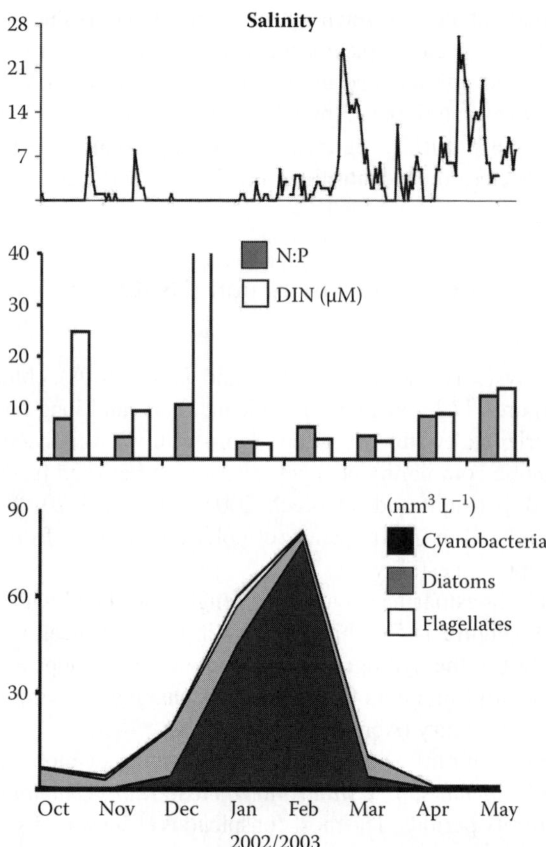

FIGURE 17.6 Daily salinity variation (upper panel) and monthly data of dissolved inorganic nitrogen (DIN, μM), the atomic ratio N:P (middle panel), and the biovolume ($mm^3 L^{-1}$) of main microalgae cyanobacteria, diatoms, and flagellates (lower panel), showing the effect of freshwater discharge in response to ENSO (October to May, 2002 to 2003) at the principal sampling site in the Patos Lagoon Estuary.

marine influence, zooplankton herbivory plays an important role in the estuary, though this activity still needs to be quantified.

17.3.1.2 *Ruppia maritima*

In coastal systems elsewhere, seagrass reductions and dieback have been mainly attributed to eutrophication, dredging, and predatory fishing, but also to global changes in temperature, precipitation, and freshwater discharge (Orth et al. 2006). Prior to 1990, 40% to 60% of the shallow estuarine bays in the Patos Lagoon were covered by high biomasses of *Ruppia maritima* (Seeliger 1997a). A significant reduction of the seagrass occurred in 1996–1997 and continued to 2000 (Figure 17.7). Shallow mid-estuarine regions, where *R. maritima* usually thrives, displayed bare substrate in the summer of 2003, with populations being present only in sheltered bays. The drastic decrease of vegetated bottoms between 2000 and 2003 coincided with increased erosion of saltmarsh margins due to extreme freshwater runoff and high water level (Costa et al., personal communication) while marsh sediment and debris buried adjacent *R. maritima* populations. Both high sediment mobility in exposed areas and high accretion rates in sheltered areas tend to limit seagrass growth (Koch 2001; Duarte et al. 2006).

Entire meadows, including belowground rhizome nets, were dislodged due to flooding conditions. As a result, only 43% of new shoots sprouted from old rhizomes in summer 2002. Additionally, the

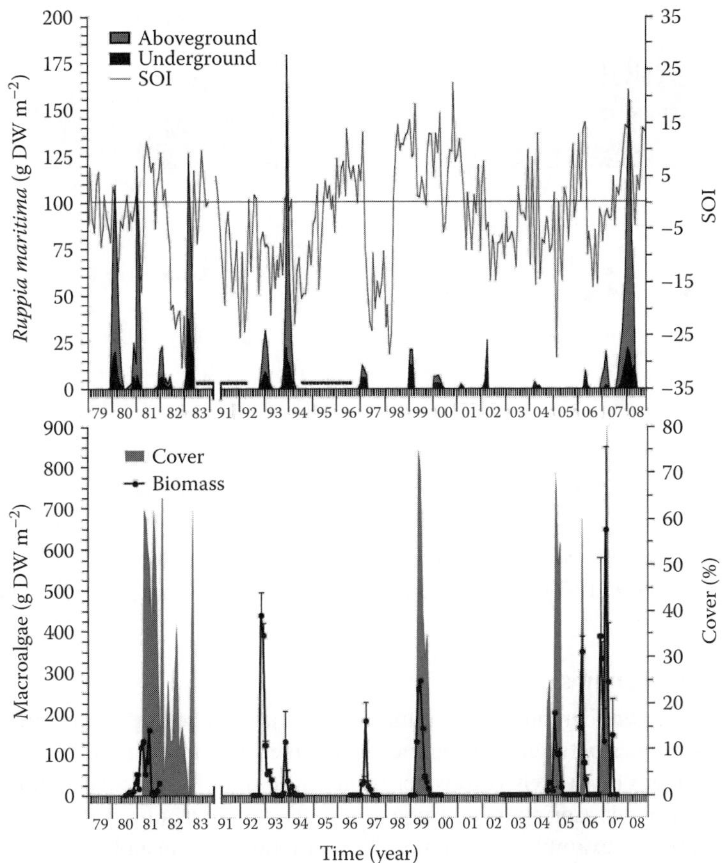

FIGURE 17.7 Monthly variability in the total biomass of seagrass *Ruppia maritima*, showing the aboveground and belowground fractions (upper panel), and drift macroalgae biomass and percentage cover (lower panel) at the principal sampling site in the Patos Lagoon Estuary. (Data for 1979 to 2008 from Coutinho 1982; Cafruni 1983; Asmus 1984; Silva 1995; Colares 1998; and Souza 2001.)

reduction of natural seed banks and low germination rates of *R. maritima* (Cordazzo 2004) further decreased meadow recovery between spring 2004 and summer 2007 despite favorable water level and transparency, salinity, and temperature for growth. *Ruppia maritima* beds visibly recovered between 2006 and 2008. Sparse meadows of small plants (maximum length of 10 cm) which developed from seedlings, appeared in the austrial summers of 2006 and 2007 (Figure 17.7). The absence of established rhizome nets rendered new meadows susceptible to seasonal hydrodynamics, leading to large numbers of floating plants at the beginning of fall. In the summer of 2008, *R. maritima* beds exhibited high shoot length and density along with high production of flowers and fruits. The rapid expansion of the meadows may have been a response of the seed bank to low water temperatures in winter of 2007 and to optimal salinities, as has been observed in culture (Koch and Seeliger 1988).

Hydrodynamics and sediment erosion and accretion appear to be key factors controlling the temporal patterns of *R. maritima* abundance in the estuary. The development, maintenance, and spatial distribution of the seagrass meadows strongly correlate with water levels that determine the amount of light penetration reaching the bottom, since *R. maritima* tends to be abundant at depth less than 1 m and is generally absent below 1.5 m depth. Estuarine salinity variations appear to have limited influence on the growth of the high salinity–tolerant *Ruppia* plant, except when there are interactions of salinity and temperature with seed germination (Knack 1986; Colares 1998).

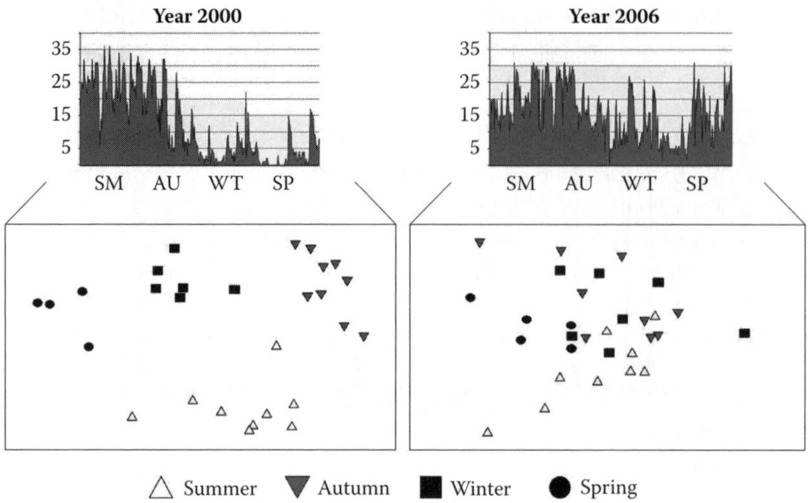

FIGURE 17.8 Daily variation of salinity in the years 2000 and 2006 at the principal sampling site in the Patos Lagoon Estuary (upper panels), and seasonal distribution of macrozoobenthos diversity in the channel (9 to 12 m depth; lower panels).

17.3.1.3 Macrozoobenthos

A comparison of macrozoobenthos in the channel area clearly shows significant differences in benthic community structure between the summer and winter/spring, with the dominance of marine species in the summer due to higher saltwater intrusion into the system. The dominance of freshwater species during winter/spring is in response to higher amounts of rainfall and thus freshwater inputs (Figure 17.8). For example, in summer 2000, a total of 39 macrozoobenthos species were observed in the channel; this number decreased to only 19 species from autumn to spring, when salinity was near zero. Greater marine influence during 2006, in contrast, resulted in a dramatic decline of seasonal macrozoobenthos variation; species richness ranged between 31 and 39.

Intense infaunal recruitment is frequently associated with high salinity and temperature during summer, which results in increased macrozoobenthos abundance (Bemvenuti 1987; Rosa and Bemvenuti 2006). Nevertheless, summers with high precipitation (i.e., El Niño of 2002–2003) decrease salinity levels, which reduce macrozoobenthic abundance in the estuary (Figure 17.9). During and immediately after El Niño events, recruitment gaps of species with direct development (e.g., *Kalliapseudes schubartii)* as well as of the bivalve *Erodona mactroides* are observed (Colling et al. 2007). Reproductive stocks of *Erodona mactroides* are mainly found in northern limnic areas of the Patos Lagoon. The high freshwater runoff during El Niño of 2002–2003 may have flushed larvae out of the lagoon into coastal waters. Intense runoff removes the nepheloid layer from the sediment surface, which reduces the organic matter content, thus causing low benthic oxygen consumption rates (Zarzur 2007). These adverse conditions negatively affect deposit feeders that are the dominant guild in estuarine shoals (Bemvenuti 1997b). During periods of greater marine influence, benthic diversity increases, owing especially to the recruitment of marine polychetes in the southern estuary channels, though inverse patterns occur during periods of higher runoff (Bemvenuti et al. 2005).

17.3.1.4 Fish

Larval fishes are most abundant and diverse during reproductive periods in spring and summer (Muelbert and Weiss 1991). The low abundance of the eggs and larvae of whitemouth croaker *Micropogonias furnieri* in spring 1982 and in summer 1993 coincided with higher than mean

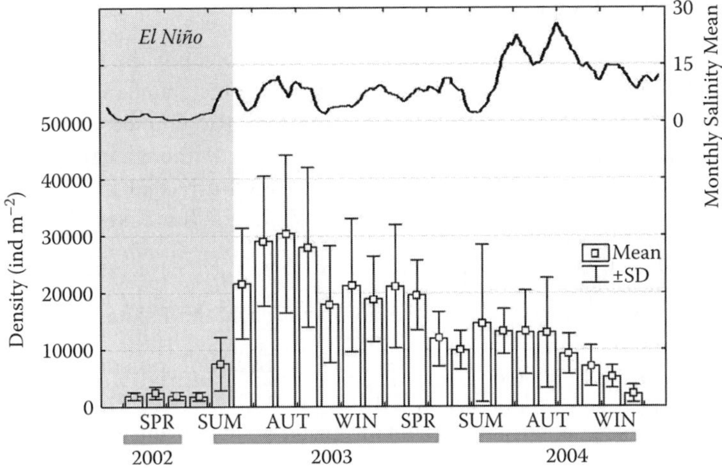

FIGURE 17.9 Monthly variation of salinity (upper line) and macrozoobenthos density (bars, mean and SD) between September 2002 and August 2004 at the principal sampling site in the Patos Lagoon Estuary.

freshwater discharge. In contrast, lower than average freshwater discharge and saltwater intrusion in summer 1992 increased the recruitment of eggs and larvae into the estuary, suggesting that the interannual variability of *M. furnieri* is related to the amount of precipitation and water discharge (A. M. Garcia et al. 2004; Bruno and Muelbert in press).

High intra- and interannual variability in relative abundance of dominant fishes (*Mugil platanus*, *M. curema*, *Atherinella brasiliensis*, and *Odontesthes argentinensis*) in the estuary tends to obscure interdecadal trends. Nevertheless, seasons or years of low abundance seem to coincide with periods of high rainfall and low salinity (Table 17.1). During El Niño years, high freshwater discharge decreases recruitment of resident and dependent fish species, though species richness increases due to an expanded range of freshwater species (i.e., *Parapimelodus nigribarbis*, Table 17.1) into the estuary (A. M. Garcia et al. 2003b). Owing to lower or higher precipitation and freshwater outflow, the abundance of marine species in the estuary increases or decreases during La Niña (1995–1996) or El Niño (1997–1998) periods, respectively (A. M. Garcia et al. 2001; A. M. Garcia and Vieira 2001).

17.3.1.5 Fisheries

Meteorological changes caused by the ENSO phenomenon are major driving forces that control the interannual variability of fish community structure and dynamics in the estuary, having great impact on fisheries (A. M. Garcia et al. 2004). Mullet fisheries serve as a good example, since both juvenile and adult mullets decline in abundance during high rainfall and near-zero salinity. A probable explanation of El Niño–induced effects is that high freshwater outflow may render ineffective the mechanisms of passive immigration of juveniles into the estuary, resulting in smaller schools of individuals. In addition, near-zero salinity in the estuarine area for several months during strong El Niño events could lead to spatial dispersion of maturing mullet during their migration offshore, resulting in smaller schools of individuals and, consequently, lower catches by artisanal fishers (Vieira et al. 2008).

Similarly, annual shrimp (*Farfantepenaeus paulensis*) catches show a strong negative correlation with freshwater discharge, which controls seawater penetration, and the transport of shrimp larvae into the estuary. High rainfall in El Niño years have generated smaller shrimp fisheries in subsequent years, causing important economic and social impacts in the region (Castello and Möller 1978; Möller et al. 2009).

17.4 HUMAN INFLUENCE

Until the arrival of European colonizers, the coastal plain surrounding the Patos Lagoon was inhabited by Chaná and Tupi-Guarani Indians, the latter exploiting the abundant fish, shrimp, and shellfish resources in the lagoon (Vieira 1983). After the town of Rio Grande was established in 1737, Portuguese colonization quickly expanded into coastal areas. Although early economic activities focused on cattle breeding, the abundance of fish supported artisanal fishery and soon became economically and socially important (Asmus 1997). Whitemouth croaker (*Micropogonias furnieri*), mullet (*Mugil platanus*), marine catfishes (*Genidens barbus*, *G. planifrons*, *G. genidens*), and black drum (*Pogonias cromis*) sustained small-scale fishery in the estuary. Today, about 3.5 million inhabitants in the cities of Porto Alegre at the northern reach and Pelotas and Rio Grande at the southern reaches of the Patos Lagoon exploit different resources. Harbor activities have become increasingly important through time. Domestic sewage and industrial effluents reach the estuary and affect various components of the system.

17.4.1 EUTROPHICATION

Near Rio Grande city, the estuarine margins of Mangueira Bay (see Figure 17.1) receive most of the effluents and therefore have muddy sediments with reducing characteristics, low faunal density, and low species diversity except for the polychetes (Bemvenuti 1987; Rosa and Bemvenuti 2006, 2007). More distant margins have nonreducing sandy sediments, as well as abundant and diverse macrozoobenthos, including crustaceans, which are sensitive to organic matter pollution (Bemvenuti and Angonesi 2009). In addition, zooplankton density and diversity are reduced in the polluted Mangueira Bay. About 14% of *Acartia tonsa* exhibit malformations that are likely caused by exposure to pollutants during several life cycles (Montú and Gloeden 1982). Compared to primary and secondary producers in more pristine areas of the estuary, isotopic tracer studies confirmed significantly higher (>3.5‰) $\delta\ ^{15}N$ values in plants, detritus, and benthic organisms in Mangueira Bay, probably related to the uptake of dissolved nitrogen in wastewater and other effluents (Abreu et al. 2006).

Despite eutrophication in part of the estuary, a long-term oligotrophication process is occurring in most areas since the concentrations of ammonium, phosphate, and silicate in shallow bays north of Rio Grande city are actually lower than the mean values recorded between 2000 and 2006 (Abreu et al. 2009). This could be due to phytoplankton, mainly heavily silicified diatoms (e.g., *Aulacoseira* spp.), cyanobacteria, and chlorophytes, which rapidly take up nutrients in the northern sector of the lagoon that receives sewage from Porto Alegre, thereby reducing nutrient levels in estuarine waters (Odebrecht et al. 2005).

The apparent reduction of nutrients in the estuary may also be associated with an increase of drift macroalgal biomass in summer when seagrass abundance declines (Figure 17.10). In some shallow-bottom generally dominated by *R. maritima* meadows, between 70% and 100% of the sediments are covered by unattached opportunistic algae, mainly *Ulva, Cladophora,* and *Rhizoclonium.* Rapid algal growth in the summers from 2004 to 2007 led to much higher macroalgal biomass than reported for previous decades (Coutinho and Seeliger 1986; Seeliger 1997b). Summer biomass peaks and productivity rates (1 to 6 g C m^{-2} day^{-1}) were comparable to eutrophic estuaries (Flindt et al. 1997; Martins and Marques 2002) possibly due to the high photosynthetic and nutrient uptake efficiency of *Ulva* (Copertino et al. 2009). Dense macroalgal growth in summer may delay or inhibit the recovery of seagrass meadows since algal filaments reduce available light and increase the drag force on *Ruppia* plants caused by winds and currents (Silva and Asmus 2001).

Drift macroalgae are important primary producers and a food source for crabs (e.g., *Neohelice granulata* and *Callinectes sapidus*), and dense algal layers over shallow bottoms may serve as refuge for larvae and juvenile fish and invertebrates (e.g., *Farfantepenaeus paulensis*). However, contrary to nursery habitats formed by well-established seagrass meadows, unstable habitats of unattached

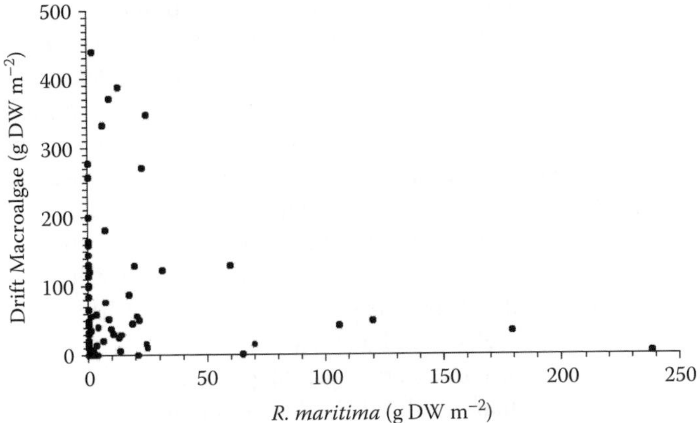

FIGURE 17.10 Relationship between the values of drift macroalgae and *R. maritima* total biomass, in spring, autumn, and summer (1979 to 2007) at the principal sampling site in the Patos Lagoon Estuary.

macroalgae and associated fauna are quickly removed by winds and currents (Souza 2001). Furthermore, at high mid-summer temperatures and low water circulation, the decomposition of drift algae causes anoxia, which leads to benthic animal mortality (Nelson et al. 2003). Finally, under high light and eutrophic condition *Ulva* may compete with phytoplankton for nutrients (Copertino et al. 2009), while *Ulva* may release compounds toxic to microalgae (Nelson et al. 2003).

17.4.2 Overfishing

Southern Brazilian estuaries have a similar composition and abundance of fish species (Ramos and Vieira 2001). Of the more than 110 fish species reported for the Patos Lagoon Estuary (Chao et al. 1982), about 26 species are numerically abundant and ecologically important (Tables 17.1 and 17.2) although only a few are commercially exploited in the local fishery (marked with an asterisk in the tables). While the landing and capture per unit effort (CPUE) data are probably underestimated (Reis et al. 1994), they indicate a reduction of several estuarine fish stocks. The use of modern transport and storage methods by both artisanal and industrial fisheries led to overfishing and to a rapid decrease of teleost catches during the last decade (Reis et al. 1994; Haimovici et al. 2006; Vieira et al. 2008). The whitemouth croaker, the king weakfish (*Macrodon ancylodon*), and marine catfishes were the most affected. Today these species only support a small-scale subsistence fishery.

The decline of artisanal fish stocks in the estuary can be attributed to inadequate fishery surveillance by government sectors, which leads to overfishing and mortality of juveniles during reproductive seasons (Chao et al. 1986; Haimovici and Umpierre 1996; Haimovici et al. 1997, 2006; Reis et al. 1994), or to decreasing water quality as a result of urban and industrial activities in the Patos Lagoon watershed (Asmus 1997; Almeida et al. 1993). However, climatic events can also mask the reasons for the decline of fish stocks in the estuary since they affect the structure of fish fauna and the recruitment of economically important species (A. M. Garcia et al. 2003b, 2004; Vieira et al. 2008; see Section 17.3.1). Indeed, intense local and oceanic fisheries seem to act synergistically with ENSO events and eventually lead to the depletion of estuarine finfish stocks like mullet (*Mugil platanus*), whose abundance depends on seawater influx to the estuary (A. M. Garcia et al. 2003b; Reis et al. 1994; Vieira et al. 2008).

17.4.3 Harbor Dredging

The main channel in the southern part of the estuary is subject to intense sedimentation and dredging of large sediment volumes (~30,000 m^3 mo^{-1}) to maintain access to Rio Grande Port (Seeliger and Costa 1997). Since estuarine sediments have been dredged for many years (von Ihering 1885), dredging impact on the biota is hard to evaluate. However, there is evidence of significant impact on macrozoobenthos species because the abundance of the dominant species *Heleobia australis* decreases in dredged areas (Bemvenuti et al. 2005). Furthermore, dredging activities in the estuary could have caused the deposition of large amounts of mud on adjacent ocean beaches (Cassino Beach, 32°13′ S; 52°15′ W) leading to changes in surf zone dynamics as well as microflora abundance and composition (Odebrecht et al. 2010). The diatom *Asterionellopsis glacialis* reaches high abundance (10^7 to 10^8cells L^{-1}), chlorophyll *a* (2 to 5 mg L^{-1}), and primary production rates (1 to 6 mg C L^{-1} h^{-1}) in the surf-zone of the exposed, gently sloping sand beaches (V. M. T. Garcia and Gianuca 1997; Odebrecht et al. 2010). During passage of the Antarctic atmospheric front, southerly winds facilitate the suspension of cells from bottom sediments which then accumulate in the surf zone. In 1998, elevated mud deposition altered surf zone characteristics by increasing viscosity and decreasing wave energy (Calliari et al. 2007). The accumulation of *A. glacialis* in the surf zone at Cassino Beach was inhibited during the following 2 years. Over a 15-year period (1992 to 2007), a significant reduction in the frequency of *A. glacialis* "blooms" occurred along with changes in phytoplankton composition (Odebrecht et al. 2010), negatively impacting the beach food web. Furthermore, the deposition of mud on surf zone diatoms may have affected mollusk and fish recruitment in the estuary. Studies should evaluate the impact of dredging on zooplankton, fish, birds, and mammals in the estuarine and coastal region.

17.4.4 Bioinvasion

Reports on bioinvasion into coastal Brazilian ecosystems are increasing, though this might also be due to more intense research efforts (Ferreira et al. 2009). High densities of nonindigenous species may potentially displace native species, leading to reduced diversity as well as food web and habitat changes. In the Patos Lagoon Estuary, the dispersal of nonindigenous marine species is largely a result of increasing navigation (ship incrustation and ballast water) and aquaculture activities. The initial bioinvasion of the Chinese freshwater golden mussel *Limnoperna fortunei* in 1991 in Argentina was apparently caused by the release of ballast water (Darrigran and Pastorino 1995). In 1998, *L. fortunei* was detected in the northern reaches of the Patos Lagoon (Mansur et al. 2003). The subsequent colonization of the estuary coincided with a low salinity period (2001 to 2003) when isolated individuals occurred near the estuarine inlet. A massive mortality of *Limnoperna fortunei* coincided with drought conditions in 2004 to 2005. Thus, population increases in the southern lagoon and estuary appear to be prevented by frequent suboptimal salinity conditions (Capítoli et al. 2008). In contrast, the establishment of the marine clam *Perna perna,* which is native to Africa and was probably introduced in Brazil in the eighteenth to nineteenth centuries (Ferreira et al. 2009), seems to be hampered by low salinities in the Patos Lagoon Estuary.

Among planktonic crustaceans, *Temora turbinate* is now one of the dominant pelagic calanoid copepods (Kaminski 2009). It was first registered in the Patos Lagoon Estuary in 1992 (Muxagata and Gloeden 1995). Prior to the invasion of this species, *T. stylifera* was the only representative of the genus at the Brazilian coast (Ferreira et al. 2009).

In recent years, nonindigenous fish species have been recorded in the Patos–Mirim system. The black catfish *Trachelyopterus lucenai* and the La Plata croaker *Pachyurus bonariensis* were first observed in the northern part of the Patos Lagoon, but recently increasing numbers were observed in the Mirim Lagoon, the São Gonçalo Channel, in tributaries of the estuary, and in the estuary itself. Furthermore, adults of four carp species (bighead carp *Aristichthys nobilis*; common carp, *Cyprinus carpio carpio*; silver carp *Hypophthalmichthys molitrix*; and grass carp, *Ctenopharyngodon idellus*) that are grown in aquaculture facilities proximate to the lagoon have been reported in the estuary, though no offspring have been found there (A. M. Garcia et al. 2004).

17.5 CONCLUDING REMARKS

The results of this study demonstrate that a large-scale and remote climatic phenomenon like the El Niño Southern Oscillation, which signals significant changes in Pacific Ocean water temperature, may influence the ecology of primary and secondary producers of an estuary in the southwestern Atlantic Ocean. Interactions between meteorology and hydrology directly affect the dynamics of plankton, benthic organisms, and fish in the Patos Lagoon Estuary, often with economic repercussions owing to the impacts of ENSO on the regional fish and shrimp fisheries. Additionally, human factors such as dredging and fishery activities can act synergistically with ENSO phenomena to affect ecological processes in the estuary. In a highly dynamic ecosystem such as the Patos Lagoon Estuary, conclusions of this nature can only be reached through long-term multidisciplinary studies as conducted within the scope of the Brazilian Long-Term Ecological Research Program (PELD – CNPq).

REFERENCES

Abreu, P. C. and J. P. Castello. 1997. Estuarine-marine interactions. Pp. 179–182, In: U. Seeliger, C. Odebrecht, and J. P. Castello (eds.), *Subtropical Convergence environments: The coast and sea in the southwestern Atlantic*. New York: Springer.

Abreu, P. C., E. Granéli, C. Odebrecht, D. Kitzmann, L. A. Proença, and C. Resgalla, Jr. 1994a. Effect of fish and mesozooplankton manipulation on phytoplankton community in the Patos Lagoon Estuary, southern Brazil. *Estuaries* 17: 575–584.

Abreu, P. C., C. Odebrecht, and A. González. 1994b. Particulate and dissolved phytoplankton production of the Patos Lagoon estuary, southern Brazil: Comparison of methods and influencing factors. *J. Plankton Res.* 16: 737–753.

Abreu, P. C., C. S. B. Costa, C. Bemvenuti, C. Odebrecht, W. Granéli, and A. M. Anésio. 2006. Eutrophication processes and trophic interactions in a shallow estuary: Preliminary results based on stable isotope analysis (∂^{13}C and (∂^{15}N). *Estuaries Coasts* 29(2): 277–285.

Abreu, P. C., M. Bergesch, L. A. Proença, and C. Odebrecht. 2010. Short- and long-term chlorophyll *a* variability in the shallow microtidal Patos Lagoon Estuary, Southern Brazil. *Estuaries Coasts.* DOI 10.1007/s12237-009-9181-9.

Almeida, M. T. A., M. G. Z. Baumgarten, and R. M. S. Rodrigues. 1993. Identificação das possíveis fontes de contaminação das águas que margeiam a cidade do Rio Grande. *Ser. Doc. Técn. Oceanografia, Rio Grande* 6: 1–63.

Asmus, M. L. 1984. Estrutura da comunidade associada a Ruppia maritima no estuário da Lagoa dos Patos, RS, Brasil. M.S. Thesis, Federal University of Rio Grande-FURG, Brazil.

Asmus, M. L. 1997. Coastal plain and Patos Lagoon. Pp. 9–12, In: U. Seeliger, C. Odebrecht, and J. P. Castello (eds.), *Subtropical Convergence environments: The coast and sea in the southwestern Atlantic.* New York: Springer.

Bemvenuti, C. E. 1987. Predation effects on a benthic community in estuarine soft sediments. *Atlântica* 9: 33–63.

Bemvenuti, C. E. 1997a. Benthic invertebrates. Pp. 43–46, In: U. Seeliger, C. Odebrecht, and J. P. Castello (eds.), *Subtropical convergence environments: The coast and the sea in the warm temperate southwestern Atlantic*. Heidelberg: Springer-Verlag.

Bemvenuti, C. E. 1997b. Unvegetated intertidal flats and subtidal bottoms. Pp. 78–82, In: U. Seeliger, C. Odebrecht, and J. P. Castello (eds.), *Subtropical convergence environments: The coast and the sea in the warm temperate southwestern Atlantic.* New York: Springer.

Bemvenuti, C. E. and L. G. Angonesi. 2009. Caracterização e comparação da estrutura da associação macrozoobentônica nas margens e parte central da enseada Saco da Mangueira. In: P. R. A. Tagliani and M. L. Asmus (eds.), *Manejo Integrado do Estuário da Lagoa dos Patos. Programa Costa Sul, Resultados, Desafios e Perspectivas.* MMA, Brasilia, DF.

Bemvenuti, C. E., S. A. Cattaneo, and S. A. Neto. 1992. Características estruturais da macrofauna bentônica em dois pontos da região estuarial da Lagoa dos Patos, RS, Brasil. *Atlântica* 14: 5–28.

Bemvenuti, C. E., L. G. Angonesi, and M. S. Gandra. 2005. Effects of dredging operations on soft bottom macrofauna in a harbor in the Patos Lagoon estuarine region of southern Brazil. *Braz. J. Biol.* 65: 29–41.

Bergesch, M., C. Odebrecht, and P. C. Abreu. 1995. Microalgas do estuário da Lagoa dos Patos: Interação entre o sedimento e a coluna de água. *Oecol. Brasil.* 1: 273–289.

Bruno, M. A. and J. H. Muelbert. In press. Distribuição espacial e variações temporais da abundância de ovos e larvas de *Micropogonias furnieri* no estuário da Lagoa dos Patos: registros históricos e forçantes ambientais. *Atlântica.*

Burns, M. D. M., A. M. Garcia, J. P. Vieira, M. A. Bemvenuti, D. M. L. M. Marques, and M. V. Condini. 2006. Evidence of habitat fragmentation affecting fish movement between the Patos and Mirim coastal lagoons in southern Brazil. *Neot. Icht.* 4: 69–72.

Cafruni, A. M. S. 1983. Estudo autoecologico de Ruppia maritima no estuario da Lagoa dos Patos. M.Sc. Thesis, Federal University of Rio Grande, Brazil.

Calliari, L. J. 1997. Geological setting. Pp. 13–18, In: U. Seeliger, C. Odebrecht, and J. P. Castello (eds.), *Subtropical Convergence environment: The coast and sea in the southwestern Atlantic*. New York: Springer.

Calliari, L. J., K. T. Holland, P. S. Pereira, R. M. C. Guedes, and R. E. Santo. 2007. The influence of mud on the inner shelf, shoreface, beach and surf zone morphodynamics, Cassino, Southern Brazil. In *Proceedings of Coastal Sediments 2007*, Vol. 2, New Orleans, Louisiana, 1455–1465.

Capítoli, R. R., L. A. Colling, and C. E. Bemvenuti. 2008. Cenários de distribuição do mexilhão dourado *Limnoperna fortunei* (Mollusca–Bivalvia) sob distintas condições de salinidade no complexo lagunar Patos-Mirim, RS, BRASIL. *Atlântica* 30: 35–44.

Castello, J. P., and O. O. Möller, Jr. 1978. On the relationship between rainfall and shrimp production in the estuary of the Patos Lagoon (Rio Grande do Sul, Brazil). *Atlântica* 3: 67–74.

Cavalli, R., W. J. Wasielesky, S. Peixoto, L. H. Poersch, M. Santos, and R. Soares. 2008. Shrimp farming as an alternative to artisanal fishermen communities: The case of Patos Lagoon, Brazil. *Braz. Arch. Biol. Techn.* 51: 991–1001.

Chao, L. N., L. E. Pereira, J. P. Vieira, M. A. Bemvenuti, and L. P. R. Cunha. 1982. Relação preliminar dos peixes estuarinos e marinhos da Lagoa dos Patos e região costeira adjacente, Rio Grande do Sul, Brasil. *Atlântica* 5: 67–75.

Chao, L.N., L.E. Pereira, and J.P. Vieira. 1985. Estuarine fish community of the Patos Lagoon, Brazil, a baseline study. Pp. 429–450, In: A. Yanez-Arancibia (ed.), *Fish community ecology in estuaries and coastal lagoons: Towards an ecosystem integration.* Mexico City: DR (R) UNAM Press.

Chao, L. N., J. P. Vieira, and L. E. Pereira. 1986. *Lagoa dos Patos as a nursery ground for shore fishes off Southern Brazil.* IOC/FAO, Oceanographic Commission Workshop Report. UNESCO 44: 143–150.

Cloern, J. E. and A. D. Jassby. 2008. Complex seasonal patterns of primary producers at the land-sea interface. *Ecol. Lett.* 11: 1–10.

Colares, I. G. 1998. Influência da luz sobre o crescimento e fotossíntese de Ruppia maritima L., no estuário da Lagoa dos Patos. Ph.D. Thesis, Federal University of Rio Grande-FURG, Rio Grande, Brazil.

Colares, I. G. and U. Seeliger. 2006. Influência da luz sobre o crescimento e biomassa de *Ruppia maritima* L. em cultivo experimental. *Acta Botan. Brasilica* 20: 31–36.

Colling, L. A., C. E. Bemvenuti, and M. S. Gandra. 2007. Seasonal variability on the structure of sublittoral macrozoobenthic association in the Patos Lagoon estuary, southern Brazil. *Iheringia, Ser. Zool.* 97: 1–6.

Copertino, M. S., T. Tormena, and U. Seeliger. 2009. Biofiltering efficiency, uptake and assimilation rates of *Ulva clathrata* (Roth) J. Agardh (Chlorophyceae) cultivated in shrimp aquaculture waste water. *J. Appl. Phycol.* 21: 31–45.

Cordazzo, C. V. 2004. Produção de sementes e estabelecimento dos fundos de *Ruppia matima* L. no estuário da Lagoa dos Patos (RS). Anais do VI Simpósio de Ecossistemas Brasileiros: patrimônio ameaçado. *Publ. ACIESP* 110: 291–298.

Coretinho, R. 1982. Taxonomia distribuicao, crescimento sazonal reproducas e biomasse, das algas bentonicas no estuario da Lagoa dos Patos (RS). M.Sc. Thesis, Federal University of Rio Grande, Brazil.

Costa, C. S. B. and U. Seeliger. 1998. Vertical distribution and biomass allocation of *Ruppia maritima* L. in a southern Brazilian estuary. *Aquat. Bot.* 33: 123–129.

Coutinho, R. and U. Seeliger. 1986. Seasonal occurrence and growth of benthic algae in the Patos Lagoon estuary, Brazil. *Estuar. Coast. Shelf Sci.* 23: 889–900.

Darrigran, G. and G. Pastorino. 1995. The recent introduction of a freshwater Asiatic bivalve *Limnoperna fortunei* (Mytilidade) into South America. *Veliger* 32 :171–175.

D'Incao, F. 1991. Pesca e biologia de *Penaeus paulensis* na Lagoa dos Patos, RS. *Atlântica* 13: 159–169.

Duarte, C. M., J. W. Fourqurean, D. Krause-Jensen, and B. Olesen. 2006. Dynamics of seagrass stability and change. Pp. 271–294, In: A. W. D. Larkum, R. J. Orth, and C. M. Duarte (eds.), *Seagrasses: Biology, ecology and conservation*. Dordrecht: Springer-Verlag.

Fernandes, E. H. L., K. R. Dyer, O. O. Möller, Jr., and L. F. H. Niencheski. 2002. The Patos Lagoon hydrodynamics during an El Niño event (1998). *Continent. Shelf Res.* 22: 1699–1713.

Ferreira, S. and U. Seeliger. 1985. The colonization process of algal epiphytes on *Ruppia maritima* L. *Bot. Mar.* 28: 245–249.

Ferreira, C. E. L., A. O. R. Junqueira, M. C. Villac, and R. M. Lopes. 2009. Marine bioinvasions in the Brazilian coast: Brief report on history of events, vectors, ecology, impacts and management of non-indigenous species. Pp. 459–477, In: G. Rilov and J. A. Crooks (eds.), *Biological invasions in marine ecosystems*. Ecological Studies 204. Berlin: Springer-Verlag.

Flindt, M., J. Salomonsen, M. Carrer, M. Bocci, and L. K. Nielsen. 1997. Loss, growth, and transport dynamics of *Chaetomorpha aerea* and *Ulva rigida* in the Lagoon of Venice. *Ecol. Model.* 102: 133–141.

Fujita, C. C. and C. Odebrecht. 2007. Short term variability of chlorophyll a and phytoplankton composition in a shallow area of Patos Lagoon estuary (southern Brazil). *Atlântica* 29: 93–106.

Garcia, A. M., M. B. Raseira, J. P. Vieira, K. O. Winemiller, and A. M. Grimm. 2003a. Spatio-temporal variation in shallow-water freshwater fish distribution and abundance in a large subtropical coastal lagoon. *Environ. Biol. Fishes* 68: 215–228.

Garcia, A. M. and J. P. Vieira. 2001. O aumento da diversidade de peixes no estuário da Lagoa dos Patos durante o episódio El Niño 1997–1998. *Atlântica* 23: 133–152.

Garcia, A. M., J. P. Vieira, and K. O. Winemiller. 2001. Dynamics of the shallow-water fish assemblage of the Patos Lagoon estuary (Brazil) during cold and warm ENSO episodes. *J. Fish Biol.* 59: 1218–1238.

Garcia, A. M., J. P. Vieira, and K. O. Winemiller. 2003b. Effects of 1997–1998 El Niño on the dynamics of the shallow-water fish assemblage of the Patos Lagoon Estuary (Brazil). *Estuar. Coast. Shelf Sci.* 57: 489–500.

Garcia, A. M., J. P. Vieira, K. O. Winemiller, and A. M. Grimm. 2004. Comparison of the 1982–1983 and 1997–1998 El Niño effects on the shallow-water fish assemblage of the Patos Lagoon Estuary (Brazil). *Estuaries* 27: 905–914.

Garcia, V. M. T. and N. Gianuca. 1997. The beach and surf zone. Pp. 166–170, In: U. Seeliger, C. Odebrecht, and J. P. Castello (eds.), *Subtropical convergence environment: The coast and sea in the southwestern Atlantic*. New York: Springer.

Grimm, A., S. E. T. Ferraz, and J. Gomes. 1998. Precipitation anomalies in southern Brazil associated with El Niño and La Niña events. *J. Climate* 11: 2863–2880.

Haimovici, M., J. P. Castello, and C. M. Vooren. 1997. *Fisheries*. Pp. 183–196, In: U. Seeliger, C. Odebrecht, and J. P. Castello. *Subtropical convergence environments: The coast and sea in the southwestern Atlantic*. New York: Springer.

Haimovici, M., V. J. Isaac, A. S. Martins, and J. M. Andrigetto. 2006. *A pesca marinha e estuarina do Brasil no início do século XXI: Recursos, tecnologias, aspectos socioeconômicos e institucionais*. Universidade Federal do Pará, Belém, Brazil.

Haimovici, M. and R. G. Umpierre. 1996. Variaciones estacionales en la estructura poblacional del efectivo pesquero de corvina blanca *Micropogonias furnieri* (Desmarest, 1823) en el extremo sur de Brasil. *Atlântica* 18: 179–203.

Kaminski, S. 2009. Variação espacial e temporal do mesozooplâncton do estuário da Lagoa dos Patos e zona costeira adjacente com ênfase nos copépodos *Acartia tonsa*, *Pseudodiaptomus richardi* e *Notodiaptomus incompositus*. Ph.D. Thesis. Federal University of Rio Grande-FURG, Brazil.

Kjerfve, B. 1986. Comparative oceanography of coastal lagoons. Pp. 63–81, In: D. A. Wolfe (ed.), *Estuarine variability*. New York: Academic Press.

Knack, R. 1986. Ecologia experimental de *Ruppia maritima* L. em condições controladas de cultivo. M.S. Thesis, Federal University of Rio Grande-FURG, Rio Grande, Brazil.

Koch, E. W. 2001. Beyond light: Physical, geological, and geochemical parameters as possible submersed aquatic vegetation habitat requirements. *Estuaries* 24: 1–17.

Koch, E. W. and U. Seeliger. 1988. Germination ecology of two *Ruppia maritima* L. populations in southern Brazil. *Aquat. Bot.* 31: 321–327.

Mansur, M. C. D., C. P. Santos, G. Darrigran, I. Heydrich, C. T. Callil, and F. R. Cardoso. 2003. Promeiros dados quail-quantitativos do "mexilhão dourado" Limnoperna fortunei (Dunker, 1857) no lago Guaíba, bacia da laguna dos Patos, Brasil e alguns aspectos de sua invasão no novo ambiente. *Rev. Bras. Zool.* 22: 75–84.

Martins, I. and J. C. Marques. 2002. A model for the growth of opportunistic macroalgae (*Enteromorpha sp.)* in tidal estuaries. *Estuar. Coast. Shelf Sci.* 55: 247–257.

Martins, I. M. S., J. M. Dias, E. H. Fernandes, and J. H. Muelbert. 2007. Numerical modelling of fish eggs dispersion at the Patos Lagoon estuary, Brazil. *J. Mar. Systems* 68: 537–555.

Möller, O. O., Jr., J. A. Lorenzzetti, J. L. Stech, and M. M. Mata. 1996. Patos Lagoon summertime circulation and dynamics. *Continent. Shelf Res.* 16: 335–351.

Möller, O. O., Jr., P. Castaing, J.-C. Salomón, and P. Lazure. 2001. The influence of local and non-local forcing effects on the subtidal circulation of Patos Lagoon. *Estuaries* 24: 297–311.

Möller, O., Jr., C. Vaz, and J. P. Castello. 2009. The effect of river discharge and winds on the interannual variability of the pink shrimp *Farfantepenaeus paulensis* production in Patos Lagoon. *Estuaries Coasts* 32: 787–796.

Montú, M. and I. Gloeden. 1982. Morphological alterations in *Acartia tonsa* (Saco da Mangueira, Lagoa dos Patos, Brasil). *Arq. Biol. Tecnol.* 25: 361–369.

Montú, M. and I. Gloeden. 1986. Atlas dos Cladocera e Copepoda (Crustacea) do estuário da Lagoa dos Patos (Rio Grande, Brasil). *Neritica* 1: 1–134.

Montú, M., A. K. Duarte, and I. Gloeden. 1997. Zooplankton. Pp. 40–43, In: U. Seeliger, C. Odebrecht, and J. P. Castello (eds.), *Subtropical convergence environments: The coast and sea in the southwestern Atlantic*. Springer: New York.

Muelbert, J. H. and G. Weiss. 1991. Abundance and distribution of fish larvae in the channel area of Patos Lagoon estuary, Brazil. In: R. Dhbyt (ed.), *Larval fish recruitment and research in the Americas*. Proceedings of the Thirteenth Annual Fish Conference, Springfield, Virginia, 95: 43–54.

Muxagata, E. and I. Gloeden. 1995. Ocorrência de *Temora turbinata* Dana, 1849 (Crustacea: Copepoda) no estuário da Lagoa dos Patos, RS, Brasil. *Nauplius* 3: 163–164.

Nelson, T. A., D. J. Lee, and B. C. Smith. 2003. Are "green tides" harmful algal blooms? Toxic properties of water-soluble extracts from two bloom-forming macroalgae, *Ulva fenestrate* and *Ulvaria obscura* (Ulvophyceae). *J. Phycol.* 39: 874–879.

Niencheski, L. F. and M. G. Baumgarten. 1997. Environmental chemistry. Pp. 20–23, In: U. Seeliger, C. Odebrecht, and J. P. Castello (eds.), *Subtropical convergence environments: The coast and sea in the southwestern Atlantic*. New York: Springer.

Niencheski, L. F. and H. L. Windom. 1994. Nutrient flux and budget in Patos Lagoon Estuary. *Sci. Total Environ.* 149: 53–60.

Odebrecht, C., P. C. Abreu, O. O. Möller, Jr., L. F. Niencheski, L. A. Proença, and L. C. Torgan. 2005. Drought effects on pelagic properties in the shallow and turbid Patos Lagoon, Brazil. *Estuaries* 28: 675–686.

Odebrecht, C., M. Bergesch, L. R. Rörig, and P. C. Abreu. 2010. Phytoplankton interannual variability at Cassino Beach, southern Brazil (1992–2007), with emphasis on the surf zone diatom *Asterionellopsis glacialis*. *Estuaries Coasts*. DOI 10.1007/s12237-009-9176-6.

Orth, R. J., T. J. B. Carruthers, W. C. Dennison, C. M. Duarte, J. W. Fourqurean, K. L. Heck, Jr., A. R. Hughes, G. A. Kendrick, W. J. Kenworthy, S. Olyarnik, F. T. Short, M. Waycott, and S. L. Williams. 2006. A global crisis for seagrass ecosystems. *BioScience* 56: 987–996.

Pereira, L. E. 1994. Variação diurna e sazonal da comunidade dos peixes demersais na barra do estuário da Lagoa dos Patos, RS. *Atlântica* 16: 5–21.

Pinedo, M. C. 1997. Marine mammals. Pp. 63–64, In: U. Seeliger, C. Odebrecht, and J. P. Castello (eds.), *Subtropical convergence environments: The coast and sea in the southwestern Atlantic*. New York: Springer.

Ramos, L. A. and J. P. Vieira. 2001. Composição específica e abundância de peixes de zonas rasas dos cinco estuários do Rio Grande do Sul, Brasil. *Bol. Inst. Pesca* 1: 109–121.

Reis, E. G., P. C. E. Vieira, and V. S. Duarte. 1994. Pesca artesanal de teleósteos no estuário da Lagoa dos Patos e costa do Rio Grande do Sul. *Atlântica* 16: 69–86.

Rosa, C. E., M. S. Souza, J. S. Yunes, L. A. O. Proença, L. E. M. Nery, and J. M. Monserrat. 2005. Cyanobacterial blooms in estuarine ecosystems: Characteristics and effects on *Laeonereis acuta* (Polychaeta, Nereididae). *Mar. Pollut. Bull.* 50: 956–964.

Rosa, L. C. and C. E. Bemvenuti. 2006. Temporal variability of the estuarine macrofauna of the Patos Lagoon. *Rev. Biol. Mar. Ocean.* 41: 1–9.

Rosa, L. C. and C. E. Bemvenuti. 2007. Seria a macrofauna bentônica de fundos não consolidados influenciada pelo aumento na complexidade estrutural do habitat? O caso do estuário da Lagoa dos Patos. *BJAST* 11: 51–56.

Seeliger, U. 1997a. Seagrass meadows. Pp. 82–85, In: U. Seeliger, C. Odebrecht, and J. P. Castello (eds.), *Subtropical convergence environments: The coast and sea in the southwestern Atlantic*. New York: Springer.

Seeliger, U. 1997b. Benthic macroalgae. Pp. 30–33, In: U. Seeliger, C. Odebrecht, and J. P. Castello (eds.), *Subtropical convergence environments: The coast and sea in the southwestern Atlantic*. New York: Springer.

Seeliger, U. 2001. *The Patos Lagoon estuary*. Pp. 167–183, In: U. Seeliger and B. Kjerfve (eds.), *Coastal marine ecosystems of Latin America*. New York: Springer.

Seeliger, U., C. Cordazzo, and F. Barbosa. 2002. *Os sites e o programa brasileiro de Pesquisas Ecológicas de Longa Duração*. Belo Horizonte, MG, Brazil.

Seeliger , U., C. Cordazzo, and L. Barcellos. 2004. *Areias do Albardão, um guia ecológico ilustrado do litoral no extremo Sul do Brasil*. Ed. Ecoscientia, Rio Grande, Brazil.

Seeliger, U. and C. S. B. Costa. 1997. *Natural and human impact*. Pp. 197–203, In: U. Seeliger, C. Odebrecht, and J. P. Castello (eds.), *Subtropical convergence environment: The coast and sea in the southwestern Atlantic*. New York: Springer.

Silva, E.T. 1995. Modelo ecológico de fundos vegetados dominados por *Ruppia maritima* (Potamogetonacea) do estuário da Lagoa dos Patos, RS. M.Sc. Thesis, Federal University of Rio Grande-FURG, Brazil.

Silva, E. T. and M. L. Asmus. 2001. A dynamic simulation model of the widgeon grass *Ruppia maritima* and its epithytes in the estuary of the Patos Lagoon, RS, Brazil. *Ecol. Model.* 137: 161–179.

Sinque, C. and J. H. Muelbert. 1997. Ichthyoplankton. Pp. 51–56, In: U. Seeliger, C. Odebrecht, and J. P. Castello (eds.), *Subtropical convergence environments: The coast and sea in the southwestern Atlantic.* New York: Springer.

Souza, R. A. 2001. Estrutura das associações macrobentônicas epifaunais associadas às macrófitas em enseadas estuarinas da Lagoa dos Patos, RS, Brasil. M.S. Thesis, Federal University of Rio Grande, Brazil.

Vaz, A. C., O. O. Möller, Jr., and T. L. Almeida. 2006. Sobre a descarga dos rios afluentes da Lagoa dos Patos. *Atlântica* 26: 13–23.

Vieira, E. F. 1983. *Rio Grande. Geografia física, humana e econômica*. Ed. Sagra, Porto Alegre, Brazil.

Vieira, J. P. 2006. Ecological analogies between estuarine bottom trawl fish assemblages from Patos Lagoon (32S), Brazil, and York River (37N), USA. *Rev. Bras. Zool.* 23: 234–247.

Vieira, J. P., and J. P. Castello. 1997. Fish fauna. Pp. 56–61, In: U. Seeliger, C. Odebrecht, and J. P. Castello (eds.), *Subtropical convergence environment: The coast and sea in the southwestern Atlantic*. New York: Springer.

Vieira, J. P., A. M. Garcia, and A. M. Grimm. 2008. Preliminary evidences of El Niño effects on the mullet fishery of Patos Lagoon estuary (Brazil). *Braz. Arch. Biol. Techn.* 51: 433–440.

Von Ihering, H. 1885. Die Lagoa dos Patos. *Dtsch. Geogr. Bl.* 8: 182–204.

Vooren, C. M. 1997. Bird fauna. Pp. 62–63, In: U. Seeliger, C. Odebrecht, and J. P. Castello (eds.), *Subtropical convergence environments: The coast and sea in the southwestern Atlantic.* New York: Springer.

Yunes, J. S., L. F. Niencheski, P. S. Salomon, M. Parise, K. A. Beattie, S. L. Ragget, and G. A. Codd. 1998. Effect of nutrient balance and physical factors on blooms of toxic cyanobacteria in the Patos Lagoon, southern Brazil. *Verh. Internat. Verein. Limnol.* 26: 1796–1800.

Zarzur, S. 2007. Regeneração bêntica de nutrientes e produção primária no estuário da Lagoa dos Patos. Ph.D. Thesis, Federal University of Rio Grande-FURG, Brazil.

18 Structure and Function of Warm Temperate East Australian Coastal Lagoons

Implications for Natural and Anthropogenic Changes

Bradley D. Eyre and Damien Maher*

CONTENTS

Abstract 458
18.1 Introduction 458
18.2 Study Area 459
18.2.1 Geomorphic Evolution 459
18.2.2 Climate 462
18.2.3 Catchment Characteristics/Land Use/Stressors 462
18.3 Water and Sediment Quality 465
18.4 Benthic Habitats 465
18.5 Carbon Budgets 469
18.6 Wild Fisheries Production 473
18.7 Implications for Natural and Anthropogenic Change 476
18.7.1 Natural Change 476
18.8 Pollutant Loads 476
18.9 Acidification, Deoxygenation, and Fish Kills 476
18.10 Climate Change Impacts 477
18.10.1 Sea Level Rise 477
18.10.2 Temperature 477
18.10.3 Rainfall and Runoff 478
18.10.4 Storminess and Wave Climate 478
18.11 Conclusions 479
Acknowledgments 479
References 479

* Corresponding author. Email: bradley.eyre@scu.edu.au

ABSTRACT

Three warm temperate east Australian coastal systems (Wallis Lake, Camden Haven, and Hastings River Estuary) serve as case studies to explore relationships between the structure and function of coastal lagoons. These three systems provide examples of bar-built or wave-dominated estuaries/coastal lagoons at different stages of maturity (infilling) on the east coast of Australia. Wallis Lake is an immature lagoon with minimal infilling and a large depositional mud basin. Camden Haven Estuary is a semimature lagoon with two partially filled shallow lakes connected by an infilled channelized river system, and the Hastings River Estuary is a mature lagoon with a highly channelized interconnected river system. As these systems evolve (mature), the areal extent of their geomorphic units (structure) changes linearly, with a decrease in the extent of deep and shallow subtidal mud shoals and an increase in extent of marine sand channel and fluvial sands. However, the function trajectory associated with the changes in the structure of these coastal lagoons as they mature show both linear and nonlinear trends. As a coastal lagoon progresses from an immature to a semimature state and the mud basin fills and shallows, there is an increase in the ratio of benthic to pelagic production due to greater light availability and an increase in net ecosystem metabolism (NEM) associated with less organic matter being trapped and respired. With further maturation, a coastal lagoon completely infills, and there is a large decrease in the benthic to pelagic ratio caused by higher turbidity (and less light) and less suitable substrate for benthic producers. This leads to a decrease in NEM due to more organic matter being trapped and respired. Total wild fisheries production (catch per fisher based on 20 years of commercial fish catch) follows a similar nonlinear trend, with maximum production in the semimature Camden Haven. In contrast, the production (catch per fisher) of higher trophic level fish species (i.e., trophic level above 2) shows a linear decrease as the system matures, most likely attributable to less organic matter availability for higher-order production. Immature lagoons are at greatest risk from pollutant loading and climate change impacts (e.g., sea level rise, higher temperatures, increased rainfall and runoff, and greater storminess and wave climate) largely because of poor flushing. In contrast, mature lagoons are at greatest risk from acidification, deoxygenation, and fish kills due to larger alluvial plains.

Key Words: structure, function, lagoon, carbon, primary production, benthic production, seagrass, benthic microalgae, net ecosystem metabolism, fisheries, trophic transfer efficiency, climate change

18.1 INTRODUCTION

The coastal zone plays an important role in the Earth's carbon budget because of the large amounts of allochthonous and autochthonous organic matter found there (Knoppers 1994; Borges et al. 2005). The amount of organic matter available for higher tropic levels (e.g., fisheries production) and export to adjacent aquatic systems (e.g., ocean, burial), or release to the atmosphere as CO_2, is determined by an ecosystem's primary production minus its respiration, or net ecosystem metabolism (NEM). Coastal lagoons usually lack large river tributary systems and freshwater inputs, which would increase the autochthonous and lateral inputs of organic matter (e.g., Santos et al. 2004; Kone et al. 2009; Eyre et al. in press). Coastal lagoons are also shallow, so light can reach much of the lagoon floor, thereby increasing the amount of primary production and respiration in the benthos (McGlathery et al. 2007; Eyre et al. in press).

There are a variety of different types of coastal lagoons shaped by the relative importance of wave action, tidal currents, and freshwater inputs (Harris and Heap 2003). The geomorphic structure of coastal lagoons also influences their function. As a coastal lagoon matures (infills), the relative significance of its main geomorphic units (i.e., marine tidal deltas, central mud basins, fluvial deltas, and riverine channel/alluvial plains) will also change (Roy et al. 2001; Harris and Heap

2003). The extent of the different benthic habitats in shallow coastal lagoons has been related to these main geomorphic units (Roy et al. 2001; Saintilan 2004).

Because much of the primary production and respiration in shallow coastal lagoons occurs in benthic habitats, it would be expected that the geomorphology of a coastal lagoon would influence higher trophic levels. For example, using multivariate analysis, Saintilan (2004) found that the degree of infilling in east Australian coastal lagoons (wave-dominated estuaries) was correlated with the type of commercial fish catches (standardized for area and effort). Multiple regression analysis indicated that the different commercial fish species caught were related to the area of seagrass in each system (Saintilan 2004). Although the area of seagrass habitat is important, especially during early life stages, differences in the type of fish landings could also be related to differences in the sources and quantity of organic matter, and perhaps differences in the transfer of this energy up the food chain. Saintilan (2004) showed by multiple regression analysis that seagrass may have been a covariate with benthic microalgae production. Clearly, a better understanding is required of how the geomorphology of coastal lagoons affects their function (productivity).

The purpose of this chapter is to examine the relationships between the geomorphology and function of coastal lagoons using case studies of three warm-temperate east Australian systems. These three systems represent typical examples of bar-built or wave-dominated estuaries/coastal lagoons at different stages of maturity (infilling) (Roy et al. 2001; Harris and Heap 2003). This work will provide insight into the roles played by these different geomorphic types of coastal lagoons in the global coastal carbon budget. About 50% of the commercial fish catch (including oyster production) in New South Wales (NSW) derives from lagoonal estuaries (NSW Fisheries Statistics 2006/07). The annual commercial fish catch in NSW estuaries is worth approximately $25 million (Australian dollars) (NSW Fisheries Statistics 2006/07) with direct follow-on economic effects (e.g., jobs, service industries, etc.) probably worth another $200 million (Australian dollars). In addition, recreational fishing in NSW estuaries adds another 180 million (Australian dollars) (Henry and Lyle 2003), giving this marine resource a total annual value of about 400 million (Australian dollars). However, similar to many other coastal lagoons around the world, the production of East Australian coastal lagoons is under increasing anthropogenic bottom-up (ecosystem changes due to nutrient enrichment, acid sulfate soils, and habitat destruction) and top-down (multisector and multispecies commercial and recreational fishing) pressures. Superimposed on these factors are anthropogenic impacts and climatic changes. Central to ecosystem-based management is an understanding of how the structure and function of the ecosystem supports a managed resource (e.g., fisheries production).

18.2 STUDY AREA

18.2.1 Geomorphic Evolution

The geomorphic evolution of the southeast Australian coastline has been intrinsically linked to the transgression of the ocean–land interface during the transition between glacial and interglacial periods, with most estuaries in this region occupying valleys previously excavated during times of lower sea level (Roy 1984). Recent evolution during the Holocene period has involved a steady infilling of these estuaries with marine and fluvial deposits, the rate of infilling being a function of tide, river discharge, and wave strength (Harris and Heap 2003). NSW estuaries have been classified into four main geomorphic types (Roy et al. 2001): (1) ocean embayment, (2) drowned river valley (tide-dominated estuaries), (3) barrier estuaries, and (4) intermittent estuaries. Type 3 and type 4 estuaries have been further divided by evolutionary stage into A, B, C, and D types; type A represents relatively immature estuaries with minimal infilling through the intermediate stages of

TABLE 18.1
Description of Geomorphic Units and Evolutionary Processes for Southeast Australian Estuaries

Geomorphic Unit	Description	Evolution
Alluvial plain/ swamp deposits	Low-lying areas on the flood plain and swamp areas surrounding estuary.	Formed by the delivery of eroded terrestrial sediments to the flood plain and infilling of the central mud basin.
Barrier sands	Marine-derived sands generally vegetated by heath, dry sclerophyll forests and, at the land–ocean edge, dune communities. Highly susceptible to storm and tidal energies.	Sands generally deposited at the start of the Holocene period along the coast and form a barrier between the ocean and the estuary. Some estuaries have several barrier systems associated with sand transgressions throughout the various glacial-interglacial cycles.
Central mud basin	Central depositional areas of low energy. Sediment comprised of fine muds and clay. Extensive macrofauna populations. Seagrass beds where light permits.	Dominant in early evolution of estuary, gradually replaced with riverine channel and alluvial plains through lateral transgression via infilling.
Fluvial delta sands	Generally located mid-estuary formed by the delivery of coarse terrestrial sediments.	Extent of facie dependent upon catchment geology, tidal energy, and river flow. Becomes proportionally more dominant as an estuary matures and sediment accumulates through evolutionary stages.
Marine tidal delta	High and low energy areas at the mouth of an estuary consisting predominantly of oceanic quartzose sands.	Generally more dominant in immature estuaries as energy from river discharge is lower due to high surface area:volume. As estuary matures, the extent of marine tidal delta is reduced as sand deposits are flushed from the lower estuary.
Riverine channel	River channel in the upper to mid-estuary.	Proportional coverage increases with maturity as the central mud basin is infilled and becomes channelized.
Rock outcrops	Areas of raised elevation with primarily volcanic geology (in the study area). Typically define the catchment boundaries, and are the source areas for the alluvial plain and fluvial delta sand facies.	Along the southeast coast of Australia, these outcrops are ancient and predate the recent Holocene estuarine evolution by many millions of years.

Source: After Roy et al. (2001).

B and C, through to mature riverine systems that are fully infilled and dominated by highly channelized river systems (D). The geomorphic facies within each estuary and catchment can be broadly divided into seven units (Table 18.1). The balance of the fundamental geomorphic units or facies changes with the evolution of an estuary. Immature estuaries are dominated by large central mud basins, and as the estuary matures, these central mud basins are infilled, being replaced by alluvial plain and mangrove/saltmarsh facies. Concurrently, the distribution and relative areal proportion of instream habitat types change (Roy et al. 2001). The changes in habitat should alter estuarine trophodynamics by shifting the relative contribution of the various primary producers and the niche availability for metazoans (e.g., a reduction in shallow mud benthic habitat as an estuary matures).

The three wave-dominated coastal lagoons investigated in this work (Wallis Lake, Camden Haven Estuary, and Hastings River) are located on the southeastern coast of Australia (Figure 18.1). They represent varying levels of infilling or maturity (Roy et al. 2001). The inlets of all three shallow systems are constricted by wave-deposited beach sand barriers, which restrict their exchange

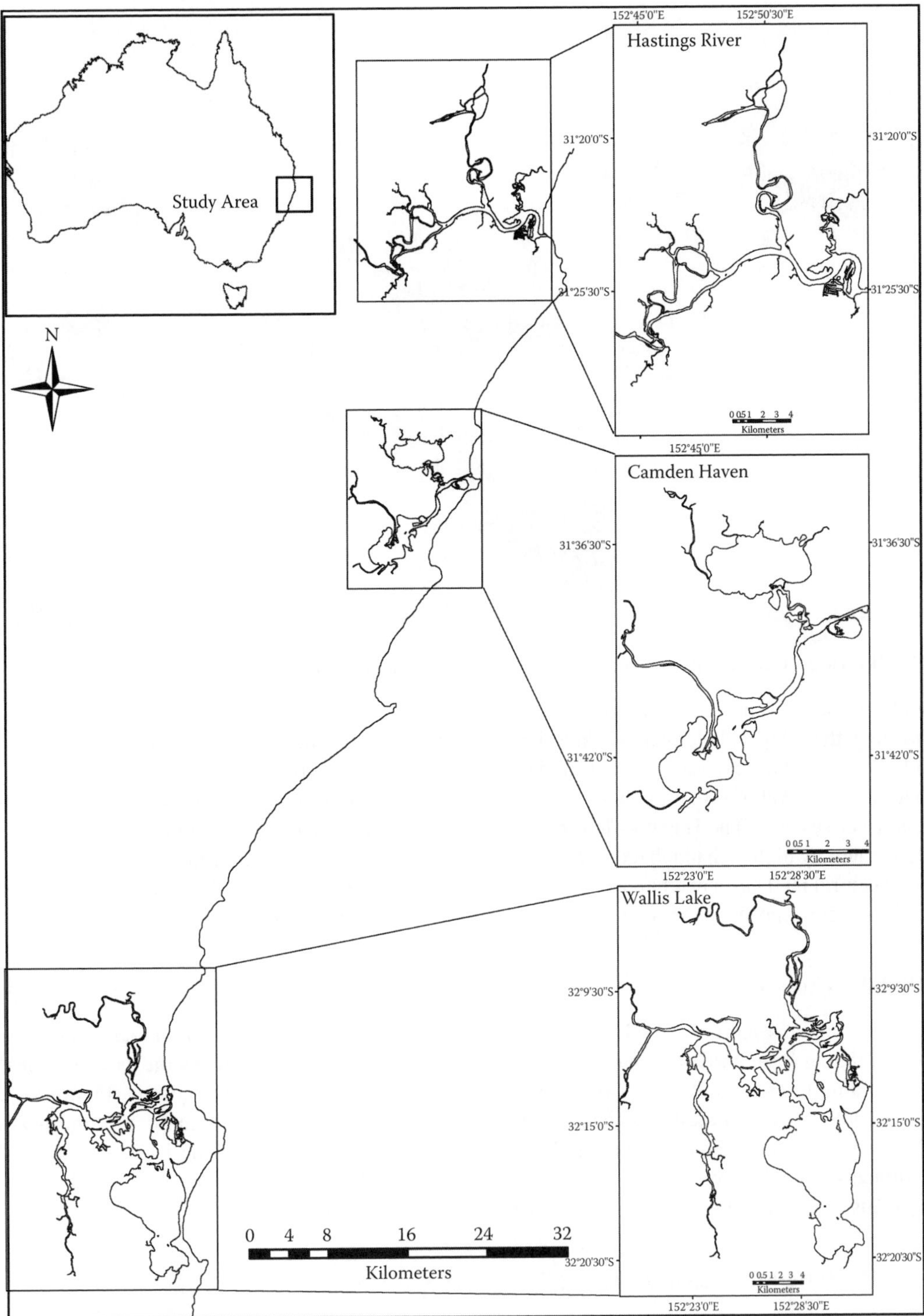

FIGURE 18.1 Location of Wallis Lake, Camden Haven, and Hastings River Estuary.

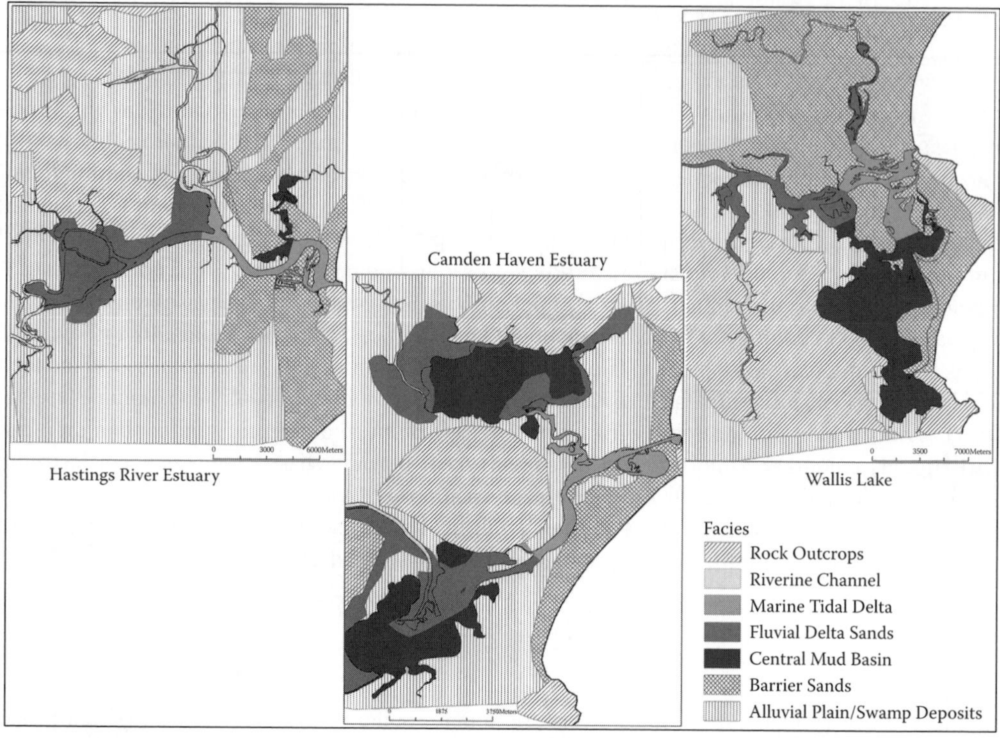

FIGURE 18.2 Geomorphic units of Wallis Lake, Camden Haven, and Hastings River Estuary.

with the ocean. The type IIIA Wallis Lake represents a wave-dominated lagoon with minimal infilling, the geomorphology of which is dominated by a large shallow (<4 m deep) lake with limited tidal flushing and freshwater input. Camden Haven Estuary is a type IIIB lagoon and represents an intermediate level of evolution, with two large shallow lakes (<2 m) connected by infilled channelized river systems. The Hastings River Estuary is a type IIID system; the geomorphology here is dominated by highly channelized interconnected river systems. The geomorphic units show a shift from central mud basin and barrier sand dominance in the immature Wallis Lake system to alluvial plain/swamp deposits in the mature Hastings River system (Figure 18.2, Table 18.2).

18.2.2 Climate

The study area is characterized by a warm temperate climate with mean maximum daily temperatures ranging from 16°C in winter to 25°C in summer. Solar radiation follows the same seasonal trend. Summer is the wettest season, while winter and spring are the driest seasons (Figure 18.3; Bom 2008). Heavy rainfall associated with east coast low pressure systems can cause significant flooding in the study area. Hence, the lagoons receive large pulses of land-derived nutrients and sediments. Flooding events can completely flush the brackish water from the estuarine basin of the east Australian lagoons (river estuaries) (Eyre and Twigg 1997; Eyre et al. 1998).

18.2.3 Catchment Characteristics/Land Use/Stressors

Table 18.3 lists the main physical and anthropogenic features of each of the three aforementioned coastal lagoon systems. All three lagoons have extensive catchment modification on the floodplain,

TABLE 18.2
The Area of Each Geomorphic Unit in Wallis Lake, Camden Haven, and Hastings River Estuary

Lagoon Geomorphic Unit	Area	%
Hastings		
Marine tidal delta	6.6	31.7
Riverine channel	6.2	29.8
Fluvial delta sands/muds	5.0	24.0
Central mud basin	3.0	14.4
Camden Haven		
Central mud basin	18.3	60.2
Fluvial delta sands/muds	7.0	23.0
Marine tidal delta	3.8	12.5
Riverine channel	1.3	4.3
Wallis Lake		
Central mud basin	48.3	44.4
Fluvial delta sands/muds	17.8	19.6
Marine tidal delta	11.8	13.0
Barrier sands	7.5	8.3
Riverine channel	5.3	5.8

Source: Modified from Maher (2010).

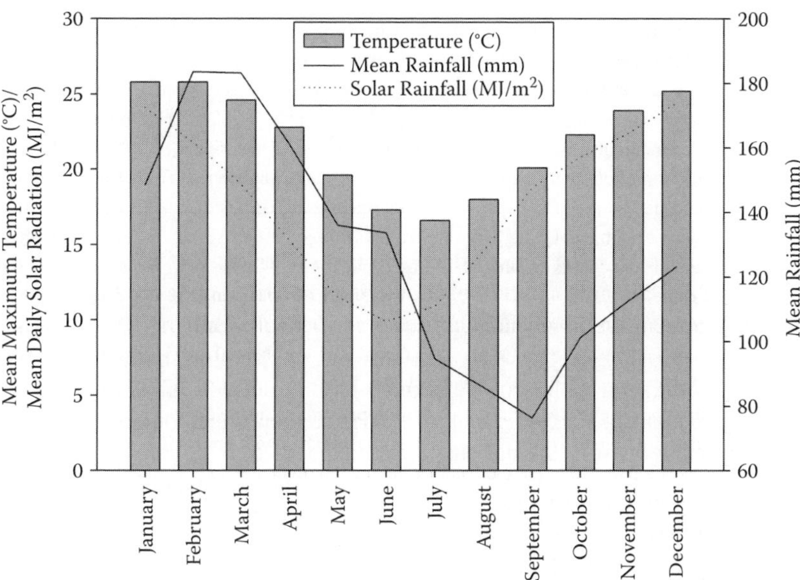

FIGURE 18.3 Mean climate data for study area. (From Bom, B. O. M. 2008. Climate dataset. Available at www.bom.gov.au/climate.)

TABLE 18.3
Physical Attributes, Catchment Characteristics, Land Use, and Stressors for Wallis Lake, Camden Haven, and Hastings River Estuary

Property	Hastings	Camden Haven	Wallis Lake
Catchment area	3,595 km^2	440 km^2	1,420 km^2
Water area	18.62 km^2	30.05 km^2	90.45 km^2
Alluvial plain area	1,315 km^2	93 km^2	477 km^2
Average depth	2.5 m	1.2 m	1.8 m
Rainfall	1,533 mm	1,540 mm	1,220 mm
Annual discharge (2006–2007)	503,471 ML	251,483 ML	147,883 ML
Mean residence time	10 days	45 days	60 days+
Land use	>70% forested, predominantly dry sclerophyll Flood plain cleared for agriculture (predominantly beef cattle)	Extensive clearing on flood plain for agriculture, upper catchment forested (dry sclerophyll)	Extensive clearing on flood plain for agriculture; forestry plantations and National Parks occur in the mid- and upper catchment
Anthropogenic stressors	2 STP plants discharge into the estuary (one at the upper tidal limit, one near the mouth) Dredging Urban pollutants Sand extraction Recreational fishing Diffuse nutrient and sediment inputs Marina/canal estate Acid sulfate soils	Fisheries (commercial and recreational) Urban pollutants STP discharge Diffuse nutrient and sediment inputs Acid sulfate soils	Fisheries (commercial and recreational) Urban pollutants STP discharge Diffuse nutrient and sediment inputs Acid sulfate soils
Geology	Upper catchment: Sedimentary rocks of the Carboniferous and Permian epochs Lower catchment: Quaternary sedimentary rocks of alluvial nature	Upper catchment: Carboniferous sedimentary (siltstone, shale, sandstone) with microgranite intrusions of Jurassic age Lower catchment: fluvial sand, silt and clay of Quaternary age	Upper catchment: Sedimentary rocks of the Carboniferous and Permian epochs Lower catchment: Quaternary sedimentary rocks of alluvial nature
Topography	Coastal plains: Lower estuary, alluvial flats less than 10 m AHD Mid-upper valley: Alluvial flats grading up to an elevation of 1200 m in the upper catchment; river characterized by steep banks and large meanders	Coastal plains: lower estuary, alluvial flats less than 10 m AHD Mid-upper valley: Steep valley banks associated with upper catchment mountains	Coastal plains: Lower estuary, alluvial flats less than 10 m AHD bounded on the seaward side by a sand barrier complex Mid-upper valley: Valley, ridge systems with relatively low maximum elevations (250 m AHD)
Fisheries catch	307 t (64% oysters)	227 t (74% oysters)	1570 t (91% oysters)
Population	48,000	21.000	34,000

due primarily to clearing for agriculture and residential development. Clearing and drainage of former wetland areas in the catchment of each lagoon have led to the exposure of acid sulfate soils (ASS). Each lagoon receives treated effluent from sewage treatment plants and stormwater runoff from urban areas.

Oyster farming is a large industry in all three lagoons, with production ranging in value from ~$1.6 million in the Hastings River and Camden Haven to ~$12 million in Wallis Lake. Commercial fishing is restricted to the Camden Haven and Wallis Lake lagoons, with the Hastings River recently declared a recreational fishing haven and all commercial licenses being revoked. Data on commercial fish catches prior to the closure of Hastings River will be used for comparative purposes. Based on the average catch data from 1990 to 1997 (Saintilan 2004), the commercial catch was 96,801 kg, 116,952 kg, and 361,006 kg for the Hastings River, Camden Haven, and Wallis Lake lagoons, respectively.

18.3 WATER AND SEDIMENT QUALITY

There was no significant difference ($p > 0.05$) in temperature between geomorphic units within the lagoons or between lagoons (Figure 18.4). Salinity was significantly lower in the riverine channel in all the systems and in the fluvial delta sands in the Hastings and Wallis lakes due to the greater river influence in these areas. The lowest salinity occurred in the Hastings, reflecting the larger freshwater inputs. Oxygen saturation across the geomorphic units and lagoons followed a similar pattern as salinity, most likely due to the input of deoxygenated water from the river and floodplain (Eyre et al. 2006). The lowest oxygen saturation occurred in the Hastings, which has the largest alluvial floodplain. Total nitrogen, ammonium, nitrate, total phosphorus and phosphate concentrations were generally elevated in the riverine channel and fluvial tidal delta reflecting a freshwater source. Nutrient concentrations were particularly elevated in the Wallis Lake riverine channel probably as a result of the longer retention time in this system. Algal biomass in the water column (chlorophyll *a* concentrations) was higher in the riverine channels of all three systems associated with high nutrient concentrations. The immature Wallis Lake had the highest chlorophyll *a* concentrations ascribable to the deeper, clearer water column. Wallis Lake and Camden Haven lagoons had higher concentrations of carbon and nitrogen in the sediment, particularly in the riverine channel, central mud basin, and fluvial delta sands. These observations most likely reflect a greater particulate trapping efficiency in these systems.

18.4 BENTHIC HABITATS

As a barrier lagoon system matures (infills), the relative coverage of various benthic habitats shifts from those associated with a central mud basin to habitats linked to the marine tidal delta and riverine channel. Large productive subtidal mud shoals in immature barrier lagoons are gradually infilled leaving mangrove/saltmarsh habitats landward with channel and remnant subtidal shoals instream (Roy et al. 2001; Harris and Heap 2003). This process is evident in the three study lagoons (Table 18.4), with the proportional coverage of subtidal mud shoals/depositional mud basins decreasing with lagoon maturity (i.e., Hastings < Camden Haven < Wallis Lake). Concurrently, marine sand channel and fluvial sands and gravel increase with maturity (Hastings > Camden Haven > Wallis Lake). These natural changes have a significant influence over the productivity and trophic dynamics of the lagoons. The relative contribution of benthic (vs. pelagic) productivity decreases as an estuary matures (see Section 18.5, Carbon Budgets, below). Along with the change in productivity, habitat structure and complexity shifts from seagrass domination in the immature Wallis Lake and Camden Haven estuaries to channel and subtidal shoal habitats in the Hastings River (Figures 18.5, 18.6, and 18.7).

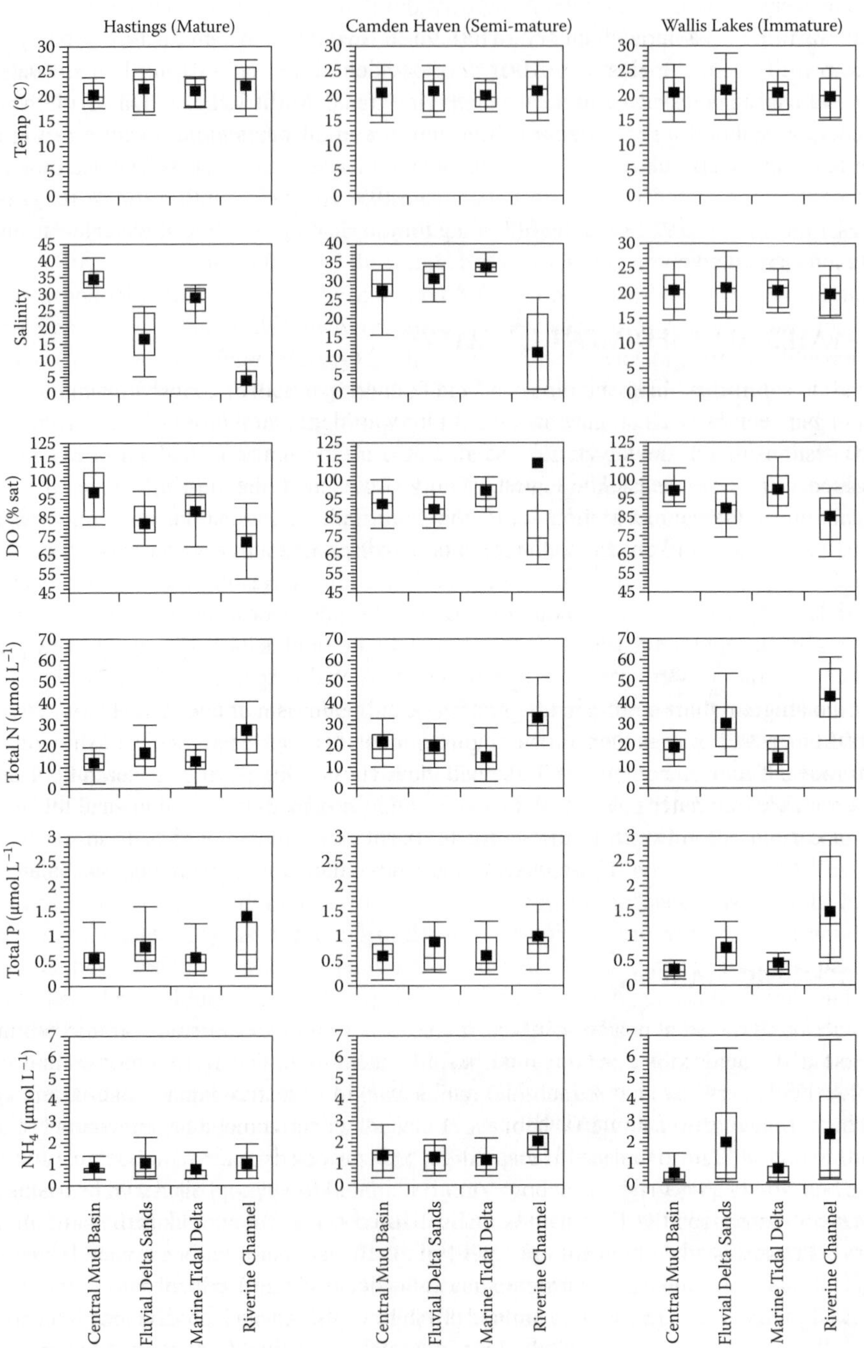

FIGURE 18.4 Water and sediment quality data for Wallis Lake, Camden Haven, and Hastings River Estuary.

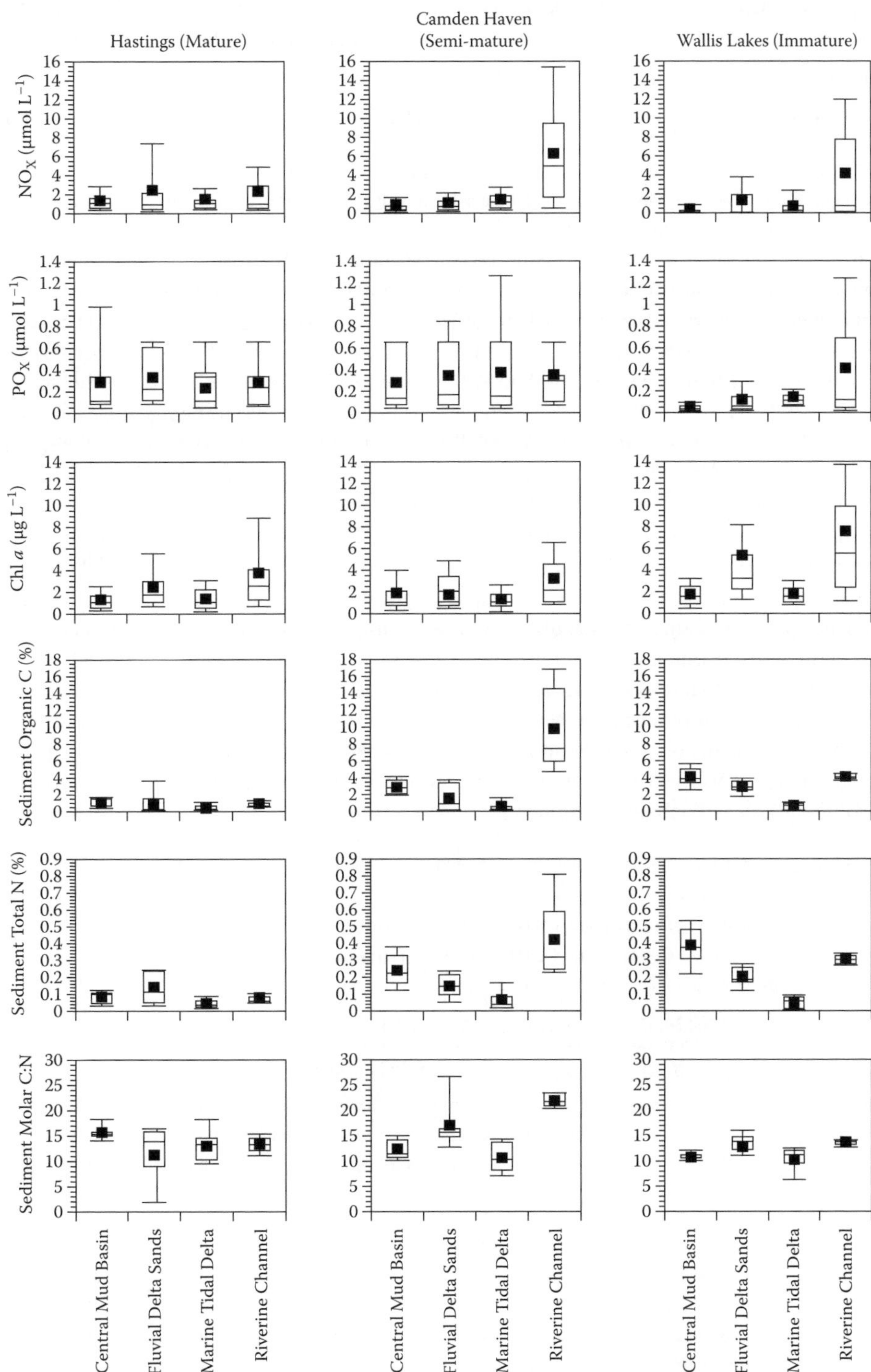

FIGURE 18.4 (continued).

TABLE 18.4
Description of the Main Benthic Habitats and Areal Contribution to Total In-Stream Habitat for Hastings River Estuary, Camden Haven, and Wallis Lake

Lagoon Benthic Habitat	Description	Area (km^2)	% of Total Area
	Hastings		
Intertidal mud shoals	Areas of fine mud sediment that are exposed and immersed throughout the tidal cycle	0.18	0.96
Halophila	Habitats where *Halophila ovalis* is the dominant vegetation type	0.38	2.05
Mangrove creek	Small tributaries dominated by mangrove community (*Avicennia marina*)	0.57	3.09
Intertidal sand shoals	Areas of sand sediment that are exposed and immersed throughout the tidal cycle	1.05	5.63
Zostera	Habitats where *Zostera capricorni* is the dominant vegetation type	1.17	6.28
Subtidal mud shoals	Areas of mud sediment that are immersed throughout the tidal cycle	1.93	10.38
Fluvial sands/muds	Upper estuarine channel habitats with coarse fluvial-derived sand and mud sediments	3.79	20.33
Subtidal sand shoals	Areas of sand habitat immersed throughout the tidal cycle	4.08	21.91
Channel marine sand	Lower estuarine channel habitat with marine-derived quartzite sands	5.47	29.36
Total		**18.62**	
	Camden Haven		
Halophila, Ruppia	Mixed *H. ovalis* and *Ruppia megacarpa* seagrass beds (*H. ovalis* dominant)	0.19	0.63
Ruppia, Zostera, Halophilia	Mixed *R. megacarpa*, *Z. capricorni,* and *H. ovalis* seagrass beds (*R. megacarpa* dominant)	0.21	0.71
Ruppia, Zostera	Mixed *R. megacarpa* and *Z. capricorni* beds (*R. megacarpa* dominant)	0.21	0.71
Intertidal mud shoals	Areas of fine mud sediment that are exposed and immersed throughout the tidal cycle	0.97	3.22
Ruppia	Habitats where *R. megacarpa* is the dominant vegetation type	1.40	4.66
Fluvial sands/muds	Upper estuarine channel habitats with coarse fluvial derived sand and mud sediments	1.52	5.05
Channel marine sand	Lower estuarine channel habitat with marine-derived quartzite sands	1.56	5.19
Halophila, Zostera	Mixed *H. ovalis* and *Z. capricorni* beds (*H. ovalis* dominant)	1.94	6.47
Intertidal sand shoals	Areas of sand sediment that are exposed and immersed throughout the tidal cycle	1.97	6.54
Halophila	Habitats where *Halophila ovalis* is the dominant vegetation type	1.99	6.61
Zostera	Habitats where *Zostera capricorni* is the dominant vegetation type	2.62	8.71
Subtidal sand shoals	Areas of sand habitat immersed throughout the tidal cycle	2.89	9.62
Zostera, Halophila	Mixed *Z. capricorni* and *H. ovalis* beds (*Z. capricorni* dominant)	3.00	9.97
Subtidal mud shoals	Areas of mud sediment that are immersed throughout the tidal cycle	9.59	31.91
Total		**30.05**	
	Wallis Lake		
Macroalgae	Habitat areas where the macroalgae *Chara* sp. and *Nitella* sp. dominate	0.75	0.83
Ruppia	Habitats where *R. megacarpa* is the dominant vegetation type	0.98	1.09
Intertidal mud shoals	Areas of fine mud sediment that are exposed and immersed throughout the tidal cycle	0.99	1.09
Channel marine sand	Lower estuarine channel habitat with marine-derived quartzite sands	1.26	1.39
Intertidal sand shoals	Areas of sand sediment that are exposed and immersed throughout the tidal cycle	2.02	2.23

TABLE 18.4 (continued)
Description of the Main Benthic Habitats and Areal Contribution to Total In-Stream Habitat for Hastings River Estuary, Camden Haven, and Wallis Lake

Lagoon Benthic Habitat	Description	Area (km^2)	% of Total Area
Zostera macroalgae	Mixed *Z. capricorni* and macroalgae beds (*Z. capricorni* dominant)	3.74	4.13
Posidonia	*Posidonia australis* seagrass beds	4.01	4.43
Subtidal mud shoals	Areas of mud sediment that are immersed throughout the tidal cycle	4.02	4.44
Lake sands	Shallow subtidal coarse quartz sands with extensive diatom coverage	4.32	4.77
Halophila	Habitats where *Halophila ovalis* is the dominant vegetation type	4.77	5.27
Subtidal sand shoals	Areas of sand habitat immersed throughout the tidal cycle	10.31	11.40
River muds	Upper estuarine habitats with fine fluvial sediments	14.92	16.49
Zostera	Habitats where *Zostera capricorni* is the dominant vegetation type	19.18	21.20
Mud basin	Depositional central basin areas with fine silt sediments and phytodetritus	19.21	21.24
Total		**90.45**	

Source: Modified from Maher (2010).

18.5 CARBON BUDGETS

Carbon budgets illustrate the distinct differences in the function of the different geomorphic types of coastal lagoons (Table 18.5). The functional differences reflect to some degree the differences in structure (see Section 18.4, Benthic Habitats, above). Carbon burial is much higher in the immature Wallis Lake (62 ± 32 t km^{-2} yr^{-1}) and semimature Camden Haven (53 ± 15 t km^{-2} yr^{-1}) compared to the mature Hastings River Estuary (28 ± 15 t km^{-2} yr^{-1}) due to the large central mud basins where material is trapped and buried. There was a slight decrease in total mean annual production as a lagoon matured (Wallis Lake, 612 ± 142 t C km^{-2} yr^{-1}; Camden Haven, 603 ± 152 t C km^{-2} yr^{-1}; Hastings, 506 ± 157 t C km^{-2} yr^{-1}), but the differences were not significant. In contrast, the benthic to pelagic production ratio showed a much larger change, with a small increase from the immature Wallis Lake (2.42 ± 0.19) to the semimature Camden Haven (3.04 ± 0.00), and then a much larger decrease in the mature Hastings River Estuary (0.84 ± 0.09) (Figure 18.8a). This pattern of benthic to pelagic production may reflect the light available to the primary producers across the three types of systems. In the semimature Camden Haven, the central mud basin is filled with sediment, and much of the system is less than 1 m deep (Table 18.3), enabling high benthic production throughout the year. A large fraction of this increased production is associated with benthic microalgae as sediment resuspension by wind in the shallow water column limits seagrass production to the more sheltered areas of the estuary. In contrast, the immature Wallis Lake has a relatively larger and deeper central mud basin and, as such, more light is attenuated in the water column and less light reaches the lagoon floor. High currents cause sediment resuspension, greater water column turbidity, and less suitable substrate conditions for benthic producers in the Hastings River Estuary, resulting in a large decrease in benthic production.

NEM, as a percentage of total production, increased from 15.3% ± 0.4% in the mature Wallis Lake to 26.9% ± 5.8% in the semimature Camden Haven, and it then decreased to 20.0% ± 6.0% in the mature Hastings River Estuary (Figure 18.8b). The differences in NEM as a percentage of total production (function) may be controlled by the geomorphology and hydrodynamics (structure) of the three systems. The immature Wallis Lake has a deep mud basin that would trap organic

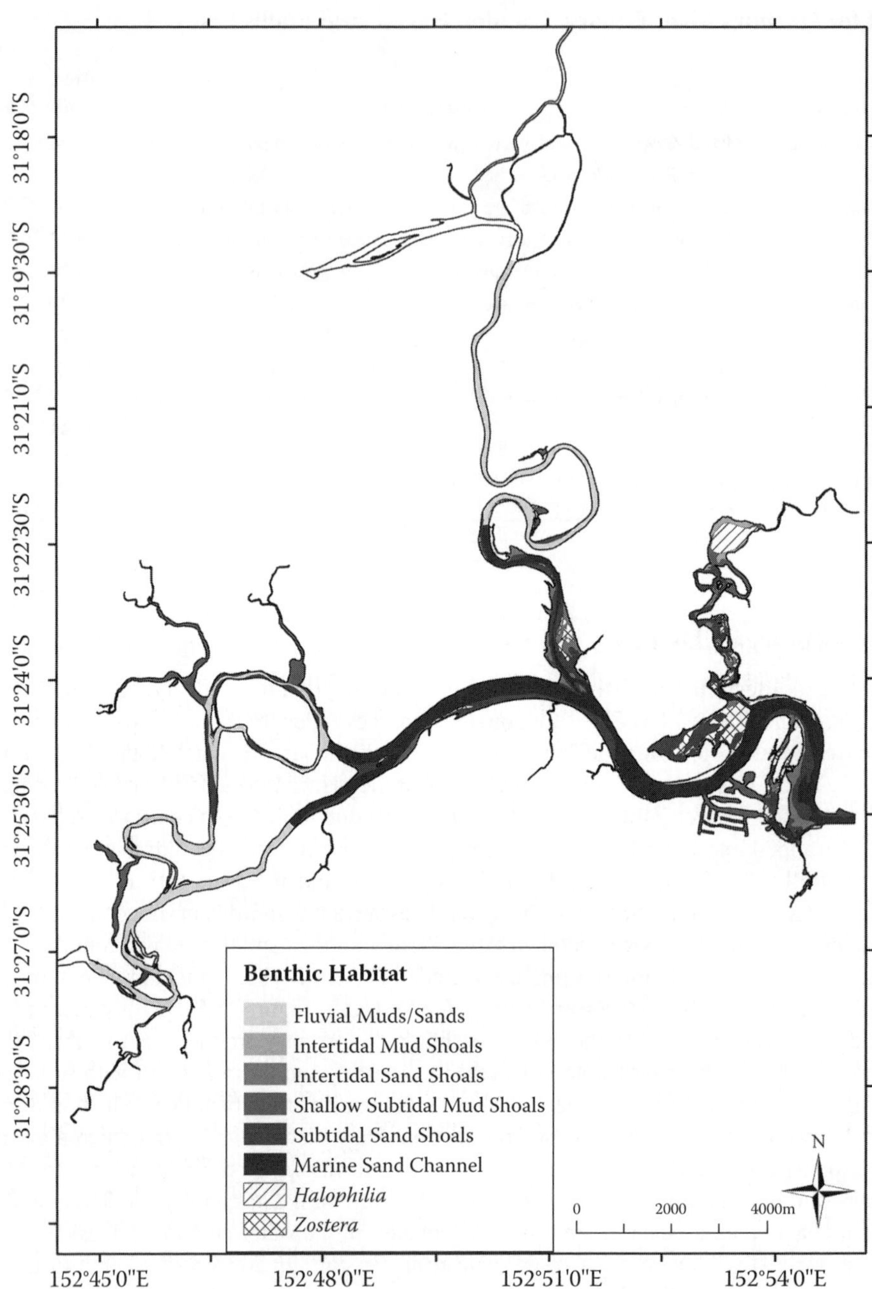

FIGURE 18.5 Benthic habitat map of the Hastings. (Modified from Maher, D. et al. 2007. *Benthic habitat mapping, primary productivity measurements and macrofauna surveys in the Camden Haven and Hastings River Estuaries.* Report to Port Macquarie Hastings Council, Centre for Coastal Biogeochemistry, Report No. 2007-5. Southern Cross University, Lismore, Australia.)

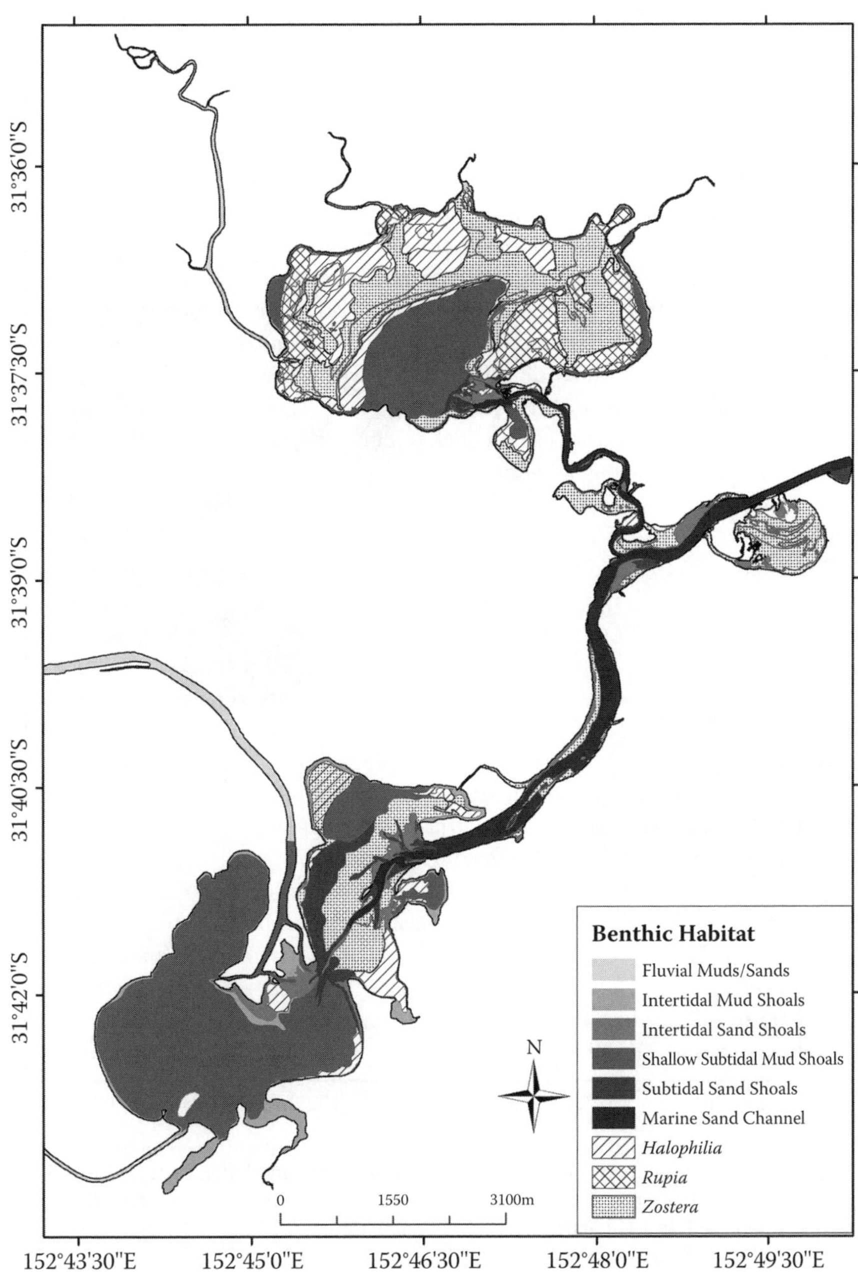

FIGURE 18.6 Benthic habitat map of the Camden Haven. (Modified from Maher, D. et al. 2007. *Benthic habitat mapping, primary productivity measurements and macrofauna surveys in the Camden Haven and Hastings River Estuaries.* Report to Port Macquarie Hastings Council, Centre for Coastal Biogeochemistry, Report No. 2007-5. Southern Cross University, Lismore, Australia.)

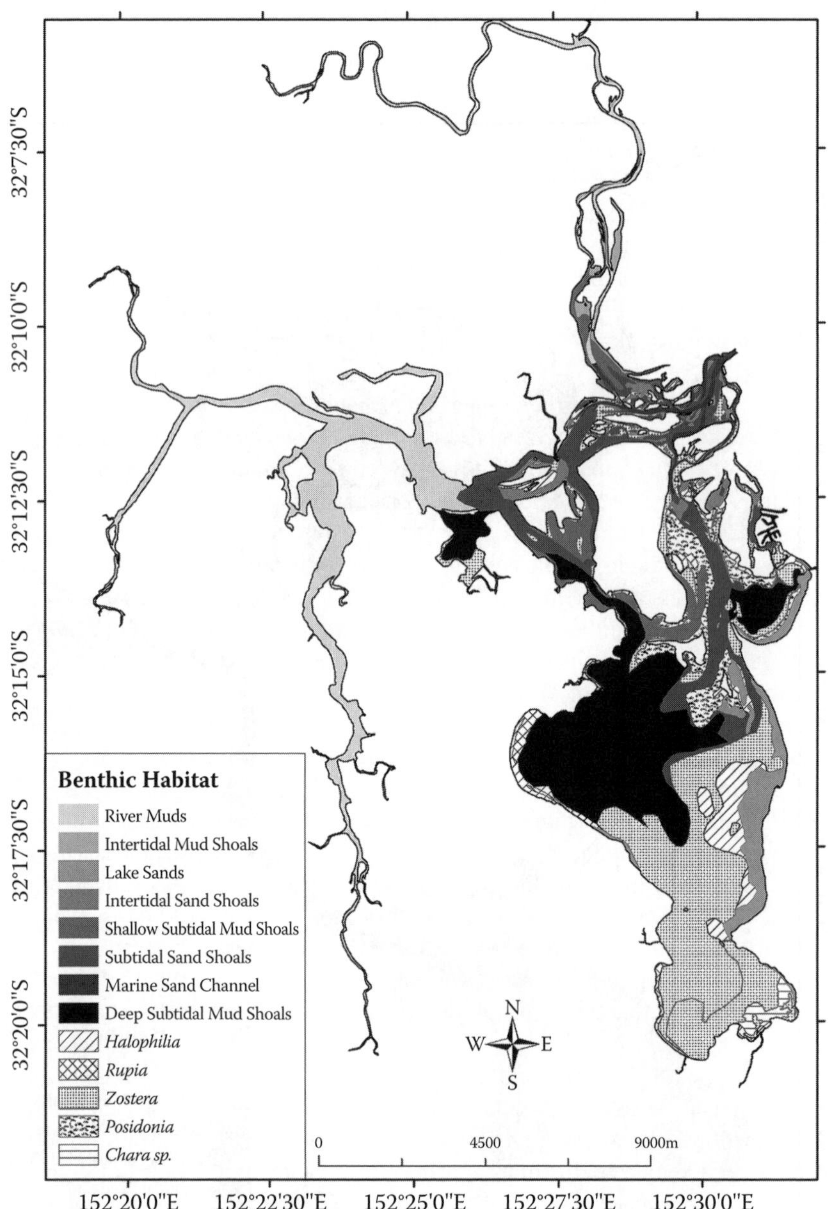

FIGURE 18.7 Benthic habitat map of the Wallis Lake. (Modified from Maher, D. et al. 2007. *Benthic habitat mapping, primary productivity measurements and macrofauna surveys in the Camden Haven and Hastings River Estuaries.* Report to Port Macquarie Hastings Council, Centre for Coastal Biogeochemistry, Report No. 2007-5. Southern Cross University, Lismore, Australia.)

matter with a concomitant increase in respiration and decrease in NEM. The mature Hastings River Estuary also has deep channels, but the whole system is more rapidly flushed (Table 18.3), which would result in more organic matter being exported to the ocean and less organic matter respired within the system than at Wallis Lake. Although the semimature Camden Haven is not as well flushed as the Hastings River Estuary, it is very shallow, particularly the areas where benthic production is highest, which probably results in organic matter being exported out of the system.

TABLE 18.5
Carbon Budgets for Hastings River Estuary, Camden Haven, and Wallis Lake

Inputs (t/C)	Hastings	Camden Haven	Wallis Lake
Diffuse (DOC)[a]	814 (±197)	806 (±98)	309 (±57)
Atmosphere[b]	32 (±32)	54 (±54)	151 (±151)
Pelagic production[c]	5138 (±1272)	3786 (±583)	16242 (±2662)
BMA production[d]	1862 (±599)	4451 (±1153)	7486 (±2823)
Mangrove production[e]	764 (±764)	261 (±261)	1485 (±1485)
Seagrass production[d]	1665 (±283)	8936 (±2026)	28091 (±5462)
Macroalgal production[d]	0	0	2017 (±420)
Outputs			
Pelagic respiration[c]	3234 (±546)	3125 (±399)	11870 (±2596)
BMA respiration[d]	2301 (±318)	3463 (±572)	6548 (±2526)
Mangrove respiration[e]	596 (±596)	204 (±204)	1158 (±1158)
Burial (non veg areas)[f]	39 (±21)	502 (±141)	2167 (±1121)
Burial seagrass[g]	129 (±69)	960 (±270)	2774 (±1435)
Burial mangrove[g]	354 (±190)	121 (±34)	688 (±356)
Total burial	521 (±280)	1582 (±445)	5629 (±2912)
Commercial Fisheries Production (2001)[h]	31 (±4)	31 (±5)	188 (±13)
Seagrass Respiration	1412 (±128)	6460 (±857)	25040 (±4419)
Macroalgae Respiration	0	0	2263 (±526)

Source: Modified from Maher (2010).
[a] Flow-weighted sampling.
[b] Volume × concentration.
[c] Directly measured (dark/light bottle) × 3 sites × 4 seasons.
[d] Combination of *in situ* chambers and cores in each habitat × 4 seasons.
[e] Allometric-based biomass vs. production scaling.
[f] Measure sedimentation rates (Pb210) and sediment C concentration.
[g] Global average value.
[h] NSW Fisheries data (including oysters) converted to C.

18.6 WILD FISHERIES PRODUCTION

The three lagoons have significantly ($p = 0.02$) different wild fisheries production (catch per fisher, based on 20 years of commercial fish catch). Camden Haven has the highest average annual catch to fisher ratio, and the Hastings River Estuary has the lowest ratio (Figure 18.9a). The wild fisheries production follows the same trend as the benthic to pelagic production ratio. However, the catch attributable to secondary consumers and higher trophic level species (i.e., trophic level above 2) shows a decrease in production with increasing system maturity (Figure 18.9b). There is also a distinct trend in the proportion of NEM transferred to fisheries production (corrected for effort on a per fisher basis) (trophic transfer efficiency, TTE), with an increase corresponding to an increase in coastal lagoon maturity (Figure 18.9c). TTE shows an inverse relationship with areal carbon burial rates that decrease with increasing maturity. This suggests that as a coastal lagoon matures, lower burial rates may increase the proportion of NEM transferred to fisheries production.

Energy appears to be more efficiently transferred into the food web in mature estuarine systems, but it is more efficiently transferred up the food chain in more immature estuaries (Figure 18.9b).

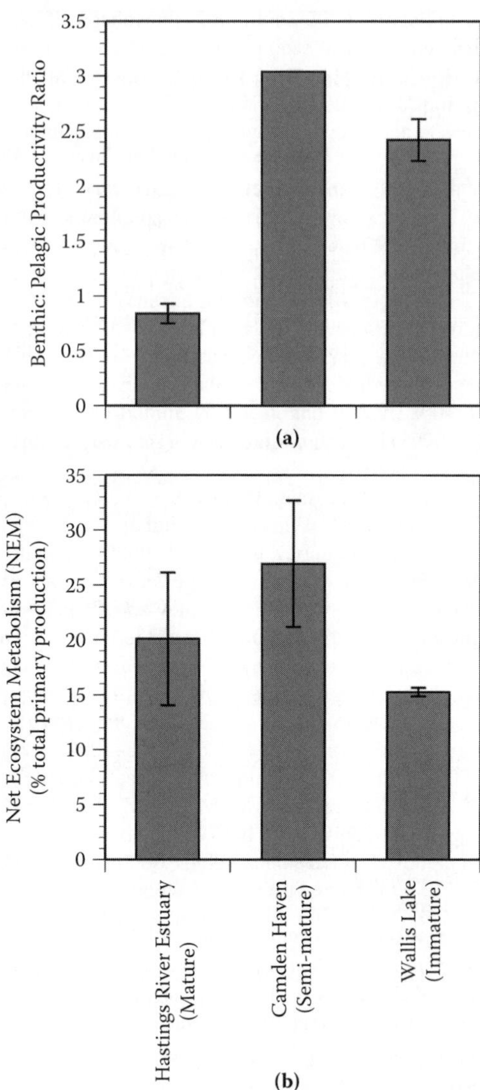

FIGURE 18.8 (a) Ratio of benthic to pelagic production and (b) net ecosystem metabolism (NEM) as a percentage of gross primary production for Hastings River Estuary, Camden Haven, and Wallis Lake.

These distinct differences in trophic dynamics are most likely due to the structure and function of the three systems. Longer residence times in the mature Wallis Lake (Table 18.3) would retain more organic matter within the system, and therefore a greater proportion of NEM is lost through burial. However, this retention of organic matter also sustains larger populations of higher-order species. Coupled with the long flushing times are the large seagrass beds in both the Wallis Lake and Camden Haven lagoons which may act as filters further increasing organic matter retention (Asmus and Asmus 2000; Papadimitriou et al. 2005). In addition, the importance of seagras habitats as nursery areas (Pollard 1984; Bell and Pollard 1989; Jenkins et al. 1997) may also have contributed to the higher-order production.

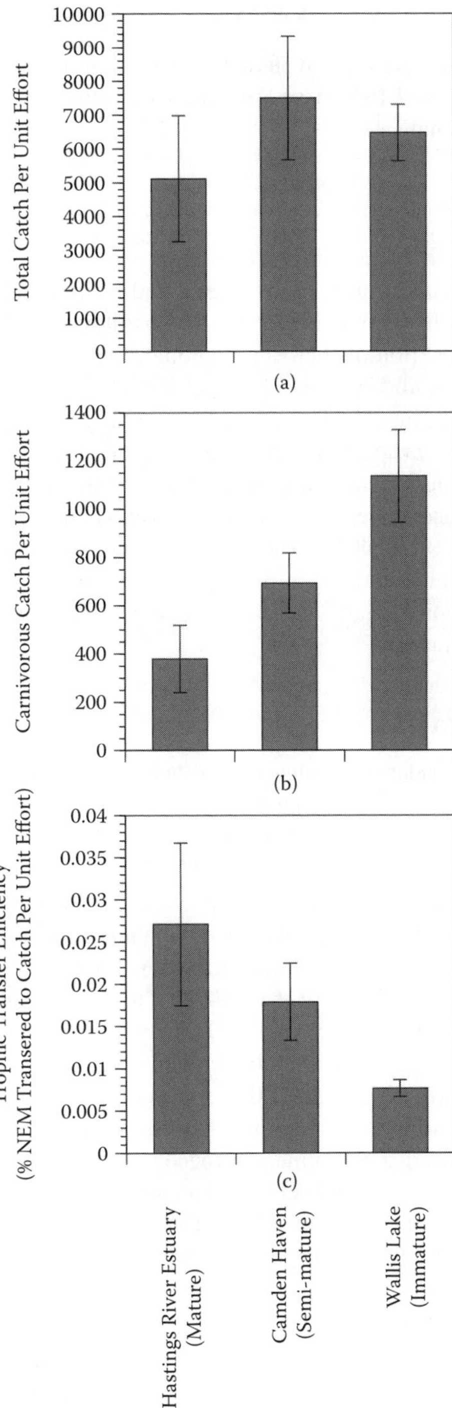

FIGURE 18.9 (a) Summary of mean effort corrected (per fisher) wild fisheries catch (oysters excluded) for the three coastal lagoons from 1984 to 2001; (b) the proportion of catch at trophic level of 2 (secondary consumer) or above for the three coastal systems for the years 1984 to 2001; (c) the percentage of NEM transferred to mean (1984 to 2001) effort corrected wild fisheries catch fish catch (trophic transfer efficiency [TTE]) for the three coastal lagoons.

18.7 IMPLICATIONS FOR NATURAL AND ANTHROPOGENIC CHANGE

The geomorphic structure of the three different types of coastal lagoons controls their function (see Table 18.5, Figures 18.8 and 18.9). This structure also influences the response of the lagoons to natural and anthropogenic change.

18.7.1 Natural Change

Bar-built or wave-dominated estuaries/coastal lagoons mature via infilling of the central basin. The rate of this infilling is controlled by the rate of sediment supply from the ocean via waves and tidal currents, and from the catchment via river inflow. As the coastal lagoons fill with sediment, the relative proportion of their main geomorphic units (i.e., marine tidal delta, central mud basin, fluvial delta, and riverine channel/alluvial plain) also changes (Roy et al. 2001; Harris and Heap 2003). Changes in the area of the main geomorphic units, in particular the central mud basin, alter the proportions of the benthic habitats. These changes, in turn, influence the benthic-pelagic coupling and the higher-order ecology (e.g., fish landing species). For example, there is a decrease in seagrass habitat and an increase in channel marine sands as coastal systems infill (see Table 18.4), resulting in a proportional decrease in the ratio of benthic to pelagic production and changes in the type of fish landed (Table 18.5, Figure 18.9). For effective management of coastal lagoons, anthropogenic change must be assessed against this background of natural evolution.

18.8 POLLUTANT LOADS

Many pollutants such as nutrients (i.e., phosphorus) and heavy metals are transported attached to sediments, and the sediments themselves may also be a pollutant (Eyre and McConchie 1993). The degree of maturity or infilling of a coastal lagoon is an important determinant of how susceptible the system will be to increased pollutant loading (Harris and Heap 2003). Much of the sediment delivered to immature coastal lagoons will be trapped within the system. In contrast, much of the sediment delivered to mature coastal lagoons will escape seaward and be delivered to the continental shelf, with interannual variations in sediment trapping efficiency dependent on the size of floods in a given year (Eyre et al. 1998; Hossain et al. 2002). The trapping efficiency of other material is also similar. For example, in the mature (type IIID) Richmond River Estuary (same coast, 300 km north of the study area) only 2.5% of the nitrogen and 5.4% of the phosphorus that is delivered to the estuary are retained (McKee et al. 2000). As such, despite a two- to threefold increase in nitrogen and phosphorus loading to the Richmond River Estuary over the past 50 years, there has been no change in water column and sediment nitrogen and phosphorus concentrations (Eyre 1997; McKee et al. 2000), illustrating its resilience to change. In contrast, in the immature Tuggerah Lakes (type IIIB), about 11% of the nitrogen and 40% of the phosphorus delivered to the lagoon is retained (Eyre and Pepperell 2001). In response to increased nutrient loads, Tuggerah Lakes show symptoms of eutrophication in the form of excess organic matter production (macroalage) (Colett et al. 1981), reflecting the higher nutrient retention.

18.9 ACIDIFICATION, DEOXYGENATION, AND FISH KILLS

Although basin infilling and low retention efficiency render mature coastal lagoons less susceptible to increased pollutant loads, they also make them more susceptible to acidification, deoxygenation, and fish kills. As a coastal lagoon fills with sediment, newly formed geomorphic units (mangroves, salt marshes, flood plains, swamps, and wetlands) will develop acid sulfate soils that are rich in pyrite (FeS_2). Alluvium will cover these soils, creating prime agricultural areas (Roy et al. 2001) that alter the vegetation types and drainage characteristics. When acid sulfate soils are exposed to air via draining for agriculture, they produce sulfuric acid that can be flushed into adjacent waterways causing a decrease in pH to as low as 2 (Ferguson and Eyre 1999). Waterway acidification results in fish mortality and diseases, impairment of oyster production, and impacts on benthic communities (Sammut et al. 1996; Roach 1997; Corfield 2000). In addition, when these modified flood plains are inundated during flooding, decomposition of the intolerant vegetation can deoxygenate the floodwaters, resulting in anoxia and fish kills in the adjacent estuary (Eyre et al. 2006). For example, following a 1:10 year return period flood in February 2001, 50 km of the Richmond River Estuary (a mature coastal lagoon, northern NSW, Australia) was closed to commercial and recreational fishing due to the complete deoxygenation of the water column (DO < 1 mg L^{-1}) and associated total fish mortality (Eyre et al. 2006). Similar deoxygenation events in the estuary, albeit not as severe, have been recorded for the last 50 years (Eyre 1997). In contrast, immature coastal lagoons have fewer geomorphic units that contain acid sulfate soils, making them less susceptible to acidification, deoxygenation, and fish kills.

18.10 CLIMATE CHANGE IMPACTS

Climate-induced changes of environmental factors that impact coastal lagoons include sea level rise, temperature, rainfall and runoff, storminess, and wave climate. The impact of changes in these environmental factors varies considerably among the different geomorphic types of coastal lagoons.

18.10.1 Sea Level Rise

Global sea level rise is predicted to range from 0.18 to 0.59 m by the end of the century (IPCC, 2007). An additional 0.12 m rise is expected along the east coast of Australia due to thermal effects of the East Australian Current (McInnes et al. 2007). Coastal lagoons will respond to sea level rise by migrating upslope along undeveloped shorelines (Bird 1994). Where shorelines have been hardened, coastal lagoons cannot migrate inland, and vegetation communities will be lost. The effect on mature and immature coastal lagoons will depend on their degree of development. Immature lagoons are most at risk because they have a higher proportion of benthic habitats such as seagrass that will be most susceptible to changes in lagoon structure (Table 18.4). Sea level rise will also modify the coastal sand supply, which would have a greater effect on mature coastal lagoons with a higher proportion of sandy benthic habitats (Table 18.4). Although previously considered to be environments of low productivity (Roy et al. 2001), sandy habitats may be highly productive with elevated rates of remineralization of organic matter (Glud et al. 2008; Eyre and Maher 2010). Immature coastal lagoons have more restricted exchange with the ocean and are therefore more likely to suffer changes in entrance conditions (i.e., mouth sand bars) than mature systems.

18.10.2 Temperature

Modeling based on medium emission predictions (IPCC scenario A1B; IPCC 2000) indicates that the temperature on the east coast of Australia is likely to rise ~1°C by 2030 (CSIRO 2007). *In situ* temperature increases are expected to be greatest in immature coastal lagoons due to more restricted circulation and flushing (Table 18.3). Temperature increases may change the phenology of a whole range of coastal lagoon processes involving organisms from phytoplankton and zooplankton to fish (Oviatt 2004; Costello et al. 2008; Mooij et al. 2008). An increase in temperature can also directly impact many marine species that live near their thermal tolerance (Somero 2002) as well as impacting marine species indirectly via associated decreases in dissolved oxygen concentrations. Oxygen depletion of the lower water column is likely to be enhanced in immature coastal lagoons that are more susceptible to stratification due to restricted flushing. If chronic oxygen depletion (hypoxia) develops in response to temperature increases, it will likely cause significant changes in benthic communities (Conley et al. 2007). In addition, immature lagoons are most at risk because they have a higher proportion of benthic habitats (e.g., seagrasses) that are sensitive to temperature increases (Blintz et al. 2003).

18.10.3 Rainfall and Runoff

An increase in summer rainfall and an increase in rainfall intensity are predicted for the southeast coast of Australia (CSIRO 2007). An increase in summer rainfall and runoff associated with climate change will increase the amount of freshwater, sediment, and nutrients delivered to coastal lagoons. The impact of these increased loads will be greater in immature systems compared to mature systems. More of the sediment delivered to immature coastal lagoons will be trapped, thereby facilitating basin infilling. Immature coastal lagoons also have a higher proportion of benthic habitats that will be susceptible to impact from increased sediment loads. The loss of seagrass habitat in these systems will change the ratio of benthic to pelagic production (Table 18.5) and the composition of fish communities. An increase in nutrient loads will also have a greater impact on immature coastal lagoons due to their higher nutrient retention efficiency, leading to enhanced eutrophication (see Section 18.8, Pollutant Loads, above). Immature coastal lagoons generally have lower salinities because of longer residence times (Table 18.3). Increases in freshwater inflow could have a dramatic effect on the structure and function of biotic communities in these systems.

Increased rainfall and runoff will have less impact on mature coastal lagoon systems that are already subject to rapid changes in flushing associated with flood events (Eyre and Twigg 1997). Any increase in the frequency and magnitude of flood events would further decrease the low sediment and nutrient retention efficiency of these systems (Eyre et al. 1998; McKee et al. 2000). The salinity gradients in mature coastal lagoons rapidly reestablish following flood events (Eyre and Twigg 1997). Therefore, increased freshwater inputs will have less of an impact than in immature coastal lagoons. An increase in rainfall and more regular flooding should decrease acid discharges in mature coastal lagoons due to higher groundwater tables. In addition, more regular flooding of the floodplain of mature coastal lagoons should decrease the buildup of organic matter that results in the deoxygenation of floodwaters (Eyre et al. 2006) and may enhance offshore fisheries production due to the greater export of organic matter.

18.10.4 Storminess and Wave Climate

Modeling of the southeastern coast of Queensland indicates an increase in the number of weather systems producing extremes in both wind and precipitation (Abbs and McInnes 2004). Any increase in the frequency and intensity of storms along the east coast of Australia will increase the probability of breaching barrier beach systems enclosing coastal lagoons. Sea level rise will further increase overwash events. The breaching of barriers will increase the flushing rate and salinity of coastal lagoons. Greater storm activity will increase the breaching of barriers enclosing both immature and mature coastal lagoons; however, the impacts would probably be greater in immature systems due to their greater proportion of more susceptible habitats (Table 18.4).

Mature coastal lagoons are hydrodynamically more strongly linked to the coastal ocean than immature coastal lagoons due to less restricted exchange and more rapid flushing. As such, any changes in ocean circulation patterns and wave climate associated with climate change that affect upwelling and delivery of nutrients will have a greater impact on mature coastal lagoons. For example, an input of nitrogen from the ocean during the dry season is evident in both the mature Richmond River and Brunswick estuaries (McKee et al. 2000; Ferguson et al. 2004). Mature east coast Australian river estuaries, in general, are nutrient limited much of the year, particularly during the dry season (Eyre 2000), and any decrease in the ocean nitrogen load would likely result in a decrease in estuarine productivity. Many fish species will expand their geographic range in a poleward shift with increasing global temperature. Changes in ocean circulation, increases in UV radiation, and changes in primary and secondary productivity are all likely to significantly impact the recruitment of fish species (Hobday et al. 2006), and in concert with fisheries harvest may lead to the collapse of some fish stocks.

18.11 CONCLUSIONS

As east Australian coastal lagoons mature, the areal extent of their geomorphic units (structure) changes linearly with a decrease in the extent of deep and shallow subtidal mud shoals and an increase in the extent of marine sand channel and fluvial sands. These structural changes, in turn, alter the function of these systems. For example, the ratio of benthic to pelagic production and NEM as a percentage of primary production and wild fisheries production (catch per fisher) all show nonlinear changes with increasing maturity, with an initial increase as the system moves from immature to semimature followed by a decrease as the system moves from semimature to mature. In contrast, the production (catch per fisher) of higher trophic level fish species shows a linear decrease with increasing system maturity, and the proportion of NEM transferred to fisheries production (trophic transfer efficiency) increases with increasing system maturity. As such, energy appears to be more efficiently transferred into food webs in mature systems, but this energy is more efficiently transferred up the food chain in more immature systems. Immature coastal lagoons most likely sustain larger populations of higher-order species due to greater retention of organic matter and larger areas of seagrass habitat. However, the higher retention capacity of immature coastal lagoons combined with their larger areas of seagrass also makes them more susceptible to pollutant loads and climate change impacts such as sea level rise, temperature increases, increased rainfall and runoff, and increased storminess and wave activity.

ACKNOWLEDGMENTS

We thank Ralph Haese and two anonymous reviewers for helpful comments. Port Macquarie Hastings Council funded much of the work in Camden Haven and the Hastings River Estuary.

REFERENCES

Abbs, D. J. and K. L. McInnes. 2004. *Impact of climate change on extreme rainfall and coastal sea levels over South-East Queensland. Part 1: Analysis of extreme rainfall and wind events in a GCM*. Report for the Gold Coast City Council, Gold Coast City, Australia.

Asmus, H. and R. Asmus. 2000. Material exchange and food web of seagrass beds in the Sylt-Rømø Bight: How significant are community changes at the ecosystem level? *Helgoland Mar. Res*. 54: 137–150.

Bell, J. D. and D. A. Pollard. 1989. Ecology of fish assemblages and fisheries associated with seagrasses. Pp. 565–609, In: A. D. W. Larkum, A. J. McComb, and S. Shepherd (eds.), *Biology of seagrasses: A treatise on the biology of seagrasses with special reference to the Australian region*. Amsterdam, the Netherlands: Elsevier.

Bird, E. C. F. 1994. Physical setting and geomorphology of coastal lagoons. Pp. 9–40, In: B. Kjerfve (ed.), *Coastal lagoons*. Amsterdam, the Netherlands: Elsevier.

Blintz, J. C., S. Nixon, B. Buckley, and S. Granger. 2003. Impacts of temperature and nutrients on coastal lagoon plant communities. *Estuaries* 26: 765–776.

Bom, B. O. M. 2008. Climate data set. Available at http:www.bom.gov.au/climate/.

Borges, A. V., B. Delille, and M. Frankignoulle. 2005. Budgeting sinks and sources of CO_2 in the coastal ocean: Diversity of ecosystem counts. *Geophys. Res. Lett.* 32 L14601. DOI: 10.1029/2005GL023053.

Colett, L. C., A. J. Collins, P. J. Gibbs, and R. J. West. 1981. Shallow dredging as a strategy for the control of sublittoral macrophytes: A case study in Tuggerah lakes, New South Wales. *Australian J. Mar. Freshw. Res.* 32: 563–571.

Conley, D. J. and others 2007. Long-term changes and impacts of hypoxia in Danish coastal waters. *Ecol. Applic*. 17: S165–S184.

Corfield, J. 2000. Macrobenthic community dynamics in the Richmond River Estuary, northern NSW, Australia. Ph.D. Thesis, Southern Cross University, Lismore, Australia.

Costello, J. H., B. K. Sullivan, and D. J. Gifford. 2008. A physical-biological interaction underlying variable phenological responses to climate change by coastal zooplankton. *J. Plankton Res*. 28: 1099–1105.

CSIRO. 2007. *Climate change in Australia: Technical report 2007*. CSIRO, Australian Bureau of Meteorology, Canberra, Australia.

Eyre, B. D. 1997. Water quality changes in an episodically flushed sub-tropical Australian estuary: A 50 year perspective. *Mar. Chem.* 59: 177–187.

Eyre, B. D. 1998. Transport, retention, and transformation of material in Australian estuaries. *Estuaries* 21: 540–551.

Eyre, B. D. 2000. Regional evaluation of nutrient transformation and phytoplankton growth in nine river-dominated sub-tropical east Australian estuaries. *Mar. Ecol. Prog. Ser*. 205: 61–83.

Eyre, B. D., A. J. P. Ferguson, A. P. Webb, D. Maher, and J. M. Oakes. In press. Metabolism of different benthic habitats and their contribution to the carbon budget of a shallow oligotrophic sub-tropical coastal system (southern Moreton Bay, Australia). *Biogeochem.*

Eyre, B. D., S. Hossain, and L. Mckee. 1998. A sediment budget for the modified sub-tropical Brisbane River estuary, Australia. *Estuar. Coastal Shelf Sci.* 47: 513–522.

Eyre, B. D., G. Kerr, and L. Sullivan. 2006. Deoxygenation potential of the lower Richmond River floodplain, northern NSW, Australia. *River Res. Applic*. 22: 981–992.

Eyre, B. D. and D. Maher. In press. Mapping ecosystem processes and function across shallow seascapes. *Contin. Shelf Res.*

Eyre, B. D., and D. McConchie. 1993. The implications of sedimentological studies for environmental pollution assessment and management: Examples from fluvial systems in North Queensland and Western Australia. *Sed. Geol.* 85: 235–252.

Eyre, B. D. and P. Pepperell. 2001. *Nitrogen and phosphorus budgets for Tuggerah Lakes, Australia*. Report to Wyong Shire Council, Centre for Coastal Management, Southern Cross University, 28 pp.

Eyre, B. D., and C. Twigg. 1997. Nutrient behavior during post-flood recovery of the Richmond River Estuary northern NSW, Australia. *Estuar. Coastal Shelf Sci.* 44: 311–326.

Ferguson, A. J. P. and B. D. Eyre. 1999. Iron and aluminium partitioning and behaviour in acidified lower Richmond River catchment waterways. *Australian Geol. Surv. Org. J.* 17: 175–191.

Ferguson, A. J. P., B. D. Eyre, and J. Gay. 2004. Nutrient cycling in the sub-tropical Brunswick Estuary, Australia. *Estuaries* 27: 1–17.

Glud, R. N., B. D. Eyre, and N. Patten. 2008. Biogeochemical responses to coral mass-spawning on the Great Barrier Reef: Effects on respiration and primary production. *Limnol. Oceanogr*. 53: 1014–1024.

Harris, P. T. and A. D. Heap. 2003. Environmental management of clastic coastal depositional environments: Inferences from an Australian geomorphic database. *Ocean Coastal Manage*. 46: 457–478.

Henry, G. and J. Lyle. 2003. *The National Recreational and Indigenous Fishing Survey.* Fisheries Research and Development Corporation, Natural Heritage Trust and NSW Fisheries, Canberra, Australia.

Hobday, A. J., T. A. Okey, E. S. Poloczanska, T. J. Kunz, and A. J. Richardson. 2006. *Impacts of climate change on Australian marine life*. CSIRO Marine and Atmosphere Research Report to the Australian Greenhouse Office.

Hossain, S., B. D. Eyre, and D. McConchie. 2002. Spatial and temporal variations of suspended sediment responses from the sub-tropical Richmond River catchment, NSW, Australia. *Australian J. Soil Res.* 40: 419–432.

IPCC. 2000. *Emissions scenarios. Special Report of the Intergovernmental Panel on Climate Change*. Cambridge: Cambridge University Press.

IPCC. 2007. *Contributions of Working Group I to the Fourth Assessment Report*. The Physical Science Basis, Cambridge: Cambridge University Press.

Jenkins, G. P., H. M. A. May, M. J. Wheatley, and M. G. Holloway. 1997. Comparison of fish assemblages associated with seagrass and adjacent unvegetated habitats of Port Phillip Bay and Corner Inlet, Victoria, Australia, with emphasis on commercial species. *Estuar. Coastal Shelf Sci.* 44: 569–588.

Kjerfve, B. 1994. Coastal lagoons. Pp. 1–8, In: B. Kjerfve (ed.), *Coastal lagoons processes*. Amsterdam, The Netherlands: Elsevier.

Knoppers, B. 1994. Aquatic primary production in coastal lagoons. Pp. 243–286, In: B. Kjerfve (ed.), *Coastal lagoon processes*. Amsterdam, the Netherlands: Elsevier.

Kone, Y. J. M., G. Abril, K. N. Kouadio, B. Delille, and A. V. Borges. 2009. Seasonal variability of carbon dioxide in the rivers and lagoons of Ivory Coast (West Africa). *Estuaries Coasts* 32: 246–260.

Maher, D. 2010. Sources of organic matter that fuel fisheries production in New South Wales estuaries. Ph.D. Thesis, Southern Cross University, Lismore, Australia, 60 pp.

Maher, D., B. D. Eyre, and P. Squire. 2007. *Benthic habitat mapping, primary productivity measurements and macrofauna surveys in the Camden Haven and Hastings River Estuaries.* Report to Port Macquarie Hastings Council, Centre for Coastal Biogeochemistry, Lismore, Australia.

McGlathery, K. J., K. Sundback, and I. C. Anderson. 2007. Eutrophication in shallow coastal bays and lagoons: The role of plants in the coastal filter. *Mar. Ecol. Prog. Ser.* 348: 1–18.

McInnes, K., D. J. Abbs, S. O'Farrell, I. Macdam, J. O'Grady, and R. Ranasinghe. 2007. *Projected changes in climate forcing for coastal erosion in NSW.* CSIRO Report prepared for NSW Department of Environment and Climate Change, Canberra, Australia.

McKee, L., B. D. Eyre, and S. Hossain. 2000. Transport and retention of nitrogen and phosphorus in the subtropical Richmond River Estuary. *Biogeochemistry* 50: 241–278.

Mooij, W. M., L. N. D. S. Domis, and S. Hulsmann. 2008. The impact of climate warming on water temperature, timing of hatching and young-of-the-year growth of fish in shallow lakes in the Netherlands. *J. Sea Res.* 60: 32–43.

NSW Fisheries Statistics 2006/2007. Unpublished data. NSW Fisheries, Cronulla, Australia.

Oviatt, C. A. 2004. The changing ecology of temperate coastal waters during a warming trend. *Estuaries* 27: 895–904.

Papadimitriou, S., H. Kennedy, D. P. Kennedy, C. M. Duarte, and N. Marba. 2005. Sources of organic matter in seagrass-colonized sediments: A stable isotope study of the silt and clay fraction from Posidonia oceanica meadows in the western Mediterranean. *Org. Geochem.* 36: 949–961.

Pollard, D. A. 1984. A review of ecological studies on seagrass-fish communities, with particular reference to recent studies in Australia. *Aquat. Botany* 18: 3–42.

Roach, A. C. 1997. The effect of acid water inflow on estuarine benthic and fish communities in the Richmond River, NSW, Australia. *Australian J. Ecotoxicol.* 3: 25–36.

Roy, P. S. 1984. New South Wales estuaries: Their origin and evolution. Pp. 99–121, In: B. G. Thom (ed.), *Coastal geomorphology in Australia.* New York: Academic Press.

Roy, P. S., R. J. Williams, A. R. Jones, I. Yassini, P. J. Gibbs, B. Coates, R. J. West, P. R. Scanes, J. P. Hudson, and S. Nichol. 2001. Structure and function of south-east Australian estuaries. *Estuar Coastal Shelf Sci.* 53: 351–384.

Saintilan, N. 2004. Relationships between estuarine geomorphology, wetland extent and fish landings in New South Wales estuaries. *Estuar. Coastal Shelf Sci.* 61: 591–601.

Sammut, J., I. White, and M. D. Melville. 1996. Acidification of an estuarine tributary in eastern Australia due to drainage of acid sulfate soils. *Mar. Freshw. Res.* 47: 669–684.

Santos, R., J. Silva, A. Alexandre, N. Navarro, C. Barron, and C. M. Duarte. 2004. Ecosystem metabolism and carbon fluxes of a tidally-dominated coastal lagoon. *Estuaries* 27: 977–985.

Somero, G. N. 2002. Thermal physiology of intertidal animals: Optima, limits, and adaptive plasticity. *Integ. Comp. Biol.* 42: 780–789.

19 Response of the Venice Lagoon Ecosystem to Natural and Anthropogenic Pressures over the Last 50 Years

Cosimo Solidoro, Vinko Bandelj, Fabrizio Aubry Bernardi, Elisa Camatti, Stefano Ciavatta, Gianpiero Cossarini, Chiara Facca, Piero Franzoi, Simone Libralato, Donata Melaku Canu, Roberto Pastres, Fabio Pranovi, Sasa Raicevich, Giorgio Socal, Adriano Sfriso, Marco Sigovini, Davide Tagliapietra, and Patrizia Torricelli*

CONTENTS

Abstract 484
19.1 Introduction 484
19.2 Historical Perspective 484
19.3 Environmental Drivers and Pressures 486
19.3.1 Subsidence and Eustatism 488
19.3.2 Freshwater Discharges 489
19.3.3 Nutrient Loads 489
19.3.4 Pollutants 489
19.3.5 Fishing Activities 490
19.3.6 Climate Change 490
19.3.7 Lagoon Governance 490
19.4 Lagoon State and Impacts 491
19.4.1 General Circulation and Confinement 491
19.4.2 Bathymetry 491
19.4.3 Dissolved Inorganic Nutrients 493
19.4.4 Sediment Nutrients 494
19.4.5 Chlorophyll a and Phytoplankton 494
19.4.6 Zooplankton 496
19.4.7 Macrophytobenthos 497
19.4.8 Macrozoobenthos 498
19.4.9 Nekton 500
19.4.10 Landings Data 500
19.5 Ecosystem Responses 501

* Corresponding author. E-mail: csolidoro@inogs.it

19.6 Concluding Remarks 505
Acknowledgments 505
References 505

ABSTRACT

Analysis of the Venice Lagoon responses to changing morphological features, freshwater discharges, nutrient loads, fishing activities, and other factors shows that this shallow coastal water body is a complex, heterogeneous, and continuously evolving dynamic system, sensitive to an array of external drivers and pressures. Both natural and anthropogenic stressors significantly affect the lagoon system. To successfully assess the impacts of anthropogenic activities on the lagoon, it is important to understand the extent of natural variability and spatial heterogeneity. Investigations of the biotic communities, biogeochemical processes, and bottom habitats demonstrate that external drivers of change can alter the structure and function of the entire ecosystem through direct and cascading effects. Therefore, responsible governance is needed to effectively manage lagoon components, and a successful resource management of the lagoon requires an integrated vision and a participatory adaptive approach.

Key Words: Venice Lagoon, human impacts, eutrophication, climate change, benthic community responses, manila clams, *Ulva*, coastal zone management.

19.1 INTRODUCTION

Venice Lagoon is a shallow coastal water body affected by an array of anthropogenic factors. To assess the significance of anthropogenic drivers of change, it is vital to understand the extent of natural variability and spatial heterogeneity of the system. In this chapter, we review studies of the lagoon responses to major external disturbances. Results of these studies indicate the following: (1) the lagoon is a complex, heterogeneous, and dynamic system that evolves continuously in response to modifications from stressors; (2) implementation of management policies and human intervention can alter ecosystem conditions; and (3) direct effects on one or several ecosystem components influence other components, leading to changes in the ecosystem structure and function. We conclude that efficient ecosystem governance is possible and may be necessary, but given the complex nature of how the ecosystem functions, great care is required in planning any intervention. Hence, effective management of the lagoon should be an outcome of an integrated vision and an adaptive approach.

This chapter provides a historical account of the Venice Lagoon showing how human activity has affected the system for centuries. In its present condition, the lagoon is not only the result of "natural" evolution, but also of continuous active management. We focus here on recent and current development and conditions of the system. The lagoon's morphological settings and drivers of change over the last 50 years are examined. Complementary information on the current state and observed changes over time for major ecosystem components is also provided, including details on the spatial variability of these components, which may be related to spatial gradients in pressures, such as distance from rivers or inlets, pollutant point sources, and areas of development. A conceptual scheme of the interactions between forcing factors and ecosystem components explains in part the recent evolution of the lagoon ecosystem.

19.2 HISTORICAL PERSPECTIVE

Venice Lagoon formed ~6000 years ago when sea level rise associated with the Holocene transgression led to the inundation of the present northern Adriatic Sea, creating a system of lagoons that extends along the entire northwestern Adriatic coast. Later, ~2000 years ago, this system was segmented into separate subsystems, including Venice Lagoon (Brambati 1992). Since the first human settlement, the lagoon has been a protective barrier against invaders and an important source of food for the local population in Venice, which grew into one of the most densely populated cities in Europe in the sixteenth century.

Human impacts on the lagoon environment also increased through time via pollution inputs, fishing activities, dredging of channels, and land use changes. However, maintaining the lagoon in a healthy state and preserving the natural resources it provides has always been an important concern of the Venetian Republic. For example, fishing was strictly controlled in 1173 by the Magistratura della Giustizia (Granzotto et al. 2001), which forbade fishing activities considered damaging to the natural resources; enforced regulations on fishing gear and net sizes; restricted the sale of fish solely to fish markets to prevent escalating prices and the sale of unhealthy seafood products (Bevilacqua 1998). Industrial activities (glass, leather, fur, dye works, etc.) were regulated as of 1465, when Provveditori alla Sanità (health superintendents) enforced strict rules against air and water pollution, with relocations from the town center to islands or the mainland, and to preclude the risk of fires (Ciriacono 1995). Starting from the late fifteenth century, senators were appointed to control the misuse of the lagoon. In 1501, a special water authority called the Magistrato alle Acque was set up, with the specific task of governance of the lagoon ecosystem, especially its morphology (Bevilacqua 1998).

Venice Lagoon is a system that has undergone constant natural and anthropogenic change, and extensive efforts have been made to stabilize and preserve its morphological and ecological features. When the Venetians settled in the area, the natural tendency was for sediment infilling of the lagoon, which was enhanced by deforestation of the mainland (Ravera 2000). To slow down this trend, major rivers (e.g., the Brenta and the Sile) were diverted from the lagoon between the thirteenth and the sixteenth centuries (Figure 19.1, sites 1 to 4). Concurrently, the number of inlets was reduced (Favero et al. 1988), and the sand bar reinforced. The Venetian Republic's interventions extended even beyond the natural boundaries of the Venetian Lagoon. For example, at the beginning of the seventeenth century (1604), flow of the Po River was diverted southward through an artificial delta mouth to prevent sediment infilling of areas close to the Venetian Lagoon (Stefani and Vincenti 2005). A number of studies and monitoring programs were conducted to assess conditions, and technical alternatives were also considered. Moreover, practical solutions were provisional and had to be tested on site. A further sign of the great attention the Venetian Republic devoted to planning is that fishermen and members of the general public were invited by law to monthly conferences on the lagoon status, since these people were recognized for their valuable experiences, insights, and views on the local environment (Caniato 2005).

In the eighteenth century, the combined effects of coastal subsidence and eustatic rise in sea level increased the flooding frequency in the lagoon, and the need to protect the city of Venice from the invading sea. Sea defenses (Murazzi) were constructed along the coastal strip at the end of the century (Figure 19.1, site 5) (Campostrini 2004). Under Austrian rule and until 1934, the shape of the Lido, Malamocco, and Chioggia inlets was altered, and outer dikes were built along the sea (Figure 19.1, sites 6 through 8).

The onset of industrialization between the end of the nineteenth and the beginning of the twentieth centuries marked a new era of major anthropogenic changes of the ecosystem. Increased urbanization and land reclamation for agriculture, aquaculture, and industry reduced the total surface of the lagoon by ~3280 ha during the period from 1924 to 1960 (Ravera 2000). A major industrial area, including chemical and oil industries, was established (Figure 19.1, sites 11 through 13), leading to the dredging of new, deep navigation channels (Figure 19.1, sites 9 and 10). The industrial zone expanded over reclaimed areas (the second industrial area; Figure 19.1, site 12). Due to growing environmental concerns, a third industrial area was never built, but a large area of the lagoon had already been converted to solid land (Figure 19.1, site 13). Throughout the twentieth century, groundwater withdrawals for industrial purposes, natural subsidence, and sea level rise led to the lowering of Venice and part of its lagoon (Carbognin et al. 2004), which further increased the

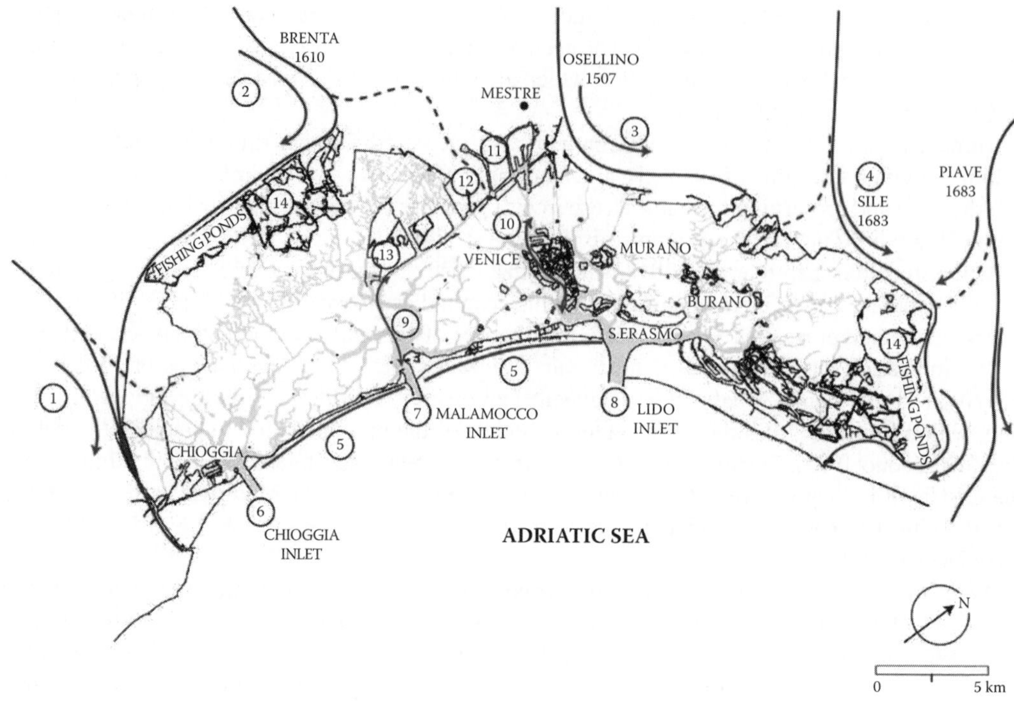

FIGURE 19.1 Schematic description of the Lagoon of Venice. Light gray indicates the network of channels, rooted at the three inlets that connect the lagoon to the sea. Areas close to water exchanges and devoted to aquaculture (fishing ponds) are in white. Dashed lines indicate original position of rivers, continuous lines the present position, after the diversion operated in the year indicated by the river name. Circles and dotted arrows indicate major anthropogenic intervention: 1 to 4, diversion of major rivers, fourteenth to seventeenth centuries; 5, fortification of coastal strip, eighteenth century; 6 to 8, modification and fortification of inlets, first half of the twentieth century; 9, navigation channel, 1960 to 1969; 10, navigation channel, 1910 to 1930; 11 to 13, land reclamation for setting up first (1917), second (1950), and third (1963) industrial areas; 14, water reclamation for aquaculture activity (vallicoltura). (Adapted from material in Quaderni Trimestrali Consorzio Venezia Nuova, www.salve.it webpage, and *Con l'acqua e contro l'acqua* CDROM.)

frequency of flooding. Concurrently, the number of fishermen and people working in the lagoon decreased sharply, and the number of city inhabitants declined to the present (~70,000) due to several causes, such as the cost of living, cost of housing, and difficulty in finding work.

19.3 ENVIRONMENTAL DRIVERS AND PRESSURES

Venice Lagoon is the largest lagoonal system in Italy, and one of the largest in the Mediterranean Sea. It surrounds the city of Venice, with ~70,000 inhabitants and an additional ~10,000,000 visitors per year. Chioggia, with ~50,000 inhabitants, and Mestre, with ~200,000 inhabitants, are also along the lagoon boundary. This is one of the most important industrial areas of Italy. Many different stressors act on the lagoon ecosystem, causing multiple environmental impacts. Among the major drivers of change there are land-based activities that deliver nutrients, heavy metals, and other pollutants; fishing (in particular clam harvesting) and aquaculture activities; other factors (e.g., groundwater extraction, subsidence, eustatism, and transportation) that affect physical and morphological features; forcing driving exchanges with the sea; and climatic conditions (Figure 19.2).

With regard to atmospheric forcing, the dominant winds are the Sirocco (from the southeast), and the Bora (from the northeast). The Bora dominates in fall and winter (Poulain and Raicich

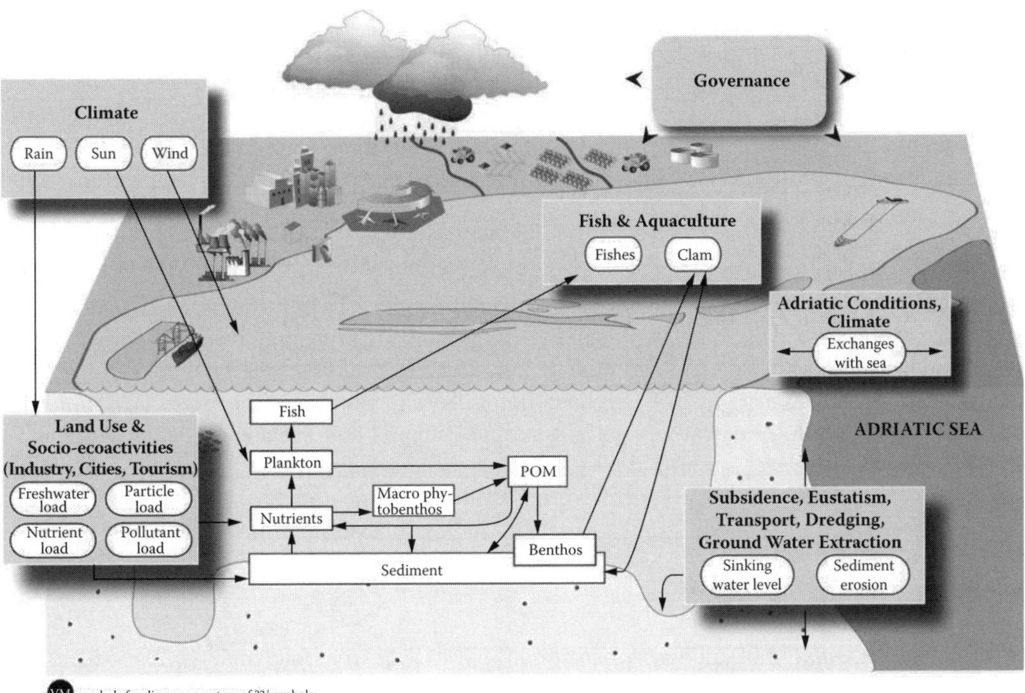

FIGURE 19.2 Conceptual scheme of major *Drivers* of changes (gray shaded boxes), related *Pressures* (white boxes within *Driver* boxes) currently acting on the lagoon ecosystem and of main relationships among lagoon ecosystem components.

2001), with speeds of over 10 m s^{-1}, though northern and northwestern winds are also observed with intensities <5 m s^{-1}. Bora events tend to decrease in spring and summer, when the Sirocco becomes more frequent, blowing up to 10 m s^{-1}. Flooding events, known worldwide as *acqua alta* (high water phenomena), mainly result from a combination of tide, seiches, and easterly winds. Spring and fall are the seasons with the most rainfall, and June, September, and October the months with the least rainfall. Winter is the driest season. Water temperatures closely follow air temperatures, with a distinct seasonal cycle; minimal values occur in January, and maximal values in July. Average monthly temperatures generally range from 3°C to 24°C, but can reach 30°C and fall to 0°C (Figure 19.3).

The lagoon covers an area of ~550 km^2, but ~150 km^2 are used as extensive aquaculture farms and are closed to natural water exchanges. Islands cover ~45 km^2 (Figure 19.1). A network of channels (65 km^2) rooted at the three inlets facilitates navigation and water exchange with the ocean. Islands, wetlands (generally located just above the mean sea level), and tidal flats (that have a lower elevation and are frequently exposed to air during low tides) are connected by channels. The average depth is only ~1 m so there is a tight coupling between pelagic and benthic environments. Decomposition processes are substantial, and biogeochemical processes accelerated.

The lagoon is characterized by a semidiurnal tidal regime with a range of about ±0.7 m. Tidal exchanges with the sea may reach 8000 m^3 s^{-1} and typically amounts to about one-third of the total lagoon volume per tidal cycle (Gačić et al. 2004). Adriatic waters entering the lagoon are typically oligotrophic, or at most mesotrophic (Bernardi-Aubry et al. 2004; Solidoro et al. 2009). The Po Delta Plume, which sustains eutrophic conditions in the Adriatic, generally flows toward the center of the Adriatic Basin or veers southward, and only marginally affects the coastal area close to Venice Lagoon.

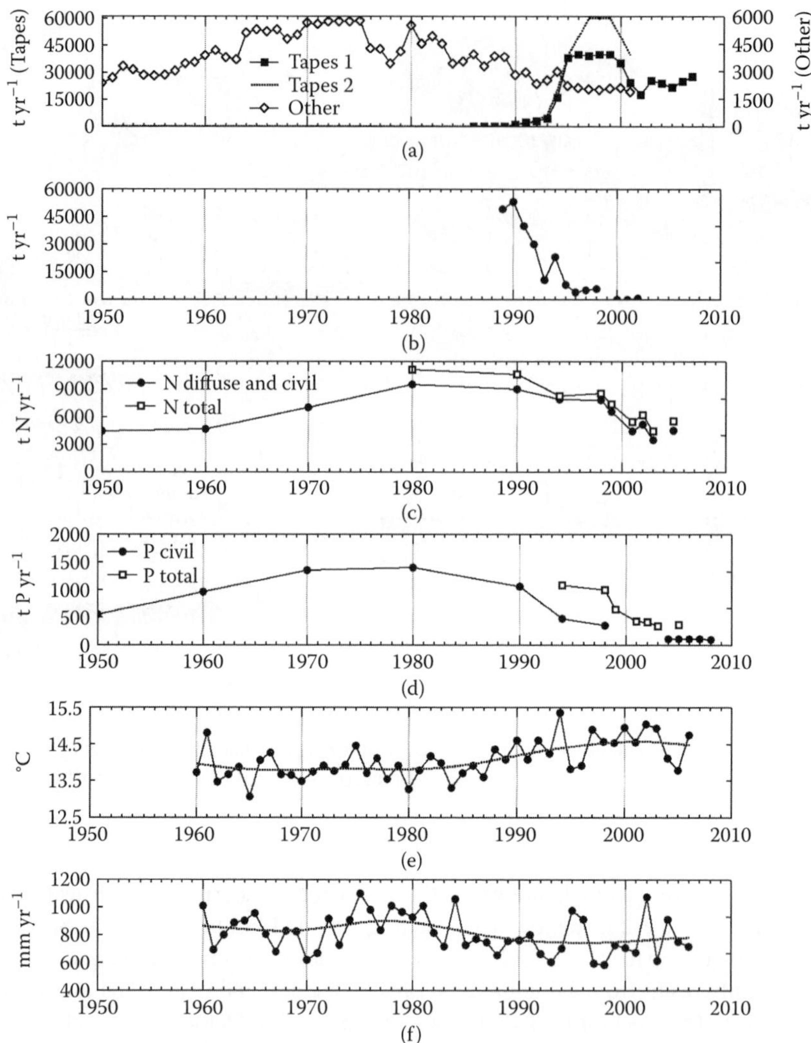

FIGURE 19.3 Multiyear time course of ecosystem drivers. (a) Two estimates of *Ruditapes philippinarum* (*R. ph.*) landing, first axis, label Tapes, and of total landing without *R. ph.* (second axis). (b) Harvested macroalgae biomass. (c) and (d) Time course of nitrogen and phosphorus loads. (e) and (f) The time evolution (and least square interpolation) of monthly mean temperature and yearly rain (elaboration of database at www.istitutoveneto.it).

19.3.1 Subsidence and Eustatism

Lowering the level of the lagoon with respect to the sea increased the frequency of flooding events (Carbognin et al. 2004) and facilitated sediment erosion (Fagherazzi et al. 2006), the loss of finer sediment (Sarretta et al. in press), and the influence of marine processes (Fagherazzi et al. 2006; Sarretta et al. in press). As a result of the diversion of major rivers from the lagoon, riverine input of particulate matter has not balanced the sinking of the lagoon, which was higher from 1930 to 1970 when groundwater and natural gas were extracted from the area. Combined with a continuous eustatic rise in sea level, subsidence has further increased relative sea level rise by ~1.5 mm yr^{-1} between 1972 and 2002 (Carbognin et al. 2004).

19.3.2 Freshwater Discharges

The lagoon receives freshwater discharges from 11 major tributaries plus several minor rivers and a number of man-regulated channels used primarily for agriculture. Freshwater discharges derive partly from rain and snow melting in the drainage basin and partly from anthropogenic control associated with agricultural use. Maximal discharges occur from October to April and are larger in the northern part of the lagoon than in the central and southern sectors. The yearly average freshwater discharge is ~40 $m^3 s^{-1}$, although pulses can be as high as 340 $m^3 s^{-1}$, with daily values >200 $m^3 s^{-1}$ (Collavini et al. 2005). The drainage basin covers 2000 km^2, mainly in the Veneto region, one of the most intensively cultivated areas in Italy; hence, rivers carry a substantial nutrient load, today roughly equivalent to 4000 t N yr^{-1} and 200 t P yr^{-1} (Cossarini et al. 2009).

19.3.3 Nutrient Loads

Nutrient loads to the lagoon are generated mainly by agriculture, industrial activity, and sewage inputs. They have a positive effect on fishing and biodiversity when delivered in moderate amounts, insofar as they sustain primary production and food web processes. However, an excess of nutrients may also fuel eutrophication, which can trigger secondary effects such as hypoxia, mortality of benthic organisms, and the loss of biodiversity (Cloern 2001). Trends of nutrient loads were reconstructed by merging databases. The time series of diffuse and civil N loads and civil P loads since 1950 shown in Figures 19.3c and 19.3d were obtained by integrating the following estimates: (1) the load estimates by Bendoricchio (1998), based on population trends and agricultural/zootechnic practices (years 1950 to 1998); (2) experimental data on river inputs in 1999 (project DRAIN; Collavini et al. 2005) and in 2001 to 2005 (Arpav 2007); (3) measurements from wastewater treatment plants in the years 2001 to 2005 (Vesta 2005a, 2005b); and (4) values from Venice City (Castellani et al. 2005). By also accounting for the industrial load estimates of Bendoricchio (1998) on N in 1980 to 1998 and P in 1994 to 1998 and the measurements at Porto Marghera by Magistrato alle Acque di Venezia in the last decade (MAV 2002), one can approximate the trend of total nutrient loads into the lagoon during the last 30 years (Figures 19.3c and 19.3d). Atmospheric deposition, which was recently estimated at ~1100 t N yr^{-1} and 44 t P yr^{-1} by MAV (2005), is not included in the estimates of total loads. These trends show that the increased use of fertilizers in agriculture and the industrialization of the drainage basin caused a threefold rise in both phosphorus and nitrogen loads from the 1950s to the 1980s. The construction of wastewater treatment plants and the total ban of phosphorus from detergents in 1989 led to a marked reversal in the trend of phosphorus load, which has been reduced to less than half of the amount recorded in 1950. Nitrogen loads have decreased only since the 1990s, and current loads are now comparable to those of the mid-1960s (Figure 19.3), due to the introduction of nitrification/denitrification processes at the major treatment plants, to the closure of fertilizer factories, and to improved fertilization practices in agriculture (Regione Veneto 2000).

19.3.4 Pollutants

Pollutants from industrial waste and other sources (e.g., drainage basins, urban areas, waste incineration, and boat engines) have been discharged to the lagoon, with a peak input during the 1960s and 1970s (Pavoni et al. 1992; Dalla Valle et al. 2005; Frignani et al. 2005), and have been accumulating in the lagoon sediments. Up to a few years ago, attention was mainly focused on heavy metals (Hg, Pb, As, Cd, Zn, and Ni), but more recently great concern has been directed toward all classes of persistent organic pollutants (POPs), such as dioxins, polychlorinated biphenyls (PCBs), hexachlorobenzene (HCB), and polycyclic aromatic hydrocarbons (PAHs). Industries are the major sources of these pollutants, which are delivered to the lagoon through point and nonpoint pollution pathways, as well as atmospheric fallout (Guerzoni et al. 2007).

19.3.5 Fishing Activities

Fishing can have significant effects on ecosystem structure, by inducing top-down controls that change communities (Pauly et al. 1998), decreasing biomass of target species, and increasing mortality of nontarget species (by-catch). Venice Lagoon has long been exploited by small-scale fishing activities using traditional methods, while semi-industrial exploitation developed in the early 1990s targeting the Manila clam (*Ruditapes philippinarum*), an allochthonous species introduced to increase lagoon production (Solidoro et al. 2000; Granzotto et al. 2004). An analysis of landings recorded over the past 50 years (Libralato et al. 2004) indicates that the total amount of catches, clams excluded, was ~2000 t yr^{-1} after World War II, which steadily increased to 5700 t yr^{-1} by the first half of the 1970s. These statistical trends illustrate the increase in catches due to the effectiveness of new, more powerful engines and fishing devices. Successive landings decreased to less than 2000 t yr^{-1} by 2001 (Figure 19.3). This might be due to a combination of changes in fish abundances and fishing activity, but also because of changes in the resources targeted.

The Manila clam, introduced in 1983, has been fished since the early 1990s. Landings data for this species show that the positive trend peaked in 1999, with a yearly production of ~40,000 t yr^{-1} (and a market price of ~2.5 euro kg^{-1}). However, in the following 5 years, landings halved, falling in 2003 to values as low as 18,000 t (Figure 19.3a; Melaku Canu et al. in press), possibly reflecting overexploitation of the resource. Fishing vessels targeting the Manila clam use heavy-impact fishing gear (Pranovi et al. 2003), and this activity is considered to be a major cause of ecological imbalance in the lagoon, since it strongly interacts with the benthic compartment (Pranovi et al. 2003; Sfriso et al. 2003b), also contributing to the process of sediment erosion and loss from the lagoon. To allay social, economic, and ecological concerns related to open-access clam fishing (Melaku Canu et al. In press), local administrators allocated ~3500 ha of the lagoon in 2005 to clam farming. However, unregulated fishing still occurs.

Aquaculture has been carried out in the Venice Lagoon for many centuries in restricted areas (mainly fishing ponds termed *valli da pesca*). This is an extensive practice that has had little impact on system trophodynamics. Semi-intensive activities began during the last century, but it was only recently, after the introduction of the Manila clam, that aquaculture became a relevant environmental issue.

19.3.6 Climate Change

Climate change is considered a major threat to the survival of Venice Lagoon mainly because of the projected increase in sea level. However, other changes can be foreseen, the most significant being the maximal warming in the summer (up to 5°C at the end of this century under the A2 IPCC scenario), increased precipitation in the fall and winter, and decreased precipitation in the spring and summer (Salon et al. 2008), which could modify system trophodynamics (Cossarini et al., 2008). Indeed, an increase in temperature and a decrease in annual rainfall have been recorded over the last 50 years. Other direct and indirect impacts due to climate change can be hypothesized, including modification of habitats and species distribution.

19.3.7 Lagoon Governance

Lagoon governance is an additional driver of change. For example, the issuance of regulations on land use or fishing has had far reaching effects on biota and habitats in the lagoon. The effects are mediated by both bottom-up and top-down processes. The harvesting of macroalgae standing crop during the 1980s is another example (Figure 19.3).

19.4 LAGOON STATE AND IMPACTS

19.4.1 General Circulation and Confinement

Water circulation in the lagoon is mainly tide driven (Gačić et al. 2004), but the actual pattern of residual transport (i.e., the actual motion after averaging over the tidal cycle) results from non-linear interaction of wind, basin morphology, inertial motion, and preexisting water levels inside the lagoon (Umgiesser et al. 2004). The typical residual circulation pattern consists of a large clockwise eddy around the Venice Islands and a net southward flux of water, and it indicates that the large flow of freshwater into the northern lagoon basin is balanced mainly by the outflow through the southern inlet (Solidoro et al. 2004a). Residence time varies over the lagoon and with meteorological conditions, but average values are on the order of a few days in areas close to the inlets and up to 40 days in the inner, more confined areas (Cucco et al. 2009). However, this pattern is significantly altered by the wind, which modifies circulation and increases both the water flux at the inlets and inside the lagoon (Solidoro et al. 2004a). Indeed, residence time varies significantly with wind velocity and direction. The spatial distribution of residence time and root mean square velocity in an idealized situation, computed using a finite element model specifically calibrated for Venice Lagoon (Umgiesser et al. 2004), are reported in Figure 19.4. Here, we define residence time as confinement *sensu* Guelorget and Perthuisot (1992), that is, the time required for each area of the lagoon to replace a given fraction of its volume with seawater, since this is a transport time scale indicator particularly suitable for lagoon systems.

A comparison of the general patterns of the spatial distribution of hydrodynamic properties obtained by considering real forcings (i.e., wind, tide, and water level at the inlets) supports the subdivision of the lagoon into at least three longitudinal sectors, whereas the comparison of the relative influence of the southern inlet, central inlets, and the two major channels originating in the northern inlet suggests a subdivision of the lagoon into four sub-basins. The combination of the two subdivisions yields the subdivision based on physical properties proposed by Solidoro et al. (2004a). Physical properties are also among the parameters considered by Guerzoni and Tagliapietra (2006) and Tagliapietra et al. (2009) in their subdivision of the lagoon.

19.4.2 Bathymetry

In the last decades, the morphological characteristics of the lagoon have changed considerably. Overall, we have observed significant erosion of sediment (Degetto and Cantaluppi 2004; Sarretta et al. in press) and a change in granulometric composition (Molinaroli et al. 2007; Sarretta et al. in press), with the loss of finer fractions. Sediment resuspension is caused by several factors, including the dredging of new large channels (Sarretta et al. in press), which altered the existing equilibrium between the channels and tidal flats, with consequent erosion of the shallow areas and infilling of the channels (Fagherazzi et al. 2006; Defina et al. 2007); an increase in the number and speed of boats, with consequent increase in waves and bottom stress (Sarretta et al. in press); a reduction in seagrass, which stabilized the sediment (Sfriso et al. 2005a,b; Marani et al. 2007); and widespread use of mechanical fishing devices for clam harvesting, which destabilized the substrate (Pranovi et al. 2004; Sfriso et al. 2005a,b). Once suspended, the finer fraction of sediment can be exported out of the lagoon (Molinaroli et al. 2007).

A comparison of the 1927, 1970, and 2002 bathymetric surveys shows that, on average, the lagoon is now ~20 cm deeper than during the 1970s and almost 90 cm deeper than during the 1930s. In addition, erosion has been more marked in the tidal flats and central basins (Sarretta et al. in press). These morphological changes are the results of sea level rise, natural and human-induced subsidence (11, 3, and 9 cm, respectively, according to Gatto and Carbogning 1981), land reclamation, dredging

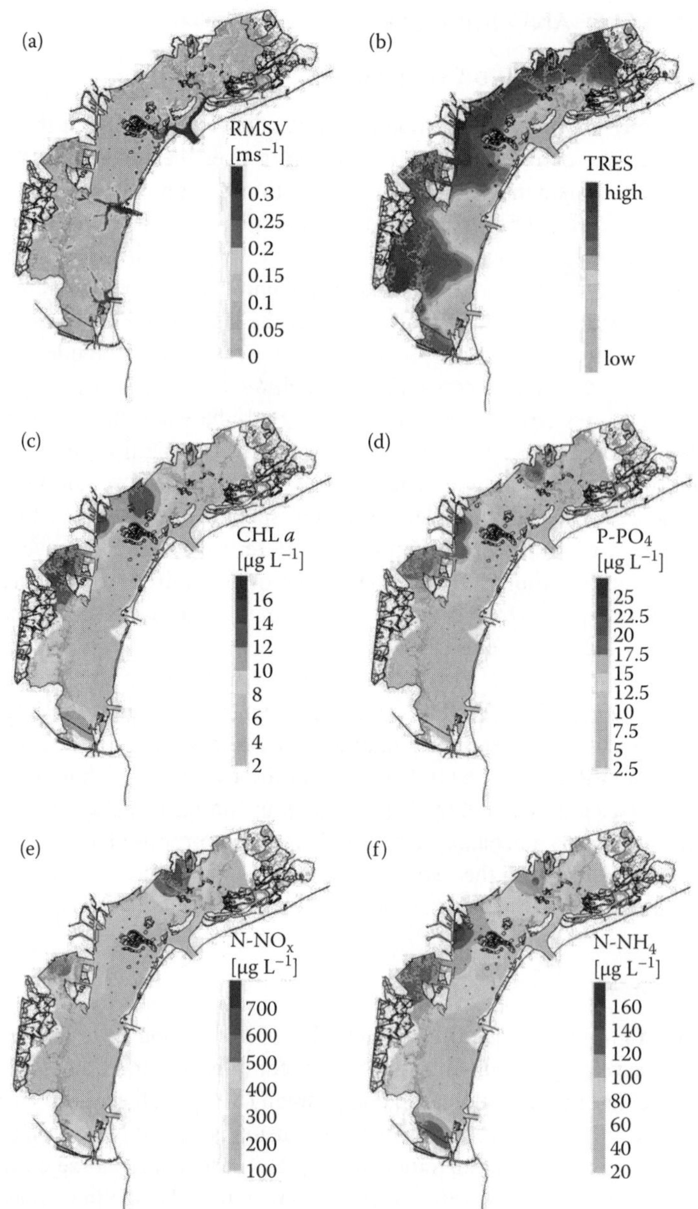

FIGURE 19.4 Spatial distribution of root mean square velocity (a), residence time (b), chlorophyll *a* (c), phosphate (d), ammonia (e), and nitrate (f). Maps (a) and (b) are obtained by numerical model simulations, (c) to (f) by spatial interpolation of experimental observation recorded in the period 2001 to 2005.

of large navigation channels including the one between the industrial area and the Malamocco inlets (site 12 in Figure 19.1) (and consequent modifications in hydrodynamic properties), and by the erosion induced by the Manila clam fishery (Sarretta et al. in press). In this regard, the increase of water level, as well as the increase of waves induced by winds and fetch, can alter the equilibrium between sediment deposition and resuspension, and promote erosion of shallow waters (Fagherazzi et al. 2006, 2007). There has been a large increase in the area covered by subtidal flats (depth between 0.75 and 2 m) from ~90 km^2 in 1927 to ~200 km^2 today. There is also a significant reduction in salt marshes, and a net loss of sediment amounting to ~0.8 mm^3 yr^{-1} (period 1970 to 2002; Sarretta et al.

in press). The emerging picture suggests the transformation from a morphologically complex and well-developed microtidal lagoon (1927) to a simpler, deeper, flatter-bottomed, and more bay-like system (Sarretta et al. in press).

19.4.3 Dissolved Inorganic Nutrients

The spatial and temporal distribution of nutrients in the lagoon results from the superimposition of nutrients derived from point and diffuse sources on land, biogeochemical transformation, and exchange with the sea. Significant interannual variation exists, but long-term trends can be deduced by comparing recent observations with experimental findings obtained for a specific area (Pastres et al. 2004). Studies close to Porto Marghera, for example, illustrate how the decrease in nutrient load to the lagoon correlated with a significant decrease in nutrient concentrations (Ciavatta 2000). Datasets obtained in five different projects (Figure 19.5) indicate that over the last 40 years ammonia decreased exponentially from 5000 to 50 μg L^{-1}, while phosphorus concentrations declined from 200 to 2 μg L^{-1}. A sharp drop in phosphorus concentrations occurred at the end of the 1980s, when phosphorus was banned in detergents. Nitrogen was relatively stable between 1965 and 1995, but also decreased during the last decade. Athough the decrease in nitrogen concentrations was less dramatic than those observed for other nutrients, the present nitrogen concentrations are now half of those observed around 2000.

The signal appears to be less clear in areas of the lagoon where pollutant point source inputs are less significant. A comparison of study periods in the late 1990s and late 1970s by Acri et al. (2004) showed a less sharp decline in phosphorus concentrations between the two periods and an increase in dissolved inorganic nitrogen in the central and north sub-basins. A statistically sophisticated analysis by Ciavatta et al. (2009) on a fairly homogeneous dataset covering different areas of the central basin for the period 1986 to 2006 revealed a more acute decrease in nutrient concentrations

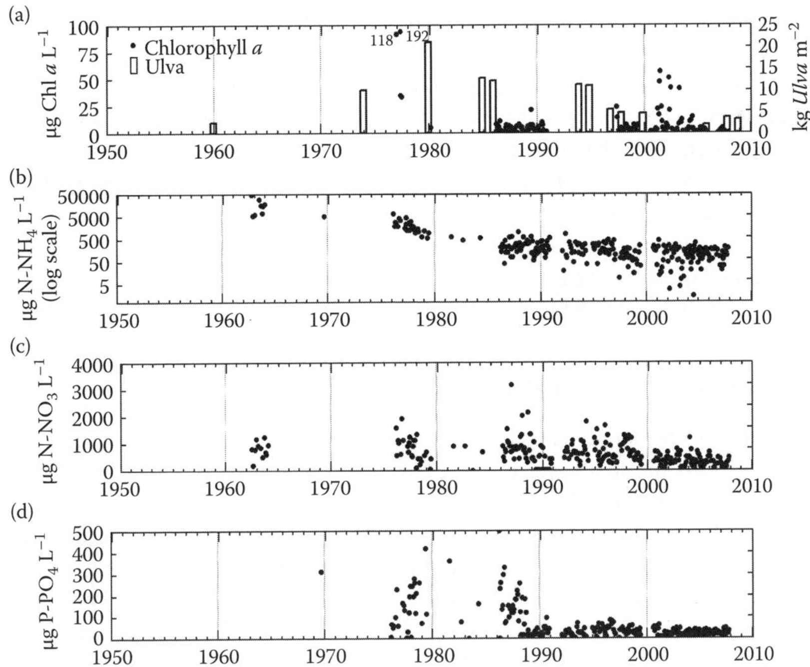

FIGURE 19.5 Multiyear time course of selected ecosystem responses. (a) Chlorophyll *a* (dots, first axis) and *Ulva* density (bar, second axis) over last 50 years. (b), (c), and (d) Evolution of concentration of ammonia (log scale), nitrate, and phosphorus, respectively.

in areas close to the nutrient point source inputs. This pattern is not as evident in areas closer to the inlets where the nutrient concentration is always lower than in the inner lagoon.

The recent nutrient status can be assessed by analysis of data collected during a 5-year-long monitoring program over 30 sampling stations during 2001 to 2003, and 20 stations during 2004 to 2005 (Solidoro et al. 2009). Figure 19.4 illustrates the spatial distribution of salinity and concentration of ammonia, nitrate, and phosphate. Maps were generated by smoothing interpolation (inverse distance to a power, quadratic law for weight computation and 0.2 smoothing parameter) of the 2001 to 2005 average values of the 20 sampling stations sampled in all periods. A decomposition of space time variability through empirical orthogonal functions shows that most of the space-time variability can be captured by a seasonal modulation of a clear spatial pattern, with a gradient of nutrients from the inner part (closer to rivers and the industrial area) toward the inlets. Indirect gradient analysis demonstrates that salinity and residence time play a major role in defining this spatial pattern (Solidoro et al. 2004b; Solidoro et al. 2009). Indeed, river signatures are clearly recognizable in nitrate and salinity distributions, indicating that tributaries represent the major sources, while industrial loads are more easily tracked from the spatial distribution of ammonia (Figure 19.4). The northern basin and the part of the southern basin close to Chioggia have the highest concentrations, while the largest fraction of the load flows into the northern basin. The concentration of phosphate today is always at a rather low level in the lagoon, possibly limiting primary productivity of the system in summer.

19.4.4 Sediment Nutrients

Today, nutrient concentrations in Venice Lagoon are significantly lower than in the past and are continuously decreasing not only in the water column, but also in bottom sediments. By analyzing the nutrient concentrations of 34 sampling sites of the central lagoon, it was observed that between 1987 and 2003, total phosphorus (TP) and total nitrogen (TN) in the top 5 cm of bottom sediments decreased from 386 ± 96 to 241 ± 85 μg cm^{-3} and from 1.21 ± 9.60 to 0.71 ± 0.27 mg cm^{-3}, respectively (Table 19.1). The concentrations of organic phosphorus (OP) in the bottom sediments were even greater (Sfriso et al. 2003b), decreasing from 104 ± 42 to 36 ± 25 μg cm^{-3}.

Anoxia events, a secondary effect of massive macroalgae blooms frequently observed during the 1980s over a large area of the lagoon, are no longer recorded. In addition, oxic conditions occur in the upper sediments (Sfriso and Facca 2007). Furthermore, continuous mixing of the upper sediments due to clam fishing activity likely promotes chemical oxidation of reduced compounds otherwise trapped in the sediments. An increase in resuspended bottom sediments is also evident based on sediment accumulation on sediment traps at rates ranging from 34 to 140 kg dry wt m^{-2} yr^{-1} in the 1980s to 300 to 1400 kg dry wt m^{-2} yr^{-1} in the early 2000s (Sfriso et al. 2005a, 2005b).

19.4.5 Chlorophyll *a* and Phytoplankton

Chlorophyll *a* (chl *a*), considered a proxy of total phytoplankton biomass, decreased from very high values during the 1970s (up to 190 μg L^{-1} in 1978) down to concentrations <25 μg L^{-1} during the 1980s, and after an increase in 2000 to 2003, even lower values were recorded in subsequent years. Figure 19.4 shows that, in most of the lagoon, average concentrations of chlorophyll *a* have been below 10 μg L^{-1} since 2001. However, concentrations over the last 3 years are even lower, mostly <7 μg L^{-1}, and maximal values are rarely higher than 15 μg L^{-1} (Solidoro et al. 2009). Chlorophyll *a* concentrations follow a seasonal pattern, being highest in summer after a minor spike in spring. The intensity of summer phytoplankton blooms mainly depends on nutrient concentrations, which may be related to river discharges and the amount of rainfall during the previous weeks. Phosphorus concentrations are often very low, possibly limiting during summer in the driest years. Indeed, significant variations can occur in yearly average values of chlorophyll *a*, even if yearly average nutrient concentrations are quite similar (Solidoro et al. 2004b). Nutrients are rapidly assimilated by

TABLE 19.1
Characteristics of Major Plant Subsystems and Sediment Nutrients in the Venice Lagoon during the 1980–2003 Period

	Past	Ulva	Ulva Decline/ Tapes Spreading	Tapes Fishing	Present	
		ha			kt	ha
Seagrasses		**1990**			**2003**	
N. noolti	—	4144	—	—	5	634
C. nodosa	—	1634	—	—	100	2946
Z. marina	—	3643	—	—	90	3443
Macroalgae central basin	**1980**	**1987**	**1993**	**1998**	**2003**	
Ulva SC (kt)	422	560	85	8,7	10,9	
Ulva NPP (kt yr^{-1})	1371	1 500	377	43	62	
Ulva GPP (kt yr^{-1})	8816	10 000	2 000	200	286	
Macroalgae lagoon	**1980**				**2003**	
Ulva SC (kt)	840	—	—	—	89	
Ulva NPP (kt yr^{-1})	2912	—	—	—	472	
Ulva GPP (kt yr^{-1})	18498	—	—	—	2335	
Sediment (top layer 5 cm)		**1987**	**1993**	**1998**	**2003**	
P tot (μg cm^{-3})	—	386	361	375	241	
N tot (μg cm^{-3})	—	1210	1140	930	710	

Note: Column heading "Past" refers to the period prior to *Ulva* massive blooms (before 1985); "Ulva" to the period of massive *Ulva* proliferation (approximately 1985–1990); "Ulva Decline" to the period ~1991–1995; "Tapes Fishing" to the period 1995–2002; and "Present" to the period after 2002.

Sources: Data from Caniglia et al. (1990); Rismondo et al. (2003); Sfriso et al. 2005a,b; Sfriso and Facca 2007; Facca et al. 2008.

phytoplankton in summer and generally enter biogeochemical cycles in the lagoon. In winter, however, when phytoplankton abundance is greatly reduced in the lagoon, far less nutrients enter biotic compartments and are more likely flushed out to sea than during the summer months (Solidoro et al. 2004b, 2005).

In Venice Lagoon, as in other large tidal-dominated coastal systems, the phytoplankton community exhibits high temporal and spatial variability. Therefore, it is difficult to provide an accurate representation of the system and to speculate about the long-term interannual trends of the community. However, general patterns can be delineated.

Voltolina (1975 and references therein) described the phytoplankton community in the lagoon prior to the 1980s. Spatial and temporal patterns of the community were reported for the area between the Burano wetlands and the Malamocco in 1971 and 1972. Phytoplankton abundance was as high as 10^8 cells L^{-1}. Frequent and abundant phytoplankton blooms were recorded, which caused brown discoloration of the water. These blooms were dominated by diatoms (mainly the centric diatom *Skeletonema marinoi,* ex *costatum*) and small green flagellates, whereas other taxonomic groups, such as euglenophyceans (*Eutreptiella*), naked dinoflagellates (*Gymnodinium*), and coccolitophorids (*Calyptrosphaera*) were observed only occasionally. In the following years, plankton abundance was generally lower, with major blooms up to 10^7 cells L^{-1}, but only in limited areas of the lagoon, usually in the inner part, where small phytoflagellates and small diatoms were commonly found. Over a 35-year period (1958 to 1992), phytoplankton primary production varied from 76 mg C m^{-3} hr^{-1} (Vatova 1960) in 1958 to 580 mg C m^{-3} hr^{-1} 20 years later (Battaglia et al. 1983). It declined to 186 mg C m^{-3} hr^{-1} in 1985 (Degobbis et al. 1986) and to 145 mg C m^{-3} hr^{-1} in 1992 (Bianchi et al. 2000). At the time of extensive macroalgal blooms in the 1980s, phytoplankton abundance was high near the industrial area and in deep channels, which were light limiting for macroalgae (Sfriso and Pavoni 1994). Phytoplankton blooms, often dominated by diatoms, were usually observed in the late winter before the increase in *Ulva* (e.g., *Skeletonema marinoi*; Socal et al. 1999), and again in May to July after the decline of *Ulva* (*Cylindrotheca closterium*; Bianchi et al. 2000). Chlorophyll *a* typically reached concentrations of ~20 mg m^{-3}, with highest concentrations exceeding 100 mg m^{-3} (Sfriso et al. 1987, 1992; Socal et al. 1999). Some investigators noted a slight increase in phytoplankton abundance just after the decline of *Ulva* blooms (Acri et al. 2004; Socal et al. 2006), but by 2002 and for the last 10 years, phytoplankton abundances have decreased, especially in shallow or more confined areas.

During the last 10 years, a distinct increase in sediment resuspension and benthic species abundance has also been recorded (Facca et al. 2002; Acri et al. 2004). A reduction in late spring-summer phytoplankton blooms has also been documented. In terms of species composition, it can be assumed that neritic diatoms and several species of large colonial diatom species (e.g., *Chaetoceros* spp.) currently dominate microalgal populations in areas near the inlets, while mixed populations composed of chlorophyceans, euglenophyceans, and brackish water diatoms develop in the inner areas, with *Skeletonema* more abundant along or close to channels and resuspended benthic diatoms and cryptophyceans observed in marshes (Bianchi et al. 2000). The most recent investigation to estimate phytoplankton primary production in Venice Lagoon was carried out at four stations in 2005. The minimal values were recorded in January (<1 mg C m^{-3} hr^{1}), and the maximal values (up to 300 mg C m^{-3} hr^{-1}) in July. An increasing gradient of values was observed moving from the inlet (16 mg C m^{-3} hr^{-1}) toward the industrial area (Pugnetti and Acri 2007).

19.4.6 Zooplankton

The ecological characteristics of the Venice Lagoon favor the occurrence of zooplankton species able to adapt to highly variable environmental conditions. Copepods account for about 80% of the total community (Camatti et al. 2006b), and *Acartia* is the most representative genus. *Cladocera* and other taxa can be found more frequently in the areas close to the inlets. A distinct seasonal pattern exists, roughly mirroring that of the phytoplankton, and there is a summer peak of ~8,000 individuals m^{-3} (Camatti et al. 2006b).

Changes have been observed in the mesozooplankton community structure. From the second half of the 1980s, the abundances of the more representative copepods in the Venice Lagoon, *Acartia margalefi* and *Acartia latisetosa*, decreased while during summer 1992, the copepod *Acartia tonsa*, a nonindigenous calanoid copepod, appeared for the first time in very high abundances (Comaschi et al. 1994; Camatti et al. 2006b). Currently, *A. tonsa* is an abundant or dominant species in regions of high particle concentration. It dominates the inner and middle part of the basin most of the year, occupying the original area of *A. latisetosa* and *A. margalefi* (Bandelj et al. 2008), whereas *A. clausi* is the most important taxon in the areas around the seaward inlets. *Acartia latisetosa*, a species specifically associated with brackish water, is becoming rare. Since the 1980s, the abundance of *A. margalefi* has decreased, and its occurrence in the lagoon is currently not as important as in the past (Camatti et al. 2006b). Currently, *A. tonsa* is the most representative taxon in the Venice mesozooplankton community (Acri et al. 2004; Camatti et al. 2006a), being the dominant species in quantitative terms, and accounting for more than 90% of the total zooplankton (Table 19.2). The increase in *A. tonsa* in Venice Lagoon and its absolute predominance during warm periods seems to

TABLE 19.2
Characteristics of Phytoplankton in the Venice Lagoon during the 1970–2002 Period

	Past	Ulva	Ulva Decline/ Tapes Spreading	Tapes Fishing
Phytoplankton abundance	**1973**		**1993**	**1998–2002**
Total lagoon (10^6 cell L^{-1})	62	—	—	22
Central sub-basin (10^6 cell L^{-1})	—	—	2.6	4.4
Central sub-basin (C μg L^{-1})	—	—	188	40
Zooplankton abundance (individuals m^{-3})	**1970**	**1980**	**1992**	**2000**
A margalefi	850	34	9	27
A clause	978	747	198	113
A. latisetosa	167	61	1	1
A. tonsa	0	0	17250	1445

Note: Column heading "Past" refers to the period prior to *Ulva* massive blooms (before 1985); "Ulva" to the period of massive *Ulva* proliferation (approximately 1985–1990); "Ulva Decline" to the period ~1991–1995; and "Tapes Fishing" to the period 1995–2002.

Sources: Data from Voltolina (1975); Facca et al. (2004a, 2004b); Camatti et al. (2006a); Comaschi et al. (1994); Sfriso and Facca (2007).

have reversed the zooplankton biomass gradient between the lagoon and sea, reflecting the greater zooplankton production in the inner area (Bressan et al. 2004).

19.4.7 Macrophytobenthos

After the 1960s, the lagoon experienced abnormal growth of Ulvaceae, which replaced seagrass populations. There are few indications of major macroalgae blooms during the 1970s, but during the 1980s, massive blooms with densities as high as 20 kg f wt m^{-2} were regularly recorded in a large area of the lagoon, with marked effects on nutrient cycles both in bottom sediments and the water column (Solidoro et al. 1997a, 1997b; Sfriso et al. 2003b). In 1980, the macroalgal standing crop (SC), and the net (NPP) and gross (GPP) production were estimated to be 840 kt f wt yr^{-1}, 2912 kt f wt yr^{-1}, and 18,498, kt f wt yr^{-1}, respectively (Table 19.1). This condition persisted until the early 1990s, then progressively decreased (Figure 19.5), likely because of a combination of factors (Sfriso and Marcomini 1996). In 2003, macroalgae SC, NPP, and GPP decreased to 89 kt f wt yr^{-1}, 472 kt f wt y^{-1}, and 2,335 kt f wt yr^{-1}, respectively. The macroalgae decline was particularly high in the central lagoon, where in 2003 the SC, NPP, and GPP decreased to 2.6%, 4.6%, and 3.4%, respectively, of the values recorded in 1980 (Sfriso and Facca 2007). The decline continued to 2005 (Miotti et al. 2007), then in successive years, some blooms were observed, especially southwest of Venice. However, the high biomasses recorded in the past were never observed again (Sfriso et al. 2007) (Table 19.1).

At present, the main producers of the lagoon are seagrasses. In 2003, *Cymodocea nodosa* accounted for 52% and 51% of the total seagrass SC and NPP, respectively (Sfriso and Facca 2007). This species also exhibited SC and NPP values higher than all the seaweeds combined (Table 19.1). *Zostera marina* had slightly lower values, whereas *Nanozostera noltii* accounted for only 5% and 3% of total SC and NPP values, respectively. The first studies of these seagrass species date back to 1990 when Caniglia et al. (1990) mapped the distribution of the three main species that colonize the lagoon (i.e., *C. nodosa, N. noltii*, and *Z. marina*). Overall, these species covered 5493 ha in 1990 and 5431 ha in 2002 (Rismondo et al. 2003). However, despite the similar coverage, the dominance of the three species changed markedly. In 1990, *N. noltii* dominated the other species, covering 4144 ha (mixed and pure beds), but in 2002 it decreased to 634 ha. In contrast, *C. nodosa* increased in areal distribution from 1634 ha in 1990 to 2946 ha in 2002. The coverage of *Z. marina* was similar both in 1990 (3643 ha) and 2002 (3443 ha). This change in dominance seems to have been caused by several factors, the most important being the improved dissolved oxygen conditions in the lagoon for *C. nodosa*, the most sensitive species (Sfriso et al. 2007), and the reduced habitat suitability and water clarity for *N. noltii* (Sfriso and Facca 2007).

A total of ~300 species of macroalgae have been recorded in the Venice Lagoon (Sfriso and Cavolo 1983; Sfriso and Curiel 2007) and the number of species appears to be increasing, although 87 taxa (54 Rhodophyceae, 25 Phaeophyceae, and 8 Chlorophyceae) have not been found in the lagoon in recent years. Some of the newly recorded species, such as *Sargassum muticum* (Yendo) Fensholt and *Undaria pinnatifida* (Harvey) Suringar, are large species of extra-Mediterranean origin introduced into the lagoon in the early 1990s. They have colonized hard substrates of the lagoon, replacing native species. Originally inhabiting cold temperate waters, these species began to grow in winter, and in spring hard substrates of the city were completely covered by algae thalli (biomass 5 to 15 kg f wt m^{-1}) which hindered boat navigation (*S. muticum* reaches 7 to 8 m). Because these species cannot grow on soft substrates, their total biomass is limited to a few thousand tonnes, and the impact on the ecosystem is much lower than that caused by *Ulva* (e.g., during the 1980s) (Curiel et al. 2010).

19.4.8 Macrozoobenthos

Analysis of available data on the macrobenthic community during three study periods over the last 70 years shows that changes occurred in the community structure between the 1990s and 2004, relative to the 1935 baseline. Species diversity sharply decreased through time, with the lowest values recorded in 1999, as indicated by Shannon index values (Table 19.3). Different trends are apparent in different sub-basins of the lagoon, with the southern sub-basin showing the most conservative pattern. The central sub-basin has exhibited the greatest changes (Pranovi et al. 2008). This pattern can be related to differences both in the structural components of the lagoon, such as the historical presence of consolidated seagrass meadows in the southern part of the water body (Caniglia et al. 1990), and the huge amount of Manila clams and the mechanical harvesting impacts, particularly in the central sub-basin (Pranovi et al. 2006). This is also reflected by an acute change in the functioning of the benthic community (Table 19.3). Data from 1935 indicate a healthier condition, with a well-diversified community, well-assorted trophic structure, and an important contribution of epifaunal species, usually characterized by high mobility. Giordani Soika and Perin (1974) referred to a remarkably impoverished benthos during the late 1950s to 1960s, with a shift toward more tolerant assemblages, a reduction in brackish species, and an increase in marine forms. The end of the 1980s was characterized by a huge biomass of benthic primary producers, dystrophic crises, and a trophic structure dominated by herbivores and detritus feeders, which by 1990 accounted for more than 55% of the total macrobenthic abundance. Then, in 1999, after the extremely successful invasion by the Manila clam, filter feeders (mainly bivalves) became dominant, accounting alone for ~50% of the macrobenthic abundance. During this period, the benthic community exhibited high secondary production, mainly due to the Manila clam (Pranovi et al. 2007, 2008).

In Venice Lagoon, as in most lagoons in the northern Adriatic Sea, distinct geomorphological zones can be recognized, including (1) inner meso-oligohaline areas close to rivers, (2) a middle mud basin characterized by low energy and sedimentation of fine sediments, and (3) marine tidal deltas around the inlets divisible into a seaward and a lagoonal component defined by ebb tidal deltas and flood tidal deltas, respectively (Tagliapietra et al. 2009; Tagliapietra and Minelli 2009).

The middle mud basin is typified by reduced hydrodynamics and more silty sediments roughly subdivided into two main parts (i.e., the shallow open waters of the central basin and the innermost part of the basin near the mainland where environmental conditions such as salinity, temperature, and dissolved oxygen are highly variable).

The near-shore areas and the ebb tidal delta are inhabited by species adapted to high salinity, high hydrodynamics, and sandy sediments with low organic content. Meadows of seagrasses, mainly *Cymodocea nodosa* and *Zostera marina*, thrive in this zone. Filter feeders dominate the benthic community here, such as the clam *Chamaelea gallina*, the razor clam *Ensis minor*, and the very abundant tellinids. Common gastropods include the muricidae *Bolinus brandaris* and *Hexaplex trunculus* and the nassarids *Nassarius mutabilis* and *N. nitidus*. Among the polychaetes, the most common species are *Owenia fusiformis* and *Arenicola marina*. Many of these species also occur in the lagoon, inhabiting the flood tidal delta area, where the razor clam *Solen marginatus* replaces *Ensis minor*, and the clam *Chamaelea gallina* declines in abundance, being progressively replaced by other venerids such as *Paphia aurea* and the Manila clam. The gastropod *Nassarius mutabilis* is replaced by other nassariids such as *Nassarius nitidus* and, to a lesser extent, by *Cyclope neritea*, which is more widespread in the inner part of the lagoon.

Filter feeders are still important in the middle basin; common bivalves are *Loripes lacteus*, the venerids *Paphia aurea* and *Ruditapes decussatus*, and the autochthonous carpet clam, which seems to have been partially replaced by the Manila clam. The Pacific oyster *Crassostrea gigas*, introduced in the lagoon around the 1960s, is very common in this zone (Mizzan 1999). Bivalves, such as *Cerastoderma glaucum* (lagoon cockle) and *Abra segmentum*, are dominant in the innermost part of the middle mud basin. Detritivores increase in abundance in the middle mud basin where the accumulation of organic matter in bottom sediments is high. Representative detritivore species include the nereid polychaetes *Neanthes succinea* and *Hediste diversicolor*. The spionid *Streblospio shrubsolii* is also characteristic of this zone and experiences wide seasonal fluctuations in abundance (Tagliapietra et al. 1998; Maggiore and Keppel 2007).

Along inner shallow areas, where rivers enter the lagoon, *Scrobicularia plana* is commonly found. In the estuarine zone of the Venice Lagoon, *S. plana* used to be very abundant, but it has declined dramatically since the middle of the twentieth century. This decline was likely due to a

TABLE 19.3
Comparison of Macrobenthic Community Characteristics in the Venice Lagoon during 1935, 1990, and 1999

	Past	Ulva	Ulva Decline/ Tapes Spreading	Tapes Fishing
Benthos	**1935**	**1990**		**1999**
Shannon	2.1	1.7	—	1.1
Composition (%)			—	
Filter feeders	28	25	—	50
Herbivores	2	17	—	1
Detritus feeders	23	39	—	17
Predators	6	3	—	4
Omnivores	41	16	—	28

Note: Column heading "Past" refers to the period prior to *Ulva* massive blooms (before 1985); "Ulva" to the period of massive *Ulva* proliferation (approximately 1985–1990); "Ulva Decline" to the period ~1991–1995; and "Tapes Fishing" to the period 1995–2002.

Source: Data from Pranovi et al. (2007).

reduction in freshwater input, although the effects of pollutants cannot be discounted. The small gastropod *Hydrobia ulvae* also occurs in high density, often together with the amphipod *Corophium orientale*. In sheltered areas removed from fish predation, red larvae of the insect *Chironomus salinarius* commonly occur. In these areas, high densities of the sedentary opportunist *Capitella capitata* and other capitellids are observed as well. *Hediste diversicolor* reaches its highest densities in the intertidal zone.

19.4.9 Nekton

A total of 79 species of fish have been identified in Venice Lagoon (Mainardi et al. 2002, 2004, 2005; Malavasi et al. 2004; Franco et al. 2006a, 2006b, 2008a, 2008b). Fish can be grouped into several ecological guilds based on species physiological tolerances to environmental conditions, type of migratory behavior, and reproductive mode. These include: (1) lagoon residents (LR), which spend their whole life cycle (or most of it) in the lagoon environment; (2) marine juvenile migrants (MJ), which spawn at sea and utilize the lagoon as a nursery; (3) marine seasonal migrants (MS), which regularly enter the lagoon environment, mainly in late spring-summer to take advantage of the high abundance of prey; (4) marine occasional visitors or stragglers (MO), which are marine spawners (stenohaline species) only occasionally found in the lagoon; (5) catadromous (MC) and anadromous (MA) migrants, which use lagoon waters mainly as pathways for ontogenetic migration between the sea and fresh water and vice versa; and (6) freshwater species (FW) (Malavasi et al. 2004; Franco et al. 2006a, 2008b).

Marine stragglers are the most abundant guild with 32 species, followed by marine migrants (i.e., MJ, MS, and MC) with 24 species, and lagoon residents with 17 species. Significant seasonal variations have been observed, including a progressive increase in species richness, total abundance, and biomass during spring as a result of both the immigration of individuals from the sea and the recruitment of resident species, a summer peak, and a decrease during late fall (Malavasi et al. 2004, 2005; Franco et al. 2006a, 2006b). On an annual basis, lagoon residents account for ~90% of the total fish abundance, whereas the euryhaline species (MJ and MS in particular) account for just 6.2%.

The fish assemblage in seagrass beds is dominated in numerical abundance by resident species such as the sand smelt *Atherina boyeri*, the grass goby *Zosterisessor ophiocephalus*, and the pipefish *Syngnathus abaster*, *S. typhle*, and *Nerophis ophidion*. The seagrass meadows provide temporal stability to the assemblage during the year, which is not characteristic of the fish fauna in other habitats (Franco et al. 2006a, 2006b).

In mudflats and tidal creeks, small-sized resident species are abundant, such as the gobies *Pomatoschistus marmoratus*, *P. canestrinii*, and *Knipowitschia panizzae*, and the pupfish *Aphanius fasciatus*. Marine migrant species (e.g., *Engraulis encrasicolus, Platichthys flesus, Solea solea, Liza saliens,* and *Pomatoschistus minutus*) are also well represented in these habitats.

In barren or sparsely vegetated sandy bottom habitat areas, the fish fauna is generally less abundant. Here, the most common species are widely distributed residents, such as *A. boyeri* and *P. marmoratus*, and marine migrants (including marine juveniles).

19.4.10 Landings Data

Data on nekton community composition over time are generally lacking, but data compiled on artisanal catches might provide a proxy for such trends. However, we must be cautious when interpreting this kind of data because not all species are fished, and because changes in landings also stem from changes in fishing effort and other activities, market values of target species, and market demand. Using market data corrected for unofficial landings for the years 1945 to 2001, it was possible to pinpoint differences in catch composition in terms of trophic guilds (Libralato et al. 2004),

TABLE 19.4
Composition of Major Biotic Groups in the Venice Lagoon Landings

	Past		Ulva	Ulva Decline/ Tapes Spreading	Tapes Fishing
Landings (%)	**1945–1960**	**1961–1984**	**1985–1989**	**1990–1995**	**1996–2001**
Predators	0.08 (0.03)	0.06 (0.03)	0.12 (0.03)	0.03 (0.05)	0.29 (0.06)
Detritivorous	0.13 (0.03)	0.14 (0.03)	0.15 (0.01)	0.16 (0.01)	0.21 (0.04)
Plankton feeders	0.16 (0.02)	0.26 (0.05)	0.22 (0.03)	0.17 (0.03)	0.09 (0.03)
Benthic feeders	0.11 (0.05)	0.13 (0.04)	0.16 (0.04)	0.16 (0.04)	0.1 (0.03)
Cephalopods	0.15 (0.04)	0.21 (0.06)	0.24 (0.04)	0.37 (0.11)	0.21 (0.03)
Shrimps	0.11 (0.03)	0.07 (0.06)	0.03 (0.01)	0.02 (0.02)	0.02 (0.01)
Crabs	0.26 (0.07)	0.12 (0.03)	0.08 (0.04)	0.09 (0.04)	0.08 (0.02)

Note: Column heading "Past" refers to the period prior to *Ulva* massive blooms (before 1985); "Ulva" to the period of massive *Ulva* proliferation (approximately 1985–1990); "Ulva Decline" to the period ~1991–1995; and "Tapes Fishing" to the period 1995–2002. Table gives average landing proportion and standard deviation (in parentheses).

Source: Data from Libralato et al. (2004).

whose averages for relevant periods are reported in Table 19.4. A steady increase in the proportion of detritivorous fishes (mainly Mugilidae) from 13% to 20% of the catches is observed. Concurrently, the proportion of scavenger crabs (mainly *Carcinus aestuarii*) and shrimps (mainly *Palaemon* sp. and *Crangon crangon*) decreased steadily from 26% to 8% and from 11% to 2%, respectively. Benthic feeder fishes, mainly Gobiidae, accounted for 11% of the landings before the 1960s, and then increased up to 16% during the *Ulva* and post-*Ulva* period. They subsequently declined to 10% of the landings during the extensive clam exploitation period (1996 to 2001). Plankton feeders, mainly *Atherina* sp. and *Engraulis encrasicolus*, accounted for 15% of the landings during the period 1945 to 1960 and then increased to 26% during the period 1961 to 1984. Plankton feeders decreased progressively from 22% to 17%, and to 8% during the following periods: 1985 to 1989, 1990 to 1995, and 1995 to 2001, respectively. Landings are also characterized by a sharp increase in piscivorous fishes (mainly *D. labrax*): 8% of the catches in 1945 to 1960, and 29% during the last period 1995 to 2001. Cephalopds (*Sepia officinalis*, *Loligo vulgaris*) comprised 15% of the landings during the period 1945 to 60, and then progressively increased up to 36% during the period 1990 to 95. This guild now accounts for ~21% of landings (Table 19.4).

19.5 ECOSYSTEM RESPONSES

The Venice Lagoon is a large, open system subjected to natural and anthropogenic stressors, and characterized by highly heterogeneous morphological structure and variable physicochemical (Molinaroli et al. 2007; Cucco et al. 2009), biogeochemical (Solidoro et al. 2004a, 2004b), and biological (Occhipinti Ambrogi et al. 1998; Malavasi et al. 2004; Franco et al. 2006b; Guerzoni and Tagliapietra 2006; Marchini and Marchini 2006; Bandelj et al. 2008, 2009; Tagliapietra et al. 2009) conditions. It can be subdivided into four sub-basins typified by internal, intermediate, and pseudo marine areas (Figure 19.6) (Solidoro et al. 2004b; Marchini and Marchini 2006; Bandelj et al. 2008, 2009; Sfriso et al. 2009; Tagliapietra et al. 2009). These sub-basins can be further subdivided into smaller geographic units, reflecting the environmental heterogeneity of a system comprised of closely interconnected, mutually interacting habitats (Tagliapietra and Volpi Ghirardini 2006; Tagliapietra et al. 2009).

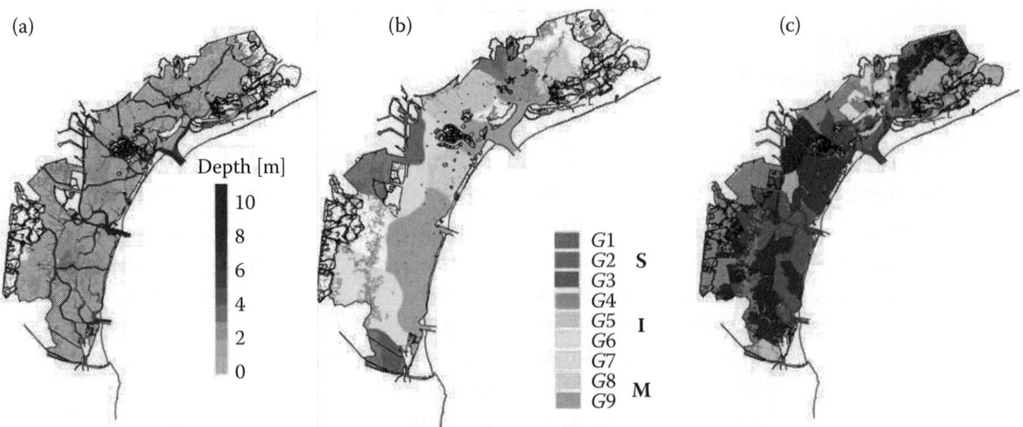

FIGURE 19.6 Lagoon partitions based on: (a) physical, (b) biogeochemical properties, and (c) morphological. Physical partition is based on confinement time and longitudinal subdivision in four sub-basins. The figure also reports bathymetry. Biogeochemical partition (b) is derived by subjectively drawing boundaries among nine groups which result from multivariate analyses (cluster and ordination methods) on data from 2001 to 2005; boundaries are drawn for visual clarity and have no objective physical meaning. (Adapted from Solidoro et al. 2004b). Landscape classification (c) subdivides lagoon based on geomorphological, hydrodynamic, and sediment properties. (Adapted from Guerzoni and Tagliapietra 2006).

The lagoon exhibits multifaceted responses to natural and anthropogenic forcing factors. The availability of a long-term time series of data also enables us to synthesize and integrate these covariations as well as to assess the causal relationships among the drivers of change on the ecosystem and their direct and indirect effects. This information also allows us to develop a conceptual scheme of ecosystem evolution over decades of time.

Industrial and agricultural activities grew exponentially from the early 1940s to the 1960s. They greatly impacted the lagoon ecosystem. For example, there was a direct impact of pollutants discharged into the lagoon, which are still present in bottom sediments and can bioaccumulate in the food chain (Pavoni et al. 1992; Dalla Valle et al. 2005; Frignani et al. 2005). Therefore, pollutant contamination and, in particular, bioaccumulation of organic pollutants remains a concern, with "hot spots" of sediment contamination in the industrial channel sediments up to 2500 ng I-TEQ kg^{-1} dw (Guerzoni and Raccanelli 2004; Guerzoni et al. 2007). Studies of contaminants in the system (Raccanelli et al. 2004; Micheletti et al. 2008) suggest that bioaccumulation of POPs in lagoon organisms may be significant. These conclusions have important implications both in terms of risk to human health (Frangipane 1999; Raccanelli et al. 2008) and an increase in natural mortality and/or energy expenditure for metabolic detoxification in lagoon species. These topics have been reviewed elsewhere (Orio and Donazzolo 1987; Guerzoni and Raccanelli 2004; Carrer et al. 2005; Guerzoni et al. 2007; Micheletti et al. 2008).

Industrial activities have also caused a number of indirect impacts on the lagoon. Aside from land reclamation and associated habitat loss, industrial activities have been responsible for dredging of large navigation channels that altered the morphological balance between the channel and tidal flats and permanently modified a large part of the central basin. They have also led to the extraction of groundwater and natural gas, causing an increase in the rate of lagoon subsidence. The diversion of fresh water for agricultural and zootechnical uses, together with the deepening of the inlets, the building of jetties, and digging of new channels, have resulted in intrusion of seawater, leading to expansion of the central mud habitat at the expense of estuarine-like oligohaline habitats, already drastically reduced by the diversion of the majority of rivers to the sea in previous centuries. These alterations of geomorphological and physical properties clearly impacted the spatial distribution, structure, and composition of the macrobenthic community in the lagoon between the late 1950s and 1960s, as evidenced by a marked impoverished community, a replacement of brackish species with more marine ones, and a shift toward assemblages that are more tolerant of eutrophic conditions (Giordani Soika and Perin 1974).

An exponential increase in the use of fertilizers, together with pollutants from industrial areas, contributed to an acute, continuous increase in nutrient loads emitted to the lagoon from the drainage basin (Figure 19.3). The increase in nutrient loads correlated with increased nutrient concentrations in the lagoon (Figures 19.2 and 19.3). Eutrophication, together with the modification of hydrodynamic and morphological features, favored the occurrence of massive macroalgae blooms dominated by Ulvacea during the 1980s. These blooms impacted nutrient cycles and the benthos. Indeed, in late spring and summer, macroalgae beds covered almost all shallow areas where water movement was low and residence time high, causing substantial depletion of the nutrient pool in the center of the beds and triggering the start of anoxic events. These events caused an increase in *Ulva* mortality and the amount of decaying organic material on the lagoon floor, resulting in additional consumption of oxygen and worsening of the anoxia, which rapidly spread from bottom sediments to the water column over large areas (Sfriso et al. 1992; Solidoro et al. 1997a). Among the more serious effects of the dystrophic crises was the release of hydrogen sulfide that adversely affected human activities in Venice and other surrounding land areas (Ravera 2000). Dystrophic crises also had significant impacts on the lagoon ecosystem, influencing biogeochemical cycles, promoting the release of large amounts of nutrients, enrichment of organic matter in bottom sediments, and secondary impacts on the benthic community (Solidoro et al. 1997a, 1997b). Subsequent decreases in shrimp and crab landings might also be coupled to this series of changes (Sfriso et al. 1992; Sfriso and Marcomini 1996). *Ulva* mats acted as traps for nutrients entering the lagoon, storing them in plant tissue. In summer, even nitrogen in waters entering from the Adriatic fueled macroalgal growth (Sfriso and Marcomini 1996).

The large biomass of *Ulva* caused a sharp decrease in other seaweeds as well as a reduction in phanerogams (Sfriso et al. 1987), while herbivorous and detritivorous species dominated the macrobenthic and nektonic communities. In addition, phytoplankton density decreased, with blooms occurring only in those areas where the macroalgae density was not very high. Typically, a spring phytoplankton bloom was triggered before the macroalgal blooms (Sfriso et al. 1992; Solidoro et al. 1997a), and other plankton blooms followed *Ulva* dystrophic crises, fostered by nutrient released during *Ulva* decomposition. The major shift in zooplankton composition, with occurrence of *A. tonsa,* can be regarded as an additional indicator of eutrophic conditions.

In order to mitigate eutrophication, several policy measures were enforced, including environmental legislation issued to specifically address environmental problems in Venice Lagoon. Harvesting of macroalgae standing crop was conducted, which removed up to 60,000 t of plant matter per year from the lagoon during the 1980s (Figure 19.3b). The effectiveness of the new regulations was evidenced by the decrease in nutrient concentrations in the system, particularly phosphorus, which dropped to very low levels after the enforcement of a total ban on phosphorus in detergents in 1989. However, the decline in *Ulva* biomass, observed during the 1990s, was probably due to a combination of factors, which triggered the system to shift to an alternative regime, then maintained by several co-occurring factors. Adverse meteorological conditions at the beginning of the 1990s probably played a very important role, with several consecutive, longer than usual cold seasons, which reduced over-wintering of the macroalgae. Other important factors were increased turbidity and sedimentation of particulate matter on *Ulva* thalli, which hindered macroalgae growth during more productive seasons (Sfriso and Marcomini 1996). The subsequent reduction in laminar free-floating thalli initiated a positive feedback by favoring wind- and tide-induced sediment resuspension, further increasing water turbidity and reducing photosynthetic activity. The drop in biomass standing crop reduced the frequency of anoxia to only episodic events, thereby enabling the establishment of zoobenthic grazer populations to exert greater control on *Ulva* biomass. Together with harvesting of macroalgae standing crops, these processes were effective in keeping the total macroalgae biomass at a much lower level than in previous years when it reached a total wet weight as high as 10^7 t.

The beginning of massive fishing of *Ruditapes philippinarum* between 1995 and 2000 (Figure 19.3), often by means of illegal fishing devices (hydraulic dredges), greatly disturbed bottom sediments, causing direct damage to the macrophytobenthos and a further increase in sediment resuspension, thereby hampering recolonization by *Ulva* and contributing to further reduction in *Ulva* biomass. As a result, the distribution of the Manila clam spread throughout the area free of macroalgae. New macrophyte recolonization appears to have been precluded by fishing activities and related impacts on the lagoon floor (Sfriso and Marcomini 1996; Sfriso and Facca 2007). Other important factors that presently control macroalgae growth are sediment resuspension due to the morphological works used to reinforce tidal lands, and the ongoing oligotrophication of the lagoon in recent years.

Following the decline in seaweed abundance, nutrients entering the lagoon were no longer readily trapped by plant uptake and could be exported to the sea without being removed by photosynthetic activity. In addition, the flux of organic matter to bottom sediments was greatly reduced. In the case of *Ulva*, therefore, biomass removal helped counterbalance sediment enrichment. In the case of *Ruditapes*, which feeds on particulate matter, harvesting promoted the effective consumption of organic matter stored in the sediments through resuspension. During the first phase of *Ruditapes* colonization, when fishing pressure was not yet very high, clam growth and metabolism facilitated the transfer of organic matter from the water column to bottom sediments, which was consequently enriched. In the following phase, however, heavy clam fishing caused the transfer of organic matter from the water column and sediments to the clams with export from the system, resulting in impoverished sediment. Clearly the demographic explosion of the Manila clam and its commercial exploitation has greatly altered the macrobenthic community, with a shift toward infaunal species and loss of more fragile or less mobile species due to the pressure exerted by mechanical fishing gear.

The decline in nutrient contents observed in lagoon bottom sediments, along with declining nutrient loads, caused a reduction in nutrient concentration, phosphorus in particular, in the water column. Chlorophyll *a* values have declined as well, probably due to a combination of bottom-up (resource limitation) and top-down (clam filtration) controls. Between 2000 and 2003, a temporary increase in chlorophyll *a* values was noted possibly because of a reduction of clam landings (Figures 19.3 and 19.5), but since 2004, only rather low levels of chlorophyll *a* have been recorded. In the last two decades, there has been a reversal in the eutrophication trend of the system.

Climate change is another major threat to the Venice Lagoon. Atmospheric temperature and sea level rise are predicted to increase significantly over the next 50 to 100 years, which will lead to greater frequency of flooding. According to Gambolati et al. (2002), extensive areas of the western part of the northern Adriatic Sea coastline could be permanently lost by 2100. In order to mitigate the impact of flooding on the city of Venice, specifically designed mobile gates (i.e., the MOSE system) are now being set in place, and will be operated to temporarily close lagoon inlets concurrently with high tide events (www.salve.it). The consequence of this intervention has been long debated because of possible adverse effects on the lagoon ecosystem. Modifications made in areas near the inlets to install the gates, and repeated temporary closures, may or may not alter the lagoon ecosystem, but it is clear that, if the frequency of closures will be so high as to significantly modify morphological and hydrodynamic properties, biogeochemical and food web processes could change significantly as well. Even considering less extreme scenarios, regional climate models indicate for the period 2070 to 2100 a significant increase in air temperature, up to 3°C (5°C in summer), and changes in the temporal pattern of rainfall over the Venice drainage basin (Salon et al. 2008).

Aside from the obvious direct effects of air temperature increase on water temperature and organism metabolism, climate change might impact the lagoon ecosystem by inducing changes in the levels and dynamics of seston and biogeochemical properties. Results obtained in a downscaling experiment confirm that the annual mean rainfall will not change appreciably in the watershed of the Venice Lagoon, whereas the seasonal patterns will likely change, with summer and spring becoming drier and winter and fall wetter (Salon et al. 2008). These changes will potentially increase winter nutrient concentrations. However, the annual planktonic primary and secondary production of the Venice Lagoon will decrease, and nutrient surplus will be exported to the Adriatic Sea (Cossarini et al. 2008). The food web structure, notably higher trophic level organisms, could be adversely affected. In addition, the success of aquaculture activity of *Ruditapes philippinarum* could be significantly impacted (Melaku Canu et al. in press).

19.6 CONCLUDING REMARKS

The Venice Lagoon is a large, complex system affected by multiple natural and anthropogenic forcing factors, and characterized by high heterogeneity in physical, biogeochemical, and biological conditions, of mutually interacting habitats. An array of stressors has altered morphological settings, biogeochemical properties, and biological components of the ecosystem either directly or through cascading effects. For example, morphological modifications of the basin, reduction of freshwater inflow, increase in nutrient loads, fishing activities, and aquaculture operations have all impacted the structure and function of the ecosystem through time. Harvesting of the Manila clam has directly affected the benthic community and also caused cascading effects such as the loss of sediment, increase in turbidity, and removal of organic matter from the system. In the past, the lagoon has been plagued by serious eutrophication problems, but in recent years, these problems have been mitigated, and the system is now in a state of progressive oligotrophication. In the future, however, rising sea level coupled to warming climatic conditions will have an increasing impact on human habitation in the region due to greater frequency of flooding and other adverse effects.

The lagoon is a continuously evolving system that responds rapidly to human activities. As such, the long-term health and viability of this important system is contingent upon sound and effective coastal zone management, which should be an outcome of an integrated vision and a participatory and adaptive approach (Suman et al. 2005). Given the complex nature of how the ecosystem functions, great care is required in planning any intervention.

ACKNOWLEDGMENTS

This chapter presents a synthesis and integration of data and knowledge collected during many different projects, mainly supported by Venice Water Authority (Magistrate alle Acque) directly or through Consorzio Venezia Nuova, and by CORILA. Authors also acknowledge contributions from CNR, Comune di Venezia, and Regione Veneto. CS gratefully acknowledges the efforts of Laura Montobbio and Alberto Giulio Bernestein (Servizio Ambiente/Consorzio Venezia Nuova) in promoting and supporting the achievement of an integrated perspective of Venice ecosystem functioning. The authors thank Valentina Mosetti for technical assistance on the manuscript. Jill Woodcock is thanked for revisions to the text. Mike Kennish greatly improved the readability of the manuscript. This paper is a contribution to Integrating Marine Biogeochemistry and Ecological Research (IMBER) and Long Term Ecological Research (LTER) projects.

REFERENCES

Acri, F., F. Bernardi-Aubry, A. Berton, F. Bianchi, A. Boldrin, E. Camatti, A. Comaschi, S. Rabitti, and G. Socal. 2004. Plankton communities and nutrients in the Venice Lagoon: Comparison between current and old data. *J. Mar. Systems.* 51: 321–329.

Arpav (Agenzia Regionale per la Prevenzione e Protezione Ambientale del Veneto). 2007. *Bacino scolante nella Laguna di Venezia, Rapporto sullo stato ambientale dei corpi idrici—Anni 2003–2004.* http://www.arpa.veneto.it/acqua/htm/pubblicazioni.asp (accessed August 26, 2009).

Bandelj, V., D. Curiel, S. Lek, A. Rismondo, and C. Solidoro. 2009. Modelling spatial distribution of hard bottom benthic communities and their functional response to environmental parameters. *Ecol. Modeling*. DOI: 10.1016/j.ecolmodel.2009.04.024.

Bandelj, V., G. Socal, Y. S. Park, L. Lek, J. Coppola, E. Camatti, E. Capuzzo, L. Milani, and C. Solidoro. 2008. Analysis of multitrophic plankton assemblages in the Lagoon of Venice. *Mar. Ecol. Prog. Ser.* 368: 23–40.

Battaglia, B., C. Datei, C. Dejak, G. Gamabarello, G. B. Guarise, G. Perin, E. Vinello, and F. Zingales. 1983. *Indagini idrotermodinamiche e biologiche per la valutazione dei riflessi ambientali del funzionamento a piena potenza della centrale termoelettrica dell'Enel di Porto Marghera.* Commissione tecnico scientifica per la sperimentazione ed i controlli periodici sulla centrale termoelettrica dell'ENEL sita in località Fusina di Porto Marghera Venezia. Venezia: Regione Veneto.

Bendoricchio, G. 1998. *Interventi per l'arresto del degrado connesso alla proliferazione delle macroalghe in Laguna di Venezia 1995–1996. Analisi dell'evoluzione recente delle immissioni di nutrienti nel periodo 1989–1995.* Relazione Finale. Venezia: Consorzio Venezia Nuova.

Bernardi Aubry, F. B., A. Berton, M. Bastianini, G. Socal, and F. Acri. 2004. Phytoplankton succession in a coastal area of the NW Adriatic, over a 10-year sampling period (1990–1999). *Continent. Shelf Res.* 24: 97–115.

Bevilacqua, P. 1998. *Venezia e le acque. Una metafora planetaria.* Rome: Donzelli Editore.

Bianchi, F., F. Acri, L. Alberighi, M. Bastianini, A. Boldrin, B. Cavalloni, F. Cioce, A. Comaschi, S. Rabitti, G. Socal, and M. M. Turchetto. 2000. Biological variability in the Venice Lagoon. Pp. 97–125, In: P. Lasserre and A. Marzollo (eds.), *The Venice Lagoon ecosystem: Inputs and interactions between land and sea*. Man and the Biosphere Series Volume 25. Paris: UNESCO and Parthenon Publishing Group.

Brambati, A. 1992. Origin and evolution of the Adriatic Sea. Pp. 327–346, In: *The Adriatic Sea.* In Marine Eutrophication and Population Dynamics, 25th European Marine Biology Symposium, Lido degli Estensi, Ferrara, Italy, September 10–15, 1990. Fredensborg, Denmark: Olsen & Olsen.

Bressan, M., E. Camatti, and A. Comaschi. 2004. Mesozooplankton at the three inlets of the Lagoon of Venice: Long- and short-term surveys 2001–2002. Pp. 389–398, In: P. Campostrini (ed.), *Scientific research and safeguarding of Venice*. CORILA Research Programme 2001–2003, 2002 Results. Venice: Corila.

Camatti, E., A. Comaschi, J. Coppola, L. Milani, M. Minocci, and G. Socal. 2006a. Analisi dei popolamenti zooplanctonici nella laguna di Venezia dal 1975 al 2004. *Biol. Mar. Mediterranea* 13: 46–53.

Camatti, E., A. Comaschi, and G. Socal. 2006b. Ciclo annuale del mesozooplancton. P. 78, In: S. Guersoni and D. Tagliapietre (eds.), *Atlante della laguna.Venezia tra terra e mare*, Eds 78. Venezia: Marsilio Editori.

Campostrini, P. 2004. Venice: The challenges of today, not only the glories of the past. *J. Mar. Systems* 51: 5–6.

Caniato, G. 2005. Between salt and freshwaters. Pp. 7–14, In: C.A. Fletcher and T. Spencer (eds.), *Flooding and environmental challenger for Venice and its lagoon: State of knowledge*. Cambridge: Cambridge University Press.

Caniglia, G., S. Borella, D. Curiel, P. Nascimbeni, F. Paloschi, A. Rispondo, F. Scarton, D. Tagliapietra, and L. Zanella. 1990. Distribuzione delle fanerogame marine *Zostera marina* L., *Zostera noltii* Hornem., *Cymodocea nodosa* (Uchria) Ascherson in laguna di Venezia. *Lav. Soc. Veneziana Sci. Nat.* 17: 137–150.

Carbognin, L., P. Teatini, and L. Tosi. 2004. Eustacy and land subsidence in the Venice Lagoon at the beginning of the new millennium. *J. Mar. Systems* 51: 345–353.

Carrer, S., G. Coffaro, M. Bocci, and A. Barbanti, 2005. Modelling partitioning and distribution of micropollutants in the Lagoon of Venice: A first step towards a comprehensive ecotoxicological model. *Ecol. Model.* 184: 83–101.

Castellani, C., A. Barbanti, and G. Colombini. 2005. Il carico di nutrienti scaricato da Venezia alla Laguna. *Acqua Aria* 3: 20–34.

Ciavatta, S., 2000. Application of mathematical models to the estimation of the Total Daily Maximum Loads (TDMLs): The Lagoon of Venice as a case study. M.S. Thesis, University Ca' Foscari of Venice, Venice, Italy.

Ciavatta, S., G. Cossarini, T. Lovato, R. Pastres, and C. Solidoro, In preparation. Exploring the long term variability of biogeochemical variables in coastal areas by using data assimilation techniques.

Ciriacono, S. 1995. Manifatture e mestieri in laguna; equilibri ambientali e sviluppo economico. Pp. 357–383, In: G. Caniato, E. Turri, and M. Zanetti (eds.), *La laguna di Venezia*. Verona: UNESCO, Cierre.

Cloern, J. E. 2001. Our evolving conceptual model for the coastal eutrophication problem. *Mar. Ecol. Prog. Ser.* 210: 223–253.

Collavini, F., C. Bettiol, L. Zaggia, and R. Zonta. 2005. Pollutant loads from the drainage basin to the Venice Lagoon (Italy). *Environ. Intern.* 31:939–947.

Comaschi, A., F. Acri, L. Alberighi, M. Bastianini, F. Bianchi, B. Cavalloni, and G. Socal. 1994. Presenza di Acartia tonsa (Copepoda: Calanoida) nella laguna di Venezia. *Biol. Mar. Mediterranea* 1: 273–274.

Cossarini, G., P. F. J. Lermusiaux, and C. Solidoro. 2009. The Lagoon of Venice ecosystem: Seasonal dynamics and environmental guidance with uncertainty analyses and error subspace data. *J. Geophys. Res.* DOI: 10.1029/2008JC005080.

Cossarini, G., S. Libralato, S. Salon, X. Gao, F. Giorgi, and C. Solidoro. 2008. Downscaling experiment for the Venice Lagoon. II. Effects of changes in precipitation on biogeochemical properties. *Climate Res.* 38: 43–59.

Cucco, A., G. Umgiesser, C. Ferrarin, A. Perilli, D. Melaku Canu, and C. Solidoro. 2009. Eulerian and lagrangian transport time scales of a tidal active coastal basin. *Ecol. Model.* 220: 913–922.

Curiel, D. and M. Marzocchi. 2010. Stato delle conoscenze nella laguna di Venezia di due /alien species: Undaria pinnatifida e Sargassum muticum. *Lav. Soc. Veneziana Sci. Nat.* 35: 93–106.

Dalla Valle, M., A. Marcomini, K. Jones, and A. J. Sweetman. 2005. Reconstruction of historical trends of PCDD/Fs and PCBs in the Venice Lagoon, Italy. *Environ. Intern.* 3: 1047–1052.

Defina, A., A. Carniello, S. Fagherazzi, and L. D'Alpaos. 2007. Self-organization of shallow basins in tidal flats and saltmarshes. *J. Geophys. Res.* 112: F03001. DOI: 10.1029/2006JF000550.

Degetto, S. and C. Cantaluppi. 2004. Radiochemical methodology for the determination of the mass balance of suspended particulate materials exchanged at the inlets of the Venice Lagoon. *J. Mar. Systems* 51: 77–94.

Degobbis, D., M. Gilmartin, and A. A. Orio. 1986. The relation of nutrient regeneration in the sediments of the northern Adriatic to eutrophication, with special reference to the Lagoon of Venice. *Sci. Total Environ.* 56: 201–210.

Facca, C., S. Ceoldo, N. Pellegrino, and A. Sfriso. 2008. Updating of sediment nutrient concentrations in the Lagoon of Venice (summer 2003). P. 102, In *9th Littoral International Conference*, November 25–28, 2008, Venice, Italy.

Facca, C., A. Sfriso, and P. F. Ghetti. 2004a. Phytoplankton community composition and distribution in a eutrophic coastal area (Venice lagoon, Italy). *Acta Adriatica* 45: 163–180.

Facca, C., A. Sfriso, and P. F. Ghetti. 2004b. Spatial and temporal distribution of phytoplankton communities in the Venice Lagoon central area. Pp. 273–277, In: P. Campostrini (ed.), *Scientific research and safeguarding of Venice*, Volume 2. Venice: Corila.

Facca, C., A. Sfriso, and G. Socal. 2002. Changes in abundance and composition of phytoplankton and microphytobenthos due to increased sediment fluxes in the Venice Lagoon, Italy. *Estuar. Coastal Shelf Sci.* 54: 773–792.

Fagherazzi, S., L. Carniello, L. D'Alpaos, and A. Defina, 2006. Critical bifurcation of shallow microtidal landforms in tidal flats and saltmarshes. *Proc. Natl. Acad. Sci. USA* 103: 8337–8341.

Fagherazzi, S., C. Palermo, M. Rulli, L. Carniello, and A. Defina. 2007. Wind waves in shallow microtidal basins and the dynamic equilibrium of tidal flats. *J. Geophys. Res.* 112: F02024. DOI: 10.1029/2006JF000572.

Favero, V., R. Parolini, and M. Scattolin. 1988. *Morfologia storia della Laguna di Venezia*. Venice: Arsenale Editrice.

Franco, A., M. Elliott, P. Franzoi, and P. Torricelli. 2008a. Life strategies of fishes in European estuaries: The functional guild approach. *Mar. Ecol. Prog. Ser.* 354: 219–228.

Franco, A., P. Franzoi, S. Malavasi, F. Riccato, P. Torricelli, and D. Mainardi. 2006a. Use of shallow water habitats by fish assemblages in a Mediterranean coastal lagoon. *Estuar. Coastal Shelf Sci.* 66: 67–83.

Franco, A., P. Franzoi, S. Malavasi, F. Riccato, and P. Torricelli. 2006b. Fish assemblages in different shallow water habitats of the Venice Lagoon. *Hydrobiologia* 555: 159–174.

Franco, A., P. Franzoi, and P. Torricelli. 2008b. Structure and functioning of Mediterranean lagoon fish assemblages: A key for the identification of water body types. *Estuar. Coastal Shelf Sci.* 79: 549–558.

Frangipane, G. 1999. *Abitudini alimentari e livelli ematici di policlorodibenzodiossine, policlorodibenzofurani e policlorobifenili in un campione di popolazione veneziana*. Venezia: Università Ca' Foscari.

Frignani, M., L. G. Bellucci, M. Favotto, and S. Albertazzi. 2005. Pollution historical trends as recorded by sediments at selected sites of the Venice Lagoon. *Environ. Intern.* 31: 1011–1022.

Gačić, M., A. Mazzoldi, V. Kovacevic, I. Mancero Mosquera, V. Cardin, F. Arena, and G. Gelsi. 2004. Temporal variations of water flow between the Venetian Lagoon and the open sea. *J. Mar. Systems* 51: 33–47.

Gambolati, G., P. Teatini, and M. Gonnella. 2002. GIS Simulations of the inundation risk in the coastal lowlands of the northern Adriatic Sea. *Math. Comp. Model* 35: 963–972.

Gatto, P. and L. Carbognin. 1981. The lagoon of Venice: Natural environmental trend and man-induced modification. *Hydrol. Sci. Bull.* 26: 379–391.

Giordani Soika, A. and G. Perin. 1974. L'inquinamento della laguna di Venezia: studio delle modificazioni chimiche e del popolamento sottobasale dei sedimenti lagunari negli ultimi vent'anni. *Bollettino del Museo Civico di Storia Naturale di Venezia* 26: 25–68.

Granzotto, A., P. Franzoi, A. Longo, F. Pranovi, and P. Torricelli. 2001. La pesca nella laguna di Venezia: un percorso di sostenibilità nel recupero delle tradizioni Lo stato dell'arte. Rapporto sullo sviluppo sostenibile 2. Fondazione Eni Enrico Mattei.

Granzotto, A., S. Libralato, F. Pranovi, S. Raicevich, and O. Giovanardi. 2004. Comparison between artisanal and industrial fisheries using ecosystem indicators. *Chem. Ecol.* 20: S435–S449.

Guelorget, O. and J. Perthuisot. 1992. Paralic ecosystems: Biological organization and functioning. *Vie et milieu* 42: 215–251.

Guerzoni, S. and S. Raccanelli. 2004. *The sick lagoon: Dioxin and other organic pollutants (POPs) in the Lagoon of Venice.* Venezia: Libreria Editrice Cafoscarina.

Guerzoni, S., P. Rossini, A. Sarretta, S. Raccanelli, G. Ferrari, and E. Molinaroli. 2007. POPs in the Lagoon of Venice: Budgets and pathways. *Chemosphere* 67: 1776–1785.

Guerzoni, S. and D. Tagliapietra (eds.). 2006. *Atlante della laguna—Venezia tra terra e mare.* Venezia: Marsilio Editori.

Libralato, S., F. Pranovi, P. Torricelli, S. Raicevich, F. Da Ponte, R. Pastres, and D. Mainardi. 2004. Ecological stages of the Venice Lagoon analysed using landing time series data. *J. Mar. Systems* 51: 331–344.

Maggiore, F. and E. Keppel. 2007. Biodiversity and distribution of polychaetes and molluscs along the Dese estuary (Lagoon of Venice, Italy). *Hydrobiologia* 588: 189–203.

Mainardi, D., R. Fiorin, A. Franco, P. Franzoi, A. Granzotto, S. Malavasi, F. Pranovi, F. Riccato, M. Zucchetta, and P. Torricelli. 2004. Seasonal distribution of fish fauna in the Venice Lagoon shallow waters: Preliminary results. Pp. 437–447, In: P. Campostrini (ed.), *Scientific research and safeguarding of Venice.* Corila Research: Program 2002 Results. Venice: Multigraf.

Mainardi, D., R. Fiorin, A. Franco, P. Franzoi, O. Giovanardi, A. Granzotto, A. Libertini, S. Malavasi, F. Pranovi, F. Riccato, P. Torricelli. 2002. *Fish diversity in the Venice lagoon: Preliminary report.* Pp. 583–594, In: P. Campostrini (eds.), *Scientific research and safeguarding of Venice.* Corila Research: Program 2001 results. Padova: La Garangola.

Mainardi, D., R. Fiorin, A. Franco, P. Franzoi, S. Malavasi, F. Pranovi, F. Riccato, M. Zucchetta, and P. Torricelli. 2005. Composition and distribution of fish assemblages in the shallow waters of the Venice Lagoon. Pp. 405–419, In: P. Campostrini (ed.), *Scientific research and safeguarding of Venice.* Corila Research: Program 2003 results. Venezia: Multigraf.

Malavasi, S., A. Franco, R. Fiorin, P. Franzoi, P. Torricelli, and D. Mainardi. 2005. The shallow water gobiid assemblage of the Venice Lagoon: Abundance, seasonal variation, and habitat partitioning. *J. Fish. Biol.* 67: 146–165.

Malavasi, S., R. Fiorin, A. Franco, P. Franzoi, A. Granzotto, F. Riccato, and D. Mainardi. 2004. Fish assemblages of Venice Lagoon shallow waters: An analysis based on species, families and functional guilds. *J. Mar. Systems* 51: 19–31.

Marani, M., A. D'Alpaos, S. Lanzoni, L. Carniello, and A. Rinaldo. 2007. Biologically-controlled multiple equilibria of tidal landforms and the fate of the Venice lagoon. *Geophys. Res. Lett.* 34: L11402. DOI: 10.1029/2007GL030178.

Marchini, A. and C. Marchini. 2006. A fuzzy logic model to recognise ecological sectors in the lagoon of Venice based on the benthic community. *Ecol. Model.* 193: 105–118.

MAV (Magistrato alle Acque di Venezia). 2002. *Relazione Sulle Caratteristiche Degli Scarichi Idrici dell'Area di Porto Marghera. Dati Relativi al 1999.* Venezia, Italy: Magistrato alle Acque di Venezia.

MAV (Magistrato alle Acque di Venezia). 2005. *Aggiornamento al 2003 delle stime dei carichi esterni ed interni e composizione del quadro relativo ai bilanci di massa degli inquinanti in laguna.* Venezia, Italy: Magistrato alle Acque di Venezia.

Melaku Canu, D., C. Solidoro, G. Cossarini, and F. Giorgi. In press. Effect of global change on bivalve rearing activity and need of implementation of adaptive management. *Climate Res.*

Micheletti, C., T. Lovato, A. Critto, R. Pastres, and A. Marcomini. 2008. Spatially distributed ecological risk for fish of a coastal food web exposed to dioxins. *Environ. Toxicol. Chem.* 27: 1217–1225.

Miotti, C., A. Pierini, A. Rismondo, and D. Curiel. 2007. Variazioni delle coperture e delle biomasse macroalgali della Laguna di Venezia: 2002–2005. *Lav. Soc. Veneta Sci. Nat.* 32: 15–24.

Mizzan, L. 1999. Le specie alloctone del macrozoobenthos della Laguna di Venezia: il punto della situazione. *Bollettino Museo civico Storia naturale di Venezia* 49: 145–177.

Molinaroli, E., S. Guerzoni, A. Sarretta, A. Cucco, and G. Umgiesser. 2007. Links between hydrology and sedimentology in the Lagoon of Venice, Italy. *J. Mar. System* 68: 303–317.

Occhipinti Ambrogi, A., T. Birkemeyer, and C. Sacchi. 1998. Indicatori ambientali in laguna di Venezia: proposta di una classificazione basata sulle comunità sessili. *Bolletino del Museo Civico di Storia Naturale di Venezia* 49:145–177.

Orio, A. A. and R. Donazzolo. 1987. Sostanze tossiche ed eutrofizzanti nella Laguna e nel Golfo di Venezia. *Memorie Istituto Veneto Scienze Lettere Arti* 11: 149–215.

Pastres, R., C. Solidoro, S. Ciavatta, A. Petrizzo, and G. Cossarini. 2004. Long-term changes of inorganic nutrients in the Lagoon of Venice (Italy). *J. Mar. Systems* 51: 179–189.

Pauly, D., V. Christensen, J. Dalsgaard, R. Froese, and F. Torres, Jr. 1998. Fishing down marine food webs. *Science* 279: 860–863.

Pavoni, B., A. Marcomini, A. Sfriso, R. Donazzolo, and A. A. Orio. 1992. Changes in an estuarine ecosystem. The Lagoon of Venice as a case-study. In: D. A. Dunnette and R. J. O'Brien (eds.), *The Science of Global Change*. ACS Symposium Series No 483. Washington, D.C.: American Chemical Society.

Poulain, P. M. and F. Raicich. 2001. *Forcings*. Pp. 45–65, In: B. Cushman-Roisin, P.-M. Poulain, M. Gačić and A. Artegiani (eds.), *Oceanography of the Adriatic Sea*. Norwell: Kluwer Academic Publishers.

Pranovi, F., F. Da Ponte, S. Raicevich, and O. Giovanardi. 2004. A multidisciplinary study of the immediate effects of mechanical clam harvesting in the Venice Lagoon. *ICES J. Mar. Sci.* 61: 43–52.

Pranovi, F., F. Da Ponte, and P. Torricelli. 2007. Application of biotic indices in the lagoon of Venice: An example by using a time series of macrobenthic community. *Mar. Pollut. Bull.* 54: 1607–1618.

Pranovi, F., F. Da Ponte, and P. Torricelli. 2008. Historical changes in the structure and functioning of the benthic community in the Lagoon of Venice. *Estuar. Coastal Shelf Sci.* 76: 753–764.

Pranovi, F., G. Franceschini, M. Casale, M. Zucchetta, O. Giovanardi, and P. Torricelli. 2006. An ecological imbalance induced by a non-native species: The Manila clam in the Venice Lagoon. *Biol. Invasion* 8: 595–609.

Pranovi, F., S. Libralato, S. Raicevich, A. Granzotto, R. Pastres, and O. Giovanardi. 2003. Mechanical clam dredging in Venice Lagoon: Ecosystem effects evaluated with a trophic mass-balance model. *Mar. Biol.* 143: 393–403.

Pugnetti, A. and F. Acri. 2007. Struttura, dinamica e caratteristiche funzionali delle comunità biologiche dominate da macrofite e da alghe planctoniche. Pp. 273–277, In: P. Campostrini (ed.), *Scientific research and safeguarding of Venice*, Volume 5. Venezia: Corila.

Raccanelli, S., S. Libralato, and M. Favotto. 2008. On the detoxification of benthic bivalve contaminated by POPs: Insights from experimental and modelling approaches. *Environ. Chem. Lett.* 6: 251–258.

Raccanelli, S., R. Pastres, M. Favotto, and P. Vio. 2004. Correlation between POPs in sediment and edible bivalve in the Lagoon of Venice and estimation of the daily intake. *Organohalogen Comp.* 66: 1823–1828.

Ravera, O. 2000. The Lagoon of Venice: The result of both natural factors and human influence. *J. Limnol.* 59: 19–30.

Regione Veneto. 2000. *Piano per la prevenzione dell'inquinamento e il risanamento delle acque del bacino idrografico immediatamente sversante nella Laguna di Venezia—Piano Direttore 2000.* http://www.serviziregionali.org/direttore2000/?pagina=i_2.html (accessed August 27, 2009).

Rismondo, A., D. Curiel, F. Scarton, D. Mion, and G. Caniglia. 2003. *A new seagrass map for the Venice Lagoon*. Pp. 843–852, In: Sixth Conference on the Mediterranean Coastal Environment. Ravenna, Italy. MEADCOAST 2.

Salon, S., G. Cossarini, S. Libralato, X. Gao , C. Solidoro, and F. Giorgi. 2008. Downscaling experiment for the Venice lagoon. I. Validation of the present-day precipitation climatology. *Climate Res.* 38: 31–41.

Sarretta, A., S. Pillon, E. Molinaroli, S. Guerzoni, and G. Fontolan. In press. Sediment budget in the Lagoon of Venice, Italy. *Cont. Shelf. Res.*

Sfriso, A. and S. Cavolo. 1983. La situazione delle alghe nella laguna di Venezia. *Ambiente Risorse Salute* 16–17: 38–39.

Sfriso, A. and D. Curiel. 2007. Check-list of marine seaweeds recorded in the last 20 years in Venice lagoon and a comparison with the previous records. *Bot. Mar.* 50: 22–58.

Sfriso, A. and C. Facca. 2007. Distribution and production of macrophytes in the Lagoon of Venice. Comparison of actual and past abundance. *Hydrobiologia* 577: 71–85.

Sfriso, A., C. Facca, S. Ceoldo, and G. Pessa. 2005a. Sedimentation rates, erosive processes, grain size and sediment density changes in the lagoon of Venice. Pp. 203–213, In: P. Campostrini (ed.) *Scientific research and safeguarding of Venice*. Corila, Venezia, Volume 3.

Sfriso, A., C. Facca, S. Ceoldo, S. Silvestri, and P. F. Ghetti. 2003a. Role of macroalgal biomass and clam fishing on spatial and temporal changes in N and P sedimentary pools in the central part of the Venice Lagoon. *Oceanol. Acta* 26: 3–13.

Sfriso, A., C. Facca, and P. F. Ghetti. 2003b. Temporal and spatial changes of macroalgae and phytoplankton in a Mediterranean coastal area: The Venice Lagoon as case study. *Mar. Environ. Res.* 56: 617–636.

Sfriso, A., C. Facca, and P. F. Ghetti. 2007. Rapid Quality Index, based mainly on macrophyte associations (R-MAQI), to assess the ecological status of the transitional environments. *Chem. Ecol.* 23: 493–503.

Sfriso, A., C. Facca, and A. Marcomini. 2005b. Sedimentation rates and erosion processes in the lagoon of Venice. *Environ. Internat.* 31: 983–992.

Sfriso, A. and A. Marcomini. 1996. Decline of *Ulva* growth in the lagoon of Venice. *Bioresource Technol.* 58: 299–307.

Sfriso, A., A. Marcomini, and B. Pavoni. 1987. Relationship between macroalgal biomass and nutrient concentrations in a hypertrophic area of the Venice lagoon. *Mar. Environ. Res.* 22: 297–312.

Sfriso, A. and B. Pavoni. 1994. Macroalgae and phytoplankton competition in the central Venice lagoon. *Environ. Technol.* 55: 1–14.

Sfriso, A., B. Pavoni, A. Marcomini, and A. A Orio. 1992. Macroalgae, nutrient cycles and pollutants in the lagoon of Venice. *Estuaries* 15: 517–528.

Sfriso A., C. Facca, and P. F. Ghetti. 2009. Validation of the Macrophyte Quality Index (MaQI) set up to assess the ecological status of Italian marine transitional environments. *Hydrobiologia* 617: 117–141.

Socal, G., F. Acri, F. Bernardi Aubry, A. Berton, F. Bianchi, E. Capuzzo, J. Coppola, C. Facca, and A. Sfriso. 2006. Analisi dei popolamenti fitoplanctonici nella laguna di Venezia dal 1977 al 2004. *Biol. Mar. Mediterranea* 13: 178–184.

Socal, G., F. Bianchi, and L. Alberghi. 1999. Effects of thermal pollution and nutrient discharges on a spring phytoplankton bloom in the industrial area of the Lagoon of Venice. *Vie et Milieu* 49: 19–31.

Solidoro, C., M. Bastianini, V. Bandelj, R. Codermatz, G. Cossarini, D. Melaku Canu, E. Ravagnan, S. Salon, and S. Trevisani. 2009. Current state, scales of variability and decadal trends of biogeochemical properties in the northern Adriatic Sea. *J. Geophys. Res.* 114: C06026. DOI: 10.1029/2008JC005080.

Solidoro, C., V. E. Brando, C. Dejak, D. Franco, R. Pastres, and G. Pecenik. 1997a. Long-term simulation of population dynamic of *Ulva r.* in the Lagoon of Venice. *Ecol. Model.* 102: 259–272.

Solidoro, C., D. Melaku Canu, A. Cucco, and G. Umgiesser. 2004a. A partition of the Venice Lagoon based on physical properties and analysis of general circulation. *J. Mar. Systems* 51: 147–160.

Solidoro, C., R. Pastres, and G. Cossarini. 2005. Nitrogen and plankton dynamics in the lagoon of Venice. *Ecol. Model.* 184: 103–124.

Solidoro, C., R. Pastres, G. Cossarini, and S. Ciavatta. 2004b. Seasonal and spatial variability of water quality parameters in the lagoon of Venice. *J. Mar. System* 51: 7–18.

Solidoro, C., R. Pastres, D. Melaku Canu, M. Pellizzato, and R. Rossi. 2000. Modelling the growth of *Tapes philippinarum* in Northern Adriatic lagoons. *Mar. Ecol. Prog. Ser.* 199: 137–148.

Solidoro, C., G. Pecenik, R. Pastres, D. Franco, and C. Dejak. 1997b. Modelling *Ulva* rigida in the Venice Lagoon: Model structure identification and first parameters estimation. *Ecol. Model.* 94: 191–206.

Stefani, M. and S. Vincenti. 2005. The interplay of eustasy, climate and human activity in the late Quaternary depositional evolution and sedimentary architecture of the Po Delta system. *Mar. Geol.* 222–223: 19–48.

Suman, D., S. Guerzoni, and E. Molinaroli. 2005. Integrated coastal management in the Venice lagoon and its watershed. *Hydrobiologia* 550: 251–269.

Tagliapietra, D. and A. Minelli. 2009. Aquatic invertebrates. In: A. Minelli (ed.), *Lagoons estuaries and deltas*. Series, Italian Habitats. Ministero dell'Ambiente e della Tutela del Territorio e del Mare. *Museo Friulano di Storia Naturale* 23: 39–61.

Tagliapietra, D., M. Pavan, and C. Wagner. 1998. Macrobenthic community changes related to eutrophication in Palude della Rosa (Venetian lagoon, Italy). *Estuar. Coastal Shelf Sci.* 47: 217–226.

Tagliapietra, D., M. Sigovini, and A. Volpi Ghirardini. 2009. A review of terms and definitions to categorize estuaries, lagoons and associated environments. *Mar. Freshw. Res.* 60: 497–509.

Tagliapietra, D. and A. Volpi Ghirardini. 2006. Notes on coastal lagoon typology in the light of the EU Water Framework Directive: Italy as a case study. *Aquat. Conserv. Mar. Freshw. Ecosystems* 16: 457–467.

Umgiesser, G., D. Melaku Canu, A. Cucco, and C. Solidoro. 2004. A finite element model for the lagoon of Venice, Italy. Development and calibration. *J. Mar. Systems* 51: 123–145.

Vatova, A. 1960. Primary production in the high Venice lagoon. *J. du Conseil Intern. pour l'Explor. de la Mer* 26: 148–155.

Vesta (Venezia Servizi Territoriali Ambientali). 2005a. *Relazione Annuale impianto di depurazione di Campalto.* Venezia, Vesta.

Vesta (Venezia Servizi Territoriali Ambientali). 2005b. *Relazione Annuale impianto di depurazione di Campalto.* Venezia, Vesta.

Voltolina, D. 1975. The phytoplankton of the lagoon of Venice: November 1971–November 1972. *Pubblicazioni della Stazione Zoologica di Napoli* 39: 206–340.

20 Effect of Freshwater Inflow on Nutrient Loading and Macrobenthos Secondary Production in Texas Lagoons

Paul A. Montagna, and Jian Li*

CONTENTS

Abstract 513
20.1 Introduction 514
20.2 Conceptual Model 514
20.3 South Texas Lagoonal Estuaries 515
 20.3.1 Lavaca-Colorado Estuary 517
 20.3.2 Guadalupe Estuary 517
 20.3.3 Nueces Estuary 518
 20.3.4 Laguna Madre Estuary 518
20.4 Materials and Methods 519
 20.4.1 Study Sites 519
 20.4.2 Databases 520
  20.4.2.1 Benthos 520
  20.4.2.2 Hydrography 520
  20.4.2.3 Predators 520
  20.4.2.4 Primary Production 520
  20.4.2.5 Hydrology 520
20.5 Modeling 520
 20.5.1 Calibration and Validation 526
20.6 Results 528
20.7 Discussion 532
Acknowledgments 537
References 537

ABSTRACT

The Texas coast is characterized by lagoons behind barrier islands, and their ecology is strongly influenced by a longitudinal ecotone where freshwater inflow and nutrients decrease from northeast to southwest. The four Texas lagoons studied here are the Lavaca-Colorado, Guadalupe, Nueces, and Laguna Madre estuaries. Each of these lagoonal estuaries consists of a primary bay (the lagoon), which is nearest to the Gulf of Mexico, and a secondary bay, which is nearest to the freshwater influent source. The hypothesis of the current study is that climatic variability and inflow differences

* Corresponding author. Email: paul.montagna@tamucc.edu

among the ecosystems alter nutrient loading, which in turn affects benthic infaunal communities and maintains secondary production. Macrobenthic biomass data from a 5-year period were used to calibrate a bioenergetic model of secondary production. Benthic organisms were divided into two trophic groups: deposit feeders (that consume detritus or sediment organic matter) and suspension feeders (that filter phytoplankton or graze on benthic diatoms). Community structure is controlled by inflow, with more suspension feeders in high inflow estuaries and more deposit feeders in low inflow estuaries. Higher inflow translates into higher productivity. Within estuaries, the production to biomass ratio (P/B, unit of 1/year) increased with proximity to the freshwater source. The P/B ratio increased with water residence time (i.e., inflow volume adjusted by the estuary volume). The one exception was Laguna Madre, with the highest P/B of 3.2, which is due to extensive seagrass habitat. Corpus Christi Bay, with little inflow, had the lowest P/B of 1.2. Results of this study show that freshwater inflow, and concomitant nutrient loads, drives benthic community structure and function.

Key Words: benthos, infauna, climate, hydrology, water resources, trophic dynamics, Matagorda Bay, San Antonio Bay, Corpus Christi Bay, Laguna Madre

20.1 INTRODUCTION

The macrobenthos in Texas estuaries are affected significantly by freshwater inflow (Kalke and Montagna 1991; Montagna and Kalke 1992, 1995). The freshwater inflows may directly affect the respiration, excretion, reproduction, natural mortality, and migration of macrobenthos through the variation of salinity and salinity tolerance of various organisms (Montagna 1989). They may also indirectly affect the consumption and assimilation of organic matter by macrobenthos, if freshwater inflow affects the primary production through the variation of nutrient input to each bay system. However, freshwater inflow is not the only important environmental factor regulating the estuarine benthic subsystem. Predation by bottom fish and crabs, temperature change, light intensity, levels of particulate organic matter (POM), autotrophic production, and microbial production are other direct or indirect driving forces. A model is a way to integrate biological processes and environmental factors to gain insight into benthic dynamics, and explore the relationship between freshwater inflow and environmental effects.

The current project had three major goals: (1) to define quantitative relationships between benthic infaunal populations and freshwater inflows; (2) to link nitrogen loading with benthos; and (3) to develop a functional relationship between freshwater inflow and secondary productivity. Year-to-year variability occurred in freshwater inflow and benthic population densities during the study. There were high-inflow conditions between 1992 and 1993 and very similar conditions between 1987 and 1988. The intervening period between 1988 and 1992 was dry. Both wet periods occurred during El Niño events, and the dry conditions occurred during La Niña periods. Nitrogen is the key element that controls estuarine primary productivity in Gulf of Mexico estuaries. Populations of estuarine benthic infauna appear to exhibit long-term variation in response to estuarine salinity changes, which are a function of historical climatic patterns (Montagna and Kalke 1995). The temporal variation of benthic infauna may be caused by multiple interactive effects of several environmental factors correlated with freshwater inflow (e.g., changes in salinity, temperature, nutrients, and possibly pollution from land). Simple, multivariate statistical methods can be used to discover these relationships (e.g., Montagna and Kalke 1992, 1995), but cannot explain the functional dependencies of these effects. A modeling study based on bioenergetic processes can be used to assess functional responses of benthos to freshwater inflow.

20.2 CONCEPTUAL MODEL

Modeling an ecosystem can start from a qualitative conceptual model, as has been created for Texas estuaries (Montagna et al. 1996). Previous studies in Texas indicate that it is easier to demonstrate freshwater effects by comparing estuaries with different freshwater inflow regimes than to study the salinity gradient within one estuary because the ecotone gradient across estuaries is greater than the salinity gradient within estuaries (Montagna and Kalke 1992, 1995; Montagna et al. 2007; Russell and Montagna 2007). Thus, a synoptic approach is needed where several estuaries are studied simultaneously to compare the biological effects of freshwater inflow. The Lavaca-Colorado and Guadalupe estuaries provide an interesting comparison. They receive about the same amount of inflow, but San Antonio Bay is smaller than Lavaca and Matagorda bays and does not have significant, direct Gulf exchange; therefore, salinity is much lower in the Guadalupe Estuary. Nueces Estuary and Laguna Madre Estuary both have much lower river inflow volumes, but Nueces has direct Gulf exchange while Upper Laguna Madre does not. Most importantly, by replicating estuaries, two with high inflow and two with low inflow, general conclusions can be deduced.

Texas estuaries differ because of differences in freshwater inflow (Longley 1994; Montagna et al. 2007). For example, the benthic macrofaunal community in the Guadalupe Estuary has a higher biomass and greater temporal variability than in the Nueces Estuary (Montagna and Kalke 1992). The difference in biomass and benthic community structure may be due to several factors. There are different amounts of rainfall and freshwater inflow (Larkin and Bomar 1983), different predators (Quast et al., 1988), and different amounts of primary production that may be transformed into food sources for benthic fauna (Montagna and Yoon 1991). Polychaetes and mollusks are the two dominant taxa in both estuaries. They compose 91% to 97% of total abundance and 91% to 98% of total biomass (Montagna and Kalke 1992). The composition of polychaetes and mollusks is different between the two estuaries. In the Guadalupe Estuary, mollusks (56% in abundance and 80% in biomass) are more dominant than polychaetes (41% in abundance and 18% in biomass). This trend is reversed in the Nueces Estuary, where polychaetes (78% in abundance and 57% in biomass) are more dominant than mollusks (13% in abundance and 34% in biomass). In general, mollusks are dominant in sandy sediments and feed on epibenthic diatoms and other food sources derived from primary production. In contrast, polychaetes are dominant in muddy sediments and feed on deposited POM. The differences between the estuaries are consistent with the concept of benthic modular components (Tenore et al. 2006) and two different trophic systems: one that has a food web based on primary production (probably dominating in the Guadalupe Estuary), and another that has a food web based on detritus or deposited POM (probably dominating in the Nueces Estuary).

There is a continuous cycle of drought and flood conditions, which can greatly influence Texas estuaries. The long-term variability of freshwater inflow cycles, which is driven by the El Niño Southern Oscillation (ENSO) cycle, results in predictable salinity changes in the estuary (Tolan 2007). These cycles regulate freshwater inflow and salinity, and therefore, directly affect the biotic communities. Past studies in Texas estuaries demonstrate the biological effects of this cycle (Kalke and Montagna 1991; Montagna and Kalke 1992, 1995). Flood conditions usually introduce nutrient-rich waters very rapidly into the estuary, which results in lower salinity. During these periods, the spatial extent of the oligohaline fauna is increased. The estuarine fauna may even replace the marine fauna. The high level of nutrients stimulates a burst of benthic productivity of predominantly oligohaline and mesohaline communities. This is followed by a transition to low inflow conditions resulting in higher salinities, lower nutrients, more marine fauna, decreased productivity and densities, and drought conditions. At first, the polyhaline fauna may respond with a burst of productivity as the remaining nutrients are utilized, but eventually nutrients are depleted, resulting in lower densities. The cycle is repeated with flooding and high freshwater inflows. This conceptual model represents a biomechanistic explanation of how the benthos responds to freshwater inflow, and why their populations fluctuate from year to year with variable inflow.

20.3 SOUTH TEXAS LAGOONAL ESTUARIES

The four lagoonal estuaries studied in South Texas are the Lavaca-Colorado, Guadalupe, Nueces, and Laguna Madre systems (Figure 20.1). Lagoons link the four estuaries. For example, Laguna

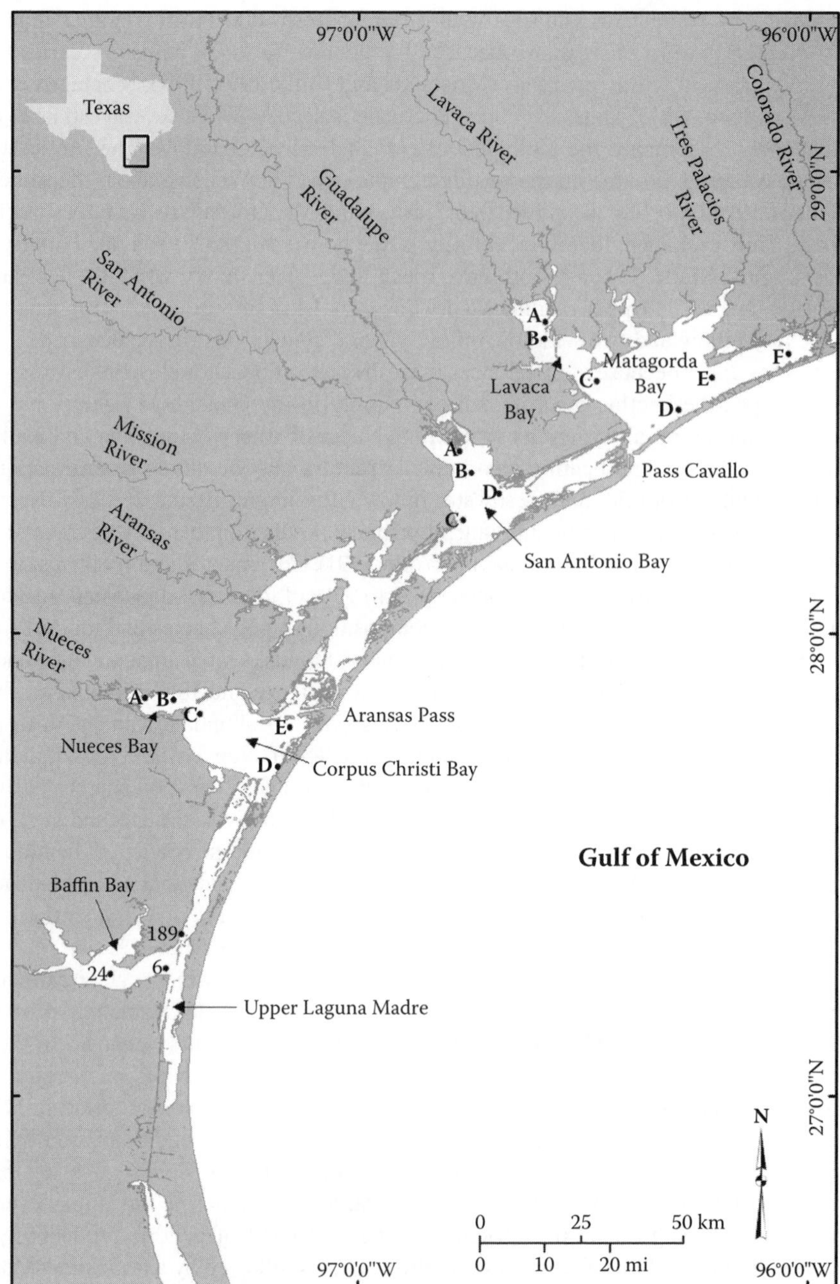

FIGURE 20.1 The Texas lagoon study area and station locations.

Madre links Baffin Bay and Corpus Christi Bay. Red Fish Bay is actually a lagoon system that links Corpus Christi Bay and Aransas Bay. Mesquite Bay is a lagoon that links Aransas Bay to San Antonio Bay, and Espiritu Santo Bay is the lagoon that links San Antonio Bay to Matagorda Bay. The series of Texas lagoons is therefore linked from the Rio Grande to the Colorado River.

The Intracoastal Waterway flows through the lower Laguna Madre to Matagorda Bay, increasing circulation among the lagoons. The bays are dominated by deeper, muddy sediments. The lagoons are shallower, narrower, and have clearer water with more seagrass beds than the bays. They also

TABLE 20.1
Comparison of Physical Characteristics of Four Texas Estuaries Examined in This Study

	Estuary			
Characteristics	**Lavaca-Colorado**	**Guadalupe**	**Nueces**	**Laguna Madre**
Size (km^2)[a]	1,158	551	433	1,139
Rainfall average (cm yr^{-1})[a]	102	91	76	69
Inflow average (10^6 m^3 yr^{-1})	3,242	2,545	509	–947
Bottom salinity average (ppt)[a]	22–22.2	13–18.6	16.6–30.6	31.3–37
Maximal monthly mean inflow (m^3 s^{-1})[a]	140 (1938–1990)	120 (1939–1989)	50 (1939–1989)	5 (1965–1987)
Average depth (m)	2	1	2	1
Maximal phytoplankton abundance (10^{-3} cell mL^{-1})[b]	4.5	19	1.1	1.6
Maximal zooplankton abundance (10^5 individuals m^{-3})[b]	0.3	0.5	500	200
Maximal benthos abundance (individuals m^{-2})[c]	9,700	35,000	7,200	13,000

[a] Orlando et al. (1993).
[b] Summarized by Armstrong (1985) and Montagna et al. (1995).
[c] Montagna and Kalke (1995), mollusks only.

have less fetch. These conditions largely account for the ecological differences observed among the four systems. The Lavaca-Colorado, Guadalupe, and Nueces estuaries are dominated by open bay and soft bottom habitat, and Laguna Madre is dominated by submerged aquatic vegetation habitat. Another significant difference is the volume of Gulf of Mexico water exchanged with estuaries through passes. The same basic habitats and processes occur in all Texas estuaries but in different proportions (Montagna et al. 2007).

20.3.1 Lavaca-Colorado Estuary

The Lavaca-Colorado is the largest of the four Texas estuaries targeted in this study (Table 20.1). It is ~1158 km^2 in area and has the largest freshwater inflow rates, 3242 × 10^6 m^3 yr^{-1}, ~1.5 times that of the Guadalupe Estuary and ~6 times that of the Nueces Estuary. The freshwater inflows are relatively constant, as are the bottom salinities. Phytoplankton abundance is less than in the Guadalupe Estuary, but four times higher than in the Nueces and Laguna Madre estuaries. Zooplankton abundance is about the same as in the Guadalupe Estuary, or 1000 times less than in the Nueces and Laguna Madre estuaries. The abundance of the benthic community is lower than in all the other estuaries.

20.3.2 Guadalupe Estuary

The Guadalupe Estuary is relatively small (551 km^2) with high freshwater inflow rates (averaging 2545 × 10^6 m^3 yr^{-1}). The small size relative to freshwater inflow results in the lowest average salinities of the four estuaries studied. Phytoplankton abundance is highest in the Guadalupe Estuary, with an average density of 19,000 cells mL^{-1}, which is 4 to 20 times higher than in other estuaries. Benthic abundance, such as mollusks, is 3 to 30 times higher than in other estuaries. The high mollusk density is probably due to high freshwater inflow into a small estuary, which results in low salinities that mollusks require, and the high densities of phytoplankton, which are the main food source for this biotic group.

20.3.3 Nueces Estuary

Nueces Estuary has a higher level of nutrient storage, kinetic storage, and oxygen storage in comparison with the Laguna Madre Estuary, due to greater inflow from rivers and creeks per unit time. However, phytoplankton primary production in Corpus Christi Bay is only 0.48 to 1.26 g C m^{-2} d^{-1} (Odum and Wilson 1962; Odum et al. 1963; Flint 1984; Stockwell, 1989). The average water depth in the Nueces Estuary is 2 m (compared to 1 m for Laguna Madre). The total area is relatively small (433 km^2), about one-third that of the Laguna Madre (see below). The amount of sun radiation per unit water volume is much lower than in the Laguna Madre. This means a lower capacity for temperature storage and lower ecological efficiency than in the Laguna Madre, which may explain the lower primary production rates and lower amount of consumer biomass.

The Nueces Estuary receives both river and creek inflow, while Laguna Madre receives only intermittent creek inflow. The Nueces Estuary has an annual inflow balance (509×10^6 m^3 y^{-1}) that is much higher than in the Laguna Madre Estuary, indicating that energy flow from the land to the bay system is higher than in the Laguna Madre. However, the energy flow from the bay to the ocean is also higher. The average rainfall decreases from the Nueces (76 cm y^{-1}) to the Laguna Madre (69 cm y^{-1}). The water residence time in the Nueces Estuary is 0.46 y. The gradient of river inflow and average rainfall accounts for the lower salinity in the Nueces Estuary (17‰ to 31‰) than in Laguna Madre Estuary (30‰ to 37‰). High salinity variation causes greater ecosystem instability than in the Laguna Madre, by affecting population aging, respiration, reproduction, and migration rates (Montagna et al. 1996). The Nueces Estuary is also more stressed, because it is the most urbanized of the four systems.

20.3.4 Laguna Madre Estuary

Laguna Madre is as large (1139 km^2) as the Lavaca-Matagorda Estuary. It has an average depth that is shallower than the other estuaries, and a larger surface area receiving sunlight (Montagna et al. 1996). Therefore, the sun radiation per unit water volume is much higher in the Laguna Madre than in the other estuaries, so the temperature storage is higher. This increases the ratio of energy flow between subsystem and storage components, by increasing all biophysiological processes, such as the synthesis, intake, decomposition, aging, respiration, migration, and reproduction rates (Montagna et al. 1995). The high energy flow in this ecosystem results in high rates of primary production. Phytoplankton primary production in Laguna Madre ranges from 2.68 to 4.78 g C m^{-2} d^{-1}, which is more than two times that of the Nueces Estuary (Odum and Wilson 1962; Odum et al. 1963). The high ecological efficiency also leads to high abundances of the higher-level consumers, such as benthic mollusks and fishes (Montagna et al. 1996). Benthic mollusk abundance in Laguna Madre (13,000 individuals m^{-2}) is twice that of the other estuaries (2500 to 7200 individuals m^{-2}). The commercial harvest of finfish in Laguna Madre (834×10^3 kg y^{-1}) is about four times higher than the others (151 to 207×10^3 kg y^{-1}). This biomass productivity is probably due to overall higher primary production in Laguna Madre.

Laguna Madre Estuary has a lower energy input from rivers compared with the other three estuaries. It has a negative inflow balance (-947×10^6 m^3 y^{-1}), which means the freshwater inflow is less than the outflow (e.g., evaporation). The negative balance also accounts for hypersaline conditions in Laguna Madre. Flow of water from the Gulf passes and the Nueces Estuary replaces the evaporating water. Water residence time in Laguna Madre, however, is very difficult to calculate because of its shallow depth, "negative" inflow, and connection to the Nueces Estuary (Montagna et al. 1996).

Energy flow from the ocean to the bay is higher for Laguna Madre than for the other estuaries (Montagna et al. 1996). The detritus storage in Laguna Madre is expected to be much higher than in the other estuaries, because of the high primary production of the extensive seagrass habitat. The consumers are dominated by deposit feeding benthos. The input of seawater and lower input of freshwater from rivers results in a stable, high salinity system with generally low nutrient concentrations. The main limitation on producers' synthesis, therefore, may be only nutrients (Montagna et al. 1996). So, Laguna Madre has higher temperature, salinity, and detritus storage, and remains a more natural ecosystem than the other three estuaries.

20.4 MATERIALS AND METHODS

As hypothesized in the conceptual model, two abiotic factors (i.e., freshwater inflow balance and physiography) are the dominant factors that drive biological processes in Texas lagoons. Thus, the overall goal of this project was to generate a quantitative model of benthic trophic dynamics to determine freshwater inflow effects on benthic productivity. A long-term dataset on the Lavaca-Colorado, Guadalupe, Nueces, and Laguna Madre estuaries was used to calibrate the model. Each estuary is subdivided into a primary and secondary bay with different levels of freshwater effects due to a river-to-sea gradient. Comparison between the two bays within each estuary represents effects of fresh water within the estuary.

20.4.1 STUDY SITES

Four to six stations were chosen in each of the four estuaries (Figure 20.1, Table 20.2). Generally, two replicate stations (A and B) were in the secondary bay where freshwater influences dominated, and two other replicate stations (C and D) were in the primary bay where marine influences predominated. By using two stations in the freshwater influenced zone and two stations in the marine influenced zone we replicated effects at the treatment level and avoided pseudoreplication. There was a diversion of the Colorado River into the east arm of Matagorda Bay in 1993; therefore, two additional stations (E and F) were located there to capture the effects of the Colorado River. The stations in Laguna Madre Estuary were located using a paired-station strategy. Two stations were located in Baffin Bay (6 and 24), and two stations in the Laguna Madre in a seagrass bed (189G) and an unvegetated sand patch (189S). The station data for each bay were pooled, enabling the characterization of eight major bays.

TABLE 20.2
Location of Sampling Stations and Sampling Periods Used in This Study

Estuary	Bay Type	Bay Name	Station	Sampling Period
Lavaca-Colorado	Secondary	Lavaca	A	1984–1995
	Secondary	Lavaca	B	1988–1995
	Primary	Matagorda	C, D	1988–1995
	Lagoon	East Matagorda	E, F	1993–1995
Guadalupe	Secondary	Upper San Antonio	A, B	1987–1995
	Primary	Lower San Antonio	C, D	1987–1995
Nueces	Secondary	Nueces	A, B	1988–1995
	Primary	Corpus Christi	C, D	1988–1995
	Primary	Corpus Christi	E	1988–1995
Laguna Madre	Secondary	Baffin Bay	6, 24	1988–1995
	Primary	Laguna Madre	189G, 189S	1988–1995

20.4.2 Databases

20.4.2.1 Benthos

Sampling began in the Lavaca-Colorado Estuary in November 1984 (Kalke and Montagna 1991). The Guadalupe Estuary and Nueces Estuary were sampled bimonthly during 1987 (Montagna and Kalke 1992). From these three studies, it was determined that long-term changes in the benthos could be characterized by sampling four stations, four times per year. A sampling program was started in July 1988 to compare the Lavaca-Colorado and Guadalupe estuaries (Table 20.2). Sampling began in Laguna Madre and Baffin Bay in 1988.

Macrobenthos were sampled using a 6.7-cm-diameter core to a depth of 10 cm. Three replicate samples were taken at each station-date combination and preserved in 5% buffered formalin, sorted using 0.5-mm sieves, identified, and counted as described in Montagna and Kalke (1992). For biomass, samples were dried for at least 24 hours at 55°C and weighed. Before drying, the mollusks were placed in 2 N HCl for 1 to 5 minutes to dissolve the carbonate shells, and then washed.

20.4.2.2 Hydrography

During each benthic sampling event, a multiparameter sonde (Hydrolab Surveyor II or YSI 6 series) was used to measure salinity, temperature, and water depth. Salinity was also measured with a refractometer. Water samples were taken to measure chlorophyll *a* (chl *a*) and nutrients. For chlorophyll *a*, the sample was filtered onto glass fiber filters and placed on ice (<0.4°C). Chlorophyll *a* was extracted overnight with methanol and read fluorometrically on a Turner Model 10-AU using a non-acidification technique (EPA method 445.0). Analysis of nutrients was performed using a Technicon Autoanalyzer (Whitledge 1989) or a LaChat QC 8000 ion analyzer.

20.4.2.3 Predators

Mobile epifauna data were obtained from Texas Park and Wildlife Department (TPWD) (McEachron and Fuls 1996). The Coastal Fisheries Division samples monthly in the four estuaries using an otter trawl and bag seine. The data are available from 1976 to present. In a study of stable isotope content and mercury bioaccumulation in different food chains, it was determined that black drum, red drum, and blue crab are the main predators on benthic infauna (Montagna and Kathmann unpublished). The average abundance for these three predators collected from otter trawls was used for all TPWD stations within a bay.

20.4.2.4 Primary Production

A range of values for primary production from previous studies is available (Table 20.3). Monthly day length for the Texas coast was obtained from Tony Amos at the University of Texas Marine Science Institute.

20.4.2.5 Hydrology

Freshwater inflow data, which includes gauged flows and modeled flows for ungauged areas, were obtained from the Texas Water Development Board (TWDB, http://www.twdb.state.tx.us/data/bays_estuaries/bays_estuary_toc.htm).

20.5 MODELING

The model consists of several mathematical equations used to calculate the variation of benthos biomass in response to the variation in environmental data. Model input was the observed long-term environmental data, and output was the simulated benthos biomass over time. The simulation of the observations is based on the parameters calibrated from the input dataset. Sensitivity analysis was performed before calibration to ensure that the output range of the simulation would extend over the full range of all observations.

TABLE 20.3
Primary Production Data Available from Previous Studies

Bay	Previous Record of Primary Production (g C m^{-2} d^{-1})	References
Lavaca	0.5–2.4	Brock (1994)
Matagorda	0.5–2.4	Brock (1994)
Upper San Antonio	0.3–1.8	Stockwell (1989)
Lower San Antonio	0.5–3.85	Stockwell (1989)
Nueces	0.35–1.7	Stockwell (1989)
Corpus Christi	0.75–4.1	Stockwell (1989)
Baffin	1.2–4.1	Odum and Wilson (1962)
Upper Laguna Madre	2.75	Odum et al. (1963)

The long-term, benthic infauna dataset was used to calibrate the model of biological processes. Energy circuit language (Odum 1971, 1972, 1983) is used to present the model structure for simulating benthos biomass (Figure 20.2). The relationship between benthic infaunal biomass and environmental factors associated with freshwater inflow is incorporated into the model. To test for inflow effects, the ideal input would be freshwater inflow as the forcing function. However, inflow rates have variable effects on salinity depending on the physiographic, inflow, and tidal characteristics of each estuary. Therefore, a physical model that predicts salinity change would be needed to provide input to the biological model. To avoid this level of complexity, the empirical salinity values measured during sampling were used as input. In this way, salinity is used as a surrogate for inflow.

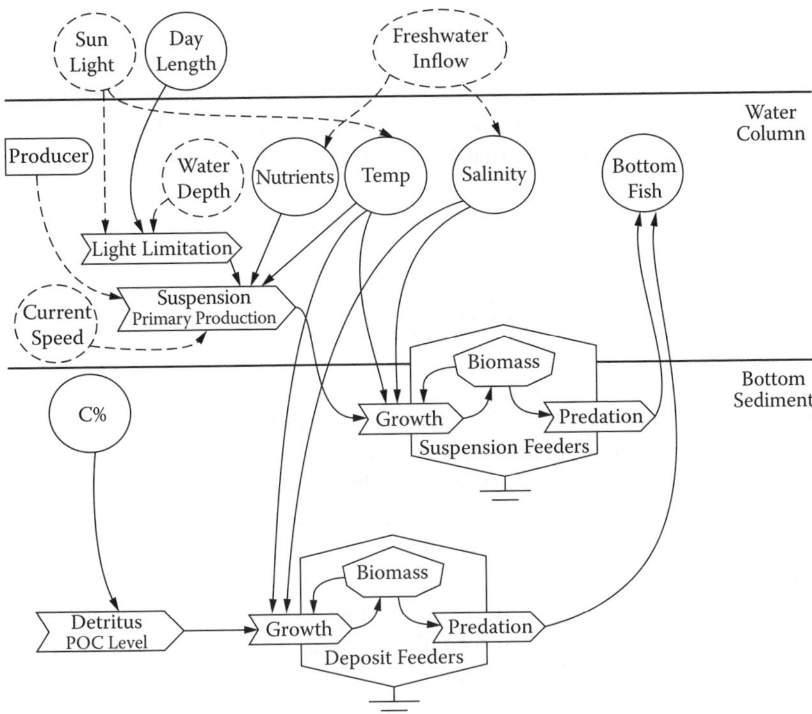

FIGURE 20.2 Energy circuit diagram for the structure of the benthic macrofaunal biomass model. Dashed lines are the parts not included in the model.

Salinity values represent the integration of all the physical characteristic of the estuary (e.g., size, inflow, outflow, oceanic exchange, and climatic variability).

The model (Figure 20.2) includes four forcing functions: salinity, temperature, food sources (bottom-up control), and predators (top-down control). The forcing functions drive the model mainly through four environmental limitations: salinity, temperature, food availability, and predation. Other forcing functions (e.g., nutrient concentrations, day length, and water depth) drive the model through the estimation of food source availability by the calculation of primary production.

The macroinfauna were divided into two trophic guilds, suspension feeders and deposit feeders, and modeled separately. The two groups represent trophic interactions in the grazing food chain and the detrital food chain (Tenore et al. 2006). Grazers utilize autotrophic production and detritivores utilize heterotrophic production. To simplify the data input, all macrobenthos were separated into one of two trophic groups: suspension feeders and deposit feeders. Suspension feeders are defined as benthos obtaining their food sources through capturing suspended particles from the sediment surface or water column, filtering phytoplankton from the water column, or grazing benthic diatoms on the sediment surface. Suspension feeding taxa include the Mollusca, Crustacea, and Chironomid larvae. Deposit feeders are defined as those organisms obtaining their food through ingestion of the sediment, predation, or omnivory. The deposit feeders include the Hemicordata, Nemertinea, Ophiuroidea, Polychaeta, and Sipunculida. Although some macrobenthos can switch between being suspension feeders and deposit feeders (Taghon et al., 1980), the simplification of categorizing each taxa in one group is necessary, and allows us to define suspension feeders as those benthos limited by autotrophic food sources, and deposit feeders as those benthos limited by heterotrophic food sources. Descriptions of the benthic community structure of these estuaries have already been published (Kalke and Montagna 1991; Montagna and Kalke 1992, 1995; Martin and Montagna 1995; Mannino and Montagna 1996; Conley 1996; Palmer et al. 2002).

Primary production (which is based mainly on plant biomass, light, nutrient concentrations, and temperature) is the principal food source for suspension feeders, that consume phytoplankton and benthic diatoms. Because primary production was not measured during every sampling period it was necessary to predict the food sources for suspension feeders in a model. Deposit feeders primarily consume POM, and this can be approximated by the concentration of total organic carbon (TOC) in sediments, which is empirically derived. The accumulation of POM and the variation of nutrient concentrations and salinity caused by temporal variations of freshwater inflows are not simulated here.

The basic mathematical formula that describes the change of benthic biomass over time (Li et al. 1996) is based on the law of conservation of mass (Crisp 1971), and in the form:

$$\frac{d(B)}{d(t)} = I \cdot A - L - D \tag{20.1}$$

where B is the benthos biomass, t is time, I is the total intake of food by benthic infauna, A is average assimilation efficiency of benthic infauna, L is the total loss due to respiration, excretion, and age-related mortality, and D is the total mortality caused by predators. Unfortunately, it is not possible to simulate B in terms of I and L, because the observed data on food source standing stocks and respiration of benthos are rare and unknown in the study area. Our approach is to substitute net growth rate for I and L.

The net growth rate is used in place of the intake rate, assimilation efficiency, respiration rate, aging mortality, and excretion rate. The formula becomes a Lotka–Volterra growth rate model (Lotka 1925) in the form above as Equation 20.2:

$$\frac{d(B)}{d(t)} = r \cdot B - g \cdot F \tag{20.2}$$

where r is the net growth rate without predation pressure. The predation loss is calculated by the feeding rate of predators, g, and the density of predatory fish, F.

Brown and Rothery (1993) have suggested that the growth rate is limited. Therefore, a logistic limitation is added to the equation in the form:

$$\frac{d(B)}{d(t)} = r \cdot B \cdot \left(1 - \frac{B}{c}\right) - g \cdot F \tag{20.3}$$

where c is the biomass carrying capacity for a population that is limited by space. The c in Equation 20.3 is only a limitation for the carrying capacity of biomass.

The limitation of a population and its biomass is also due to many other environmental effects. Our model is based on Equation 20.3, and has been modified to include other factors of environmental limitation. The new equation contains a parameter to reduce the maximal growth rate (r) and maximal predation rate (g) by the effects of environmental limitation (E). The values of E are between 0 and 1. When $E = 1$, there is no environmental limitation, and the benthic population reaches maximal growth rate, or the predators reach maximal feeding rate. When $E = 0$, environmental factors reach maximal limitation; benthic populations do not grow, or predators do not consume benthos. Because there is more than one predator, the final equation for the model becomes:

$$\frac{d(B_{(i,j)})}{d(t)} = \frac{r_{(i)}}{12} \cdot E_{ben(i,j)} \cdot B_{(i,j)} \cdot \left(1 - \frac{B_{(i,j)}}{c_{(i)}}\right) - 30 \cdot E_{fish(i,j)} \cdot \sum_k g_{(i,j,k)} \cdot F_{(j,k)} \tag{20.4}$$

where $i = 1$ or 2 for deposit feeders or suspension feeders; $j = 1$ to 8 for eight bay systems; $k = 1$ to 3 for three different predators: red drum, black drum, and blue crab. The net growth rate is $r_{(i)}$. The environmental limitation for benthos biomass growth is $E_{ben(i,j)}$. The biomass carrying capacity levels for the two feeding groups is $c_{(i)}$. The predation rate by fish k in bay j to prey benthos i is $g_{(i,j,k)}$. The term $E_{fish(i,j)}$ is the environmental limitation for predation. The different benthos species have different biomasses in each bay, computed by their r, E, c, and g. Predator abundances are also different in the eight bays. The benthos should have the same r and c in all eight bays. However, the dominant species in the deposit-feeding group and suspension feeding group differ in the different bays. Therefore, it is necessary to run the model separately for each of the eight bays.

The term $E_{ben(i,j)}$ includes three limitation effects: temperature, salinity, and food.

$$E_{ber(i,j)} = E_{tem(j)} \cdot E_{sal(i,j)} \cdot E_{food(i,j)} \tag{20.5}$$

An exponential equation was used for the temperature effect (Carrada 1983):

$$E_{tem(j)} = \frac{1}{\dfrac{\left|T_{(j)} - T_{\max}\right|}{e^{P(1)}}} \tag{20.6}$$

where $E_{tem(j)}$ is the temperature limitation, T is the temperature, and $T_{\max}$ is the most suitable temperature, which is fixed at the highest temperature recorded at each location. When T is close to $p_{(1)}$, $E_{tem(j)} = 1$, and there is no temperature limitation. Therefore, $p_{(1)}$ is a parameter that describes the weighting due to temperature limitation. The higher $p_{(1)}$ is, the higher the sensitivity to temperature.

The salinity effects are the most interesting environmental factor, being directly correlated with freshwater inflow. All invertebrates have suitable salinity ranges, at which the population growth is maximal, that is, the highest metabolism rates (Wohlschlag et al. 1977). An exponential equation is used to model salinity limitation. The equation is similar in form to that used for temperature limitation:

$$E_{sal(i,j)} = \frac{1}{\dfrac{\left|s_{(i,j)} - p_{(i,3)}\right|}{e^{p_{(2)}}}} \tag{20.7}$$

where $E_{sal(i,j)}$ is the salinity limitation, $S_{(j)}$ is salinity, $p_{(i,3)}$ is the optimal salinity for a population, and $P_{(2)}$ is a parameter that describes the weight of the salinity limitation. There is no salinity effect when $P_{(2)} = \infty$. Salinity limitation has a centralized optimum, with greater affects at high and low salinities. The greater the salinity tolerance range, the higher the $P_{(2)}$ value.

Michaelis–Menten kinetics is used to describe the food source limitation (see a review, Keen and Spain 1992):

$$E_{food(i,j)} = \frac{F_{(i,j)}}{F_{(i,j)} + p_{(i,4)}} \tag{20.8}$$

where $E_{food(i,j)}$ is the food limitation, $F_{(i,j)}$ is the concentration of food source for the infauna, which is sedimentary POC for deposit feeders and primary production for suspension feeders, and $p_{(i,4)}$ is a parameter at which the food concentration is at half the maximum level of the population growth rate.

Because two feeding groups (deposit feeders and suspension feeders) were simulated, two different food sources were considered: detritus in sediment and organic matter in the water. Sedimentary POM was used as food sources for deposit feeders, and expected primary production was used for suspension feeders. Increased consumer biomass ($B_{(i,j)}$) can increase food limitation. Therefore, Equation 20.8 transforms to the ratio as a function of benthic biomass:

$$E_{food(i,j)} = \frac{\dfrac{F_{(i,j)}}{B_{(i,j)}}}{\dfrac{F_{(i,j)}}{B_{(i,j)}} + p_{(i,4)}} \tag{20.9}$$

The sedimentary POM is expected to be a constant level in each bay. We used eight parameters as POM levels, one for each of the eight bays. The POM levels were precalibrated by the observed carbon concentration ($C\%_{(j)}$) in the sediment ($j = 1$ to 8 for eight bays):

$$C\%_{(j)} = \frac{p_{pom(j)}}{p_{sed}} \cdot 100\% \tag{20.10}$$

where $p_{pom(j)}$ is the sedimentary POM level for each bay, and p_{sed} is a parameter for the average dry weight of whole sediment. The POM levels for each bay are the food sources for deposit feeders ($F_{(i,j)}$) in each bay:

$$\left(F_{(i,j)}\right) = p_{pom(j)} \tag{20.11}$$

Primary production is expected to be the main food source for suspension feeders. It is simulated as a function of day length, temperature, nutrient concentration, and water depth. Primary production was precalibrated using data from previous studies (Armstrong 1985; Stockwell 1989) using the following formula:

$$F_{(2,j)} = D_{(j)} \cdot p_{mic(2)} \cdot \frac{1}{\dfrac{|T_{(j)} - T_{\max}|}{e_{pmic(3)}}} \cdot L_{(t)} \cdot E_{nut(j)} \tag{20.12}$$

where $F_{(2j)}$ is the available food source for suspension feeders, $p_{mic(2)}$ is the maximal monthly primary production rate, $p_{mic(3)}$ is the temperature limitation for primary production, $L_{(t)}$ is the day length that represents light limitation, and $E_{nut(j,t)}$ is the nutrient limitation for photosynthesis that includes concentrations of nitrogen (N), silica (Si), and phosphorus (P). Because suspension feeders only use the available food source 10 cm above the sediment surface, the following adjustment must be taken into consideration:

$$D_{(j)} = \frac{100.030}{0.42.10 \cdot d_{(j)}} \tag{20.13}$$

Nutrient limitation ($E_{nut(j)}$) for photosynthesis was modeled according to the Redfield ratio of 106:16:15:1, which assumes that producers use C, N, Si, and P proportionally by weight (Redfield 1934; Parsons et al. 1961):

$$E_{nut(j)} = MIN\left(\frac{\dfrac{[N]_{(j)}}{[N]_{(j)} + p_{mic(1)}}}{16}, \frac{\dfrac{[P]_{(j)}}{[P]_{(j)} + p_{mid(1)}}}{1}, \frac{\dfrac{[Si]_{(j)}}{[Si]_{(j)} + p_{mic(1)}}}{15} \right) \tag{20.14}$$

where $[N]_{(j)}$, $[P]_{(j)}$, and $[Si]_{(j)}$ are concentrations of inorganic nitrogen, phosphorus, and silica.

Photosynthesis is limited by light, which varies seasonally. A sine function is used to simulate the seasonal cycle of day length:

$$L_{(t)} = p_{avg} + p_{amp} \cdot \sin\left(\frac{2\pi \cdot (t)}{12} - p_{pha} \right) \tag{20.15}$$

where $L_{(t)}$ is day length at time t, p_{avg} is the average day length over a year, p_{amp} is the amplitude of the seasonal fluctuation, and p_{pha} is the correction factor for starting the phase of the sine cycle at a given time.

Predation can be limited by temperature, salinity, prey abundance, and predator density. In this study, we considered only the benthic prey abundance, because predation rates on benthos are strongly related to the benthos biomass. A complete ecosystem model would also include fish bioenergetics. Predator limitation, $E_{fish(i,j)}$, may be different for different predators, but in this study, the same $E_{fish(i,j)}$ was used for all predators, including black drum, red drum, and blue crab. However, $E_{fish(i,j)}$ is different for deposit feeders and suspension feeders because of the different vertical distribution of these two groups within the soft bottom habitat.

In addition to standing stock, a second characteristic of prey is its distribution. The feeding rate of predators is expected to increase exponentially when the prey are aggregated in time or space. An "S"-shaped curve is used to simulate this effect (Montagna et al. 1993):

$$E_{fist_{(i,j)}} = 1 - e^{-P_{(s)} - B_{(i,j)}} \quad (20.16)$$

where $B_{(i,j)}$ is the biomass of the prey benthos (i = 1 or 2 for deposit feeders or suspension feeders, and j = 1 to 8 for the eight bays), and $p_{(5)}$ is a new parameter for the aggregation effect. When biomass ($B_{(i,j)}$) is at a very low level, the value of term $E_{fish(i,j)}$ is close to 0, and limitation due to the aggregation effect is nil. When the predator reaches its maximal grazing rate and $B_{(i,j)}$ is very high, the term $E_{fish(i,j)}$ is close to 1, and the limitation due to aggregation is at the maximal level.

The model was constructed using the FORTRAN 77 language and facilitated by the PC software package SENECA (Simulation ENvironment for ECological Application) (de Hoop et al. 1989). SENECA is a PC/DOS software package produced by the Netherlands Institute of Ecology. It simplifies model setup, supports techniques for calibration of the model (i.e., estimating the best fit parameter values according to a goodness of fit test), and links programs to a FORTRAN Compiler. The FORTRAN programs created by SENECA are listed in Montagna and Li (1996).

20.5.1 Calibration and Validation

The model was calibrated for the eight bay systems for two periods: a short-term period from January 1991 to December 1995, where the dataset was balanced and complete for all measurements in all eight bays, and a longer period from January 1987 to December 1995, where there were some missing data. The initial ranges for the 17 parameters were set as the same for each bay. For each case, there were over 10,000 calibration runs, and all parameter ranges were reduced to less than 50% of the initial ranges (Table 20.4).

Model validation was judged using three methods: (1) goodness of fit test values, (2) biological meaning of the estimated parameters, and (3) comparison of different simulation periods. Two periods were defined: (1) a short-term synoptic period (1991 to 1995), which was used for calibration; and (2) a long-term period (1987 to 1995), which included all data available.

The goodness of fit test compares each simulation against the observed values in each bay–date combination. The model is valid when the variance of the goodness of fit test is close to zero. The goodness of fit values for all simulations or parameters calibrated by short-term datasets ranged from 0.59 to 3.25 (Table 20.5). The simulation of Corpus Christi Bay has the best goodness of fit values, followed by Lower San Antonio Bay, Matagorda Bay, Laguna Madre, Baffin Bay, Nueces Bay, and Upper San Antonio Bay. Lavaca Bay has the worst goodness of fit values. The short-term simulation has a better fit than the long-term simulation. The short-term simulation was not better using parameters calibrated from the long-term database than those from the short-term database. Surprisingly, the long-term simulation was better using the parameters calibrated from the short-term database than parameters calibrated using the long-term database. This was true for both benthic feeding types in Lavaca Bay (2.39 < 3.25 and 1.45 < 1.63) and Lower San Antonio Bay (0.84 < 0.89 and 0.95 < 1.05), but only true for suspension feeders in Matagorda Bay (1.00 < 1.08) and deposit feeders in Corpus Christi Bay (0.84 < 0.92). In general, the short-term synoptic data simulation is more appropriate for use in this study than the long-term data simulation.

Biological responses are indicated by the different response of the two trophic groups. Suspension feeders, which rely on food from the overlying water column, should have lower suitable salinity ranges than deposit feeders in each bay. Therefore, the model should predict higher production to biomass (P/B) ratios for suspension feeders in the secondary bays than in the primary bays. The predicted annual P/B ratios are similar for deposit feeders in all eight bays. In contrast, the annual P/B ratios for suspension feeders are always higher in the secondary bays than in the primary bays (Table 20.6). Therefore, the model structure correctly predicts the expected biological responses based on knowledge of the feeding mechanisms. An invalid model structure would predict a biological response that is counterintuitive.

TABLE 20.4
Best Fit Parameter Values from the Calibration of Eight Bay Systems for the Continuous Four-Year Database: 1991–1995

Parameter	Best Fit Values for Each Bay								Initial Range
	LB	MB	US	LS	NB	CC	BB	LM	
$p_{(1)}$	44.73	42.42	44.57	41.43	40.24	44.20	42.55	42.34	20, 45
$p_{(2)}$	19.66	17.14	17.78	39.32	17.91	17.73	32.67	35.27	20, 45
$p_{(1,3)}$	34.89	37.77	27.05	34.99	29.95	29.62	35.80	32.04	20, 40
$p_{(1,4)}$	96.11	62.42	2.55	44.03	74.83	4.95	93.46	85.88	0, 100
$g_{(1,j,1)}$	—	0.553	3.944	3.096	3.176	0.620	0.110	4.910	0, 5
$g_{(1,j,2)}$	—	3.471	4.976	4.823	4.744	4.973	2.614	4.959	0, 5
$g_{(1,j,3)}$	0.6814	1.2477	0.1388	0.0743	0.4850	1.6649	1.4502	0.7158	0, 5
$p_{(5)}$	0.002059	0.003724	0.001323	0.009225	0.004171	0.002859	0.002285	0.002087	0.001, 0.01
$r_{(2)}$	5.806	6.670	5.075	5.077	6.316	6.260	5.101	6.541	5, 20
$c_{(1)}$	68.88	30.31	69.81	51.29	42.27	30.62	46.34	34.76	30, 70
$p_{(2,3)}$	16.75	5.11	8.11	19.25	7.11	9.87	17.58	14.17	20, 40
$p_{(2,4)}$	6.46	15.13	89.89	40.02	81.46	44.78	85.17	64.64	0, 100
$g_{(2,j,1)}$	—	3.705585	0.8315539	4.858335	3.312729	1.334923	0.0976517	3.821	0, 5
$g_{(2,j,2)}$	—	0.4145367	0.5318332	4.515656	4.939147	0.400967	0.635004	4.975	0, 5
$g_{(2,j,3)}$	0.9929386	1.195466	0.4320632	0.09369156	0.4154133	1.625607	1.46157	0.5934	0, 5
$r_{(2)}$	5.343904	8.033044	5.942176	5.186666	7.632288	7.890732	6.403854	6.628	5, 20
$c_{(2)}$	50.24743	99.65412	89.25981	52.67889	87.83488	97.43611	52.60339	76.38	50, 100

Note: Abbreviations: LB = Lavaca Bay, MB = Matagorda Bay, US = Upper San Antonio Bay, LS = Lower San Antonio Bay, NB = Nueces Bay, CC = Corpus Christi Bay, BB = Baffin Bay, LM = Upper Laguna Madre. LB did not have red drum and black drum observations, so the parameters $g_{(i,1,1)}$ and $g_{(i,1,2)}$ were not computed. The parameters are defined in Equations 20.4–20.15 and presented to four significant digits.

TABLE 20.5
The Goodness of Fit Values for Simulation and Model Validation

Bay	Simulation				Validation				Average of All Tests
	Short Term		Long Term		Short Term		Long Term		
	Deposit	Suspension	Deposit	Suspension	Deposit	Suspension	Deposit	Suspension	
LB	1.94	1.06	3.25	1.63	3.25	2.04	2.39	1.45	2.28
MB	0.96	0.81	1.04	1.08	0.93	1.39	1.21	1.00	1.23
US	0.91	0.75	0.94	1.01	1.79	2.28	1.15	1.23	1.61
LS	0.75	1.07	0.89	1.05	1.06	1.34	0.84	0.95	1.05
NB	0.72	0.59	0.91	1.13	1.03	2.76	1.26	1.14	1.55
CC	0.63	0.63	0.92	0.87	0.96	1.29	0.84	0.92	1.00
BB	0.81	0.70	0.73	1.15	1.03	1.19	1.18	1.62	1.26
LM	0.80	0.75	1.47	0.79	1.26	1.08	1.49	1.14	1.24

Note: Abbreviations: LB = Lavaca Bay, MB = Matagorda Bay, US = Upper San Antonio Bay, LS = Lower San Antonio Bay, NB = Nueces Bay, CC = Corpus Christi Bay, BB = Baffin Bay, LM = Upper Laguna Madre. Simulation is performed using the parameters calibrated from the same period, and validation is performed using parameters calibrated from different periods. The short-term synoptic period was 1991–1995, and the long-term period (which is different for each bay) was 1988–1995. Test values are given for both deposit and suspension feeders.

TABLE 20.6
The Average Levels of Simulated Biomass (Standing Stock), Production, and Annual P/B for the Period 1991 to 1995

	Standing Stocks (g dw m^{-2})			Production (g dw m^{-2} y^{-1})			Annual P/B (y^{-1})		
Bays	Deposit Feeders	Suspension Feeders	Total	Deposit Feeders	Suspension Feeders	Total	Deposit Feeders	Suspension Feeders	Total
LB	2.2	1.7	3.9	3.0	4.4	7.4	1.4	2.6	1.9
MB	5.0	0.4	5.4	8.4	0.9	9.3	1.7	2.0	1.7
US	1.4	5.2	6.6	2.2	14.4	16.6	1.6	2.8	2.5
LS	2.1	5.6	7.7	4.4	13.8	18.2	2.1	2.5	2.4
NB	4.4	4.7	9.1	11.7	11.0	22.7	2.7	2.4	2.5
CC	11.1	0.7	11.8	22.9	1.4	24.3	2.1	2.0	1.2
BB	0.6	0.6	1.2	1.7	1.7	3.4	3.0	3.0	2.8
LM	5.5	1.2	6.7	18.3	3.2	21.5	3.3	2.7	3.2

Note: Abbreviations: LB = Lavaca Bay, MB = Matagorda Bay, US = Upper San Antonio Bay, LS = Lower San Antonio Bay, NB = Nueces Bay, CC = Corpus Christi Bay, BB = Baffin Bay, LM = Upper Laguna Madre.

The simulations with different calibration periods were similar (Table 20.5). The simulation of the 1991 to 1995 period using the calibration results of the data from the 1988 to 1995 period is stable for Matagorda, Nueces, Corpus Christi, Upper and Lower San Antonio, and Baffin bays and upper Laguna Madre. The simulation of the 1988 to 1995 period using the calibration results of the data from the 1991 to 1995 period are also stable for Matagorda, Corpus Christi, Upper and Lower San Antonio, and Baffin bays, and upper Laguna Madre. The worst simulation is for both deposit feeders and suspension feeders biomass in Lavaca Bay.

20.6 RESULTS

Predators are an important driver of top-down processes. The blue crab was the dominant species in trawls (83.4% of the three predators), followed by black drum (16.4%), and red drum (0.2%). Blue crabs have strong seasonal migration patterns and have peak abundance each spring (Figure 20.3a). There are differences in the average density of blue crabs among the bay systems (Figure 20.3b). They are most common in upper and lower San Antonio Bay, averaging two to three times the abundance in the other systems.

The simulations of benthic biomass are based on the best fit parameters from the calibration of the period 1991 to 1995 (Table 20.4). The average values of biomass over the entire simulation period are the expected values of the benthos standing stocks. Production, annual production to biomass ratio (P/B), production efficiency, and environmental limitation are all based on the calibrations and simulations of the period 1991 to 1995.

The simulations of benthos biomass for deposit feeders and suspension feeders were partly successful (Figures 20.4 through 20.7). The simulations fit well for most bays, except for Lavaca Bay (Table 20.5, Figure 20.4). In contrast, successful simulations for both feeding types with goodness of fit values less than 1 include Matagorda Bay, upper San Antonio Bay, Nueces Bay, Corpus Christi Bay, Baffin Bay, and upper Laguna Madre. In lower San Antonio Bay, the goodness of fit value for the simulation of deposit feeders is less than 1, while it is 1.07 for suspension feeders. In contrast, both simulations of the two feeding types in Lavaca Bay are higher than 1, and near 2 for deposit feeders.

Suspension feeders have higher simulated biomass variation through the 1991 to 1995 period over all eight bays than do deposit feeders (Figures 20.4 through 20.7). The primary bays closest to the sea, such as Matagorda Bay, lower San Antonio Bay, Corpus Christi Bay, and Laguna Madre

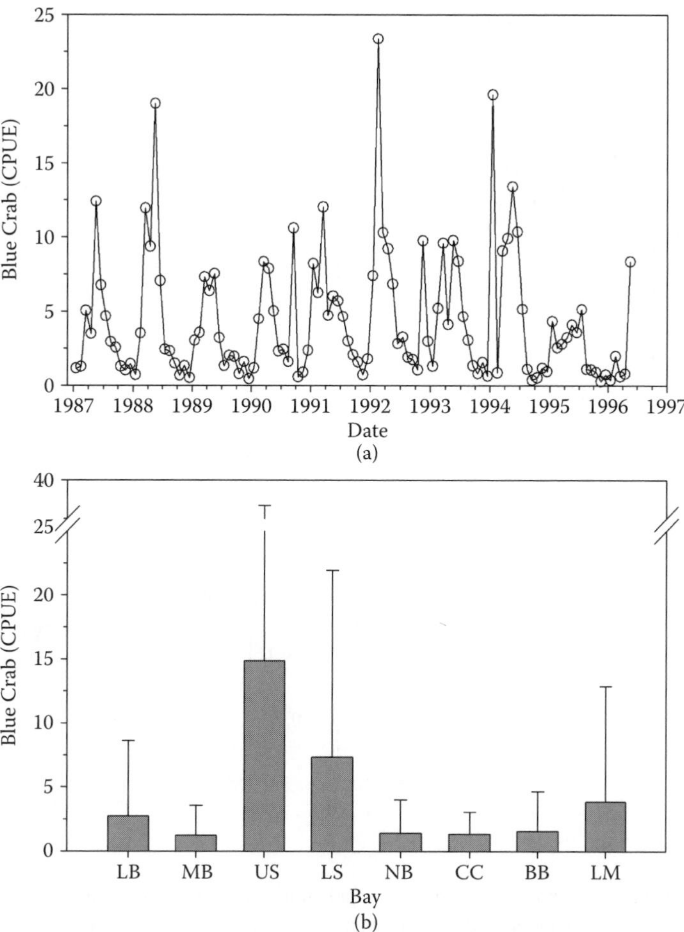

FIGURE 20.3 Catch per unit effort (CPUE) in 10 minute otter trawls for blue crab, *Callinectes sapidus*. (a) Average all bays by month. (b) Average all dates by bay. Abbreviations: LB = Lavaca Bay, MB = Matagorda Bay, US = upper San Antonio Bay, LS = lower San Antonio Bay, NB = Nueces Bay, CC = Corpus Christi Bay, BB = Baffin Bay, and LM = Laguna Madre.

have more stable biomass for both feeding types than do the secondary bays close to the freshwater sources, such as Lavaca, lower San Antonio, Nueces, and Baffin bays.

The average simulated benthic biomass is the predicted standing stock of benthos in the eight bays during the period 1991 to 1995. The standing stocks in the Nueces Estuary, 9.1 to 11.8 g dw m^{-2}, is higher than in other estuaries (Table 20.6). It is followed by the Guadalupe Estuary (6.6 to 7.7 g dw m^{-2}). Upper Laguna Madre has a median level standing stock of 6.7 g dw m^{-2}. Lavaca-Colorado Estuary has a lower level at 3.9 to 5.4 g dw m^{-2} and Baffin Bay has the lowest level in the study area at only 1.2 g dw m^{-2}. Overall, it appears that the secondary bays close to the freshwater source have higher levels of suspension feeder standing stocks, while the primary bays near the sea have higher standing stock levels of deposit feeders.

The average production rate of benthos in Texas estuaries ranges from 3.4 to 24.3 g dw m^{-2} y^{-1} during the simulation period 1991 to 1995 (Table 20.6). Nueces Estuary has the highest productivity due to both deposit feeders and suspension feeders, which have annual production rates of over 11 g dw m^{-2}. A much higher production rate of deposit feeders (22.9 g dw m^{-2} y^{-1}) occurs in Corpus Christi Bay. Guadalupe Estuary has a medium level of production in the range of 16.6 to 18.2 g

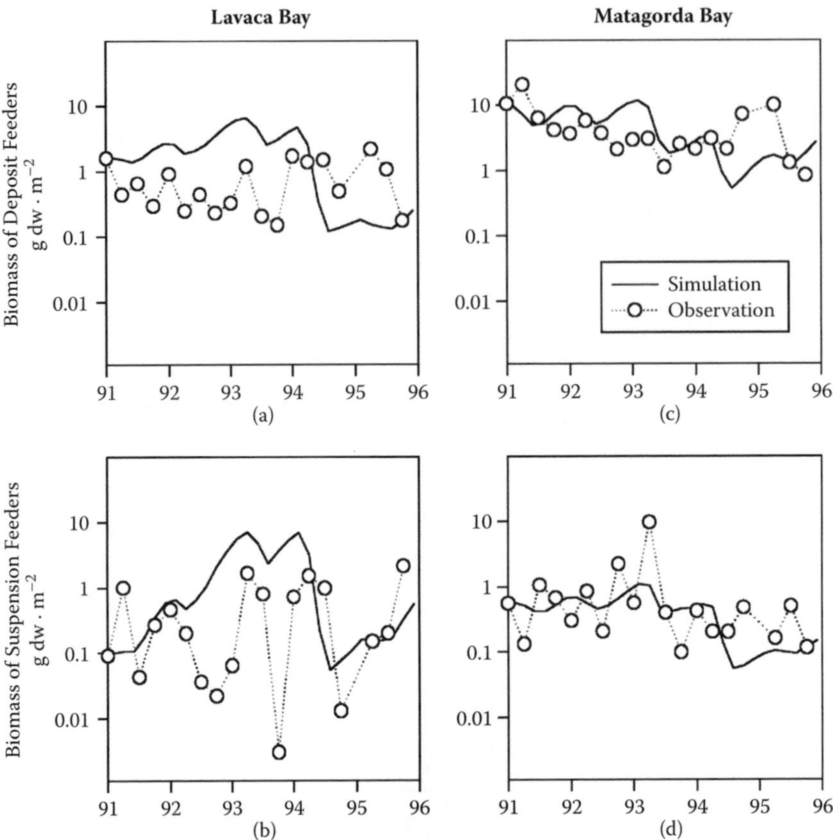

FIGURE 20.4 Simulation of deposit feeders and suspension feeders biomass in the Lavaca-Colorado Estuary for the period 1991–1995. (a) Deposit feeders in Lavaca Bay. (b) Suspension feeders in Lavaca Bay. (c) Deposit feeders in Matagorda Bay. (d) Suspension feeders in Matagorda Bay.

dw m^{-2} y^{-1} due to a higher level of production by suspension feeders (13.8 to 14.4). Laguna Madre Estuary has a large difference of production between upper Laguna Madre and Baffin Bay. In upper Laguna Madre, the production of deposit feeders is as high as 18.3 g dw m^{-2} y^{-1}, which is second only to Corpus Christi Bay, and 20 times higher than in Baffin Bay.

Monthly production is more stable for deposit feeders than for suspension feeders in Nueces Bay, but the reverse is true in the upper Laguna Madre. The monthly production rate decreased for both feeding types in 1994 for Lavaca Bay and in 1992 for lower San Antonio Bay.

The annual P/B ratio is the annual turnover rate of the benthos, because the units are y^{-1}. The P/B ratios of benthos in the Texas estuaries range from 1.2 to 3.2 y^{-1} during the period 1991 to 1995 (Table 20.6). The P/B ratios for the two trophic groups were similar, ranging from 1.4 to 3.3 y^{-1} for deposit feeders and 2.0 to 3.0 y^{-1} for suspension feeders. During the period 1991 to 1995, the monthly turnover rates (i.e., monthly P/B) changed seasonally based on the environmental limitations. Variation of turnover is different from variation of biomass because of predation. A high monthly P/B ratio indicates that the benthos biomass is able to rapidly recover standing stock levels following consumption by predators. A low monthly P/B ratio means that the standing stock falls to a lower level and does not recover following predation mortality.

The variance of monthly P/B ratio of both feeding types was different in all eight bays (Figure 20.8). Overall, the suspension feeders had higher variance than deposit feeders. Deposit feeders in the primary bays close to the sea (Matagorda Bay, lower San Antonio Bay, Corpus Christi Bay, and Laguna Madre) had higher variation than in the secondary bays close to the freshwater

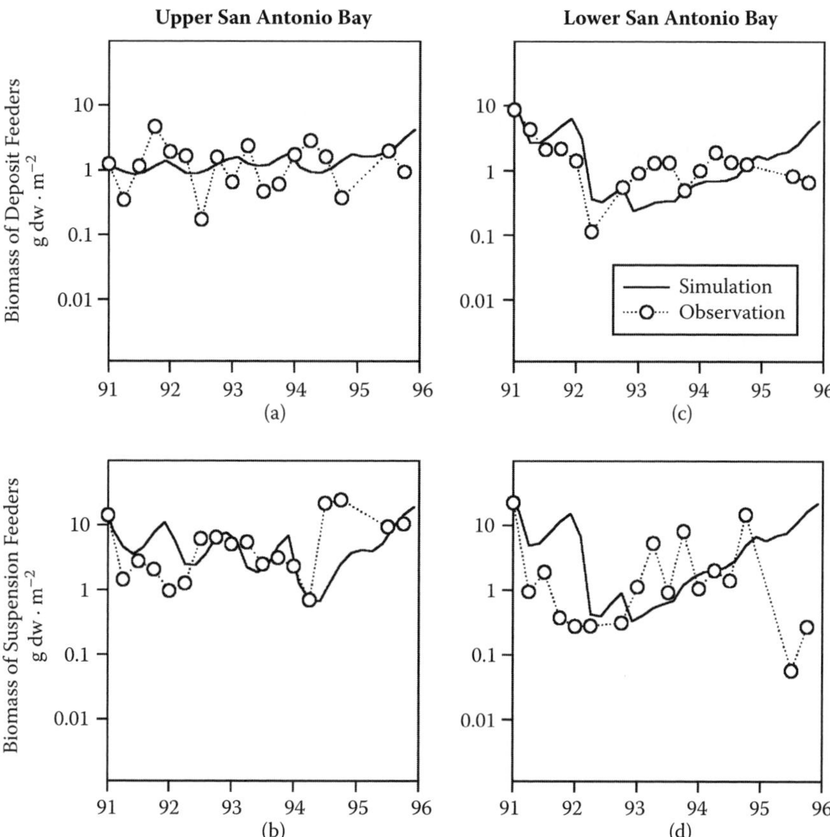

FIGURE 20.5 Simulation of deposit feeders and suspension feeders biomass in the Guadalupe Estuary for the period 1991–1995. (a) Deposit feeders in Upper San Antonio Bay. (b) Suspension feeders in Upper San Antonio. (c) Deposit feeders in Lower San Antonio Bay. (d) Suspension feeders in Lower San Antonio Bay.

sources (Lavaca Bay, upper San Antonio Bay, Nueces Bay, and Baffin Bay). The opposite is true for suspension feeders. In the upper San Antonio Bay, particularly, suspension feeders are the most variable, while deposit feeders are the most stable relative to other bays.

The temporal variation of the monthly P/B ratio correlated with the variation of environmental limitation factors. Temperature affects physiological rates, thus the production rate of both feeding types in each bay had seasonal fluctuation. However, there is not a large difference in temperature limitation between feeding types or among the eight bays. Salinity is the major limiting environmental factor. Salinity variation has a greater effect on the production rate than does temperature. Deposit feeders are more salinity limited than suspension feeders in secondary bays close to freshwater inflow sources (e.g., in the Lavaca-Colorado, Guadalupe, and Nueces estuaries). Suspension feeders are more salinity limited in primary bays close to the sea (e.g., in the Lavaca-Colorado, Nueces, and Laguna Madre estuaries). Food availability does not appear to limit either feeding type in all eight bays. The food limitation factor is close to 1 for both groups in all bays, indicating little or no effect due to food limitation.

The varibility of benthic biomass can be affected by the variability of predation rates. The variation of predation mortality caused by mobile epibenthos ranges from 0 to more than 15 g dw m^{-2} mo^{-1}. Predation limitation by crab and fish was low for deposit feeders in Lavaca Bay, lower San Antonio Bay, upper San Antonio Bay, and Baffin Bay, and suspension feeders in Lavaca Bay, Corpus Christi Bay, Baffin Bay and Laguna Madre. High rates of predation on both feeding groups occur in

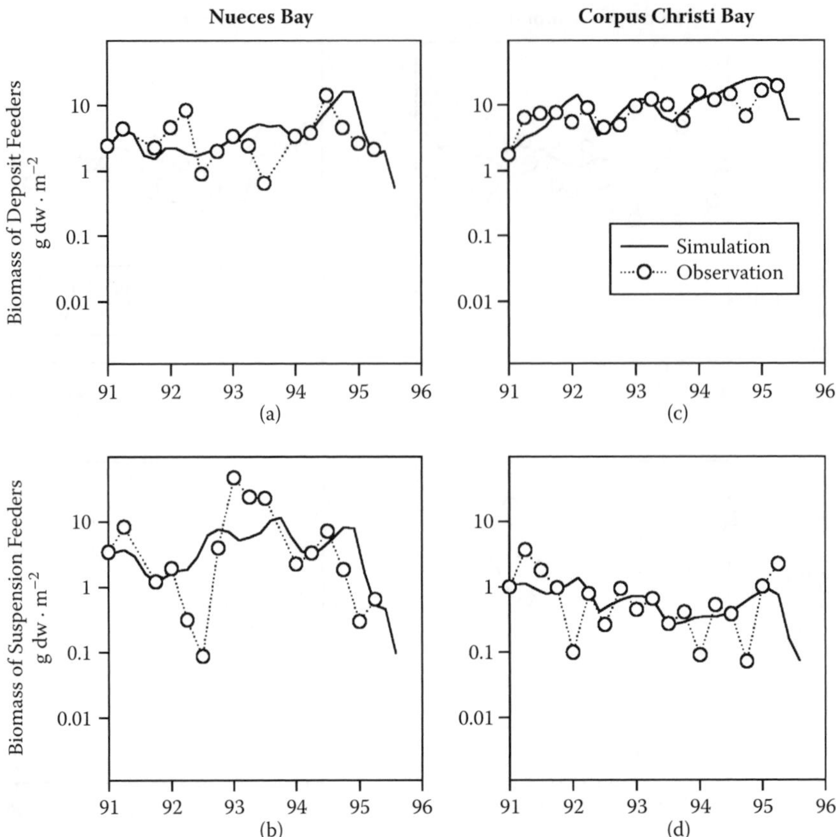

FIGURE 20.6 Simulation of deposit feeders and suspension feeders biomass in the Nueces Estuary for the period 1991–1995. (a) Deposit feeders in Nueces Bay. (b) Suspension feeders in Nueces Bay. (c) Deposit feeders in Corpus Christi Bay. (d) Suspension feeders in Corpus Christi Bay.

Matagorda Bay and Nueces Bay only. A strong reduction of benthos biomass occurs after the peaks of predation rates; such is the case for both feeding types in Lavaca Bay in 1994, lower San Antonio Bay in 1992, and Corpus Christi Bay in 1992.

20.7 DISCUSSION

Biomass change is an indicator of secondary productivity (Banse and Mosher 1980). Within Texas estuaries, the mean annual P/B ratio increased with proximity to the freshwater inflow source. The range of P/B values indicates that macrofauna in Texas bays turns over one to three times per year. Low turnover times are associated with bays that have low inflow rates and other anthropogenic disturbances. Corpus Christi Bay had the lowest P/B of 1.2 y^{-1}, and this is probably due to several forms of natural and anthropogenic disturbance, including low inflow, poor circulation and exchange with the Gulf of Mexico, disturbance due to shrimp trawling, and extensive coastal development. Laguna Madre had the highest P/B ratio (3.2 y^{-1}), and is associated with the presence of extensive seagrass habitat. Therefore, turnover rates appear to be enhanced by the habitats present within the estuaries.

Estuary physiography is also important because salinity change is a function of inflow being diluted by the volume of water in the estuary. The estuarine water residence time is calculated by the fraction of freshwater diluted by oceanic water: $f \times (V \div Q)$; where f = (bay salinity – Gulf salinity) ÷ Gulf salinity, V = volume of the estuary, and Q = the net inflow (Armstrong 1982). Residence

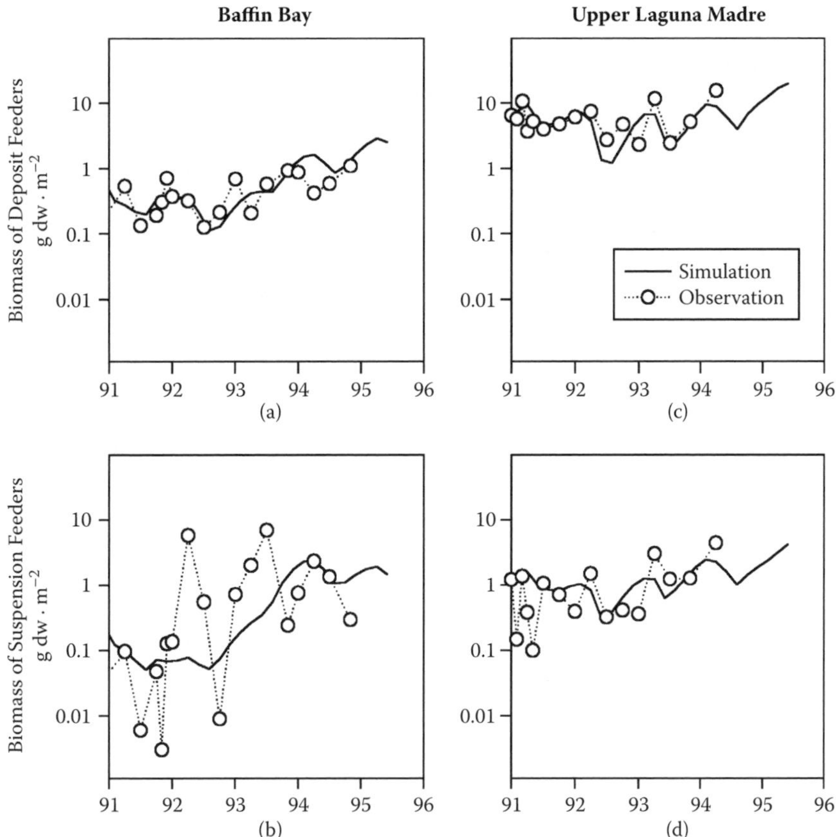

FIGURE 20.7 Simulation of deposit feeders and suspension feeders biomass in the Laguna Madre Estuary for the period 1991–1995. (a) Deposit feeders in Baffin Bay. (b) Suspension feeders in Baffin Bay. (c) Deposit feeders in Laguna Madre. (d) Suspension feeders in Laguna Madre.

time is calculated on an estuarine-wide basis; therefore, the standing stocks and productivity for each estuary were calculated by adding the values of the estuarine components from Table 20.7. The P/B ratio for deposit feeders increases with water residence time (Figure 20.9, $r = 0.98$, $P = 0.018$). As water residence time increases, the estuary is increasingly depositional, and thus better habitat for deposit feeders. In contrast, suspension feeder turnover times should increase as the water turnover rate increases, meaning that the water residence time decreases, because they depend on nutrients stimulating plankton. On first inspection, it appears that this is not the case (Figure 20.9, $r = 0.42$, $P = 0.58$). Laguna Madre has a high turnover time and a long water residence time. However, Laguna Madre benthos is dominated by suspension feeders associated with the base of seagrass blades, which is specific to the submerged aquatic habitat. When only the bays with the soft-bottom muddy habitat are included (GE, LC, and NC), the trend is negative as predicted, but not statistically significant due to the low sample sizes (Figure 20.9, $r = -0.96$, $P = 0.17$). There are two implications to this finding. First, suspension feeders, particularly mollusks, are the best indicators of the importance of freshwater inflow to maintaining the productivity of an ecosystem, which is consistent with previous findings (Montagna and Kalke 1995; Montagna et al. 2008). The second implication is that communities in bays where inflow is reduced will change in dominance patterns, from a dominant suspension feeder community to a dominant deposit feeder community. This change from a mollusk- and crustacean-dominated community to a polychaete-dominated community is consistent with an emerging paradigm on how benthos changes in response to anthropogenic disturbance (Peterson et al. 1996).

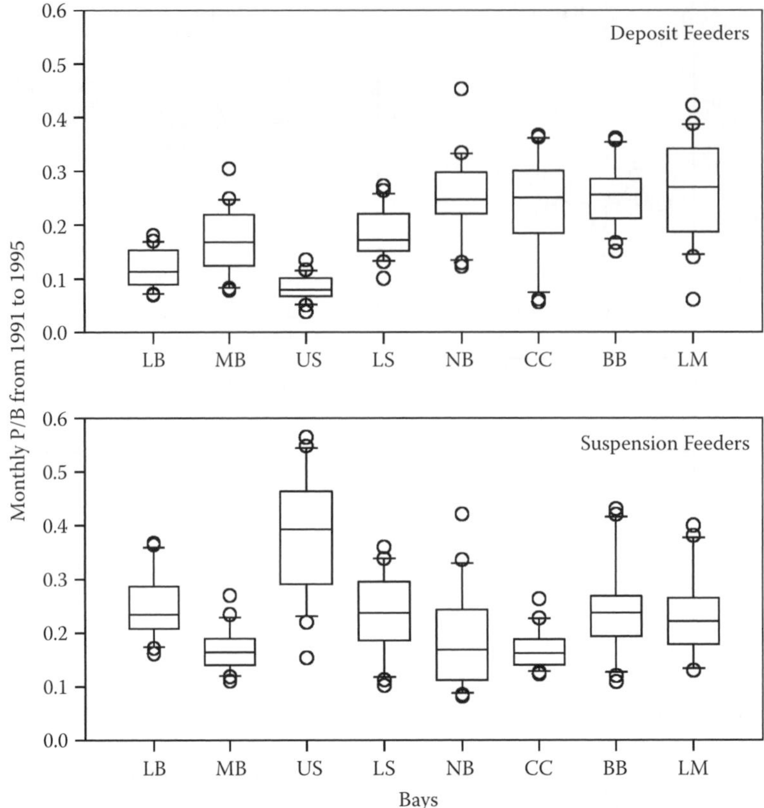

FIGURE 20.8 Tukey box plots for the distribution of the monthly P/B for two feeding types in eight bays. The box encompasses the 25th through 50th percentile points of the data. The horizontal lines mark the 10th, 50th, and 90th percentiles, and the symbols mark the 5th and 95th percentiles. Abbreviations: LB = Lavaca Bay, MB = Matagorda Bay, US = upper San Antonio Bay, LS = lower San Antonio Bay, NB = Nueces Bay, CC = Corpus Christi Bay, BB = Baffin Bay, and LM = Laguna Madre.

TABLE 20.7
Mean Monthly Freshwater Inflow (FWI), Salinity (Sal.), Total Nitrogen (N), Total Phosphorus (P), and Total Organic Carbon Volume Loading (C) (g m^{-3} yr^{-1}) for the Texas Estuaries

Bay	FWI (10^6 m^3 month^{-1})	Sal. (ppt)	Loading (g m^{-3} y^{-1}) N	P	C	N:P	C:N	Res. Time (y)	Res. Time Weighted N
LB/MB	100	13.17	3.18	0.48	19.6	6.63	6.16	0.21	0.66
US/LS	241	11.94	10.8	2.25	34.2	4.80	3.17	0.19	2.09
NB/CC	65	21.49	2.10	0.43	6.3	4.88	3.00	0.46	0.97

Note: Abbreviations: LB = Lavaca Bay, MB = Matagorda Bay, US = Upper San Antonio Bay, LS = Lower San Antonio Bay, NB = Nueces Bay, CC = Corpus Christi Bay, BB = Baffin Bay, LM = Upper Laguna Madre. Residence (Res.) times (y) influence nitrogen availability (g m^{-3} y^{-1}).

Source: Adapted from Longley (1994).

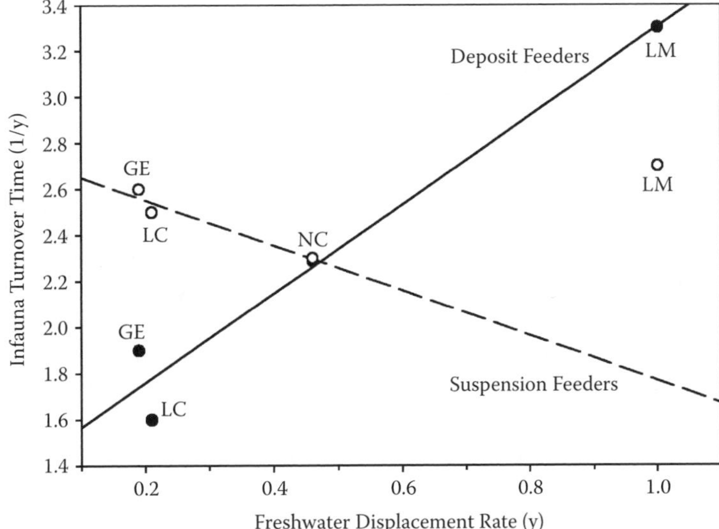

FIGURE 20.9 Trophic group response in terms of P/B (y^{-1}) to freshwater residence time (y). Each point represents the measurement from one estuary, which is placed on the abscissa at its water residence time. Suspension feeder regression line without Laguna Madre.

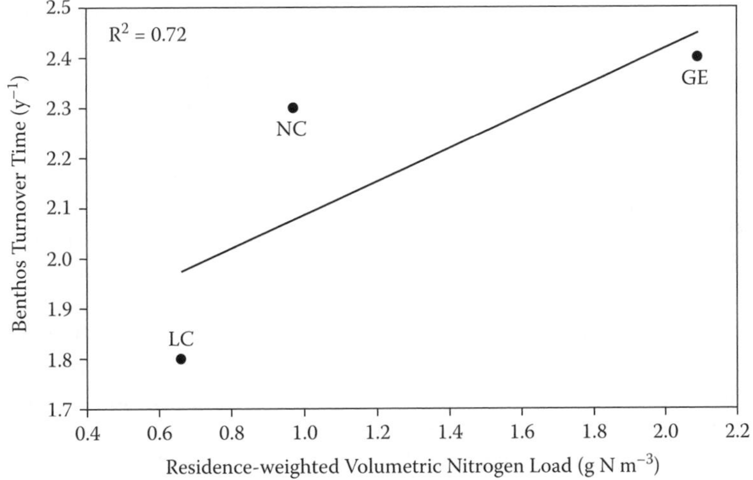

FIGURE 20.10 Trophic group response in terms of P/B (y^{-1}) to total nitrogen load weighted by the residence time (y).

Nitrogen loads have been estimated by Longley (1994) (Table 20.7). In general, nitrogen loading increases as freshwater inflow to the estuary increases. However, the effect that nitrogen load can have is affected by the size of the receiving water. The same loading rate can have different effects if it is flowing into a small system as opposed to being diluted in a system with a larger volume of water. For this reason, the appropriate nitrogen loading value to compare with benthic production is the loading weighted by the water residence time. The benthic turnover time is positively correlated with nitrogen loading rate (Figure 20.10). The implication is that ecosystems with more nitrogen will produce more benthic biomass.

A conceptual modeling study (Montagna et al. 1996) predicted that Laguna Madre would have a higher rate of energy flow than other Texas lagoons. This hypothesis is based on two characteristics unique to Laguna Madre: (1) its large size and (2) the extensive seagrass habitat. In the present quantitative modeling study, we estimated that benthos of Laguna Madre has the highest annual P/B ratio (3.2 y^{-1}) in comparison with other bays, which range from 1.2 to 2.5 y^{-1}. The higher production is mainly due to production of the deposit feeders, which is up to 18.3 g dw m^{-2} y^{-1}. These deposit feeders are probably exploiting a detrital food web enhanced by seagrass detritus. The conceptual model also predicted that higher inflow would yield high productivity of suspension feeders, because river-borne nutrients would stimulate primary production, which in turn would fuel benthic productivity. San Antonio Bay has the highest nutrient load levels, and the highest production of suspension feeders of all the bays. Suspension feeder production in San Antonio Bay is 14 g dw m^{-2} y^{-1} in comparison with a range of 0.9 to 11 g dw m^{-2} y^{-1} found in the other bays. However, the annual P/B ratio in San Antonio Bay, which is 2.5 y^{-1}, is not highest in comparison to the range of P/B values recorded in other Texas bays, 2.0 to 3.0 y^{-1}.

As might be expected, modeling applications do not always yield perfect results. In general, one can expect to encounter at least two main problems: (1) a lack of realism or completeness of the model and (2) a lack of data to calibrate the model. While these problems also occurred in this study, the most interesting result was the poor fit of any model with deposit feeders and suspension feeders biomass in Lavaca Bay. Why would fit be poor in just Lavaca Bay? Perhaps it is related to the lack of predator data or pollution influence in Lavaca Bay. There were no red drum and black drum in trawls from Lavaca Bay reported by the Texas Park and Wildlife Department during the simulation period. These predatory fish exist in the bay, but were not recorded for the period. Predation mortality is very important in limiting biomass. This lack of predator data means that the simulation of mortality is underestimated; therefore, biomass was generally overestimated (Figure 20.4). The second reason is that Lavaca Bay may have a high level of pollutant input that can affect benthos productivity, and pollutants are not a limitation factor in this model. Lavaca Bay is a U.S. EPA Superfund Site with high levels of mercury contamination. The P/B ratio in Lavaca Bay is very low, in the range of 1.7 to 1.9 y^{-1}. This value is in the lower range of all the Texas bays. Productivity is probably much higher, because it is being consumed by fish. Unfortunately, we do not have sufficient data to get a better estimate of productivity at this time.

The simulation of the Nueces and Laguna Madre estuaries is much better than the simulations of the Lavaca-Colorado and Guadalupe Estuaries. Therefore, the model fits estuaries with low inflow (Nueces and Laguna Madre estuaries) better than estuaries with high inflow (Lavaca-Colorado and Guadalupe estuaries). This may be due to a larger effect of interannual variability of inflow in the high inflow estuaries than in the low inflow estuaries. The estuaries with low inflow generally have a smaller salinity range from year to year than the high inflow estuaries, and this could be causing the observed biological effects.

For most cases, predation should decrease benthos biomass. However, where there are multiple environmental stressors (such as the presence of pollution and unsuitable salinity ranges), respiration and natural mortality could be higher than other losses or growth. In this case, benthos biomass may be lost before predation pressure can have an effect, and there may be a net negative growth rate. The current limitation equation (Equation 20.5) assumes that the minimal growth rate is a function of just salinity, temperature, and food, and can only be as low as zero. Additional stressor equations can be added to the model. In the future, it may be useful to model the effect of brown tide, which may cause a decline in benthos in Laguna Madre (Conley 1996), or model the effect of mercury contamination in Lavaca Bay, as has been done for zinc in the Gulf of Mexico (Montagna and Li 1997). Inclusion of a multiple stressor term could improve model performance, especially in Lavaca Bay.

Overall, the results of the current modeling study indicate that inflow differences among the Texas lagoons alter nutrient loading, which in turn affects benthic infaunal communities and maintains secondary production of benthos.

ACKNOWLEDGMENTS

Funding for the modeling study was provided by the Texas Water Development Board's (TWDB) Water Research and Planning Fund, authorized under Texas Water code Sections 15.402 and 16.058(e), and administered by the Department under Interagency Cooperative Contract No. 96-483-132. Benthic data used to calibrate the simulations described in this chapter were obtained as a result of many different research projects, but the primary source of funding was through the TWDB. The research in the Lavaca-Colorado, Guadalupe, and Nueces estuaries was partially funded through the TWDB Water Research and Planning Fund, Department under Interagency Cooperative Contracts Nos. 8-483-607, 9-483-705, 9-483-706, 93-483-352, 94-483-003, and 95-483-068. Other partial support for work in Matagorda Bay was supplied by the lower Colorado River Authority (LCRA). Work in the Nueces Estuary was partially funded by Institutional Grant NA16RG0445-01 to the Texas A&M University Sea Grant Program from the National Sea Grant Office, National Oceanic and Atmospheric Administration, U.S. Department of Commerce. Partial support was provided for work in Laguna Madre and Baffin Bay by the Texas Higher Education Coordinating Board, Advanced Technology Program under Grant Nos. 4541 and 3658-264.

The authors especially thank Mr. Rick Kalke for playing a vital role in the collection and analysis of the benthic biomass data. Mr. Kalke also supervised many technicians and graduate and undergraduate students over the years, including Mary Conley, Landon Ward, Antonio Mannino, Chris Martin, Rob Rewolinsky, Amy Rutter, and Greg Street. The authors thank Ms. Carol Simanek for playing a vital role in data management.

REFERENCES

Armstrong, N. E. 1982. Responses of Texas estuaries to freshwater inflows. Pp. 103–120, In: Kennedy, V. S. (ed.), *Estuarine Comparisons*. New York: Academic Press.

Armstrong, N. E. 1985. *The ecology of open-bay bottoms of Texas: A community profile*. Fish and Wildlife Service, U.S. Department of the Interior, Biological Report, 85 (7.12).

Banse, K. and S. Mosher. 1980. Adult body mass and annual production/biomass relationships of field populations. *Ecol. Monogr.* 50: 355–379.

Brock, D. A. 1994. Estuarine, phytoplankton, primary productivity, and freshwater inflows. Pp. 74–92, In: Longley, W. L. (ed.), *Freshwater inflows to Texas bays and estuaries: Ecological relationships and methods for determination of needs*. Austin: Texas Water Development Board and Texas Parks and Wildlife Department.

Brown, D. and P. Rothery. 1993. *Models in biology: Mathematics, statistics and computing*. New York: John Wiley & Sons.

Carrada, G. C. 1983. Modeling of the Gulf of Naples. Pp. 80–153, In: G. Carrada, T. Hopkins, L. Jeftie, and S. Morcos (eds.), *Quantitative analysis and simulation of Mediterranean coastal ecosystems: The Gulf of Naples, a case study*. UNESCO Reports in Marine Science 20. Paris: UNESCO.

Conley, M. F. 1996. Effect of a persistent brown tide bloom on macroinfaunal communities in Baffin Bay and Laguna Madre, Texas. M.A. Thesis, University of Texas at Austin, Texas.

Crisp, D. J. 1971. Energy flow measurements. In: N. A. Hole and A. D. McIntyre (eds.), *Methods for the study of marine benthos*. I. B. P. Handbook. Oxford: Blackwell Press.

de Hoop, B. J., P. M. J. Herman, H. Scholten, and K. Soetaert. 1989. *SENECA 1.5: A simulation environment for ecological application*. Yerseke, Netherlands. Netherlands Institute of Ecology Center for Estuarine and Coastal Ecology.

Flint, R. W. 1984. Phytoplankton production in the Corpus Christi Bay Estuary. *Continent. Mar. Sci.* 27: 65–83.

Keen, R. E. and J. D. Spain. 1992. Computer simulation in biology: A BASIC introduction. New York: Wiley-Liss.

Larkin, T. J. and G. W. Bomar. 1983. *Climatic atlas of Texas*. Austin: Texas Department of Water Resources.

Li, J., M. Vincx, and P. M. J. Hermen. 1996. A model for nematode dynamics in the Westerschelde estuary. *Ecol. Model.* 90: 271–284.

Longley, W. L. (ed.). 1994. *Freshwater inflows to Texas bays and estuaries: Ecological relationships and methods for determination of needs*. Austin: Texas Water Development Board and Texas Parks and Wildlife Department.

Lotka, A. J. 1925. *Elements of physical biology*. Baltimore, Maryland: Williams & Wilkins.

Kalke, R. D. and P. A. Montagna. 1991. The effect of freshwater inflow on macrobenthos in the Lavaca River Delta and upper Lavaca Bay, Texas. *Contrib. Mar. Sci.* 32: 49–72.

Mannino, A. and P. A. Montagna. 1996. Small scale spatial variation of macrobenthic community structure. *Estuaries* 20: 159–173.

Martin, C. M. and P. A. Montagna. 1995. Environmental assessment of La Quinta Channel, Corpus Christi Bay, Texas. *Texas J. Sci.* 47: 203–222.

McEachron, L. W. and B. Fuls. 1996. *Trends in relative abundance and size of selected finfishes and shellfishes along the Texas coast.* Coastal Fisheries Division, Management Data Series, Texas Parks and Wildlife Department, Austin, Texas.

Montagna, P. A. 1989. *Nitrogen process studies (Nips): The effect of freshwater inflow on benthos communities and dynamics.* Technical Report No. TR/89-011, Marine Science Institute, University of Texas, Port Aransas, Texas.

Montagna, P.A. 1991. *Predicting long-term effects of freshwater inflow on macrobenthos in the Lavaca-Colorado and Guadalupe Estuaries.* Final Report to Texas Water Development Board. Technical Report No. TR/91-004, Marine Science Institute, University of Texas, Port Aransas, Texas.

Montagna, P.A. 1994. *Inflow needs assessment: Effect of the Colorado River diversion on benthic communities.* Final Report to the Lower Colorado River Authority. Technical Report No. TR/94-001, Marine Science Institute, University of Texas, Port Aransas, Texas.

Montagna, P. A., E. D. Estevez, T. A. Palmer, and M. S. Flannery. 2008. Meta-analysis of the relationship between salinity and molluscs in tidal river estuaries of southwest Florida, U.S.A. *Am. Malacol. Bull.* 24: 101–115.

Montagna, P. A., J. C. Gibeaut, and J. W. Tunnell, Jr. 2007. South Texas climate 2100: Coastal impacts. Pp. 57–77, In: J. Norwine and K. John (eds.), *South Texas Climate 2100: Problems and prospects, impacts and implications.* CREST-RESSACA. Texas A&M University–Kingsville, Texas.

Montagna, P. A. and R. D. Kalke. 1992. The effect of freshwater inflow on meiofaunal and macrofaunal populations in the Guadalupe and Nueces Estuaries, Texas. *Estuaries* 15: 307–326.

Montagna, P. A. and R. D. Kalke. 1995. Ecology of Mollusca in south Texas estuaries. *Am. Malacol. Bull.* 11: 163–175.

Montagna, P. A. and J. Li. 1996. *Modeling and monitoring long-term change in macrobenthos in Texas estuaries.* Final Report to the Texas Water Development Board. University of Texas at Austin, Marine Science Institute, Technical Report No. TR/96-001, Port Aransas, Texas. http://harteresearchinstitute.org/montagna/docs/TWDB1996.pdf.

Montagna, P. A. and J. Li. 1997. Modeling contaminant effects on deposit feeding nematodes near Gulf of Mexico production platforms. *Ecol. Model.* 98: 151–162.

Montagna, P. A., J. Li, and G. T. Street. 1996. *A conceptual ecosystem model of the Corpus Christi Bay National Estuary Program study area.* Publication CCBNEP-08, Texas Natural Resource Conservation Commission, Austin, Texas.http://cbbep.org/publications/virtuallibrary/ccbnep08.pdf.

Montagna, P. A., D. A. Stockwell, and R. D. Kalke. 1993. Dwarf surfclam *Mulina lateralis* (Say, 1822) populations and feeding during the Texas brown tide event. *J. Shellfish Res.* 12: 433–442.

Montagna, P. A. and W. B.Yoon. 1991. The effect of freshwater inflow on meiofaunal consumption of sediment bacteria and microphytobenthos in San Antonio Bay, Texas USA. *Estuar. Coastal Shelf Sci.* 33: 529–547.

Odum, H. T., 1971. *Environment, Power, and Society.* Wiley Interscience, New York. 331 p.

Odum, H. T., 1972. An energy circuit language. In: B. C. Patten (ed.). *Systems Analysis and Simulation in Ecology*, vol. 2. Academic Press, New York: 139–211.

Odum, H. T., 1983. *Systems Ecology.* Wiley Interscience, New York, 644 p.

Odum, H. T., R. P. Cuzon, R. J. Beyers, and C. Allbaugh. 1963. Diurnal metabolism, total phosphorus, Ohle anomaly, and zooplankton diversity of abnormal marine ecosystems of Texas. *Pub. Inst. Mar. Sci. Univ. Texas* 8: 404–453.

Odum, H. T. and R. Wilson. 1962. Further studies on re-aeration and metabolism of Texas Bays, 1958–1960. *Pub. Inst. Mar. Sci. Univ. Texas* 8: 23–55.

Orlando, S. P., Jr., L. P. Rozas, G. H. Ward, and C. J. Klein. 1993. *Salinity characteristics of Gulf of Mexico estuaries.* Silver Spring, Maryland: National Oceanic and Atmospheric Administration, Office of Ocean Resources Conservation and Assessment.

Palmer, T. A., P. A. Montagna, and R. D. Kalke. 2002. Downstream effects of restored freshwater inflow to Rincon Bayou, Nueces Delta, Texas, USA. *Estuaries* 25: 1448–1456.

Parsons, T. R., K. Stevens, and J. D. H. Strickland. 1961. On the chemical composition of eleven species of marine phytoplankton. *J. Fish. Res. Board Canada* 18: 1001–1016.

Peterson, C. H., M. C. Kennicutt II, R. Green, P. Montagna, E. Powell, and P. Roscigno. 1996. Ecological consequences of environmental perturbations associated with offshore hydrocarbon production: A perspective from study of long-term exposures in the Gulf of Mexico. *Canadian J. Fish. Aquat. Sci.* 53: 2637–2654.

Quast, W. D., T. S. Searcy, and H. R. Osborn. 1988. *Trends in Texas commercial fishery landings, 1977–1987*. Management Data Services, No. 149. Austin: Texas Parks and Wildlife Department, Coastal Fisheries Branch.

Redfield, A. C. 1934. On the proportions of organic derivations in sea water and their relation to the composition of plankton. Pp. 177–192, In: R. J. Daniel (ed.), *James Johnstone memorial volume*. Liverpool: University Press of Liverpool.

Russell, M .J. and P. A. Montagna. 2007. Spatial and temporal variability and drivers of net ecosystem metabolism in Western Gulf of Mexico Estuaries. *Estuaries Coasts* 30: 137–153.

Stockwell, D. A. 1989. *Effects of freshwater inflow on the primary production of a Texas coastal bay system.* Final Report, Data Synthesis Study (NIPS), University of Texas Marine Science Institute, Technical Report No. TR/89-010, Aransas, Texas.

Taghon, G. L., A. R. M. Nowell, and P. A. Jumars. 1980. Induction of suspension feeding in spionid polychaetes by high particulate fluxes. *Science* 210: 562–564

Tenore, K. R., R. N. Zajac, J. Terwin, F. Andrade, J. Blanton, W. Boynton, D. Carey, R. Diaz, A. F. Holland, E. Lopez, P. Jamar, P. Montagna, F. Nichols, R. Rosenberg, H. Queiroga, M. Sprung, and R. B. Whitlatch. 2006. Characterizing the role benthos plays in large coastal seas and estuaries: A modular approach. *J. Exp. Mar. Biol. Ecol.* 330: 392–402.

Tolan, J. M. 2007. El Niño-Southern Oscillation impacts translated to the watershed scale: Estuarine salinity patterns along the Texas Gulf coast, 1982 to 2004. *Estuar. Coastal Shelf Sci.* 72: 247–260.

Tumbiolo, M. L. and J. A. Downing. 1994. An empirical model for the prediction of secondary production in marine benthic invertebrate populations. *Mar. Ecol. Prog. Ser.* 114: 165–174.

Whitledge, T. E. 1989. *Nutrient distributions and dynamics in Lavaca, San Antonio and Nueces/Corpus Christi Bays in relation to freshwater inflow*. Final Report, Data Synthesis Study (NIPS), University of Texas Marine Science Institute, Technical Report No. TR/89-007, Aransas, Texas.

Wohlschlag, D. E., J. M. Wakeman, R. Vetter, and R. J. Ilg. 1977. *Analysis of freshwater inflow effects on metabolic stresses of South Texas Bay and estuarine fishes: Continuation and extension.* Report to Texas Department of Water Resources, by Marine Science Institute, University of Texas, Austin, Texas.

Index

A

Abiotic conditions, 404–407
Aboveground biomass, 176–179
Abra segmentum, 499
Abundance, 121
Acanthocyclops sp., 274
Acartia clausi, 496
Acartia latisetosa, 496
Acartia margalefi, 496
Acartia tonsa
  Patos Lagoon, 440, 448
  Venice Lagoon, 496, 503
Acidification, 476–477
Agardhiella sp., 48
Agriculture intensification
  discussion, 158–162
  drainage control structures, 141
  drivers of change, water quality, 148–151
  estuarine water quality, 143, 157–158
  excess concentrations, 142
  fundamentals, 11, 136–137
  groundwater piezometers, 141–142
  hydrology, 139
  improvement, water quality, 159–162
  methods, 139–143
  N and P agricultural reductions, 151–155, 157
  nontidal monitoring stations, 147–148
  regional hydrology, 145–146
  results, 145–158
  status, water quality, 158–159
  study site, 137, 139
  surface water chemistry, 139, 141
  trends, water quality, 147–148
  water quality, 147–151, 158–162
Algal physiology, 102–103
Allochthonous nitrogen sources, 49–55
Alona sp., 440
Alternative stable states, 234, 236
Ampelisca abdita, 124, 127, 129
Amphora sp., 314
Anabaena oumiana, 443
Anabaenopsis sp., 443
Anguilla anguilla, 276
Anthropogenic pressures, *see also* Human impacts
  acidification, 476–477
  anthropogenic change, 476
  bathymetry, 491–493
  benthic habitats, 465
  carbon budgets, 469, 472
  catchment characteristics, 462, 465
  chlorophyll *a,* 494–496
  circulation, 491
  climate and climate change, 490
  climate and climate changes, 462, 477–478
  confinement, 491
  deoxygenation, 476–477
  dissolved inorganic nutrients, 493–494
  ecosystem responses, 501–505
  environmental drivers and pressures, 486–490
  eustatism, 488
  fisheries, 473–474
  fishing activities, 490
  fish kills, 476–477
  freshwater discharges, 489
  geomorphic evolution, 459–462
  governance, 490
  historical developments, 484–486
  Hong Kong waters, 371–377
  impacts, 7–9
  lagoon state and impacts, 491–501
  landings data, 500–501
  land use, 462, 465
  macrophytobenthos, 497–498
  macrozoobenthos, 498–500
  natural changes, 476
  nekton, 500
  nutrient loads, 489
  phytoplankton, 494–496
  pollutants, 476, 489
  rainfall, 478
  Ria Formosa, 343, 344–345
  runoff, 478
  sea level rise, 477
  sediment nutrients, 494
  sediment quality, 465
  storminess and wave climate, 478
  stressors, 462, 465
  study area, 459–465
  subsidence, 488
  temperature, 477
  water quality, 465
  zooplankton, 496
Aphanius fasciatus, 500
Aphanizomenon aphanizomeniodes, 443
Arenicola marina
  Venice Lagoon, 499
  Wadden Sea, 421
Argopecten irradians, 168–169, 190–192, 195
Argyrosomus regius, 276
Arundo sp., 273
Ascención Bay, 325–326
Assessment, ecosystem health
  biotic characteristics, 4
  discussion, 83–86
  ecosystem health framework, 84–86
  fundamentals, 10, 73–76, 86–88
  methods, 76–83
  network analysis, 82–83
  network construction, 77, 79, 82
  results, 83–86
  study sites, 76–77

trophic status, 83–84
uncertainty sensitivity, 82–83
Assessment, human and climatic effects
biogeochemical considerations and time scales, 21–22, 34–36
climatic impacts, 22–34
climatological time scales, 21–22
eutrophication changes, 9
fundamentals, 7–21, 34
geological time scales, 21–22
historic perspectives, 22–34
human impacts, 22
hurricanes, 9, 23–33
integrated assessment response program, 36–38
large-scale storm events response, 36–38
lessons learned, 34–36
smaller storms, 33–34
socioeconomic aspects and considerations, 36
tropic considerations, 34–36
Assessment, subtropical karstic lagoons
Ascención Bay, 325–326
Bojorquez Lagoon, 325
Celestún Lagoon, 314–315
Chacmochuk Lagoon, 321, 324
characterization of lagoons, 312–326
Chelem Lagoon, 315–317
Chetumal Bay, 326
context, 309–311
Dzilam Lagoon, 317–318
fundamentals, 307–309, 329
Holbox Lagoon, 320–321
lessons learned, 326–329
monitoring experience, 311–312
Nichupte Lagoon, 324–325
phytoplankton, 328
research, 311–312
Rio Lagartos Lagoon, 318–319
status of lagoons, 312–326
submerged aquatic vegetation, 328–329
water quality characteristics, 326–328
Assimilation, 58–60
Asterionellopsis glacialis, 450
Atherina boyeri, 500
Atherina sp., 501
Atherinella brasiliensis, 441, 447
Atmospheric deposition, 54–55
Aulacoseira spp., 448
Aureococcus anophagefferens
algal physiology, 102
blooms, 99, 105
Laguna Madre, 116, 129
Maryland coastline, 46
New Jersey lagoons, 170, 190, 192, 197
Aureococcus lagunensis
blooms, 105
Laguna Madre, 115, 116
Australia, *see* East Australian coastal lagoons
Autochthonouse nitrogen sources, 56–58
Avicennia germinans
Celestún Lagoon, 314
Rio Lagartos Lagoon, 319
Azolla filiculoides, 273

B

Babrus bayad, 276
Bacillaria paxillifer, 439
Balanus improvisus, 440
Barnegat Bay-Little Egg Harbor Estuary
classification, 2
eutrophication changes, 8–9
Bathymetry, 491–493
Batophora sp., 291, 299
Belowground biomass, 179–184
Benthic habitats, 465
Benthic invertebrates, 274–275
Benthic macro- and microalgae, 61–63
Benthic macrophyte communities
bottom-up controls, 204–206
case studies, 207–219
function, 204
fundamentals, 11, 203–204, 219–221
Jobos Bay, 215–219
structure, 204
top-down controls, 204–206
Waquoit Bay, 201–215
Benthic primary production, 61
Benthos, 520
Biogeochemical considerations and time scales, 21–22, 34–36
Bioinvasion, 450
Biological assemblages, 340
Biological characteristics, 257–268
Biological components, 439–442
Biomass, 122
Biotic characteristics, 4–6
Biotic factors, 407–410
Biotic indicators, 175–190
Biotic responses, 442–447
bioinvasion, 450
biological components, 439–442
changes, 442–447
classification, 2
dissolved inorganic nutrients, 437–438
El Niño Southern Oscillation, 443–447
estuarine habits, 439–442
eutrophication, 448–449
fish, 446–447
fisheries, 447
fundamentals, 13, 434
geological evolution, 434–435
harbor dredging, 449–450
historical developments, 434–435
human influence, 448–450
macrozoobenthos, 446
natural protection areas, 435
overfishing, 449
phytoplankton, 443
primary producers, 439–440
processes, 439–442
Ruppia maritima, 444–445
salinity, 436–437
secondary producers, 440–442
temperature, 436–437
water exchange, 436–437

Blooms, *see also* Laguna Madre (Texas)
  algal physiology, 102–103
  data resources, 94–97
  differential nutrient use, 102–103
  dominant algal species, composition and size, 99–100
  fundamentals, 10, 91–93, 106–108
  intensity, 97–99
  nutrient composition and sources, 100–102
  "poised on the edge," 103–106
  seasonality, 97–99
  site descriptions, 94–97
Bojorquez Lagoon, 325
Bolinus brandaris, 499
Bonnemaisonia hamifera, 189–190, 197
Boodleopsis pusilla, 291, 295
Bottom-up controls
  benthic macrophyte communities, 204–206
  Jobos Bay, 215–219
  Waquoit Bay, 207–215
Bottom-up forcing by nutrients, 236–238
Brachionus angularis, 274
Brachionus calyciflorus, 274
Brachionus plicatilis, 274
Brachionus urceolaris, 274
Branchioasychis americana, 122, 130
Brania furcelligera, 126–127
Branta bernicla, 58
Brevoortia pectinata, 441
Brown tide, 190
Brown tide, Laguna Madre
  abundance, 121
  biomass, 122
  community structure, 124, 126–128
  discussion, 128–132
  diversity, 123–124
  freshwater inflow, 518–519
  fundamentals, 11, 115–117
  macrofauna, 121–124
  materials and methods, 117–119
  results, 119–128
  water column, 119–120
Buffering capacities, comparisons
  anthropogenic influences, 371–377
  climate change effects, 388–389
  Deep Bay, 377, 387–388
  dissolved oxygen, 383
  drivers of change, 369, 371
  eutrophication processes, 377, 380–386
  fundamentals, 12, 368–369, 389–391
  geographic locations, 369
  geomorphologic setting, 369
  management strategy and perspectives, 386–388
  nutrient stoichiometry, 383–386
  oceanographic setting, 369, 371
  physical processes, 371–377
  Port Shelter, 371–372, 387
  Tolo Harbour, 372, 377, 386
Burullus Lagoon, 258–260

C

Caecum pulchellum, 124, 126
Calibration, 526, 528
Callinectes sapidus
  New Jersey lagoons, 168–169, 195
  Patos Lagoon, 448
Calyptrosphaera sp., 495
Capitella capitata
  Laguna Madre, 124
  Venice Lagoon, 500
Carbon budgets, 469, 472
Carcinus aestuarii, 501
Carpobrotus edulis, 347
Carteria sp., 272
Case studies
  benthic macrophyte communities, 207–219
  Jobos Bay, 215–219
  Southern California lagoons, 241–245
  Waquoit Bay, 201–215
Catchment characteristics, 462, 465
Caulerpa mexicana, 218–219
Caulerpa verticillata, 291
Celestún Lagoon, 314–315
Centropages velificatus, 440
Ceramium cimbricum, 187
Ceramium deslongchampsii, 187, 189
Ceramium sp., 295
Ceramium strictum, 187
Ceramium (unidentifiable), 189
Cerastoderma edule, 416, 421
Cerastoderma glaucum, 499
Ceratophyllum demersum, 273
Ceriodaphnia cornuta, 440
Cerithidia californica, 240
Cerstoderma edule, 409
Chacmochuk Lagoon, 321, 324
Chaetoceros spp.
  blooms, 99
  Chelem Lagoon, 316
  Venice Lagoon, 496
Chaetoceros subtilis, 443
Chaetomorpha sp., 48
Chamaelea gallina, 499
Champia parvula, 187, 189, 197
Changes, impact of continuous, 404–407
Characterization of lagoons, 312–326
Chara fibrosa, 314
Chara sp., 284, 291, 295, 299
Chelem Lagoon, 315–317
Chemical characteristics, 257–268
Chetumal Bay, 326
Chincoteague Bay, 9
Chinese lagoons, *see* Hong Kong waters
Chione cancellata, 126
Chironomid sp., 274
Chironomus salinarius, 500
Chlorophyll *a*
  Hong Kong waters, 380–381, 383
  Venice Lagoon ecosystem, 494–496
Chocked lagoons, 2
Choptank Basin (Delmarva Peninsula)
  discussion, 158–162
  drainage control structures, 141
  drivers of water quality change, 148–151
  estuarine water quality, 143, 157–158

excess concentrations, 142
fundamentals, 11, 136–137
groundwater piezometers, 141–142
hydrology, 139
improvement, water quality, 159–162
methods, 139–143
N and P agricultural reductions, 151–155, 157
nontidal monitoring stations, 147–148
regional hydrology, 145–146
results, 145–158
status, water quality, 158–159
study site, 137, 139
surface water chemistry, 139, 141
trends, water quality, 147–148
water quality, 147–151, 158–162
Chroomonas sp., 314
Circulation, 491
Cladocera sp., 496
Cladophora sp.
Patos Lagoon, 448
Waquoit Bay, 210
Yucatan Peninsula, 284, 286, 291, 295, 298–299
Cladophora spp., 440
Cladophora vagbunda, 214
Clarias Lazera, 270, 276
Classification, 2–3
Climate and climate changes, *see also* Pamlico Sound (North Carolina)
Australian coastal lagoons, 462, 477–478
Hong Kong waters, 388–389
Ria Formosa, 339, 343–344
Southern California lagoons, 229–230, 245–246
Venice Lagoon ecosystem, 490
Wadden Sea, 419–422
Climate and climate changes, management implications
biogeochemical considerations and time scales, 21–22, 34–36
climatic impacts, 22–34
climatological time scales, 21–22
eutrophication changes, 9
fundamentals, 7–21, 34
geological time scales, 21–22
historic perspectives, 22–34
human impacts, 22
hurricanes, 9, 23–33
integrated assessment response program, 36–38
large-scale storm events response, 36–38
lessons learned, 34–36
smaller storms, 33–34
socioeconomic aspects and considerations, 36
tropic considerations, 34–36
Climatological time scales, 21–22
Closed lagoon systems, 233
Coastal lagoons
biotic characteristics, 4–6
classification, 2–3
ecological and economic value, 6–7
fundamentals, 1–2
human impacts, 7–9
natural stressors, 9
physiochemical characteristics, 3–4
Codium fragile, 189
Codium isthmocladum, 299
Codium sp., 191
Community shift impact, 240–245
Community stability, *see also* Texas lagoons
abundance, 121
biomass, 122
community structure, 124, 126–128
discussion, 128–132
diversity, 123–124
freshwater inflow, 518–519
fundamentals, 11, 115–117
macrofauna, 121–124
materials and methods, 117–119
results, 119–128
water column, 119–120
Community structure, 124, 126–128
Conceptual model, 514–515
Confinement, 491
Continuous changes, impact
abiotic conditions, 404–407
biotic factors, 407–410
classification, 3
climate change effects, 419–422
ecosystem function, 419–422
fundamentals, 13, 399–400
future research, 422–423
habitat change and destruction, 416–417
higher trophic levels, 409–410
historical drivers, 410, 416–419
invasive species, 418–419
major changes, 403–410
nutrients, 417–418
overexploitation, 416
physiography, 400, 403
pollution, 417
primary producers and productivity, 407, 409
secondary producers and productivity, 409
Controls, benthic macrophyte communities
bottom-up controls, 204–206
case studies, 207–219
function, 204
fundamentals, 11, 203–204, 219–221
Jobos Bay, 215–219
structure, 204
top-down controls, 204–206
Waquoit Bay, 201–215
Core sampling, 173
Corophium orientale, 500
Corophium orientalis, 274
Corophium volutator, 274
Crangon crangon
Venice Lagoon, 501
Wadden Sea, 416
Crassostrea gigas
Venice Lagoon, 499
Wadden Sea, 418, 422
Crepidula fornicata, 418
Crucigenia tetrapedia, 273
Ctenopharyngodon idellus, 450
Cyclope neritea, 499
Cyclotella sp., 316
Cygnus melancorhyphus, 442

Cylindrothea closterium
Dzilam Lagoon, 317
Patos Lagoon, 443
Venice Lagoon, 496
Cylindrotheca closterium, 439
Cymodocea nodosa
Ria Formosa, 340
Venice Lagoon, 497–499
Cymodocea sp., 6
Cynoscion nebulosus, 168
Cynoscion regalis, 169
Cyperus sp., 273
Cyprinus carpio carpio, 450

D

Databases, 520
Data resources, 94–97
Dataset, 340–343
Deep Bay
management strategy and perspectives, 387–388
physical processes and anthropogenic influences, 377
Delmarva Peninsula, *see* Choptank Basin
Deoxygenation, 476–477
Derbesia sp., 291
Diaphanosoma sarsi, 440
Dicentrarchus labrax
Nile Delta lagoons, 276
Venice Lagoon, 501
Differential nutrient use, 102–103
Direct groundwater discharge, 52, 54
Discussions
Choptank Basin, 158–162
Hog Island Bay, 83–86
Laguna Madre, 128–132
New Jersey coastal lagoons, 190–196
Sacca di Goro, 83–86
Texas lagoons, 532–536
Yucatan Peninsula, 291–299
Dissolved inorganic nutrients
Patos Lagoon Estuary, 437–438
Venice Lagoon ecosystem, 493–494
Dissolved oxygen, 383
Diversity, 123–124
Dominant algal species, composition and size, 99–100
Drainage control structures, 141
Drivers of change, 369, 371
Dynophysis spp., 345
Dzilam Lagoon, 317–318

E

East Australian coastal lagoons
acidification, 476–477
anthropogenic change, 476
benthic habitats, 465
carbon budgets, 469, 472
catchment characteristics, 462, 465
climate and climate changes, 462, 477–478
deoxygenation, 476–477
fisheries, 473–474
fish kills, 476–477
fundamentals, 13, 458–459, 479
geomorphic evolution, 459–462
land use, 462, 465
natural changes, 476
pollutant loads, 476
rainfall, 478
runoff, 478
sea level rise, 477
sediment quality, 465
storminess and wave climate, 478
stressors, 462, 465
study area, 459–465
temperature, 477
water quality, 465
Ecological value, 6–7
Economic value, 6–7
Ecosystem function, 419–422
Ecosystem health assessment
biotic characteristics, 4
discussion, 83–86
ecosystem health framework, 84–86
fundamentals, 10, 73–76, 86–88
methods, 76–83
network analysis, 82–83
network construction, 77, 79, 82
results, 83–86
study sites, 76–77
trophic status, 83–84
uncertainty sensitivity, 82–83
Ecosystem health framework
Hog Island Bay, 84–86
Sacca di Goro, 84–86
Ecosystem responses
Venice Lagoon ecosystem, 501–505
Edku Lagoon, 260–262
Egretta thula, 442
Eichornia carssipes, 273
Elasmopus sp., 126, 127
El Niño Southern Oscillation
Patos Lagoon Estuary, 442–447, 451
Southern California lagoons, 246
Eneromorpha intestinalis, 187
Engraulis encrasicolus, 500–501
Ensis americanus, 418
Ensis minor, 499
Enteromorpha spp.
New Jersey lagoons, 187, 191, 193
Ria Formosa, 340
Environmental drivers and pressures, 486–490
Environmental setting, 347–350, 354–356
Epinephelus spp., 276
Erodona mactroides, 440, 446
Escherichia coli
Tolo Harbour, 386
Yucatan Peninsula, 296
Estuarine habits, 439–442, *see also* Blooms
Estuarine water quality, 143, 157–158
Eubosmina tubicen, 440
Eustatism, 488
Euterpina acutrifrons, 440
Eutreptiella sp., 495
Eutrophication processes
aboveground biomass, 176–179
belowground biomass, 179–184

biotic indicators, 175–190
blade length, 186–187
brown tide, 190
composition, macroalgae, 187, 189–190
core sampling, 173
cover, 184–185
density, 184
discussion, 158–162, 190–196
distribution, 175
drainage control structures, 141
drivers of change, water quality, 148–151
estuarine water quality, 143, 157–158
excess concentrations, 142
fundamentals, 11, 136–137, 168–170, 196–198
groundwater piezometers, 141–142
Hong Kong waters, 377, 380–386
hydrology, 139
improvement, water quality, 159–162
macroalgae, 173, 184–185, 187, 189–190
materials and methods, 170–173
methods, 139–143
N and P agricultural reductions, 151–155, 157
nontidal monitoring stations, 147–148
Patos Lagoon Estuary, 448–449
physicochemical conditions, 174–175
quadrat sampling, 172–173
regional hydrology, 145–146
results, 145–158, 174–175
Ria Formosa, 345–346
sampling, 170–173
sediment sampling, 173
status, water quality, 158–159
study sites, 137, 139, 170
surface water chemistry, 139, 141
trends, water quality, 147–148
videographic imaging, 173
water quality, 147–151, 158–162
water quality sampling, 173
Eutrophication processes, comparisons
anthropogenic influences, 371–377
climate change effects, 388–389
Deep Bay, 377, 387–388
dissolved oxygen, 383
drivers of change, 369, 371
eutrophication processes, 377, 380–386
fundamentals, 12, 368–369, 389–391
geographic locations, 369
geomorphologic setting, 369
management strategy and perspectives, 386–388
nutrient stoichiometry, 383–386
oceanographic setting, 369, 371
physical processes, 371–377
Port Shelter, 371–372, 387
Tolo Harbour, 372, 377, 386
Excess concentrations, 142
Exogone sp., 127, 128
Exploitation, biological production, 346–347

F

Farfantepenaeus paulensis, 447–448
Fate of nitrogen, 61–63
Feedback mechanisms, 238–240
Fish, 446–447
Fisheries
Australian coastal lagoons, 473–474
Nile Delta, 275–278
Patos Lagoon Estuary, 447
Fishing activities, 490
Fish kills, 476–477
Frequently closed lagoon systems, 233
Freshwater discharges, 489
Freshwater inflow
benthos, 520
calibration, 526, 528
conceptual model, 514–515
databases, 520
discussion, 532–536
fundamentals, 13, 513–514
Guadalupe Estuary, 517
hydrography, 520
hydrology, 520
lagoonal estuaries, 515–519
Laguna Madre Estuary, 518–519
Lavaca-Colorado Estuary, 517
materials and methods, 519–520
modeling, 520–528
Nueces Estuary, 518
predators, 520
primary production, 520
results, 528–532
study sites, 519
validation, 526, 528
Fulica armillata, 442
Future challenges and directions
Nile Delta, 278
Southern California lagoons, 245–246
Virginia coastal bays, 63–66
Wadden Sea, 422–423

G

Gambierdiscus sp., 99
Gambierdiscus toxicus, 316
Gammis locusta, 274
Gammis oceanicus, 274
Genidens barbus, 448
Genidens genidens, 448
Genidens planifrons, 448
Geographic locations, 369
Geological evolution, 434–435
Geological time scales, 21–22
Geology, 229–230
Geomorphic evolution, 459–462
Geomorphologic setting, 369
Gloeocapsa sp., 319
Gloethece sp., 319
Governance, 490
Gracilaria sp.
biotic characteristics, 6
Delmarva Peninsula, 48
New Jersey lagoons, 191
nitrogen assimilation, benthic micro/macroalgae, 62

Sacca di Goro locations, 76
Waquoit Bay, 210
Gracilaria tikvahiae
Jobos Bay, 218
New Jersey lagoons, 187, 189, 197
Waquoit Bay, 211–215
Grandidierella bonnieroides, 124
Great South Bay, 9
Groundwater piezometers, 141–142
Guadalupe Estuary, 517
Gymnodinium catenatum, 345
Gymnodinium sp., 495

H

Habitat alterations, 346
Habitat change and destruction, 416–417
Halichoerus grypus, 410
Halodule wrightii
Ascención Bay, 325
Bojorquez Lagoon, 325
Celestún Lagoon, 314
Characterization and status, 324–325
Chelem Lagoon, 316
Dzilam Lagoon, 318
Holbox Lagoon, 321
Jobos Bay, 216
Laguna Madre, 130
Rio Lagartos Lagoon, 319
Yucatan Peninsula, 286–287, 290–291, 298–299, 328
Hannea sp., 316
Harbor dredging, 449–450
Hediste diversicolor, 499–500
Heleobia australis, 440, 450
Heteromastus similis, 440
Heterotrophic bacterioplankton, 352–354
Hexaplex trunculus, 499
Higher trophic levels, 409–410
Historical developments
Pamlico Sounds, 22–34
Patos Lagoon Estuary, 434–435
Venice Lagoon ecosystem, 484–486
Historical drivers, 410, 416–419
Hog Island Bay (HIB-Virginia)
biotic characteristics, 4
discussion, 83–86
ecosystem health framework, 84–86
fundamentals, 10, 73–76, 86–88
methods, 76–83
network analysis, 82–83
network construction, 77, 79, 82
results, 83–86
study sites, 76–77
trophic status, 83–84
uncertainty sensitivity, 82–83
Holbox Lagoon, 320–321
Hong Kong waters
anthropogenic influences, 371–377
climate change effects, 388–389
Deep Bay, 377, 387–388
dissolved oxygen, 383
drivers of change, 369, 371
eutrophication processes, 377, 380–386
fundamentals, 12, 368–369, 389–391
geographic locations, 369
geomorphologic setting, 369
management strategy and perspectives, 386–388
nutrient stoichiometry, 383–386
oceanographic setting, 369, 371
physical processes, 371–377
Port Shelter, 371–372, 387
Tolo Harbour, 372, 377, 386
Human impacts, *see also* Anthropogenic pressures
coastal lagoons, 7–9
Pamlico Sound, 22
Patos Lagoon Estuary, 448–450
Human impacts, assessing response
biogeochemical considerations and time scales, 21–22, 34–36
climatic impacts, 22–34
climatological time scales, 21–22
eutrophication changes, 9
fundamentals, 7–21, 34
geological time scales, 21–22
historic perspectives, 22–34
human impacts, 22
hurricanes, 9, 23–33
integrated assessment response program, 36–38
large-scale storm events response, 36–38
lessons learned, 34–36
smaller storms, 33–34
socioeconomic aspects and considerations, 36
tropic considerations, 34–36
Human impacts, biotic responses
bioinvasion, 450
biological components, 439–442
biotic responses to change, 442–447
classification, 2
dissolved inorganic nutrients, 437–438
El Niño Southern Oscillation, 443–447
estuarine habits, 439–442
eutrophication, 448–449
fish, 446–447
fisheries, 447
fundamentals, 13, 434
geological evolution, 434–435
harbor dredging, 449–450
historical developments, 434–435
human influence, 448–450
macrozoobenthos, 446
natural protection areas, 435
overfishing, 449
phytoplankton, 443
primary producers, 439–440
processes, 439–442
Ruppia maritima, 444–445
salinity, 436–437
secondary producers, 440–442
temperature, 436–437
water exchange, 436–437
Human impacts, mesotidal lagoon
anthropogenic changes, 343, 344–345
approach, 340–343

biological assemblages, 340
climate and climate changes, 339, 343–344
dataset, 340–343
environmental setting, 347–350, 354–356
eutrophication, 345–346
exploitation, biological production, 346–347
fundamentals, 12, 336–337
habitat alterations, 346
heterotrophic bacterioplankton, 352–354
hydrodynamics, 339–340, 346
hydrologic setting, 339
interannual dynamics, planktonic microbes, 354–360
introduced species, 347
morphology, 339–340
oceanographic setting, 339
overview, 339–340
phytoplankton, 351–352, 356–360
seasonal dynamics, 347–352
sedimentary dynamics, 346
water quality, 345–346
Hurricanes, *see also* Storminess and wave climate
natural stressors, 9
Pamlico Sound, 23–33
Hydrobia ulvae, 500
Hydrodynamics, 339–340, 346
Hydrography, 520
Hydrologic setting, 339
Hydrology
Choptank Basin, 139
Southern California lagoons, 230, 232–234
Texas lagoons, 520
Hypophthalmichthys molitrix, 450

I

Inorganic nitrogen, origins and fate
discussion, 291–299
fundamentals, 12, 283–286
methods, 286–288
nutrient impacts, 298–299
nutrient sources, 295–298
plant isotope values, 291
results, 288–291
sample collection and analysis, 286–288
seagrass biomass, 290
statistical analysis, 288
study area, 286
tissue nutrient content, 291
water chemistry, 288, 290
water column processes, 291–295
Integrated assessment response program, 36–38
Intensity of blooms, 97–99
Interannual dynamics, planktonic microbes, *see* Seasonal and interannual variations
Introduced species, 347, *see also specific type*
Invasive species, 418–419, *see also specific type*
Isochrysis galbana, 129

J

Jenynsia multidentata, 441
Jobos Bay
bottom-up controls, 215–219
case studies, 215–219
top-down controls, 215–219

K

Kalliapseudes schubartii, 440, 446
Karenia brevis, 99
Karstic assessment, subtropical lagoons
Ascención Bay, 325–326
Bojorquez Lagoon, 325
Celestún Lagoon, 314–315
Chacmochuk Lagoon, 321, 324
characterization of lagoons, 312–326
Chelem Lagoon, 315–317
Chetumal Bay, 326
context, 309–311
Dzilam Lagoon, 317–318
fundamentals, 307–309, 329
Holbox Lagoon, 320–321
lessons learned, 326–329
monitoring experience, 311–312
Nichupte Lagoon, 324–325
phytoplankton, 328
research, 311–312
Rio Lagartos Lagoon, 318–319
status of lagoons, 312–326
submerged aquatic vegetation, 328–329
water quality characteristics, 326–328
Knipowitschia panizzae, 500

L

Labarinthula sp., 58
Labyrinthula zosterae, 191, 195
Laeonereis acuta, 440
Lagao dos Patos, *see* Patos Lagoon Estuary (Brazil)
Lagoonal estuaries, 515–519
Lagoon state and impacts, 491–501
Laguna Madre (Texas), *see also* Texas lagoons
abundance, 121
biomass, 122
community structure, 124, 126–128
discussion, 128–132
diversity, 123–124
freshwater inflow, 518–519
fundamentals, 11, 115–117
macrofauna, 121–124
materials and methods, 117–119
results, 119–128
water column, 119–120
Laguncularia racemosa, 314, 317
Laminaria sp., 6
Landings data, 500–501
Land use, 462, 465
La Niña periods, 443
Large-scale storm events response, 36–38
Larus maculipennis, 442
Lavaca-Colorado Estuary, 517
Leaky lagoons, 2–3
Lessons learned
karstic ecosystem, 326–329
Pamlico Sound, 34–36
phytoplankton, 328

submerged aquatic vegetation, 328–329
water quality characteristics, 326–328
Yucatan Peninsula, 326–329
Limanda limanda, 410
Limited riverine/oceanic influence, 233–234
Limnoperna fortunei, 450
Liza auratus, 276
Liza ramada, 276
Liza saliens
Nile Delta lagoons, 276
Venice Lagoon, 500
Loligo vulgaris, 501
Lomentaria baileyana, 189–190
Loripes lacteus, 499

M

Macoma balthica, 409, 420–421
Macroalgae
composition, 187, 189–190
cover, 184–185
sampling, 173
Macrobenthos secondary production
benthos, 520
calibration, 526, 528
conceptual model, 514–515
databases, 520
discussion, 532–536
fundamentals, 13, 513–514
Guadalupe Estuary, 517
hydrography, 520
hydrology, 520
lagoonal estuaries, 515–519
Laguna Madre Estuary, 518–519
Lavaca-Colorado Estuary, 517
materials and methods, 519–520
modeling, 520–528
Nueces Estuary, 518
predators, 520
primary production, 520
results, 528–532
study sites, 519
validation, 526, 528
Macrodon ancylodon, 449
Macroinfaunal diversity, *see also* Texas lagoons
abundance, 121
biomass, 122
community structure, 124, 126–128
discussion, 128–132
diversity, 123–124
freshwater inflow, 518–519
fundamentals, 11, 115–117
macrofauna, 121–124
materials and methods, 117–119
results, 119–128
water column, 119–120
Macrophytobenthos, 497–498
Macrozoobenthos
Patos Lagoon Estuary, 446
Venice Lagoon ecosystem, 498–500
Magelona pettiboneae, 122, 130
Management implications, human and climatic effects
biogeochemical considerations and time scales, 21–22, 34–36
climatic impacts, 22–34
climatological time scales, 21–22
eutrophication changes, 9
fundamentals, 7–21, 34
geological time scales, 21–22
historic perspectives, 22–34
human impacts, 22
hurricanes, 9, 23–33
integrated assessment response program, 36–38
large-scale storm events response, 36–38
lessons learned, 34–36
smaller storms, 33–34
socioeconomic aspects and considerations, 36
tropic considerations, 34–36
Management strategy and perspectives, 386–388
Manzalah Lagoon, 262–265
Maryut Lagoon, 266–268
Materials and methods
Laguna Madre, 117–119
New Jersey coastal lagoons, 170–173
Texas lagoons, 519–520
Maya arenaria, 418, 421
Mediomastus ambiseta, 127
Melania tuberculata, 274–275
Melanoides tuberculata, 275
Mercenaria mercenaria, 169, 190, 192, 195
Mercierella enigmatica, 275
Merismopedia tenussima, 273
Mesocyclops annulatus, 440
Mesotidal lagoons, seasonal/interannual variations
anthropogenic changes, 343, 344–345
approach, 340–343
biological assemblages, 340
climate and climate changes, 339, 343–344
dataset, 340–343
environmental setting, 347–350, 354–356
eutrophication, 345–346
exploitation, biological production, 346–347
fundamentals, 12, 336–337
habitat alterations, 346
heterotrophic bacterioplankton, 352–354
hydrodynamics, 339–340, 346
hydrologic setting, 339
interannual dynamics, planktonic microbes, 354–360
introduced species, 347
morphology, 339–340
oceanographic setting, 339
overview, 339–340
phytoplankton, 351–352, 356–360
seasonal dynamics, 347–352
sedimentary dynamics, 346
water quality, 345–346
Methods
Choptank Basin, 139–143
Hog Island Bay, 76–83
Sacca di Goro, 76–83
Yucatan Peninsula, 286–288
Microcystis sp., 439
Micropogonias furnieri, 441, 446, 448
Mircocoleus lyngbyaceus, 284
Mnemiopsos leidyi, 419

Modeling, 520–528
Moina micrura
Nile Delta lagoons, 274
Patos Lagoon, 440
Monitoring experience, 311–312
Morphology, 339–340
Mugil cephalus
Nile Delta lagoons, 276
Patos Lagoon, 447
Mugil curema, 441
Mugil gaimardianus, 441
Mugil platanus, 441, 447–449
Mulinia lateralis, 122, 124, 128–130
Mya arenaria, 409
Mytilicola intestinalis, 416
Mytilus edulis
New Jersey lagoons, 168–169, 195
Wadden Sea, 416, 421, 422

N

N agricultural reductions, 151–155, 157
Nanozostera noltii, 497–498
Nassarius mutabilis, 499
Nassarius nitidus, 499
Natural changes, East Australia
acidification, 476–477
anthropogenic change, 476
benthic habitats, 465
carbon budgets, 469, 472
catchment characteristics, 462, 465
climate and climate changes, 462, 477–478
deoxygenation, 476–477
fisheries, 473–474
fish kills, 476–477
fundamentals, 13, 458–459, 479
geomorphic evolution, 459–462
land use, 462, 465
natural changes, 476
pollutant loads, 476
rainfall, 478
runoff, 478
sea level rise, 477
sediment quality, 465
storminess and wave climate, 478
stressors, 462, 465
study area, 459–465
temperature, 477
water quality, 465
Natural changes, Venice
bathymetry, 491–493
chlorophyll *a,* 494–496
circulation, 491
climate change, 490
confinement, 491
dissolved inorganic nutrients, 493–494
ecosystem responses, 501–505
environmental drivers and pressures, 486–490
eustatism, 488
fishing activities, 490
freshwater discharges, 489
fundamentals, 13, 484, 505
governance, 490
historical developments, 484–486
lagoon state and impacts, 491–501
landings data, 500–501
macrophytobenthos, 497–498
macrozoobenthos, 498–500
nekton, 500
nutrient loads, 489
phytoplankton, 494–496
pollutants, 489
sediment nutrients, 494
subsidence, 488
zooplankton, 496
Natural protection areas, 435
Natural stressors, 9
Nature of change, 234–240
Navicula sp., 314
Neanthes succinea, 499
Negative effects, 240–241
Nekton, 500
Neohelice granulata, 448
Neosiphonia harveyi, 187
Nephtys fluviatilis, 440
Nereis diversicolor
Nile Delta lagoons, 274
Wadden Sea, 421
Nereis sp., 274
Nereis virens, 421
Nerophis ophidion, 500
Network analysis
Hog Island Bay, 82–83
Sacca di Goro, 82–83
Network construction
Hog Island Bay, 77, 79, 82
Sacca di Goro, 77, 79, 82
New Jersey coastal lagoons
aboveground biomass, 176–179
belowground biomass, 179–184
biotic indicators, 175–190
blade length, 186–187
brown tide, 190
composition, macroalgae, 187, 189–190
core sampling, 173
cover, 184–185
density, 184
discussion, 190–196
distribution, 175
fundamentals, 11, 168–170, 196–198
macroalgae, 173, 184–185, 187, 189–190
materials and methods, 170–173
physicochemical conditions, 174–175
quadrat sampling, 172–173
results, 174–175
sampling, 170–173
sediment sampling, 173
study area, 170
videographic imaging, 173
water quality sampling, 173
Newport Bay, 9
Nichupte Lagoon, 324–325
Nile Delta
benthic invertebrates, 274–275
biological characteristics, 257–268
Burullus Lagoon, 258–260

chemical characteristics, 257–268
Edku Lagoon, 260–262
fisheries, 275–278
fundamentals, 12, 253–257
future directions, 278
Manzalah Lagoon, 262–265
Maryut Lagoon, 266–268
nutrient dynamics, 270–272, 277–278
physical characteristics, 257–268
pollutants, 268–270
primary producers, 272–274
substrates, 268
zooplankton, 274–275
Nistzchia longissima, 319
Nitrogen, sources and fates
allochthonous nitrogen sources, 49–55
assimilation, 58–60
atmospheric deposition, 54–55
autochthonouse nitrogen sources, 56–58
benthic macro- and microalgae, 61–63
benthic primary production, 61
direct groundwater discharge, 52, 54
fate of nitrogen, 61–63
fundamentals, 10, 43–50
future trends and predictions, 63–66
nitrogen fixation, 56
nutrient gradient, 49
physical description, 48
primary producers, 58–60
remineralization, 57–58
restoration of seagrass, 65–66
seagrasses, 58, 65–66
seasonality, 58
sinks of nitrogen, 58–61
sources of, 48–58
stream flow, 50–52
Nitrogen cycling, ecosystem health assessment
biotic characteristics, 4
discussion, 83–86
ecosystem health framework, 84–86
fundamentals, 10, 73–76, 86–88
methods, 76–83
network analysis, 82–83
network construction, 77, 79, 82
results, 83–86
study sites, 76–77
trophic status, 83–84
uncertainty sensitivity, 82–83
Nitrogen fixation, 56
Nitrogen (inorganic), origins and fate
discussion, 291–299
fundamentals, 12, 283–286
methods, 286–288
nutrient impacts, 298–299
nutrient sources, 295–298
plant isotope values, 291
results, 288–291
sample collection and analysis, 286–288
seagrass biomass, 290
statistical analysis, 288
study area, 286
tissue nutrient content, 291
water chemistry, 288, 290
water column processes, 291–295
Nitzschia sp., 272
Noctiluca scantillans, 381
Nontidal monitoring stations, 147–148
Notodiaptomus carteri, 440
Notodiaptomus incompositus, 440
Nueces Estuary, 518
Nutrients
composition and sources, 100–102
dynamics, 270–272, 277–278
gradient, 49
historical drivers, 417–418
impacts, 298–299
loads, 489
retention and recycling, 241–245
sources, 295–298
stoichiometry, 383–386
supply, 230, 232–234

O

Obelia geniculata, 409
Oceanic influence, limited, 233–234
Oceanographic setting
Hong Kong waters, 369, 371
Ria Formosa, 339
Odontesthes argentinensis, 441, 447
Oncea conifera, 440
Open estuarine/lagoon systems, 232–233
Oreochromis aureus, 276
Oreochromis niloticus, 276
Ostrea edulis, 417
Otaria flavescens, 442
Overexploitation, 416
Overfishing, 449
Owenia fusiformis, 499

P

Pachyurus bonariensis, 450
P agricultural reductions, 151–155, 157
Palaemon sp., 501
Pamlico Sound (North Carolina)
biogeochemical considerations and time scales, 21–22, 34–36
climatic impacts, 22–34
climatological time scales, 21–22
eutrophication changes, 9
fundamentals, 7–21, 34
geological time scales, 21–22
historic perspectives, 22–34
human impacts, 22
hurricanes, 9, 23–33
integrated assessment response program, 36–38
large-scale storm events response, 36–38
lessons learned, 34–36
smaller storms, 33–34
socioeconomic aspects and considerations, 36
tropic considerations, 34–36
Paphia aurea, 499
Paralichthys orbignyanus, 442
Parapimelodus nigribarbis, 447

Parona signata, 442
Patos Lagoon Estuary (Brazil)
  bioinvasion, 450
  biological components, 439–442
  biotic responses to change, 442–447
  classification, 2
  dissolved inorganic nutrients, 437–438
  El Niño Southern Oscillation, 443–447
  estuarine habits, 439–442
  eutrophication, 448–449
  fish, 446–447
  fisheries, 447
  fundamentals, 13, 434
  geological evolution, 434–435
  harbor dredging, 449–450
  historical developments, 434–435
  human influence, 448–450
  macrozoobenthos, 446
  natural protection areas, 435
  overfishing, 449
  phytoplankton, 443
  primary producers, 439–440
  processes, 439–442
  Ruppia maritima, 444–445
  salinity, 436–437
  secondary producers, 440–442
  temperature, 436–437
  water exchange, 436–437
Penilia avirostris, 440
Peprilus paru, 442
Perkinsus sp., 347
Permanently closed lagoon systems, 233
Perna perna, 450
Pfiesteria sp., 99
Phaeocystis sp., 409
Phalacrocorax olivaceus, 442
Phase shifts, 234, 236
Phoca vitulina, 410, 417
Phocoena phocoena, 410
Phragmites australis, 273
Phragmites communis, 273
Physical characteristics, 48, 257–268
Physical processes, 371–377
Physicochemical characteristics, 3–4, 174–175
Physiography, 400, 403
Phytoplankton
  lessons learned, 328
  Patos Lagoon Estuary, 443
  Ria Formosa, 351–352, 356–360
  Venice Lagoon ecosystem, 494–496
  Yucatan Peninsula, 328
Planktonic microbe variability
  anthropogenic changes, 343, 344–345
  approach, 340–343
  biological assemblages, 340
  climate and climate changes, 339, 343–344
  dataset, 340–343
  environmental setting, 347–350, 354–356
  eutrophication, 345–346
  exploitation, biological production, 346–347
  fundamentals, 12, 336–337
  habitat alterations, 346
  heterotrophic bacterioplankton, 352–354
  hydrodynamics, 339–340, 346
  hydrologic setting, 339
  interannual dynamics, planktonic microbes, 354–360
  introduced species, 347
  morphology, 339–340
  oceanographic setting, 339
  overview, 339–340
  phytoplankton, 351–352, 356–360
  seasonal dynamics, 347–352
  sedimentary dynamics, 346
  water quality, 345–346
Planktospaeria gelatinosa, 273
Plant isotope values, 291
Platanichthys platana, 441
Platichthys flesus, 500
Pleopys polyphemoides, 440
Pleuronectes platessa, 410
Podon intermedius, 440
Pogonias cromis, 448
"Poised on the edge," 103–106
Pollution
  Australian coastal lagoons, 476
  Nile Delta, 268–270
  Venice Lagoon ecosystem, 489
  Wadden Sea, 417
Polydora ligni, 128
Polysiphonia sp., 291
Polysiphonia spp., 187
Pomatoschistus canestrinii, 500
Pomatoschistus marmoratus, 500
Pomatoschistus minutus, 500
Porichthys porosissimus, 442
Port Shelter
  management strategy and perspectives, 387
  physical processes and anthropogenic influences, 371–372
Potamocorbula amurensis, 44
Potamogeton pectinatus, 273
Predators, 520
Primary production
  benthic, Virginia coastal bays, 61
  biotic responses, natural and anthropogenic impacts, 439–440
  continuous changes, impact, 407, 409
  freshwater inflow, 520
  human impacts, biotic responses, 439–440
  macrobenthos secondary production, 520
  Nile Delta, 272–274
  Patos Lagoon Estuary, 439–440
  responses, natural and anthropogenic pressures, 439–440
  Southern California lagoons, 236–238
  Texas lagoons, 520
  Virginia coastal bays, 58–60
  Wadden Sea, 407, 409
Prionospio heterobranchia, 116, 121, 124, 127–128
Priontus punctatus, 442
Prorocentrum minimum, 99
Prorocentrum sp., 318
Protoperidinium sp., 318
PS, *see* Pamlico Sound (North Carolina)
Pseudonitzschia spp., 345

Q

Quadrat sampling, 172–173

R

Rainfall, 478
Ramnogaster arcuata, 441
Regional hydrology, 145–146
Rehoboth Bay, 9
Remineralization, 57–58
Responses, eutrophication
aboveground biomass, 176–179
belowground biomass, 179–184
biotic indicators, 175–190
blade length, 186–187
brown tide, 190
composition, macroalgae, 187, 189–190
core sampling, 173
cover, 184–185
density, 184
discussion, 190–196
distribution, 175
fundamentals, 11, 168–170, 196–198
macroalgae, 173, 184–185, 187, 189–190
materials and methods, 170–173
physicochemical conditions, 174–175
quadrat sampling, 172–173
results, 174–175
sampling, 170–173
sediment sampling, 173
study area, 170
videographic imaging, 173
water quality sampling, 173
Responses, human and climatic effects assessment
biogeochemical considerations and time scales, 21–22, 34–36
climatic impacts, 22–34
climatological time scales, 21–22
eutrophication changes, 9
fundamentals, 7–21, 34
geological time scales, 21–22
historic perspectives, 22–34
human impacts, 22
hurricanes, 9, 23–33
integrated assessment response program, 36–38
large-scale storm events response, 36–38
lessons learned, 34–36
smaller storms, 33–34
socioeconomic aspects and considerations, 36
tropic considerations, 34–36
Responses, natural and anthropogenic pressures
bathymetry, 491–493
bioinvasion, 450
biological components, 439–442
biotic responses to change, 442–447
chlorophyll *a,* 494–496
circulation, 491
classification, 2
climate change, 490
confinement, 491
dissolved inorganic nutrients, 437–438, 493–494
ecosystem responses, 501–505
El Niño Southern Oscillation, 443–447
environmental drivers and pressures, 486–490
estuarine habits, 439–442
eustatism, 488
eutrophication, 448–449
fish, 446–447
fisheries, 447
fishing activities, 490
freshwater discharges, 489
fundamentals, 13, 434, 484, 505
geological evolution, 434–435
governance, 490
harbor dredging, 449–450
historical developments, 434–435, 484–486
human influence, 448–450
lagoon state and impacts, 491–501
landings data, 500–501
macrophytobenthos, 497–498
macrozoobenthos, 446, 498–500
natural protection areas, 435
nekton, 500
nutrient loads, 489
overfishing, 449
phytoplankton, 443, 494–496
pollutants, 489
primary producers, 439–440
processes, 439–442
Ruppia maritima, 444–445
salinity, 436–437
secondary producers, 440–442
sediment nutrients, 494
subsidence, 488
temperature, 436–437
water exchange, 436–437
zooplankton, 496
Restricted lagoons, 2
Results
Choptank Basin, 145–158
Hog Island Bay, 83–86
Laguna Madre, 119–128
New Jersey coastal lagoons, 174–175
Sacca di Goro, 83–86
Texas lagoons, 528–532
Yucatan Peninsula, 288–291
Rhizoclonium riparium, 440
Rhizoclonium sp., 448
Rhizophora mangle
Celestún Lagoon, 314
Chelem Lagoon, 315
Dzilam Lagoon, 317
Rio Lagartos Lagoon, 319
Rhizosolenia spp., 99
Ria Formosa (Portugal)
anthropogenic changes, 343, 344–345
approach, 340–343
biological assemblages, 340
climate and climate changes, 339, 343–344
dataset, 340–343
environmental setting, 347–350, 354–356
eutrophication, 345–346
exploitation, biological production, 346–347
fundamentals, 12, 336–337
habitat alterations, 346

heterotrophic bacterioplankton, 352–354
hydrodynamics, 339–340, 346
hydrologic setting, 339
interannual dynamics, planktonic microbes, 354–360
introduced species, 347
morphology, 339–340
oceanographic setting, 339
overview, 339–340
phytoplankton, 351–352, 356–360
seasonal dynamics, 347–352
sedimentary dynamics, 346
water quality, 345–346
Rio Lagartos Lagoon, 318–319
Riverine influence, limited, *see also* Blooms; Freshwater discharges and inflow
Southern California lagoons, 233–234
Ruditapes decussatus
Ria Formosa, 340
Venice Lagoon, 499
Ruditapes philippinarum
Ria Formosa, 347
Venice Lagoon, 490, 504–505
Ruditapes sp., 504
Runoff, 478
Ruppia maritima
Ascención Bay, 325
Celestún Lagoon, 314
Characterization and status, 324
Dzilam Lagoon, 318
New Jersey lagoons, 175
Patos Lagoon, 439, 441, 448
Patos Lagoon Estuary, 444–445
Rio Lagartos Lagoon, 319
Wadden Sea, 421
Yucatan Peninsula, 328
Rynchops niger, 442

S

Sabellaria spinulosa, 417
Sabellidae (unidentified), 126
Sacca di Goro (Italy)
discussion, 83–86
ecosystem health framework, 84–86
fundamentals, 10, 73–76, 86–88
methods, 76–83
network analysis, 82–83
network construction, 77, 79, 82
results, 83–86
study sites, 76–77
trophic status, 83–84
uncertainty sensitivity, 82–83
Sagitta enflata, 440
Sagitta tenuis, 440
Salinity, 436–437
Sampling
New Jersey coastal lagoons, 170–173
Yucatan Peninsula, 286–288
Sargassum muticum
Ria Formosa, 347
Venice Lagoon, 498
Sarotherodon galilaeus, 276
Scolopolos rubra, 122, 130
Scripsiella trochoidea, 319
Scrobhicularia plana, 499
Seagrass
density, 184
distribution, 175
New Jersey coastal lagoons, 184–187
restoration, 65–66
Virginia coastal bays, 58
Yucatan Peninsula, 290
Seagrass decline, eutrophication
aboveground biomass, 176–179
belowground biomass, 179–184
biotic indicators, 175–190
blade length, 186–187
brown tide, 190
composition, macroalgae, 187, 189–190
core sampling, 173
cover, 184–185
density, 184
discussion, 190–196
distribution, 175
fundamentals, 11, 168–170, 196–198
macroalgae, 173, 184–185, 187, 189–190
materials and methods, 170–173
physicochemical conditions, 174–175
quadrat sampling, 172–173
results, 174–175
sampling, 170–173
sediment sampling, 173
study area, 170
videographic imaging, 173
water quality sampling, 173
Sea level rise, 477
Seasonal and interannual variations
anthropogenic changes, 343, 344–345
approach, 340–343
biological assemblages, 340
climate and climate changes, 339, 343–344
dataset, 340–343
environmental setting, 347–350, 354–356
eutrophication, 345–346
exploitation, biological production, 346–347
fundamentals, 12, 336–337
habitat alterations, 346
heterotrophic bacterioplankton, 352–354
hydrodynamics, 339–340, 346
hydrologic setting, 339
interannual dynamics, planktonic microbes, 354–360
introduced species, 347
morphology, 339–340
oceanographic setting, 339
overview, 339–340
phytoplankton, 351–352, 356–360
seasonal dynamics, 347–352
sedimentary dynamics, 346
water quality, 345–346
Seasonality
blooms, 97–99
Ria Formosa, 347–352
Virginia coastal bays, 58
Secondary production
continuous changes, impact, 409
human impacts, biotic responses, 440–442

Secondary production, macrobenthos
benthos, 520
calibration, 526, 528
conceptual model, 514–515
databases, 520
discussion, 532–536
fundamentals, 13, 513–514
Guadalupe Estuary, 517
hydrography, 520
hydrology, 520
lagoonal estuaries, 515–519
Laguna Madre Estuary, 518–519
Lavaca-Colorado Estuary, 517
materials and methods, 519–520
modeling, 520–528
Nueces Estuary, 518
predators, 520
primary production, 520
results, 528–532
study sites, 519
validation, 526, 528
Sediment
nutrients, 494
quality, 465
sampling, 173
Sedimentary dynamics, 346
Semi-enclosed embayments
anthropogenic influences, 371–377
climate change effects, 388–389
Deep Bay, 377, 387–388
dissolved oxygen, 383
drivers of change, 369, 371
eutrophication processes, 377, 380–386
fundamentals, 12, 368–369, 389–391
geographic locations, 369
geomorphologic setting, 369
management strategy and perspectives, 386–388
nutrient stoichiometry, 383–386
oceanographic setting, 369, 371
physical processes, 371–377
Port Shelter, 371–372, 387
Tolo Harbour, 372, 377, 386
Sepia officinalis, 501
Simosa sp., 440
Sinepuxent Bay, 9
Sinks of nitrogen, 58–61
Sites, *see* Study sites
Skeletonema costatum, 99
Skeletonema marinoi, 495–496
Skeletonema spp., 443
Smaller storms, 33–34
Socioeconomic aspects and considerations, 36
Solea solea
Venice Lagoon, 500
Wadden Sea, 410
Solea spp., 276
Solen marginatus, 499
Southern California lagoons
alternative stable states, 234, 236
bottom-up forcing by nutrients, 236–238
case study, 241–245
climate and climate changes, 229–230, 245–246
community shift impact, 240–245
feedback mechanisms, 238–240
frequently/permanently closed lagoon systems, 233
fundamentals, 11–12, 227–229, 247
future challenges, 245–246
geology, 229–230
hydrology, 230, 232–234
limited riverine/oceanic influence, 233–234
nature of change, 234–240
negative effects, 240–241
nutrient retention and recycling, 241–245
nutrient supply, 230, 232–234
open estuarine/lagoon systems, 232–233
phase shifts, 234, 236
primary producer communities, 236–238
trophic support, 240–241
Spartina alterniflora, 59
Spartina densiflora, 347
Spartina maritima, 340
Sparus aurata, 276
Spyridia filamentosa, 187, 189, 197
Statistical analysis, 288
Status of lagoons, 312–326
Sterna hirundo, 442
Sterna trudeaui, 442
Storminess and wave climate, 478, *see also* Hurricanes
Stream flow, 50–52
Streblospio benedicti, 116, 121, 123–124, 127–131
Streblospio shrubsolii, 499
Stressors, 462, 465, *see also* Natural stressors
Structure, 204
Structure and function, temperate lagoons
acidification, 476–477
anthropogenic change, 476
benthic habitats, 465
carbon budgets, 469, 472
catchment characteristics, 462, 465
climate and climate changes, 462, 477–478
deoxygenation, 476–477
fisheries, 473–474
fish kills, 476–477
fundamentals, 13, 458–459, 479
geomorphic evolution, 459–462
land use, 462, 465
natural changes, 476
pollutant loads, 476
rainfall, 478
runoff, 478
sea level rise, 477
sediment quality, 465
storminess and wave climate, 478
stressors, 462, 465
study area, 459–465
temperature, 477
water quality, 465
Study sites
Australian coastal lagoons, 459–465
blooms, 94–97
Choptank Basin, 137, 139
Hog Island Bay, 76–77
New Jersey coastal lagoons, 170
Sacca di Goro, 76–77
Texas lagoons, 519
Yucatan Peninsula, 286

Submerged aquatic vegetation, 328–329
Subsidence, 488
Substrates, 268
Subtropical karstic assessment
  Ascención Bay, 325–326
  Bojorquez Lagoon, 325
  Celestún Lagoon, 314–315
  Chacmochuk Lagoon, 321, 324
  characterization of lagoons, 312–326
  Chelem Lagoon, 315–317
  Chetumal Bay, 326
  context, 309–311
  Dzilam Lagoon, 317–318
  fundamentals, 307–309, 329
  Holbox Lagoon, 320–321
  lessons learned, 326–329
  monitoring experience, 311–312
  Nichupte Lagoon, 324–325
  phytoplankton, 328
  research, 311–312
  Rio Lagartos Lagoon, 318–319
  status of lagoons, 312–326
  submerged aquatic vegetation, 328–329
  water quality characteristics, 326–328
Subtropical semi-enclosed embayments
  anthropogenic influences, 371–377
  climate change effects, 388–389
  Deep Bay, 377, 387–388
  dissolved oxygen, 383
  drivers of change, 369, 371
  eutrophication processes, 377, 380–386
  fundamentals, 12, 368–369, 389–391
  geographic locations, 369
  geomorphologic setting, 369
  management strategy and perspectives, 386–388
  nutrient stoichiometry, 383–386
  oceanographic setting, 369, 371
  physical processes, 371–377
  Port Shelter, 371–372, 387
  Tolo Harbour, 372, 377, 386
Surface water chemistry, 139, 141
Surirella spp., 439
Syllis cornuta, 124, 126–128
Synechococcus sp.
  algal physiology, 102
  blooms, 99, 104, 105
Syngnathus abaste, 500
Syringodium filiforme
  Ascención Bay, 325
  Characterization and status, 324
  Dzilam Lagoon, 318
  Holbox Lagoon, 321
  Jobos Bay, 216
  Yucatan Peninsula, 286–287, 290–291, 298, 328

T

Tapes philippinarum, 77
Temora longicornis, 409
Temora stylifera, 450
Temora turbinate, 450
Temperate and tropical estuaries
  bottom-up controls, 204–206
  case studies, 207–219
  function, 204
  fundamentals, 11, 203–204, 219–221
  Jobos Bay, 215–219
  structure, 204
  top-down controls, 204–206
  Waquoit Bay, 201–215
Temperate lagoons, structure and function
  acidification, 476–477
  anthropogenic change, 476
  benthic habitats, 465
  carbon budgets, 469, 472
  catchment characteristics, 462, 465
  climate and climate changes, 462, 477–478
  deoxygenation, 476–477
  fisheries, 473–474
  fish kills, 476–477
  fundamentals, 13, 458–459, 479
  geomorphic evolution, 459–462
  land use, 462, 465
  natural changes, 476
  pollutant loads, 476
  rainfall, 478
  runoff, 478
  sea level rise, 477
  sediment quality, 465
  storminess and wave climate, 478
  stressors, 462, 465
  study area, 459–465
  temperature, 477
  water quality, 465
Temperature
  Australian coastal lagoons, 477
  Patos Lagoon Estuary, 436–437
Texas lagoons, *see also* Laguna Madre (Texas)
  benthos, 520
  calibration, 526, 528
  conceptual model, 514–515
  databases, 520
  discussion, 532–536
  fundamentals, 13, 513–514
  Guadalupe Estuary, 517
  hydrography, 520
  hydrology, 520
  lagoonal estuaries, 515–519
  Laguna Madre Estuary, 518–519
  Lavaca-Colorado Estuary, 517
  materials and methods, 519–520
  modeling, 520–528
  Nueces Estuary, 518
  predators, 520
  primary production, 520
  results, 528–532
  study sites, 519
  validation, 526, 528
Thalassia testudinum
  Ascención Bay, 325–326
  Bojorquez Lagoon, 325
  Characterization and status, 324–325

Chelem Lagoon, 316
Dzilam Lagoon, 318
Holbox Lagoon, 321
Jobos Bay, 215–216, 219
Rio Lagartos Lagoon, 319
Yucatan Peninsula, 286–288, 290–291, 298–299, 328
Thalassiosira issima, 316
Thalassiosira sp., 314
Tilapia zillii, 276
Tissue nutrient content, 291
Tolo Harbour
management strategy and perspectives, 386
physical processes and anthropogenic influences, 372, 377
Top-down controls
benthic macrophyte communities, 204–206
Jobos Bay, 215–219
Waquoit Bay, 207–215
Trachelyopterus lucenai, 450
Trophic status
Hog Island Bay, 83–84
Sacca di Goro, 83–84
Trophic support, 240–241
Tropical climate, *see* Temperate and tropical estuaries
Tropic considerations, 34–36
Tursiops truncatus, 442
Tyhpa australis, 273

U

Ulothrix flacca, 187, 190
Ulva lactuca
Jobos Bay, 218–219
New Jersey lagoons, 168, 187, 189, 191, 194, 197
Waquoit Bay, 211–214
Ulva sp.
Delmarva Peninsula, 48
New Jersey lagoons, 191, 193
Patos Lagoon, 448–449
Venice Lagoon, 496
Ulva spp.
Patos Lagoon, 440
Ria Formosa, 340
Sacca di Goro locations, 76–77
Venice Lagoon, 501, 503–504
Waquoit Bay, 210
Uncertainty sensitivity
Hog Island Bay, 82–83
Sacca di Goro, 82–83
Undaria pinnatifida, 498
Urbanization, slow
discussion, 158–162
drainage control structures, 141
drivers of change, water quality, 148–151
estuarine water quality, 143, 157–158
excess concentrations, 142
fundamentals, 11, 136–137
groundwater piezometers, 141–142
hydrology, 139
improvement, water quality, 159–162
methods, 139–143
N and P agricultural reductions, 151–155, 157
nontidal monitoring stations, 147–148
regional hydrology, 145–146
results, 145–158
status, water quality, 158–159
study site, 137, 139
surface water chemistry, 139, 141
trends, water quality, 147–148
water quality, 147–151, 158–162

V

Validation, 526, 528
Venice Lagoon ecosystem
bathymetry, 491–493
chlorophyll *a,* 494–496
circulation, 491
climate change, 490
confinement, 491
dissolved inorganic nutrients, 493–494
ecosystem responses, 501–505
environmental drivers and pressures, 486–490
eustatism, 488
fishing activities, 490
freshwater discharges, 489
fundamentals, 13, 484, 505
governance, 490
historical developments, 484–486
lagoon state and impacts, 491–501
landings data, 500–501
macrophytobenthos, 497–498
macrozoobenthos, 498–500
nekton, 500
nutrient loads, 489
phytoplankton, 494–496
pollutants, 489
sediment nutrients, 494
subsidence, 488
zooplankton, 496
Videographic imaging, 173
Virginia coastal bays
allochthonous nitrogen sources, 49–55
assimilation, 58–60
atmospheric deposition, 54–55
autochthonouse nitrogen sources, 56–58
benthic macro- and microalgae, 61–63
benthic primary production, 61
direct groundwater discharge, 52, 54
fate of nitrogen, 61–63
fundamentals, 10, 43–50
future trends and predictions, 63–66
nitrogen fixation, 56
nutrient gradient, 49
physical description, 48
primary producers, 58–60
remineralization, 57–58
restoration of seagrass, 65–66
seagrasses, 58, 65–66
seasonality, 58
sinks of nitrogen, 58–61
sources of, 48–58
stream flow, 50–52

W

Wadden Sea
  abiotic conditions, 404–407
  biotic factors, 407–410
  classification, 3
  climate change effects, 419–422
  ecosystem function, 419–422
  fundamentals, 13, 399–400
  future research, 422–423
  habitat change and destruction, 416–417
  higher trophic levels, 409–410
  historical drivers, 410, 416–419
  invasive species, 418–419
  major changes, 403–410
  nutrients, 417–418
  overexploitation, 416
  physiography, 400, 403
  pollution, 417
  primary producers and productivity, 407, 409
  secondary producers and productivity, 409
Waquoit Bay
  bottom-up controls, 207–215
  case studies, 201–215
  top-down controls, 207–215
Water chemistry, 288, 290
Water column, 119–120, 291–295
Water exchange, 436–437
Water quality
  Australian coastal lagoons, 465
  Choptank Basin, 147–151, 158–162
  lessons learned, 326–328
  New Jersey coastal lagoons, 173
  Ria Formosa, 345–346
  Yucatan Peninsula, 326–328

Y

Yucatan Peninsula (Mexico)
  Ascención Bay, 325–326
  Bojorquez Lagoon, 325
  Celestún Lagoon, 314–315
  Chacmochuk Lagoon, 321, 324
  characterization of lagoons, 312–326
  Chelem Lagoon, 315–317
  Chetumal Bay, 326
  context, 309–311
  discussion, 291–299
  Dzilam Lagoon, 317–318
  fundamentals, 12, 283–286, 307–309, 329
  Holbox Lagoon, 320–321
  lessons learned, 326–329
  methods, 286–288
  monitoring experience, 311–312
  Nichupte Lagoon, 324–325
  nutrient impacts, 298–299
  nutrient sources, 295–298
  phytoplankton, 328
  plant isotope values, 291
  research, 311–312
  results, 288–291
  Rio Lagartos Lagoon, 318–319
  sample collection and analysis, 286–288
  seagrass biomass, 290
  statistical analysis, 288
  status of lagoons, 312–326
  study area, 286
  submerged aquatic vegetation, 328–329
  tissue nutrient content, 291
  water chemistry, 288, 290
  water column processes, 291–295
  water quality characteristics, 326–328

Z

Zooplankton
  Nile Delta, 274–275
  Venice Lagoon ecosystem, 496
Zostera marina
  Delmarva Peninsula, 47–48
  Hog Island Bay, 58
  Jobos Bay, 219
  New Jersey lagoons, 168–170, 175–179, 181–187, 191–196
  Ria Formosa, 340
  Venice Lagoon, 497–499
  Wadden Sea, 407
  Waquoit Bay, 209–210, 215
Zostera noltii
  Ria Formosa, 340
  Wadden Sea, 407, 421
Zosterisessor ophiocephalus, 500